Ann Xavier Gantert

AMSCO SCHOOL PUBLICATIONS, INC.
315 HUDSON STREET, NEW YORK, N.Y. 10013

Dedication

This book is dedicated to Edward Keenan who left a profound influence on mathematics education in New York State and on the development of Amsco texts.

Ann Xavier Gantert

This author has been associated with mathematics education in New York State as a teacher and an author throughout the many changes of the past fifty years. She has worked as a consultant to the Mathematics Bureau of the Department of Education in the development and writing of Sequential Mathematics and has been a coauthor of Amsco's *Integrated Mathematics* series, which accompanied that course of study.

Reviewers:

Steven Balasiano
Assistant Principal, Mathematics
Canarsie High School
Brooklyn, NY

Debbie Calvino
Mathematics Chairperson
Valley Central High School
Montgomery, NY

Donna Getchell
Mathematics Teacher
Garden City High School
Garden City, NY

Sal Sutera
Mathematics Teacher
New Utrecht High School
Brooklyn, New York

Text Designer: Nesbitt Graphics, Inc.
Compositor: Compset, Inc.
Cover Designer by Meghan J. Shupe
Cover Art by Brand X Pictures

Please visit our Web site at: www.amscopub.com

When ordering this book, please specify:
R 239 P *or* INTEGRATED ALGEBRA 1, *Paperback*
or
R 239 H *or* INTEGRATED ALGEBRA 1, *Hardbound*

ISBN 978-1-56765-584-1 (Paperback edition)
NYC Item 56765-584-0 (Paperback edition)
ISBN 978-1-56765-585-8 (Hardbound edition)
NYC Item 56765-585-7 (Hardbound edition)

Printed in the United States of America

4 5 6 7 8 9 10 11 10 09 08 07 (Paperback edition)
11 12 13 14 11 10 09 08 (Hardbound edition)

PREFACE

Integrated Algebra 1 is a new text for high school algebra that continues the approach that has made Amsco a leader in presenting mathematical ideas in a contemporary, integrated manner. Over the past decades, this approach has undergone numerous changes and refinements to keep pace with ever changing technology.

This Amsco book uses an integrated approach to the teaching of high school mathematics that is promoted by the National Council of Teachers of Mathematics in its *Principles and Standards for School Mathematics* and mandated by the New York State Board of Regents in the *New York State Mathematics Core Curriculum*. This text presents a range of materials and explanations that are guidelines for achieving a high level of excellence in the study of mathematics.

In this book:

✔ **The graphing calculator** is introduced and used throughout the book as a routine tool in the study of mathematics. Underlying mathematical concepts and procedures are clearly presented, stressing calculator use as a learning and computational aid.

✔ **The real number system** is fully developed, to help students understand and correctly interpret technological limitations such as the calculator displays of rational approximations. The role of precision and accuracy, in determining acceptable computational results, is carefully explained and illustrated.

✔ **Application** of algebra to the solution of problems from geometry, probability, statistics, finance, and other real-world applications is developed throughout the text.

✔ **Enrichment** is stressed throughout the text and in the Teacher's Manual where multiple suggestions are given for teaching strategies, for further explorations of related topics, and for alternative assessment. The text, as well as the Manual, includes opportunities for cooperative learning, hands-on activities, extended tasks, and independent investigation. Reproducible *Enrichment Activities* for each chapter provide both material for review and reinforcement as well as for in-depth study.

- ✔ **Exercises** are divided into three categories. *Writing About Mathematics* provides questions in which students are asked to contrast, compare, evaluate, and justify their own ideas or those of others. These questions help students incorporate the tools of the performance indicators—investigate, explore, discover, conjecture, reason, justify, explain, prove, and apply—into their study of mathematics. These questions also provide a valuable source of material for classroom discussion or for inclusion in a student portfolio. *Developing Skills* provides routine practice exercises that enable the student and the teacher to evaluate the student's ability to both manipulate mathematical symbols and understand mathematical relationships. *Applying Skills* provides exercises in which the new ideas of each section, together with previously learned skills, are used to solve problems that reflect real-life situations.

- ✔ **Conceptual understanding**, **procedural fluency**, and **problem solving**, which are the primary goals of the *Core Curriculum* are addressed throughout the text. General concepts and principles are fully addressed and developed in detail, then further explored in the examples and exercise sections. The *Procedures* throughout the text explain how to perform both arithmetic and geometric processes. The *Examples* given in each section demonstrate problem solving approaches, often presenting alternative strategies for solution. Both routine and non-routine problems are presented.

The material in this text is intended to present basic algebra and its relationship to other branches of mathematics. The text aims at developing mathematics as a unified whole in which each branch of mathematics is integrally related. Many of the concepts presented in this text have been introduced in previous mathematics courses. The text provides the opportunity for students to review familiar material that is the foundation for the development of new topics, and presents all the material needed to develop the skills and achieve the goals suggested in the New York State *Core Curriculum* for Integrated Algebra.

An intent of the author was to make this text of greatest service to the average student. However, the materials for reinforcement and for enrichment that the text contains make it appropriate for varying abilities. Specifically:

- ✔ Concepts are carefully developed using appropriate language and mathematical symbolism.

- ✔ General principles and procedures are stated clearly and concisely.

- ✔ Numerous solved examples serve as models for students, with detailed step-by-step explanations.

- ✔ Abundant and varied exercises develop skills and test understanding. Additional enrichment activities challenge the most capable student.

This text is offered so that teachers may effectively continue to help students to comprehend, master, and enjoy mathematics.

CONTENTS

Chapter 3

ALGEBRAIC EXPRESSIONS AND OPEN SENTENCES 88

Chapter 4

FIRST DEGREE EQUATIONS AND INEQUALITIES IN ONE VARIABLE 116

Chapter 5

OPERATIONS WITH ALGEBRAIC EXPRESSIONS 167

Chapter 11
SPECIAL PRODUCTS AND FACTORS 442

Chapter 12
OPERATIONS WITH RADICALS 469

Chapter 13
QUADRATIC RELATIONS AND FUNCTIONS 502

Chapter 14

ALGEBRAIC FRACTIONS, AND EQUATIONS AND INEQUALITIES INVOLVING FRACTIONS 539

Chapter 15

PROBABILITY 575

STATISTICS 660

CHAPTER

1

NUMBER SYSTEMS

- The athletic department needs to transport 125 students, including the basketball team and supporters, to a playoff game. If each bus can accommodate 48 students, how many buses will be needed for the trip?
- The distance from the school to the game is 125 miles. If the bus travels at an average rate of 48 miles per hour, how long will the trip take?
- Students are having a recycling drive to help pay for the trip. One group of students collected 125 cans that will be placed in cases of 48 cans each. Only full cases can be returned to the distributor for a deposit refund. How many cases can be returned?

Each of these is a simple problem. How are the three problems alike? Why are their answers different?

In this chapter you will review the real numbers system and its subsets, use estimation skills and rational approximations to interpret calculator results, and begin to integrate the different areas of mathematics through the study of numbers, number lines, graphs, and geometric figures.

CHAPTER TABLE OF CONTENTS

1-1 THE INTEGERS

Mathematics is the study of numbers, shapes, arrangements, relationships, and reasoning. Mathematics is both a science and an art that can be used to describe the world in which we live, to solve problems, and to create new ideas.

Numbers, which are a basic part of mathematics, help us to understand algebra, to measure geometric objects, and to make predictions using probability and statistics. In this chapter we will study numbers such as those shown below:

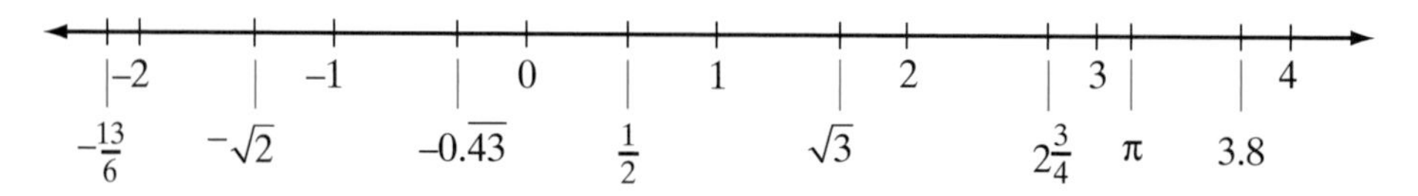

Every point on this number line corresponds to a **real number**. What are real numbers? What is meant by values such as $\sqrt{3}$ and $-0.\overline{43}$? Let us begin with simpler numbers that we know.

Symbols for Numbers

A **number** is really an idea: it is something that we can talk about and think about. We represent numbers in writing by using the symbols 1, 2, 3, 4, and so on. These symbols, called **numerals**, are not numbers but are used to represent numbers.

Counting Numbers or Natural Numbers

The **counting numbers**, which are also called **natural numbers**, are represented by the symbols

$$1, 2, 3, 4, 5, 6, 7, 8, 9, 10, 11, 12, \ldots$$

The three dots after the 12 indicate that the numbers continue *in the same pattern* without end. The smallest counting number is 1. Every counting number has a **successor** that is 1 more than that number. The successor of 1 is 2, the successor of 2 is 3, and so on. Since this process of counting is endless, there is no last counting number.

On the number line, the points associated with counting numbers are highlighted and an arrow shows that these numbers continue without end.

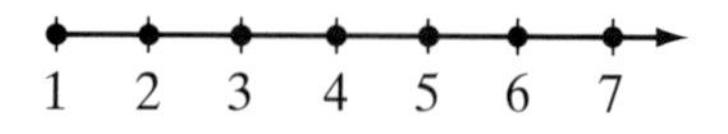

The Set of Whole Numbers

Zero is not a counting number. By combining 0 with all the counting numbers, we form the set of **whole numbers**. The whole numbers are represented by the symbols

$$0, 1, 2, 3, 4, 5, 6, 7, 8, 9, 10, 11, 12, \ldots$$

The smallest whole number is 0. There is no largest whole number. Notice that the number line has been extended to include the number 0.

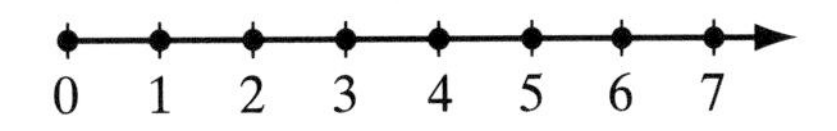

A **set** is a collection of distinct objects or elements. A set is usually indicated by enclosing the names or symbols for its elements within a pair of braces, { }. For example, the set of whole numbers can be written as $\{0, 1, 2, 3, 4, \ldots\}$.

Types of Sets

A **finite set** is a set whose elements can be counted. For example, the set of **digits** consists of only ten symbols, 0 through 9, that are used to write our numerals:

$$\{0, 1, 2, 3, 4, 5, 6, 7, 8, 9\}$$

An **infinite set** is a set whose elements cannot be counted because there is no end to the set. For example, the counting numbers and the whole numbers are both infinite sets.

The **empty set** or **null set** is a set that has no elements, written as $\{\}$ or $\varnothing$. For example, the set of months with 32 days is empty, and the set of counting numbers between 1 and 2 is also empty.

Numerical Expressions

A **numerical expression** is a way of writing a number in symbols. The expression can be a single numeral, or it can be a collection of numerals with one or more operation symbols. For example:

$6 + 2$	$18 - 10$	4×2	$640 \div 80$
$2 \times 2 \times 2$	$2 + 2 + 2 + 2$	$1 \times 7 + 1$	8

Each of these expressions is a symbol for the number 8. In general, to **simplify** a numerical expression means to find the single number that is its value.

A calculator can be used to find the value of a numerical expression. The primary purpose of any calculator is to perform arithmetic operations, in particular, the four basic operations: addition, subtraction, multiplication, and division.

In this book, we will show the keys used on a TI-83+/84+ graphing calculator. However, here, and in many of the calculator examples throughout the book, the keys listed, or similar keys, can be used on any graphing calculator.

Add 6 to the product of 3 and 9.

ENTER: 3 [×] 9 [+] 6 [ENTER]

DISPLAY:

```
3*9+6
                    33
```

Answer: 33

From the quotient of 10 and 2, subtract 1.

ENTER: 10 [÷] 2 [−] 1 [ENTER]

DISPLAY:

```
10/2-1
                     4
```

Answer: 4

Note that a scientific calculator uses [=] in place of [ENTER].

The Set of Integers

The temperature on a winter day may be below 0 degrees, or someone may write a check for an amount that cannot be covered by funds in the checking account. Both situations describe **negative numbers**.

Just as the number line was extended to include 0, we can again extend it to include negative numbers. A number that is 1 less than 0 is -1; a number 2 less than 0 is -2; and so on.

Imagine that a mirror is placed at the number 0, and the counting numbers (which are thought of as *positives*) are reflected in the mirror to show *negative* numbers. Our new number line extends forever in two directions. It has no beginning and no end.

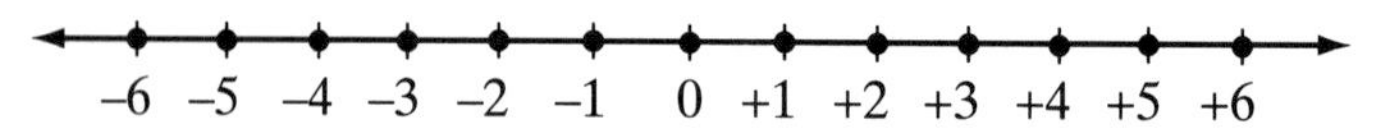

Each positive number can be paired with a negative number that is the same distance from 0 but on the opposite side of 0. The numbers of each pair are called **opposites**.

- The opposite of $+1$ is -1, and the opposite of -1 is $+1$.
- The opposite of $+2$ is -2, and the opposite of -2 is $+2$, and so on.

Notice that 0 is neither positive nor negative. 0 is considered its own opposite.

The set that contains the counting numbers, 0, and the opposites of the counting numbers is the **set of integers**. The most common way to write the set of integers is to write them from the smallest to largest, in the order in which they occur on the number line. Since there is no smallest integer, the list begins with three dots to indicate that there are an infinite number of integers that are smaller than the first integer that is named. Since there is no largest integer, the list ends with three dots to indicate that there are an infinite number of integers that are larger than the last integer that is named.

$$\{\ldots, -3, -2, -1, 0, +1, +2, +3, \ldots\}$$

Subsets of the Integers

Set A is called a **subset** of set B, written $A \subseteq B$, if every element of set A is also an element of set B. If A is a subset of B and there is at least one element in B that is not an element of A, then $A \subset B$.

Using this definition, we know that the whole numbers and the counting numbers are subsets of the integers. Counting numbers also form a subset of the whole numbers. These subsets can be illustrated in a diagram, as shown to the right.

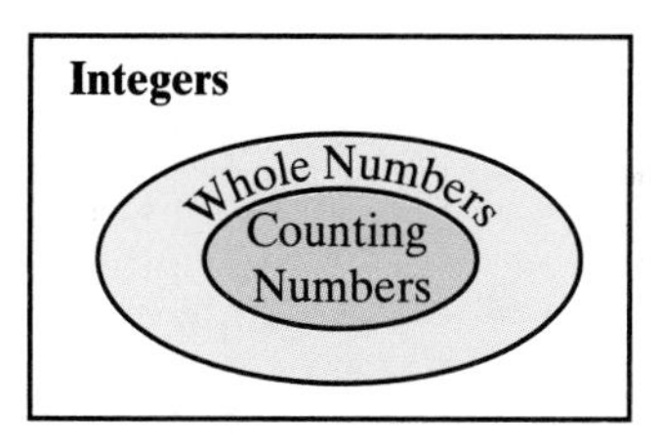

Counting numbers $\subset$ Whole numbers

Whole numbers $\subset$ Integers

Counting numbers $\subset$ Integers

There are many other subsets of the integers, such as:

1. Odd whole numbers $\{1, 3, 5, 7, 9, 11, \ldots\}$

2. Odd integers $\{\ldots, -5, -3, -1, 1, 3, 5, \ldots\}$

3. Even whole numbers $\{0, 2, 4, 6, 8, 10, \ldots\}$

4. Even integers $\{\ldots, -6. -4. -2, 0, 2, 4, 6, \ldots\}$

5. One-digit whole numbers $\{0, 1, 2, 3, 4, \ldots, 9\}$

Ordering the Integers

A **number line** can be used to show the numbers of a set in their relationship to each other. Each number is represented by a point on the line called the **graph** of the number.

There are two standard forms of the number line that we use. One a vertical number line (as pictured left), such as one seen on a thermometer, the higher up we go, the greater will be the number or the higher the temperature. Just as 4 is greater than 3, so is 3 greater than 0, and 0 greater than -1. It follows that -1 is greater than -2, and -2 is greater than -20.

On a horizontal number line, positive numbers appear to the right of 0, and the negative numbers to the left of 0. The greater of any two numbers will be the one to the right and the smaller of any two numbers the one to the left. We will call this number line the **standard number line**. To build the standard number line:

1. Draw a horizontal line. Label one point on the line 0. Choose a point to the right of 0 and label it 1. The distance from 0 to 1 is called the **unit measure**.
2. Place arrowheads at the ends of the line that you drew to show that this is just a part of a line that extends without end in both directions.
3. Use the unit measure to mark off equally spaced points to the right of 1 and label these points 2, 3, 4, and so on.
4. Use the unit measure to mark off equally spaced points to the left of 0 and label these points $-1, -2, -3$, and so on.

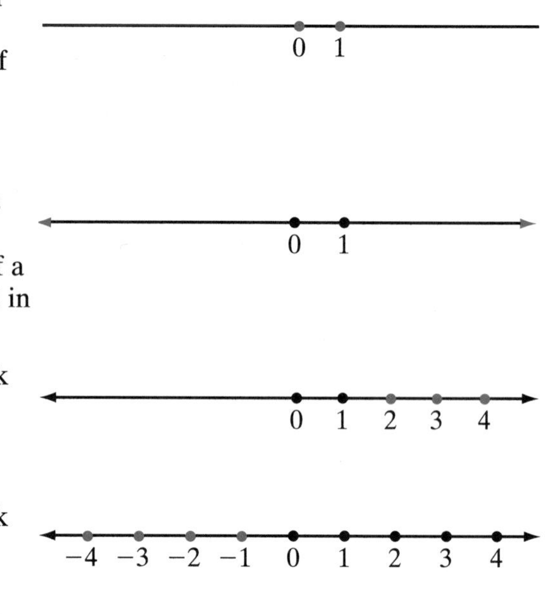

From the number line, we see that 2 is greater than 0 and -1 is greater than -3.

Absolute Value

In every pair of nonzero opposites, the positive number is the greater. On a standard horizontal number line, the positive number is always to the right of the negative number that is its opposite. For example, 10 is greater than its opposite, -10. On a number line, 10 is to the right of -10.

The greater of a nonzero number and its opposite is called the **absolute value** of the number. The absolute value of 0 is 0.

The absolute value of a number, a, is symbolized as $|a|$. Since 10 is the greater of the two numbers 10 and -10, the absolute value of 10 is 10 and the absolute value of -10 is 10.

$$|10| = 10$$
$$|-10| = 10$$
$$|10| = |-10|$$

▶ **The absolute value of a positive number is the number itself; the absolute value of a negative number is the opposite of the number.**

The absolute value of a number can also be thought of as the distance between 0 and the graph of that number on the real number line. For example, $|3| = 3$, the distance between 0 and P, the graph of 3 on the real number line shown below. Also, $|-3| = 3$, the distance between 0 and S, the graph of -3 on the real number line.

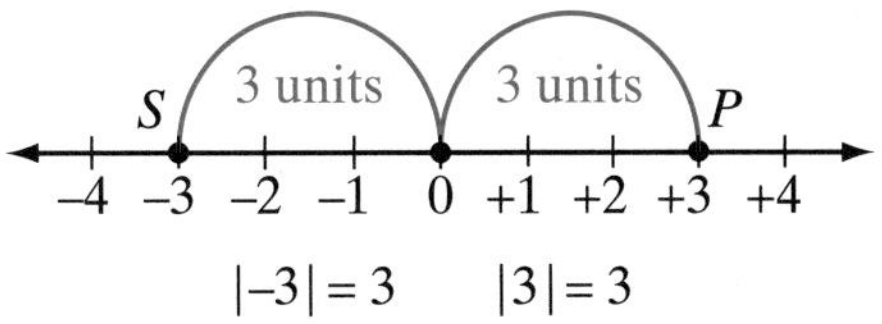

Symbols of Inequality

In our daily lives, we are often asked to compare quantities. Which is cheaper? Which weighs more? Who is taller? Which will last longer? Are two objects the same size? The answers to these questions are given by comparing quantities that are stated in numerical terms.

If two numbers are not equal, the relationship between them can be expressed as an **inequality** that can be written in several different ways.

Symbol	Example	Read
$>$	$9 > 2$	9 is greater than 2.
$<$	$2 < 9$	2 is less than 9.
$\geq$	$9 \geq 2$	9 is greater than or equal to 2.
$\leq$	$2 \leq 9$	2 is less than or equal to 9.
$\neq$	$9 \neq 2$	9 is not equal to 2.

Notice that in an inequality, the symbols $>$ and $<$ point to the smaller number.

EXAMPLE 1

Find the value of each expression.

a. $|12| + |-3|$ **b.** $|12 - 3|$

Solution **a.** Since $|12| = 12$, and $|-3| = 3$, $|12| + |-3| = 12 + 3 = 15$ *Answer*

b. First, evaluate the expression inside the absolute value symbol. Then, find the absolute value.

$$|12 - 3| = |9| = 9 \quad \textit{Answer}$$

EXAMPLE 2

Tell whether each statement is true or false.

a. $-3 > -5$ *Answer:* True **b.** $0 < -4$ *Answer:* False

c. $12 - 5 > 2$ *Answer:* True **d.** $(2)(7) \leq 14$ *Answer:* True

EXAMPLE 3

Use the symbol < to order the numbers -4, 2, and -7.

Solution

–7 –6 –5 –4 –3 –2 –1 0 1 2 3

On the number line, -7 is to the left of -4 and -4 is to the left of 2. Therefore, $-7 < -4$ and $-4 < 2$.

Answer $-7 < -4 < 2$

EXAMPLE 4

Write, in each case, at least three true statements to compare the numbers in the order given.

a. 8 and 2 *Answer:* $8 > 2$; $8 \neq 2$; $8 \geq 2$

b. 12 and 12 *Answer:* $12 = 12$; $12 \geq 12$; $12 \leq 12$

EXERCISES

Writing About Mathematics

1. Olga said that the absolute value of any real number is always greater than or equal to the number. Do you agree with Olga? Explain your answer.

2. A number is represented by a and its opposite by b. If $a < b$, which letter represents a positive number and which represents a negative number. Explain your answer.

Developing Skills

In 3–12: **a.** Give the absolute value of each given number. **b.** Give another number that has the same absolute value.

3. 10.4 **4.** -7 **5.** $3 + 18$ **6.** -13 **7.** -20

8. $1\frac{1}{2}$ **9.** $-3\frac{3}{4}$ **10.** -1.45 **11.** $+2.7$ **12.** -0.02

In 13–20, state whether each sentence is true or false.

13. $|20| = 20$ **14.** $|-13| = 13$ **15.** $|-15| = -15$ **16.** $|-9| = |9|$

17. $|-7| < |7|$ **18.** $|-10| > |3|$ **19.** $|8| < |-19|$ **20.** $|-21| \geq 21$

In 21–30, find the value of each expression.

21. $|9| + |3|$ **22.** $|+8| - |+2|$ **23.** $|-6| + |4|$ **24.** $|-10| - |-5|$

25. $|4.5| - |-4.5|$ **26.** $|+6| + |-4|$ **27.** $|+6 - 4|$ **28.** $|7 - 2|$

29. $|15 - 15|$ **30.** $|-8| + |8 - 2|$

In 31–34, state whether each sentence is true or false. Give a reason for each answer.

31. $+5 > +2$ **32.** $-3 < 0$ **33.** $-7 > -1$ **34.** $-2 > -10$

In 35–40, write each inequality using the symbol $>$ or the symbol $<$.

35. +8 is greater than +6. **36.** –8 is less than 0.

37. –5 is less than −2. **38.** –5 is greater than −25.

39. The sum of 16 and 3 is greater than the product of 9 and 2.

40. The product of 6 and 7 is less than the quotient of 100 divided by 2.

In 41–44, express each inequality in words.

41. $+7 > -7$ **42.** $-20 < -3$ **43.** $-4 < 0$ **44.** $-9 \geq -90$

In 45–48, use the symbol $<$ to order the numbers.

45. $-4, +8, -5$ **46.** $-3, -6, +3, +6$ **47.** $+3, -2, -4, 0$ **48.** $-2, +8, 0, -8$

In 49–52, write, in each case, three true statements to compare the numbers, using the order in which they are given.

49. 8 and 14 **50.** 9 and 3 **51.** 15 and 15 **52.** 6 and -2

53. In Column I, sets of numbers are described in words. In Column II, the sets are listed using patterns and dots. Match the patterns in Column II with their correct sets in Column I.

Column I	*Column II*
1. Counting numbers	a. $0, 1, 2, \ldots, 9$
2. Whole numbers	b. $0, 1, 2, \ldots$
3. Even whole numbers	c. $0, 2, 4, 6, \ldots$
4. Odd whole numbers	d. $0, 2, 4, 6, 8$
5. Even counting numbers	e. $0, 2, -2, 4, -4, 6, -6, \ldots$
6. Odd integers	f. $1, 2, 3, 4, \ldots$
7. Even integers	g. $1, 2, 3, \ldots, 9$
8. One-digit whole numbers	h. $1, 3, 5, 7, \ldots$
9. One-digit counting numbers	i. $1, 3, 5, 7, 9$
10. Odd whole numbers less than 10	j. $2, 4, 6, 8, \ldots$
11. Even whole numbers less than 10	k. $-2, -1, 0, 1, 2, 3, 4, \ldots$
12. Integers greater than -3	l. $1, -1, 3, -3, 5, -5, \ldots$

Applying Skills

For 54 and 55, read the problem carefully, solve the problem, and check the solution.

54. The athletic department of a school wants to transport 151 students to a basketball game. Some buses that seat 25 passengers and others that seat 34 passengers are available.

a. How many buses of each size should be scheduled for the trip so that the smallest number of buses will be used and the smallest number of seats will be empty?

b. Based on your answer to part a, how many empty seats will there be?

55. A shopkeeper has a bag of rice that he wants to divide into smaller bags. He has a container that holds 3 pounds and another that holds 4 pounds of rice. How can he use these containers to measure 5 pounds of rice?

56. Give three examples in which a negative number can be used in describing a measurement or an event.

1-2 THE RATIONAL NUMBERS

In earlier years, you worked with many numbers other than integers, such as fractions, decimals, and mixed numbers. These numbers from arithmetic, which can be located on the real number line, behave in a special way. Consider the following examples:

$$\frac{3}{5} \qquad 3\frac{1}{2} = \frac{7}{2} \qquad -\frac{41}{100}$$

$$0.9 = \frac{9}{10} \qquad 8.1 = 8\frac{1}{10} = \frac{81}{10} \qquad 0.25 = \frac{25}{100} = \frac{1}{4}$$

Each of the numbers shown here is written in the form of a fraction. In fact, every integer can be written as a fraction by writing the integer with a denominator of 1:

$$5 = \frac{5}{1} \qquad 0 = \frac{0}{1} \qquad -12 = \frac{-12}{1}$$

In general, any integer n can be written as $\frac{n}{1}$, which is a quotient of two integers.

The Set of Rational Numbers

The **rational numbers** are all numbers that can be expressed in the form $\frac{a}{b}$ where a and b are integers and $b \neq 0$.

Notice that the first five letters of the word *rational* form the word *ratio*, which means a comparison of two quantities by division.

The counting numbers, the whole numbers, and the integers are all subsets of the set of rational numbers, as illustrated in the diagram.

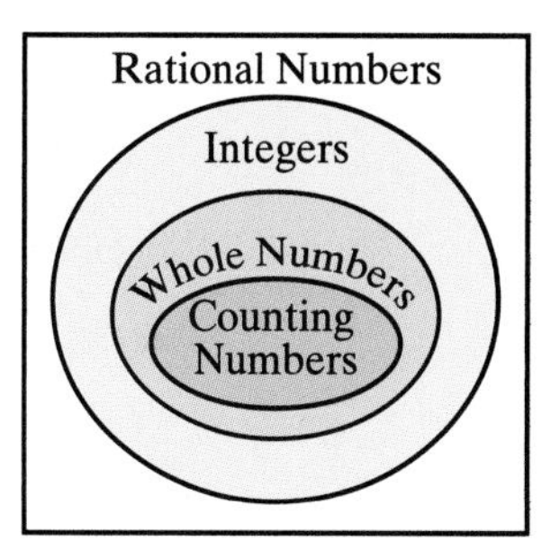

The Rational Number Line

Every rational number can be associated with a point on the real number line. For example, $\frac{1}{2}$ is midway between 0 and 1, and -2.25 $\left(\text{or } -2\frac{1}{4}\right)$ is one-quarter of the way between -2 and -3, closer to -2.

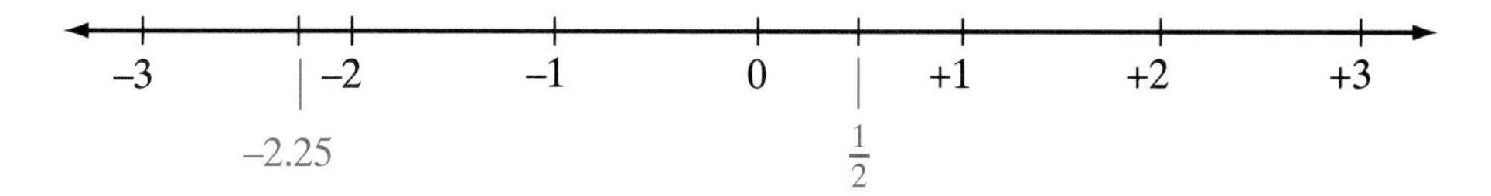

The rational numbers, like other sets studied earlier, can be ordered. In other words, given any two unequal rational numbers, we can tell which one is greater. For example, we know that $\frac{1}{2} > -1$ because, on the standard number line, $\frac{1}{2}$ is to the right of -1. There are also other ways to determine which of two rational numbers is greater, as shown in the following example.

EXAMPLE 1

Which is the greater of the numbers $\frac{7}{9}$ and $\frac{8}{11}$?

Solution **METHOD 1.** Express the numbers as equivalent fractions with a common denominator, and compare the numerators.

$$\frac{7}{9} = \frac{7}{9} \times \frac{11}{11} = \frac{77}{99}$$
$$\frac{8}{11} = \frac{8}{11} \times \frac{9}{9} = \frac{72}{99}$$

Since $\frac{77}{99} > \frac{72}{99}$, then $\frac{7}{9} > \frac{8}{11}$.

METHOD 2. Change the fractions to decimals by dividing each numerator by its denominator to see which is greater. The answers here are from a calculator that shows ten places in each display.

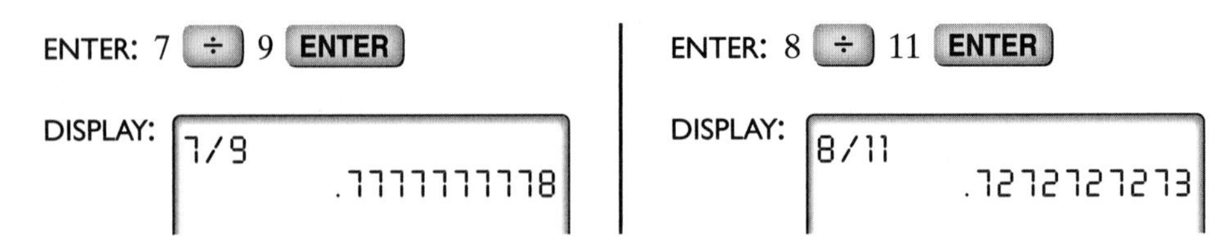

Compare the numbers in the first two decimal places.
Since $0.77 > 0.72$, $\frac{7}{9} > \frac{8}{11}$.

Answer $\frac{7}{9} > \frac{8}{11}$

A Property of the Rational Numbers

► **The set of rational numbers is everywhere dense.**

In other words, given two unequal rational numbers, it is always possible to find a rational number that lies between them.

For example, some rational numbers between 1 and 2 are $1\frac{1}{2}$, $1\frac{1}{8}$, $1\frac{2}{3}$, and $1\frac{9}{10}$. In fact, there is an infinite number of rational numbers between two rational numbers.

One way to find a rational number between two rational numbers is to find their average, called the *mean*. To find the mean of two numbers, add the numbers and divide by 2. The mean (or average) of 3 and 5 is $(3 + 5) \div 2 = 8 \div 2 = 4$, a number that is between 3 and 5 on a number line. In the same way, a rational number between $\frac{1}{4}$ and $\frac{3}{4}$ can be found as follows:

$$\left(\frac{1}{4} + \frac{3}{4}\right) \div 2 = 1 \div 2 = \frac{1}{2}$$

A calculator can be used to do this.

ENTER: (1 ÷ 4 + 3 ÷ 4) ÷ 2 ENTER

DISPLAY:
```
(1/4+3/4)/2
             .5
```

Note that, on a calculator, the rational number $\frac{1}{2}$ is written as 0.5.

Expressing a Rational Number as a Decimal

Every rational number that is not an integer can be written as a fraction. A **common fraction** is written with a numerator and denominator, for example, $\frac{3}{4}$. In a **decimal fraction** or decimal, the numerator is written after the decimal point and the denominator is indicated by the place value of the last digit. For example, the decimal fraction 0.75 has a numerator of 75 and a denominator of 100. To express as a decimal fraction a rational number named as a common fraction, we simply perform the indicated division. For example, express the following as decimals: $\frac{1}{2}, \frac{3}{4}, \frac{1}{16}$.

ENTER: 1 ÷ 2 ENTER

DISPLAY:
```
1/2
             .5
```

ENTER: 3 ÷ 4 ENTER

DISPLAY:
```
3/4
            .75
```

ENTER: 1 ÷ 16 ENTER

DISPLAY:
```
1/16
          .0625
```

In each of the examples, $\frac{1}{2}$, $\frac{3}{4}$, and $\frac{1}{16}$, when we perform the division, we reach a point at which the division has no remainder, that is, a remainder of 0. If we were to continue the division with paper and pencil, we would continually obtain only zeros in the quotient. Decimals that result from such divisions, for example, 0.5, 0.75, and 0.0625, are called **terminating decimals**.

Not all rational numbers can be expressed as terminating decimals, as shown in the following examples.

Express the following as decimals: $\frac{1}{3}$, $\frac{2}{11}$, $\frac{1}{6}$.

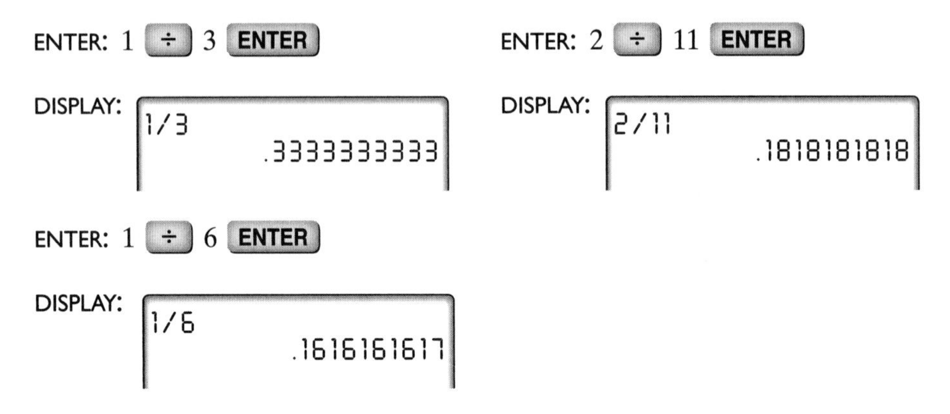

In each of the above examples, when we perform the division, we find, in the quotient, that the same digit or group of digits is continually repeated in the same order. The calculator prints as many digits as possible and rounds the digit in the last decimal place that can be displayed. Decimals that keep repeating endlessly are called **repeating decimals** or **periodic decimals**.

A repeating decimal may be written in abbreviated form by placing a bar (–) over the group of digits that is to be continually repeated. For example:

$$0.333333\ldots = 0.\overline{3} \qquad 0.181818\ldots = 0.\overline{18} \qquad 0.166666\ldots = 0.1\overline{6}$$

The examples above illustrate the truth of the following statement:

► **Every rational number can be expressed as either a terminating decimal or a repeating decimal.**

Note that the equalities $0.5 = 0.5\overline{0}$ and $0.75 = 0.75\overline{0}$ illustrate the fact that every terminating decimal can be expressed as a repeating decimal that, after a point, repeats with all 0's. Then since every terminating decimal can be expressed as a repeating decimal, we will henceforth regard terminating decimals as repeating decimals. Therefore, we may say:

► **Every rational number can be expressed as a repeating decimal.**

Expressing a Decimal as a Rational Number

Use the following steps to change a terminating decimal to a fraction:

STEP 1. Read it (using place value): 0.8 is read as 8 tenths.

STEP 2. Write it as a fraction, using the same words as in step 1: $\frac{8}{10}$ is also 8 tenths

STEP 3. Reduce it (if possible): $\frac{8}{10} = \frac{4}{5}$

EXAMPLE 2

Express each decimal as a fraction:

a. 0.3 **b.** 0.37 **c.** 0.139 **d.** 0.0777

Answers **a.** $0.3 = \frac{3}{10}$ **b.** $0.37 = \frac{37}{100}$ **c.** $0.139 = \frac{139}{1,000}$ **d.** $0.0777 = \frac{777}{10,000}$

EXERCISES

Writing About Mathematics

1. Bennie used his calculator to find the decimal value of $\frac{1}{17}$. The number in the display was 0.0588235294. Bennie knows that this is not a terminating decimal equivalent to $\frac{1}{17}$ because $\frac{588,235,294}{10,000,000,000} \neq \frac{1}{17}$. Therefore, Bennie concluded that $\frac{1}{17}$ is a rational number that is a *nonrepeating* decimal. Explain why Bennie's conclusion is incorrect.

2. Explain how you know that there is not a smallest positive rational number.

Developing Skills

In 3–12, write each rational number in the form $\frac{a}{b}$ where a and b are integers, and $b \neq 0$.

3. 0.7 **4.** 0.18 **5.** –0.21 **6.** 9 **7.** –3

8. 0 **9.** $5\frac{1}{2}$ **10.** $-3\frac{1}{3}$ **11.** 0.007 **12.** –2.3

In 13–22, state, in each case, which of the given numbers is the greater.

13. $\frac{5}{2}, \frac{7}{2}$ **14.** $-\frac{9}{3}, \frac{11}{3}$ **15.** $\frac{5}{6}, -\frac{13}{6}$ **16.** $\frac{1}{5}, -5$ **17.** $\frac{5}{2}, \frac{7}{4}$

18. $-\frac{10}{3}, -\frac{13}{6}$ **19.** $\frac{13}{6}, \frac{15}{10}$ **20.** $-\frac{5}{8}, -\frac{5}{12}$ **21.** 1.275, 1.2 **22.** $0.\overline{6}, 0.6$

In 23–32, find a rational number between each pair of given numbers.

23. 5, 6 **24.** –4, –3 **25.** –1, 0 **26.** $\frac{1}{4}, \frac{1}{2}$ **27.** $\frac{1}{2}, \frac{7}{8}$

28. $-\frac{3}{4}, -\frac{2}{3}$ **29.** –2.1, –2.2 **30.** $2\frac{1}{2}, 2\frac{5}{8}$ **31.** $-1\frac{1}{3}, -1\frac{1}{4}$ **32.** $3.05, 3\frac{1}{10}$

In 33–42, write each rational number as a repeating decimal. (Hint: Every terminating decimal has a repeating zero, for example, $0.3 = 0.3\overline{0}$.)

33. $\frac{5}{8}$ **34.** $\frac{9}{4}$ **35.** $-5\frac{1}{2}$ **36.** $\frac{13}{8}$ **37.** $-\frac{7}{12}$

38. $\frac{5}{3}$ **39.** $\frac{7}{9}$ **40.** $\frac{2}{11}$ **41.** $\frac{5}{99}$ **42.** $-\frac{5}{6}$

In 43–52, find a common fraction that names the same rational number as each decimal fraction.

43. 0.5 **44.** 0.555 **45.** -0.2 **46.** 0.12 **47.** 0.111

48. $0.125\overline{0}$ **49.** 0.2525 **50.** 0.07 **51.** 0.99875 **52.** -0.3

In 53–59, tell whether each statement is true or false, and give a reason for each answer.

53. Every integer is a rational number.

54. Whole numbers can be negative.

55. On a standard horizontal number line, the greater of two numbers is always the number farther to the right.

56. Every rational number can be written as a repeating decimal.

57. Between 0 and 1, there are an infinite number of fractions.

58. There are an infinite number of numbers between -2 and -1.

59. For every rational number, there is another rational number that is larger than the given number.

Applying Skills

For each of the following, read the problem carefully and then solve it.

60. Jacob baked some cookies. For every two cookies that he kept for his family, he gave three away to his friends. What fractional part of the cookies did he give away?

61. Margarita took part in a walk to raise money for a food pantry. After every forty-five minutes of walking, she rested for five minutes. What fractional part of the total time that she took to complete the walk was spent resting?

62. Hannah walked $\frac{3}{4}$ of the way from school to her home.

a. What fractional part of the distance from school to her home does she have left to walk?

b. The remaining distance is what fractional part of the distance she has already walked?

63. Josh is 72 inches tall. Ruben is $\frac{9}{10}$ as tall as Josh. John is $\frac{11}{12}$ as tall as Ruben.

a. What is Ruben's height in inches?

b. What fractional part of Josh's height is John?

64. Brendan has a strip of paper that is gray on the front and white on the back. The strip can be divided into three squares of the same size. He folds the paper along the diagonal of the middle square, as shown in the diagram.

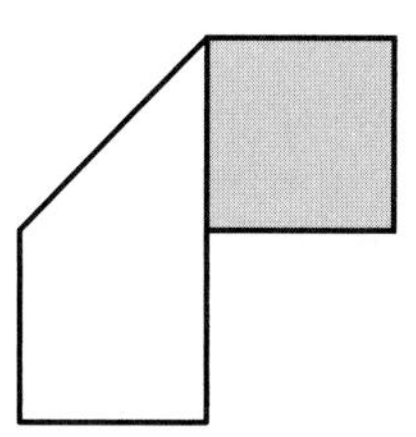

a. What fractional part of the white side of the paper is now showing?

b. What fractional part of the area showing is gray?

1-3 THE IRRATIONAL NUMBERS

We have learned that on the real number line, there is one point for every rational number. We also know that there is an infinite number of rational numbers and, in turn, an infinite number of points assigned to these numbers. When we draw a number line, the dots that represent these points appear to be so dense and crowded together that the line appears to be complete. However, there are still points on the real number line that are not associated with rational numbers.

The Set of Irrational Numbers

Recall that every rational number is a repeating decimal. This includes terminating decimals, where 0 is repeated. There are infinitely many decimals, however, that do not terminate and are nonrepeating. Here is one example of such a decimal:

$$0.03003000300003000003\ldots$$

Observe that, in this number, only the digits 0 and 3 appear. First, there is a zero to the left of the decimal point, and a 0 to the right of the decimal point, then a 3 followed by two 0's, a 3 followed by three 0's, a 3 followed by four 0's and so on. If this pattern of digits continues with the number of 0's between the 3's always increasing, the number is not a repeating decimal. It is not a rational number.

An infinite, nonrepeating decimal is an **irrational number**. An irrational number cannot be expressed in the form $\frac{a}{b}$ where a and b are integers and $b \neq 0$.

When writing an irrational number, we use three dots (. . .) after a series of digits to indicate that the number does not terminate. The dots do *not* indicate a pattern, and no raised bar can be placed over any digits. In an irrational number, we are never certain what the next digit will be when these dots (. . .) are used.

In this section, we will see more examples of irrational numbers, both positive and negative. First, however, we need to review a few terms you learned in earlier mathematics courses.

Squares and Square Roots

To **square** a number means to multiply the number by itself. For example:

The square of 3 is 9. $3^2 = 3 \cdot 3 = 9$

The square of 4 is 16. $4^2 = 4 \cdot 4 = 16$

Calculators have a special key, x^2, that will square a number.

ENTER: 5 x^2 ENTER

DISPLAY:

```
5²
                25
```

To find a **square root** of a number means to find a number that, when multiplied by itself, gives the value under the **radical sign**, $\sqrt{\ }$. For example:

$\sqrt{9} = 3$ A square root of 9 equals 3 because $3 \cdot 3 = 9$.

$\sqrt{16} = 4$ A square root of 16 equals 4 because $4 \cdot 4 = 16$.

Calculators also have a key, $\sqrt{\ }$, that will display the square root of a number. This key is often the second function of the x^2 key. For example:

ENTER: 2nd $\sqrt{\ }$ 25 ENTER

DISPLAY:

```
√(25
                 5
```

When the square root key is pressed, the calculator displays a square root sign followed by a left parenthesis. It is not necessary to close the parentheses if the entire expression that follows is under the radical sign. However, when other numbers and operations follow that are not part of the expression under the radical sign, the right parenthesis must be entered to indicate the end of the radical expression.

More Irrational Numbers

When a square measures 1 unit on every side, its diagonal measures $\sqrt{2}$ units. You can use a ruler to measure the diagonal and then show the placement of $\sqrt{2}$ on a number line.

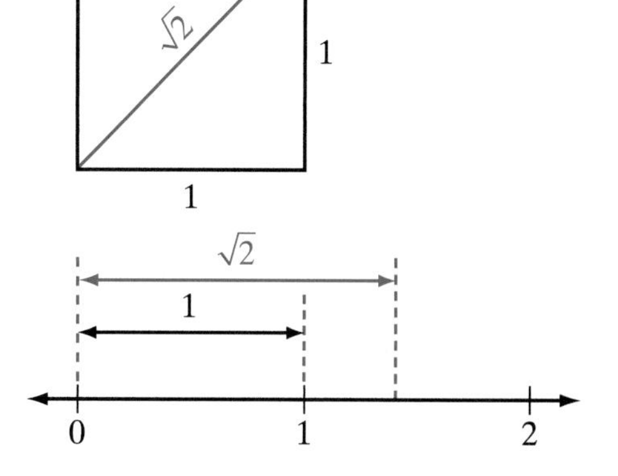

What is the value of $\sqrt{2}$? Can we find a decimal number that, when multiplied by itself, equals 2? We expect $\sqrt{2}$ to be somewhere between 1 and 2.

Use a calculator to find the value.

ENTER: **2nd** **√** 2 **ENTER**

DISPLAY:

Check this answer by multiplying:

$$1.414213562 \times 1.414213562 = 1.999999999, \text{ too small.}$$
$$1.414213563 \times 1.414213563 = 2.000000002, \text{ too large.}$$

Note that if, instead of rewriting the digits displayed on the screen, we square the answer using **2nd** **ANS**, the graphing calculator will display 2 because in that case it uses the value of $\sqrt{2}$ that is stored in the memory of the calculator, which has more decimal places than are displayed on the screen.

No matter how many digits can be displayed on a calculator, no terminating decimal, nor any repeating decimal, can be found for $\sqrt{2}$ because

► **$\sqrt{2}$ is an irrational number.**

In the same way, an infinite number of square roots are irrational numbers, for example:

$$\sqrt{3} \qquad \sqrt{5} \qquad \sqrt{3.2} \qquad \sqrt{0.1} \qquad -\sqrt{2} \qquad -\sqrt{3}$$

The values displayed on a calculator for irrational square roots are called rational approximations. A **rational approximation** for an irrational number is a rational number that is *close to*, but *not equal to*, the value of the irrational number.

The symbol $\approx$ means *approximately equal to*. Therefore, it is not correct to write $\sqrt{3} = 1.732$, but it is correct to write $\sqrt{3} \approx 1.732$.

Another interesting number that you have encountered in earlier courses is π, read as "pi." Recall that π equals the circumference of a circle divided by its diameter, or $\pi = \frac{C}{d}$.

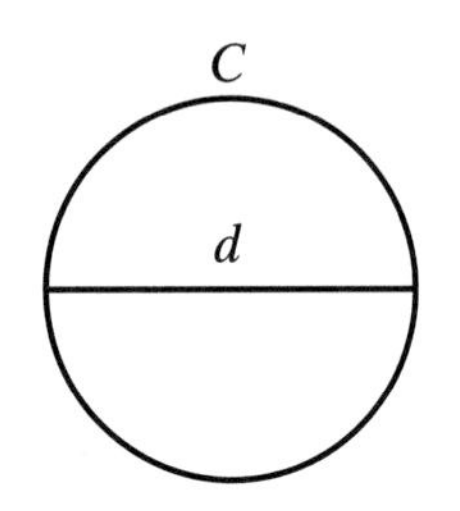

▶ **π is an irrational number.**

There are many rational approximations for π, including:

$$\pi \approx 3.14 \qquad \pi \approx \frac{22}{7} \qquad \pi \approx 3.1416$$

If π is doubled, or divided by two, or if a rational number is added to or subtracted from π, the result is again an irrational number. There are infinitely many such irrational numbers, for example:

$$2\pi \qquad \frac{\pi}{2} \qquad \pi + 7 \qquad \pi - 3$$

Approximation

Scientific calculators have a key that, when pressed, will place in the display a rational approximation for π that is more accurate than the ones given above. On a graphing calculator, when the [π] key is accessed, the screen shows the symbol π but a rational approximation is used in the calculation.

On a graphing calculator:

ENTER: [2nd] [π] [ENTER]

DISPLAY:

```
π
            3.141592654
```

With a calculator, however, you must be careful how you interpret and use the information given in the display. At times, the value shown is exact, but, more often, displays that fill the screen are rational approximations. To write a rational approximation to a given number of decimal places, **round** the number.

Procedure

To round to a given decimal place:

1. Look at the digit in the place at the immediate right of the decimal place to which you are rounding the number.
2. If the digit being examined is *less than* 5, drop that digit and all digits to the right. (Example: 3.1415927 . . . rounded to two decimal places is 3.14 because the digit in the third decimal place, 1, is less than 5.)
3. If the digit being examined is *greater than or equal to* 5, add 1 to the digit in the place to which you are rounding and then drop all digits to the right. (Example: 3.1415927 . . . rounded to four decimal places is 3.1416 because the digit in the fifth decimal place, 9, is greater than 5.)

EXAMPLE 1

True or False: $\sqrt{5} + \sqrt{5} = \sqrt{10}$? Explain why.

Solution Use a calculator.

ENTER: [2nd] [√] 5 [)] [+] [2nd] [√] 5 [ENTER]

DISPLAY:

```
√(5)+√(5
          4.472135955
```

ENTER: [2nd] [√] 10 [ENTER]

DISPLAY:

```
√(10
          3.16227766
```

Use these rational approximations to conclude that the values are not equal.

Answer False. $\sqrt{5} + \sqrt{5} \neq \sqrt{10}$ because $\sqrt{5} + \sqrt{5} > 4$ while $\sqrt{10} < 4$.

EXAMPLE 2

Find a rational approximation for each irrational number, to the *nearest hundredth*. **a.** $\sqrt{3}$ **b.** $\sqrt{0.1}$

Solution Use a calculator.

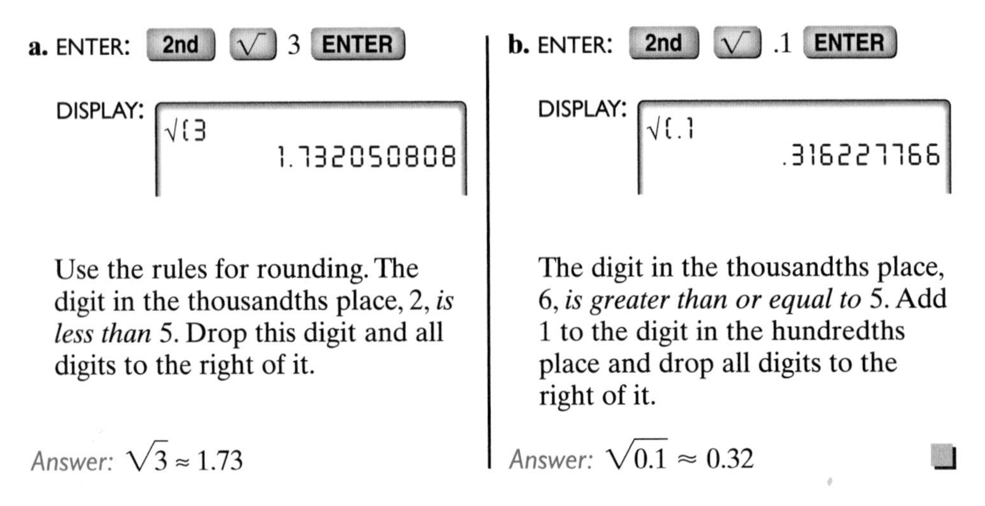

a. Use the rules for rounding. The digit in the thousandths place, 2, *is less than* 5. Drop this digit and all digits to the right of it.

Answer: $\sqrt{3} \approx 1.73$

b. The digit in the thousandths place, 6, *is greater than or equal to* 5. Add 1 to the digit in the hundredths place and drop all digits to the right of it.

Answer: $\sqrt{0.1} \approx 0.32$

EXAMPLE 3

The circumference C of a circle with a diameter d is found by using the formula $C = \pi d$.

a. Find the exact circumference of a circle whose diameter is 8.

b. Find, to the nearest thousandth, a rational approximation of the circumference of this circle.

Solution **a.** $C = \pi d$

$C = \pi \cdot 8$ or 8π

b. Use a calculator.

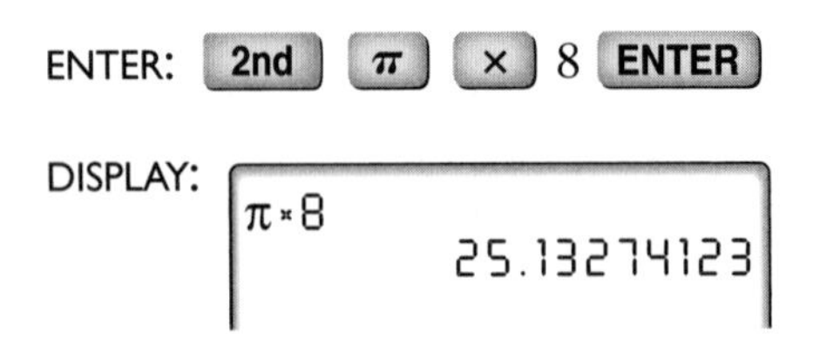

Round the number in the display to three decimal places: 25.133.

Answers **a.** 8π is the *exact* circumference, an *irrational* number.

b. 25.133 is the *rational approximation* of the circumference, to the *nearest thousandth.*

EXAMPLE 4

Which of the following four numbers is an irrational number? In each case, the ... that follows the last digit indicates that the established pattern of digits repeats.

(1) 0.12
(2) 0.12121212 ...
(3) 0.12111111 ...
(4) 0.12112111211112 ...

Solution Each of the first three numbers is a repeating decimal. Choice (1) is a terminating decimal that can be written with a repeating zero. Choice (2) repeats the pair of digits 12 from the first decimal place and choice (3) repeats the digit 1 from the third decimal place. In choice (4), the pattern increases the number of times the digit 1 occurs after each 2. Therefore, (4) is not a repeating decimal and is irrational.

Answer (4) 0.12112111211112 ... is irrational.

EXERCISES

Writing About Mathematics

1. Erika knows that the sum of two rational numbers is always a rational number. Therefore, she concludes that the sum of two irrational numbers is always an irrational number. Give some examples that will convince Erika that she is wrong.

2. Carlos said that 3.14 is a better approximation for π than $\frac{22}{7}$. Do you agree with Carlos? Explain your answer.

Developing Skills

In 3–22, tell whether each number is rational or irrational.

3. 0.36
4. 0.36363636 ...
5. $0.3\overline{6}$
6. 0.363363336 ...
7. $\sqrt{8}$
8. 10π
9. 0.12131415 ...
10. $\sqrt{16}$
11. 0.989989998 ...
12. 0.725
13. $\sqrt{121}$
14. $\pi + 30$
15. -5.28
16. 0.14141414 ...
17. $-\sqrt{5}$
18. $-\pi$
19. $\sqrt{48}$
20. $\sqrt{49}$
21. $0.246\overline{82}$
22. $\pi - 2$

23. Determine which of the following irrational numbers are between 1 and 4.

(1) $\frac{\pi}{2}$ (2) $\sqrt{5}$ (3) $\frac{\sqrt{2}}{4}$ (4) $\sqrt{11}$ (5) $-\sqrt{3}$

In 24–43 write the rational approximation of each given number:

a. as shown on a calculator display,

b. rounded to the *nearest thousandth* (three decimal places)

c. rounded to the *nearest hundredth* (two decimal places).

24. $\sqrt{5}$ **25.** $\sqrt{7}$ **26.** $\sqrt{19}$ **27.** $\sqrt{75}$ **28.** $\sqrt{63}$

29. $\sqrt{90}$ **30.** $-\sqrt{14}$ **31.** $-\sqrt{22}$ **32.** $\sqrt{0.2}$ **33.** $\sqrt{0.3}$

34. $\sqrt{12}$ **35.** $\sqrt{16}$ **36.** $\frac{\sqrt{17}}{3}$ **37.** $\frac{\pi}{3}$ **38.** $\sqrt{0.17}$

39. $-\sqrt{82}$ **40.** $\sqrt{6.5}$ **41.** $-\sqrt{55}$ **42.** $\sqrt{1,732}$ **43.** $\sqrt{241}$

44. A rational approximation for $\sqrt{3}$ is 1.732.

a. Multiply 1.732 by 1.732. **b.** Which is larger, $\sqrt{3}$ or 1.732?

45. a. Find $(3.162)^2$. **b.** Find $(3.163)^2$.

c. Is 3.162 or 3.163 a better approximation for $\sqrt{10}$? Explain why.

In 46–50, use the formula $C = \pi d$ to find, in each case, the circumference C of a circle when the diameter d is given. **a.** Write the exact value of C by using an irrational number. **b.** Find a rational approximation of C to the nearest hundredth.

46. $d = 7$ **47.** $d = 15$ **48.** $d = 72$ **49.** $d = \frac{1}{2}$ **50.** $d = 3\frac{1}{3}$

51. True or False: $\sqrt{4} + \sqrt{4} = \sqrt{8}$? Explain why or why not.

52. True or False: $\sqrt{18} + \sqrt{18} = \sqrt{36}$? Explain why or why not.

Hands-On Activity

Cut two squares, each of which measures 1 foot on each side. Cut each square along a diagonal (the line joining opposite corners of the square). Arrange the four pieces of the squares into a larger square.

a. What is the area of each of the two squares that you cut out?

b. What is the area of the larger square formed by using the pieces of the smaller squares?

c. What should be the length of each side of the larger square? Is this length rational or irrational?

d. Measure the length of each side of the larger square? Is this measurement rational or irrational?

e. Should the answers to parts **c** and **d** be the same? Explain your answer.

1-4 THE REAL NUMBERS

Recall that rational numbers can be written as repeating decimals, and that irrational numbers are decimals that do not repeat. Taken together, rational and irrational numbers make up the set of all numbers that can be written as decimals.

The **set of real numbers** is the set that consists of all rational numbers and all irrational numbers.

The accompanying diagram shows that the rational numbers are a subset of the real numbers, and the irrational numbers are also a subset of the real numbers. Notice, however, that the rationals and the irrationals take up different spaces in the diagram because they have no numbers in common. Together, these two sets of numbers form the real numbers. The cross-hatched shaded portion in the diagram contains no real numbers. The cross-hatched shading indicates that no other numbers except the rationals and irrationals are real numbers.

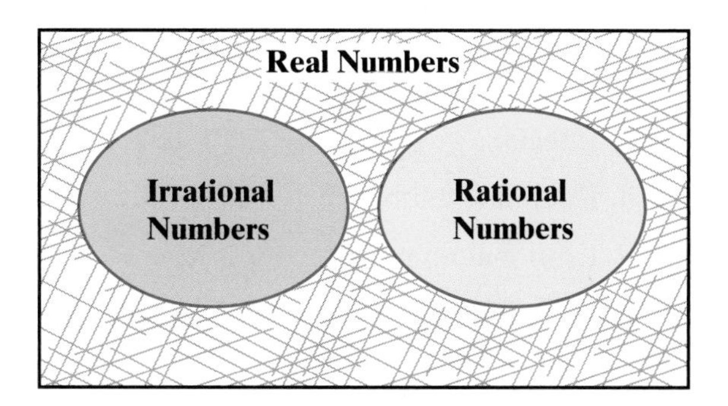

We have seen that there are an infinite number of rational numbers and an infinite number of irrationals. For every rational number, there is a corresponding point on the number line, and, for every irrational number, there is a corresponding point on the number line. All of these points, taken together, make up the **real number line**. Since there are no more holes in this line, we say that the real number line is now complete. The **completeness property of real numbers** may be stated as follows:

► **Every point on the real number line corresponds to a real number, and every real number corresponds to a point on the real number line.**

Ordering Real Numbers

There are two ways in which we can order real numbers:

1. Use a *number line*. On the standard horizontal real number line, the graph of the greater number is always to the right of the graph of the smaller number.

2. Use *decimals*. Given any two real numbers that are not equal, we can express them in decimal form (even using rational approximations) to see which is greater.

EXAMPLE 1

The number line that was first seen in Section 1-1 is repeated below.

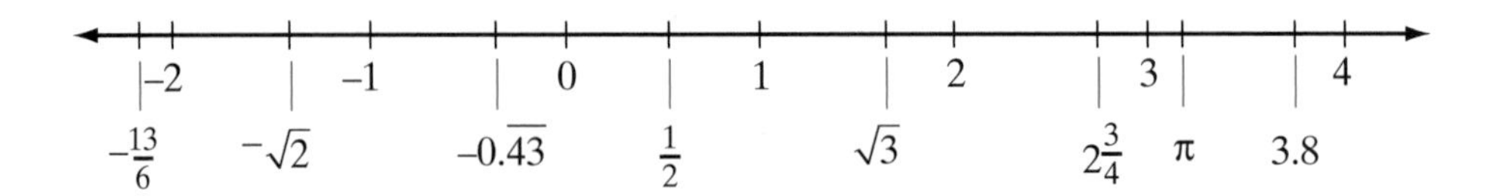

Of the numbers shown here, tell which are: **a.** counting numbers **b.** whole numbers **c.** integers **d.** rational numbers **e.** irrational numbers **f.** real numbers.

Solution **a.** Counting numbers: $1, 2, 3, 4$
b. Whole numbers: $0, 1, 2, 3, 4$
c. Integers: $-2, -1, 0, 1, 2, 3, 4$
d. Rational Numbers: $-\frac{13}{6}, -2, -1, -0.\overline{43}, 0, \frac{1}{2}, 1, 2, 2\frac{3}{4}, 3, 3.8, 4$
e. Irrational numbers $-\sqrt{2}, \sqrt{3}, \pi$
f. Real numbers: All: $-\frac{13}{6}, -2, -\sqrt{2}, -1, -0.\overline{43}, 0, \frac{1}{2}, 1, \sqrt{3}, 2, 2\frac{3}{4}, 3, \pi, 3.8, 4$

EXAMPLE 2

Order these real numbers from least to greatest, using the symbol <.

$0.3 \qquad \sqrt{0.3} \qquad 0.\overline{3}$

Solution **STEP 1.** Write each real number in decimal form:

$$0.3 = 0.3000000\ldots$$

$$\sqrt{0.3} \approx 0.547722575 \text{ (a rational approximation, displayed on a calculator)}$$

$$0.\overline{3} = 0.3333333\ldots$$

STEP 2. Compare these decimals: $0.3000000\ldots < 0.3333333\ldots < 0.547722575$

STEP 3. Replace each decimal with the number in its original form:

$$0.3 < 0.\overline{3} < \sqrt{0.3}$$

Answer $0.3 < 0.\overline{3} < \sqrt{0.3}$

EXERCISES

Writing About Mathematics

1. There are fewer than 6 persons in my family.
The board is less than 6 feet long.
Each of the given statements can be designated by the inequality $x < 6$. How are the numbers that make the first statement true different from those that make the second statement true? How are they the same?

2. Dell said that it is impossible to decide whether π is larger or smaller than $\sqrt{10}$ because the calculator gives only rational approximations for these numbers. Do you agree with Dell? Explain.

3. The decimal form of a real number consists of two digits that repeat for the first one-hundred decimal places. The digits in the places that follow the one-hundredth decimal place are random, form no pattern, and do not terminate. Is the number rational or irrational? Explain.

Developing Skills

4. Twelve numbers have been placed on a number line as shown here.

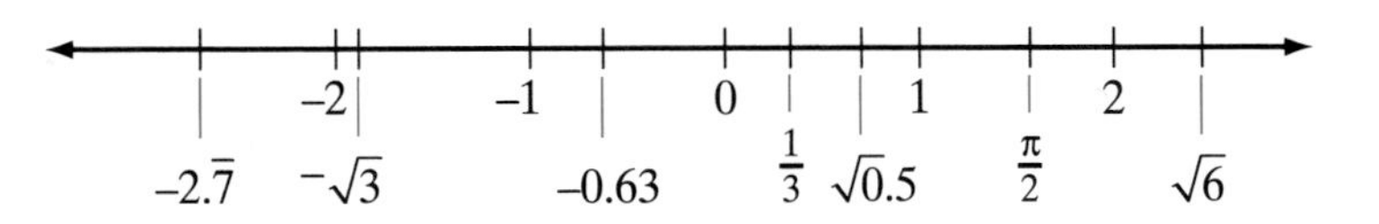

Of these numbers, tell which are:

a. counting numbers **b.** whole numbers **c.** integers
d. rational numbers **e.** irrational numbers **f.** real numbers

5. Given the following series of numbers:

$\sqrt{0}, \sqrt{1}, \sqrt{2}, \sqrt{3}, \sqrt{4}, \sqrt{5}, \sqrt{6}, \sqrt{7}, \sqrt{8}, \sqrt{9}$

Of these ten numbers, tell which is (are): **a.** rational **b.** irrational **c.** real

6. Given the following series of numbers: $\pi, 2\pi, 3\pi, 4\pi, 5\pi$
Of these five numbers, tell which is (are): **a.** rational **b.** irrational **c.** real

In 7–18, determine, for each pair, which is the greater number.

7. 2 or 2.5 **8.** 8 or $\sqrt{8}$ **9.** $0.\overline{2}$ or 0.22 **10.** $0.\overline{2}$ or 0.23

11. 0.7 or $0.\overline{7}$ **12.** -5.6 or -5.9 **13.** 0.43 or -0.431 **14.** $0.\overline{21}$ or $0.\overline{2}$

15. 3.14 or π **16.** 0.5 or $\sqrt{0.5}$ **17.** $\sqrt{2}$ or 1.414 **18.** π or $\frac{22}{7}$.

In 19–24, order the numbers in each group from least to greatest by using the symbol <.

19. $0.202, 0.\overline{2}, 0.2022$ **20.** $0.\overline{4}, 0.45, 0.4499$ **21.** $0.\overline{67}, 0.\overline{6}, 0.\overline{667}$

22. $-\sqrt{2}, -\sqrt{3}, -1.5$ **23.** $0.5, 0.\overline{5}, \sqrt{0.3}$ **24.** $\pi, \sqrt{10}, 3.\overline{15}$

In 25–34, tell whether each statement is true or false.

25. Every real number is a rational number.

26. Every rational number is a real number.

27. Every irrational number is a real number.

28. Every real number is an irrational number.

29. Every rational number corresponds to a point on the real number line.

30. Every point on the real number line corresponds to a rational number.

31. Every irrational number corresponds to a point on the real number line.

32. Every point on the real number line corresponds to an irrational number.

33. Some numbers are both rational and irrational.

34. Every repeating decimal corresponds to a point on the real number line.

Hands-On Activity

a. Using a cloth or paper tape measure, find, as accurately as you can, the distance across and the distance around the top of a can or other object that has a circular top. If you do not have a tape measure, fit a narrow strip of paper around the circular edge and measure the length of the strip with a yardstick.

b. Divide the measure of the circumference, the distance around the circular top, by the measure of the diameter, the distance across the circular top at its center.

c. Repeat steps **a** and **b** for other circular objects and compare the quotients obtained in step **b**. Compare your results from step **b** with those of other members of your class. What conclusions can you draw?

1-5 NUMBERS AS MEASUREMENTS

In previous sections, we defined the subsets of the real numbers. When we use a counting number to identify the number of students in a class or the number of cars in the parking lot, these numbers are exact. However, to find the length of a block of wood, we must use a ruler, tape measure, or some other measuring instrument. The length that we find is dependent upon the instrument we use to measure and the care with which we make the measurement.

For example, in the diagram, a block of wood is placed along the edge of a ruler that is marked in tenths of an inch. We might say that the block of wood is 2.7 inches in length but is this measure exact?

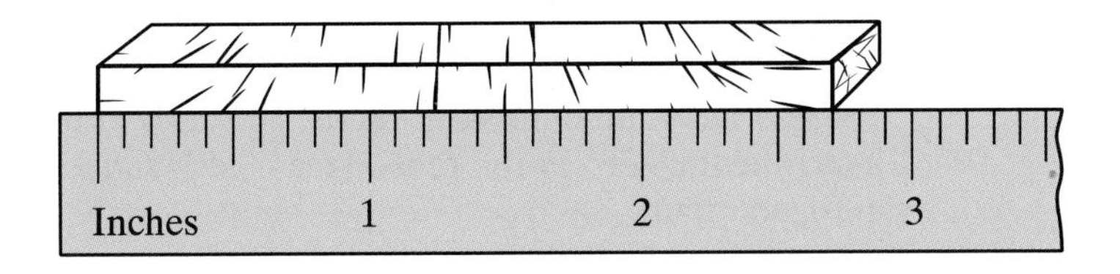

All measurements are approximate. When we say that the length of the block of wood is 2.7 inches, we mean that it is closer to 2.7 inches than it is to 2.6 inches or to 2.8 inches. Therefore, the true measure of the block of wood whose length is given as 2.7 inches is between 2.65 and 2.75 inches. In other words, the true measure is less than 0.05 inches from 2.7 and can be written as 2.7 ± 0.05 inches. The value 0.05 is called the **greatest possible error (GPE)** of measurement and is half of the place value of the last digit.

Significant Digits

The accuracy of measurement is often indicated in terms of the number of significant digits. **Significant digits** are those digits used to determine the measure and excludes those zeros that are used as place holders at the beginning of a decimal fraction and at the end of an integer.

Rules for Determining Significant Digits

RULE 1 All nonzero digits are significant.

135.6 has four significant digits. All digits are significant.

RULE 2 All zeros between significant digits are significant.

130.6 has four significant digits. The zero is significant because it is between significant digits.

RULE 3 All zeros at the end of a decimal fraction are significant.

135.000 has six significant digits. The three zeros at the end of the decimal fraction are significant.

RULE 4 Zeros that precede the first nonzero digit in a decimal fraction are not significant.

0.00424 has three significant digits. The zeros that precede the nonzero digits in the decimal fraction are placeholders and are not significant.

RULE 5 Zeros at the end of an integer may or may nor be significant. Sometimes a dot is placed over a zero if it is significant.

4,500 has two significant digits. Neither zero is significant.

$4{,}5\dot{0}0$ has three significant digits. The zero in the tens place is significant but the zero in the ones place is not.

$4{,}50\dot{0}$ has four significant digits. The zero in the ones place is significant. Therefore, the zero in the tens place is also significant because it is between significant digits.

In any problem that uses measurement, the rules of greatest possible error and significant digits are used to determine how the answer should be stated. We can apply these rules to problems of perimeter and area. Recall the formulas for perimeter and area that you learned in previous courses.

Let P represent the perimeter of a polygon, C the circumference of a circle, and A the area of any geometric figure.

Triangle	$P = a + b + c$	$A = \frac{1}{2}bh$
Rectangle	$P = 2l + 2w$	$A = lw$
Square	$P = 4s$	$A = s^2$
Circle	$C = \pi d$ or $C = 2\pi r$	$A = \pi r^2$

Precision

The **precision** of a measurement is the place value of the last significant digit in the number. The greatest possible error of a measurement is one-half the place value of the last significant digit. In the measurement 4,500 feet, the last significant digit is in the hundreds place. Therefore, the greatest possible error is $\frac{1}{2} \times 100 = 50$. We can write the measurement as $4{,}500 \pm 50$ feet. One number is said to be more precise than another if the place value of its last significant digit is smaller. For example, 3.40 is more precise than 3.4 because 3.40 is correct to the nearest hundredth and 3.4 is correct to the nearest tenth.

When measures are added, the sum can be no more precise than the least precise number of the given values. For example, how should the perimeter of a triangle be stated if the measures of the sides are 34.2 inches, 27.52 inches, and 29 inches?

$$P = a + b + c$$

$$P = 34.2 + 27.52 + 29 = 90.72$$

Since the least precise measure is 29 which is precise to the nearest integer, the perimeter of the triangle should be given to the nearest integer as 91 inches.

Accuracy

The **accuracy** of a measure is the number of significant digits in the measure. One number is said to be more accurate than another if it has a larger number of significant digits. For example, 0.235 is more accurate than 0.035 because 0.235 has three significant digits and 0.035 has two, but 235 and 0.235 have the same degree of accuracy because they both have three significant digits.

When measures are multiplied, the product can be no more accurate than the least accurate of the given values. For example, how should the area of a triangle be stated if the base measures 0.52 meters and the height measures 0.426 meters?

$$A = \tfrac{1}{2}bh$$
$$A = \tfrac{1}{2}(0.52)(0.426) = 0.5(0.52)(0.426) = 0.11076$$

Since the less accurate measure is 0.52, which has two significant digits, the area should be written with two significant digits as 0.11 square meters. Note that the $\frac{1}{2}$ or 0.5 is not a measurement but an exact value that has been determined by counting or by reasoning and therefore is not used to determine the accuracy of the answer.

One last important note: when doing multi-step calculations, *make sure to keep at least one more significant digit in intermediate results than is needed in the final answer*. For example, if a computation requires three significant digits, then use at least four significant digits in your calculations. Otherwise, you may encounter what is known as **round-off error**, which is the phenomena that occurs when you discard information contained in the extra digit, skewing your calculations.

In this text, you will often be asked to find the answer to an exercise in which the given numbers are thought of as exact values and the answers are given as exact values. However, in certain problems that model practical applications, when the given data are approximate measurements, you may be asked to use the precision or accuracy of the data to determine how the answer should be stated.

EXAMPLE 1

State the precision and accuracy of each of the following measures.

a. 5.042 cm **b.** 12.0 ft **c.** 93,000,000 mi

Solution

	Precision	*Accuracy*
a. 5.042 cm	thousandths	4 significant digits
b. 12.0 ft	tenths	3 significant digits
c. 93,000,000 mi	millions	2 significant digits

EXAMPLE 2

Of the measurements 125 feet and 6.4 feet, **a.** which is the more precise? **b.** which is the more accurate?

Solution The measurement 125 feet is correct to the nearest foot, has an error of 0.5 feet, and has three significant digits.

The measurement 6.4 feet is correct to the nearest tenth of a foot, has an error of 0.05 feet, and has two significant digits.

Answers **a.** The measure 6.4 feet is more precise because it has the smaller error.
b. The measure 125 feet is more accurate because it has the larger number of significant digits.

EXAMPLE 3

The length of a rectangle is 24.3 centimeters and its width is 18.76 centimeters. Using the correct number of significant digits in the answer, express **a.** the perimeter **b.** the area.

Solution **a.** Use the formula for the perimeter of a rectangle.

$$P = 2l + 2w$$
$$P = 2(24.3) + 2(18.76)$$
$$P = 86.12$$

Perimeter is a sum since $2l$ means $l + l$ and $2w$ means $w + w$. The answer should be no more precise than the least precise measurement. The least precise measurement is 24.3, given to the nearest tenth. The perimeter should be written to the nearest tenth as 86.1 centimeters.

b. To find the area of a rectangle, multiply the length by the width.

$$A = lw$$
$$A = (24.3)(18.76)$$
$$A = 455.868$$

Area is a product and the answer should be no more accurate than the least accurate of the given dimensions. Since there are three significant digits in 24.3 and four significant digits in 18.76, there should be three significant digits in the answer. Therefore, the area should be written as 456 square centimeters.

Answers **a.** 86.1 cm **b.** 456 sq cm

EXERCISES

Writing about Mathematics

1. If 12.5 = 12.50, explain why a measure of 12.50 inches is more accurate and more precise than a measurement of 12.5 inches.

2. A circular track has a radius of 63 meters. Mario rides his bicycle around the track 10 times. Mario multiplied the radius of the track by 2π to find the circumference of the track. He said that he rode his bicycle 4.0 kilometers. Olga said that it would be more correct to say that he rode his bicycle 4 kilometers. Who is correct? Explain your answer.

Developing Skills

In 3–10, for each of the given measurements, find **a.** the accuracy **b.** the precision **c.** the error.

3. 24 in. **4.** 5.05 cm **5.** 2,400 ft **6.** 454 lb

7. 0.0012 kg **8.** 1.04 yd **9.** 1.005 m **10.** 900 mi

In 11–14, for each of the following pairs, select the measure that is **a.** the more precise **b.** the more accurate.

11. 57 in. and 4,250 in. **12.** 2.50 ft and 2.5 ft

13. 0.0003 g and 32 g **14.** 500 cm and 0.055 m

Applying Skills

In 15–18, express each answer to the correct number of significant digits.

15. Alicia made a square pen for her dog using 72.4 feet of fencing.

a. What is the length of each side of the pen?

b. What is the area of the pen?

16. Corinthia needed 328 feet of fencing to enclose her rectangular garden. The length of the garden is 105 feet.

a. Find the width of the garden.

b. Find the area of the garden.

17. Brittany is making a circular tablecloth. The diameter of the tablecloth is 10.5 inches. How much lace will she need to put along the edge of the tablecloth?

18. The label on a can of tomatoes is a rectangle whose length is the circumference of the can and whose width is the height of the can. If a can has a diameter of 7.5 centimeters and a height of 10.5 centimeters, what is the area of the label?

CHAPTER SUMMARY

A **set** is a collection of distinct objects or elements.

The **counting numbers** or **natural numbers** are $\{1, 2, 3, 4, \ldots\}$.

The **whole numbers** are $\{0, 1, 2, 3, 4, \ldots\}$.

The **integers** are $\{\ldots, -4, -3, -2, -1, 0, 1, 2, 3, 4, \ldots\}$.

These sets of numbers form the basis for a **number line**, on which the length of a segment from 0 to 1 is called the **unit measure** of the line.

The **rational numbers** are all numbers that can be expressed in the form $\frac{a}{b}$ where a and b are integers and $b \neq 0$. Every rational number can be expressed as a repeating decimal or as a terminating decimal (which is actually a decimal in which 0 is repeated).

The **irrational numbers** are decimal numbers that do not terminate and do not repeat. On calculators and in the solution of many problems, *rational approximations* are used to show values that are close to, but not equal to, irrational numbers.

The **real numbers** consist of all rational numbers and all irrational numbers taken together. On a **real number line**, every point represents a real number and every real number is represented by a point.

The **precision** of a measurement is determined by the place value of the last significant digit. The **accuracy** of a measurement is determined by the number of significant digits in the measurement.

VOCABULARY

1-1 Mathematics • Real number • Number • Numeral • Counting numbers • Natural numbers • Successor • Whole numbers • Set • Finite set • Digit • Infinite set • Empty set • Null set • Numerical expression • Simplify • Negative numbers • Opposites • Integers • Subset • Number line • Graph • Standard number line • Unit measure • Absolute value • Inequality

1-2 Rational numbers • Everywhere dense • Common fraction • Decimal fraction • Terminating decimal • Repeating decimal • Periodic decimal

1-3 Irrational numbers • Square • Square root • Radical sign • Rational approximation • Pi (π) • Round

1-4 Real numbers • Real number line • Completeness property of real numbers

1-5 Greatest possible error (GPE) • Significant digits • Precision • Accuracy

REVIEW EXERCISES

In 1–5, use a calculator to evaluate each expression and round the result to the *nearest hundredth*.

1. 29.73×14.6 **2.** $38 \div 9$ **3.** 12.23^2 **4.** $\sqrt{216}$ **5.** $\pi \times 12$

6. Order the numbers -5, 3, and -1 using the symbol $>$.

In 7–10, state whether each sentence is true of false.

7. $7 > -8$ **8.** $-7 > |-2|$ **9.** $4 < -8$ **10.** $9 \leq -9$

In 11–16, write each rational number in the form $\frac{a}{b}$, where a and b are integers and $b \neq 0$.

11. 0.9 **12.** 0.45 **13.** $8\frac{1}{2}$ **14.** 14 **15.** $0.\overline{3}$ **16.** -63

17. Find a rational number between 19.9 and 20.

In 18–22, tell whether each number is rational or irrational.

18. $0.\overline{64}$ **19.** $\sqrt{6}$ **20.** $\sqrt{64}$ **21.** π

22. 0.040040004 . . .

In 23–27, write a rational approximation of each given number: **a.** as shown on a calculator display **b.** rounded to the *nearest hundredth*.

23. $\sqrt{11}$ **24.** $\sqrt{0.7}$ **25.** $\sqrt{905}$ **26.** $\sqrt{1,599}$ **27.** π

In 28–32, determine which is the greater number in each pair.

28. 5 or $\sqrt{20}$ **29.** $|12 - 8|$ or $|12| + |-8|$ **30.** 3.2 or π

31. 0.41 or $0.\overline{4}$ **32.** $0.\overline{12}$ or 0.121

In 33–37, tell whether each statement is true or false.

33. Every integer is a real number.

34. Every rational number is an integer.

35. Every whole number is a counting number.

36. Every irrational number is a real number.

37. Between 0 and 1, there is an infinite number of rational numbers.

38. Draw a number line, showing the graphs of these numbers: 0, 1, 4, -3, -1.5, and π.

In 39 and 40, use the given number line where the letters are equally spaced.

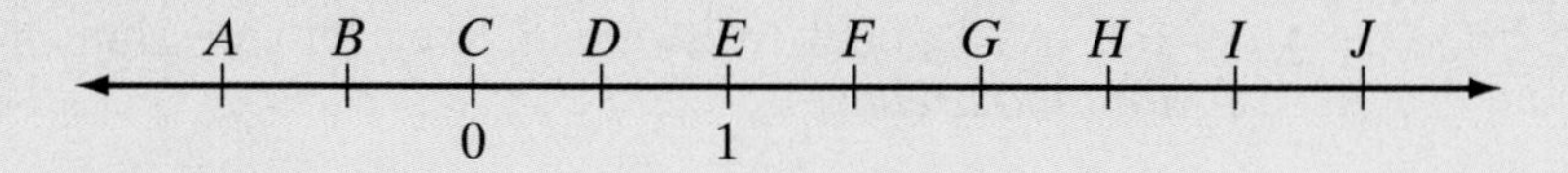

39. Find the real number that corresponds to each point indicated by a letter shown on the number line when $C = 0$ and $E = 1$.

40. Between what two consecutive points on this number line is the graph of:

a. 1.8 **b.** -0.6 **c.** $\sqrt{2}$ **d.** π **e.** $\sqrt{6}$

41. The distance across a circular fountain (the diameter of the fountain) is 445 centimeters. The distance in centimeters around the fountain (the circumference of the fountain) can be found by multiplying 445 by π.

a. Find the circumference of the fountain in centimeters. Round your answer to the nearest ten centimeters.

b. When the circumference is rounded to the nearest ten centimeters, are the zeros significant?

Exploration

Using only the digits 5 and 6, and without using a radical sign:

a. Write an irrational number.

b. Write three irrational numbers that are between 5 and 6 in increasing order.

c. Write three irrational numbers that are between 0.55 and 0.56 in increasing order.

d. Write three irrational numbers that are between $0.5\overline{56}$ and $0.55\overline{6}$ in increasing order.

CHAPTER

2

OPERATIONS AND PROPERTIES

Jesse is fascinated by number relationships and often tries to find special mathematical properties of the five-digit number displayed on the odometer of his car. Today Jesse noticed that the number on the odometer was a palindrome and an even number divisible by 11, with 2 as three of the digits. What was the five-digit reading? (Note: A palindrome is a number, word, or phrase that is the same read left to right as read right to left, such as 57375 or Hannah.)

In this chapter you will review basic operations of arithmetic and their properties. You will also study operations on sets.

CHAPTER TABLE OF CONTENTS

2-1 ORDER OF OPERATIONS

The Four Basic Operations in Arithmetic

Bicycles have two wheels. Bipeds walk on two feet. Biceps are muscles that have two points of origin. Bilingual people can speak two languages. What do these *bi*-words have in common with the following examples?

$$6.3 + 0.9 = 7.2 \qquad 21.4 \times 3 = 64.2$$

$$11\tfrac{3}{7} - 2\tfrac{1}{7} = 9\tfrac{2}{7} \qquad 9 \div 2 = 4\tfrac{1}{2}$$

The prefix *bi*- means "two." In each example above, an operation or rule was followed to replace *two* rational numbers with a *single* rational number. These familiar operations of addition, subtraction, multiplication, and division are called **binary operations**. Each of these operations can be performed with any pair of rational numbers, except that division by zero is meaningless and is not allowed.

In every binary operation, two elements from a set are replaced by exactly one element from the same set. There are some important concepts to remember when working with binary operations:

1. A *set* must be identified, such as the set of whole numbers or the set of rational numbers. When no set is identified, use the set of all real numbers.
2. The *rule* for the binary operation must be clear, such as the rules you know for addition, subtraction, multiplication, and division.
3. The order of the elements is important. Later in this chapter, we will use the notation (a, b) to indicate an *ordered pair* in which a is the first element and b is the second element. For now, be aware that answers may be different depending on which element is first and which is second. Consider subtraction. If 8 is the first element and 5 is the second element, then: $8 - 5 = 3$. But if 5 is the first element and 8 is the second element, then: $5 - 8 = -3$
4. Every problem using a binary operation must have an *answer*, and there must be *only one* answer. We say that each answer is *unique*, meaning there is *one and only one* answer.

DEFINITION

A **binary operation** in a set assigns to every ordered pair of elements from the set a unique answer from the set.

Note that, even when we find the sum of three or more numbers, we still add only two numbers at a time, indicating the binary operation:

$$4 + 9 + 7 = (4 + 9) + 7 = 13 + 7 = 20$$

Factors

When two or more numbers are multiplied to give a certain product, each number is called a **factor** of the product. For example:

- Since $1 \times 16 = 16$, then 1 and 16 are factors of 16.
- Since $2 \times 8 = 16$, then 2 and 8 are factors of 16.
- Since $4 \times 4 = 16$, then 4 is a factor of 16.
- The numbers 1, 2, 4, 8, and 16 are all factors of 16.

Prime Numbers

A **prime number** is a whole number greater than 1 that has no whole number factors other than itself and 1. The first seven prime numbers are 2, 3, 5, 7, 11, 13, 17. Whole numbers greater than 1 that are not prime are called **composite numbers**. Composite numbers have three or more whole number factors. Some examples of composite numbers are 4, 6, 8, 9, 10.

Bases, Exponents, Powers

When the same number appears as a factor many times, we can rewrite the expression using *exponents*. For example, the exponent 2 indicates that the factor appears twice. In the following examples, the repeated factor is called a *base*.

$4 \times 4 = 16$ can be written as $4^2 = 16$.

4^2 is read as "4 squared," or "4 raised to the second power," or "the second power of 4."

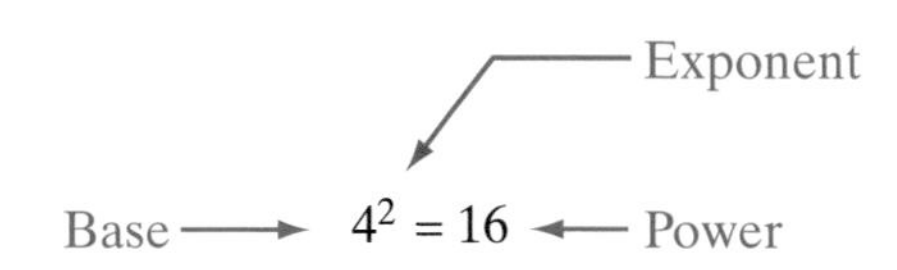

The exponent 3 indicates that a factor is used three times.

$4 \times 4 \times 4 = 64$ can be written as $4^3 = 64$.

4^3 is read as "4 cubed," or "4 raised to the third power," or "the third power of 4."

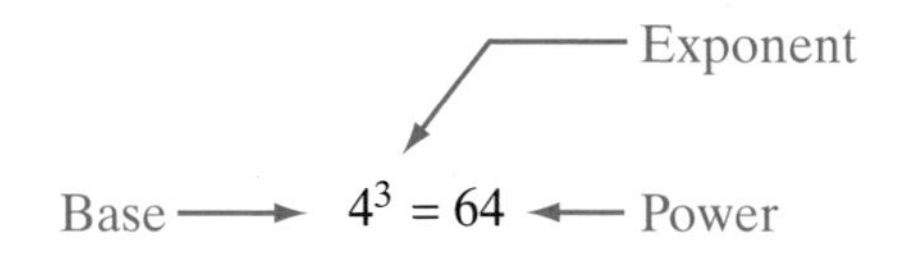

The examples shown above lead to the following definitions:

DEFINITION

A **base** is a number that is used as a factor in the product.

An **exponent** is a number that tells how many times the base is to be used as a factor. The exponent is written, in a smaller size, to the upper right of the base.

A **power** is a number that is a product in which all of its factors are equal.

A number raised to the first power is equal to the number itself, as in $6^1 = 6$. Also, when no exponent is shown, the exponent is 1, as in $9 = 9^1$.

EXAMPLE 1

Compute the value of 4^5.

Solution $4 \times 4 \times 4 \times 4 \times 4 = 1{,}024$

Calculator Solution Use the exponent key, **^**, on a calculator.

ENTER: 4 **^** 5 **ENTER**

DISPLAY:

```
4^5
                1024
```

Answer 1,024

EXAMPLE 2

Find $\left(\frac{2}{3}\right)^3$ **a.** as an exact value **b.** as a rational approximation.

Solution **a.** The exact value is a fraction:

$$\left(\frac{2}{3}\right)^3 = \frac{2}{3} \times \frac{2}{3} \times \frac{2}{3} = \frac{8}{27}$$

b. Use a calculator.

ENTER: **(** 2 **÷** 3 **)** **^** 3 **ENTER**

DISPLAY:

```
(2/3)^3
        .2962962963
```

Note: The exact value is a rational number that can also be written as the repeating decimal $0.\overline{296}$.

Answers **a.** $\frac{8}{27}$ **b.** 0.2962962963

Computations With More Than One Operation

When a numerical expression involves two or more different operations, we need to agree on the order in which they are performed. Consider this example:

$$11 - 3 \times 2$$

Suppose that one person multiplied first.

$$\begin{aligned} 11 - 3 \times 2 &= 11 - 6 \\ &= 5 \end{aligned}$$

Suppose another person subtracted first.

$$\begin{aligned} 11 - 3 \times 2 &= 8 \times 2 \\ &= 16 \end{aligned}$$

Who is correct?

In order that there will be one and only one correct answer to problems like this, mathematicians have agreed to follow this **order of operations**:

1. Simplify powers (terms with exponents).

2. Multiply and divide, from left to right.

3. Add and subtract, from left to right.

Therefore, we multiply before we subtract, and $11 - 3 \times 2 = 11 - 6 = 5$ is correct.

A different problem involving powers is solved in this way:

1. Simplify powers: $5 \times 2^3 + 3 = 5 \times 8 + 3$

2. Multiply and divide: $= 40 + 3$

3. Add and subtract: $= 43$

Expressions with Grouping Symbols

In mathematics, **parentheses ( )** act as grouping symbols, giving different meanings to expressions. For example, $(4 \times 6) + 7$ means "add 7 to the product of 4 and 6," while $4 \times (6 + 7)$ means "multiply the sum of 6 and 7 by 4."

When simplifying any numerical expression, always perform the operations within parentheses first.

$$(4 \times 6) + 7 = 24 + 7 = 31 \qquad 4 \times (6 + 7) = 4 \times 13 = 52$$

Besides parentheses, other symbols are used to indicate grouping, such as **brackets [ ]**. The expressions $2(5 + 9)$ and $2[5 + 9]$ have the same meaning: 2 is multiplied by the sum of 5 and 9. A *bar*, or *fraction line*, also acts as a symbol of grouping, telling us to perform the operations in the numerator and/or denominator first.

$$\frac{20 - 8}{3} = \frac{12}{3} = 4 \qquad \frac{6}{3 + 1} = \frac{6}{4} = \frac{3}{2} = 1\tfrac{1}{2}$$

However, when entering expressions such as these into a calculator, the line of the fraction is usually entered as a division and a numerator or denominator that involves an operation must be enclosed in parentheses.

ENTER: (20 − 8) ÷ 3 ENTER

DISPLAY: (20−8)/3
4

ENTER: 6 ÷ (3 ÷ 1) ENTER

DISPLAY: 6/(3+1)
1.5

When there are *two or more* grouping symbols in an expression, we perform the operations on the numbers in the *innermost symbol first.* For example:

$$
\begin{aligned}
&5 + 2[6 + (3 - 1)^3] \\
&= 5 + 2[6 + 2^3] \\
&= 5 + 2[6 + 8] \\
&= 5 + 2[14] \\
&= 5 + 28 \\
&= 33
\end{aligned}
$$

Procedure

To simplify a numerical expression, follow the correct order of operations:

1. Simplify any numerical expressions within parentheses or within other grouping symbols, starting with the innermost.
2. Simplify any powers.
3. Do all multiplications and divisions in order from left to right.
4. Do all additions and subtractions in order from left to right.

EXAMPLE 3

Simplify the numerical expression $80 - 4(7 - 5)$.

Solution Remember that, in the given expression, $4(7 - 5)$ means 4 *times* the value in the parentheses.

How to Proceed

(1) Write the expression:	$80 - 4(7 - 5)$
(2) Simplify the value within the parentheses:	$= 80 - 4(2)$
(3) Multiply:	$= 80 - 8$
(4) Subtract:	$= 72$

Calculator Solution ENTER: 80 [−] 4 [(] 7 [−] 5 [)] [ENTER]

DISPLAY:

```
80-4(7-5)
                72
```

Answer 72

EXERCISES

Writing About Mathematics

1. Explain why 2 is the only even prime.

2. Delia knows that every number except 2 that ends in a multiple of 2 is composite. Therefore, she concludes that every number except 3 that ends in a multiple of 3 is composite. Is Delia correct? Explain how you know.

Developing Skills

In 3–10, state the meaning of each expression in part **a** and in part **b**, and simplify the expression in each part.

3. a. $20 + (6 + 1)$ **b.** $20 + 6 + 1$

4. a. $18 - (4 + 3)$ **b.** $18 - 4 + 3$

5. a. $12 - (3 - 0.5)$ **b.** $12 - 3 - 0.5$

6. a. $15 \times (2 + 1)$ **b.** $15 \times 2 + 1$

7. a. $(12 + 8) \div 4$ **b.** $12 + 8 \div 4$

8. a. $48 \div (8 - 4)$ **b.** $48 \div 8 - 4$

9. a. $7 + 5^2$ **b.** $(7 + 5)^2$

10. a. 4×3^2 **b.** $(4 \times 3)^2$

11. Noella said that since the line of a fraction indicates division, $\frac{10 \times 15}{5 \times 3}$ is the same as $10 \times 15 \div 5 \times 3$. Do you agree with Noella? Explain why or why not.

In 12–15: **a.** Find, in each case, the value of the three given powers. **b.** Name, in each case, the expression that has the greatest value.

12. $5^2, 5^3, 5^4$ **13.** $(0.5)^2, (0.5)^3, (0.5)^4$ **14.** $(0.5)^2, (0.6)^2, (0.7)^2$ **15.** $(1.1)^2, (1.2)^2, (1.3)^2$

In 16–23: **a.** List all of the whole numbers that are factors of each of the given numbers. **b.** Is the number prime, composite, or neither?

16. 82 **17.** 101 **18.** 71 **19.** 15

20. 1 **21.** 808 **22.** 67 **23.** 397

Applying Skills

In 24–28, write a numerical expression for each of the following and find its value to answer the question.

24. What is the cost of two chocolate chip and three peanut butter cookies if each cookie costs 28 cents?

25. What is the cost of two chocolate chip cookies that cost 30 cents each and three peanut butter cookies that cost 25 cents each?

26. How many miles did Ms. McCarthy travel if she drove 30 miles per hour for $\frac{3}{4}$ hour and 55 miles per hour for $1\frac{1}{2}$ hours?

27. What is the cost of two pens at \$0.38 each and three notebooks at \$0.69 each?

28. What is the cost of five pens at \$0.29 each and three notebooks at \$0.75 each if ordered from a mail order company that adds \$1.75 in postage and handling charges?

In 29–30, use a calculator to find each answer.

29. The value of \$1 invested at 6% for 20 years is equal to $(1.06)^{20}$. Find, to the nearest cent, the value of this investment after 20 years.

30. The value of \$1 invested at 8% for n years is equal to $(1.08)^n$. How many years will be required for \$1 invested at 8% to double in value? (*Hint:* Guess at values of n to find the value for which $(1.08)^n$ is closest to 2.00.)

31. In each box insert an operational symbol +, –, ×, ÷, and then insert parentheses if needed to make each of the following statements true.

a. 3 □ 2 □ 1 = 4 **b.** 1 □ 3 □ 1 = 4 **c.** 1 □ 2 □ 3 □ 4 = 5

d. 4 □ 3 □ 2 □ 1 = 5 **e.** 6 □ 6 □ 6 □ 6 = 5 **f.** 6 □ 6 □ 6 □ 6 = 6

2-2 PROPERTIES OF OPERATIONS

When numbers behave in a certain way for an operation, we describe this behavior as a **property**. You are familiar with these operations from your study of arithmetic. As we examine the properties of operations, no proofs are given, but the examples will help you to see that these properties make sense and to identify the sets of numbers for which they are true.

The Property of Closure

A set is said to be **closed** under a binary operation when every pair of elements from the set, under the given operation, yields an element from that set.

1. Add any two numbers.

$23 + 11 = 34$	The sum of two whole numbers is a whole number.
$7.8 + 4.8 = 12.6$	The sum of two rational numbers is a rational number.
$\pi + (-\pi) = 0$	The sum of π and its opposite, $-\pi$, two irrational numbers, is 0, a rational number.

Even though the sum of two irrational numbers is usually an irrational number, the set of irrational numbers is not closed under addition. There are some pairs of irrational numbers whose sum is not an irrational number. However, π, $-\pi$, 0, and each of the other numbers used in these examples are real numbers and the sum of two real numbers is a real number.

▶ **The sets of whole numbers, rational numbers, and real numbers are each closed under addition.**

2. Multiply any two numbers.

$(2)(4) = 8$	The product of two whole numbers is a whole number.
$\frac{3}{4} \times \frac{1}{2} = \frac{3}{8}$	The product of two rational numbers is a rational number.
$\sqrt{2} \times \sqrt{2} = \sqrt{4} = 2$	The product of $\sqrt{2}$ and $\sqrt{2}$, two irrational numbers, is 2, a rational number.

Though the product of two irrational numbers is usually an irrational number, there are some pairs of irrational numbers whose product is not an irrational number. The set of irrational numbers is not closed under multiplication. However, $\sqrt{2}$, 2, and each of the other numbers used in these examples are real numbers and the product of two real numbers is a real number.

▶ **The sets of whole numbers, rational numbers, and real numbers are each closed under multiplication.**

3. Subtract any two numbers.

$7 - 12 = -5$	The difference of two whole numbers is not a whole number, but these whole numbers are also integers and the difference between two integers is an integer.
$12.7 - 8.2 = 4.5$	The difference of two rational numbers is a rational number.
$\sqrt{3} - \sqrt{3} = 0$	The difference of $\sqrt{3}$ and $\sqrt{3}$, two irrational numbers, is 0, a rational number.

Even though the difference of two irrational numbers is usually an irrational number, there are some pairs of irrational numbers whose difference is not an irrational number. The set of irrational numbers is not closed under subtraction. However, $\sqrt{3}$, 0, and each of the other numbers used in these examples are real numbers and the difference of two real numbers is a real number.

► **The sets of integers, rational numbers, and real numbers are each closed under subtraction.**

4. Divide any two numbers by a nonzero number. (Remember that division by 0 is not allowed.)

$9 \div 2 = 4.5$	The quotient of two whole numbers or two integers is not always a whole number or an integer.
$\frac{3}{4} \div \frac{2}{3} = \frac{3}{4} \times \frac{3}{2} = \frac{9}{8}$	The quotient of two rational numbers is a rational number.
$\sqrt{5} \div \sqrt{5} = 1$	The quotient of $\sqrt{5}$ and $\sqrt{5}$, two irrational numbers, is 1, a rational number.

Though the quotient of two irrational numbers is usually an irrational number, there are some pairs of irrational numbers whose quotient is not an irrational number. The set of irrational numbers is not closed under division. However, $\sqrt{5}$, 1, and each of the other numbers used in these examples are real numbers and the quotient of two nonzero real numbers is a nonzero real number.

► **The sets of nonzero rational numbers, and nonzero real numbers are each closed under division.**

Later in this book, we will study operations with signed numbers and operations with irrational numbers in greater detail. For now, we will simply make these observations:

► **The set of whole numbers is closed under the operations of addition and multiplication.**

- **The set of integers is closed under the operations of addition, subtraction, and multiplication.**

- **The set of rational numbers is closed under the operations of addition, subtraction, and multiplication, and the set of nonzero rational numbers is closed under division.**

- **The set of real numbers is closed under the operations of addition, subtraction, and multiplication, and the set of nonzero real numbers is closed under division.**

Commutative Property of Addition

When we add rational numbers, we assume that we can change the *order* in which two numbers are added without changing the sum.

For example, $4 + 5 = 5 + 4$ and $\frac{1}{2} + \frac{1}{4} = \frac{1}{4} + \frac{1}{2}$. These examples illustrate the **commutative property of addition.**

In general, we assume that for every number a and every number b:

$$a + b = b + a$$

Commutative Property of Multiplication

In the same way, when we multiply rational numbers, we assume that we can change the *order* of the factors without changing the product.

For example, $5 \times 4 = 4 \times 5$, and $\frac{1}{2} \times \frac{1}{4} = \frac{1}{4} \times \frac{1}{2}$. These examples illustrate the **commutative property of multiplication.**

In general, we assume that for every number a and every number b:

$$a \times b = b \times a$$

Subtraction and *division* are not commutative, as shown by the following counterexample.

$12 - 7 \neq 7 - 12$	$12 \div 3 \neq 3 \div 12$
$5 \neq -5$	$4 \neq \frac{3}{12}$

Associative Property of Addition

Addition is a binary operation; that is, we add two numbers at a time. If we wish to add three numbers, we find the sum of two and add that sum to the third. For example:

$$2 + 5 + 8 = (2 + 5) + 8 = 7 + 8 = 15 \quad \text{or} \quad 2 + 5 + 8 = 2 + (5 + 8) = 2 + 13 = 15$$

The way in which we *group* the numbers to be added does not change the sum. Therefore, we see that $(2 + 5) + 8 = 2 + (5 + 8)$. This example illustrates the **associative property of addition.**

In general, we assume that for every number a, every number b, and every number c:

$$(a + b) + c = a + (b + c)$$

Associative Property of Multiplication

In a similar way, to find a product that involves three factors, we first multiply any two factors and then multiply this result by the third factor. We assume that we do not change the product when we change the grouping. For example:

$$5 \times 4 \times 2 = (5 \times 4) \times 2 = 20 \times 2 = 40 \quad \text{or} \quad 5 \times 4 \times 2 = 5 \times (4 \times 2) = 5 \times 8 = 40$$

Therefore, $(5 \times 4) \times 2 = 5 \times (4 \times 2)$. This example illustrates the **associative property of multiplication.**

In general, we assume that for every number a, every number b, and every number c:

$$a \times (b \times c) = (a \times b) \times c$$

Subtraction and *division* are not associative, as shown in the following counterexamples.

$(15 - 4) - 3 \neq 15 - (4 - 3)$	$(8 \div 4) \div 2 \neq 8 \div (4 \div 2)$
$11 - 3 \neq 15 - 1$	$2 \div 2 \neq 8 \div 2$
$8 \neq 14$	$1 \neq 4$

The Distributive Property

We know $4(3 + 2) = 4(5) = 20$, and also $4(3) + 4(2) = 12 + 8 = 20$. Therefore, we see that $4(3 + 2) = 4(3) + 4(2)$.

This result can be illustrated geometrically. Recall that the area of a rectangle is equal to the product of its length and its width.

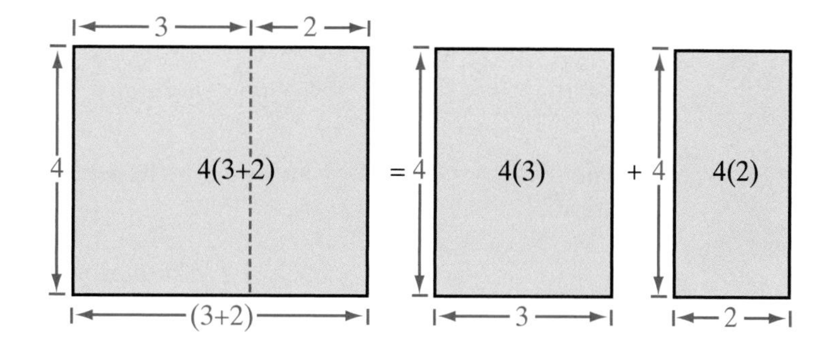

This example illustrates the **distributive property of multiplication over addition**, also called the **distributive property**. This means that the product of one number times the sum of a second and a third number equals the product of the first and second numbers plus the product of the first and third numbers.

In general, we assume that for every number a, every number b, and every number c:

$$a(b + c) = ab + ac \qquad \text{and} \qquad (a + b)c = ac + bc$$

The distributive property is also true for multiplication over subtraction:

$$a(b - c) = ab - ac \qquad \text{and} \qquad (a - b)c = ac - bc$$

The distributive property can be useful for mental computations. Observe how we can use the distributive property to find each of the following products as a sum:

1. $6 \times 23 = 6(20 + 3) = 6 \times 20 + 6 \times 3 = 120 + 18 = 138$

2. $9 \times 3\frac{1}{3} = 9\left(3+\frac{1}{3}\right) = 9 \times 3 + 9 \times \frac{1}{3} = 27 + 3 = 30$

3. $6.5 \times 8 = (6 + 0.5)8 = 6 \times 8 + 0.5 \times 8 = 48 + 4 = 52$

Working backward, we can also use the distributive property to change the form of an expression from a sum or a difference to a product:

1. $5(12) + 5(8) = 5(12 + 8) = 5(20) = 100$

2. $7(14) - 7(4) = 7(14 - 4) = 7(10) = 70$

Addition Property of Zero and the Additive Identity Element

The equalities $5 + 0 = 5$ and $0 + 2.8 = 2.8$ are true. They illustrate that the sum of a rational number and zero is the number itself. These examples lead us to observe that:

1. The **addition property of zero** states that for every number a:

$$a + 0 = a \quad \text{and} \quad 0 + a = a$$

2. The **identity element of addition**, or the **additive identity**, is 0. Thus, for any number a:

If $a + x = a$, or if $x + a = a$, it follows that $x = 0$.

Additive Inverses (Opposites)

When we first studied integers, we learned about *opposites*. For example, the opposite of 4 is -4, and the opposite of -10 is 10.

Every rational number a has an *opposite*, $-a$, such that their sum is 0, the identity element in addition. The opposite of a number is called the **additive inverse** of the number.

In general, for every rational number a and its opposite $-a$:

$$a + (-a) = 0$$

On a calculator, the (-) key, is used to enter the opposite of a number. The following example shows that the opposite of -4.5 is 4.5.

ENTER: (-) (-) 4.5 ENTER

DISPLAY:

```
--4.5
                4.5
```

Multiplication Property of One and the Multiplicative Identity Element

The sentences $5 \times 1 = 5$ and $1 \times 4.6 = 4.6$ are true. They illustrate that the product of a number and one is the number itself. These examples lead us to observe that:

1. The **multiplication property of one** states that for every number a:

$$a \cdot 1 = a \quad \text{and} \quad 1 \cdot a = a$$

2. The **identity element of multiplication**, or the **multiplicative identity**, is **1**.

Multiplicative Inverses (Reciprocals)

When the product of two numbers is 1 (the identity element for multiplication), then each of these numbers is called the **multiplicative inverse** or **reciprocal** of the other. Consider these examples:

$4 \cdot \frac{1}{4} = 1$	$\left(2\frac{1}{2}\right)\left(\frac{2}{5}\right) = \left(\frac{5}{2}\right)\left(\frac{2}{5}\right) = 1$
The reciprocal of 4 is $\frac{1}{4}$.	The multiplicative inverse of $2\frac{1}{2}$ or $\frac{5}{2}$ is $\frac{2}{5}$.
The reciprocal of $\frac{1}{4}$ is 4.	The multiplicative inverse of $\frac{2}{5}$ is $\frac{5}{2}$ or $2\frac{1}{2}$.

Since there is no number that, when multiplied by 0, gives 1, the number 0 has no reciprocal, or no multiplicative inverse.

In general, for every *nonzero* number a, there is a unique number $\frac{1}{a}$ such that:

$$a \cdot \frac{1}{a} = 1$$

On the calculator, a special key, x^{-1} displays the reciprocal. For example, if each of the numbers shown above is entered and the reciprocal key is pressed, the reciprocal appears in decimal form.

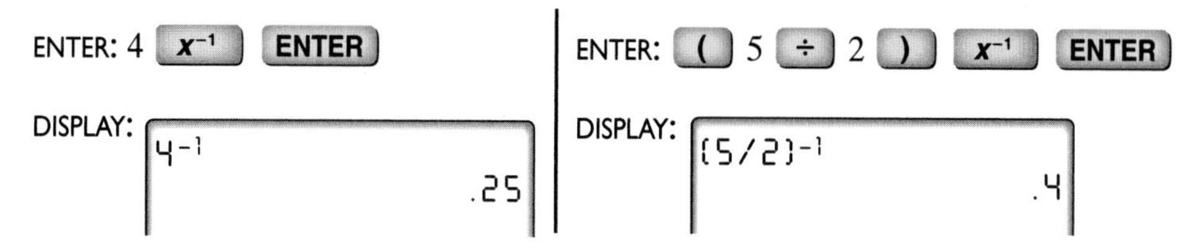

Note: Parentheses must be used when calculating the reciprocal of a fraction.

For many other numbers, however, the decimal form of the reciprocal is not shown in its entirety in the display. For example, we know that the reciprocal of 6 is $\frac{1}{6}$, but what appears is a rational approximation of $\frac{1}{6}$.

ENTER: 6 x^{-1} ENTER

DISPLAY:

6⁻¹

.1666666667

The display shows the rational approximation of $\frac{1}{6}$ rounded to the last decimal place displayed by the calculator. A calculator stores more decimal places in its operating system than it has in its display. The decimal displayed times the original number will equal 1.

To display the reciprocal of 6 from the example on the previous page as a common fraction, we use 6 [x⁻¹] [MATH] [ENTER] [ENTER].

Multiplication Property of Zero

The sentences $7 \times 0 = 0$ and $0 \times \frac{3}{4} = 0$ are true. They illustrate that the product of a rational number and zero is zero. This property is called the **multiplication property of zero**:

In general, for every number a:

$$a \cdot 0 = 0 \quad \text{and} \quad 0 \cdot a = 0$$

EXAMPLE 1

Write, in simplest form, the opposite (additive inverse) and the reciprocal (multiplicative inverse) of each of the following: **a.** 7 **b.** $-\frac{3}{8}$ **c.** $1\frac{1}{5}$ **d.** 0.2 **e.** π

Solution

Number	*Opposite*	*Reciprocal*
a. 7	-7	$\frac{1}{7}$
b. $-\frac{3}{8}$	$\frac{3}{8}$	$-\frac{8}{3} = -2\frac{2}{3}$
c. $1\frac{1}{5} = \frac{6}{5}$	$-1\frac{1}{5} = -\frac{6}{5}$	$\frac{5}{6}$
d. 0.2	-0.2	5
e. π	$-\pi$	$\frac{1}{\pi}$

EXAMPLE 2

Express $6t + t$ as a product and give the reason for each step.

Solution

	Step	*Reason*
(1)	$6t + t = 6t + 1t$	Multiplication property of 1.
(2)	$= (6 + 1)t$	Distributive property.
(3)	$= 7t$	Addition.

Answer $7t$

EXERCISES

Writing About Mathematics

1. If x and y represent real numbers and $xy = x$:

a. What is the value of y if the equation is true for all x? Explain your answer.

b. What is the value of x if the equation is true for all y? Explain your answer.

2. Cookies and brownies cost \$0.75 each. In order to find the cost of 2 cookies and 3 brownies Lindsey added 2 + 3 and multiplied the sum by \$0.75. Zachary multiplied \$0.75 by 2 and then \$0.75 by 3 and added the products. Explain why Lindsey and Zachary both arrived at the correct cost of the cookies and brownies.

Developing Skills

3. Give the value of each expression.

a. $9 + 0$ **b.** 9×0 **c.** 9×1 **d.** $\frac{2}{3} \times 0$ **e.** $0 + \frac{2}{3}$ **f.** $1 \times \frac{2}{3}$

g. $(\pi)\left(\frac{1}{\pi}\right)$ **h.** 1×4.5 **i.** $0 + \sqrt{7}$ **j.** $1.\overline{63} \times 0$ **k.** $1 \times \sqrt{5}$ **l.** $0 \times \sqrt{225}$

In 4–13: **a.** Replace each question mark with the number that makes the sentence true. **b.** Name the property illustrated in each sentence that is formed when the replacement is made.

4. $8 + 6 = 6 + ?$

5. $17 \times 5 = ? \times 17$

6. $(3 \times 9) \times 15 = 3 \times (9 \times ?)$

7. $6(5 + 8) = 6(5) + ?(8)$

8. $(0.5 + 0.2) + 0.7 = 0.5 + (? + 0.7)$

9. $4 + 0 = ?$

10. $(3 \times 7) + 5 = (? \times 3) + 5$

11. $(?)(8 + 2) = (8 + 2)(9)$

12. $7(4 + ?) = 7(4)$

13. $?x = x$

In 14–25: **a.** Name the additive inverse (opposite) of each number. **b.** Name the multiplicative inverse (reciprocal) of each number.

14. 17 **15.** 1 **16.** –10 **17.** 2.5 **18.** –1.8 **19.** $\frac{1}{9}$

20. $2\frac{1}{3}$ **21.** $-\pi$ **22.** $\frac{3}{7}$ **23.** $1.78\overline{0}$ **24.** $-\frac{1}{11}$ **25.** $3\frac{5}{71}$

In 26–31, state whether each sentence is a correct application of the distributive property. If you believe that it is not, state your reason.

26. $6(5 + 8) = 6(5) + 6(8)$

27. $10\left(\frac{1}{2} + \frac{1}{5}\right) = 10 \times \frac{1}{2} + \frac{1}{5}$

28. $5 + (8 \times 6) = (5 + 8) \times (5 + 6)$

29. $3(x + 5) = 3x + 3 \times 5$

30. $14a - 4a = (14 - 4)a$

31. $18(2.5) = 18(2) + 18(0.5)$

In 32–35: **a.** Tell whether each sentence is true or false. **b.** Tell whether the commutative property holds for the given operation.

32. $357 + 19 = 19 + 357$

33. $2 \div 1 = 1 \div 2$

34. $25 - 7 = 7 - 25$

35. $18(3.6) = 3.6(18)$

In 36–39: **a.** Tell whether each sentence is true of false. **b.** Tell whether the associative property holds for the given operation

36. $(73 \times 68) \times 92 = 73 \times (68 \times 92)$

37. $(24 \div 6) \div 2 = 24 \div (6 \div 2)$

38. $(19 - 8) - 5 = 19 - (8 - 5)$

39. $9 + (0.3 + 0.7) = (9 + 0.3) + 0.7$

40. Insert parentheses to make each statement true.

a. $3 \times 2 + 1 \div 3 = 3$

b. $4 \times 3 \div 2 + 2 = 3$

c. $8 + 8 \div 8 - 8 \times 8 = 8$

d. $3 \div 3 + 3 \times 3 - 3 = 1$

e. $3 \div 3 + 3 \times 3 - 3 = 0$

f. $0 \times 12 \times 3 - 16 \div 8 = 0$

Applying Skills

41. Steve Heinz wants to give a 15% tip to the taxi driver. The fare was $12. He knows that 10% of $12 is $1.20 and that 5% would be half of $1.20. Explain how this information can help Steve calculate the tip. What mathematical property is he using to determine the tip?

42. Juana rides the bus to and from work each day. Each time she rides the bus the fare is $1.75. She works five days a week. To find what she will spend on bus fare each week, Juana wants to find the product 2(1.75)(5). Juana rewrote the product as 2(5)(1.75).

a. What property of multiplication did Juana use when she changed 2(1.75)5 to 2(5)(1.75)?

b. What is her weekly bus fare?

2-3 ADDITION OF SIGNED NUMBERS

Adding Numbers That Have the Same Signs

The number line can be used to find the sum of two numbers. Start at 0. To add a positive number, move to the right. To add a negative number, move to the left.

EXAMPLE 1

Add +3 and +2.

Solution Start at 0 and move 3 units to the right to +3; then move 2 more units to the right, arriving at +5.

Calculator Solution ENTER: 3 [+] 2 [ENTER]

DISPLAY:

```
3+2
                5
```

Answer 5

The sum of two positive integers is the same as the sum of two whole numbers. The sum +5 is a number whose absolute value is the sum of the absolute values of +3 and +2 and whose sign is the same as the sign of +3 and +2.

EXAMPLE 2

Add -3 and -2.

Solution Start at 0 and move 3 units to the left to -3: then move 2 more units to the left, arriving at -5.

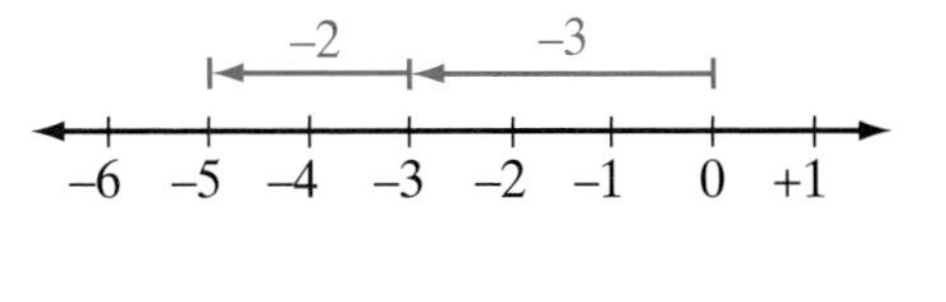

Calculator Solution ENTER: [(-)] 3 [+] [(-)] 2 [ENTER]

DISPLAY:

```
-3+-2
               -5
```

Answer -5

The sum -5 is a number whose absolute value is the sum of the absolute values of -3 and -2 and whose sign is the same as the sign of -3 and -2.

Examples 1 and 2 illustrate that the sum of two numbers with the same sign is a number whose absolute value is the sum of the absolute values of the numbers and whose sign is the sign of the numbers.

Procedure

To add two numbers that have the same sign:

1. Find the sum of the absolute values.
2. Give the sum the common sign.

Adding Numbers That Have Opposite Signs

EXAMPLE 3

Add: +3 and −2.

Solution Start at 0 and move 3 units to the right to +3; then move 2 units to the left, arriving at +1.

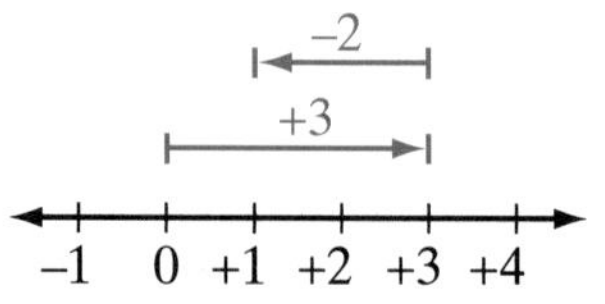

Calculator Solution ENTER: 3 [+] [(-)] 2 [ENTER]

DISPLAY:

Answer 1

This sum can also be found by using properties. In the first step, substitution is used, replacing (+3) with the sum (+1) + (+2).

$(+3) + (-2) = [(+1) + (+2)] + (-2)$	Substitution
$= (+1) + [(+2) + (-2)]$	Associative property
$= +1 + 0$	Addition property of opposites
$= +1$	Addition property of zero

The sum +1 is a number whose absolute value is the difference of the absolute values of +3 and −2 and whose sign is the same as the sign of +3, the number with the greater absolute value.

EXAMPLE 4

Add: −3 and +2.

Solution Start at 0 and move 3 units to the left to −3; then move 2 units to the right, arriving at −1.

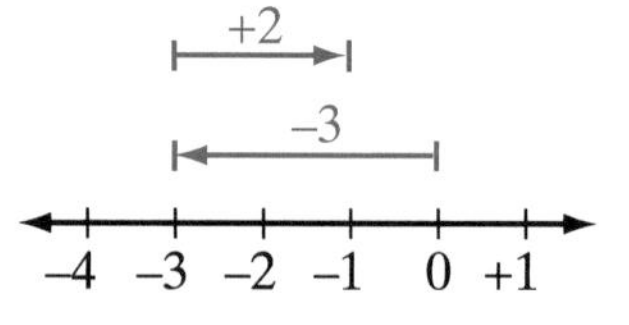

Calculator Solution ENTER: [(-)] 3 [+] 2 [ENTER]

DISPLAY:

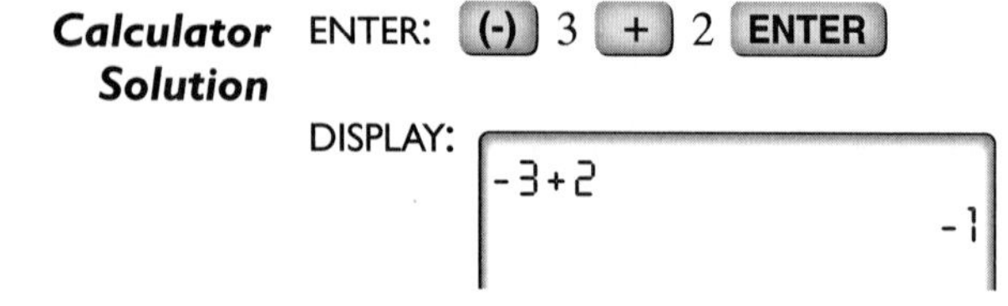

Answer −1

This sum can also be found by using properties. In the first step, substitution is used, replacing (-3) with the sum $(-1) + (-2)$.

$$\begin{aligned} (-3) + (+2) &= [(-1) + (-2)] + (+2) && \text{Substitution} \\ &= (-1) + [(-2) + (+2)] && \text{Associative property} \\ &= -1 + 0 && \text{Addition property of opposites} \\ &= -1 && \text{Addition property of zero} \end{aligned}$$

The sum -1 is a number whose absolute value is the difference of the absolute values of -3 and $+2$ and whose sign is the same as the sign of -3, the number with the greater absolute value.

Examples 3 and 4 illustrate that the sum of a positive number and a negative number is a number whose absolute value is the difference of the absolute values of the numbers and whose sign is the sign of the number having the larger absolute value.

Procedure

To add two numbers that have different signs:

1. Find the difference of the absolute values of the numbers.
2. Give this difference the sign of the number that has the greater absolute value.
3. The sum is 0 if both numbers have the same absolute value.

EXAMPLE 5

Find the sum of $-3\frac{3}{4}$ and $1\frac{1}{4}$.

Solution

How to Proceed

(1) Since the numbers have different signs, find the difference of their absolute values: $3\frac{3}{4} - 1\frac{1}{4} = 2\frac{2}{4}$

(2) Give the difference the sign of the number with the greater absolute value: $-2\frac{2}{4} = -2\frac{1}{2}$ *Answer*

Calculator Solution The number $3\frac{3}{4}$ is the sum of the whole number 3 and the fraction $\frac{3}{4}$. The opposite of $3\frac{3}{4}$, $-3\frac{3}{4}$, is the sum of -3 and $-\frac{3}{4}$.

Enclose the absolute value of the sum of 3 and $\frac{3}{4}$ in parentheses.

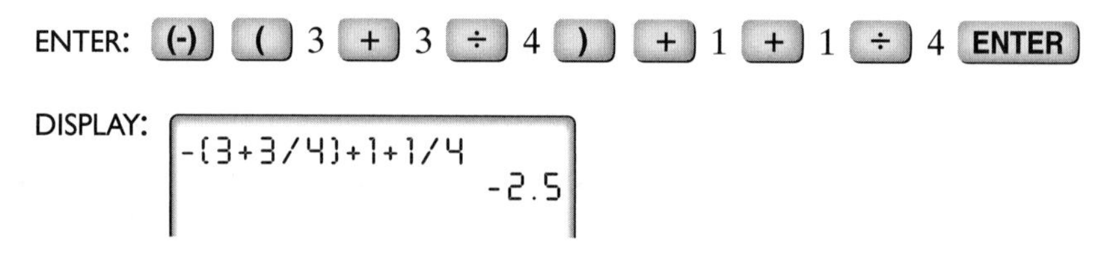

Answer -2.5 or $-2\frac{1}{2}$

The method used in Example 5 to enter the negative mixed number $-3\frac{3}{4}$ illustrate the following property:

Property of the Opposite of a Sum For all real numbers a and b:

$$-(a + b) = (-a) + (-b)$$

When adding more than two signed numbers, the commutative and associative properties allow us to arrange the numbers in any order and to group them in any way. It may be helpful to add positive numbers first, add negative numbers next, and then add the two results.

EXERCISES

Writing About Mathematics

1. The sum of two numbers is positive. One of the numbers is a positive number that is larger than the sum. Is the other number positive or negative? Explain your answer.

2. The sum of two numbers is positive. One of the numbers is negative. Is the other number positive or negative? Explain your answer.

3. The sum of two numbers is negative. One of the numbers is a negative number that is smaller than the sum. Is the other number positive or negative? Explain your answer.

4. The sum of two numbers is negative. One of the numbers is positive. Is the other number positive or negative? Explain your answer.

5. Is it possible for the sum of two numbers to be smaller than either of the numbers? If so, give an example.

Developing Skills

In 6–10, find each sum or difference.

6. $|-6| + |4|$ **7.** $|-10| - |-5|$ **8.** $|4.5| - |-4.5|$ **9.** $|+6| + |-4|$ **10.** $|+6 - 4|$

In 11–27, add the numbers. Use a calculator to check your answer.

11. $-17 + (-28)$ **12.** $+23 + (-35)$ **13.** $-87 + (+87)$ **14.** $-2.06 + 1.37$

15. $-33\frac{1}{3} + 19\frac{2}{3}$ **16.** $-5\frac{3}{4} + 8\frac{1}{2}$ **17.** $+7 + \left(-6\frac{2}{7}\right)$ **18.** $(-3.72) + (-5.28)$

19. $(-47) + (-35) + (+47)$ **20.** $34\frac{3}{8} + (-73)$ **21.** $-73\frac{1}{2} + 86$

22. $-14\frac{3}{4} + \left(-17\frac{3}{4}\right)$ **23.** $|-13| + (-13)$ **24.** $|-42| + (-|43|)$

25. $-6.25 + (-0.75)$ **26.** $(-12.4) + 13.0$ **27.** $|-12.4| + 13.0$

Applying Skills

28. In 1 hour, the temperature rose 4° Celsius and in the next hour it dropped 6° Celsius. What was the net change in temperature during the two-hour period?

29. An elevator started on the first floor and rose 30 floors. Then it came down 12 floors. At which floor was it at that time?

30. A football team gained 7 yards on the first play, lost 2 yards on the second, and lost 8 yards on the third. What was the net result of the three plays?

31. Fay has $250 in a bank. During the month, she made a deposit of $60 and a withdrawal of $80. How much money did Fay have in the bank at the end of the month?

32. During a four-day period, the dollar value of a share of stock rose $1.50 on the first day, dropped $0.85 on the second day, rose $0.12 on the third day, and dropped $1.75 on the fourth day. What was the net change in the stock during this period?

2-4 SUBTRACTION OF SIGNED NUMBERS

In arithmetic, to subtract 3 from 7, we find the number that, when added to 3, gives 7. We know that $7 - 3 = 4$ because $3 + 4 = 7$. Subtraction is the inverse operation of addition.

DEFINITION

In general, for every number c and every number b, the expression $c - b$ is the number a such that $b + a = c$.

We use this definition in order to subtract signed numbers. To subtract -2 from $+3$, written as $(+3) - (-2)$, we must find a number that, when added to -2, will give $+3$. We write:

$$(-2) + (?) = +3$$

We can use a number line to find the answer to this open sentence. From a point 2 units to the left of 0, move to the point that represents +3, that is, 3 units to the right of 0. We move 5 units to the right, a motion that represents +5.

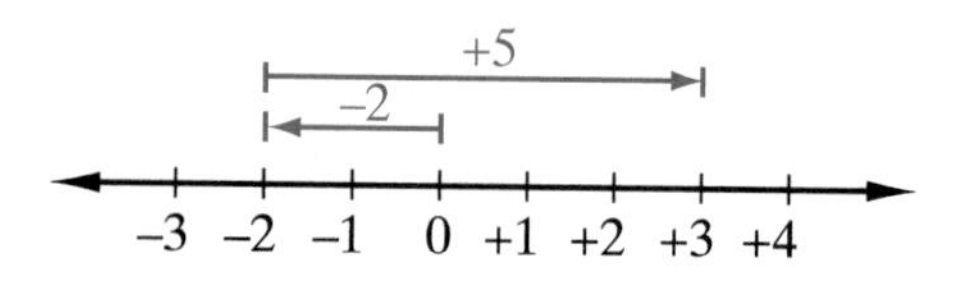

Therefore, $(+3) - (-2) = +5$ because $(-2) + (+5) = +3$.

Notice that $(+3) - (-2)$ can also be represented as the directed distance from −2 to +3 on the number line.

Subtraction can be written vertically as follows:

(+3)	or	*Subtract:*	(+3)	minuend
(−2)			(−2)	subtrahend
+5			+5	difference

When you first learned to subtract numbers, you learned to check your answer. The answer (the difference) plus the number being subtracted (the subtrahend) must be equal to the number from which you are subtracting (the minuend).

Check each of the following examples using:

subtrahend + difference = minuend:

Subtract:	+9	−7	+5	−3
	+6	−2	−2	+1
	+3	−5	+7	−4

Now, consider another way in which addition and subtraction are related. In each of the following examples, compare the result obtained when a signed number is subtracted with the result obtained when the opposite of that signed number is added.

Subtract	*Add*	*Subtract*	*Add*	*Subtract*	*Add*	*Subtract*	*Add*
+9	+9	−7	−7	+5	+5	−3	−3
+6	−6	−2	+2	−2	+2	+1	−1
+3	+3	−5	−5	+7	+7	−4	−4

Observe that, in each example, adding the opposite (the additive inverse) of a signed number gives the same result as subtracting that signed number. It therefore seems reasonable to define subtraction as follows:

DEFINITION

If a is any signed number and b is any signed number, then:

$$a - b = a + (-b)$$

Procedure

To subtract one signed number from another, add the opposite (additive inverse) of the subtrahend to the minuend.

Uses of the Symbol −

We have used the symbol − in three ways:

1. To indicate that a number is negative: -2 Negative 2

2. To indicate the opposite of a number: $-(-4)$ Opposite of negative 4

$-a$ Opposite of a

3. To indicate subtraction: $4 - (-3)$ Difference between 4 and −3

Note that an arithmetic expression such as $+3 - 7$ can mean either the difference between $+3$ and $+7$ or the sum of $+3$ and -7.

$$+3 - 7 = +3 - (+7) = +3 + (-7)$$

When writing an arithmetic expression, we use the same sign for both a negative number and subtraction. On a calculator, the key for subtraction, [−], is not the same key as the key for a negative number, [(-)]. Using the wrong key will result in an error message.

EXAMPLE 1

Perform the indicated subtractions.

a. $(+30) - (+12)$ **b.** $(-19) - (-7)$ **c.** $(-4) - (0)$ **d.** $0 - 8$

Answers **a.** $+18$ **b.** -12 **c.** -4 **d.** -8

EXAMPLE 2

From the sum of −2 and +8, subtract −5.

Solution $(-2 + 8) - (-5) = +6 + (+5) = +11$

Calculator Solution ENTER: [(-)] 2 [+] 8 [−] [(-)] 5 [ENTER]

DISPLAY:

```
-2+8--5
             11
```

Answer 11

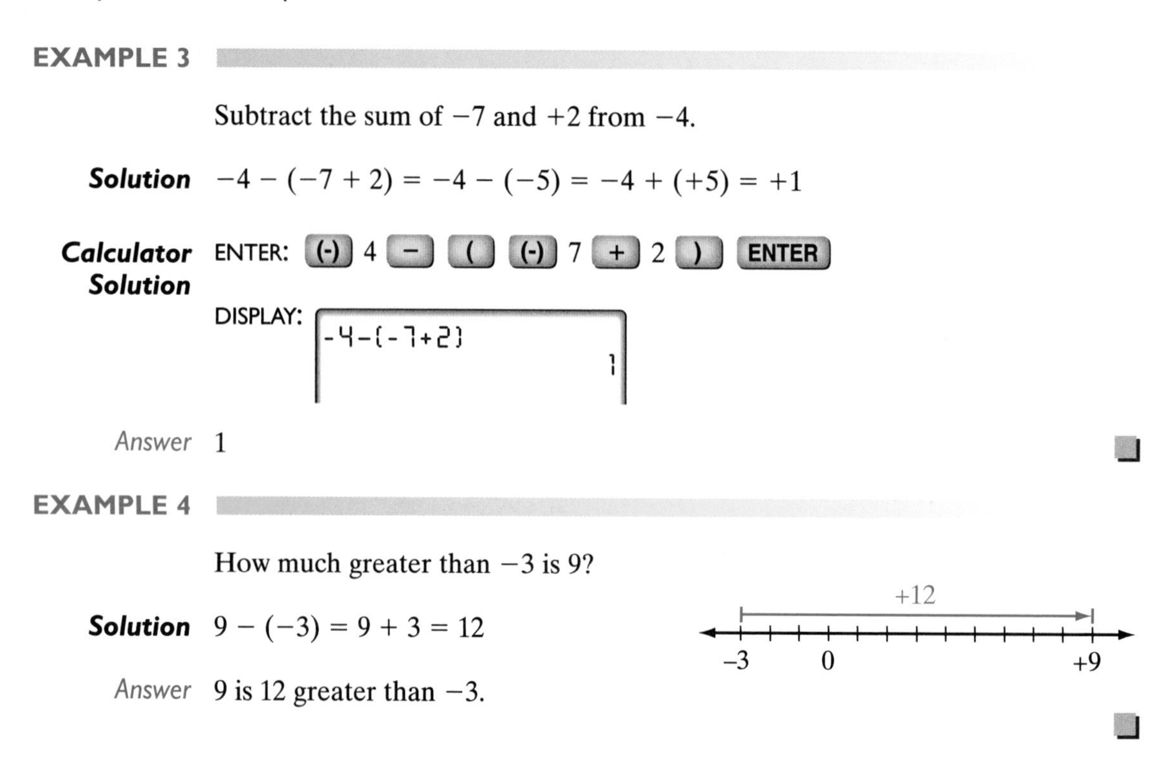

EXAMPLE 3

Subtract the sum of -7 and $+2$ from -4.

Solution $-4 - (-7 + 2) = -4 - (-5) = -4 + (+5) = +1$

Calculator Solution ENTER: (-) 4 − ((-) 7 + 2) ENTER

DISPLAY:
```
-4-(-7+2)
                1
```

Answer 1

EXAMPLE 4

How much greater than -3 is 9?

Solution $9 - (-3) = 9 + 3 = 12$

Answer 9 is 12 greater than -3.

Note that parentheses need not be entered in the calculator in Example 2 since the additions and subtractions are to be done in the order in which they occur in the expression. Parentheses are needed, however, in Example 3 since the sum of -7 and $+2$ is to be found first, before subtracting this sum from -4.

EXERCISES

Writing About Mathematics

1. Does $+8 - 12$ mean the difference between $+8$ and 12 or the sum of $+8$ and -12? Explain your answer.
2. How is addition used to check subtraction?

Developing Skills

In 3–10, perform each indicated subtraction. Check your answers using a calculator.

3. $-23 - (-35)$ **4.** $-87 - (+87)$ **5.** $+5.4 - (-8.6)$ **6.** $-8.8 - (-3.7)$

7. $-2.06 - (+1.37)$ **8.** $-33\frac{1}{3} - \left(+19\frac{2}{3}\right)$ **9.** $-5\frac{3}{4} - \left(-8\frac{1}{2}\right)$ **10.** $+7 - \left(6\frac{2}{7}\right)$

In 11–18, Find the value of each given expression.

11. $(+18) - (-14)$ **12.** $(-3.72) - (-5.28)$ **13.** $(-12) + (-57 - 12)$

14. $(-47) - (-35 + 47)$ **15.** $-73\frac{1}{2} - 86$ **16.** $-14\frac{3}{4} - \left(-17\frac{3}{4}\right)$

17. $-34\frac{3}{8} - \left(-73 - 72\frac{3}{8}\right)$ **18.** $|-32| - (-32)$

19. How much is 18 decreased by -7?

20. How much greater than -15 is 12?

21. How much greater than -4 is -1?

22. What number is 6 less than -6?

23. Subtract 8 from the sum of -6 and -12.

24. Subtract -7 from the sum of $+18$ and -10.

25. State whether each of the following sentences is true or false:

a. $(+5) - (-3) = (-3) - (+5)$ **b.** $(-7) - (-4) = (-4) - (-7)$

26. If x and y represent real numbers:

a. Does $x - y = y - x$ for all replacements of x and y? Justify your answer.

b. Does $x - y = y - x$ for any replacements of x and y? For which values of x and y?

c. What is the relation between $x - y$ and $y - x$ for all replacements of x and y?

d. Is the operation of subtraction commutative? In other words, for all signed numbers x and y, does $x - y = y - x$?

27. State whether each of the following sentences is true or false:

a. $(15 - 9) - 6 = 15 - (9 - 6)$ **b.** $[(-10) - (+4)] - (+8) = (-10) - [(+4) - (+8)]$

28. Is the operation of subtraction associative? In other words, for all signed numbers x, y, and z, does $(x - y) - z = x - (y - z)$? Justify your answer.

Applying Skills

29. Express as a signed number the increase or decrease when the Celsius temperature changes from:

a. $+5°$ to $+8°$ **b.** $-10°$ to $+18°$ **c.** $-6°$ to $-18°$ **d.** $+12°$ to $-4°$

30. Find the change in altitude when you go from a place that is 15 meters below sea level to a place that is 95 meters above sea level.

31. In a game, Sid was 35 points "in the hole." How many points must he make in order to have a score of 150 points?

32. The record high Fahrenheit temperature in New City is 105°; the record low is $-9°$. Find the difference between these temperatures.

33. At one point, the Pacific Ocean is 0.50 kilometers in depth; at another point it is 0.25 kilometers in depth. Find the difference between these depths.

2-5 MULTIPLICATION OF SIGNED NUMBERS

Four Possible Cases in the Multiplication of Signed Numbers

We will use a common experience to illustrate the various cases that can arise in the multiplication of signed numbers.

1. Represent a gain in weight by a positive number and a loss of weight by a negative number.
2. Represent a number of months in the future by a positive number and a number of months in the past by a negative number.

CASE 1 *Multiplying a Positive Number by a Positive Number*
If a boy gains 2 pounds each month, 4 months from now he will be 8 pounds heavier than he is now. Using signed numbers, we may write:

$$(+2)(+4) = +8$$

The product of the two positive numbers is a positive number.

CASE 2 *Multiplying a Negative Number by a Positive Number*
If a boy loses 2 pounds each month, 4 months from now he will be 8 pounds lighter than he is now. Using signed numbers, we may write:

$$(-2)(+4) = -8$$

The product of the negative number and the positive number is a negative number.

CASE 3 *Multiplying a Positive Number by a Negative Number*
If a girl gained 2 pounds each month, 4 months ago she was 8 pounds lighter than she is now. Using signed numbers, we may write:

$$(+2)(-4) = -8$$

The product of the positive number and the negative number is a negative number.

CASE 4 *Multiplying a Negative Number by a Negative Number*
If a girl lost 2 pounds each month, 4 months ago she was 8 pounds heavier than she is now. Using signed numbers, we may write:

$$(-2)(-4) = +8$$

The product of the two negative numbers is a positive number.

In all four cases, the absolute value of the product, 8, is equal to the product of the absolute values of the factors, 4 and 2.

Using the Properties of Real Numbers to Multiply Signed Numbers

You know that the sum of any real number and its inverse is 0, $a + (-a) = 0$, and that for all real numbers a, b, and c, $a(b + c) = ab + ac$.

These two facts will enable us to demonstrate the rules for multiplying signed numbers.

$$\begin{aligned} 4[7 + (-7)] &= 4(7) + 4(-7) \\ 4(0) &= 28 + 4(-7) \\ 0 &= 28 + 4(-7) \end{aligned}$$

In order for this to be a true statement, $4(-7)$ must equal -28, the opposite of 28.

Since addition is a commutative operation, $4(-7) = -7(4)$. Then

$$\begin{aligned} -7[4 + (-4)] &= -7(4) + (-7)(-4) \\ -7(0) &= -28 + (-7)(-4) \\ 0 &= -28 + (-7)(-4) \end{aligned}$$

In order for this to be a true statement, $-7(-4)$ must be 28, the additive inverse of -28.

Rules for Multiplying Signed Numbers

RULE 1 The product of two positive numbers or of two negative numbers is a positive number whose absolute value is the product of the absolute values of the numbers.

RULE 2 The product of a positive number and a negative number is a negative number whose absolute value is the product of the absolute values of the numbers.

In general, if a and b are both positive or are both negative, then:

$$ab = |a| \cdot |b|$$

If one of the numbers, a or b, is positive and the other is negative, then:

$$ab = -(|a| \cdot |b|)$$

Procedure

To multiply two signed numbers:

1. Find the product of the absolute values.
2. Write a plus sign before this product when the two numbers have the same sign.
3. Write a minus sign before this product when the two numbers have different signs.

EXAMPLE 1

Find the product of each of the given pairs of numbers.

	Answers		Answers
a. $(+12)(+4)$	$= +48$	**b.** $(-13)(-5)$	$= +65$
c. $(+18)(-3)$	$= -54$	**d.** $(-15)(+6)$	$= -90$
e. $(+3.4)(-3)$	$= -10.2$	**f.** $\left(-7\frac{1}{8}\right)(-3)$	$= +21\frac{3}{8}$

EXAMPLE 2

Use the distributive property of multiplication over addition to find the product $8(-72)$.

Solution

$$\begin{aligned} 8(-72) &= 8[(-70) + (-2)] \\ &= 8(-70) + 8(-2) \\ &= (-560) + (-16) \\ &= -576 \text{ Answer} \end{aligned}$$

EXAMPLE 3

Find the value of $(-2)^3$.

Solution

$$(-2)^3 = (-2)(-2)(-2) = +4(-2) = -8$$

Answer -8

Note: The product of an odd number (3) of negative factors is negative.

EXAMPLE 4

Find the value of $(-3)^4$.

Solution

$$(-3)^4 = (-3)(-3)(-3)(-3) = [(-3)(-3)][(-3)(-3)] = (+9)(+9) = +81$$

Answer $+81$

Note: The product of an even number (4) of negative factors is positive.

In this example, the value of $(-3)^4$ was found to be $+81$. This is not equal to -3^4, which is the opposite of 3^4 or $-1(3^4)$. To find the value of -3^4, first find the value of 3^4, which is 81, and then write the opposite of this power, -81. Thus, $(-3)^4 = 81$ and $-3^4 = -81$.

On a calculator, evaluate $(-3)^4$.

On a calculator, evaluate -3^4.

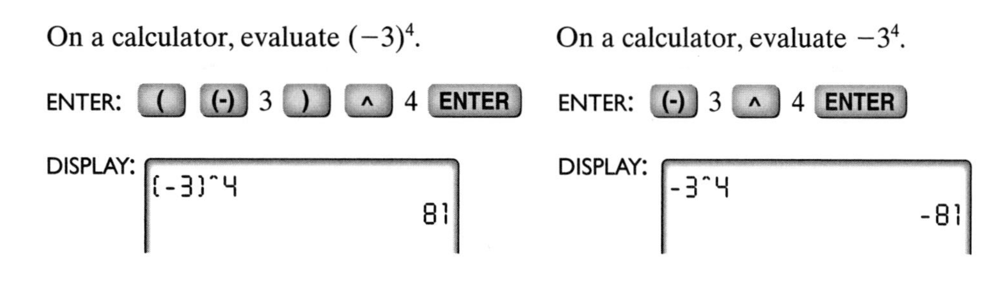

EXERCISES

Writing About Mathematics

1. Javier said that $(-5)(-4)(-2) = +40$ because the product of numbers with the same sign is positive. Explain to Javier why he is wrong.

2. If $-a(b)$ means the opposite of ab, explain how knowing that $3(-4) = -12$ can be used to show that $-3(-4) = +12$.

Developing Skills

In 3–12, find the product of each pair of numbers. Check your answers using a calculator.

3. $-17(-6)$ **4.** $+27(-6)$ **5.** $-9(+27)$ **6.** $-23(-15)$ **7.** $+4(-4)$

8. $+5.4(-0.6)$ **9.** $-2.6(+0.05)$ **10.** $-3\frac{1}{3}\left(+1\frac{2}{3}\right)$ **11.** $-5\frac{3}{4}\left(-\frac{1}{2}\right)$ **12.** $-6\frac{2}{3}(7)$

In 13–18, find the value of each expression.

13. $(+18)(-4)(-5)$ **14.** $(-3.72)(-0.5)(+0.2)$ **15.** $(-4)(-35 + 7)$

16. $(-12)(-7 - 3)$ **17.** $-8\left(-73 + 72\frac{3}{8}\right)$ **18.** $-4\left|-\frac{3}{4} + \frac{1}{4}\right|$

In 19–28, find the value of each power.

19. $(-3)^2$ **20.** -3^2 **21.** $(-5)^3$ **22.** $(+2)^4$ **23.** $(-2)^4$

24. -2^4 **25.** $(0.5)^2$ **26.** $(-0.5)^2$ **27.** -0.5^2 **28.** $-(-3)^3$

In 29–34, name the property that is illustrated in each statement.

29. $2(-1 + 3) = 2(-1) + 2(3)$

30. $(-2) + (-1) = (-1) + (-2)$

31. $(-1 + 2) + 3 = -1 + (2 + 3)$

32. $-2 + 0 = -2$

33. $(-5)(7) = (7)(-5)$

34. $3 \times (2 \times -3) = (3 \times 2) \times (-3)$

2-6 DIVISION OF SIGNED NUMBERS

Using the Inverse Operation in Dividing Signed Numbers

Division may be defined as the inverse operation of multiplication, just as subtraction is defined as the inverse operation of addition. To divide 6 by 2 means to find a number that, when multiplied by 2, gives 6. That number is 3 because $3(2) = 6$. Thus, $\frac{6}{2} = 3$, or $6 \div 2 = 3$. The number 6 is the dividend, 2 is the divisor, and 3 is the quotient.

It is impossible to divide a signed number by 0; that is, division by 0 is undefined. For example, to solve $(-9) \div 0$, we would have to find a number that, when multiplied by 0, would give -9. There is no such number since the product of any signed number and 0 is 0.

▶ **In general, for all signed numbers a and b ($b \neq 0$), $a \div b$ or $\frac{a}{b}$ means to find the unique number c such that $cb = a$**

In dividing nonzero signed numbers, there are four possible cases. Consider the following examples:

CASE 1 $\frac{+6}{+3} = ?$ implies $(?)(+3) = +6$. Since $(+2)(+3) = +6$, $\frac{+6}{+3} = +2$

CASE 2 $\frac{-6}{-3} = ?$ implies $(?)(-3) = -6$. Since $(+2)(-3) = -6$, $\frac{-6}{-3} = +2$

CASE 3 $\frac{-6}{+3} = ?$ implies $(?)(+3) = -6$. Since $(-2)(+3) = -6$, $\frac{-6}{+3} = -2$

CASE 4 $\frac{+6}{-3} = ?$ implies $(?)(-3) = +6$. Since $(-2)(-3) = +6$, $\frac{+6}{-3} = -2$

In the preceding examples, observe that:

1. When the dividend and divisor are both positive or both negative, the quotient is positive.
2. When the dividend and divisor have opposite signs, the quotient is negative.
3. In all cases, the absolute value of the quotient is the absolute value of the dividend divided by the absolute value of the divisor.

Rules for Dividing Signed Numbers

RULE 1 The quotient of two positive numbers or of two negative numbers is a positive number whose absolute value is the absolute value of the dividend divided by the absolute value of the divisor.

RULE 2 The quotient of a positive number and a negative number is a negative number whose absolute value is the absolute value of the dividend divided by the absolute value of the divisor.

In general, if a and b are both positive or are both negative, then:

$$a \div b = |a| \div |b| \quad \text{or} \quad \frac{a}{b} = \frac{|a|}{|b|}$$

If one of the numbers, a or b, is positive and the other is negative, then:

$$a \div b = -(|a| \div |b|) \quad \text{or} \quad \frac{a}{b} = -\left(\frac{|a|}{|b|}\right)$$

Procedure

To divide two signed numbers:

1. Find the quotient of the absolute values.
2. Write a plus sign before this quotient when the two numbers have the same sign.
3. Write a minus sign before this quotient when the two numbers have different signs.

Rule for Dividing Zero by a Nonzero Number

If the expression $0 \div (-5)$ or $\frac{0}{-5} = ?$, then $(?)(-5) = 0$. Since 0 is the only number that can replace ? and result in a true statement, $0 \div (-5)$ or $\frac{0}{-5} = 0$. This illustrates that 0 divided by any nonzero number is 0.

In general, if a is a nonzero number ($a \neq 0$), then

$$\mathbf{0 \div a = \frac{0}{a} = 0.}$$

EXAMPLE 1

Perform each indicated division, if possible.

a. $\frac{+60}{+15}$ **b.** $\frac{+10}{-90}$ **c.** $\frac{-27}{-3}$ **d.** $(-45) \div 9$ **e.** $0 \div (-9)$ **f.** $-3 \div 0$

Answers **a.** $+4$ **b.** $-\frac{1}{9}$ **c.** $+9$ **d.** -5 **e.** 0 **f.** Undefined

Using the Reciprocal in Dividing Signed Numbers

In Section 2-2, we learned that for every nonzero number a, there is a unique number $\frac{1}{a}$, called the *reciprocal* or *multiplicative inverse*, such that $a \cdot \frac{1}{a} = 1$.

Using the reciprocal of a number, we can define division in terms of multiplication as follows:

For all numbers a and b ($b \neq 0$):

$$a \div b = \frac{a}{b} = a \cdot \frac{1}{b} \ (b \neq 0)$$

Procedure

To divide a signed number by a nonzero signed number, multiply the dividend by the reciprocal of the divisor.

Notice that we exclude division by 0. The set of nonzero real numbers is closed with respect to division because every nonzero real number has a unique reciprocal, and multiplication by this reciprocal is always possible.

EXAMPLE 2

Perform each indicated division by using the reciprocal of the divisor.

	Answers
a. $\frac{+30}{+2}$	$= (+30)\left(+\frac{1}{2}\right) = +15$
b. $\frac{-30}{+90}$	$= (-30)\left(+\frac{1}{90}\right) = -\frac{1}{3}$
c. $(-54) \div 6$	$= (-54)\left(\frac{1}{6}\right) = -9$
d. $-27 \div \left(-\frac{1}{3}\right)$	$= -27 \times \left(-\frac{3}{1}\right) = +81$
e. $(+3) \div \left(-\frac{3}{5}\right)$	$= +3\left(-\frac{5}{3}\right) = -5$
f. $0 \div (-9)$	$= 0\left(-\frac{1}{9}\right) = 0$

EXERCISES

Writing About Mathematics

1. If x and y represent nonzero numbers, what is the relationship between $x \div y$ and $y \div x$?

2. If $x \neq y$, are there any values of x and y for which $x \div y = y \div x$?

Developing Skills

In 3–10, name the reciprocal (the multiplicative inverse) of each given number.

3. $+6$ **4.** -5 **5.** 1 **6.** -1

7. $\frac{1}{2}$ **8.** $-\frac{1}{10}$ **9.** $-\frac{3}{4}$ **10.** x if $x \neq 0$

In 11–26, find the indicated quotients or write "undefined" if no quotient exists.

11. $\frac{-63}{-9}$ **12.** $\frac{-48}{+16}$ **13.** $\frac{-10}{+10}$ **14.** $\frac{0}{-13}$

15. $\frac{-15}{-45}$ **16.** $\frac{+3.6}{-0.12}$ **17.** $\frac{-0.25}{-2.5}$ **18.** $\frac{+0.01}{-0.001}$

19. $(+100) \div (-2.5)$ **20.** $(-75) \div (0)$

21. $(+0.5) \div (-0.25)$ **22.** $(-1.5) \div (-0.03)$

23. $(+12) \div \left(-\frac{1}{3}\right)$ **24.** $\left(-\frac{3}{4}\right) \div (+6)$

25. $\left(+\frac{7}{8}\right) \div \left(-\frac{21}{32}\right)$ **26.** $\left(-1\frac{1}{4}\right) \div \left(-2\frac{1}{2}\right)$

In 27–28, state whether each sentence is true or false:

27. a. $[(+16) \div (+4)] \div (+2) = (+16) \div [(+4) \div (+2)]$

b. $[(-36) \div (+6)] \div (-2) = (-36) \div [(+6) \div (-2)]$

c. Division is associative.

28. a. $(12 + 6) \div 2 = 12 \div 2 + 6 \div 2$

b. $[(+25) + (-10)] \div (-5) = (+25) \div (-5) + (-10) \div (-5)$

c. $2 \div (3 + 5) = 2 \div 3 + 2 \div 5$

d. Division is distributive over addition and subtraction.

2-7 OPERATIONS WITH SETS

Recall that a set is simply a collection of distinct objects or elements, such as a set of numbers in arithmetic or a set of points in geometry. And, just as there are operations in arithmetic and in geometry, there are operations with sets. Before we look at these operations, we need to understand one more type of set.

The **universal set**, or the **universe**, is the set of all elements under consideration in a given situation, usually denoted by the letter U. For example:

1. Some universal sets, such as all the numbers we have studied, are infinite. Here, U = {real numbers}.

2. In other situations, such as the scores on a classroom test, the universal set can be finite. Here, using whole-number grades, $U = \{0, 1, 2, 3, \ldots, 100\}$.

Three operations with sets are called *intersection*, *union*, and *complement*.

Intersection of Sets

The **intersection of two sets**, A and B, denoted by $A \cap B$, is the set of all elements that belong to both sets, A and B. For example:

1. When $A = \{1, 2, 3, 4, 5\}$, and $B = \{2, 4, 6, 8, 10\}$, then $A \cap B$ is $\{2, 4\}$.

2. In the diagram, two lines called $\overleftrightarrow{AB}$ and $\overleftrightarrow{CD}$ intersect. Each line is an infinite set of points although only three points are marked on each line. The intersection is a set that has one element, point E, the point that is on line AB ($\overleftrightarrow{AB}$) and on line CD ($\overleftrightarrow{CD}$). We write the intersection of the lines in the example shown as $\overleftrightarrow{AB} \cap \overleftrightarrow{CD} = E$.

3. Intersection is a binary operation, $F \cap G = H$, where sets F, G, and H are subsets of the universal set and $\cap$ is the operation symbol. For example:

$$\begin{aligned} U &= \text{set of natural numbers} = \{1, 2, 3, 4, 5, 6, 7, 8, 9, 10, 11, 12, \ldots\} \\ F &= \text{multiples of } 2 = \{2, 4, 6, 8, 10, 12, \ldots\} \\ G &= \text{multiples of } 3 = \{3, 6, 9, 12, 15, 18, \ldots\} \\ F \cap G &= \text{multiples of } 6 = \{6, 12, 18, 24, 30, \ldots\} \end{aligned}$$

4. Two sets are **disjoint sets** if their intersection is the **empty set** ($\varnothing$ or $\{\}$); that is, if they do not have a common element. For example, when $K = \{1, 3, 5, 7, 9, 11, 13\}$ and $L = \{2, 4, 6, 8\}$, $K \cap L = \varnothing$. Therefore, K and L are disjoint sets.

Union of Sets

The **union of two sets**, A and B, denoted by $A \cup B$, is the set of all elements that belong to set A or to set B, or to both set A and set B. For example:

1. If $A = \{1, 2, 3, 4\}$ and $B = \{2, 4, 6\}$, then $A \cup B = \{1, 2, 3, 4, 6\}$. Note that an element is not repeated in the union of two sets even if it is an element of each set.

2. In the diagram, both region R (gray shading) and region S (light color shading) represent sets of points. The shaded parts of both regions represents $R \cup S$, and the dark color shading where the regions overlap represents $R \cap S$.

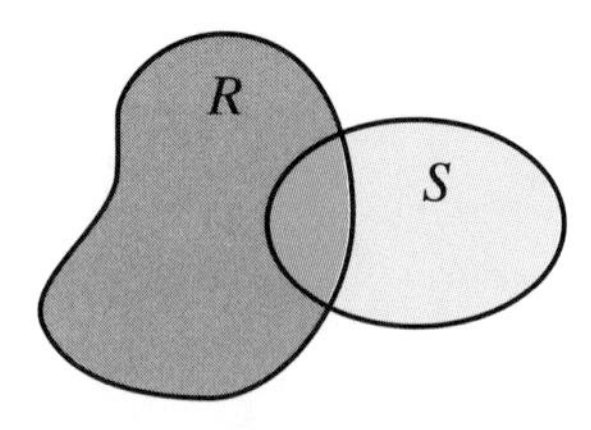

3. If $A = \{1, 2\}$ and $B = \{1, 2, 3, 4, 5\}$, then the union of A and B is $\{1, 2, 3, 4, 5\}$. We can write $A \cup B = \{1, 2, 3, 4, 5\}$, or $A \cup B = B$. Once again we have an example of a binary operation, where the elements are taken from a universal set and where the operation here is union.

4. The union of the set of all rational numbers and the set of all irrational numbers is the set of real numbers.

$$\{\text{real numbers}\} = \{\text{rational numbers}\} \cup \{\text{irrational numbers}\}$$

Complement of a Set

The **complement of a set A**, denoted by $\overline{A}$, is the set of all elements that belong to the universe U but do not belong to set A. Therefore, before we can determine the complement of A, we must know U. For example:

1. If $A = \{3, 4, 5\}$ and $U = \{1, 2, 3, 4, 5\}$, then $\overline{A} = \{1, 2\}$ because 1 and 2 belong to the universal set U but do not belong to set A.
2. If the universe is {whole numbers} and A = {even whole numbers} then $\overline{A}$ = {odd whole numbers} because the odd whole numbers belong to the universal set but do not belong to set A.

Although it seems at first that only one set is being considered in writing the complement of A as $\overline{A}$, actually there are two sets. This fact suggests a binary operation, in which the universe U and the set A are the pair of elements, *complement* is the operation, and the unique result is $\overline{A}$. The complement of any universe is the empty set. Note that the complement can also be written as $U \setminus A$ to emphasize that it is a binary operation.

EXAMPLE 1

If $U = \{1, 2, 3, 4, 5, 6, 7\}$, $A = \{6, 7\}$, and $B = \{3, 5, 7\}$, determine $\overline{A} \cap \overline{B}$.

Solution

$$U = \{1, 2, 3, 4, 5, 6, 7\}$$

Since $A = \{6, 7\}$, then $\overline{A} = \{1, 2, 3, 4, 5\}$.

Since $B = \{3, 5, 7\}$, then $\overline{B} = \{1, 2, 4, 6\}$.

Since 1, 2, and 4 are elements in both $\overline{A}$ and $\overline{B}$, we can write:

$$\overline{A} \cap \overline{B} = \{1, 2, 4\} \text{ Answer}$$

EXAMPLE 2

Using sets U, A, and B given for Example 1, find the complement of the set $A \cup B$, that is, determine $\overline{A \cup B}$.

Solution Since $A \cup B$ contains all of the elements that are common to A and B, $A \cup B = \{3, 5, 6, 7\}$. Therefore,

$$\overline{A \cup B} = \{1, 2, 4\} \text{ Answer}$$

EXERCISES

Writing About Mathematics

1. A line is a set of points. Can the intersection of two lines be the empty set? Explain.

2. Is the union of the set of prime numbers and the set of composite numbers equal to the set of counting numbers? Explain.

Developing Skills

In 3–10, $A = \{1, 2, 3\}$, $B = \{3, 4, 5, 6\}$, and $C = \{1, 3, 4, 6\}$. In each case, perform the given operation and list the element(s) of the resulting set.

3. $A \cap B$ **4.** $A \cup B$ **5.** $A \cap C$ **6.** $A \cup C$

7. $B \cap C$ **8.** $B \cup C$ **9.** $B \cup \varnothing$ **10.** $B \cap \varnothing$

11. Using the sets A, B, and C given for Exercises 3–10, list the element(s) of the smallest possible universal set of which A, B, and C are all subsets.

In 12–19, the universe $U = \{1, 2, 3, 4, 5\}$, $A = \{1, 5\}$, $B = \{2, 5\}$, and $C = \{2\}$. In each case, perform the given operation and list the element(s) of the resulting set.

12. $\overline{A}$ **13.** $\overline{B}$ **14.** $\overline{C}$ **15.** $A \cup B$

16. $A \cap B$ **17.** $A \cup \overline{B}$ **18.** $\overline{A} \cap \overline{B}$ **19.** $\overline{A} \cup \overline{B}$

20. If $U = \{2, 4, 6, 8\}$ and $\overline{A} = \{6\}$, what are the elements of A?

21. If $U = \{2, 4, 6, 8\}$, $A = \{2\}$, and $\overline{B} = \{2, 4\}$, what are the elements of $A \cup B$?

22. If $U = \{2, 4, 6, 8\}$, $A = \{2\}$, and $\overline{B} = \{2, 4\}$, what is the set $A \cap B$?

23. Suppose that the set A has two elements and the set B has three elements.

a. What is the greatest number of elements that $A \cup B$ can have?

b. What is the least number of elements that $A \cup B$ can have?

c. What is the greatest number of elements that $A \cap B$ can have?

d. What is the least number of elements that $A \cap B$ can have?

24. Let the universe $U = \{2, 4, 6, 8, 10, 12\}$, $A = \{2, 8, 12\}$ and $B = \{4, 10\}$.

a. $\overline{A}$ **b.** $\overline{\overline{A}}$ (the complement of $\overline{A}$) **c.** $\overline{B}$ **d.** $\overline{\overline{B}}$ (the complement of $\overline{B}$)

25. Let the universe $U = \{1, 2, 3, 4, 5, 6, 7, 8\}$.

a. Find the elements of $A \cap B$, $\overline{A} \cap \overline{B}$, and $\overline{A} \cup \overline{B}$ when A and B are equal to:

(1) $A = \{1, 2, 3, 4\}$; $B = \{5, 6, 7, 8\}$ (2) $A = \{2, 4\}$; $B = \{6, 8\}$

(3) $A = \{1, 3, 5, 7\}$; $B = \{2, 4, 6, 8\}$ (4) $A = \{2\}$; $B = \{4\}$

b. When A and B are disjoint sets, describe, in words, the set $\overline{A} \cap \overline{B}$.

c. If A and B are disjoint sets, what is the set $\overline{A} \cup \overline{B}$? Explain.

2-8 GRAPHING NUMBER PAIRS

Even though we know that the surface of the earth is approximately the surface of a sphere, we often model the earth by using maps that are plane surfaces. To locate a place on a map, we choose two reference lines, the equator and the prime meridian. The location of a city is given in term of east or west *longitude* (distance from the prime meridian) and north or south *latitude* (distance from the equator). For example, the city of Lagos in Nigeria is located at 3° east longitude and 6° north latitude, and the city of Dakar in Senegal is located 17° west longitude and 15° north latitude.

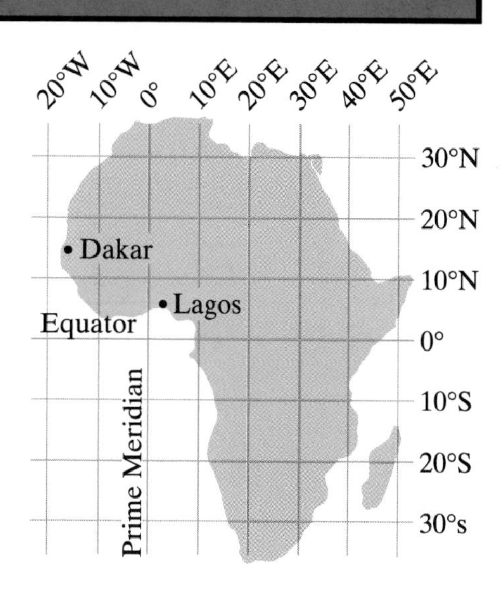

Points on a Plane

The method used to locate cities on a map can be used to locate any point on a plane. The reference lines are a horizontal number line called the **x-axis** and a vertical number line called the **y-axis**. These two number lines, which have the same scale and are drawn perpendicular to each other, are called the **coordinate axes**. The plane determined by the axes is called the **coordinate plane**.

In a coordinate plane, the intersection of the two axes is called the **origin** and is indicated as point O. This point of intersection is assigned the value 0 on both the x- and y-axes.

Moving to the right and moving up are regarded as movements in the positive direction. In the coordinate plane, points to the right of O on the x-axis and on lines parallel to the x-axis and points above O on the y-axis and on lines parallel to the y-axis are assigned positive values.

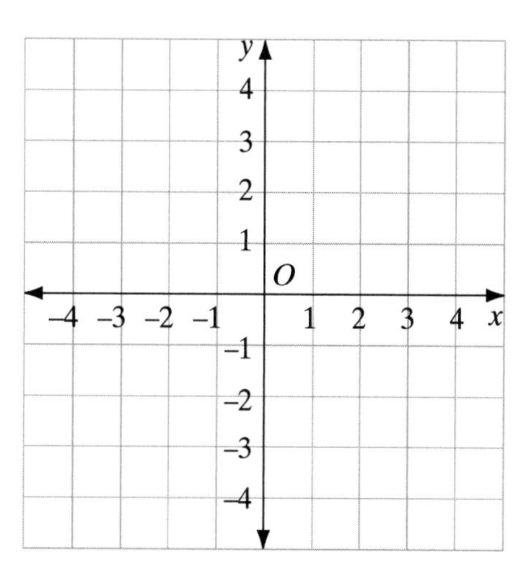

Moving to the left and moving down are regarded as movements in the negative direction. In the coordinate plane, points to the left of O on the x-axis and on lines parallel to the x-axis and points below O on the y-axis and on lines parallel to the y-axis are assigned negative values.

The x-axis and the y-axis separate the plane into four regions called **quadrants**. These quadrants are numbered I, II, III, and IV in a counterclockwise order, beginning at the upper right, as shown in the accompanying diagram. The points on the axes are not in any quadrant.

Coordinates of a Point

Every point on the plane can be described by two numbers, called the **coordinates** of the point, usually written as an **ordered pair**. The first number in the pair is called the **x-coordinate** or the **abscissa**. The second number is the **y-coordinate** or the **ordinate**. In general, the coordinates of a point are represented as **(x, y)**.

In the graph at the right, point A, which is the **graph of the ordered pair** $(+2, +3)$, lies in quadrant I. Here, A lies a distance of 2 units to the right of the origin (in a positive direction along the x-axis) and then up a distance of 3 units (in a positive direction parallel to the y-axis).

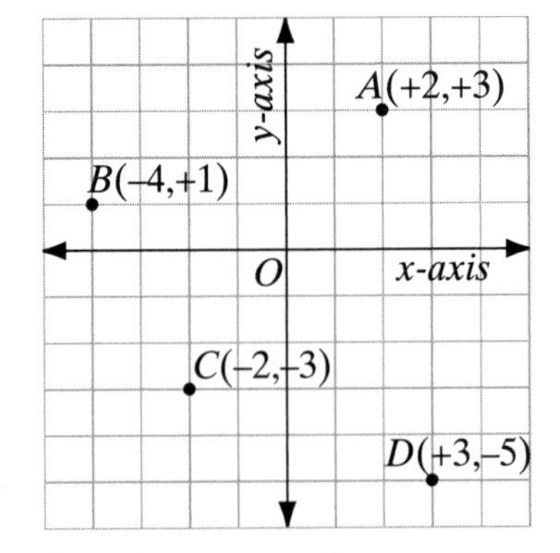

Point B, the graph of $(-4, +1)$ in quadrant II, lies a distance of 4 units to the left of the origin (in a negative direction along the x-axis) and then up 1 unit (in a positive direction parallel to the y-axis).

In quadrant III, every ordered pair (x, y) consists of two negative numbers. For example, C, the graph of $(-2, -3)$, lies 2 units to the left of the origin (in the negative direction along the x-axis) and then down 3 units (in the negative direction parallel to the y-axis).

Point D, the graph of $(+3, -5)$ in quadrant IV, lies 3 units to the right of the origin (in a positive direction along the x-axis) and then down 5 units (in a negative direction parallel to the y-axis).

Point O, the origin, has the coordinates $(0, 0)$.

Locating a Point on the Coordinate Plane

An ordered pair of signed numbers uniquely determines the location of a point.

Procedure

To find the location of a point on the coordinate plane:

1. Starting from the origin O, move along the x-axis the number of units given by the x-coordinate. Move to the right if the number is positive or to the left if the number is negative. If the x-coordinate is 0, there is no movement along the x-axis.
2. Then, from the point on the x-axis, move parallel to the y-axis the number of units given by the y-coordinate. Move up if the number is positive or down if the number is negative. If the y-coordinate is 0, there is no movement in the y direction.

To locate the point $A(-3, -4)$, from O, move 3 units to the left along the x-axis, then 4 units down, parallel to the y-axis.

To locate the point $B(4, 0)$, from O, move 4 units to the right along the x-axis. There is no movement parallel to the y-axis.

To locate the point $C(0, -5)$, there is no movement along the x-axis. From O, move 5 units down along the y-axis.

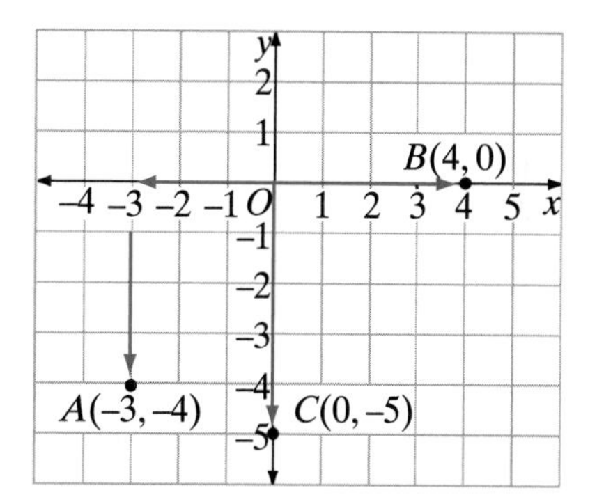

Finding the Coordinates of a Point on a Plane

The location of a point on the coordinate plane uniquely determines the coordinates of the point.

Procedure

To find the coordinates of a point:

1. From the point, move along a vertical line to the x-axis. The number assigned to that point on the x-axis is the x-coordinate of the point.
2. From the point, move along a horizontal line to the y-axis. The number assigned to that point on the y-axis is the y-coordinate of the point.

To find the coordinates of point R, from point R, move in the vertical direction to 5 on the x-axis and in the horizontal direction to -6 on the y-axis. The coordinates of R are $(5, -6)$

To find the coordinates of point S, from point S, move in the vertical direction to -2 on the x-axis and in the horizontal direction to 4 on the y-axis. The coordinates of S are $(-2, 4)$

Point T, lies at -5 on the x-axis. Therefore, the x-coordinate is -5 and the y-coordinate is 0. The coordinates of T are $(-5, 0)$.

Note that if a point lies on the y-axis, the x-coordinate is 0. If a point lies on the x-axis, the y-coordinate is 0.

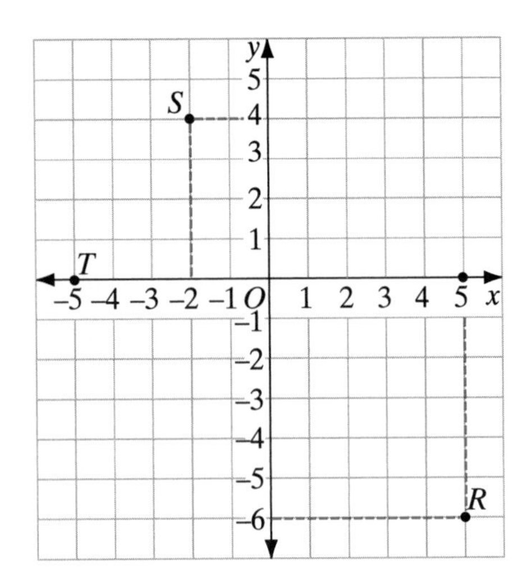

Graphing Polygons

A **polygon** is a closed figure whose sides are line segments. A **quadrilateral** is a polygon with four sides. The endpoints of the sides are called **vertices**. A quadrilateral can be represented in the coordinate plane by locating its vertices and then drawing the sides, connecting the vertices in order. The graph at the right shows the rectangle $ABCD$. The vertices are $A(3, 2)$, $B(-3, 2)$, $C(-3, -2)$ and $D(3, -2)$. From the graph, note the following:

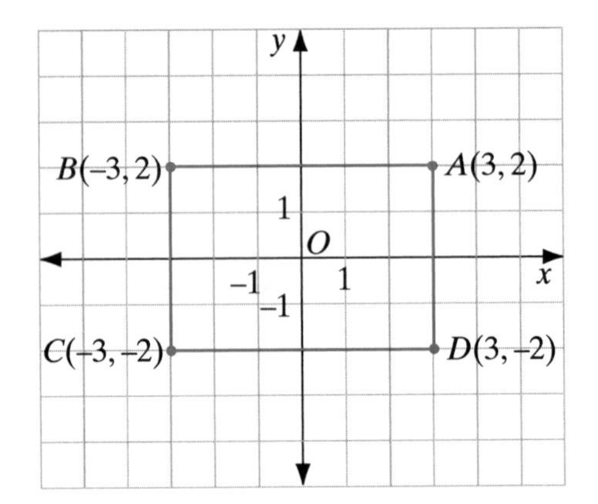

1. Points A and B have the same y-coordinate and are on a line parallel to the x-axis.
2. Points C and D have the same y-coordinate and are on a line parallel to the x-axis.
3. Lines parallel to the x-axis are parallel to each other.
4. Lines parallel to the x-axis are perpendicular to the y-axis.
5. Points B and C have the same x-coordinate and are on a line parallel to the y-axis.

6. Points A and D have the same x-coordinate and are on a line parallel to the y-axis.
7. Lines parallel to the y-axis are parallel to each other.
8. Lines parallel to the y-axis are perpendicular to the x-axis.

Now, we know that $ABCD$ is a rectangle, because it is a parallelogram with right angles. From the graph, we can find the dimensions of this rectangle. To find the length of the rectangle, we can count the number of units from A to B or from C to D. $AB = CD = 6$. Because points on the same horizontal line have the same y-coordinate, we can also find AB and CD by subtracting their x-coordinates.

$$AB = CD = 3 - (-3) = 3 + 3 = 6$$

To find the width of the rectangle, we can count the number of units from B to C or from D to A. $BC = DA = 4$. Because points on the same vertical line have the same x-coordinate, we can find BC and DA by subtracting their y-coordinates.

$$BC = DA = 2 - (-2) = 2 + 2 = 4$$

EXAMPLE 1

Graph the following points: $A(4, 1)$, $B(1, 5)$, $C(-2, 1)$. Then draw $\triangle ABC$ and find its area.

Solution The graph at the right shows $\triangle ABC$.

To find the area of the triangle, we need to know the lengths of the base and of the altitude drawn to that base. The base of $\triangle ABC$ is $\overline{AC}$.

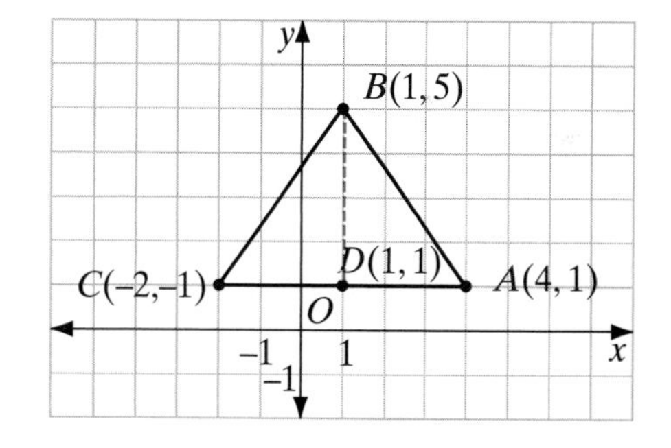

$$AC = 4 - (-2) = 4 + 2 = 6$$

The line segment drawn from B perpendicular to $\overline{AC}$ is the altitude $\overline{BD}$.

$$BD = 5 - 1 = 4$$

$$\begin{aligned} \text{Area} &= \tfrac{1}{2}(AC)(BD) \\ &= \tfrac{1}{2}(6)(4) \\ &= 12 \end{aligned}$$

Answer The area of $\triangle ABC$ is 12 square units.

EXERCISES

Writing About Mathematics

1. Mark is drawing triangle ABC on the coordinate plane. He locates points $A(-2, -4)$ and $C(5, -4)$. He wants to make $AC = BC$, $\angle C$ a right angle, and point B lie in the first quadrant. What must be the coordinates of point B? Explain how you found your answer.

2. Phyllis graphed the points $D(3, 0)$, $E(0, 5)$, $F(-2, 0)$, and $G(0, -4)$ on the coordinate plane and joined the points in order. Explain how Phyllis can find the area of this polygon, then find the area.

Developing Skills

3. Write as ordered number pairs the coordinates of points A, B, C, D, E, F, G, H, and O in the graph.

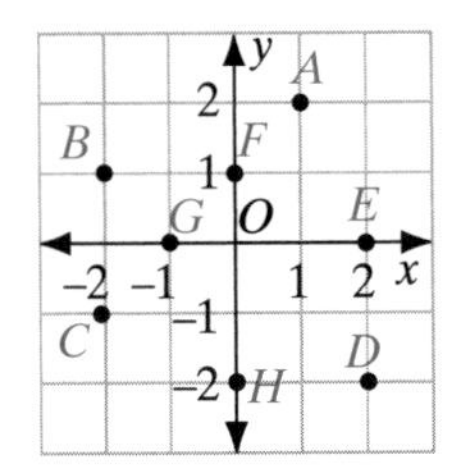

In 4–15, draw a pair of coordinate axes on graph paper and locate the point associated with each ordered number pair. Label each point with its coordinates.

4. $(5, 7)$ **5.** $(-3, 2)$ **6.** $(2, -6)$ **7.** $(-4, -5)$

8. $(1, 6)$ **9.** $(-8, 5)$ **10.** $(4, -4)$ **11.** $(5, 0)$

12. $(-3, 0)$ **13.** $(0, 4)$ **14.** $(0, -6)$ **15.** $(0, 0)$

In 16–20, name the quadrant in which the graph of each point described appears.

16. $(5, 7)$ **17.** $(-3, -2)$ **18.** $(-7, 4)$ **19.** $(1, -3)$ **20.** $(|-2|, |-3|)$

21. Graph several points on the x-axis. What is the value of the y-coordinate for every point in the set of points on the x-axis?

22. Graph several points on the y-axis. What is the value of the x-coordinate for every point in the set of points on the y-axis?

23. What are the coordinates of the origin in the coordinate plane?

Applying Skills

In 24–33: **a.** Graph the points and connect them with straight lines in order, forming a polygon. **b.** Identify the polygon. **c.** Find the area of the polygon.

24. $A(1, 1)$, $B(8, 1)$, $C(1, 5)$ **25.** $P(0, 0)$, $Q(5, 0)$, $R(5, 4)$, $S(0, 4)$

26. $C(8, -1)$, $A(9, 3)$, $L(4, 3)$, $F(3, -1)$ **27.** $H(-4, 0)$, $O(0, 0)$, $M(0, 4)$, $E(-4, 4)$

28. $H(5, -3), E(5, 3), N(-2, 0)$

29. $F(5, 1), A(5, 5), R(0, 5), M(-2, 1)$

30. $B(-3, -2), A(2, -2), R(2, 2), N(-3, 2)$

31. $P(-3, 0), O(0, 0), N(2, 2), D(-1, 2)$

32. $R(-4, 2), A(0, 2), M(0, 7)$

33. $M(-1, -1), I(3, -1), L(3, 3), K(-1, 3)$

34. Graph points $A(1, 1)$, $B(5, 1)$, and $C(5, 4)$. What must be the coordinates of point D if $ABCD$ is a rectangle?

35. Graph points $P(-1, -4)$ and $Q(2, -4)$. What are the coordinates of R and S if $PQRS$ is a square? (Two answers are possible.)

36. a. Graph points $S(3, 0)$, $T(0, 4)$, $A(-3, 0)$, and $R(0, -4)$, and draw the quadrilateral $STAR$.

b. Find the area of $STAR$ by adding the areas of the triangles into which the axes divide it.

37. a. Graph points $P(2, 0)$, $L(1, 1)$, $A(-1, 1)$, $N(-2, 0)$, $E(-1, -1)$, and $T(1, -1)$. Draw $PLANET$, a six-sided polygon called a hexagon.

b. Find the area of $PLANET$. (*Hint:* Use the x-axis to separate the hexagon into two parts.).

CHAPTER SUMMARY

A **binary operation** in a set assigns to every ordered pair of elements from the set a unique answer from that set. The general form of a binary operation is $a * b = c$, where a, b, and c are elements of the set and $*$ is the operation symbol. Binary operations exist in arithmetic, in geometry, and in sets.

Operations in arithmetic include addition, subtraction, multiplication, division, and raising to a power.

Powers are the result of repeated multiplication of the same **factor**, as in $5^3 = 125$. Here, the **base** 5 with an **exponent** of 3 equals $5 \times 5 \times 5$ or 125, the power.

Numerical expressions are simplified by following a clear **order of operations**:

(1) simplify within parentheses or other grouping symbols;

(2) simplify powers;

(3) multiply and divide from left to right;

(4) add and subtract from left to right.

Many **properties** are used in operations with real numbers, including:

- **closure** under addition, subtraction, and multiplication;
- **commutative** properties for addition and multiplication, $a * b = b * a$;
- **associative** properties for addition and multiplication, $(a * b) * c = a * (b * c)$;

- **distributive** property of multiplication over addition or subtraction, $a \cdot (b \pm c) = a \cdot b \pm a \cdot c$ and $(a \pm b) \cdot c = a \cdot c \pm b \cdot c$;
- **additive identity (0)**;
- **additive inverses** (opposites), called $-a$ for every element a;
- **multiplicative identity (1)**;
- **multiplicative inverses** (reciprocals), $\frac{1}{a}$ for every nonzero element a; the nonzero real numbers are closed under division.

Operations with sets include the **intersection** of sets, the **union** of sets, and the **complement** of a set.

Basic operations with signed numbers:

- To add two numbers that have the same sign, find the sum of the absolute values and give this sum the common sign.
- To add two numbers that have different signs, find the difference of the absolute values of the numbers. Give this difference the sign of the number that has the greater absolute value. The sum is 0 if both numbers have the same absolute value.
- To subtract one signed number from another, add the opposite (additive inverse) of the subtrahend to the minuend.
- To multiply two signed numbers, find the product of the absolute values. Write a plus sign before this product when the two numbers have the same sign. Write a minus sign before this product when the two numbers have different signs.
- To divide two signed numbers, find the quotient of the absolute values. Write a plus sign before this quotient when the two numbers have the same sign. Write a minus sign before this quotient when the two numbers have different signs. When 0 is divided by any number, the quotient is 0. Division by 0 is not defined.

The location of a point on a plane is given by an **ordered pair** of numbers that indicate the distance of the point from two reference lines, a horizontal line and a vertical line, called **coordinate axes**. The horizontal line is called the ***x*-axis**, the vertical line is the ***y*-axis**, and their intersection is the **origin**. The pair of numbers that are used to locate a point on the plane are called the **coordinates** of the point. The first number in the pair is called the ***x*-coordinate** or **abscissa**, and the second number is the ***y*-coordinate** or **ordinate**. The coordinates of a point are represented as (x, y).

VOCABULARY

2-1 Binary operation • Factor • Prime • Composite • Base • Exponent • Power • Order of operations • Parentheses • Brackets

2-2 Property • Closure • Commutative property • Associative property • Distributive property of multiplication over addition and subtraction • Addition property of zero • Additive identity • Additive inverse (opposite) • Multiplication property of one • Multiplicative inverse (reciprocal) • Multiplication property of zero

2-3 Property of the opposite of a sum

2-7 Universal set • Intersection • Disjoint sets • Empty set • Union • Complement

2-8 x-axis • y-axis • Coordinate axes • Coordinate plane • Origin • Quadrants • Coordinates • Ordered pair • x-coordinate • Abscissa • y-coordinate • Ordinate • Graph of the ordered pair

2-9 Polygon • Quadrilateral • Vertices

REVIEW EXERCISES

In 1–8, simplify each numerical expression.

1. $20 - 3 \times 4$ **2.** $(20 - 3) \times (-4)$ **3.** $-8 + 16 \div 4 - 2$

4. $(8 - 16) \div (4 - 2)$ **5.** $(0.16)^2$ **6.** $6^2 + 8^2$

7. $(6 + 8)^2$ **8.** $7(9 - 7)^3$

In 9–14: **a.** Replace each question mark with a number that makes the sentence true. **b.** Name the property illustrated in each sentence that is formed when the replacement is made.

9. $8 + (2 + 9) = 8 + (9 + ?)$ **10.** $8 + (2 + 9) = (8 + ?) + 9$

11. $3(?) = 3$ **12.** $3(?) = 0$

13. $5(7 + 4) = 5(7) + ?(4)$ **14.** $5(7 + 4) = (7 + 4)?$

In 15–24, the universe $U = \{1, 2, 3, 4, 5, 6\}$, set $A = \{1, 2, 4, 5\}$, and set $B = \{2, 4, 6\}$. In each case, perform the given operation and list the element(s) of the resulting set.

15. $A \cap B$ **16.** $A \cup B$ **17.** $\overline{A}$ **18.** $\overline{A} \cap B$ **19.** $A \cap \overline{A}$

20. $A \cup \overline{A}$ **21.** $A \cap \overline{B}$ **22.** $A \cup \overline{B}$ **23.** $\overline{A} \cap \overline{B}$ **24.** $\overline{A} \cup \overline{B}$

In 25–32, find each sum or difference.

25. $|6 - 6|$ **26.** $|6| + |-6|$ **27.** $|-3.2| - |-4.5|$ **28.** $|4| - |-5|$

29. $|-23| + |0|$ **30.** $|-54 + 52|$ **31.** $|100| + |-25|$ **32.** $|0 - 7|$

In 33–40, to each property named in Column I, match the correct application of the property found in Column II.

Column I	Column II
33. Associative property of multiplication	**a.** $3 + 4 = 4 + 3$
34. Associative property of addition	**b.** $3 \times 1 = 3$
35. Commutative property of addition	**c.** $0 \times 4 = 0$
36. Commutative property of multiplication	**d.** $3 + 0 = 3$
37. Identity element of multiplication	**e.** $3 \times 4 = 4 \times 3$
38. Identity element of addition	**f.** $3(4 + 5) = 3(4) + 3(5)$
39. Distributive property	**g.** $3(4 \times 5) = (3 \times 4)5$
40. Multiplication property of zero	**h.** $(3 + 4) + 5 = 3 + (4 + 5)$

41. a. Find the number on the odometer of Jesse's car described in the chapter opener on page 37.

b. What would the number be if it were an odd number?

42. a. Rectangle $ABCD$ is drawn with two sides parallel to the x-axis. The coordinates of vertex A are $(-2, -4)$ and the coordinates of C are $(3, 5)$. Find the coordinates of vertices B and D.

b. What is the area of rectangle $ABCD$?

43. Maurice answered all of the 60 questions on a multiple-choice test. The test was scored by using the formula $S = R - \frac{W}{4}$ where S is the score on the test, R is the number right, and W is the number wrong.

a. What is the lowest possible score?

b. How many answers did Maurice get right if his score was -5?

c. Is it possible for a person who answers all of the questions to get a score of -4? Explain why not or find the numbers of right and wrong answers needed to get this score.

d. Is it possible for a person who does not answer all of the questions to get a score of -4? Explain why not or find the numbers of right and wrong answers needed to get this score.

44. Solve the following problem using signed numbers.

Doug, the team's replacement quarterback, started out on his team's 30-yard line. On the first play, one of his linemen was offsides for a loss of 5 yards. On the next play, Doug gave the ball to the runningback who made a gain of 8 yards. He then made a 17-yard pass. Then Doug was tackled, for a loss of 3 yards. Where was Doug on the field after he was tackled?

Exploration

In this activity, you will derive a rule to determine if a number is divisible by 3. We will start our exploration with the number 23,568. For steps 1–4, fill in the blanks with a digit that will make the equality true.

STEP 1. Write the number as a sum of powers of ten.

$$23{,}568 = 20{,}000 + 3{,}000 + 500 + 60 + 8$$
$$= \square \times 10{,}000 + \square \times 1{,}000 + \square \times 100 + \square \times 10 + \square$$

STEP 2. Rewrite the powers of ten as a multiple of 9, plus 1.

$$\square \times 10{,}000 + \square \times 1{,}000 + \square \times 100 + \square \times 10 + \square$$
$$= \square \times (9{,}999 + 1) + \square \times (999 + 1) + \square \times (99 + 1) + \square \times (9 + 1) + \square$$

STEP 3. Use the distributive property to expand the product terms. Do not multiply out products involving the multiples of 9.

$$\square \times (9{,}999 + 1) + \square \times (999 + 1) + \square \times (99 + 1) + \square \times (9 + 1) + \square$$
$$= (\square \times 9{,}999 + \square) + (\square \times 999 + \square) + (\square \times 99 + \square) + (\square \times 9 + \square) + \square$$

STEP 4. Group the products involving the multiples of 9 first and then the remaining digit terms.

$$(\square \times 9{,}999 + \square) + (\square \times 999 + \square) + (\square \times 99 + \square) + (\square \times 9 + \square) + \square$$
$$= (\square \times 9{,}999 + \square \times 999 + \square \times 99 + \square \times 9) + (\square + \square + \square + \square + \square)$$

STEP 5. Compare the expression involving the digit terms with the original number. What do they have in common?

STEP 6. The expression involving the multiples of 9 is divisible by 3. Why?

STEP 7. If the expression involving the digit terms is divisible by 3, will the entire expression be divisible by 3? Explain.

STEP 8. Based on steps 1–7 write the rule to determine if a number is divisible by 3.

CUMULATIVE REVIEW CHAPTERS 1–2

Part I

Answer all questions in this part. Each correct answer will receive 2 credits. No partial credit will be allowed.

1. Which number is not an integer?

(1) 7 (2) -2 (3) 0.2 (4) $\sqrt{9}$

2. Which inequality is false?

(1) $4 > 3$ (2) $3 \geq 3$ (3) $-4 > -3$ (4) $+3 > -3$

3. What is the opposite of -4?

(1) $-\frac{1}{4}$ (2) $\frac{1}{4}$ (3) $+4$ (4) -4

4. Which of the following numbers has the greatest value?

(1) $\frac{5}{7}$ (2) 0.7 (3) $\frac{\pi}{5}$ (4) $\frac{25}{36}$

5. Under which operation is the set of integers *not* closed?

(1) Addition (2) Subtraction (3) Multiplication (4) Division

6. Which of the following numbers has exactly three factors?

(1) 1 (2) 2 (3) 9 (4) 15

7. The graph of the ordered pair $(-3, 5)$ is in which quadrant?

(1) I (2) II (3) III (4) IV

8. Put the following numbers in order, starting with the smallest:

$-\frac{1}{3}, -\frac{1}{5}, -\frac{2}{7}$

(1) $-\frac{1}{3} < -\frac{2}{7} < -\frac{1}{5}$ (3) $-\frac{1}{5} < -\frac{1}{3} < -\frac{2}{7}$

(2) $-\frac{1}{3} < -\frac{1}{5} < -\frac{2}{7}$ (4) $-\frac{1}{5} < -\frac{2}{7} < -\frac{1}{3}$

9. When rounded to the nearest hundredth, $\sqrt{3}$ is approximately equal to

(1) 1.732 (2) 1.730 (3) 1.73 (4) 1.74

10. For which value of t is $\frac{1}{t} > t > t^2$?

(1) 1 (2) 2 (3) 0 (4) $\frac{1}{2}$

Part II

Answer all questions in this part. Each correct answer will receive 2 credits. Clearly indicate the necessary steps, including appropriate formula substitutions, diagrams, graphs, charts, etc. For all questions in this part, a correct numerical answer with no work shown will receive only 1 credit.

11. Mrs. Ling spends more than \$4.90 and less than \$5.00 for meat for tonight's dinner. Write the set of all possible amounts that she could have paid for the meat. Is this a finite or an infinite set? Explain.

12. In a basketball league, 100 students play on 8 teams. Each team has at least 12 players. What is the largest possible number of players on any one team?

Part III

Answer all questions in this part. Each correct answer will receive 3 credits. Clearly indicate the necessary steps, including appropriate formula substitutions, diagrams, graphs, charts, etc. For all questions in this part, a correct numerical answer with no work shown will receive only 1 credit.

13. A teacher wrote the sequence 2, 4, 6, . . . and asked the class what the next number could be. Three students each gave a different answer. The teacher said that each of the answers was correct.

a. Josie said 8. Explain the rule that she used.

b. Emil said 10. Explain the rule that he used.

c. Ross said 12. Explain the rule that he used.

14. Evaluate the following expression without using a calculator. Show each step in your computation.

$$4(-7 + 3) - 8 \div (-2 + 6)^2$$

Part IV

Answer all questions in this part. Each correct answer will receive 4 credits. Clearly indicate the necessary steps, including appropriate formula substitutions, diagrams, graphs, charts, etc. For all questions in this part, a correct numerical answer with no work shown will receive only 1 credit.

15. The vertices of triangle ABC are $A(-2, -3)$, $B(5, -3)$, and $C(0, 4)$. Draw triangle ABC on the coordinate plane and find its area.

16. A survey to which 250 persons responded found that 140 persons said that they watch the news on TV at 6 o'clock, 120 persons said that they watch the news on TV at 11 o'clock and 40 persons said that they do not watch the news on TV at any time.

a. How many persons from this group watch the news both at 6 and at 11?

b. How many persons from this group watch the news at 6 but not at 11?

c. How many persons from this group watch the news at 11 but not at 6?

CHAPTER

3

ALGEBRAIC EXPRESSIONS AND OPEN SENTENCES

CHAPTER TABLE OF CONTENTS

An express delivery company will deliver a letter or package locally, within two hours. The company has the following schedule of rates. In addition to the basic charge of \$25, the cost is \$3 per mile or part of a mile for the first 10 miles or less and \$4.50 per mile or part of a mile for each additional mile over 10.

Costs such as those described, that vary according to a schedule, are often shown by a formula or set of formulas. Formulas can be used to solve many different problems.

In this chapter, you will learn to write algebraic expressions and formulas, to use algebraic expressions and formulas to solve problems, and to determine the solution set of an open sentence.

3-1 USING LETTERS TO REPRESENT NUMBERS

Eggs are usually sold by the dozen, that is, 12 in a carton. Therefore, we know that:

In 1 carton, there are 12×1 or 12 eggs.
In 2 cartons, there are 12×2 or 24 eggs.
In 3 cartons, there are 12×3 or 36 eggs.
In n cartons, there are $12 \times n$ or $12n$ eggs.

Here, n is called a **variable** or a **placeholder** that can represent different numbers from the set of whole numbers, $\{1, 2, 3, \ldots\}$. The set of numbers that can replace a variable is called the **domain** or the **replacement set** of that variable.

Recall that a numerical expression contains only numbers. An **algebraic expression**, such as $12n$, however, is an expression or phrase that contains one or more variables.

In this section, you will see how verbal phrases are translated into algebraic expressions, using letters as variables and using symbols to represent operations.

Verbal Phrases Involving Addition

The algebraic expression $a + b$ may be used to represent several different verbal phrases, such as:

a plus b	a added to b	a is increased by b
the sum of a and b	b is added to a	b more than a

The word *exceeds* means "is more than." Thus, the number that exceeds 5 by 2 can be written as "2 *more than* 5" or $5 + 2$.

Compare the numerical and algebraic expressions shown below.

A numerical expression: The number that exceeds 5 by 2 is $5 + 2$, or 7.
An algebraic expression: The number that exceeds a by b is $a + b$.

Verbal Phrases Involving Subtraction

The algebraic expression $a - b$ may be used to represent several different verbal phrases, such as:

a minus b	a decreased by b	b less than a
b subtracted from a	a diminished by b	a reduced by b

the difference between a and b

Verbal Phrases Involving Multiplication

The algebraic expressions $a \times b$, $a \cdot b$, $(a)(b)$ and ab may be used to represent several different verbal phrases, such as:

a times b the product of a and b b multiplied by a

The preferred form to indicate multiplication in algebra is ab. Here, a product is indicated by using *no symbol* between the variables being multiplied.

The multiplication symbol $\times$ is avoided in algebra because it can be confused with the letter or variable x. The raised dot, which is sometimes mistaken for a decimal point, is also avoided. Parentheses are used to write numerical expressions: $(3)(5)(2)$ or $3(5)(2)$. Note that all but the first number must be in parentheses. In algebraic expressions, parentheses may be used but they are not needed: $3(b)(h) = 3bh$.

Verbal Phrases Involving Division

The algebraic expressions $a \div b$ and $\frac{a}{b}$ may be used to represent several different verbal phrases, such as:

a divided by *b*   the quotient of *a* and *b*

The symbols $a \div 4$ and $\frac{a}{4}$ mean one-fourth of *a* as well as *a* divided by 4.

Phrases and Commas

In some verbal phrases, using a comma can prevent misreading. For example, in "the product of x and y, decreased by 2," the comma after y makes it clear that the x and y are to be multiplied before subtracting 2 and can be written as $(xy) - 2$ or $xy - 2$. Without the comma, the phrase, "the product of x and y decreased by 2," would be written $x(y - 2)$.

EXAMPLE 1

Use mathematical symbols to translate the following verbal phrases into algebraic language:

	Answers
a. w more than 3	$3 + w$
b. w less than 3	$3 - w$
c. r decreased by 2	$r - 2$
d. the product of $5r$ and s	$5rs$
e. twice x, decreased by 10	$2x - 10$
f. 25, diminished by 4 times n	$25 - 4n$
g. the sum of t and u, divided by 6	$\frac{t+u}{6}$
h. 100 decreased by twice $(x + 5)$	$100 - 2(x + 5)$

EXERCISES

Writing About Mathematics

1. Explain why the sum of a and 4 can be written as $a + 4$ or as $4 + a$.

2. Explain why 3 less than a can be written as $a - 3$ but not as $3 - a$.

Developing Skills

In 3–20, use mathematical symbols to translate the verbal phrases into algebraic language.

3. y plus 8
4. 4 minus r
5. 7 times x
6. x times 7
7. x divided by 10
8. 10 divided by x
9. c decreased by 6
10. one-tenth of w
11. the product of x and y
12. 5 less than d
13. 8 divided by y
14. y multiplied by 10
15. t more than w
16. one-third of z
17. twice the difference of p and q
18. a number that exceeds m by 4
19. 5 times x, increased by 2
20. 10 decreased by twice a

In 21–30, using the letter n to represent "number," write each verbal phrase in algebraic language.

21. a number increased by 2
22. 20 more than a number
23. 8 increased by a number
24. a number decreased by 6
25. 2 less than a number
26. 3 times a number
27. three-fourths of a number
28. 4 times a number, increased by 3
29. 3 less than twice a number
30. 10 times a number, decreased by 2

In 31–34, use the given variable(s) to write an algebraic expression for each verbal phrase.

31. the number of baseball cards, if b cards are added to a collection of 100 cards

32. Hector's height, if he was h inches tall before he grew 2 inches

33. the total cost of n envelopes that cost $0.39 each

34. the cost of one pen, if 12 pens cost d dollars

3-2 TRANSLATING VERBAL PHRASES INTO SYMBOLS

A knowledge of arithmetic is important in algebra. Since the variables represent numbers that are familiar to you, it will be helpful to solve each problem by first using a simpler related problem; that is, relate similar arithmetic problems to the given algebraic one.

Procedure

To write an algebraic expression involving variables:

1. Think of a similar problem in arithmetic.
2. Write an expression for the arithmetic problem, using numbers.
3. Write a similar expression for the problem, using letters or variables.

EXAMPLE 1

Represent each phrase by an algebraic expression.

a. a distance that is 20 meters shorter than x meters

b. a bill for n baseball caps, each costing d dollars

c. a weight that is 40 pounds heavier than p pounds

d. an amount of money that is twice d dollars

Solution **a.**

How to Proceed	
(1) Think of a similar problem in arithmetic:	Think of a distance that is 20 meters shorter than 50 meters.
(2) Write an expression for this arithmetic problem:	$50 - 20$
(3) Write a similar expression using the letter x in place of 50.	$x - 20$ *Answer*

b.

How to Proceed	
(1) Think of a similar problem in arithmetic:	Think of a bill for 6 caps, each costing 5 dollars.
(2) Write an expression for this arithmetic problem. Multiply the number of caps by the cost of one cap:	$6(5)$
(3) Write a similar expression using n and d:	nd *Answer*

Note: After some practice, you will be able to do steps (1) and (2) mentally.

c. $(p + 40)$ pounds or $(40 + p)$ pounds

d. $2d$ dollars

Answers **a.** $(x - 20)$ meters **b.** nd dollars

c. $(p + 40)$ or $(40 + p)$ pounds **d.** $2d$ dollars

EXAMPLE 2

Brianna paid 17 dollars for batteries and film for her camera. If the batteries cost x dollars, express the cost of the film in terms of x.

Solution If Brianna had spent 5 dollars for the batteries, the amount that was left is found by subtracting the 5 dollars from the 17 dollars, $(17 - 5)$ dollars. This would have been the cost of the film. If Brianna spent x dollars for the batteries, then the difference, $(17 - x)$ dollars would have been the cost of the film.

Answer $(17 - x)$ dollars

Note: In general, if we know the sum of two quantities, then we can let x represent one of these quantities and (the sum $- x$) represent the other.

EXERCISES

Writing About Mathematics

1. a. Represent the number of pounds of grapes you can buy with d dollars if each pound costs b dollars.

b. Does the algebraic expression in part **a** always represent a whole number? Explain your answer by showing examples using numbers.

2. a. If x apples cost c cents, represent the cost of one apple.

b. If x apples cost c cents, represent the cost of n apples.

c. Do the algebraic expressions in parts **a** and **b** always represent whole numbers? Explain your answer.

Developing Skills

In 3–18, represent each answer in algebraic language, using the variable mentioned in the problem.

3. The number of kilometers traveled by a bus is represented by x. If a train traveled 200 kilometers farther than the bus, represent the number of kilometers traveled by the train.

4. Mr. Gold invested \$1,000 in stocks. If he lost d dollars when he sold the stocks, represent the amount he received for them.

5. The cost of a mountain bike is 5 times the cost of a skateboard. If the skateboard costs x dollars, represent the cost of the mountain bike.

6. The length of a rectangle is represented by l. If the width of the rectangle is one-half of its length, represent its width.

7. After 12 centimeters had been cut from a piece of lumber, c centimeters were left. Represent the length of the original piece of lumber.

8. Paul and Martha saved 100 dollars. If the amount saved by Paul is represented by x, represent the amount saved by Martha.

9. A ballpoint pen sells for 39 cents. Represent the cost of x pens.

10. Represent the cost of t feet of lumber that sells for g cents a foot.

11. If Hilda weighed 45 kilograms, represent her weight after she had lost x kilograms.

12. Ronald, who weighs c pounds, is d pounds overweight. Represent the number of pounds Ronald should weigh.

13. A woman spent $250 for jeans and a ski jacket. If she spent y dollars for the ski jacket, represent the amount she spent for the jeans.

14. A man bought an article for c dollars and sold it at a profit of $25. Represent the amount for which he sold it.

15. The width of a rectangle is represented by w meters. Represent the length of the rectangle if it exceeds the width by 8 meters.

16. The width of a rectangle is x centimeters. Represent the length of the rectangle if it exceeds twice the width by 3 centimeters.

17. If a plane travels 550 kilometers per hour, represent the distance it will travel in h hours.

18. If a car traveled for 5 hours at an average rate of r kilometers per hour, represent the distance it traveled.

19. **a.** Represent the total number of days in w weeks and 5 days.

b. Represent the total number of days in w weeks and d days.

Applying Skills

20. An auditorium with m rows can seat a total of c people. If each row in the auditorium has the same number of seats, represent the number of seats in one row.

21. Represent the total number of calories in x peanuts and y potato chips if each peanut contains 6 calories and each potato chip contains 14 calories.

22. The charges for a telephone call are $0.45 for the first 3 minutes and $0.09 for each additional minute or part of a minute. Represent the cost of a telephone call that lasts m minutes when m is greater than 3.

23. A printing shop charges a 75-cent minimum for the first 8 photocopies of a flyer. Additional copies cost 6 cents each. Represent the cost of c copies if c is greater than 8.

24. A utility company measures gas consumption by the hundred cubic feet, CCF. The company has a three-step rate schedule for gas customers. First, there is a minimum charge of $5.00 per month for up to 3 CCF of gas used. Then, for the next 6 CCF, the charge is $0.75 per CCF. Finally, after 9 CCF, the charge is $0.55 per CCF. Represent the cost of g CCF of gas if g is greater than 9.

3-3 ALGEBRAIC TERMS AND VOCABULARY

Terms

A **term** is a number, a variable, or any *product* or *quotient* of numbers and variables. For example: 5, x, $4y$, $8ab$, $\frac{k^2}{5}$, and $\frac{5}{c}$ are terms.

An algebraic expression that is written as a *sum* or a *difference* has more than one term. For example, $4a + 2b - 5c$ has three terms. These terms, $4a$, $2b$, and $5c$, are separated by $+$ and $-$ signs.

Factors of a Term

If a term contains two or more numbers or variables, then each number, each variable, and each product of numbers and variables is called a **factor** of the term, or factor of the product. For example, the factors of $3xy$ are 1, 3, x, y, $3x$, $3y$, xy, and $3xy$. When we factor whole numbers, we write only factors that are integers.

Any factor of an algebraic term is called the **coefficient** of the remaining factor, or product of factors, of that term. For example, consider the algebraic term $3xy$:

3 is the coefficient of xy $\quad$ $3x$ is the coefficient of y
$3y$ is the coefficient of x $\quad$ xy is the coefficient of 3

When an algebraic term consists of a number and one or more variables, the number is called the **numerical coefficient** of the term. For example:

In $8y$, the numerical coefficient is 8.
In $4abc$, the numerical coefficient is 4.

When the word *coefficient* is used alone, it usually means a numerical coefficient. Also, since x names the same term as $1x$, the coefficient of x is understood to be 1. This is true of all terms that contain only variables. For example:

7 is the coefficient of b in the term $7b$. $\quad$ 1 is the coefficient of b in the term b.
2.25 is the coefficient of gt in the term $2.25gt$. $\quad$ 1 is the coefficient of gt in the term gt.

Bases, Exponents, and Powers

You learned in Chapter 2 that a *power* is the product of equal factors. A power has a base and an exponent.

The *base* is one of the equal factors of the power.

The *exponent* is the number of times the base is used as a factor. (If a term is written without an exponent, the exponent is understood to be 1.)

$4^2 = 4(4)$:	base = 4	exponent = 2	power = $4^2 = 16$
$x^3 = x(x)(x)$:	base = x	exponent = 3	power = x^3
$35m$:	base = m	exponent = 1	power = m^1 or m
$5d^2 = 5(d)(d)$:	base = d	exponent = 2	power = d^2

An exponent refers only to the number or variable that is directly to its left, as seen in the last example, where 2 refers only to the base d. To show the product $5d$ as a base (or to show any sum, difference, product, or quotient as a base), we must enclose the base in parentheses.

$(5d)^2 = (5d)(5d)$:	base = $5d$	exponent = 2	power = $(5d)^2$
$(a + 4)^2 = (a + 4)(a + 4)$:	base = $(a + 4)$	exponent = 2	power = $(a + 4)^2$

Note that $-d^4$ is not the same as $(-d)^4$.

$-d^4 = -1(d)(d)(d)(d)$ is always a negative number.
$(-d)^4 = 1(-d)(-d)(-d)(-d)$ is always a positive number since the exponent is even.

EXAMPLE 1

For each term, name the coefficient, base, and exponent.

Answers

a. $4x^5$	coefficient = 4	base = x	exponent = 5
b. $-w^8$	coefficient = –1	base = w	exponent = 8
c. $2\pi r$	coefficient = 2π	base = r	exponent = 1

Note: Remember that coefficient means *numerical coefficient*, and that 2π is a real number.

EXERCISES

Writing About Mathematics

1. Does squaring distribute over multiplication, that is, does $(ab)^2 = (a^2)(b^2)$? Write $(ab)^2$ as $(ab)(ab)$ and use the associative and commutative properties of multiplication to justify your answer.

2. Does squaring distribute over addition, that is, does $(a + b)^2 = a^2 + b^2$? Substitute values for a and b to justify your answer.

Developing Skills

In 3–6, name the factors (other than 1) of each product.

3. xy **4.** $3a$ **5.** $7mn$ **6.** $1st$

In 7–14, name, in each case, the numerical coefficient of x.

7. $8x$ **8.** $(5 + 2)x$ **9.** $\frac{1}{2}x$ **10.** x

11. $-1.4x$ **12.** $2 + 7x$ **13.** $3.4x$ **14.** $-x$

In 15–22, name, in each case, the base and exponent of the power.

15. m^2 **16.** $-s^3$ **17.** t **18.** $(-a)^4$

19. 10^6 **20.** $(5y)^4$ **21.** $(x + y)^5$ **22.** $12c^3$

In 23–29, write each expression, using exponents.

23. $b \cdot b \cdot b \cdot b \cdot b$ **24.** $\pi \cdot r \cdot r$ **25.** $a \cdot a \cdot a \cdot a \cdot b \cdot b$

26. $7 \cdot r \cdot r \cdot r \cdot s \cdot s$ **27.** $(6a)(6a)(6a)$ **28.** $(a - b)(a - b)(a - b)$

29. the fourth power of $(m + 2n)$

In 30–33, write each term as a product without using exponents.

30. r^6 **31.** $5x^4$ **32.** $4a^4b^2$ **33.** $(3y)^5$

In 34–41, name, for each given term, the coefficient, base, and exponent.

34. $-3k$ **35.** $-k^3$ **36.** πr^2 **37.** $(ax)^5$

38. $\sqrt{2}y$ **39.** $0.0004t^{12}$ **40.** $\frac{3}{2}a^4$ **41.** $(-b)^3$

Applying Skills

42. If x represents the cost of a can of soda, what could $5x$ represent?

43. If r represents the speed of a car in miles per hour, what could $3r$ represent?

44. If n represents the number of CDs that Alice has, what could $n - 5$ represent?

45. If d represents the number of days until the end of the year, what could $\frac{d}{7}$ represent?

46. If s represents the length of a side of a square, what could $4s$ represent?

47. If r represents the measure of the radius of a circle, what could $2r$ represent?

48. If w represents the number of weeks in a school year, what could $\frac{w}{4}$ represent?

49. If d represents the cost of one dozen bottles of water, what could $\frac{d}{12}$ represent?

50. If q represents the point value of one field goal, what could $7q$ represent?

3-4 WRITING ALGEBRAIC EXPRESSIONS IN WORDS

In Section 1 of this chapter, we listed the words that can be represented by each of the four basic operations. We can use these same lists to write algebraic expressions in words and to write problems that can be represented by a given algebraic expression.

For an algebraic expression such as $2n - 3$, n could be any real number. That is, associated with any real number n, there is exactly one real number that is the value of $2n - 3$. However, if n and $2n - 3$ represent the number of cans of tuna that two customers buy, then n must be a whole number greater than or equal to 2 in order for both n and $2n - 3$ to be whole numbers.

For this situation, the domain or replacement set would be the set of whole numbers.

EXAMPLE 1

If n represents the number of points that Hradish scored in a basketball game and $2n - 3$ represents the number of points that his friend Brad scored, describe in words the number of points that Brad scored. What is a possible domain for the variable n?

Solution The number of points scored is always a whole number. In order that $2n - 3$ be a whole number, n must be at least 2.

Answer The number of points that Brad scored is 3 less than twice the number that Hradish scored. A possible domain for n is the set of whole numbers greater than or equal to 2.

EXAMPLE 2

Molly earned d dollars in July and $\frac{1}{2}d + 10$ dollars in August. Describe in words the number of dollars that Molly earned in August.

Answer In August, Molly earned 10 more than half the number of dollars that she earned in July.

EXAMPLE 3

Describe a situation in which x and $12 - x$ can be used to represent variable quantities. List the domain or replacement set for the answer.

Solution If x eggs are used from a full dozen of eggs, there will be $12 - x$ eggs left.

Answer The domain or replacement set is the set of whole numbers less than or equal to 12.

Another Solution The distance from my home to school is 12 miles. On my way to school, after I have traveled x miles, I have $12 - x$ miles left to travel.

Answer The domain or replacement set is the set of non-negative real numbers that are less than or equal to 12.

Many other answers are possible.

EXERCISES

Writing About Mathematics

1. a. If $4 + n$ represents the number of books Ken read in September and $4 - n$ represents the number of books he read in October, how many books did he read in these two months?

b. What is the domain of the variable n?

2. Pedro said that the replacement set for the amount that we pay for any item is the set of rational numbers of the form $0.01x$ where x is a whole number. Do you agree with Pedro? Explain why or why not.

Developing Skills

In 3–14: **a.** Write in words each of the given algebraic expressions. **b.** Describe a possible domain for each variable.

3. By one route, the distance that Ian walks to school is d miles. By a different route, the distance is $d - 0.2$ miles.

4. Juan pays n cents for a can of soda at the grocery store. When he buys soda from a machine, he pays $n + 15$ cents.

5. Yesterday Alexander spent a minutes on leisurely reading and $3a + 10$ minutes doing homework.

6. The width of a rectangle is w meters and the length is $2w + 8$ meters.

7. During a school day, Abby spends h hours in class, $\frac{h}{6}$ hours at lunch and $\frac{h}{3}$ hours on sports.

8. Jen spends d hours at work and $\frac{d}{12}$ hours driving to and from work.

9. Alicia's score for 18 holes of golf was g and her son's score was $10 + g$.

10. Tom paid d cents for a notebook and $5d + 30$ cents for a pen.

11. Seema's essay for English class had w words and Dominic's had $\frac{3}{4}w + 80$ words.

12. Virginia read r books last month and Anna read $3r - 5$ books.

13. Mario and Pete are playing a card game where it is possible to have a negative score. Pete's score is s and Mario's score is $s - 220$.

14. In the past month, Agatha has increased the time that she walks each day from m minutes to $3m - 10$ minutes.

3-5 EVALUATING ALGEBRAIC EXPRESSIONS

Benjamin has 1 more tape than 3 times the number of tapes that Julia has. If Julia has n tapes, then Benjamin has $3n + 1$ tapes.

The algebraic expression $3n + 1$ represents an unspecified number. Only when the variable n is replaced by a specific number does $3n + 1$ become a specific number. For example:

If $n = 10$, then $3n + 1 = 3(10) + 1 = 30 + 1 = 31$.
If $n = 15$, then $3n + 1 = 3(15) + 1 = 45 + 1 = 46$.

Since in this example, n represents the number of tapes that Julia has, only whole numbers are reasonable replacements for n. Therefore, the replacement set is the set of whole numbers or some subset of the set of whole numbers.

When we substitute specific values for the variables in an algebraic expression and then determine the value of the resulting expression, we are **evaluating the algebraic expression.**

When we determine the number that an algebraic expression represents for specific values of its variables, we are **evaluating the algebraic expression.**

Procedure

To evaluate an algebraic expression, replace the variables by the given values, and then follow the rules for the order of operations.

1. Replace the variables by the given values.
2. Evaluate the expression within the grouping symbols such as parentheses, always simplifying the expressions in the innermost groupings first.
3. Simplify all powers and roots.
4. Multiply and divide, from left to right.
5. Add and subtract, from left to right.

EXAMPLE 1

Evaluate $50 - 3x$ when $x = 7$.

Solution *How to Proceed*

(1) Write the expression: $50 - 3x$

(2) Replace the variable by its given value:	$50 - 3(7)$
(3) Multiply:	$50 - 21$
(4) Subtract:	29

Answer 29

EXAMPLE 2

Evaluate $2x^2 - 5x + 4$ when: **a.** $x = -7$ **b.** $x = 1.2$

Solution *How to Proceed*

a. (1) Write the expression:	$2x^2 - 5x + 4$
(2) Replace the variable by the value -7:	$2(-7)^2 - 5(-7) + 4$
(3) Evaluate the power:	$2(49) - 5(-7) + 4$
(4) Multiply:	$98 + 35 + 4$
(5) Add:	137
b. (1) Write the expression:	$2x^2 - 5x + 4$
(2) Replace the variable by the value 1.2:	$2(1.2)^2 - 5(1.2) + 4$
(3) Evaluate the power:	$2(1.44) - 5(1.2) + 4$
(4) Multiply:	$2.88 - 6 + 4$
(5) Add and subtract:	0.88

Answers **a.** 137 **b.** 0.88

EXAMPLE 3

Evaluate $\frac{2a}{5} + (n - 1)d$ when $a = -4, n = 10$, and $d = 3$.

Solution *How to Proceed*

(1) Write the expression:	$\frac{2a}{5} + (n - 1)d$
(2) Replace the variables with their given values:	$\frac{2(-4)}{5} + (10 - 1)(3)$
(3) Simplify the expressions grouped by parentheses or fraction bar:	$\frac{-8}{5} + (9)(3)$
(4) Multiply and divide:	$-1\frac{3}{5} + 27$
(5) Add:	$-1\frac{3}{5} + 26\frac{5}{5}$
	$25\frac{2}{5}$ *Answer*

Calculator Solution The values given for the variables can be stored in the calculator.

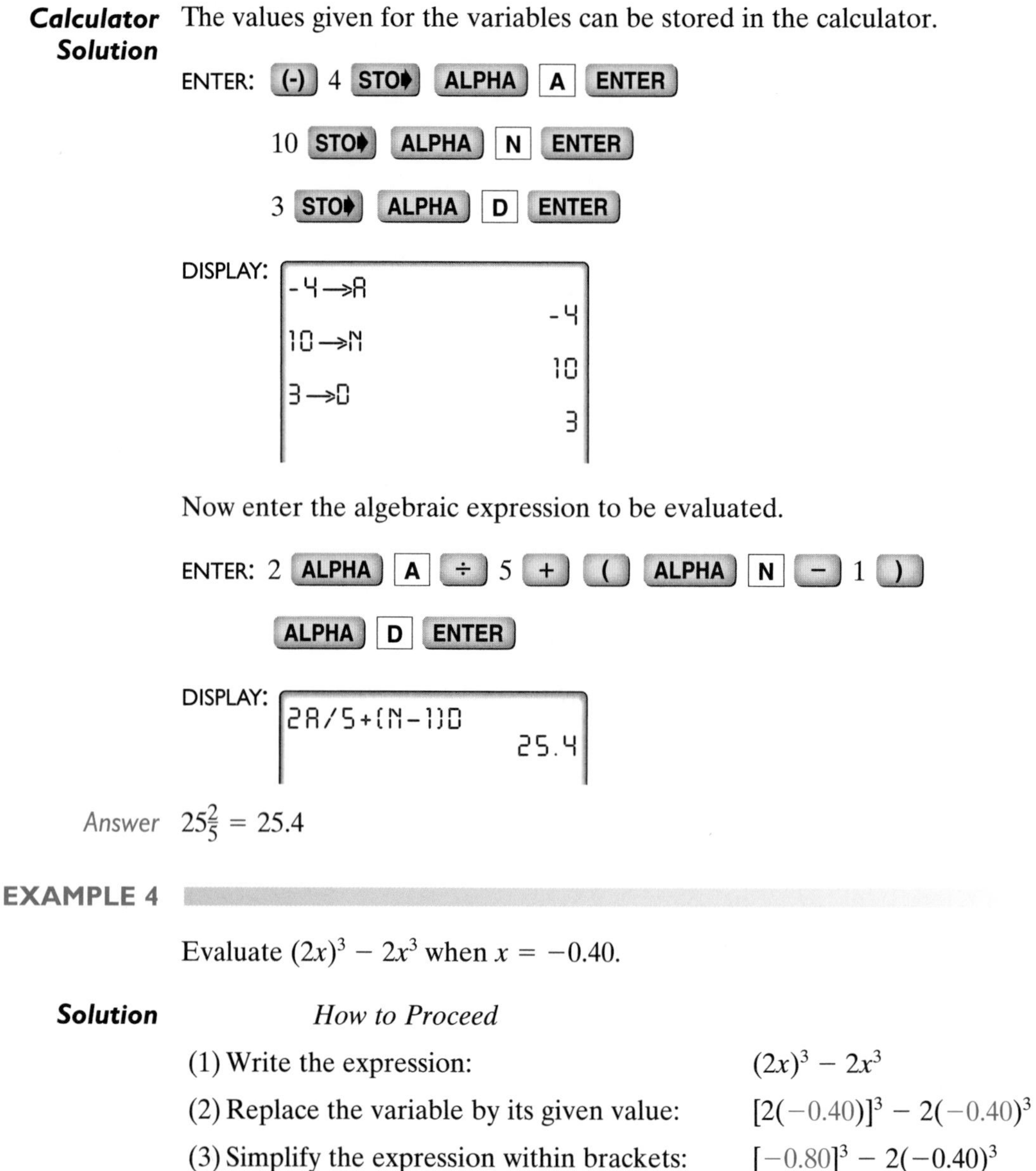

Now enter the algebraic expression to be evaluated.

Answer $25\frac{2}{5} = 25.4$

EXAMPLE 4

Evaluate $(2x)^3 - 2x^3$ when $x = -0.40$.

Solution

How to Proceed	
(1) Write the expression:	$(2x)^3 - 2x^3$
(2) Replace the variable by its given value:	$[2(-0.40)]^3 - 2(-0.40)^3$
(3) Simplify the expression within brackets:	$[-0.80]^3 - 2(-0.40)^3$
(4) Evaluate the powers:	$-0.512 - 2(-0.064)$
(5) Multiply:	$-0.512 + 0.128$
(6) Subtract:	-0.384

Answer -0.384

EXERCISES

Writing About Mathematics

1. Explain why, in an algebraic expression such as $12ab$, 12 is called a constant and a and b are called variables?

2. Explain why, in step 2 of Example 1, parentheses were needed when x was replaced by its value.

Developing Skills

To understand this topic, you should first evaluate the expressions in Exercises 3 to 27 without a calculator. Then, store the values of the variables in the calculator and enter the given algebraic expressions to check your work.

In 3–27, find the numerical value of each expression. Use $a = 8, b = -6, d = 3, x = -4$, and $y = 0.5$.

3. $5a$ **4.** $\frac{1}{2}x$ **5.** $0.3y$ **6.** $a + 3$

7. $b - 2$ **8.** ax^2 **9.** $\frac{3bd}{9}$ **10.** $5x - 2y$

11. $7xy^3$ **12.** $ab - dx$ **13.** $\frac{2}{5}a + \frac{1}{5}b$ **14.** $0.2d + 0.3b$

15. $\frac{3}{4}x^3$ **16.** $(3y)^2$ **17.** $\frac{1}{4}x^2y$ **18.** $a^2 + 3d^2$

19. $(ay)^3$ **20.** $x(y - 2)$ **21.** $4(2x + 3y)$ **22.** $\frac{1}{2}x(y + 0.1)^2$

23. $3y - (x - d)$ **24.** $2(x + y) - 5$ **25.** $(x - d)^5$

26. $(2a - 5d)^2$ **27.** $(2a)^2 - (5d)^2$

Applying Skills

28. At one car rental agency, the cost of a car for one day can be determined by using the algebraic expression \$32.00 + \$0.10m where m represents the number of miles driven. Determine the cost of rental for each of the following:

a. Mike Baier drove the car he rented for 35 miles.

b. Dana Morse drove the car he rented for 435 miles.

c. Jim Szalach drove the car he rented for 102 miles.

29. The local pottery co-op charges \$40.00 a year for membership and \$0.75 per pound for firing pottery pieces made by the members. The algebraic expression $40 + 0.75p$ represents the yearly cost to a member who brings p pounds of pottery to be fired. Determine the yearly cost for each of the following:

a. Tiffany is an amateur potter who fired 35 pounds of work this year.

b. Nia sells her pottery in a local craft shop and fired 485 pounds of work this year.

30. If a stone is thrown down into a deep gully with an initial velocity of 30 feet per second, the distance it has fallen, in feet, after t seconds can be found by using the algebraic expression $16t^2 + 30t$. Find the distance the stone has fallen:

a. after 1 second. **b.** after 2 seconds. **c.** after 3 seconds.

31. The Parkside Bread Company sells cookies and scones as well as bread. Bread (b) costs \$4.50 a loaf, cookies (c) cost \$1.10 each, and scones (s) cost \$1.50 each. The cost of a bakery order can be represented by $4.50b + 1.10c + 1.50s$. Determine the cost of each of the following orders:

a. six cookies and two scones

b. three loaves of bread and one cookie

c. one loaf of bread, a dozen cookies, and a half-dozen scones

32. A Green Thumb volunteer can plant shrubbery at a rate of 6 shrubs per hour and a Friendly Garden volunteer can plant shrubbery at a rate of 8 shrubs per hour. The total number of shrubs that g Green Thumb volunteers and f Friendly Garden volunteers can plant in h hours is given by the algebraic expression $6gh + 8fh$. Determine the number of shrubs planted:

a. in 3 hours by 2 Green Thumb and 1 Friendly Garden volunteers.

b. in 2 hours by 4 Green Thumb and 4 Friendly Garden volunteers.

3-6 OPEN SENTENCES AND SOLUTION SETS

In this chapter, you learned how to translate words into algebraic expressions. The value of an algebraic expression depends on the value of the variables. When the values of the variables change, the value of the algebraic expression changes. For example, $x + 6$ is an algebraic expression. The value of $x + 6$ depends on the value of x.

If one value is assigned to an algebraic expression, an algebraic sentence is formed. These sentences may be formulas, equations, or inequalities. For example, when the value 9 is assigned to the algebraic expression $x + 6$, we can write the sentence "Six more than x is 9." This sentence can be written in symbols as $x + 6 = 9$.

Every sentence that contains a variable is called an **open sentence**.

$x + 6 = 9 \qquad 3y = 12 \qquad 2n > 0 \qquad x + 5 \leq 8$

An open sentence is neither true nor false.

The sentence will be true or false only when the variables are replaced by numbers from a domain or a replacement set, such as {0, 1, 2, 3}.

The numbers from the domain that make the sentence *true* are the elements of the **solution set** of the open sentence. A solution set, as seen below, can contain one or more numbers or, at times, no numbers at all, from the replacement set.

EXAMPLE 1

Using the domain {0, 1, 2, 3}, find the solution set of each open sentence: **a.** $x + 6 = 9$ **b.** $2n > 0$

Solution **a.** *Procedure:* Replace x in the open sentence with numbers from the domain $\{0, 1, 2, 3\}$.

$x + 6 = 9$

Let $x = 0$.
Then $0 + 6 = 9$ is false.

Let $x = 1$.
Then $1 + 6 = 9$ is false.

Let $x = 2$.
Then $2 + 6 = 9$ is false.

Let $x = 3$.
Then $3 + 6 = 9$ is true.

Here, only when $x = 3$ does the open sentence become a true sentence.

Answer: **a.** Solution set = $\{3\}$.

b. *Procedure:* Replace n in the open sentence with numbers from the domain $\{0, 1, 2, 3\}$.

$2n > 0$

Let $n = 0$.
Then $2(0) > 0$ or $0 > 0$ is false.

Let $n = 1$.
Then $2(1) > 0$ or $2 > 0$ is true.

Let $n = 2$.
Then $2(2) > 0$ or $4 > 0$ is true.

Let $n = 3$.
Then $2(3) > 0$ or $6 > 0$ is true.

Here, three elements of the domain make the open sentence true.

Answer: **b.** Solution set = $\{1, 2, 3\}$

EXAMPLE 2

Find the solution set for the open sentence $3y = 12$ using:
a. the domain = $\{3, 5, 7\}$ **b.** the domain = {whole numbers}

Solution **a.** *Procedure:* Replace y with 3, 5, and 7.

If $y = 3$, then $3(3) = 12$ is false.

If $y = 5$, then $3(5) = 12$ is false.

If $y = 7$, then $3(7) = 12$ is false.

When y is replaced by each of the numbers from the domain, no true statement is found. The solution set *for this domain* is the empty set or the null set, written in symbols as { } or as $\varnothing$.

Answer: **a.** The solution set is { } or $\varnothing$.

b. *Procedure:* Of course, you cannot replace y with every whole number, but you can use multiplication facts learned previously.

You know that $3(4) = 12$. Let $y = 4$. Then $3(4) = 12$ is true. No other whole number would make the open sentence $3y = 12$ a true sentence.

Answer: **b.** The solution set is $\{4\}$.

EXERCISES

Writing About Mathematics

1. For the open sentence $x + 7 > 12$, write a domain for which the solution set is the empty set.

2. For the open sentence $x + 7 > 12$, write a domain for which the solution set has only one element.

3. For the open sentence $x + 7 > 12$, write a domain for which the solution set is an infinite set.

Developing Skills

In 4–11, tell whether each is an open sentence, a true sentence, or an algebraic expression.

4. $2 + 3 = 5 + 0$ **5.** $x + 10 = 14$ **6.** $y - 4$ **7.** $3 + 7 = 2(5)$

8. $n > 7$ **9.** $3 + 2 < 10$ **10.** $3r + 2$ **11.** $2x - 7 = 15$

In 12–15, name the variable in each open sentence.

12. $x + 5 = 9$ **13.** $4y = 20$ **14.** $r - 6 > 2$ **15.** $7 < 3 + a$

In 16–23, using the domain $\{-5, -4, -3, -2, -1, 0, 1, 2, 3, 4, 5\}$, find the solution set for each open sentence.

16. $n + 3 = 7$ **17.** $x - x = 0$ **18.** $5 - n = 2$ **19.** $n + 3 > 9$

20. $2n + 1 < 8$ **21.** $\frac{2n+1}{3} = 4$ **22.** $\frac{3x}{2} < x$ **23.** $2x < -4$

Applying Skills

24. Pencils sell for \$0.19 each. Torry wants to buy at least one but not more than 10 pencils and has \$1.50 in his pocket.

a. Use the number of pencils that Torry wants to buy to write a domain for this problem.

b. The number of pencils that Torry might buy, x, can be found using the open sentence $0.19x \leq 1.50$. Find the solution set of this open sentence using the domain from part **a**.

c. How many pencils can Torry buy?

25. The local grocery store has frozen orange juice on sale for \$0.99 a can but limits the number of cans that a customer may buy at the sale price to no more than 5.

a. The domain for this problem is the number of cans of juice that a customer may buy at the sale price. Write the domain.

b. If Mrs. Dajhon does not want to spend more than \$10, the number of cans that she might buy at the sale price, y, is given by the equation $0.99y \leq 10$. Find the solution set of this equation using the domain from part **a**.

c. How many cans can Mrs. Dajhon buy if she does not want to spend more than \$10?

26. Admission to a recreation park is \$17.50. This includes all rides except for a ride called *The Bronco* that costs \$1.50 for each ride. Ian has \$25 to spend.

a. Find the domain for this problem, the number of times a person might ride *The Bronco*.

b. The number of times Ian might ride *The Bronco*, z, can be found using the open sentence $17.50 + 1.50z \leq 25$. Find the solution set of this open sentence using the domain from part **a**.

c. How many times can Ian ride *The Bronco*?

3-7 WRITING FORMULAS

A **formula** uses mathematical language to express the relationship between two or more variables. Some formulas are found by the strategy of looking for patterns. For example, how many square units are shown in the rectangle on the left? This rectangle, measuring 4 units in length and 3 units in width, contains a total of 12 square units of area.

Many such examples led to the conclusion that the area of a rectangle is equal to the product of its length and width. This relationship is expressed by the formula $A = lw$ where A, l, and w are variables that represent, respectively, the area, the length, and the width of a rectangle.

A formula is an open sentence that states that two algebraic expressions are equal. In formulas, the word *is* is translated into the symbol =.

EXAMPLE 1

Write a formula for each of the following relationships.

a. The perimeter P of a square is equal to 4 times the length of one side.

b. The total cost C of an article is equal to its price p plus an 8% tax on the price.

c. The sum S of the measures of the interior angles of an n-sided polygon is 180 times 2 less than the number of sides.

Solution **a.** Let s represent the length of each side of a square.

$$P = 4s \text{ Answer}$$

b. 8% (or 8 percent) means 8 hundredths, written as 0.08 or $\frac{8}{100}$.

$$C = p + 0.08p \text{ or } C = p + \frac{8}{100}p \text{ Answer}$$

c. "2 less than the number of sides" means $(n - 2)$.

$$S = 180(n - 2) \text{ Answer}$$

EXAMPLE 2

Write a formula that expresses the number of months m that are in y years.

Solution Look for a pattern.

In 1 year, there are 12 months.

In 2 years, there are 12(2) or 24 months.

In 3 years, there are 12(3) or 36 months.

In y years, there are $12(y)$ or $12y$ months. This equals m, the number of months.

Answer $m = 12y$

EXAMPLE 3

The Short Stop Diner pays employees \$6.00 an hour for working 40 hours a week or less. For working overtime, an employee is paid \$9.00 for each hour over 40 hours. Write a formula for the wages, W, of an employee who works h hours in a week.

Solution Two formulas are needed, one for $h \leq 40$ and the other for $h > 40$.

If $h \leq 40$, the wage is 6.00 times the number of hours, h. $W = 6.00h$

If $h > 40$, the employee has worked 40 hours at \$6.00 an hour and the remaining hours, $h - 40$, at \$9.00 an hour. $W = 6.00(40) + 9.00(h - 40)$.

Answer $W = 6h$ if $h \leq 40$ and $W = 6(40) + 9(h - 40)$ if $h > 40$.

(Note that the formula for $h > 40$ may also be given as $W = 240 + 9(h - 40)$.)

EXERCISES

Writing About Mathematics

1. Fran said that a recipe is a type of formula. Do you agree or disagree with Fran? Explain your answer.

2. a. Is an algebraic expression a formula? Explain why or why not.

b. Is a formula an open sentence? Explain why or why not.

Developing Skills

In 3–17, write a formula that expresses each relationship.

3. The total length l of 10 pieces of lumber, each m meters in length, is 10 times the length of each piece of lumber.

4. An article's selling price S equals its cost c plus the margin of profit m.

5. The perimeter P of a rectangle is equal to the sum of twice its length l and twice its width w.

6. The average m of three numbers, a, b, and c is their sum divided by 3.

7. The area A of a triangle is equal to one-half the length of the base b multiplied by the length of the altitude h.

8. The area A of a square is equal to the square of the length of a side s.

9. The volume V of a cube is equal to the cube of the length of an edge e.

10. The surface area S of a cube is equal to 6 times the square of the length of an edge e.

11. The surface area S of a sphere is equal to the product of 4π and the square of the radius r.

12. The average rate of speed r is equal to the distance that is traveled d divided by the time spent on the trip t.

13. The Fahrenheit temperature F is 32° more than nine-fifths of the Celsius temperature C.

14. The Celsius temperature C is equal to five-ninths of the difference between the Fahrenheit temperature F and 32°.

15. The dividend D equals the product of the divisor d and the quotient q plus the remainder r.

16. A sales tax T that must be paid when an article is purchased is equal to 8% of the price of the article v.

17. A salesman's weekly earnings F is equal to his weekly salary s increased by 2% of his total volume of sales v.

Applying Skills

18. A ferry takes cars, drivers, and passengers across a body of water. The total ferry charge C in dollars is $20.00 for the car and driver, plus d dollars for each passenger.

    a. Write a formula for C in terms of d and the number of passengers, n.

    b. Find the cost of the ferry for a car if $d = \$15$ and there are 5 persons in the car.

    c. Find the cost of the ferry for a car with only the driver.

19. The cost C in cents of an internet telephone call lasting m minutes is x cents for the first 3 minutes and y cents for each additional minute.

a. Write two formulas for C, one for the cost of calls lasting 3 minutes or less ($m \leq 3$), and another for the cost of calls lasting more than 3 minutes ($m > 3$).

b. Find the cost of a 2.5 minute telephone call if $x = \$0.25$ and $y = \$0.05$.

c. Find the cost of a 10 minute telephone call if $x = \$0.25$ and $y = \$0.05$.

20. The cost D in dollars of sending a fax of p pages is a dollars for sending the first page and b dollars for each additional page.

a. Write two formulas for D, one for the cost of faxing 1 page ($p = 1$), and another for the cost of faxing more than 1 page ($p > 1$).

b. Find the cost of faxing 1 page if $a = \$1.00$ and $b = \$0.60$.

c. Find the cost of faxing 5 pages if $a = \$1.00$ and $b = \$0.60$.

21. A gasoline dealer is allowed a profit of 12 cents a gallon for each gallon sold. If more than 25,000 gallons are sold in a month, an additional profit of 3 cents for every gallon over that number is given.

a. Write two formulas for the gasoline dealer's profit, P, one for when the number of gallons sold, n, is not more than 25,000 ($n \leq 25{,}000$), and another for when more than 25,000 gallons are sold ($n > 25{,}000$).

b. Find P when 21,000 gallons of gasoline are sold in one month.

c. Find P when 30,000 gallons of gasoline are sold in one month.

22. Gabriel earns a bonus of $25 for each sale that he makes if the number of sales, s, in a month is 20 or less. He earns an extra $40 for each additional sale if he makes more than 20 sales in a month.

a. Write a formula for Gabriel's bonus, B, when $s \leq 20$.

b. Write a formula for B when $s > 20$.

c. In August, Gabriel made 18 sales. Find his bonus for August.

d. In September, Gabriel made 25 sales. Find his bonus for September.

23. Mrs. Lucy is selling cookies at a local bake sale. If she sells exactly 3 dozen cookies, the cost of ingredients will equal her earnings. If she sells more than 3 dozen cookies, Mrs. Lucy will make a profit of 25 cents for each cookie sold.

a. Write a formula for Mrs. Lucy's earnings, E, when the number of cookies sold, c, is equal to 36.

b. Write a formula for E when $c > 36$.

c. Find the number of cookies Mrs. Lucy sold if she makes a profit of $2.00.

CHAPTER SUMMARY

An **algebraic expression**, such as $x + 6$, is an expression or a phrase that contains one or more variables, such as x. The **variable** is a placeholder for numbers. To evaluate an expression, replace each variable with a number and follow the order of operations.

A **term** is a number, a variable, or any product or quotient of numbers and variables. In the term $6by$, 6 is the **numerical coefficient**. In the term n^3, the **base** is n, the **exponent** is 3, and the **power** is n^3. The power n^3 means that base n is used as a factor 3 times.

An **open sentence**, which can be an equation or an inequality, contains a variable. When the variable is replaced by numbers from a **domain**, the numbers that make the open sentence true are the elements of the **solution set** of the sentence.

A **formula** is a sentence that shows the relationship between two or more variables.

VOCABULARY

3-1 Variable • Placeholder • Domain • Replacement set • Algebraic expression

3-3 Term • Factor • Coefficient • Numerical coefficient

3-5 Evaluating an algebraic expression

3-6 Open sentence • Solution set

3-7 Formula

REVIEW EXERCISES

1. Explain the difference between an algebraic expression and an open sentence.

2. Explain the difference between $2a^2$ and $(2a)^2$.

In 3–6, use mathematical symbols to translate the verbal phrases into algebraic language.

3. x divided by b

4. 4 less than r

5. q decreased by d

6. 3 more than twice g

In 7–14, find the value of each expression.

7. $6ac - d$ when $a = 10, c = 8$, and $d = 5$

8. $4b^2$ when $b = 2.5$

9. $3b + c$ when $b = 7$ and $c = 14$

10. $km - 9$ when $k = 15$ and $m = 0.6$

11. $\frac{bc}{a}$ when $a = 5, b = 3$, and $c = 12$

12. $2a^2 - 2a$ when $a = \frac{1}{4}$

13. $(2a)^2 - 2a$ when $a = \frac{1}{4}$

14. $a(b + c)$ when $a = 2.5, b = 1.1$, and $c = 8.9$

15. Write an algebraic expression for the total number of cents in n nickels and q quarters.

16. In the term $2xy^3$ what is the coefficient?

17. In the term $2xy^3$ what is the exponent of y?

18. In the term $2xy^3$, what is the base that is used 3 times as a factor?

19. If distance is the product of rate and time, write a formula for distance, d, in terms of rate, r, and time, t.

20. What is the smallest member of the solution set of $19.4 \leq y - 29$ if the domain is $\{46.25, 47.9, 48, 48.5, 49, 50, 51.\overline{3}\}$?

21. What is the smallest member of the solution set of $19.4 \leq y - 29$ if the domain is the set of whole numbers?

22. What is the smallest member of the solution set of $19.4 \leq y - 29$ if the domain is the set of real numbers?

23. In a baseball game, the winning team scored n runs and the losing team scored $2n - 5$ runs.

a. Describe in words the number of runs that the losing team scored.

b. What could have been the score of the game? Is there more than one answer?

c. What are the possible values for n?

24. A mail order book club offers books for \$8.98 each plus \$3.50 for shipping and handling on each order. The cost of Bethany's order, which totaled less than \$20, can be expressed as $8.98b + 3.50 < 20$ where b represents the number of books Bethany ordered.

a. What could be the domain for this problem?

b. What is the solution set for this open sentence?

25. Write two algebraic expressions to represent the cost of sending an express delivery based on the rates given in the chapter opener on page 88, the first if the delivery distance is 10 miles or less, and the second if the delivery distance is more than 10 miles.

26. A list of numbers that follows a pattern begins with the numbers 2, 5, 8, 11, ...

a. Find the next number in the list.

b. Write a rule or explain how the next number is determined.

c. What is the 25th number in the list?

27. Each of the numbers given below is different from the others, that is, it belongs to a set of numbers to which the others do not belong. Explain why each is different.

3 6 9 35

28. Two oranges cost as much as five bananas. One orange costs the same as a banana and an apple. How many apples cost the same as three bananas?

Exploration

STEP 1. Write a three-digit multiple of 11 by multiplying any whole number from 10 to 90 by 11. Add the digits in the hundreds and the ones places. If the sum is greater than or equal to 11, subtract 11. Compare this result to the digit in the tens place. Repeat the procedure for other three-digit multiples of 11.

STEP 2. Write a three-digit number that is not a multiple of 11 by adding any counting number less than 11 to a multiple of 11 used in step 1. Add the digits in the hundreds and the ones places. If the sum is greater than or equal to 11, subtract 11. Compare this result to the digit in the tens place. Repeat the procedure with another number.

STEP 3. Based on steps 1 and 2, can you suggest a way of determining whether or not a three-digit number is divisible by 11?

STEP 4. Write a four-digit multiple of 11 by multiplying any whole number from 91 to 909 by 11. Add the digits in the hundreds and ones places. Add the digits in the thousands and tens places. If one sum is greater than or equal to 11, subtract 11. Compare these results. Repeat the procedure for another four-digit multiple of 11.

STEP 5. Write a four-digit number that is not a multiple of 11 by adding any counting number less than 11 to a multiple of 11 used in step 4. Add the digits in the hundreds place and ones place. Add the digits in the thousands place and tens place. If one sum is greater than or equal to 11, subtract 11. Compare these results. Repeat the procedure starting with another number.

STEP 6. Based on steps 4 and 5, can you suggest a way of determining whether or not a four-digit number is divisible by 11?

STEP 7. Write a rule for determining whether or not any whole number is divisible by 11.

CUMULATIVE REVIEW CHAPTERS 1–3

Part I

Answer all questions in this part. Each correct answer will receive 2 credits. No partial credit will be allowed.

1. Which of the following is the set of negative integers greater than –3?
(1) $\{-4, -5, -6, -7, \ldots\}$
(2) $\{-3, -4, -5, -6, \ldots\}$
(3) $\{-3, -2, -1\}$
(4) $\{-2, -1\}$

2. The exact value of the rational number $\frac{5}{3}$ can be written as
(1) 1.6 (2) $1.\overline{6}$ (3) 1.666666667 (4) 1.666666666

3. Rounded to the nearest hundredth, $\sqrt{5}$ is approximately equal to
(1) 2.23 (2) 2.236 (3) 2.24 (4) 2.240

4. Which of the following numbers is rational?
(1) π (2) $\sqrt{2}$ (3) $1.4\overline{2}$ (4) $\sqrt{0.4}$

5. Which of the following inequalities is a true statement?
(1) $0.026 < 0.25 < 0.2$
(2) $0.2 < 0.026 < 0.25$
(3) $0.2 < 0.25 < 0.026$
(4) $0.026 < 0.2 < 0.25$

6. The length of a rectangle is given as 30.02 yards. This measure has how many significant digits?
(1) 1 (2) 2 (3) 3 (4) 4

7. Which of the following is not a prime?
(1) 7 (2) 23 (3) 37 (4) 51

8. The arithmetic expression $8 - 5(-0.2)^2 \div 10$ is equal to
(1) 0.9 (2) 0.78 (3) 7.98 (4) 8.02

9. Which of the following is a correct application of the distributive property?
(1) $4(8 + 0.2) = 4(8) + 4(0.2)$
(2) $6(3 - 1) = 6(-1 + 3)$
(3) $8(5 + 4) = 8(5) + 4$
(4) $8(5)(4) = 8(5) + 8(4)$

10. The additive inverse of 7 is
(1) –7 (2) 0 (3) $\frac{1}{7}$ (4) $|7|$

Part II

Answer all questions in this part. Each correct answer will receive 2 credits. Clearly indicate the necessary steps, including appropriate formula substitutions, diagrams, graphs, charts, etc. For all questions in this part, a correct numerical answer with no work shown will receive only 1 credit.

11. Of the 80 students questioned about what they had read in the past month, 35 had read nonfiction, 55 had read fiction, and 22 had read neither fiction nor nonfiction. How many students had read both fiction and nonfiction?

12. What is the largest number that is the product of three different two-digit primes?

Part III

Answer all questions in this part. Each correct answer will receive 3 credits. Clearly indicate the necessary steps, including appropriate formula substitutions, diagrams, graphs, charts, etc. For all questions in this part, a correct numerical answer with no work shown will receive only 1 credit.

13. The formula for the area of an equilateral triangle (a triangle with three sides of equal measure) is $A = \frac{\sqrt{3}}{4}s^2$. Find the area of an equilateral triangle if the measure of one side is 12.6 centimeters. Express your answer to the number of significant digits determined by the given data.

14. A teacher wrote the sequence 1, 2, 4, . . . and asked what the next number could be. Three students each gave a different answer and the teacher said that all three answers were correct.

a. Adam said 7. Explain what rule Adam used.

b. Bette said 8. Explain what rule Bette used.

c. Carlos said 5. Explain what rule Carlos used.

Part IV

Answer all questions in this part. Each correct answer will receive 4 credits. Clearly indicate the necessary steps, including appropriate formula substitutions, diagrams, graphs, charts, etc. For all questions in this part, a correct numerical answer with no work shown will receive only 1 credit.

15. If $a = 7$, $b = -5$, and $c = \frac{1}{3}$, evaluate the expression $\frac{a-b}{c} + 3(b - 2)$. Do not use a calculator. Show each step in your calculation.

16. Michelle bought material to make a vest and skirt. She used half of the material to make the skirt and two-thirds of what remained to make the vest. She had $1\frac{1}{4}$ yards of material left.

a. How many yards of material did she buy?

b. How many yards of material did she use for the vest?

c. How many yards of material did she use for the skirt?

CHAPTER

4

CHAPTER TABLE OF CONTENTS

FIRST DEGREE EQUATIONS AND INEQUALITIES IN ONE VARIABLE

An equation is an important problem-solving tool. A successful business person must make many decisions about business practices. Some of these decisions involve known facts, but others require the use of information obtained from equations based on expected trends.

For example, an equation can be used to represent the following situation. Helga sews hand-made quilts for sale at a local craft shop. She knows that the materials for the last quilt that she made cost \$76 and that it required 44 hours of work to complete the quilt. If Helga received \$450 for the quilt, how much did she earn for each hour of work, taking into account the cost of the materials?

Most of the problem-solving equations for business are complex. Before you can cope with complex equations, you must learn the basic principles involved in solving any equation.

4-1 SOLVING EQUATIONS USING MORE THAN ONE OPERATION

Some Terms and Definitions

An **equation** is a sentence that states that two algebraic expressions are equal. For example, $x + 3 = 9$ is an equation in which $x + 3$ is called the **left side**, or **left member**, and 9 is the **right side**, or **right member**.

An equation may be a *true* sentence such as $5 + 2 = 7$, a *false* sentence such as $6 - 3 = 4$, or an *open* sentence such as $x + 3 = 9$. The number that can replace the variable in an open sentence to make the sentence true is called a **root**, or a **solution**, of the equation. For example, 6 is a root of $x + 3 = 9$.

As discussed in Chapter 3, the replacement set or domain is the set of possible values that can be used in place of the variable in an open sentence. If no replacement set is given, the replacement set is the set of real numbers. The set consisting of all elements of the replacement set that are solutions of the open sentence is called the **solution set** of the open sentence. For example, if the replacement set is the set of real numbers, the solution set of $x + 3 = 9$ is $\{6\}$. If no element of the replacement set makes the open sentence true, the solution set is the empty or null set, $\varnothing$ or $\{\}$. If every element of the domain satisfies an equation, the equation is called an **identity**. Thus, $5 + x = x - (-5)$ is an identity when the domain is the set of real numbers because every element of the domain makes the sentence true.

Two equations that have the same solution set are **equivalent equations**. To **solve an equation** is to find its solution set. This is usually done by writing simpler equivalent equations.

If not every element of the domain makes the sentence true, the equation is called a **conditional equation**, or simply an equation. Therefore, $x + 3 = 9$ is a conditional equation.

Properties of Equality

When two numerical or algebraic expressions are equal, it is reasonable to assume that if we change each in the same way, the resulting expressions will be equal. For example:

$$5 + 7 = 12$$
$$(5 + 7) + 3 = 12 + 3$$
$$(5 + 7) - 8 = 12 - 8$$
$$-2(5 + 7) = -2(12)$$
$$\frac{5 + 7}{3} = \frac{12}{3}$$

These examples suggest the following properties of equality:

Properties of Equality

1. **The addition property of equality.** If equals are added to equals, the sums are equal.
2. **The subtraction property of equality.** If equals are subtracted from equals, the differences are equal.
3. **The multiplication property of equality.** If equals are multiplied by equals, the products are equal.
4. **The division property of equality.** If equals are divided by nonzero equals, the quotients are equal.
5. **The substitution principle.** In a statement of equality, a quantity may be substituted for its equal.

To solve an equation, you need to work backward or "undo" what has been done by using inverse operations. To undo the addition of a number, add its opposite. For example, to solve the equation $x + 7 = 19$, use the addition property of equality. Add the opposite of 7 to both sides.

$$\begin{aligned} x + 7 &= 19 \\ -7 \quad & \quad -7 \\ \hline x \quad &= 12 \end{aligned}$$

The variable x is now alone on one side and it is easy to read the solution, $x = 12$.

To solve an equation in which the variable has been multiplied by a number, either divide by that number or multiply by its reciprocal. (Remember multiplying by the reciprocal is the same as dividing by the number.) To solve $6x = 24$, divide both sides by 6 or multiply both sides by $\frac{1}{6}$.

$$\begin{aligned} 6x &= 24 \\ \frac{6x}{6} &= \frac{24}{6} \\ x &= 4 \end{aligned} \qquad \text{or} \qquad \begin{aligned} 6x &= 24 \\ \tfrac{1}{6}(6x) &= \tfrac{1}{6}(24) \\ x &= 4 \end{aligned}$$

To solve $\frac{x}{3} = 5$, multiply each side by the reciprocal of $\frac{1}{3}$ which is 3.

$$\begin{aligned} \frac{x}{3} &= 5 \\ (3)\frac{x}{3} &= (3)5 \\ x &= 15 \end{aligned}$$

In the equation $2x + 3 = 15$, there are two operations in the left side: multiplication and addition. In forming the left side of the equation, x was first multiplied by 2, and then 3 was added to the product. To solve this equation, we must undo these operations by using the inverse elements in the reverse order. Since the last operation was to add 3, the first step in solving the equation is to add its opposite, -3, to both sides of the equation or subtract 3 from both sides of the equation. Here we are using either the addition or the subtraction property of equality.

$$\begin{aligned} 2x + 3 &= 15 \\ 2x + 3 + (-3) &= 15 + (-3) \\ 2x &= 12 \end{aligned} \quad \text{or} \quad \begin{array}{r} 2x + 3 = 15 \\ \underline{-3 \quad -3} \\ 2x \quad = 12 \end{array}$$

Now we have a simpler equation that has the same solution set as the original and includes only multiplication by 2. To solve this simpler equation, we multiply both sides of the equation by $\frac{1}{2}$, the reciprocal of 2, or divide both sides of the equation by 2. Here we can use either the multiplication or the division property of equality.

$$\begin{aligned} 2x &= 12 \\ \tfrac{1}{2}(2x) &= \tfrac{1}{2}(12) \\ x &= 6 \end{aligned} \quad \text{or} \quad \begin{aligned} 2x &= 12 \\ \tfrac{2x}{2} &= \tfrac{12}{2} \\ x &= 6 \end{aligned}$$

After an equation has been solved, we **check** the equation, that is, we verify that the solution does in fact make the given equation true by replacing the variable with the solution and performing any computations.

Check:

$$\begin{aligned} 2x + 3 &= 15 \\ 2(6) + 3 &= 15 \\ 12 + 3 &= 15 \\ 15 &= 15 \checkmark \end{aligned}$$

To find the solution of the equation $2x + 3 = 15$, we used several properties of the four basic operations and of equality. The solution below shows the mathematical principle that we used in each step.

$2x + 3 = 15$	Given
$(2x + 3) + (-3) = 15 + (-3)$	Addition property of equality
$2x + [3 + (-3)] = 15 + (-3)$	Associative property of addition
$2x + 0 = 12$	Additive inverse property
$2x = 12$	Additive identity property
$\frac{1}{2}(2x) = \frac{1}{2}(12)$	Multiplication property of equality
$\left[\frac{1}{2}(2)\right]x = \frac{1}{2}(12)$	Associative property of multiplication
$1x = 6$	Multiplicative inverse property
$x = 6$	Multiplicative identity property

These steps and properties are necessary to justify the solution of an equation of this form. However, when solving an equation, we do not need to write each of the steps, as shown in the examples that follow.

EXAMPLE 1

Solve and check: $7x + 15 = 71$

Solution *How to Proceed*

(1) Write the equation: $7x + 15 = 71$

(2) Add -15, the opposite of $+15$ to each side: $\begin{array}{r} -15 \quad -15 \\ \hline 7x = 56 \end{array}$

(3) Since multiplication and division are inverse operations, divide each side by 7: $\frac{7x}{7} = \frac{56}{7}$, $x = 8$

(4) Check the solution. Write the solution in place of x and perform the computations:

$$7x + 15 = 71$$
$$7(8) + 15 \stackrel{?}{=} 71$$
$$56 + 15 \stackrel{?}{=} 71$$
$$71 = 71 \checkmark$$

Answer $x = 8$

Note: The check is based on the substitution principle.

EXAMPLE 2

Find the solution set and check: $\frac{3}{5}x - 6 = -18$

Solution

$$\frac{3}{5}x - 6 = -18$$
$$\underline{+6 \quad +6} \quad \text{Addition property of equality}$$
$$\frac{3}{5}x = -12$$
$$\frac{5}{3}\left(\frac{3}{5}x\right) = \frac{5}{3}(-12) \quad \text{Multiplication property of equality}$$
$$x = -20$$

Check

$$\frac{3}{5}x - 6 = -18$$
$$\frac{3}{5}(\mathbf{-20}) - 6 \stackrel{?}{=} -18$$
$$-12 - 6 \stackrel{?}{=} -18$$
$$-18 = -18 \checkmark$$

Answer The solution set is $\{-20\}$.

EXAMPLE 3

Solve and check: $7 - x = 9$

Solution **METHOD 1.** Think of $7 - x$ as $7 + (-1x)$.

$$\begin{aligned} 7+(-x) &= 9 \\ -7 \quad & \quad -7 \\ \hline -x &= 2 \end{aligned}$$ Addition property of equality

$$\frac{-1x}{-1} = \frac{2}{-1}$$ Division property of equality

$$x = -2$$

Check

$$7 - x = 9$$
$$7 - (-2) \stackrel{?}{=} 9$$
$$7 + 2 \stackrel{?}{=} 9$$
$$9 = 9 \checkmark$$

METHOD 2. Add x to both sides of the equation so that the variable has a positive coefficient.

How to Proceed

(1) Write the equation: $7 - x = 9$

(2) Add x to each side of the equation: $7 - x + x = 9 + x$

$7 = 9 + x$

(3) Add -9 to each side of the equation: $-9 + 7 = -9 + 9 + x$

$-2 = x$

The check is the same as for Method 1.

Answer $\{-2\}$ or $x = -2$

EXERCISES

Writing About Mathematics

1. Is it possible for the equation $2x + 5 = 0$ to have a solution in the set of positive real numbers? Explain your answer.

2. Max wants to solve the equation $7x + 15 = 71$. He begins by multiplying both sides of the equation by $\frac{1}{7}$, the reciprocal of the coefficient of x.

a. Is it possible for Max to solve the equation if he begins in this way? If so, what would be the result of multiplying by $\frac{1}{7}$ and what would be his next step?

b. In this section you learned to solve the equation $7x + 15 = 71$ by first adding the opposite of 15, -15, to both sides of the equation. Which method do you think is better? Explain your answer.

Developing Skills

In 3 and 4, write a complete solution for each equation, listing the property used in each step.

3. $3x + 5 = 35$

4. $\frac{1}{2}x - 1 = 15$

In 5–32, solve and check each equation.

5. $55 = 6a + 7$ **6.** $17 = 8c - 7$ **7.** $9 - 1x = 7$ **8.** $11 = 15t + 16$

9. $15 - a = 3$ **10.** $11 = -6d - 1$ **11.** $8 - y = 1$ **12.** $\frac{3a}{8} = 12$

13. $\frac{2}{3}x = -8$ **14.** $12 = \frac{3}{4}y$ **15.** $\frac{5t}{4} = \frac{45}{2}$ **16.** $-\frac{3}{5}m = 30$

17. $7.2 = \frac{4m}{5}$ **18.** $\frac{a}{4} + 9 = 5$ **19.** $-2 = \frac{y}{5} + 3$ **20.** $9d - \frac{1}{2} = 17\frac{1}{2}$

21. $4a + 0.2 = 5$ **22.** $4 = 3t - 0.2$ **23.** $\frac{1}{4}x + 11 = 5$ **24.** $13 = 5 - \frac{2}{3}y$

25. $\frac{4}{5}t + 7 = 47$ **26.** $0.04c + 1.6 = 0$ **27.** $15x + 14 = 19$ **28.** $8 = 18c - 1$

29. $\frac{1}{7} = 14 - x$ **30.** $0.8r + 19 = 20$ **31.** $\frac{1}{3}w + 6 = -2$ **32.** $842 - 162m = -616$

Applying Skills

33. The formula $F = \frac{9}{5}C + 32$ gives the relationship between the Fahrenheit temperature F and the Celsius temperature C. Solve the equation $59 = \frac{9}{5}C + 32$ to find the temperature in degrees Celsius when the Fahrenheit temperature is 59°.

34. When Kurt orders from a catalog, he pays \$3.50 for shipping and handling in addition to the cost of the goods that he purchases. Kurt paid \$33.20 when he ordered six pairs of socks. Solve the equation $6x + 3.50 = 33.20$ to find x, the price of one pair of socks.

35. When Mattie rents a car for one day, the cost is \$29.00 plus \$0.20 a mile. On her last trip, Mattie paid \$66.40 for the car for one day. Find the number of miles, m, that Mattie drove by solving the equation $29 + 0.20x = 66.40$.

36. On his last trip to the post office, Hal paid \$4.30 to mail a package and bought some 39-cent stamps. He paid a total of \$13.66. Find s, the number of stamps that he bought, by solving the equation $0.39s + 4.30 = 13.66$.

4-2 SIMPLIFYING EACH SIDE OF AN EQUATION

An equation is often written in such a way that one or both sides are not in simplest form. Before starting to solve the equation by using additive and multiplicative inverses, you should simplify each side by removing parentheses if necessary and adding like terms.

Recall that an algebraic expression that is a number, a variable, or a product or quotient of numbers and variables is called a *term*. **First-degree equations in one variable** contain two kinds of terms, terms that are constants and terms that contain the variable to the first power only.

Like and Unlike Terms

Two or more terms that contain the same variable or variables, with corresponding variables having the same exponents, are called **like terms** or **similar terms**. For example, the following pairs are like terms.

$6k$ and k $\quad$ $5x^2$ and $-7x^2$ $\quad$ $9ab$ and $0.4ab$ $\quad$ $\frac{9}{2}x^2y^3$ and $-\frac{11}{3}x^2y^3$

Two terms are **unlike terms** when they contain different variables, or the same variable or variables with different exponents. For example, the following pairs are unlike terms.

$3x$ and $4y$ $\quad$ $5x^2$ and $5x^3$ $\quad$ $9ab$ and $0.4a$ $\quad$ $\frac{8}{3}x^3y^2$ and $\frac{4}{7}x^2y^3$

To add like terms, we use the distributive property of multiplication over addition.

$$9x + 2x = (9 + 2)x = 11x$$
$$-16d + 3d = (-16 + 3)d = -13d$$

Note that in the above examples, when like terms are added:

1. The sum has the same variable factor as the original terms.
2. The numerical coefficient of the sum is the sum of the numerical coefficients of the terms that were added.

The sum of like terms can be expressed as a single term. The sum of unlike terms cannot be expressed as a single term. For example, the sum of $2x$ and 3 cannot be written as a single term but is written $2x + 3$.

EXAMPLE 1

Solve and check: $2x + 3x + 4 = -6$

Solution

How to Proceed		*Check*
(1) Write the equation:	$2x + 3x + 4 = -6$	$2x + 3x + 4 = -6$
(2) Simplify the left side by combining like terms:	$5x + 4 = -6$	$2(-2) + 3(-2) + 4 \stackrel{?}{=} -6$
(3) Add -4, the additive inverse of $+4$, to each side:	$\underline{\quad -4 \quad -4}$ $5x \quad = -10$	$-4 - 6 + 4 \stackrel{?}{=} -6$ $-6 = -6$ ✔
(4) Multiply by $\frac{1}{5}$, the multiplicative inverse of 5:	$\frac{1}{5}(5x) = \frac{1}{5}(-10)$	
(5) Simplify each side.	$x = -2$	

Answer -2

Note: When solving equations, remember to check the answer in the original equation and not in the simplified one.

The algebraic expression that is on one side of an equation may contain parentheses. Use the distributive property to remove the parentheses solving the equation. The following examples illustrate how the distributive and associative properties are used to do this.

EXAMPLE 2

Solve and check: $27x - 3(x - 6) = 6$

Solution Since $-3(x - 6)$ means that $(x - 6)$ is to be multiplied by -3, we will use the distributive property to remove parentheses and then combine like terms. Note that for this solution, in the first three steps the left side is being simplified. These steps apply only to the left side and only change the form but not the numerical value. The next two steps undo the operations of addition and multiplication that make up the expression $24x + 18$. Since adding -18 and dividing by 24 will change the value of the left side, the right side must be changed in the same way to retain the equality.

How to Proceed

(1) Write the equation: $27x - 3(x - 6) = 6$

(2) Use the distributive property: $27x - 3x + 18 = 6$

(3) Combine like terms: $24x + 18 = 6$

(4) Use the addition property of equality. Add -18, the additive inverse of $+18$, to each side:
$$\begin{array}{rr} 24x + 18 = & 6 \\ -18 & -18 \\ \hline 24x = & -12 \end{array}$$

(5) Use the division property of equality. Divide each side by 24: $\frac{24x}{24} = \frac{-12}{24}$

(6) Simplify each side: $x = -\frac{1}{2}$

Check

(1) Write the equation: $27x - 3(x - 6) = 6$

(2) Replace x by $-\frac{1}{2}$: $27\left(-\frac{1}{2}\right) - 3\left(-\frac{1}{2} - 6\right) \stackrel{?}{=} 6$

(3) Perform the indicated computation:

$$27\left(-\frac{1}{2}\right) - 3\left(-6\frac{1}{2}\right) \stackrel{?}{=} 6$$
$$-\frac{27}{2} + 18\frac{3}{2} \stackrel{?}{=} 6$$
$$-\frac{27}{2} + \frac{39}{2} \stackrel{?}{=} 6$$
$$\frac{12}{2} \stackrel{?}{=} 6$$
$$6 = 6 \checkmark$$

Answer $x = -\frac{1}{2}$

Representing Two Numbers with the Same Variable

Problems often involve finding two or more different numbers. It is useful to express these numbers in terms of the same variable. For example, if you know the sum of two numbers, you can express the second in terms of the sum and the first number.

- If the sum of two numbers is 12 and one of the numbers is 5, then the other number is $12 - 5$ or 7.
- If the sum of two numbers is 12 and one of the numbers is 9, then the other number is $12 - 9$ or 3.
- If the sum of two numbers is 12 and one of the numbers is x, then the other number is $12 - x$.

A problem can often be solved algebraically in more than one way by writing and solving different equations, as shown in the example that follows. The methods used to obtain the solution are different, but both use the facts stated in the problem and arrive at the same solution.

EXAMPLE 3

The sum of two numbers is 43. The larger number minus the smaller number is 5. Find the numbers.

Solution This problem states two facts:

FACT 1 The sum of the numbers is 43.

FACT 2 The larger number minus the smaller number is 5. In other words, the larger number is 5 more than the smaller.

(1) Represent each number in terms of the same variable using Fact 1: the sum of the numbers is 43.
Let x = the larger number.
Then, $43 - x$ = the smaller number.

(2) Write an equation using Fact 2:

$$\underbrace{\text{The larger number}}_{x}\ \underbrace{\text{minus}}_{-}\ \underbrace{\text{the smaller number}}_{(43 - x)}\ \underbrace{\text{is 5.}}_{= 5}$$

(3) Solve the equation.

(a) Write the equation: $x - (43 - x) = 5$

(b) To subtract $(43 - x)$, add its opposite: $x + (-43 + x) = 5$

(c) Combine like terms: $2x - 43 = 5$

(d) Add the opposite of -43 to each side:

$$\begin{array}{rcr} 2x - 43 &=& 5 \\ +43 && +43 \\ \hline 2x &=& 48 \end{array}$$

(e) Divide each side by 2:

$$\frac{2x}{2} = \frac{48}{2}$$
$$x = 24$$

(4) Find the numbers.

The larger number $= x = 24$.

The smaller number $= 43 - x = 43 - 24 = 19$.

Check A word problem is checked by comparing the proposed solution with the facts stated in the original wording of the problem. Substituting numbers in the equation is not sufficient since the equation formed may not be correct.

The sum of the numbers is 43: $24 + 19 = 43$.

The larger number minus the smaller number is 5: $24 - 19 = 5$.

Alternate Solution Reverse the way in which the facts are used.

(1) Represent each number in terms of the same variable using Fact 2: the larger number is 5 more than the smaller.

Let x = the smaller number.

Then, $x + 5$ = the larger number.

(2) Write an equation using the first fact.

The sum of the numbers is 43.

$$x + (x + 5) = 43$$

(3) Solve the equation.

(a) Write the equation: $x + (x + 5) = 43$

(b) Combine like terms: $2x + 5 = 43$

(c) Add the opposite of 5 to each side:

$$\begin{array}{rcr} 2x + 5 &=& 43 \\ -5 && -5 \\ \hline 2x &=& 38 \end{array}$$

(d) Divide each side by 2:

$$\frac{2x}{2} = \frac{38}{2}$$
$$x = 19$$

(4) Find the numbers.

The smaller number $= x = 19$.

The larger number $= x + 5 = 19 + 5 = 24$.

(5) *Check.* (See the first solution.)

Answer The numbers are 24 and 19.

EXERCISES

Writing About Mathematics

1. Two students are each solving a problem that states that the difference between two numbers is 12. Irene represents one number by x and the other number by $x + 12$. Henry represents one number by x and the other number by $x - 12$. Explain why both students are correct.

2. A problem states that the sum of two numbers is 27. The numbers can be represented by x and $27 - x$. Is it possible to determine which is the larger number and which is the smaller number? Explain your answer.

Developing Skills

In 3–28, solve and check each equation.

3. $x + (x - 6) = 20$
4. $x - (12 - x) = 38$
5. $(15x + 7) - 12 = 4$
6. $(14 - 3c) + 7c = 94$
7. $x + (4x + 32) = 12$
8. $7x - (4x - 39) = 0$
9. $5(x + 2) = 20$
10. $3(y - 9) = 30$
11. $8(2c - 1) = 56$
12. $6(3c - 1) = -42$
13. $30 = 2(10 - y)$
14. $4(c + 1) = 32$
15. $25 - 2(t - 5) = 19$
16. $18 = -6x + 4(2x + 3)$
17. $55 = 4 + 3(m + 2)$
18. $5(x - 3) - 30 = 10$
19. $3(2b + 1) - 7 = 50$
20. $5(3c - 2) + 8 = 43$
21. $7r - (6r - 5) = 7$
22. $8b - 4(b - 2) = 24$
23. $5m - 2(m - 5) = 17$
24. $28y - 6(3y - 5) = 40$
25. $3(a - 5) - 2(2a + 1) = 0$
26. $0.04(2r + 1) - 0.03(2r - 5) = 0.29$
27. $0.3a + (0.2a - 0.5) + 0.2(a + 2) = 1.3$
28. $\frac{3}{4}(8 + 4x) - \frac{1}{3}(6x + 3) = 9$

Applying Skills

In 29–33, write and solve an equation for each problem. Follow these steps:

a. List two facts in the problem.

b. Choose a variable to represent one of the numbers to be determined.

c. Use one of the facts to write any other unknown numbers in terms of the chosen variable.

d. Use the second fact to write an equation.

e. Solve the equation.

f. Answer the question.

g. Check your answer using the words of the problem.

29. Sandi bought 6 yards of material. She wants to cut it into two pieces so that the difference between the lengths of the two pieces will be 1.5 yards. What should be the length of each piece?

30. The Tigers won eight games more than they lost, and there were no ties. If the Tigers played 78 games, how many games did they lose?

31. This month Erica saved \$20 more than last month. For the two months, she saved a total of \$70. How much did she save each month?

32. On a bus tour, there are 100 passengers on three buses. Two of the buses each carry four fewer passengers than the third bus. How many passengers are on each bus?

33. For a football game, $\frac{4}{5}$ of the seats in the stadium were filled. There were 31,000 empty seats at the game. What is the stadium's seating capacity?

4-3 SOLVING EQUATIONS THAT HAVE THE VARIABLE IN BOTH SIDES

A variable represents a number. As you know, any number may be added to both sides of an equation without changing the solution set. Therefore, the same variable (or the same multiple of the same variable) may be added to or subtracted from both sides of an equation without changing the solution set.

For instance, to solve $8x = 30 + 5x$, write an equivalent equation that has only a constant in the right side. To do this, eliminate $5x$ from the right side by adding its opposite, $-5x$, to each side of the equation.

METHOD 1

$$\begin{array}{rcr} 8x & = & 30 + 5x \\ \underline{-5x} & & \underline{\ -5x} \\ 3x & = & 30 \\ x & = & 10 \end{array}$$

METHOD 2

$$\begin{aligned} 8x &= 30 + 5x \\ 8x + (-5x) &= 30 + 5x + (-5x) \\ 3x &= 30 \\ x &= 10 \end{aligned}$$

Check

$$\begin{aligned} 8x &= 30 + 5x \\ 8(\mathbf{10}) &\stackrel{?}{=} 30 + 5(\mathbf{10}) \\ 80 &\stackrel{?}{=} 30 + 50 \\ 80 &= 80 \checkmark \end{aligned}$$

Answer: $x = 10$

To solve an equation that has the variable in both sides, transform it into an equivalent equation in which the variable appears in only one side. Then, solve the equation.

EXAMPLE 1

Solve and check: $7x = 63 - 2x$

Solution *How to Proceed*

(1) Write the equation: $7x = 63 - 2x$

(2) Add $2x$ to each side of the equation:

$$\begin{array}{rcl} 7x &=& 63 - 2x \\ +2x & & +\,2x \\ \hline 9x &=& 63 \end{array}$$

(3) Divide each side of the equation by 9: $\frac{9x}{9} = \frac{63}{9}$

(4) Simplify each side: $x = 7$

Check

$$\begin{aligned} 7x &= 63 - 2x \\ 7(\mathbf{7}) &\stackrel{?}{=} 63 - 2(\mathbf{7}) \\ 49 &\stackrel{?}{=} 63 - 14 \\ 49 &= 49 \checkmark \end{aligned}$$

Answer $x = 7$

To solve an equation that has both a variable and a constant in both sides, first write an equivalent equation with only a variable term on one side. Then solve the simplified equation. The following example shows how this can be done.

EXAMPLE 2

Solve and check: $3y + 7 = 5y - 3$

Solution

METHOD 1

$$\begin{array}{rcr} 3y + 7 &=& 5y - 3 \\ -5y & & -5y \\ \hline -2y + 7 &=& -3 \\ -7 & & -7 \\ \hline -2y &=& -10 \end{array}$$

$$\frac{-2y}{-2} = \frac{-10}{-2}$$

$$y = 5$$

METHOD 2

$$\begin{array}{rcr} 3y + 7 &=& 5y - 3 \\ -3y & & -3y \\ \hline 7 &=& 2y - 3 \\ +3 & & +3 \\ \hline 10 &=& 2y \end{array}$$

$$\frac{10}{2} = \frac{2y}{2}$$

$$y = 5$$

Check

$$\begin{aligned} 3y + 7 &= 5y - 3 \\ 3(\mathbf{5}) + 7 &\stackrel{?}{=} 5(\mathbf{5}) - 3 \\ 15 + 7 &\stackrel{?}{=} 25 - 3 \\ 22 &= 22 \checkmark \end{aligned}$$

Answer $y = 5$

A graphing calculator can be used to check an equation. The calculator can determine whether a given statement of equality or inequality is true or false. If the statement is true, the calculator will display 1; if the statement is false, the calculator will display 0. The symbols for equality and inequality are found in the TEST menu.

To check that $y = 5$ is the solution to the equation $3y + 7 = 5y - 3$, first store 5 as the value of y. then enter the equation to be checked.

ENTER: 5 **STO▸** **ALPHA** **Y** **ENTER**

3 **ALPHA** **Y** **+** 7 **2nd** **TEST** **ENTER** 5 **ALPHA** **Y** **−** 3 **ENTER**

DISPLAY:

```
5→Y
                5
3Y+7=5Y-3
                1
```

The calculator displays 1 which indicates that the statement of equality is true for the value that has been stored for y.

EXAMPLE 3

The larger of two numbers is 4 times the smaller. If the larger number exceeds the smaller number by 15, find the numbers.

Note: When s represents the smaller number and $4s$ represents the larger number, "the larger number *exceeds* the smaller by 15" has the following meanings. Use any one of them.

1. The larger equals 15 more than the smaller, written as $4s = 15 + s$.
2. The larger decreased by 15 equals the smaller, written as $4s - 15 = s$.
3. The larger decreased by the smaller is 15, written as $4s - s = 15$.

Solution Let s = the smaller number.

Then $4s$ = the larger number.

The larger is 15 more than the smaller.

$$4s \quad = 15 \quad + \quad s$$

$$4s = 15 + s$$

$$\begin{array}{rcl} 4s & = & 15 + s \\ -s & & -s \\ \hline 3s & = & 15 \\ s & = & 5 \\ 4s & = & 4(5) = 20 \end{array}$$

Check The larger number, 20, is 4 times the smaller number, 5. The larger number, 20, exceeds the smaller number, 5, by 15.

Answer The larger number is 20; the smaller number is 5.

EXAMPLE 4

In his will, Uncle Clarence left \$5,000 to his two nieces. Emma's share is to be \$500 more than Clara's. How much should each niece receive?

Solution (1) Use the fact that the sum of the two shares is \$5,000 to express each share in terms of a variable.
Let x = Clara's share.
Then $5{,}000 - x$ = Emma's share.

(2) Use the fact that Emma's share is \$500 more than Clara's share to write an equation.

Emma's share is \$500 more than Clara's share.

$$5{,}000 - x \quad = \quad 500 \quad + \quad x$$

(3) Solve the equation to find Clara's share.

$$\begin{aligned} 5{,}000 - x &= 500 + x \\ +x \quad & \quad\;\; +x \\ \hline 5{,}000 &= 500 + 2x \\ -500 \quad & \quad -500 \\ \hline 4{,}500 &= 2x \\ 2{,}250 &= x \end{aligned}$$

Clara's share is x = \$2,250.

(4) Find Emma's share: $5{,}000 - x = 5{,}000 - 2{,}250$ = \$2,750.

Alternate Solution (1) Use the fact that Emma's share is \$500 more than Clara's share to express each share in terms of a variable.
Let x = Clara's share.
Then $x + 500$ = Emma's share.

(2) Use the fact that the sum of the two shares is \$5,000 to write an equation.

Clara's share plus Emma's share is \$5,000.

$$x \quad + \quad (x + 500) \quad = \quad 5{,}000$$

(3) Solve the equation to find Clara's share

$$\begin{aligned} x + (x + 500) &= 5{,}000 \\ 2x + 500 &= 5{,}000 \\ -500 \quad & \quad -500 \\ \hline 2x &= 4{,}500 \\ x &= 2{,}250 \end{aligned}$$

Clara's share is x = \$2,250.

(4) Find Emma's share: $x + 500 = 2250 + 500 = \$2,750$.

Check $2,750 is $500 more than $2,250, and $2,750 + $2,250 = $5,000.

Answer Clara's share is $2,250, and Emma's share is $2,750.

EXERCISES

Writing About Mathematics

1. Milus said that he finds it easier to work with integers than with fractions. Therefore, in order to solve the equation $\frac{3}{4}a - 7 = \frac{1}{2}a + 3$, he began by multiplying both sides of the equation by 4.

$$4\left(\frac{3}{4}a - 7\right) = 4\left(\frac{1}{2}a + 3\right)$$
$$3a - 28 = 2a + 12$$

Do you agree with Milus that this is a correct way of obtaining the solution? If so, what mathematical principle is Milus using?

2. Katie said that Example 3 could be solved by letting $\frac{x}{4}$ equal the smaller number and x equal the larger number. Is Katie correct? If so, what equation would she write to solve the problem?

Developing Skills

In 3–36, solve and check each equation.

3. $7x = 10 + 2x$
4. $9x = 44 - 2x$
5. $5c = 28 + c$
6. $y = 4y + 30$
7. $2d = 36 + 5d$
8. $2\frac{1}{4}y = 1\frac{1}{4}y - 8$
9. $0.8m = 0.2m + 24$
10. $8y = 90 - 2y$
11. $2.3x + 36 = 0.3x$
12. $2\frac{3}{4}x + 24 = 3x$
13. $5a - 40 = 3a$
14. $5c = 2c - 81$
15. $x = 9x - 72$
16. $0.5m - 30 = 1.1m$
17. $4\frac{1}{4}c = 9\frac{3}{4}c + 44$
18. $7r + 10 = 3r + 50$
19. $4y + 20 = 5y + 9$
20. $7x + 8 = 6x + 1$
21. $x + 4 = 9x + 4$
22. $9x - 3 = 2x + 46$
23. $y + 30 = 12y - 14$
24. $c + 20 = 55 - 4c$
25. $2d + 36 = -3d - 54$
26. $7y - 5 = 9y + 29$
27. $3m - (m + 1) = 6m + 1$
28. $x - 3(1 - x) = 47 - x$
29. $3b - 8 = 10 + (4 - 8b)$
30. $\frac{2}{3}t - 11 = 4(16 - t) - \frac{1}{3}t$
31. $18 - 4n = 8 - 2(1 + 8n)$
32. $8c + 1 = 7c - 2(7 + c)$
33. $8a - 3(5 + 2a) = 85 - 3a$
34. $4(3x - 5) = 5x + 2(x + 15)$
35. $3m - 5m - 12 = 7m - 88 - 5$
36. $5 - 3(a + 6) = a - 1 + 8a$

In 37–42, **a.** write an equation to represent each problem, and **b.** solve the equation to find each number.

37. Eight times a number equals 35 more than the number. Find the number.

38. Six times a number equals 3 times the number, increased by 24. Find the number.

39. If 3 times a number is increased by 22, the result is 14 less than 7 times the number. Find the number.

40. The greater of two numbers is 1 more than twice the smaller. Three times the greater exceeds 5 times the smaller by 10. Find the numbers.

41. The second of three numbers is 6 more than the first. The third number is twice the first. The sum of the three numbers is 26. Find the three numbers.

42. The second of three numbers is 1 less than the first. The third number is 5 less than the second. If the first number is twice as large as the third, find the three numbers.

Applying Skills

In 43–50, use an algebraic solution to solve each problem.

43. It took the Gibbons family 2 days to travel 925 miles to their vacation home. They traveled 75 miles more on the first day than on the second. How many miles did they travel each day?

44. During the first 6 month of last year, the interest on an investment was $130 less than during the second 6 months. The total interest for the year was $1,450. What was the interest for each 6-month period?

45. Gemma has 7 more five-dollar bills than ten-dollar bills. The value of the five-dollar bills equals the value of the ten-dollar bills. How many five-dollar bills and ten-dollar bills does she have?

46. Leonard wants to save $100 in the next 2 months. He knows that in the second month he will be able to save $20 more than during the first month. How much should he save each month?

47. The ABC Company charges $75 a day plus $0.05 a mile to rent a car. How many miles did Mrs. Kiley drive if she paid $92.40 to rent a car for one day?

48. Kesha drove from Buffalo to Syracuse at an average rate of 48 miles per hour. On the return trip along the same road she was able to travel at an average rate of 60 miles per hour. The trip from Buffalo to Syracuse took one-half hour longer than the return trip. How long did the return trip take?

49. Carrie and Crystal live at equal distances from school. Carries walks to school at an average rate of 3 miles per hour and Crystal rides her bicycle at an average rate of 9 miles per hour. It takes Carrie 20 minutes longer than Crystal to get to school. How far from school do Crystal and Carrie live?

50. Emmanuel and Anthony contributed equal amounts to the purchase of a gift for a friend. Emmanuel contributed his share in five-dollar bills and Anthony gave his share in one-dollar bills. Anthony needed 12 more bills than Emmanuel. How much did each contribute toward the gift?

4-4 USING FORMULAS TO SOLVE PROBLEMS

To solve for the subject of a formula, substitute the known values in the formula and perform the required computation. For example, to find the area of a triangle when $b = 4.70$ centimeters and $h = 3.20$ centimeters, substitute the given values in the formula for the area of a triangle:

$$\begin{aligned} A &= \tfrac{1}{2}bh && \text{A is the subject of the formula.} \\ &= \tfrac{1}{2}(4.70 \text{ cm})(3.20 \text{ cm}) \\ &= 7.52 \text{ cm}^2 \end{aligned}$$

Now that you can solve equations, you will be able to find the value of any variable in a formula when the values of the other variables are known. To do this:

1. Write the formula.
2. Substitute the given values in the formula.
3. Solve the resulting equation.

The values assigned to the variables in a formula often have a unit of measure. It is convenient to solve the equation without writing the unit of measure, but the answer should always be given in terms of the correct unit of measure.

EXAMPLE 1

The perimeter of a rectangle is 48 centimeters. If the length of the rectangle is 16 centimeters, find the width to the nearest centimeter.

Solution You know that the **perimeter** of a geometric figure is the sum of the lengths of all of its sides. When solving a perimeter problem, it is helpful to draw and label a figure to model the region. Use the formula $P = 2l + 2w$.

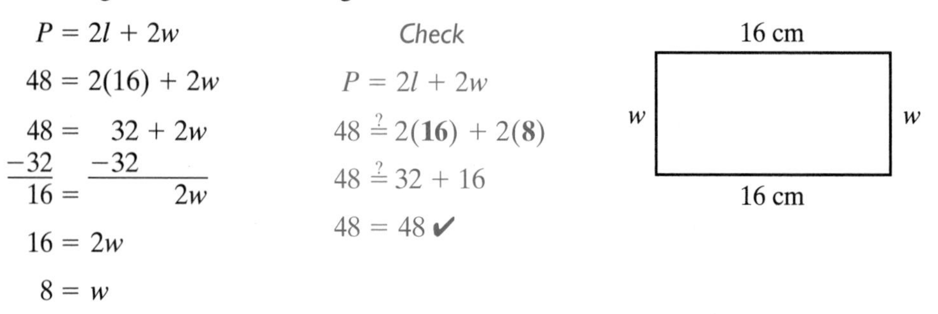

$$\begin{aligned} P &= 2l + 2w \\ 48 &= 2(16) + 2w \\ 48 &= 32 + 2w \\ -32 &\quad -32 \\ 16 &= 2w \\ 16 &= 2w \\ 8 &= w \end{aligned}$$

Check

$$\begin{aligned} P &= 2l + 2w \\ 48 &\stackrel{?}{=} 2(\mathbf{16}) + 2(\mathbf{8}) \\ 48 &\stackrel{?}{=} 32 + 16 \\ 48 &= 48 \checkmark \end{aligned}$$

Answer 8 centimeters

EXAMPLE 2

A garden is in the shape of an isosceles triangle, a triangle that has two sides of equal measure. The length of the third side of the triangle is 2 feet greater than the length of each of the equal sides. If the perimeter of the garden is 86 feet, find the length of each side of the garden.

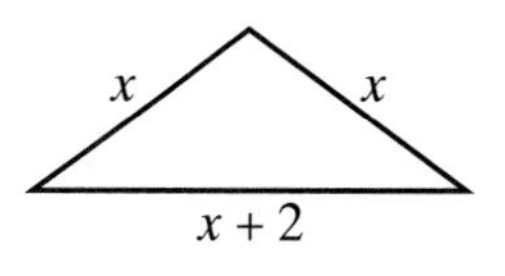

Solution Let x = the length of each of the two equal sides.

Then, $x + 2$ = the length of the third side.

The perimeter is the sum of the lengths of the sides.

$$86 = x + x + (x + 2)$$

$$86 = 3x + 2$$

$$84 = 3x$$

$$28 = x$$

The length of each of the equal sides $= x = 28$.

The length of the third side $= x + 2 = 28 + 2 = 30$.

Check Perimeter $= 28 + 28 + 30 = 86$ ✔

Answer The length of each of the equal sides is 28 feet. The length of the third side (the base) is 30 feet.

EXAMPLE 3

The perimeter of a rectangle is 52 feet. The length is 2 feet more than 5 times the width. Find the dimensions of the rectangle.

Solution Use the formula for the perimeter of a rectangle, $P = 2l + 2w$, to solve this problem.

Let w = the width, in feet, of the rectangle.

Then $5w + 2$ = the length, in feet, of the rectangle.

	Check
$P = 2l + 2w$	The length, 22, is 2 more than 5 times the width, 4. ✔
$52 = 2(5w + 2) + 2w$	$P = 2l + 2w$
$52 = 10w + 4 + 2w$	$52 \stackrel{?}{=} 2(\mathbf{22}) + 2(\mathbf{4})$
$52 = 12w + 4$	$52 \stackrel{?}{=} 44 + 8$
$48 = 12w$	$52 = 52$ ✔
$4 = w$	

Answer The width is 4 feet; the length is $5(w) + 2 = 5(4) + 2 = 22$ feet.

EXAMPLE 4

Sabrina drove from her home to her mother's home which is 150 miles away. For the first half hour, she drove on local roads. For the next two hours she drove on an interstate highway and increased her average speed by 15 miles per hour. Find Sabrina's average speed on the local roads and on the interstate highway.

Solution List the facts stated by this problem:

FACT 1 Sabrina drove on local roads for $\frac{1}{2}$ hour or 0.5 hour.

FACT 2 Sabrina drove on the interstate highway for 2 hours.

FACT 3 Sabrina's rate or speed on the interstate highway was 15 mph more than her rate on local roads.

This problem involves rate, time, and distance. Use the **distance formula**, $d = rt$, where r is the rate, or speed, in miles per hour, t is time in hours, and d is distance in miles.

(1) Represent Sabrina's speed for each part of the trip in terms of r.

Let r = Sabrina's speed on the local roads.

Then $r + 15$ = Sabrina's speed on the interstate highway.

(2) Organize the facts in a table, using the distance formula.

	Rate	× Time	= Distance
Local Roads	r	0.5	$0.5r$
Interstate highway	$r + 15$	2	$2(r + 15)$

(3) Write an equation.

The distance on the local roads plus the distance on the highway is 150 miles.

$$0.5r \quad + \quad 2(r + 15) \quad = \quad 150$$

(4) Solve the equation.

(a) Write the equation: $0.5r + 2(r + 15) = 150$

(b) Use the distributive property: $0.5r + 2r + 30 = 150$

(c) Combine like terms: $2.5r + 30 = 150$

(d) Add −30, the opposite of +30 to each side of the equation: $\underline{-30 \quad -30}$

$2.5r = 120$

(e) Divide each side by 2.5: $\frac{2.5r}{2.5} = \frac{120}{2.5}$

$r = 48$

(5) Find the average speed for each part of the trip.
Sabrina's speed on local roads $= r = 48$ mph.
Sabrina's speed on the highway $= r + 15 = 48 + 15 = 63$ mph.

Check On local roads: $0.5(48) = 24$ miles

On the interstate highway: $2(63) = \underline{126}$ miles

The total distance traveled: 150 miles ✔

Answer Sabrina traveled at an average speed of 48 miles per hour on local roads and 63 miles per hour on the interstate highway.

EXERCISES

Writing About Mathematics

1. In Example 4, step 4 uses the equivalent equation $2.5r + 30 = 150$. Explain what each term in the left side of this equation represents in the problem.

2. Antonio solved the problem in Example 4 by letting r represent Sabrina's rate of speed on the interstate highway. To complete the problem correctly, how should Antonio represent her rate of speed on the local roads? Explain your answer.

Developing Skills

In 3–19, state the meaning of each formula, tell what each variable represents, and find the required value. In 3–12, express each answer to the correct number of significant digits.

3. If $P = a + b + c$, find c when $P = 85$ in., $a = 25$ in., and $b = 12$ in.

4. If $P = 4s$, find s when $P = 32$ m.

5. If $P = 4s$, find s when $P = 6.8$ ft.

6. If $P = 2l + 2w$, find w when $P = 26$ yd and $l = 8$ yd.

7. If $P = 2a + b$, find b when $P = 80$ cm and $a = 30$ cm.

8. If $P = 2a + b$, find a when $P = 18.6$ m and $b = 5.8$ m.

9. If $A = bh$, find b when $A = 240 \text{ cm}^2$ and $h = 15$ cm.

10. If $A = bh$, find h when $A = 3.6 \text{ m}^2$ and $b = 0.90$ m.

11. If $A = \frac{1}{2}bh$, find h when $A = 24$ sq ft and $b = 8.0$ ft.

12. If $V = lwh$, find w when $V = 72 \text{ yd}^3$, $l = 0.75$ yd, and $h = 12$ yd.

13. If $d = rt$, find r when $d = 120$ mi and $t = 3$ hr.

14. If $I = prt$, find the principal, p, when the interest, I, is \$135, the yearly rate of interest, r, is 2.5%, and the time, t, is 3 years.

15. If $I = prt$, find the rate of interest, r, when $I = \$225$, $p = \$2{,}500$, and $t = 2$ years.

16. If $T = nc$, find the number of items purchased, n, if the total cost, T, is \$19.80 and the cost of one item, c, is \$4.95.

17. If $T = nc$, find the cost of one item purchased, c, if $T = \$5.88$ and $n = 12$.

18. If $S = nw$, find the hourly wage, w, if the salary earned, S, is \$243.20 and the number of hours worked, n, is 38.

19. If $S = nw$, find the number of hours worked, n, if $S = \$315.00$ and $w = \$8.40$.

In 20–32, **a.** write a formula that can be used to solve each problem, **b.** use the formula to solve each problem and check the solution. All numbers may be considered to be exact values.

20. Find the length of a rectangle whose perimeter is 34.6 centimeters and whose width is 5.7 centimeters.

21. The length of the second side of a triangle is 2 inches less than the length of the first side. The length of the third side is 12 inches more than the length of the first side. The perimeter of the triangle is 73 inches. Find the length of each side of the triangle.

22. Two sides of a triangle are equal in length. The length of the third side exceeds the length of one of the other sides by 3 centimeters. The perimeter of the triangle is 93 centimeters. Find the length of each side of the triangle.

23. The length of a rectangle is 5 meters more than its width. The perimeter is 66 meters. Find the dimensions of the rectangle.

24. The width of a rectangle is 3 yards less than its length. The perimeter is 130 yards. Find the length and the width of the rectangle.

25. The length of each side of an equilateral triangle is 5 centimeters more than the length of each side of a square. The perimeters of the two figures are equal. Find the lengths of the sides of the square and of the triangle.

26. The length of each side of a square is 1 centimeter more than the width of a rectangle. The length of the rectangle is 1 centimeter less than twice its width. The perimeters of the two figures are equal. Find the dimensions of the rectangle.

27. The area of a triangle is 36 square centimeters. Find the measure of the altitude drawn to the base when the base is 8 centimeters.

28. The altitude of a triangle is 4.8 meters. Find the length of the base of the triangle if the area is 8.4 square meters.

29. The length of a rectangle is twice the width. If the length is increased by 4 inches and the width is decreased by 1 inch, a new rectangle is formed whose perimeter is 198 inches. Find the dimensions of the original rectangle.

30. The length of a rectangle exceeds its width by 4 feet. If the width is doubled and the length is decreased by 2 feet, a new rectangle is formed whose perimeter is 8 feet more than the perimeter of the original rectangle. Find the dimensions of the original rectangle.

31. A side of a square is 10 meters longer than the side of an equilateral triangle. The perimeter of the square is 3 times the perimeter of the triangle. Find the length of each side of the triangle.

32. The length of each side of a hexagon is 4 inches less than the length of a side of a square. The perimeter of the hexagon is equal to the perimeter of the square. Find the length of a side of the hexagon and the length of a side of the square.

Applying Skills

33. The perimeter of a rectangular parking lot is 146 meters. Find the dimensions of the lot, using the correct number of significant digits, if the length is 7.0 meters less than 4 times the width.

34. The perimeter of a rectangular tennis court is 228 feet. If the length of the court exceeds twice its width by 6.0 feet, find the dimensions of the court using the correct number of significant digits.

In 35–48, make a table to organize the information according to the formula to be used. All numbers may be considered to be exact values.

35. Rahul has 25 coins, all quarters and dimes. Copy the table given below and organize the facts in the table using the answers to **a** through **c**.

	(Number of coins in one denomination) ×	(Value of one coin) =	Total value of the coins of that denomination
	Number of Coins	**Value of One Coin**	**Total Value**
Dimes			
Quarters			

a. If x is the number of dimes Rahul has, express, in terms of x, the number of quarters he has.

b. Express the value of the dimes in terms of x.

c. Express the value of the quarters in terms of x.

d. If the total value of the dimes and quarters is $4.90, write and solve an equation to find how many dimes and how many quarters Rahul has.

e. Check your answer in the words of the problem.

36. If the problem had said that the total value of Rahul's 25 dimes and quarters was $5.00, what conclusion could you draw?

37. When Ruth emptied her bank, she found that she had 84 coins, all nickels and dimes. The value of the coins was $7.15. How many dimes did she have? (Make a table similar to that given in exercise 35.)

38. Adele went to the post office to buy stamps and postcards. She bought a total of 25 stamps, some 39-cent stamps and the rest 23-cent postcards. If she paid $8.47 altogether, how many 39-cent stamps did she buy?

39. Carlos works Monday through Friday and sometimes on Saturday. Last week Carlos worked 38 hours. Copy the table given below and organize the facts in the table using the answers to **a** through **c**.

	Hours Worked	Wage Per Hour	Earnings
Monday–Friday			
Saturday			

a. If x is the total number of hours Carlos worked Monday through Friday, express, in terms of x, the number of hours he worked on Saturday.

b. Carlos earns $8.50 an hour when he works Monday through Friday. Express, in terms of x, his earnings Monday through Friday.

c. Carlos earns $12.75 an hour when he works on Saturday. Express, in terms of x, his earnings on Saturday.

d. Last week Carlos earned $340. How many hours did he work on Saturday?

40. Janice earns $6.00 an hour when she works Monday through Friday and $9.00 an hour when she works on Saturday. Last week, her salary was $273 for 42 hours of work. How many hours did she work on Saturday? (Make a table similar to that given in exercise 39.)

41. Candice earns $8.25 an hour and is paid every two weeks. Last week she worked 4 hours longer than the week before. Her pay for these two weeks, before deductions, was $594. How many hours did she work each week?

42. Akram drove from Rochester to Albany, a distance of 219 miles. After the first 1.5 hours of travel, it began to snow and he reduced his speed by 26 miles per hour. It took him another 3 hours to complete the trip. Copy the table given below and fill in the entries using the answers to **a** through **c**.

	Rate	Time	Distance
First part of the trip			
Last part of the trip			

a. If r is the average speed at which Akram traveled for the first part of the trip, express, in terms of r, his average speed for the second part of the trip.

b. Express, in terms of r, the distance that Akram traveled in the first part of the trip.

c. Express, in terms of r, the distance that Akram traveled in the second part of the trip.

d. Find the speed at which Akram traveled during each part of the trip.

43. Vera walked from her home to a friend's home at a rate of 3 miles per hour. She rode to work with her friend at an average rate of 30 miles per hour. It took Vera a total of 50 minutes $\left(\frac{5}{6}\text{ of an hour}\right)$ to walk to her friend's home and to get to work, traveling a total distance of 16 miles. How long did she walk and how long did she ride with her friend to get to work? (Make a table similar to that given in exercise 42.)

44. Peter drove a distance of 189 miles. Part of the time he averaged 65 miles per hour and for the remaining time, 55 miles per hour. The entire trip took 3 hours. How long did he travel at each rate?

45. Shelly and Jack left from the same place at the same time and drove in opposite directions along a straight road. Jack traveled 15 miles per hour faster than Shelly. After 3 hours, they were 315 miles apart. Find the rate at which each traveled.

46. Carla and Candice left from the same place at the same time and rode their bicycles in the same direction along a straight road. Candice bicycled at an average speed that was three-quarters of Carla's average speed. After 2 hours they were 28 miles apart. What was the average speed of Carla and Candice?

47. Nolan walked to the store from his home at the rate of 5 miles per hour. After spending one-half hour in the store, his friend gave him a ride home at the rate of 30 miles per hour. He arrived home 1 hour and 5 minutes $\left(1\frac{1}{12}\text{ hours}\right)$ after he left. How far is the store from Nolan's home?

48. Mrs. Dang drove her daughter to school at an average rate of 45 miles per hour. She returned home by the same route at an average rate of 30 miles per hour. If the trip took one-half hour, How long did it take to get to school? How far is the school from their home?

4-5 SOLVING FOR A VARIABLE IN TERMS OF ANOTHER VARIABLE

An equation may contain more than one variable. For example, the equation $ax + b = 3b$ contains the variables a, b, and x. To solve this equation for x means to express x in terms of the other variables.

To plan the steps in the solution, it is helpful to use the strategy of using a simpler related problem, that is, to compare the solution of this equation with the solution of a simpler equation that has only one variable. In Example 1, the solution of $ax + b = 3b$ is compared with the solution of $2x + 5 = 15$. The same operations are used in the solution of both equations.

EXAMPLE 1

Solve for x in $ax + b = 3b$.

Solution Compare with $2x + 5 = 15$.

$$\begin{aligned} 2x + 5 &= 15 \\ -5 &\quad -5 \\ 2x &= 10 \\ \frac{2x}{2} &= \frac{10}{2} \\ x &= 5 \end{aligned}$$

$$\begin{aligned} ax + b &= 3b \\ -b &\quad -b \\ ax &= 2b \\ \frac{ax}{a} &= \frac{2b}{a} \\ x &= \frac{2b}{a} \end{aligned}$$

Check

$$\begin{aligned} ax + b &= 3b \\ a\left(\frac{2b}{a}\right) + b &\stackrel{?}{=} 3b \\ 2b + b &\stackrel{?}{=} 3b \\ 3b &= 3b \checkmark \end{aligned}$$

Answer $x = \frac{2b}{a}$

EXAMPLE 2

Solve for x in $x - a = b$.

Solution Compare with $x - 5 = 9$.

$$\begin{aligned} x - 5 &= 9 \\ +5 &\quad +5 \\ x &= 14 \end{aligned}$$

$$\begin{aligned} x - a &= b \\ +a &\quad +a \\ x &= b + a \end{aligned}$$

Check

$$\begin{aligned} x - a &= b \\ b + a - a &\stackrel{?}{=} b \\ b &= b \checkmark \end{aligned}$$

Answer $x = b + a$

EXAMPLE 3

Solve for x in $2ax = 10a^2 - 3ax$ $(a \neq 0)$.

Solution Compare with $2x = 10 - 3x$.

$2x = 10 - 3x$	$2ax = 10a^2 - 3ax$
$\underline{+3x \quad + 3x}$	$\underline{+3ax \quad + 3ax}$
$5x = 10$	$5ax = 10a^2$
$\frac{5x}{5} = \frac{10}{5}$	$\frac{5ax}{5a} = \frac{10a^2}{5a}$
$x = 2$	$x = 2a$

Check

$2ax = 10a^2 - 3ax$

$2a(\mathbf{2a}) \stackrel{?}{=} 10a^2 - 3a(\mathbf{2a})$

$4a^2 \stackrel{?}{=} 10a^2 - 6a^2$

$4a^2 = 4a^2$ ✔

Answer $x = 2a$

EXERCISES

Writing About Mathematics

1. Write a simpler related equation in one variable that can be used to suggest the steps needed to solve the equation $a(x + b) = 4ab$ for x.
2. Write a simpler related equation in one variable that can be used to suggest the steps needed to solve the equation $5cy = d + 2cy$ for y.

Developing Skills

In 3–24, solve each equation for x or y and check.

3. $5x = b$ **4.** $sx = 8$ **5.** $ry = s$ **6.** $hy = m$

7. $x + 5r = 7r$ **8.** $x + a = 4a$ **9.** $y + c = 9c$ **10.** $4 + x = k$

11. $d + y = 9$ **12.** $3x - q = 5q$ **13.** $3x - 8r = r$ **14.** $cy - d = 4d$

15. $ax + b = 3b$ **16.** $dx - 5c = 3c$ **17.** $r + sy = t$ **18.** $m = 2(x + n)$

19. $bx = 9b^2$ **20.** $cx + c^2 = 5c^2 - 7cx$ **21.** $rsx - rs^2 = 0$

22. $m^2x - 3m^2 = 12m^2$ **23.** $9x - 24a = 6a + 4x$ **24.** $8ax - 7a^2 = 19a^2 - 5ax$

4-6 TRANSFORMING FORMULAS

A formula is an equation that contains more than one variable. Sometimes you want to solve for a variable in the formula that is different from the subject of the formula. For example, the formula for distance, d, in terms of rate, r, and time, t, is $d = rt$. Distance is the subject of the formula, but you might want to rewrite the formula so that it expresses time in terms of distance and rate. You do this by solving the equation $d = rt$ for t in terms of d and r.

EXAMPLE 1

a. Solve the formula $d = rt$ for t.
b. Use the answer obtained in part **a** to find the value of t when $d = 200$ miles and $r = 40$ miles per hour.

Solution **a.** $d = rt$

$$\frac{d}{r} = \frac{rt}{r}$$

$$\frac{d}{r} = t$$

b. $t = \frac{d}{r}$

$$= \frac{200}{40}$$

$$= 5$$

Answers **a.** $t = \frac{d}{r}$ **b.** $t = 5$ hours

Note that the rate is 40 miles per hour, that is, $\frac{40 \text{ miles}}{1 \text{ hour}}$. Therefore,

$$200 \text{ miles} \div \frac{40 \text{ miles}}{1 \text{ hour}} = 200 \cancel{\text{miles}} \times \frac{1 \text{ hour}}{40 \cancel{\text{miles}}} = 5 \text{ hours}$$

We can think of canceling miles in the numerator and the denominator of the fractions being multiplied.

EXAMPLE 2

a. The formula for the volume of a cone is $V = \frac{1}{3}Bh$. Solve this formula for h.
b. Find the height of a cone that has a volume of 92.0 cubic centimeters and a circular base with a radius of 2.80 centimeters. Express the answer using the correct number of significant digits.

Solution **a.**

$$V = \tfrac{1}{3}Bh$$

$$3V = 3\left(\tfrac{1}{3}Bh\right)$$

$$3V = Bh$$

$$\frac{3V}{B} = \frac{Bh}{B}$$

$$\frac{3V}{B} = h$$

b. Find B, the area of the base of the cone. Since the base is a circle, its area is π times the square of the radius, r.

$$B = \pi r^2$$

$$= \pi(2.80)^2$$

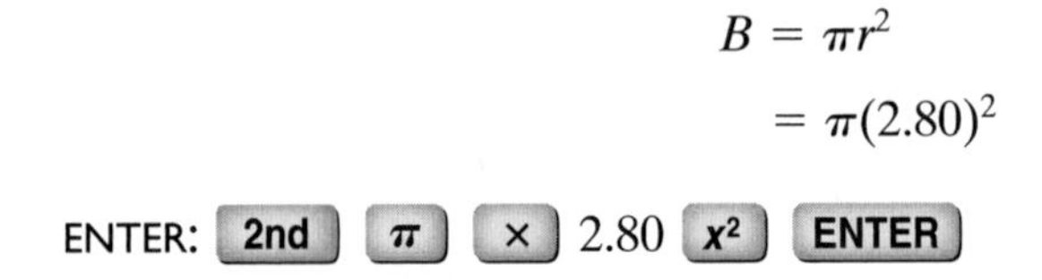

DISPLAY:

```
π*2.80²
          24.6300864
```

Now use the answer to part **a** to find h: $h = \frac{3V}{B} \approx \frac{3(92.0)}{24.63}$

ENTER: 3 [×] 92.0 [÷] 24.63 [ENTER]

DISPLAY:

```
3*92.0/24.63
         11.20584653
```

Since each measure is given to three significant digits, round the answer to three significant digits.

Answers **a.** $h = \frac{3V}{B}$ **b.** The height of the cone is 11.2 centimeters.

EXERCISES

Developing Skills

In 1–14, transform each given formula by solving for the indicated variable.

1. $P = 4s$ for s

2. $A = bh$ for h

3. $d = rt$ for r

4. $V = lwh$ for l

5. $P = br$ for r

6. $I = prt$ for t

7. $A = \frac{1}{2}bh$ for h

8. $V = \frac{1}{3}Bh$ for B

9. $s = \frac{1}{2}gt$ for g

10. $P = 2a + b$ for b

11. $P = 2a + b$ for a

12. $P = 2l + 2w$ for w

13. $F = \frac{9}{5}C + 32$ for C

14. $2S = n(a + l)$ for a

Applying Skills

15. The concession stand at a movie theater wants to sell popcorn in containers that are in the shape of a cylinder. The volume of the cylinder is given by the formula $V = \pi r^2 h$, where V is the volume, r is the radius of the base, and h is the height of the container.

a. Solve the formula for h.

b. If the container is to hold 1,400 cubic centimeters of popcorn, find, to the *nearest tenth*, the height of the container if the radius of the base is:
(1) 4.0 centimeters (2) 5.0 centimeters (3) 8.0 centimeters

c. The concession stand wants to put an ad with a height of 20 centimeters on the side of the container. Which height from part **b** do you think would be the best for the container? Why?

16. A bus travels from Buffalo to Albany, stopping at Rochester and Syracuse. At each city there is a 30-minute stopover to unload and load passengers and baggage. The driving distance from Buffalo to Rochester is 75 miles, from Rochester to Syracuse is 85 miles, and from Syracuse to Albany is 145 miles. The bus travels at an average speed of 50 miles per hour.

a. Solve the formula $d = rt$ for t to find the time needed for each part of the trip.

b. Make a schedule for the times of arrival and departure for each city if the bus leaves Buffalo at 9:00 A.M.

4-7 PROPERTIES OF INEQUALITIES

The Order Property of Real Numbers

If two real numbers are graphed on the number line, only one of the following three situations can be true:

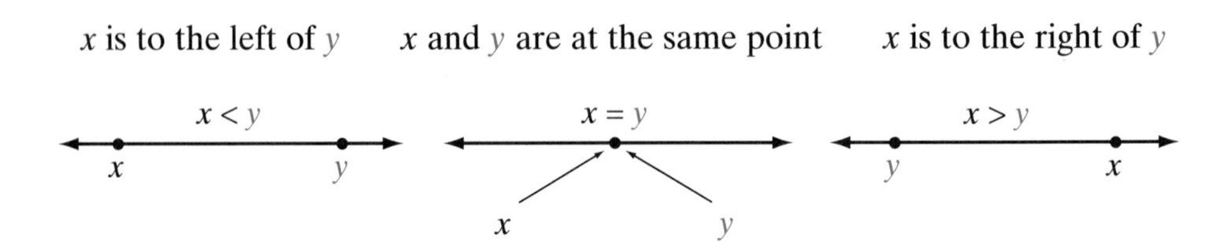

These three cases illustrate the **order property of real numbers**:

▶ **If x and y are two real numbers, then one and only one of the following can be true:**

$$x < y \text{ or } x = y \text{ or } x > y$$

Let y be a fixed point, for example, $y = 3$. Then y separates the real numbers into three sets. For any real number, one of the following must be true: $x < 3$, $x = 3$, $x > 3$.

The real numbers, x, that make the inequality $x < 3$ true are to the left of 3 on the number line. The circle at 3 indicates that 3 is the boundary value of the set. The circle is not filled in, indicating that 3 does not belong to this set.

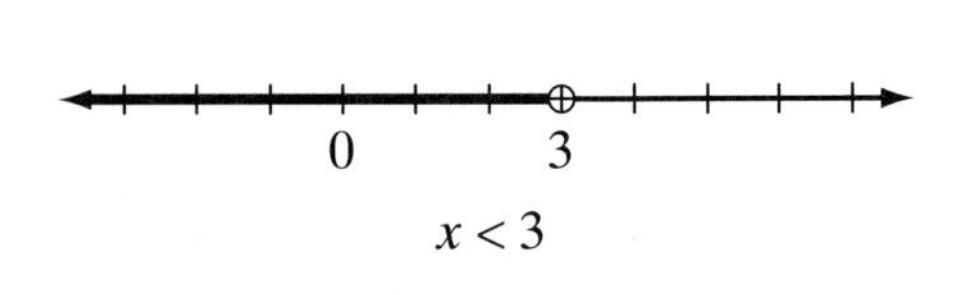

The real number that makes the corresponding equality, $x = 3$, true is a single point on the number line. This point, $x = 3$, is also the boundary between the values of x that make $x < 3$ true and the values of x that make $x > 3$ true. The circle is filled in, indicating that 3 belongs to this set. Here, 3 is the only element of the set.

The real numbers, x, that make $x > 3$ true are to the right of 3 on the number line. Again, the circle at 3 indicates that 3 is the boundary value of the set. The circle is not filled in, indicating that 3 does not belong to this set.

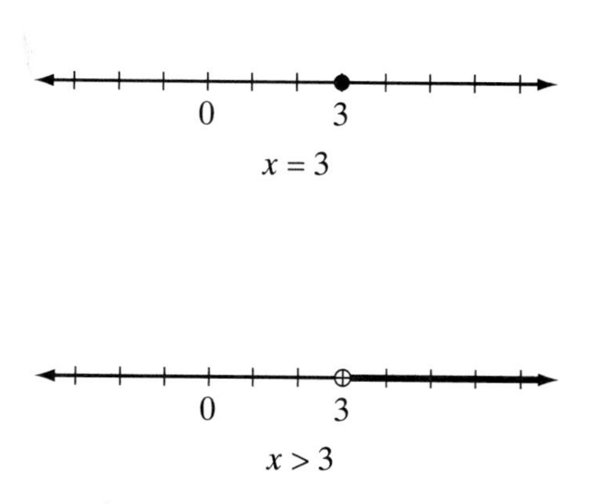

The Transitive Property of Inequality

From the graph at the right, you can see that, if x lies to the left of y, and y lies to the left of z, then x lies to the left of z.

The graph illustrates the **transitive property of inequality**:

▶ **For the real numbers x, y, and z:**

If $x < y$ and $y < z$, then $x < z$; and if $z > y$ and $y > x$, then $z > x$.

The Addition Property of Inequality

The following table shows the result of adding a number to both sides of an inequality.

True Sentence	Number to Add to Both Sides	Result
$9 > 2$ Order is "greater than."	$9 + 3 \ ?\ 2 + 3$ Add a positive number.	$12 > 5$ Order is unchanged.
$9 > 2$ Order is "greater than."	$9 + (-3) \ ?\ 2 + (-3)$ Add a negative number.	$6 > -1$ Order is unchanged.
$2 < 9$ Order is "less than."	$2 + 3 \ ?\ 9 + 3$ Add a positive number.	$5 < 12$ Order is unchanged
$2 < 9$ Order is "less than."	$2 + (-3) \ ?\ 9 + (-3)$ Add a negative number.	$-1 < 6$ Order is unchanged

The table illustrates the **addition property of inequality**:

▶ **For the real numbers x, y, and z:**

$$\text{If } x < y, \text{ then } x + z < y + z; \text{ and if } x > y, \text{ then } x + z > y + z.$$

Since subtracting the same number from both sides of an inequality is equivalent to adding the additive inverse to both sides of the inequality, the following is true:

▶ **When the same number is added to or subtracted from both sides of an inequality, the order of the new inequality is the same as the order of the original one.**

EXAMPLE 1

Use the inequality $5 < 9$ to write a new inequality:

a. by adding 6 to both sides

b. by adding -9 to both sides

Solution **a.** $5 + 6 < 9 + 6$

$11 < 15$

b. $5 + (-9) < 9 + (-9)$

$-4 < 0$

Answers **a.** $11 < 15$ **b.** $-4 < 0$

The Multiplication Property of Inequality

The following table shows the result of multiplying both sides of an inequality by the same number.

True Sentence	Number to Multiply Both Sides	Result
$9 > 2$ Order is "greater than."	$9(3) \overset{?}{} 2(3)$ Multiply by a positive number	$27 > 6$ Order is unchanged.
$5 < 9$ Order is "less than."	$5(3) \overset{?}{} 9(3)$ Multiply by a positive number.	$15 < 27$ Order is unchanged.
$9 > 2$ Order is "greater than."	$9(-3) \overset{?}{} 2(-3)$ Multiply by a negative number	$-27 < -6$ Order is changed.
$5 < 9$ Order is "less than."	$5(-3) \overset{?}{} 9(-3)$ Multiply by a negative number.	$-15 > -27$ Order is changed.

The table illustrates that the order does not change when both sides are multiplied by the same positive number, but does change when both sides are multiplied by the same negative number.

In general terms, the **multiplication property of inequality** states:

▶ **For the real numbers x, y, and z:**

If z is positive ($z > 0$) and $x < y$, then $xz < yz$.

If z is positive ($z > 0$) and $x > y$, then $xz > yz$.

If z is negative ($z < 0$) and $x < y$, then $xz > yz$.

If z is negative ($z < 0$) and $x > y$, then $xz < yz$.

Dividing both sides of an inequality by a number is equivalent to multiplying both sides by the multiplicative inverse of the number. A number and its multiplicative inverse always have the same sign. Therefore, the following is true:

▶ **When both sides of an inequality are multiplied or divided by the same *positive* number, the order of the new inequality is the same as the order of the original one.**

▶ **When both sides of an inequality are multiplied or divided by the same *negative* number, the order of the new inequality is the opposite of the order of the original one.**

EXAMPLE 2

Use the inequality $6 < 9$ to write a new inequality:

a. by multiplying both sides by 2.

b. by multiplying both sides by $-\frac{1}{3}$.

Solution

a.

$$6 < 9$$
$$6(2) \ ?\ 9(2)$$
$$12 < 18$$

b.

$$6 < 9$$
$$6\left(-\tfrac{1}{3}\right) \ ?\ 9\left(-\tfrac{1}{3}\right)$$
$$-2 > -3$$

Answers **a.** $12 < 18$ **b.** $-2 > -3$

EXERCISES

Writing About Mathematics

1. Sadie said that if $5 > 4$, then it must be true that $5x > 4x$. Do you agree with Sadie? Explain why or why not.
2. Lucius said that if $x > y$ and $a > b$ then $x + a > y + b$. Do you agree with Lucius? Explain why or why not.
3. Jason said that if $x > y$ and $a > b$ then $x - a > y - b$. Do you agree with Jason? Explain why or why not.

Developing Skills

In 4–31, replace each question mark with the symbol > or < so that the resulting sentence will be true.

4. Since $8 > 2, 8 + 1 \ ? \ 2 + 1$.
5. Since $-6 < 2, -6 + (-4) \ ? \ 2 + (-4)$.
6. Since $9 > 5, 9 - 2 \ ? \ 5 - 2$.
7. Since $-2 > -8, -2 - \left(\frac{1}{4}\right) \ ? \ -8 - \left(\frac{1}{4}\right)$.
8. Since $7 > 3, \frac{2}{3}(7) \ ? \ \frac{2}{3}(3)$.
9. Since $-8 < 4, (-8) \div (4) \ ? \ (4) \div (4)$.
10. Since $9 > 6, 9 \div \left(-\frac{1}{3}\right) \ ? \ 6 \div \left(-\frac{1}{3}\right)$.
11. If $5 > x$, then $5 + 7 \ ? \ x + 7$.
12. If $y < 6$, then $y - 2 \ ? \ 6 - 2$.
13. If $20 > r$, then $4(20) \ ? \ 4(r)$.
14. If $t < 64$, then $t \div 8 \ ? \ 64 \div 8$.
15. If $x > 8$, then $-2x \ ? \ -2(8)$.
16. If $y < 8$, then $y \div (-4) \ ? \ 8 \div (-4)$.
17. If $x + 2 > 7$, the $x + 2 + (-2) \ ? \ 7 + (-2)$ or $x \ ? \ 5$.
18. If $y - 3 < 12$, then $y - 3 + 3 \ ? \ 12 + 3$ or $y \ ? \ 15$.
19. If $a + 5 < 14$, then $a + 5 - 5 \ ? \ 14 - 5$ or $a \ ? \ 9$.
20. If $2x > 8$, then $\frac{2x}{2} \ ? \ \frac{8}{2}$ or $x \ ? \ 4$.
21. If $\frac{1}{3}y < 4$, then $3\left(\frac{1}{3}y\right) \ ? \ 3(4)$ or $y \ ? \ 12$.
22. If $-3x < 36$, then $\frac{-3x}{-3} \ ? \ \frac{36}{-3}$ or $x \ ? \ -12$.
23. If $-2x > 6$, then $-\frac{1}{2}(-2x) \ ? \ -\frac{1}{2}(6)$ or $x \ ? \ -3$.
24. If $x < 5$ and $5 < y$, then $x \ ? \ y$.
25. If $m > -7$ and $-7 > a$, then $m \ ? \ a$.
26. If $3 < 7$, then $7 \ ? \ 3$.
27. If $-4 > -12$, then $-12 \ ? \ -4$.
28. If $9 > x$, then $x \ ? \ 9$.
29. If $-7 < a$, then $a \ ? \ -7$.
30. If $x < 10$ and $10 < z$, then $x \ ? \ z$.
31. If $a > b$ and $c < b$, then $a \ ? \ c$.

4-8 FINDING AND GRAPHING THE SOLUTION OF AN INEQUALITY

When an inequality contains a variable, the **domain** or **replacement set of the inequality** is the set of all possible numbers that can be used to replace the variable. When an element from the domain is used in place of the variable, the inequality may be true or it may be false. The **solution set of an inequality** is the set of numbers from the domain that make the inequality true. Inequalities that have the same solution set are **equivalent inequalities**.

To find the solution set of an inequality, solve the inequality by methods similar to those used in solving an equation. Use the properties of inequalities to transform the given inequality into a simpler equivalent inequality whose solution set is evident.

In Examples 1–5, the domain is the set of real numbers.

EXAMPLE 1

Find and graph the solution of the inequality $x - 4 > 1$.

Solution *How to Proceed*

(1) Write the inequality: $x - 4 > 1$

(2) Use the addition property of inequality. Add 4 to each side: $\begin{array}{r} x - 4 > 1 \\ \underline{+4 \quad +4} \\ x > 5 \end{array}$

–2 –1 0 1 2 3 4 5 6 7 8

The graph above shows the solution set. The circle at 5 indicates that 5 is the boundary between the numbers to the right, which belong to the solution set, and the numbers to the left, which do not belong. Since 5 is not included in the solution set, the circle is not filled in.

Check (1) Check one value from the solution set, for example, 7. This value will make the inequality true. $7 - 4 > 1$ is true.

(2) Check the boundary value, 5. This value, which separates the values that make the inequality true from the values that make it false, will make the corresponding equality, $x - 4 = 1$, true. $5 - 4 = 1$ is true.

Answer $x > 5$

An alternative method of expressing the solution set is **interval notation**. When this notation is used, the solution set is written as $(5, \infty)$. The first number, 5 names the lower boundary. The symbol ∞, often called infinity, indicates that there is no upper boundary, that is, that the set of real numbers continues without end. The parentheses indicate that the boundary values are not elements of the set.

EXAMPLE 2

Find and graph the solution of $5x + 4 < 11 - 2x$.

Solution The solution set of $5x + 4 \leq 11 - 2x$ includes all values of the domain for which either $5x + 4 < 11 - 2x$ is true or $5x + 4 = 11 - 2x$ is true.

How to Proceed

(1) Write the inequality: $5x + 4 \leq 11 - 2x$

(2) Add $2x$ to each side: $\underline{+2x \qquad\quad + 2x}$

$7x + 4 \leq 11$

(3) Add -4 to each side: $\underline{\quad - 4 \quad -4}$

$7x \leq 7$

(4) Divide each side by 7: $\frac{7x}{7} \leq \frac{7}{7}$

$x \leq 1$

The solution set includes 1 and all of the real numbers less than 1. This is shown on the graph below by filling in the circle at 1 and drawing a heavy line to the left of 1.

Answer $x \leq 1$

–4 –3 –2 –1 0 1 2 3

The solution set can also be written in interval notation as $(-\infty, 1]$. The symbol $-\infty$, often called negative infinity, indicates that there is no lower boundary, that is, all negative real numbers less than the upper boundary are included. The number, 1, names the upper boundary. The right bracket indicates that the upper boundary value is an element of the set.

EXAMPLE 3

Find and graph the solution set: $2(2x - 8) - 8x \leq 0$.

Solution *How to Proceed*

(1) Write the inequality: $2(2x - 8) - 8x \leq 0$

(2) Use the distributive property: $4x - 16 - 8x \leq 0$

(3) Combine like terms in the left side: $-4x - 16 \leq 0$

(4) Add 16 to each side: $\underline{\quad + 16 \quad +16}$

$-4x \leq 16$

(5) Divide both sides by -4. Dividing by a negative number reverses the inequality: $\frac{-4x}{-4} \geq \frac{16}{-4}$

$x \geq -4$

(6) The graph of the solution set includes -4 and all of the real numbers to the right of -4 on the number line:

–5 –4 –3 –2 –1 0 1 2 3

Answer $x \geq -4$ or $[-4, \infty)$

Graphing the Intersection of Two Sets

The inequality $3 < x < 6$ is equivalent to $(3 < x)$ and $(x < 6)$. This statement is true when both simple statements are true and false when one or both statements are false. The solution set of this inequality consists of all of the numbers that are in the solution set of both simple inequalities. The graph of $3 < x < 6$ can be drawn as shown below.

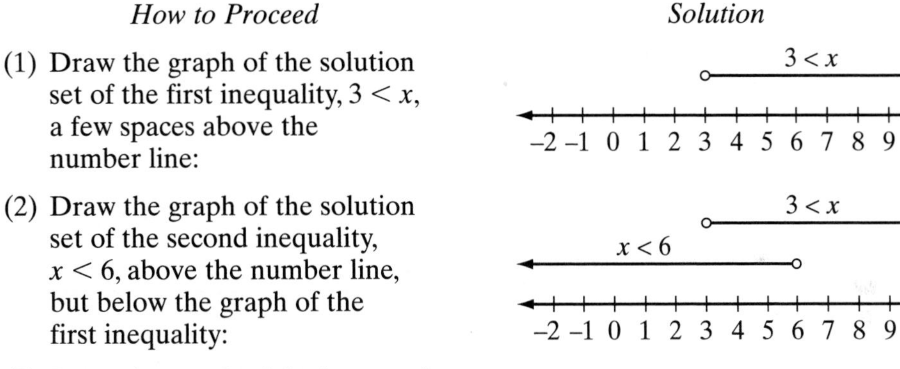

(3) Draw the graph of the intersection of these two sets by shading, on the number line, the points that belong to the solution set of both simple inequalities:

$3 < x < 6$

–2 –1 0 1 2 3 4 5 6 7 8 9

Since 3 is in the solution set of $x < 6$ but not in the solution set of $3 < x$, 3 is not in the intersection of the two sets. Also, since 6 is in the solution set of $3 < x$ but not in the solution set of $x < 6$, 6 is not in the intersection of the two sets. Therefore, the circles at 3 and 6 are not filled in, indicating that these boundary values are not elements of the solution set of $3 < x < 6$.

This set can also be written as $(3, 6)$, a pair of numbers that list the left and right boundaries of the set. The parentheses indicate that the boundary values do not belong to the set. Similarly, the set of numbers $3 \leq x \leq 6$ can be written as $[3, 6]$. The brackets indicate that the boundary values do belong to the set.

Although this notation is similar to that used for an ordered pair that names a point in the coordinate plane, the context in which the interval or ordered pair is used will determine the meaning.

EXAMPLE 4

Solve the inequality and graph the solution set: $-7 < x - 5 < 0$.

Solution *How to Proceed*

(1) First solve the inequalities for x:

$$\begin{array}{rcl} -7 < & x - 5 & < 0 \\ +5 & +5 & +5 \\ \hline -2 < & x & < 5 \end{array}$$

(2) Draw the graphs of $-2 < x$ and $x < 5$ above the number line:

$-2 < x$

$x < 5$

–4 –3 –2 –1 0 1 2 3 4 5 6 7 8

(3) Draw the graph of all points that are common to the graphs of $-2 < x$ and $x < 5$:

$-2 < x < 5$

–4 –3 –2 –1 0 1 2 3 4 5 6 7 8

Answer $-2 < x < 5$ or $(-2, 5)$

Graphing the Union of Two Sets

The inequality $(x > 3)$ or $(x > 6)$ is true when one or both of the simple statements are true. It is false when both simple statements are false. The solution set of the inequality consists of the union of the solution sets of the two simple statements. The graph of the solution set can be drawn as shown below.

How to Proceed *Solution*

(1) Draw the graph of the solution set of the first inequality a few spaces above the number line:

(2) Draw the graph of the solution set of the second inequality above the number line, but below the graph of the first inequality:

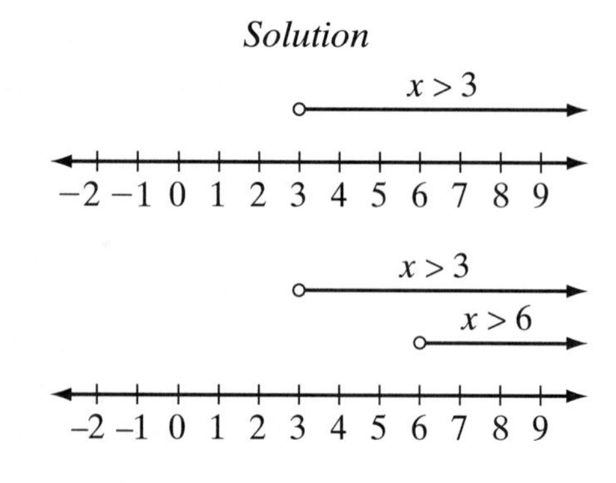

(3) Draw the graph of the union by shading, on the number line, the points that belong to the solution of one or both of the simple inequalities:

Since 3 is not in the solution set of either inequality, 3 is not in the union of the two sets. Therefore, the circle at 3 is not filled in.

Answer: $x > 3$ or $(3, \infty)$

EXAMPLE 5

Solve the inequality and graph the solution set: $(x + 2 < 0)$ or $(x - 3 > 0)$.

Solution

How to Proceed

(1) Solve each inequality for x:

$$\begin{array}{rr} x + 2 < & 0 \\ -2 & -2 \\ \hline x < & -2 \end{array} \quad \text{or} \quad \begin{array}{rr} x - 3 > & 0 \\ +3 & +3 \\ \hline x > & 3 \end{array}$$

(2) Draw the graphs of $x < -2$ and $x > 3$ above the number line:

(3) Draw the graph of all points of the graphs of $x < -2$ or $x > 3$:

Answer $(x < -2)$ or $(x > +3)$ or $(\infty, -2)$ or $(+3, \infty)$

Note: Since the solution is the union of two sets, the answer can also be expressed using set notation: $(x < -2) \cup (x > 3)$ or $(-\infty, -2) \cup (3, \infty)$.

EXERCISES

Writing About Mathematics

1. Give an example of a situation that can be modeled by the inequality $x > 5$ in which **a.** the solution set has a smallest value, and **b.** the solution set does not have a smallest value.

2. Abram said that the solution set of $(x < 4)$ or $(x > 4)$ is the set of all real numbers. Do you agree with Abram? Explain why or why not.

Developing Skills

In 3–37, find and graph the solution set of each inequality. The domain is the set of real numbers.

3. $x - 2 > 4$ **4.** $z - 6 < 4$ **5.** $y - \frac{1}{2} > 2$ **6.** $x - 1.5 < 3.5$

7. $x + 3 > 6$ **8.** $19 < y + 17$ **9.** $d + \frac{1}{4} > 3\frac{1}{4}$ **10.** $-3.5 > c + 0.5$

11. $y - 4 \geq 4$ **12.** $25 \leq d + 22$ **13.** $3t > 6$ **14.** $2x \leq 12$

15. $15 \leq 3y$ **16.** $-10 \leq 4h$ **17.** $-6y < 24$ **18.** $27 > -9y$

19. $-10x > -20$ **20.** $12 \leq -1.2r$ **21.** $\frac{1}{3}x > 2$ **22.** $-\frac{2}{3}z \geq 6$

23. $\frac{x}{2} > 1$ **24.** $-10 \geq 2.5z$ **25.** $2x - 1 > 5$ **26.** $3y - 6 \geq 12$

27. $5y + 3 \geq 13$ **28.** $5x - 4 > 4 - 3x$ **29.** $8y - 1 \leq 3y + 29$ **30.** $6x + 2 < 8x + 14$

31. $8m \geq 2(2m + 3)$ **32.** $0 < x + 3 < 6$ **33.** $-5 \leq x + 2 < 7$

34. $0 < 2x - 4 \leq 6$ **35.** $(x - 1 > 3)$ or $(x + 1 > 9)$ **36.** $(2x < 2)$ or $(x + 5 \geq 10)$

37. $(x - 5 < 2x)$ or $(x + 8 > 3x)$

38. Which of the following is equivalent to $y + 4 \geq 9$?

(1) $y > 5$ (2) $y \geq 5$ (3) $y \geq 13$ (4) $y = 13$

39. Which of the following is equivalent to $4x < 5x + 6$?

(1) $x > -6$ (2) $x < -6$ (3) $x > 6$ (4) $x < 6$

40. The smallest member of the solution set of $3x - 7 \geq 8$ is

(1) 3 (2) 4 (3) 5 (4) 6

41. The largest member of the solution set of $4x \leq 3x + 2$ is

(1) 1 (2) 2 (3) 3 (4) 4

In 42–47, write an inequality for each graph using interval notation.

42. –4 –3 –2 –1 0 1 2 3 4

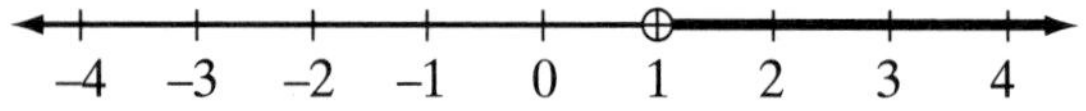

43. –4 –3 –2 –1 0 1 2 3 4

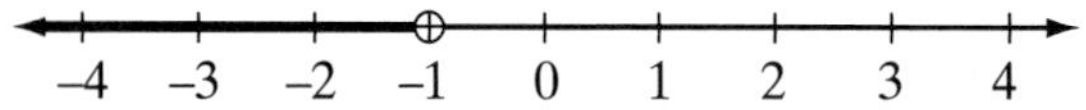

44. –4 –3 –2 –1 0 1 2 3 4

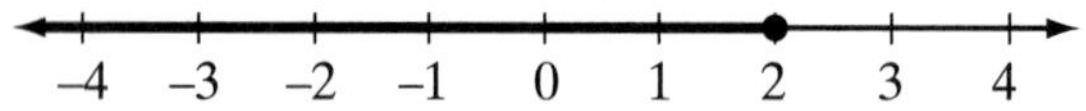

45. –4 –3 –2 –1 0 1 2 3 4

46. –4 –3 –2 –1 0 1 2 3 4

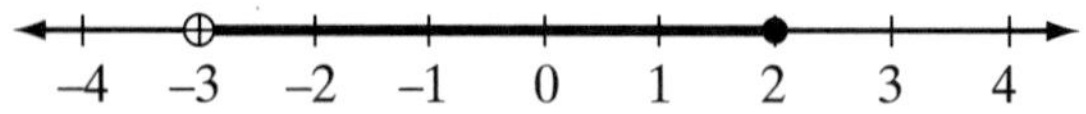

47. –4 –3 –2 –1 0 1 2 3 4

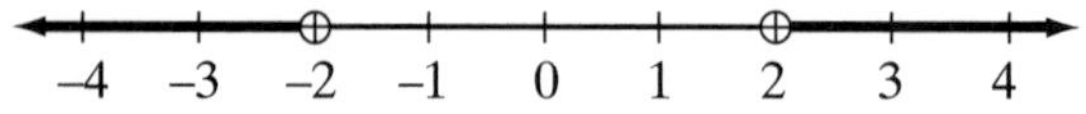

48. **a.** Graph the inequality $(x > 2)$ or $(x = 2)$.

b. Write an inequality equivalent to $(x > 2)$ or $(x = 2)$.

4-9 USING INEQUALITIES TO SOLVE PROBLEMS

Many problems can be solved by writing an inequality that describes how the numbers in the problem are related and then solving the inequality. An inequality can be expressed in words in different ways. For example:

$x > 12$	$x \geq 12$
A number is more than 12. A number exceeds 12. A number is greater than 12. A number is over 12.	A number is at least 12. A number has a minimum value of 12. A number is not less than 12. A number is not under 12.
$x < 12$	$x \leq 12$
A number is less than 12. A number is under 12.	A number is at most 12. A number has a maximum value of 12. A number is not greater than 12. A number does not exceed 12. A number is not more than 12.

Procedure

To solve a problem that involves an inequality:

1. Choose a variable to represent one of the unknown quantities in the problem.
2. Express other unknown quantities in terms of the same variable.
3. Choose an appropriate domain for the problem.
4. Write an inequality using a relationship given in the problem, a previously known relationship, or a formula.
5. Solve the inequality.
6. Check the solution using the words of the problem.
7. Use the solution of the inequality to answer the question in the problem.

EXAMPLE 1

Serafina has \$53.50 in her pocket and wants to purchase shirts at a sale price of \$14.95 each. How many shirts can she buy?

Solution (1) Choose a variable to represent the number of shirts Serafina can buy and the cost of the shirts.

Let x = the number of shirts that she can buy.
Then, $14.95x$ = the cost of the x shirts.

The domain is the set of whole numbers, since she can only buy a whole number of shirts.

(2) Write an inequality using a relationship given in the problem.

The cost of the shirts is less than or equal to \$53.50.

$$14.95x \quad \leq \quad \$53.50$$

(3) Solve the inequality.

$$14.95x \leq 53.50$$

$$\frac{14.95x}{14.95} \leq \frac{53.50}{14.95}$$

Use a calculator to complete the computation.

ENTER: 53.50 **÷** 14.95 **ENTER**

DISPLAY:
```
53.50/14.95
            3.578595318
```

Therefore, $x \leq 3.578595318$. Since the domain is the set of whole numbers, the solution set is $\{x : x \text{ is a counting number less than or equal to } 3\}$ or $\{0, 1, 2, 3\}$.

(4) Check the solution in the words of the problem.
0 shirt costs \$14.95(0) = \$0
1 shirt costs \$14.95(1) = \$14.95
2 shirts cost \$14.95(2) = \$29.90
3 shirts cost \$14.95(3) = \$44.85
4 or more shirts cost more than \$14.95 · (4) or more than \$59.80

Answer Serafina can buy 0, 1, 2, or 3 shirts.

EXAMPLE 2

The length of a rectangle is 5 centimeters more than its width. The perimeter of the rectangle is at least 66 centimeters. Find the minimum measures of the length and width.

Solution If the perimeter is at least 66 centimeters, then the sum of the measures of the four sides is either equal to 66 centimeters or is greater than 66 centimeters.

Let x = the width of the rectangle.

Then, $x + 5$ = the length of the rectangle.

The domain is the set of positive real numbers.

$$\underbrace{\text{The perimeter of the rectangle}}_{x + (x + 5) + x + (x + 5)} \; \underbrace{\text{is at least}}_{\geq} \; \underbrace{\text{66 centimeters.}}_{66}$$

$$4x + 10 \geq 66$$

$$4x + 10 - 10 \geq 66 - 10$$

$$4x \geq 56$$

$$x \geq 14$$

The width can be any real number that is greater than or equal to 14 and the length is any real number that is 5 more than the width. Since we are looking for the minimum measures, the smallest possible width is 14 and the smallest possible length is 14 + 5 or 19.

Answer The minimum width is 14 centimeters, and the minimum length is 19 centimeters.

EXERCISES

Writing About Mathematics

1. If there is a number x such that $-x < -3$, is it true that $x > -3$? Explain why or why not.

2. Is the solution set of $-x < -3$ the same as the solution set of $x > -3$? Explain why or why not.

Developing Skills

In 3–12, represent each sentence as an algebraic inequality.

3. x is less than or equal to 15.

4. y is greater than or equal to 4.

5. x is at most 50.

6. x is more than 50.

7. The greatest possible value of $3y$ is 30.

8. The sum of $5x$ and $2x$ is at least 70.

9. The maximum value of $4x - 6$ is 54.

10. The minimum value of $2x + 1$ is 13.

11. The product of $3x$ and $x + 1$ is less than 35.

12. When x is divided by 3 the quotient is greater than 7.

In 13–19, in each case write and solve the inequality that represents the given conditions. Use n as the variable.

13. Six less than a number is less than 4.

14. Six less than a number is greater than 4.

15. Six times a number is less than 72.

16. A number increased by 10 is greater than 50.

17. A number decreased by 15 is less than 35.

18. Twice a number, increased by 6, is less than 48.

19. Five times a number, decreased by 24, is greater than 3 times the number.

Applying Skills

In 20–29, in each case write an inequality and solve the problem algebraically.

20. Mr. Burke had a sum of money in a bank. After he deposited an additional sum of $100, he had at least $550 in the bank. At least how much money did Mr. Burke have in the bank originally?

21. The members of a club agree to buy at least 250 tickets for a theater party. If they expect to buy 80 fewer orchestra tickets than balcony tickets, what is the least number of balcony tickets they will buy?

22. Mrs. Scott decided that she would spend no more than $120 to buy a jacket and a skirt. If the price of the jacket was $20 more than 3 times the price of the skirt, find the highest possible price of the skirt.

23. Three times a number increased by 8 is at most 40 more than the number. Find the greatest value of the number.

24. The length of a rectangle is 8 meters less than 5 times its width. If the perimeter of the rectangle is at most 104 meters, find the greatest possible width of the rectangle.

25. The length of a rectangle is 12 centimeters less than 3 times its width. If the perimeter of the rectangle is at most 176 centimeters, find the greatest possible length of the rectangle.

26. Mrs. Diaz wishes to save at least $1,500 in 12 months. If she saves $300 during the first 4 months, what is the least possible average amount that she must save in each of the remaining 8 months?

27. Two consecutive even numbers are such that their sum is greater than 98 decreased by twice the larger. Find the smallest possible values for the integers.

28. Minou wants $29 to buy music online. Her father agrees to pay her $6 an hour for gardening in addition to her $5 weekly allowance for helping around the house. What is the minimum number of hours Minou must work at gardening to receive at least $29 this week?

29. Allison has more than 2 but less than 3 hours to spend on her homework. She has work in math, English, and social studies. She plans to spend equal amounts of time studying English and studying social studies, and to spend twice as much time studying math as in studying English.

a. What is the minimum number of minutes she can spend on English homework?

b. What is the maximum number of minutes she can spend on social studies?

c. What is the maximum number of minutes she can devote to math?

CHAPTER SUMMARY

The properties of equality allow us to write equivalent equations to solve an equation.

1. **The addition property of equality:** If equals are added to equals the sums are equal.
2. **The subtraction property of equality:** If equals are subtracted from equals the differences are equal.
3. **The multiplication property of equality:** If equals are multiplied by equals the products are equal.
4. **The division property of equality:** If equals are divided by nonzero equals the quotients are equal.
5. **The substitution property:** In any statement of equality, a quantity may be substituted for its equal.

Before solving an equation, simplify each side if necessary. To solve an equation that has the variable in both sides, transform it into an equivalent equation in which the variable appears in only one side. Do this by adding the opposite of the variable term on one side to both sides of the equation. Use the properties of equality.

Any equation or formula containing two or more variables can be transformed so that one variable is expressed in terms of all other variables. To do this, think of solving a simpler but related equation that contains only one variable.

Order property of numbers: For real numbers x and y, one and only one of the following can be true: $x < y$, $x = y$, or $x > y$.

Transitive property of inequality: For real numbers x, y, and z, if $x < y$ and $y < z$, then $x < z$, and if $x > y$ and $y > z$, then $x > z$.

Addition property of inequality: When the same number is added to or subtracted from both sides of an inequality, the order of the new inequality is the same as the order of the original one.

Multiplication property of inequality: When both sides of an inequality are multiplied or divided by the same positive number, the order of the new inequality is the same as the order of the original one. When both sides of an inequality are multiplied or divided by the same negative number, the order of the new inequality is the opposite of the order of the original one.

The **domain** or **replacement set** of an inequality is the set of all possible numbers that can be used to replace the variable. The **solution set** of an inequality is the set of numbers from the domain that make the inequality true. Inequalities that have the same solution set are **equivalent inequalities**.

VOCABULARY

4-1 Equation • Left side • Left member • Right side • Right member • Root • Solution • Solution set • Identity • Equivalent equations • Solve an equation • Conditional equation • Check

4-2 First degree equation in one variable • Like terms • Similar terms • Unlike terms

4-4 Perimeter • Distance formula

4-7 Order property of the real numbers • Transitive property of inequality • Addition property of inequality • Multiplication property of inequality

4-8 Domain of an inequality • Replacement set of an inequality • Solution set of an inequality • Equivalent inequalities • Interval notation

REVIEW EXERCISES

1. Compare the properties of equality with the properties of inequality. Explain how they are alike and how they are different.

In 2–9, solve for the variable and check.

2. $8w = 60 - 4w$ **3.** $8w - 4w = 60$

4. $4h + 3 = 23 - h$ **5.** $5y + 3 = 2y$

6. $8a - (6a - 5) = 1$ **7.** $2(b - 4) = 4(2b + 1)$

8. $3(4x - 1) - 2 = 17x + 10$ **9.** $(x + 2) - (3x - 2) = x + 3$

In 10–15, solve each equation for x in terms of a, b, and c.

10. $a + x = bc$ **11.** $cx + a = b$ **12.** $bx - a = 5a + c$

13. $\frac{a+c}{2}x = b$ **14.** $\frac{ax}{b} = c$ **15.** $ax + 2b = c$

16. a. Solve $A = \frac{1}{2}bh$ for h in terms of A and b.

b. Find h when $A = 5.4$ and $b = 0.9$.

17. If $P = 2l + 2w$, find w when $P = 17$ and $l = 5$.

18. If $F = \frac{9}{5}C + 32$, find C when $F = 68$.

In 19–26, find and graph the solution set of each inequality.

19. $6 + x > 3$ **20.** $2x - 3 \geq -5$ **21.** $\frac{1}{3}x < 1$

22. $-x \geq 4$ **23.** $-3 < x - 1 \leq 2$ **24.** $(x + 2 \geq 5)$ and $(2x < 14)$

25. $(-x \geq 2)$ or $(x > 0)$

26. $(x - 4 \geq 1)$ and $(-2x > -18)$

In 27–30, tell whether each statement is sometimes, always, or never true. Justify your answer by stating a property of inequality or by giving a counterexample.

27. If $x > y$, then $a + x > a + y$.

28. If $x > y$, then $ax > ay$.

29. If $x > y$ and $y > z$, then $x > z$.

30. If $x > y$, then $-x > -y$.

In 31–33, select the answer choice that correctly completes the statement or answers the question.

31. An inequality that is equivalent to $4x - 3 > 5$ is

(1) $x > 2$ (2) $x < 2$ (3) $x > \frac{1}{2}$ (4) $x < \frac{1}{2}$

32.

–1 0 1 2 3 4

The solution set of which inequality is shown in the graph above?

(1) $x - 2 \geq 0$ (2) $x - 2 > 0$ (3) $x - 2 < 0$ (4) $x - 2 \leq 0$

33.

–5 –4 –3 –2 –1 0 1 2

The above graph shows the solution set of which inequality?

(1) $-4 < x < 1$ (2) $-4 \leq x < 1$ (3) $-4 < x \leq 1$ (4) $-4 \leq x \leq 1$

34. The figure on the right consists of two squares and two isosceles right triangles. Express the area of the figure in terms of s, the length of one side of a square.

35. Express in terms of w the number of days in w weeks and 4 days.

36. The length of a rectangular room is 5 feet more than 3 times the width. The perimeter of the room is 62 feet. Find the dimensions of the room.

37. A truck must cross a bridge that can support a maximum weight of 24,000 pounds. The weight of the empty truck is 1,500 pounds, and the driver weighs 190 pounds. What is the weight of a load that the truck can carry?

38. In an apartment building there is one elevator, and the maximum load that it can carry is 2,000 pounds. The maintenance supervisor wants to move a replacement part for the air-conditioning unit to the roof. The part weighs 1,600 pounds, and the mechanized cart on which it is being moved weighs 250 pounds. When the maintenance supervisor drives the cart onto the elevator, the alarm sounds to signify that the elevator is overloaded.

a. How much does the maintenance supervisor weigh?

b. How can the replacement part be delivered to the roof if the part cannot be disassembled?

39. A mail-order digital photo developer charges 8 cents for each print plus a \$2.98 shipping fee. A local developer charges 15 cents for each print. How many digital prints must be ordered in order that:

a. the local developer offers the lower price?

b. the mail-order developer offers the lower price?

40. a. What is an appropriate replacement set for the problem in the chapter opener on page 116 of this chapter?

b. Write and solve the equations suggested by this problem.

c. Write the solution set for this problem.

Exploration

The figure at the right shows a circle inscribed in a square. Explain how this figure shows that $\pi r^2 < (2r)^2$.

CUMULATIVE REVIEW CHAPTERS 1–4

Part I

Answer all questions in this part. Each correct answer will receive 2 credits. No partial credit will be allowed.

1. The rational numbers are a subset of
(1) the integers
(2) the counting numbers
(3) the whole numbers
(4) the real numbers

2. If $x - 12.6 = 8.4 + 0.7x$, then x equals
(1) 0.07 (2) 0.7 (3) 7 (4) 70

3. The solution set of $2x - 4 = 5x + 14$ is
(1) $\left\{\frac{-10}{3}\right\}$ (2) $\left\{\frac{10}{3}\right\}$ (3) $\{6\}$ (4) $\{-6\}$

4. Which of the following inequalities is false?
(1) $\frac{2}{3} \leq 0.6$ (2) $\frac{2}{3} \leq 0.\overline{6}$ (3) $\frac{2}{3} \neq 0.6$ (4) $\frac{2}{3} \geq 0.6$

5. Which of the following identities is an illustration of the commutative property of addition?

(1) $(x + 3) + 2 = x + (3 + 2)$
(2) $x + 3 = 3 + x$
(3) $5(x + 3) = 5x + 15$
(4) $x + 0 = x$

6. Which of the following sets is closed under division?

(1) nonzero whole numbers
(2) nonzero integers
(3) nonzero even integers
(4) nonzero rational numbers

7. The measure of one side of a rectangle is 20.50 feet. This measure is given to how many significant digits?

(1) 1 (2) 2 (3) 3 (4) 4

8. In the coordinate plane, the vertices of quadrilateral $ABCD$ have the coordinates $A(-2, 0)$, $B(7, 0)$, $C(7, 5)$, and $D(0, 5)$. The quadrilateral is

(1) a rhombus
(2) a rectangle
(3) a parallelogram
(4) a trapezoid

9. One element of the solution set of $(x \leq -3)$ or $(x > 5)$ is

(1) -4 (2) -2 (3) 5
(4) None of the above. The solution set is the empty set.

10. When $x = -3$, $-x^2$ is

(1) -6 (2) 6 (3) -9 (4) 9

Part II

Answer all questions in this part. Each correct answer will receive 2 credits. Clearly indicate the necessary steps, including appropriate formula substitutions, diagrams, graphs, charts, etc. For all questions in this part, a correct numerical answer with no work shown will receive only 1 credit.

11. A quadrilateral has four sides. Quadrilateral $ABCD$ has three sides that have equal measures. The measure of the fourth side is 8.0 cm longer than each of the other sides. If the perimeter of the quadrilateral is 28.0 m, find the measure of each side using the correct number of significant digits.

12. To change degrees Fahrenheit, F, to degrees Celsius, C, subtract 32 from the Fahrenheit temperature and multiply the difference by five-ninths.

a. Write an equation for C in terms of F.

b. Normal body temperature is 98.6° Fahrenheit. What is normal body temperature in degrees Celsius?

Part III

Answer all questions in this part. Each correct answer will receive 3 credits. Clearly indicate the necessary steps, including appropriate formula substitutions, diagrams, graphs, charts, etc. For all questions in this part, a correct numerical answer with no work shown will receive only 1 credit.

13. Is it possible for the remainder to be 2 when a prime number that is greater than 2 is divided by 4? Explain why or why not.

14. A plum and a pineapple cost the same as three peaches. Two plums cost the same as a peach. How many plums cost the same as a pineapple?

Part IV

Answer all questions in this part. Each correct answer will receive 4 credits. Clearly indicate the necessary steps, including appropriate formula substitutions, diagrams, graphs, charts, etc. For all questions in this part, a correct numerical answer with no work shown will receive only 1 credit.

15. A trapezoid is a quadrilateral with only one pair of parallel sides called the bases of the trapezoid. The formula for the area of a trapezoid is $A = \frac{h}{2}(b_1+b_2)$ where h represents the measure of the altitude to the bases, and b_1 and b_2 represent the measures of the bases. Find the area of a trapezoid if $h = 5.25$ cm, $b_1 = 12.75$ cm, and $b_2 = 9.50$ cm. Express your answer to the number of significant digits determined by the given data.

16. Fred bought three shirts, each at the same price, and received less than \$12.00 in change from a \$50.00 bill.

a. What is the minimum cost of one shirt?

b. What is the maximum cost of one shirt?

CHAPTER

5

OPERATIONS WITH ALGEBRAIC EXPRESSIONS

Marvin is planning two rectangular gardens that will have the same width. He wants one to be 5 feet longer than it is wide and the other to be 8 feet longer than it is wide. How can he express the area of each of the gardens and the total area of the two gardens in terms of w, the width of each?

Problems like this often occur in many areas of business, science and technology as well as every day life. When we use variables and the rules for adding and multiplying expressions involving variables, we can often write general expressions that help us investigate many possibilities in the solution of a problem. In this chapter, you will learn to add, subtract, multiply, and divide algebraic expressions.

CHAPTER TABLE OF CONTENTS

5-1 ADDING AND SUBTRACTING ALGEBRAIC EXPRESSIONS

Recall that an algebraic expression that is a number, a variable, or a product or quotient of numbers and variables is called a **term**. Examples of terms are:

$$7 \qquad a \qquad -2b \qquad -\tfrac{4}{7}y^2 \qquad 0.7ab^5 \qquad -\tfrac{5}{w}$$

Two or more terms that contain the same variable or variables with corresponding variables having the same exponents, are called **like terms** or **similar terms**. For example, the following pairs are like terms.

$$6k \text{ and } k \qquad 5x^2 \text{ and } -7x^2 \qquad 9ab \text{ and } 0.4ab \qquad \tfrac{9}{2}x^2y^3 \text{ and } -\tfrac{11}{3}x^2y^3$$

Two terms are **unlike terms** when they contain different variables, or the same variable or variables with different exponents. For example, the following pairs are unlike terms.

$$3x \text{ and } 4y \qquad 5x^2 \text{ and } 5x^3 \qquad 9ab \text{ and } 0.4a \qquad \tfrac{8}{3}x^3y^2 \text{ and } \tfrac{4}{7}x^2y^3$$

To add or subtract like terms, we use the distributive property of multiplication over addition or subtraction.

$$9x + 2x = (9 + 2)x = 11x$$
$$-16cd + 3cd = (-16 + 3)cd = -13cd$$
$$18y^2 - 5y^2 = (18 - 5)y^2 = 13y^2$$
$$7ab - ab = 7ab - 1ab = (7 - 1)ab = 6ab$$

Since the distributive property is true for any number of terms, we can express the sum or difference of any number of like terms as a single term.

$$-3ab^2 + 4ab^2 - 2ab^2 = (-3 + 4 - 2)ab^2 = -1\,ab^2 = -\,ab^2$$
$$x^3 + 11x^3 - 8x^3 - 4x^3 = (1 + 11 - 8 - 4)x^3 = 0x^3 = 0$$

Recall that when like terms are added:

1. The sum or difference has the same variable or variables as the original terms.
2. The numerical coefficient of the sum or difference is the sum or difference of the numerical coefficients of the terms that were added.

The sum of unlike terms cannot be expressed as a single term. For example, the sum of $2x$ and $3y$ cannot be written as a single term but is written $2x + 3y$.

EXAMPLE 1

Add: *Answers*

a. $+3a + (-8a)$ $= [3 + (-8)]a = -5a$

b. $-12b^2 - (-5b^2)$ $= [-12 - (-5)]b^2 = [-12 + 5]b^2 = -7b^2$

c. $-15abc + 6abc$ $= (-15 + 6)abc = -9abc$

d. $8x^2 y - x^2y$ $= (8 - 1)x^2y = 7x^2y$

e. $-9y + 9y$ $= (-9 + 9)y = 0y = 0$

f. $2(a + b) + 6(a + b)$ $= (2 + 6)(a + b) = 8(a + b)$

EXAMPLE 2

An isosceles triangle has two sides that are equal in length. The length of each of the two equal sides of an isosceles triangle is twice the length of the third side of the triangle. If the length of the third side is represented by n, represent in simplest form the perimeter of the triangle.

Solution n represents the length of the base.

$2n$ represents the length of one of the equal sides.

$2n$ represents the length of the other equal side.

$$\text{Perimeter} = n + 2n + 2n = (1 + 2 + 2)n = 5n.$$

Note that the length of a side of a geometric figure is a positive number. Therefore, the variable n must represent a positive real number, that is, the replacement set for n must be the set of positive real numbers.

Answer $5n$

Monomials and Polynomials

A term that has no variable in the denominator is called a **monomial**. For example, 5, $-5w$, and $\frac{3w^2}{5}$ are monomials, but $\frac{5}{w}$ is *not* a monomial.

A monomial or the sum of monomials is called a **polynomial**. A polynomial may have one or more terms. Some polynomials are given special names to indicate the number of terms.

- A monomial such as $4x^2$ may be considered to be a polynomial of one term. (*Mono-* means "one"; *poly-* means "many.")
- A polynomial of two unlike terms, such as $10a + 12b$, is called a **binomial**. (*Bi-* means "two.")

- A polynomial of three unlike terms, such as $x^2 + 3x + 2$, is called a **trinomial**. (*Tri-* means "three.")
- A polynomial such as $5x^2 + (-2x) + (-4)$ is usually written as $5x^2 - 2x - 4$.

A polynomial has been simplified or is in **simplest form** when it contains no like terms. For example, $5x^3 + 8x^2 - 5x^3 + 7$, when expressed in simplest form, becomes $8x^2 + 7$.

A polynomial is said to be in **descending order** when the exponents of a particular variable decrease as we move from left to right. The polynomial $x^3 + 5x^2 - 4x + 9$ is in a descending order of powers of x.

A polynomial is said to be in **ascending order** when the exponents of a particular variable increase as we move from left to right. The polynomial $4 + 5y + y^2$ is in an ascending order of powers of y.

To add two polynomials, we use the commutative, associative, and distributive properties to combine like terms.

EXAMPLE 3

Simplify: $3ab + 5b - ab + 4ab - 2b$

Solution

How to Proceed

(1) Write the expression: $3ab + 5b - ab + 4ab - 2b$

(2) Group like terms together by using the commutative and associative properties: $3ab - ab + 4ab + 5b - 2b$
$(3ab - ab + 4ab) + (5b - 2b)$

(3) Use the distributive property: $(3 - 1 + 4)ab + (5 - 2)b$

(4) Simplify the numerical expressions that are in parentheses: $6ab + 3b$

Answer $6ab + 3b$

EXAMPLE 4

Find the sum: $(3x^2 + 5) + (6x^2 + 8)$

Solution

How to Proceed

(1) Write the expression: $(3x^2 + 5) + (6x^2 + 8)$

(2) Use the associative property: $3x^2 + (5 + 6x^2) + 8$

(3) Use the commutative property: $3x^2 + (6x^2 + 5) + 8$

(4) Use the associative property: $(3x^2 + 6x^2) + (5 + 8)$

(5) Add like terms: $9x^2 + 13$

Answer $9x^2 + 13$

The sum of polynomials can also be arranged vertically, placing like terms under one another. The sum of $3x^2 + 5$ and $6x^2 + 8$ can be arranged as shown at the right.

$$\begin{array}{r} 3x^2 + 5 \\ \underline{6x^2 + 8} \\ 9x^2 + 13 \end{array}$$

Addition can be checked by substituting any convenient value for the variable and evaluating each polynomial and the sum.

Check Let $x = 4$

$$\begin{array}{l} 3x^2 + 5 = 3(4)^2 + 5 = 53 \\ \underline{6x^2 + 8 = 6(4)^2 + 8 = 104} \\ 9x^2 + 13 = 9(4)^2 + 13 = 157 \checkmark \end{array}$$

EXAMPLE 5

Simplify: $6a + [5a + (6 - 3a)]$

Solution When one grouping symbol appears within another, first simplify the expression within the innermost grouping symbol.

How to Proceed

(1) Write the expression	$6a + [5a + (6 - 3a)]$
(2) Use the commutative property:	$6a + [5a + (-3a + 6)]$
(3) Use the associative property:	$6a + [(5a - 3a) + 6]$
(4) Combine like terms:	$6a + [2a + 6]$
(5) Use the associative property:	$(6a + 2a) + 6$
(6) Combine like terms:	$8a + 6$

Answer $8a + 6$

EXAMPLE 6

Express the difference $(4x^2 + 2x - 3) - (2x^2 - 5x - 3)$ in simplest form.

Solution

How to Proceed

(1) Write the subtraction problem:	$(4x^2 + 2x - 3) - (2x^2 - 5x - 3)$
(2) To subtract, add the opposite of the polynomial to be subtracted:	$(4x^2 + 2x - 3) + (-2x^2 + 5x + 3)$
(3) Use the commutative and associative properties to group like terms:	$(4x^2 - 2x^2) + (2x + 5x) + (-3 + 3)$
(4) Add like terms:	$2x^2 + 7x + 0$
	$2x^2 + 7x$

Answer $2x^2 + 7x$

EXERCISES

Writing About Mathematics

1. Christopher said that $3x - x = 3$.

a. Use the distributive property to show Christopher that his answer is incorrect.

b. Substitute a numerical value of x to show Christopher that his answer is incorrect.

2. Explain how the procedure for adding like terms is similar to the procedure for adding fractions.

Developing Skills

In 3–27, write each algebraic expression in simplest form.

3. $(+8c) + (+7c)$

4. $(-4a) + (-6a)$

5. $(-20r) + (5r)$

6. $(-7w) + (+7w)$

7. $(5ab) + (-9ab)$

8. $(+6x) + (-4x) + (-5x) + (10x)$

9. $-5y + 6y + 9y - 14y$

10. $4m + 9m - 12m - m$

11. $(+8x^2) + (-x^2) + (-12x^2) + (+2x^2)$

12. $4a + (9a + 3)$

13. $7b + (4b - 6)$

14. $8c + (7 - 9c)$

15. $(-6x - 4) + 5x$

16. $r + (s + 2r)$

17. $8d^2 + (6d^2 - 4d)$

18. $(5x + 3) - (6x - 5)$

19. $-9y + [7 + (6y - 7)]$

20. $(5 - 6y) + (-9y + 2)$

21. $5a + [3b - (-2a + 4b)]$

22. $(5x^2 - 4) + (-3x^2 - 9)$

23. $3y^2 + [6y^2 + (3y - 4)]$

24. $(x^3 + 3x^2) - (-2x^2 - 9)$

25. $-d^2 + [9d + (2 - 4d^2)]$

26. $(x^2 + 5x - 24) + (-x^2 - 4x + 9)$

27. $(x^3 + 9x - 5) - (-4x^2 - 12x - 5)$

In 28–31, state whether each expression is a monomial, a binomial, a trinomial, or none of these.

28. $8x + 3$

29. $7y$

30. $-2a^2 + 3a - 6$

31. $x^3 + 2x^2 + x - 7$

32. a. Give an example of the sum of two binomials that is a binomial.

b. Give an example of the sum of two binomials that is a monomial.

c. Give an example of the sum of two binomials that is a trinomial.

d. Give an example of the sum of two binomials that has four terms.

e. Can the sum of two binomials have more than four terms?

Applying Skills

In 33–41, write each answer as a polynomial in simplest form.

33. A cheeseburger costs 3 times as much as a soft drink, and an order of fries costs twice as much as a soft drink. If a soft drink costs s cents, express the total cost of a cheeseburger, an order of fries, and a soft drink in terms of s.

34. Jack deposited some money in his savings account in September. In October he deposited twice as much as in September, and in November he deposited one-half as much as in September. If x represents the amount of money deposited in September, represent, in terms of x, the total amount Jack deposited in the 3 months.

35. On Tuesday, Melita read 3 times as many pages as she read on Monday. On Wednesday she read 1.5 times as many pages as on Monday, and on Thursday she read half as many pages as on Monday. If Melita read p pages on Monday, represent in terms of p, the total number of pages she read in the 4 days.

36. The cost of 12 gallons of gas is represented by $12x$, and the cost of a quart of oil is represented by $2x - 30$. Represent the cost of 12 gallons of gas and a quart of oil.

37. In the last basketball game of the season, Tom scored $2x$ points, Tony scored $x + 5$ points, Walt scored $3x + 1$ points, Dick scored $4x - 7$ points, and Dan scored $2x - 2$ points. Represent the total points scored by these five players.

38. Last week, Greg spent twice as much on bus fare as he did on lunch, and 3 dollars less on entertainment than he did on bus fare. If x represents the amount, in dollars, spent on lunch, express in terms of x the total amount Greg spent on lunch, bus fare, and entertainment.

39. The cost of a chocolate shake is 40 cents less than the cost of a hamburger. If h represents the cost, *in cents*, of a hamburger, represent in terms of h the cost of a hamburger and a chocolate shake *in dollars*.

40. Rosie spent 12 dollars more for fabric for a new dress than she did for buttons, and 1 dollar less for thread than she did for buttons. If b represents the cost, in dollars, of the buttons, represent in terms of b the total cost of the materials needed for the dress.

41. The length of a rectangle is $7z^2 + 3$ inches and the width is $9z^2 + 2$ inches. Represent the perimeter of the rectangle.

5-2 MULTIPLYING POWERS THAT HAVE THE SAME BASE

Finding the Product of Powers

We know that y^2 means $y \cdot y$ and y^3 means $y \cdot y \cdot y$. Therefore,

$$y^2 \cdot y^3 = \overbrace{(y \cdot y)}^{2}\overbrace{(y \cdot y \cdot y)}^{3} = \overbrace{(y \cdot y \cdot y \cdot y \cdot y)}^{5} = y^5$$

Similarly,

$$c^2 \cdot c^4 = \overbrace{(c \cdot c)}^{2}\overbrace{(c \cdot c \cdot c \cdot c)}^{4} = \overbrace{(c \cdot c \cdot c \cdot c \cdot c \cdot c)}^{6} = c^6$$

and

$$x \cdot x^3 = \overbrace{(x)}^{1}\overbrace{(x \cdot x \cdot x)}^{3} = \overbrace{(x \cdot x \cdot x \cdot x)}^{4} = x^4$$

The exponent in each product is the sum of the exponents in the factors, as shown in these examples.

In general, when x is a real number and a and b are positive integers:

$$x^a \cdot x^b = x^{a+b}$$

EXAMPLE 1

Simplify each of the following products:

a. $x^5 \cdot x^2$ **b.** $a^7 \cdot a$ **c.** $3^2 \cdot 3^4$

Answers **a.** $x^5 \cdot x^2 = x^{5+2} = x^7$ **b.** $a^7 \cdot a = a^{7+1} = a^8$ **c.** $3^2 \cdot 3^4 = 3^{2+4} = 3^6$

Note: When we multiply powers with like bases, we do not actually perform the operation of multiplication but rather *count up* the number of times that the base is to be used as a factor to find the product. In Example 1c above, the answer does not give the value of the product but indicates only the number of times that 3 must be used as a factor to obtain the product. We can use the power key ^ to evaluate the products $3^2 \cdot 3^4$ and 3^6 to show that they are equal.

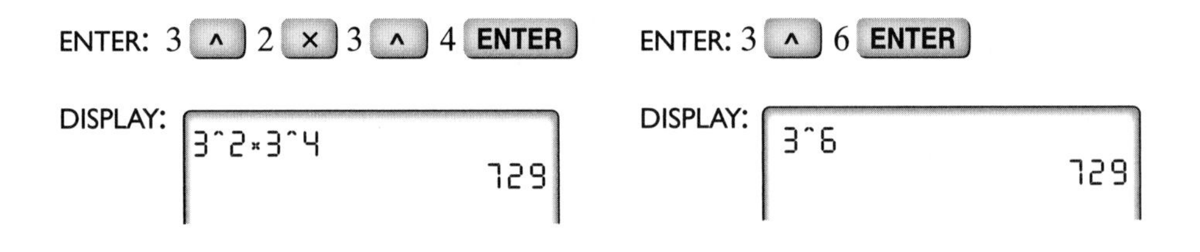

Finding a Power of a Power

Since $(x^3)^4 = x^3 \cdot x^3 \cdot x^3 \cdot x^3$, then $(x^3)^4 = x^{12}$. The exponent 12 can be obtained by addition: $3 + 3 + 3 + 3 = 12$ or by multiplication: $4 \times 3 = 12$.

In general, when x is a real number and a and c are positive integers:

$$(x^a)^c = x^{ac}$$

An expression such as $(x^5y^2)^3$ can be simplified by using the commutative and associative properties:

$$\begin{aligned}(x^5y^2)^3 &= (x^5y^2)(x^5y^2)(x^5y^2)\\ &= (x^5 \cdot x^5 \cdot x^5)(y^2 \cdot y^2 \cdot y^2)\\ &= x^{15}y^6\end{aligned}$$

When the base is the product of two or more factors, we apply the rule for the power of a power to each factor.

$$(x^5y^2)^3 = (x^5)^3\,(y^2)^3 = x^{5(3)}y^{2(3)} = x^{15}y^6$$

Thus,

$$(x^ay^b)^c = (x^a)^c(y^b)^c = x^{ac}y^{bc}$$

The expression $(5 \cdot 4)^3$ can be evaluated in two ways.

$$(5 \cdot 4)^3 = 5^3 \cdot 4^3 = 125 \cdot 64 = 8{,}000$$
$$(5 \cdot 4)^3 = 20^3 = 8{,}000$$

EXAMPLE 2

Simplify each expression in two ways.

a. $(a^2)^3$ **b.** $(ab^2)^4$ **c.** $(3^2 \cdot 4^2)^3$

Solution **a.**

$$(a^2)^3 = a^2 \cdot a^2 \cdot a^2 = a^{2+2+2} = a^6$$

or

$$(a^2)^3 = a^{2(3)} = a^6$$

Answer a^6

b. $$(ab^2)^4 = ab^2 \cdot ab^2 \cdot ab^2 \cdot ab^2 = (a \cdot a \cdot a \cdot a)(b^2 \cdot b^2 \cdot b^2 \cdot b^2) = a^4b^8$$

or

$$(ab^2)^4 = a^{1(4)}b^{2(4)} = a^4b^8$$

Answer a^4b^8

c. $$(3^2 \cdot 4^2)^3 = (3^2)^3 \cdot (4^2)^3 = 3^6 \cdot 4^6 = (3 \cdot 4)^6 = 12^6$$

or

$$(3^2 \cdot 4^2)^3 = ((3 \cdot 4)^2)^3 = (12^2)^3 = 12^6$$

Answer 12^6

To evaluate the expression in Example 2c, use a calculator.

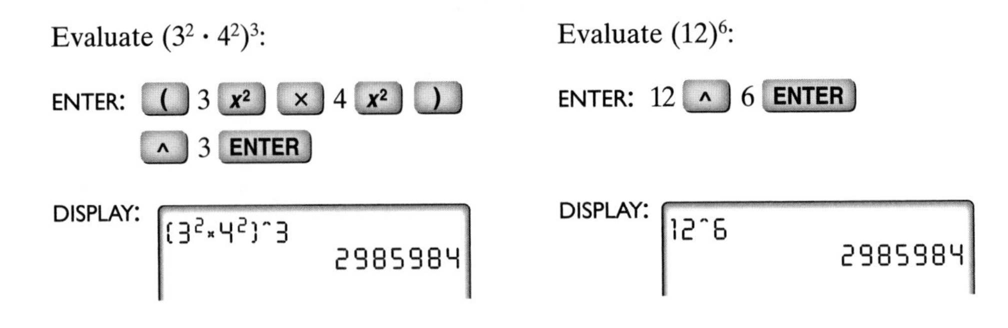

EXERCISES

Writing About Mathematics

1. Does $-5^3 \cdot 5^3 = -25^3$? Use the commutative and associative properties of multiplication to explain why or why not.

2. Does $-2^4 \cdot 4 = -2^6$? Use the commutative and associative properties of multiplication to explain why or why not.

Developing Skills

In 3–26, multiply in each case.

3. $a^2 \cdot a^3$
4. $b^3 \cdot b^4$
5. $r^2 \cdot r^4 \cdot r^5$
6. $r^3 \cdot r^3$
7. $z^3 \cdot z^3 \cdot z^5$
8. $t^8 \cdot t^4 \cdot t^2$
9. $x \cdot x$
10. $a^2 \cdot a$
11. $e^4 \cdot e^5 \cdot e$
12. $2^3 \cdot 2^2$
13. $3^4 \cdot 3^3$
14. $5^2 \cdot 5^4$
15. $4^3 \cdot 4$
16. $2^4 \cdot 2^5 \cdot 2$
17. $(x^3)^2$
18. $(a^4)^2$
19. $(z^3)^2 \cdot (z^4)^2$
20. $(x^2y^3)^2$
21. $(ab^2)^4$
22. $(rs)^3$
23. $(2^2 \cdot 3^2)^3$
24. $(5 \cdot 2^3)^4$
25. $(100^2 \cdot 10^3)^5$
26. $(a^2)^5 \cdot a$

In 27–31, multiply in each case. (All exponents are positive integers.)

27. $x^a \cdot x^{2a}$
28. $y^c \cdot y^2$
29. $c^r \cdot c^2$
30. $x^m \cdot x$
31. $(3y)^a \cdot (3y)^b$

In 32–39, state whether each sentence is true or false.

32. $10^4 \cdot 10^3 = 10^7$
33. $2^4 \cdot 2^2 = 2^8$
34. $3^3 \cdot 2^2 = 6^5$
35. $14^{80} \cdot 14^{10} = 14^{90}$
36. $3^3 \cdot 2^2 = 6^6$
37. $5^4 \cdot 5 = 5^5$
38. $(2^2)^3 = 2^5$
39. $(6^3)^4 = (6^4)^3$

Applying Skills

40. Two students attended the first meeting of the Chess Club. At that meeting, they decided that each person would bring one additional person to the next meeting, doubling the membership. At the second meeting, they again decided that each person would bring one additional person to the next meeting, again doubling the membership. If this plan was carried out for n meetings, the membership would equal 2^n persons.

a. How many persons attended the fifth meeting?

b. At which meeting would the membership be twice as large as at the fifth meeting?

41. In the metric system, 1 meter $= 10^2$ centimeters and 1 kilometer $= 10^3$ meters. How many centimeters equal one kilometer?

5-3 MULTIPLYING BY A MONOMIAL

Multiplying a Monomial by a Monomial

We know that the commutative property of multiplication makes it possible to arrange the factors of a product in any order and that the associative property of multiplication makes it possible to group the factors in any combination. For example:

$$(5x)(6y) = (5)(6)(x)(y) = (5 \cdot 6)(x \cdot y) = 30xy$$

$$(3x)(7x) = (3)(7)(x)(x) = (3 \cdot 7)\ (x \cdot x) = 21x^2$$

$$(-2x^2)(+5x^4) = (-2)(x^2)(+5)(x^4) = [(-2)(+5)]\ [(x^2)(x^4)] = -10x^6$$

$$(-3a^2b^3)(-4a^4b) = (-3)(a^2)\ (b^3)(-4)(a^4)\ (b)$$

$$= [(-3)(-4)][(a^2)(a^4)][(b^3)(b)] = 12a^6b^4$$

In the preceding examples, the factors may be rearranged and grouped mentally.

Procedure

To multiply a monomial by a monomial:

1. Use the commutative and associative properties to rearrange and group the factors. This may be done mentally.
2. Multiply the numerical coefficients.
3. Multiply powers with the same base by adding exponents.
4. Multiply the products obtained in Steps 2 and 3 and any other variable factors by writing them with no sign between them.

EXAMPLE 1

Multiply:

	Answers		Answers
a. $(8xy)(3z)$	$= 24xyz$	**b.** $(-4a^3)(-5a^5)$	$= 20a^8$
c. $(-6y^3)(y)$	$= -6y^4$	**d.** $(3a^2b^3)(4a^3b^4)$	$= 12a^5b^7$
e. $(-5x^2y^3)(-2xy^2)$	$= 10x^3y^5$	**f.** $(6c^2d^4)(-0.5d)$	$= -3c^2d^5$

g. $(-3x^2)^3$ $= (-3x^2)(-3x^2)(-3x^2) = -27x^6$

or

$= (-3)^3(x^2)^3 = -27x^6$

EXAMPLE 2

Represent the area of a rectangle whose length is $3x$ and whose width is $2x$.

Solution

How to Proceed

(1) Write the area formula: $A = lw$

(2) Substitute the values of l and w: $= (3x)(2x)$

(3) Perform the multiplication: $= (3 \cdot 2)(x \cdot x)$

$= 6x^2$

Answer $6x^2$

Multiplying a Polynomial by a Monomial

The distributive property of multiplication over addition is used to multiply a polynomial by a monomial. Therefore,

$$a(b + c) = ab + ac$$

$$x(4x + 3) = x(4x) + x(3)$$

$$= 4x^2 + 3x$$

This result can be illustrated geometrically. Let us separate a rectangle, whose length is $4x + 3$ and whose width is x, into two smaller rectangles such that the length of one rectangle is $4x$ and the length of the other is 3.

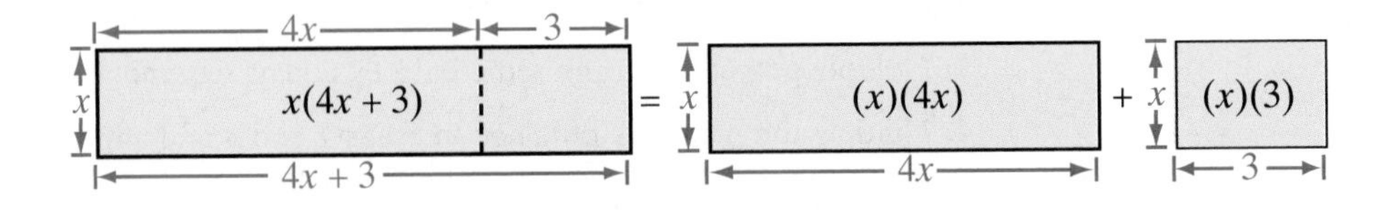

Since the area of the largest rectangle is equal to the sum of the areas of the two smaller rectangles:

$$x(4x + 3) = x(4x) + x(3) = 4x^2 + 3x$$

Procedure

To multiply a polynomial by monomial, use the distributive property: Multiply each term of the polynomial by the monomial and write the result as the sum of these products.

Multiplication and Grouping Symbols

When an algebraic expression involves grouping symbols such as parentheses, we follow the general order of operations and perform operations with algebraic terms.

In the example at the right, first simplify the expression within parentheses:	$8y - 2(7y - 4y) + 5$ $8y - 2(3y) + 5$
Next, multiply:	$8y - 6y + 5$
Finally, combine like terms by addition or subtraction:	$2y + 5$

In many expressions, however, the terms within parentheses cannot be combined because they are unlike terms. When this happens, we use the distributive property to clear parentheses and then follow the order of multiplying before adding.

Here, clear the parentheses by using the distributive property:	$3 + 7(2x + 3)$ $3 + 7(2x) + 7(3)$
Next, multiply:	$3 + 14x + 21$
Finally, combine like terms by addition:	$24 + 14x$

The multiplicative identity property states that $a = 1 \cdot a$. By using this property, we can say that $5 + (2x - 3) = 5 + 1(2x - 3)$ and then follow the procedures shown above.

$$5 + (2x - 3) = 5 + 1(2x - 3) = 5 + 1(2x) + 1(-3) = 5 + 2x - 3 = 2 + 2x$$

Also, since $-a = -1 \cdot a$, we can use this property to simplify expressions in which a parentheses is preceded by a negative sign:

$$6y - (9 - 7y) = 6y - 1(9 - 7y) = 6y - 1(9) - 1(-7y) = 6y - 9 + 7y = 13y - 9$$

EXAMPLE 3

Multiply:

	Answers
a. $5(r - 7)$	$= 5r - 35$
b. $8(3x - 2y + 4z)$	$= 24x - 16y + 32z$
c. $-5x(x^2 - 2x + 4)$	$= -5x^3 + 10x^2 - 20x$
d. $-3a^2b^2(4ab^2 - 3b^2)$	$= -12a^3b^4 + 9a^2b^4$

EXAMPLE 4

Simplify: **a.** $-5x(x^2 - 2) - 7x$ **b.** $3a - (5 - 7a)$

a. *How to Proceed*

(1) Write the expression:	$-5x(x^2 - 2) - 7x$
(2) Use the distributive property:	$-5x(x^2) - 5x(-2) - 7x$
(3) Multiply:	$-5x^3 + 10x - 7x$
(4) Add like terms:	$-5x^3 + 3x$

Answer $-5x^3 + 3x$

b. *How to Proceed*

(1) Write the expression:	$3a - (5 - 7a)$
	$3a - 1(5 - 7a)$
(2) Use the distributive property:	$3a - 5 + 7a$
(3) Add like terms:	$10a - 5$

Answer $10a - 5$

EXERCISES

Writing About Mathematics

1. In an algebraic term, how do you show the product of a constant times a variable or the product of different variables?

2. In the expression $2 + 3(7y)$, which operation is performed first? Explain your answer.

3. In the expression $(2 + 3)(7y)$, which operation is performed first? Explain your answer.

4. In the expression $5y(y + 3)$, which operation is performed first? Explain your answer.

5. Can the sum $x^2 + x^3$ be written in simpler form? Explain your answer.

6. Can the product $x^2(x^3)$ be written in simpler form? Explain your answer.

Developing Skills

In 7–29, find each product.

7. $(-4b)(-6b)$ **8.** $(+5)(-2y)(-3y)$ **9.** $(4a)(5b)$

10. $(-8r)(-2r)$ **11.** $(+7x)(-2y)(3z)$ **12.** $(+6x)(-0.5y)$

13. $\left(-\frac{3}{4}a\right)\left(+8b\right)$ **14.** $\left(-6x\right)\left(\frac{1}{2}y\right)\left(-\frac{1}{3}z\right)$ **15.** $(+5ab)(-3c)$

16. $(-7r)(5st)$ **17.** $(-2)(+6cd)(-e)$ **18.** $(+9xy)(-2x)$

19. $(3s)(-4s)(5t)$ **20.** $(+5a^2)(-4a^2)$ **21.** $(-6x^4)(-3x^3)$

22. $(20y^3)(-7y^2)$ **23.** $(18r^5)(-5r^2)$ **24.** $(+3z^2)(4z)$

25. $(-8y^5)(5y)$ **26.** $(-9z)(8z^4)(z^3)$ **27.** $(+6x^2y^3)(-4x^4y^2)$

28. $(-7a^3b)(+5a^2b^2)$ **29.** $(+4ab^2)(-2a^2b^3)$

In 30–47, write each product as a polynomial.

30. $3(6c + 3d)$ **31.** $-5(4m - 6n)$ **32.** $-2(8a + 6b)$

33. $10(2x - 0.2y)$ **34.** $12\left(\frac{2}{3}m - 4n\right)$ **35.** $-8\left(4r - \frac{1}{4}s\right)$

36. $-16\left(\frac{3}{4}c - \frac{5}{8}d\right)$ **37.** $4x(5x + 6)$ **38.** $5d(d^2 - 3d)$

39. $-5c^2(15c - 4c)$ **40.** $mn(m + n)$ **41.** $-ab(a - b)$

42. $3ab(5a^2 - 7b^2)$ **43.** $-r^3s^3(-2r^4s - 3s^4)$ **44.** $10d(2a - 3c + 4b)$

45. $-8(2x^2 - 3x - 5)$ **46.** $3xy(x^2 + xy + y^2)$ **47.** $5r^2s^2(-2r^2 + 3rs - 4s^2)$

In 48–50, represent the area of each rectangle whose length l and width w are given.

48. $l = 5y, w = 3y$ **49.** $l = 3x, w = 5y$ **50.** $l = 3c, w = 8c - 2$

51. The dimensions of the outer rectangle pictured at the right are $4x$ by $3x - 6$. The dimensions of the inner rectangle are $2x$ by $x + 2$.

a. Express the area of the outer rectangle in terms of x.

b. Express the area of the inner rectangle in terms of x.

c. Express as a polynomial in simplest form the area of the shaded region.

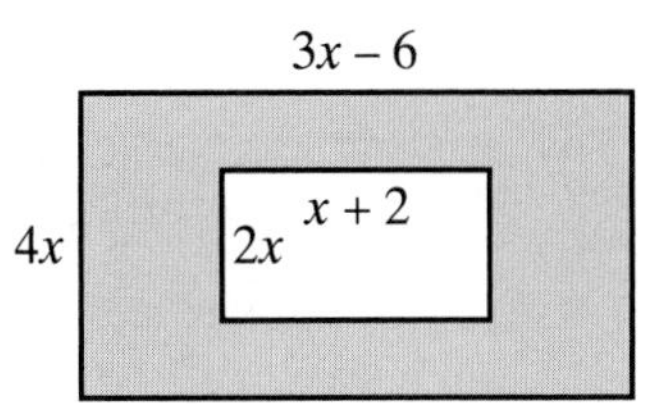

In 52–73, simplify each expression.

52. $5(d + 3) - 10$ **53.** $3(2 - 3c) + 5c$ **54.** $7 + 2(7x - 5)$

55. $2(x - 1) + 6$ **56.** $4(3 - 6a - 8s)$ **57.** $5 - 4(3e - 5)$

58. $8 + (4e - 2)$

59. $a + (b - a)$

60. $(6b + 4) - 2b$

61. $9 - (5t + 6)$

62. $4 - (2 - 8s)$

63. $-(6x - 7) + 14$

64. $5x(2x - 3) + 9x$

65. $12y - 3y(2y - 4)$

66. $7x + 3(2x - 1) - 8$

67. $7c - 4d - 2(4c - 3d)$

68. $3a - 2a(5a - a) + a^2$

69. $(a + 3b) - (a - 3b)$

70. $4(2x + 5) - 3(2 - 7x)$

71. $3(x + y) + 2(x - 3y)$

72. $5x(2 - 3x) - x(3x - 1)$

73. $y(y + 4) - y(y - 3) - 9y$

Applying Skills

In 74–86, write each answer as a polynomial in simplest form.

74. If 1 pound of grass seed costs $25x$ cents, represent in terms of x the cost of 7 pounds of seed.

75. If a bus travels at the rate of $10z$ miles per hour for 4 hours, represent in terms of z the distance traveled.

76. If Lois has $2n$ nickels, represent in terms of n the number of cents she has.

77. If the cost of a notebook is $2x - 3$, express the cost of five notebooks.

78. If the length of a rectangle is $5y - 7$ and the width is $3y$, represent the area of the rectangle.

79. If the measure of the base of a triangle is $3b + 2$ and the height is $4b$, represent the area of the triangle.

80. Represent the distance traveled in 3 hours by a car traveling at $3x - 7$ miles per hour.

81. Represent in terms of x and y the amount saved in $3y$ weeks if $x - 2$ dollars are saved each week.

82. The length of a rectangular skating rink is 2 less than 3 times the width. If w represents the width of the rink, represent the area in terms of w.

83. An internet bookshop lists used books for $3x - 5$ dollars each. The cost for shipping and handling is 2 dollars for five books or fewer. Represent the total cost of an order for four used books.

84. A store advertises skirts for $x - 5$ dollars and allows an additional 10-dollar reduction on the total purchase if three or more skirts are bought. Represent the cost of five skirts.

85. A store advertises skirts for $x - 5$ dollars and allows an additional two-dollar reduction on each skirt if three or more skirts are purchased. Represent the cost of five skirts.

86. A store advertises skirts for $x - 5$ dollars and tops for $2x - 3$ dollars. Represent the cost of two skirts and three tops.

5-4 MULTIPLYING POLYNOMIALS

As discussed in Section 5-3, to find the product $(x + 4)(a)$, we use the distributive property of multiplication over addition:

$$(x + 4)(a) = x(a) + 4(a)$$

Now, let us use this property to find the product of two binomials, for example, $(x + 4)(x + 3)$.

$$\begin{aligned}(x + 4)\quad(a)\quad &= x\quad(a)\quad + 4\quad(a)\\ (x + 4)(x + 3) &= x(x + 3) + 4(x + 3)\\ &= x^2 + 3x + 4x + 12\\ &= x^2 + 7x + 12\end{aligned}$$

This result can also be illustrated geometrically.

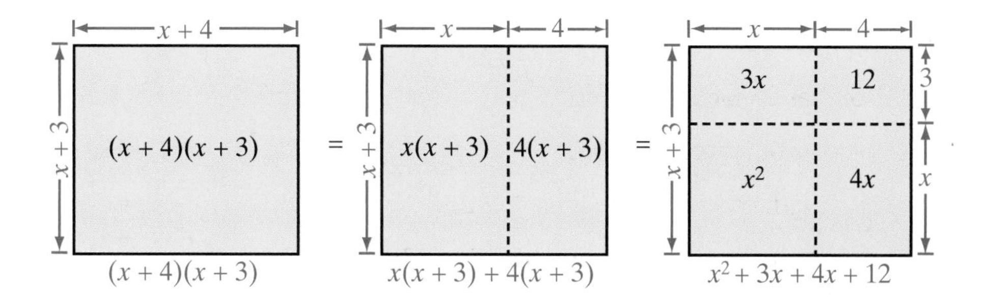

In general, for all a, b, c, and d:

$$\begin{aligned}\mathbf{(a + b)(c + d)} &= \mathbf{a(c + d) + b(c + d)}\\ &= \mathbf{ac + ad + bc + bd}\end{aligned}$$

Notice that each term of the first polynomial multiplies each term of the second.

At the right is a convenient vertical arrangement of the preceding multiplication, similar to the arrangement used in arithmetic multiplication. *Note that multiplication is done from left to right.*

$$\begin{array}{rr} & x + 3\\ & \underline{x + 4}\\ x(x + 3) \rightarrow & x^2 + 3x\\ 4(x + 3) \rightarrow & \underline{+ 4x + 12}\\ & x^2 + 7x + 12\end{array}$$

The word **FOIL** serves as a convenient way to remember the steps necessary to multiply two binomials.

$$\begin{array}{rcccccccc} & & \text{First} & & \text{Outside} & & \text{Inside} & & \text{Last}\\ (2x - 5)(x + 4) & = & 2x(x) & + & 2x(+4) & - & 5(x) & - & 5(+4)\\ & = & 2x^2 & + & 8x & - & 5x & - & 20\\ & = & 2x^2 & & & + & 3x & - & 20\end{array}$$

Procedure

To multiply a polynomial by a polynomial, first arrange each polynomial in descending or ascending powers of the same variable. Then use the distributive property: multiply each term of the first polynomial by each term of the other.

EXAMPLE 1

Simplify: $(3x - 4)(4x + 5)$

Solution

METHOD 1

$$\begin{aligned}(3x - 4)(4x + 5) &= 3x(4x + 5) - 4(x + 5)\\ &= 12x^2 + 15x - 16x - 20\\ &= 12x^2 - x - 20\end{aligned}$$

METHOD 2

$$\begin{array}{rrr} 3x & -4 & \\ 4x & +5 & \\ \hline 12x^2 & -16x & \\ & +15x & -20 \\ \hline 12x^2 & -\ x & -20 \end{array}$$

Answer $12x^2 - x - 20$

EXAMPLE 2

Simplify: $(x^2 + 3xy + 9y^2)(x - 3y)$

Solution

$$\begin{aligned}(x^2 + 3xy + 9y^2)(x - 3y) &= x^2(x - 3y) + 3xy(x - 3y) + 9y^2(x - 3y)\\ &= x^3 - 3x^2y + 3x^2y - 9xy^2 + 9xy^2 - 27y^3\\ &= x^3 + 0x^2y + 0xy^2 - 27y^3\end{aligned}$$

Answer $x^3 - 27y^3$

EXAMPLE 3

Simplify: $(2x - 5)^2 - (x - 3)$

Solution

$$\begin{aligned}(2x - 5)^2 - (x - 3) &= (2x - 5)(2x - 5) - (x - 3)\\ &= 2x(2x) + 2x(-5) - 5(2x) - 5(-5) + (-x) + (+3)\\ &= 4x^2 - 10x - 10x + 25 - x + 3\\ &= 4x^2 - 21x + 28\end{aligned}$$

Answer $4x^2 - 21x + 28$

EXERCISES

Writing About Mathematics

1. The product of two binomials in simplest form can have four terms, three terms, or two terms.

a. When does the product of two binomials have four terms?

b. When is the product of two binomials a trinomial?

c. When is the product of two binomials a binomial?

2. Burt wrote $(a + 3)^2$ as $a^2 + 9$. Prove to Burt that he is incorrect.

Developing Skills

In 3–35, write each product as a polynomial.

3. $(a + 2)(a + 3)$
4. $(x - 5)(x - 3)$
5. $(d + 9)(d - 3)$
6. $(x - 7)(x - 2)$
7. $(m + 3)(m - 7)$
8. $(t + 15)(t - 6)$
9. $(b - 8)(b - 10)$
10. $(6 + y)(5 + y)$
11. $(8 - e)(6 - e)$
12. $(12 - r)(6 + r)$
13. $(x + 5)(x - 5)$
14. $(2y + 7)(2y - 7)$
15. $(5a + 9)(5a - 9)$
16. $(2x + 1)(x - 6)$
17. $(5y - 2)(3y - 1)$
18. $(2x + 3)(2x - 3)$
19. $(3d + 8)(3d - 8)$
20. $(x + y)(x + y)$
21. $(a - b)(a + b)$
22. $(a + b)(a + b)$
23. $(a + b)^2$
24. $(x - 4y)(x + 4y)$
25. $(x - 4y)^2$
26. $(9x - 5y)(2x + 3y)$
27. $(r^2 + 5)(r^2 - 2)$
28. $(x^2 - y^2)(x^2 + y^2)$
29. $(x + 2)(x^2 + 3x + 5)$
30. $(2c + 1)(2c^2 - 3c + 1)$
31. $(3 - 2a - a^2)(5 - 2a)$
32. $(2x + 1)(3x - 4)(x + 3)$
33. $(x + 4)(x + 4)(x + 4)$
34. $(a + 5)^3$
35. $(x - y)^3$

In 36–43, simplify each expression.

36. $(x + 7)(x - 2) - x^2$
37. $2(3x + 1)(2x - 3) + 14x$
38. $r(r - 2) - (r - 5)$
39. $8x^2 - (4x + 3)(2x - 1)$
40. $(x + 4)(x + 3) - (x - 2)(x - 5)$
41. $(3y + 5)(2y - 3) - (y + 7)(5y - 1)$
42. $(y + 4)^2 - (y - 3)^2$
43. $a[(a + 2)(a - 2) - 4]$

Applying Skills

In 44–46, use grouping symbols to write an algebraic expression that represents the answer. Then, express each answer as a polynomial in simplest form.

44. The length of a rectangle is $2x - 5$ and its width is $x + 7$. Express the area of the rectangle as a trinomial.

45. The dimensions of a rectangle are represented by $11x - 8$ and $3x + 5$. Represent the area of the rectangle as a trinomial.

46. A train travels at a rate of $(15x + 100)$ kilometers per hours.

a. Represent the distance it can travel in $(x + 3)$ hours as a trinomial.

b. If $x = 2$, how fast does the train travel?

c. If $x = 2$, how far does it travel in $(x + 3)$ hours?

5-5 DIVIDING POWERS THAT HAVE THE SAME BASE

We know that any nonzero number divided by itself is 1. Therefore, $x \div x = 1$ and $y^3 \div y^3 = 1$.

In general, when $x \neq 0$ and a is a positive integer:

$$x^a \div x^a = 1$$

Therefore,

$$\frac{x^5}{x^3} = \frac{x^2 \cdot x^3}{x^3} = x^2 \cdot 1 = x^2$$

$$\frac{y^9}{y^4} = \frac{y^5 \cdot y^4}{y^4} = y^5 \cdot 1 = y^5$$

$$\frac{c^5}{c} = \frac{c^4 \cdot c}{c} = c^4 \cdot 1 = c^4$$

These same results can be obtained by using the relationship between division and multiplication:

If $a \cdot b = c$, then $c \div b = a$.

- Since $x^2 \cdot x^3 = x^5$, then $x^5 \div x^3 = x^2$.
- Since $y^5 \cdot y^4 = y^9$, then $y^9 \div y^4 = y^5$.
- Since $c^4 \cdot c = c^5$, then $c^5 \div c^1 = c^4$.

Observe that the exponent in each quotient is the difference between the exponent of the dividend and the exponent of the divisor.

In general, when $x \neq 0$ and a and b are positive integers with $a > b$:

$$x^a \div x^b = x^{a-b}$$

Procedure

To divide powers of the same base, find the exponent of the quotient by subtracting the exponent of the divisor from the exponent of the dividend. The base of the quotient is the same as the base of the dividend and of the divisor.

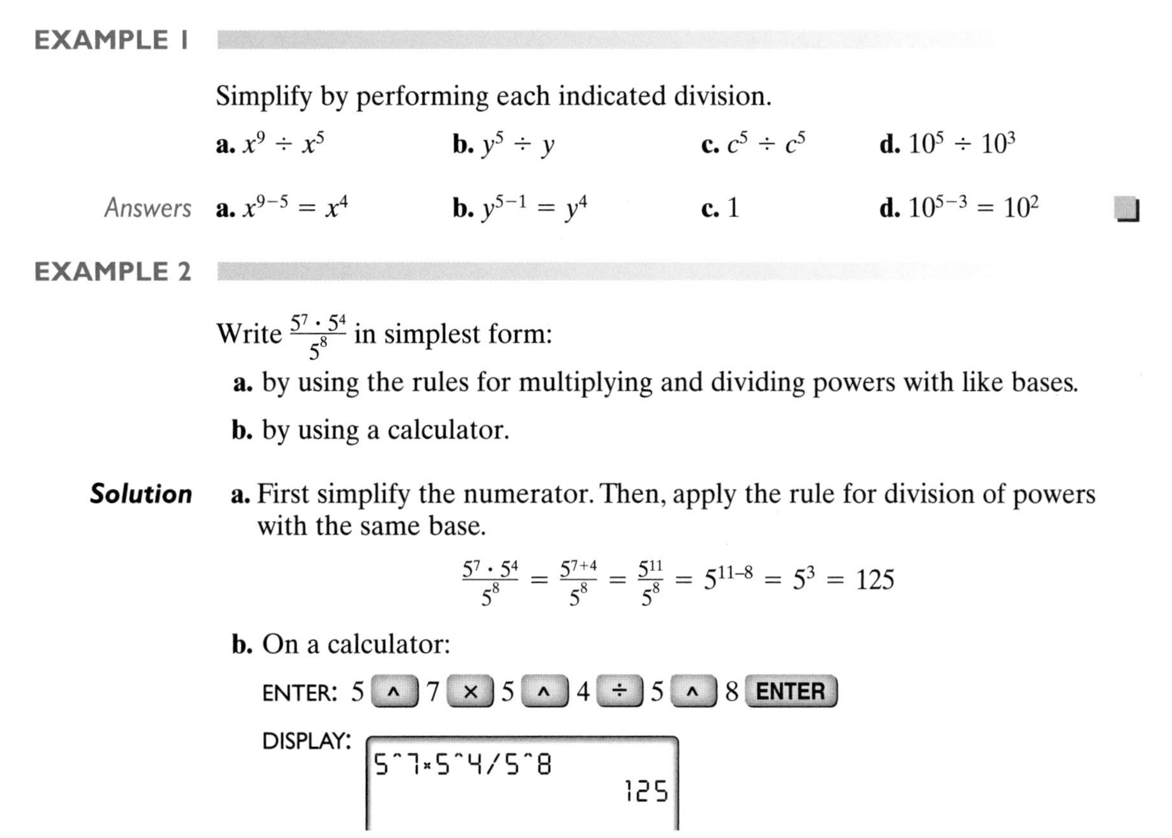

EXAMPLE 1

Simplify by performing each indicated division.

a. $x^9 \div x^5$ **b.** $y^5 \div y$ **c.** $c^5 \div c^5$ **d.** $10^5 \div 10^3$

Answers **a.** $x^{9-5} = x^4$ **b.** $y^{5-1} = y^4$ **c.** 1 **d.** $10^{5-3} = 10^2$

EXAMPLE 2

Write $\frac{5^7 \cdot 5^4}{5^8}$ in simplest form:

a. by using the rules for multiplying and dividing powers with like bases.

b. by using a calculator.

Solution **a.** First simplify the numerator. Then, apply the rule for division of powers with the same base.

$$\frac{5^7 \cdot 5^4}{5^8} = \frac{5^{7+4}}{5^8} = \frac{5^{11}}{5^8} = 5^{11-8} = 5^3 = 125$$

b. On a calculator:

ENTER: 5 [^] 7 [×] 5 [^] 4 [÷] 5 [^] 8 [ENTER]

DISPLAY: 5^7*5^4/5^8
125

Answer 125

EXERCISES

Writing About Mathematics

1. Coretta said that $5^4 \div 5 = 1^4$. Do you agree with Coretta? Explain why or why not.

2. To evaluate the expression $\frac{3^8}{3^5 \cdot 3^2}$,

a. in what order should the operations be performed? Explain your answer.

b. does $\frac{3^8}{3^5 \cdot 3^2} = 3^8 \div 3^5 \cdot 3^2$? Explain why or why not.

Developing Skills

In 3–18, divide in each case.

3. $x^8 \div x^2$ **4.** $a^{10} \div a^5$ **5.** $c^5 \div c^4$ **6.** $x^7 \div x^7$

7. $\frac{e^9}{e^3}$ **8.** $\frac{m^{12}}{m^4}$ **9.** $\frac{n^{10}}{n^9}$ **10.** $\frac{r^6}{r^6}$

11. $x^8 \div x$ **12.** $z^{10} \div z$ **13.** $t^5 \div t$ **14.** $2^5 \div 2^2$

15. $10^6 \div 10^4$ **16.** $3^4 \div 3^2$ **17.** $5^3 \div 5$ **18.** $10^4 \div 10$

In 19–24, divide in each case. (All exponents are positive integers.)

19. $x^{5a} \div x^{2a}$ **20.** $y^{10b} \div y^{2b}$ **21.** $r^c \div r^d$ $(c > d)$

22. $s^x \div s^2$ $(x > 2)$ **23.** $a^b \div a^b$ **24.** $2^a \div 2^b$ $(a > b)$

In 25–32: **a.** Simplify each expression by using the rules for multiplying and dividing powers with like bases. **b.** Evaluate the expression using a calculator. Compare your answers to parts **a** and **b**.

25. $\frac{2^3 \cdot 2^4}{2^2}$ **26.** $\frac{5^8}{5^4 \cdot 5}$ **27.** $\frac{10^2 \cdot 10^3}{10^4}$ **28.** $\frac{3^3 \cdot 3^2}{3^2}$

29. $\frac{10^6}{10^2 \cdot 10^4}$ **30.** $\frac{10^8 \cdot 10^2}{(10^5)^2}$ **31.** $\frac{6^4 \cdot 6^9}{6^2 \cdot 6^3}$ **32.** $\frac{4^5 \cdot 4^5}{(4^2)^4}$

In 33–35, tell whether each sentence is true or false.

33. $100^{99} \div 10^{98} = 100^2$ **34.** $a^6 \div a^2 = a^4$ $(a \neq 0)$ **35.** $450^{45} \div 450^{40} = 1^5$

5-6 POWERS WITH ZERO AND NEGATIVE EXPONENTS

Integers that are negative or zero, such as 0, −1, and −2 can also be used as exponents. We will define powers having zero and negative integral exponents in such a way that the properties that were valid for positive integral exponents will also be valid for zero and negative integral exponents. In other words, the following properties will be true when the exponents a and b are positive integers, negative integers, or 0:

$$x^a \cdot x^b = x^{a+b} \qquad x^a \div x^b = x^{a-b} \qquad (x^a)^b = x^{ab}$$

The Zero Exponent

We know that, for $x \neq 0$, $\frac{x^3}{x^3} = 1$. If $\frac{x^3}{x^3} = x^{3-3} = x^0$ is to be a meaningful statement, we must let $x^0 = 1$ since x^0 and 1 are each equal to $\frac{x^3}{x^3}$. This leads to the following definition:

DEFINITION

$x^0 = 1$ if x is a number such that $x \neq 0$.

It can be shown that all the laws of exponents remain valid when x^0 is defined as 1. For example:

- Using the definition $10^0 = 1$, we have $10^3 \cdot 10^0 = 10^3 \cdot 1 = 10^3$

- Using the law of exponents, we have $10^3 \cdot 10^0 = 10^{3+0} = 10^3$.

The two procedures result in the same product.

- Using the definition $10^0 = 1$, we have $10^3 \div 10^0 = 10^3 \div 1 = 10^3$
- Using the law of exponents, we have $10^3 \div 10^0 = 10^{3-0} = 10^3$.

The two procedures result in the same quotient.

The definition $x^0 = 1$ $(x \neq 0)$ permits us to say that the zero power of any number except 0 equals 1.

$$4^0 = 1 \qquad (-4)^0 = 1 \qquad (4x)^0 = 1 \qquad (-4x)^0 = 1$$

A calculator will return this value. For example, to evaluate 4^0:

ENTER: 4 [^] 0 [ENTER]

DISPLAY:

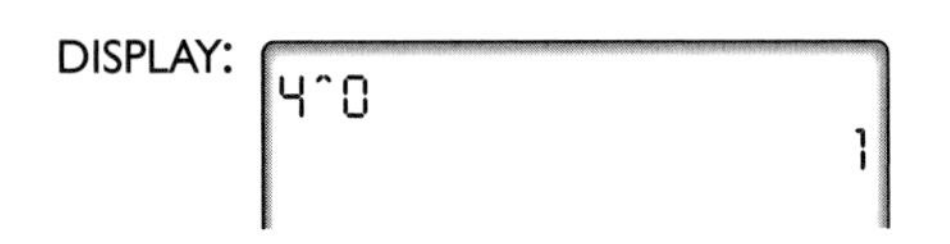

Note that $4x^0 = 4^1 \cdot x^0 = 4 \cdot 1 = 4$ but $(4x)^0 = 4^0 \cdot x^0 = 1 \cdot 1 = 1$.

The Negative Integral Exponent

We know that, for $x \neq 0$,

$$\frac{x^3}{x^5} = \frac{1 \cdot x^3}{x^2 \cdot x^3} = \frac{1}{x^2} \cdot \frac{x^3}{x^3} = \frac{1}{x^2} \cdot 1 = \frac{1}{x^2}.$$

If $\frac{x^3}{x^5} = x^{3-5} = x^{-2}$ is to be a meaningful statement, we must let $x^{-2} = \frac{1}{x^2}$ since x^{-2} and $\frac{1}{x^2}$ are each equal to $\frac{x^3}{x^5}$. This leads to the following definition:

DEFINITION

$x^{-n} = \frac{1}{x^n}$ if $x \neq 0$.

A graphing calculator will return equal values for x^{-2} and $\frac{1}{x^2}$. For example, let $x = 5$.

Evaluate 5^{-2}.

ENTER: 5 [^] [(-)] 2 [ENTER]

DISPLAY:

5^-2
.04

Evaluate $\frac{1}{5^2}$.

ENTER: 1 [÷] 5 [^] 2 [ENTER]

DISPLAY:

1/5^2
.04

It can be shown that all the laws of exponents remain valid if x^{-n} is defined as $\frac{1}{x^n}$. For example:

- Using the definition $2^{-4} = \frac{1}{2^4}$, we have $2^2 \cdot 2^{-4} = 2^2 \cdot \frac{1}{2^4} = \frac{2^2}{2^4} = \frac{1}{2^2} = 2^{-2}$.
- Using the law of exponents, we have $2^2 \cdot 2^{-4} = 2^{2+(-4)} = 2^{-2}$.

The two procedures give the same result.

Now we can say that, for all integral values of a and b,

$$\frac{x^a}{x^b} = x^{a-b}\ (x \neq 0)$$

EXAMPLE 1

Transform each given expression into an equivalent one with a positive exponent.

	Answers
a. 4^{-3}	$= \frac{1}{4^3}$
b. 10^{-1}	$= \frac{1}{10^1} = \frac{1}{10}$
c. $\frac{1}{2^{-5}}$	$= 1 \div 2^{-5} = 1 \div \frac{1}{2^5} = 1 \times \frac{2^5}{1} = 2^5$
d. $\left(\frac{5}{3}\right)^{-2}$	$= \frac{5^{-2}}{3^{-2}} = \frac{1}{5^2} \div \frac{1}{3^2} = \frac{1}{5^2} \times \frac{3^2}{1} = \frac{3^2}{5^2} = \left(\frac{3}{5}\right)^2$

EXAMPLE 2

Compute the value of each expression.

	Answers
a. 3^0	$= 1$
b. 10^{-2}	$= \frac{1}{10^2} = \frac{1}{100}$
c. $(-5)^0 + 2^{-4}$	$= 1 + \frac{1}{2^4} = 1 + \frac{1}{16} = 1\frac{1}{16}$
d. $6(3^{-3})$	$= 6\left(\frac{1}{3^3}\right) = 6\left(\frac{1}{27}\right) = \frac{6}{27} = \frac{2}{9}$

EXAMPLE 3

Use the laws of exponents to perform the indicated operations.

	Answers		*Answers*
a. $2^7 \cdot 2^{-3}$	$= 2^{7+(-3)} = 2^4$	**b.** $3^{-6} \div 3^{-2}$	$= 3^{-6-(-2)} = 3^{-6+2} = 3^{-4}$
c. $(x^4)^{-3}$	$= x^{4(-3)} = x^{-12}$	**d.** $(y^{-2})^{-4}$	$= y^{-2(-4)} = y^8$

EXERCISES

Writing About Mathematics

1. Sasha said that for all $x \neq 0$, x^{-2} is a positive number less than 1. Do you agree with Sasha? Explain why or why not.

2. Brandon said that, when n is a whole number, the number 10^n when written in ordinary decimal notation uses $n + 1$ digits. Do you agree with Brandon? Explain why or why not.

Developing Skills

In 3–7, transform each given expression into an equivalent expression involving a positive exponent.

3. 10^{-4} **4.** 2^{-1} **5.** $\left(\frac{2}{3}\right)^{-2}$ **6.** $m^{-6}, m \neq 0$ **7.** $r^{-3}, r \neq 0$

In 8–19, compute each value using the definitions of zero and negative exponents. Compare your answers with the results obtained using a calculator.

8. 10^0 **9.** $(-4)^0$ **10.** -4^0 **11.** 3^{-2}

12. 10^{-1} **13.** 10^{-2} **14.** 10^{-3} **15.** $4(10)^{-2}$

16. $1.5(10)^{-3}$ **17.** $7^0 + 6^{-2}$ **18.** $\left(\frac{1}{2}\right)^0 + 3^{-3}$ **19.** $-2 \cdot 4^{-1}$

In 20–27, use the laws of exponents to perform each indicated operation.

20. $10^{-2} \cdot 10^5$ **21.** $3^{-4} \cdot 3^{-2}$ **22.** $10^{-3} \div 10^{-5}$ **23.** $(4^{-2})^2 \div 4^{-4}$

24. $3^4 \div 3^0$ **25.** $(4^{-1})^2$ **26.** $(3^{-3})^{-2}$ **27.** $2^0 \cdot 2^{-5}$

28. Find the value of $7x^0 - (6x)^0, (x \neq 0)$.

29. Find the value of $5x^0 + 2x^{-1}$ when $x = 4$.

5-7 SCIENTIFIC NOTATION

Scientists and mathematicians often work with numbers that are very large or very small. In order to write and compute with such numbers more easily, these workers use **scientific notation**. A number is expressed in scientific notation when it is written as the product of two quantities: the first is a number greater than or equal to 1 but less than 10, and the second is a power of 10. In other words, a number is in scientific notation when it is written as

$$a \times 10^n$$

where $1 \leq a < 10$ and n is an integer.

Writing Numbers in Scientific Notation

To write a number in scientific notation, first write it as the product of a number between 1 and 10 times a power of 10. Then express the power of 10 in exponential form.

The table at the right shows some integral powers of 10. When the exponent is a positive integer, the power can be written as 1 followed by the number of 0's equal to the exponent of 10. When the exponent is a negative integer, the power can be written as a decimal value with the number of decimal places equal to the absolute value of the exponent of 10.

Powers of 10
$10^5 = 100{,}000$
$10^4 = 10{,}000$
$10^3 = 1{,}000$
$10^2 = 100$
$10^1 = 10$
$10^0 = 1$
$10^{-1} = \frac{1}{10^1} = \frac{1}{10} = 0.1$
$10^{-2} = \frac{1}{10^2} = \frac{1}{100} = 0.01$
$10^{-3} = \frac{1}{10^3} = \frac{1}{1{,}000} = 0.001$
$10^{-4} = \frac{1}{10^4} = \frac{1}{10{,}000} = 0.0001$

$$3{,}000{,}000 = 3 \times 1{,}000{,}000 = 3 \times 10^6$$
$$780 = 7.8 \times 100 = 7.8 \times 10^2$$
$$3 = 3 \times 1 = 3 \times 10^0$$
$$0.025 = 2.5 \times 0.01 = 2.5 \times 10^{-2}$$
$$0.0003 = 3 \times 0.0001 = 3 \times 10^{-4}$$

When writing a number in scientific notation, keep in mind the following:

- A number equal to or greater than 10 has a positive exponent of 10.
- A number equal to or greater than 1 but less than 10 has a zero exponent of 10.
- A number between 0 and 1 has a negative exponent of 10.

EXAMPLE 1

The distance from the earth to the sun is approximately 93,000,000 miles. Write this number in scientific notation.

Solution *How to Proceed*

(1) Write the number, placing a decimal point after the last digit. — 93,000,000.

(2) Place a caret (^) after the first nonzero digit so that replacing the caret with a decimal point will give a number between 1 and 10. — $9_{\wedge}3{,}000{,}000.$

(3) Count the number of digits between the caret and the decimal point. This is the exponent of 10 in scientific notation. The exponent is positive because the given number is greater than 10. — $9_{\wedge}\underbrace{3{,}000{,}000}_{7}.$

(4) Write the number in the form $a \times 10^n$, where where a is found by replacing the caret with a decimal point and n is the exponent found in Step 3. — 9.3×10^7

Answer 9.3×10^7

EXAMPLE 2

Express 0.0000029 in scientific notation.

Solution Since the number is between 0 and 1, the exponent will be negative.

Place a caret after the first nonzero digit to indicate the position of the decimal point in scientific notation.

Answer $0.0000029 = 0.\underbrace{000002}_{6}{}_{\wedge}9 = 2.9 \times 10^{-6}$

Graphing calculators can be placed in scientific notation mode and will return the results shown in Examples 1 and 2 when the given numbers are entered.

ENTER: MODE ▸ ENTER CLEAR

.0000029 ENTER

DISPLAY:

```
.0000029
              2.9E-6
```

This display is read as 2.9×10^{-6}, where the integer following "E" is the exponent to the base 10 used to write the number in scientific notation.

Changing to Ordinary Decimal Notation

We can change a number that is written in scientific notation to ordinary decimal notation by expanding the power of 10 and then multiplying the result by the number between 1 and 10.

EXAMPLE 3

The approximate population of the United States is 2.81×10^8. Find the approximate number of people in the United States.

Solution *How to Proceed*

(1) Evaluate the second factor, which is a power of 10: $2.81 \times 10^8 = 2.81 \times 100{,}000{,}000$

(2) Multiply the factors: $2.81 \times 10^8 = 281{,}000{,}000$

Answer 281,000,000 people

Note: We could have multiplied 2.81 by 10^8 quickly by moving the decimal point in 2.81 eight places to the right.

EXAMPLE 4

The diameter of a red blood corpuscle is expressed in scientific notation as 7.5×10^{-4} centimeters. Write the number of centimeters in the diameter as a decimal fraction.

Solution *How to Proceed*

(1) Evaluate the second factor, which is a power of 10: $7.5 \times 10^{-4} = 7.5 \times 0.0001$

(2) Multiply the factors: $7.5 \times 10^{-4} = 0.00075$

Answer 0.00075 cm

Note: We could have multiplied 7.5 by 10^{-4} quickly by moving the decimal point in 7.5 four places to the left.

EXAMPLE 5

Use a calculator to find the product: $45{,}000 \times 570{,}000$.

Calculator Solution ENTER: 45000 [×] 570000 [ENTER] DISPLAY:

```
45000*570000
        2.565E10
```

A calculator will shift to scientific notation when the number is too large or too small for the display. The number in this display can be changed to decimal notation by using the procedure shown in Examples 3 and 4.

Answer $2.565 \times 10^{10} = 2.565 \times 10{,}000{,}000{,}000 = 25{,}650{,}000{,}000$

EXAMPLE 6

Use a calculator to find the mass of 2.70×10^{15} hydrogen atoms if the mass of one hydrogen atom is 1.67×10^{-24} grams. Round the answer to three significant digits.

Solution Multiply the mass of one hydrogen atom by the number of hydrogen atoms.

$$\begin{aligned}&(1.67 \times 10^{-24}) \times (2.70 \times 10^{15})\\&= (1.67 \times 2.70) \times (10^{-24} \times 10^{15})\\&= (1.67 \times 2.70) \times (10^{-24+15})\\&= 4.509 \times 10^{-9}\end{aligned}$$

Round 4.509 to 4.51, which has three significant digits.

Answer $4.51 \times 10^{-9} = 4.51 \times 0.000000001 = 0.00000000451$ grams

Calculator Solution Use a calculator to multiply the mass of one hydrogen atom by the number of hydrogen atoms. Enter the numbers in scientific notation.

ENTER: 1.67 [2nd] [EE] [(-)] 24 [×] 2.7 [2nd] [EE] 15 [ENTER]

DISPLAY:
```
1.67E-24*2.7E15
          4.509E-9
```

Round 4.509 to three significant digits.

Answer $4.51 \times 10^{-9} = 4.51 \times 0.000000001 = 0.00000000451$ grams

EXERCISES

Writing About Mathematics

1. Jared said that when a number is in scientific notation, $a \times 10^n$, the number of digits in a is the number of significant digits. Do you agree with Jared? Explain why or why not.

2. When Corey wanted to enter 2.54×10^{-5} into his calculator, he used this sequence of keys: 2.54 [×] [2nd] [EE] [−] 5 [ENTER]. Is this a correct way to enter the number? Explain why or why not.

Developing Skills

In 3–8, write each number as a power of 10.

3. 100
4. 10,000
5. 0.01
6. 0.0001
7. 1,000,000,000
8. 0.0000001

In 9–20, find the number that is expressed by each numeral.

9. 10^7
10. 10^{10}
11. 10^{-3}
12. 10^{-5}
13. 3×10^5
14. 4×10^8
15. 6×10^{-1}
16. 9×10^{-7}
17. 1.3×10^4
18. 8.3×10^{-10}
19. 1.27×10^3
20. 6.14×10^{-2}

In 21–32, find the value of n that will make each resulting statement true.

21. $120 = 1.2 \times 10^n$
22. $9{,}300 = 9.3 \times 10^n$
23. $5{,}280 = 5.28 \times 10^n$
24. $0.00161 = 1.61 \times 10^n$
25. $0.0000760 = 7.60 \times 10^n$
26. $52{,}000 = 5.2 \times 10^n$
27. $0.00000000375 = 3.75 \times 10^n$
28. $872{,}000{,}000 = 8.72 \times 10^n$
29. $0.800 = 8.00 \times 10^n$
30. $2.54 = 2.54 \times 10^n$
31. $0.00456 = 4.56 \times 10^n$
32. $7{,}123{,}000 = 7.123 \times 10^n$

In 33–44, express each number in scientific notation.

33. 8,400
34. 27,000
35. 54,000,000
36. 320,000,000
37. 0.00061
38. 0.0000039
39. 0.0000000140
40. 0.156
41. 453,000
42. 0.00381
43. 375,000,000
44. 0.0000763

In 45–48, compute the result of each operation. Using the correct number of significant digits: **a.** write the result in scientific notation, **b.** write the result in ordinary decimal notation.

45. $(2.9 \times 10^3)(3.0 \times 10^{-3})$
46. $(2.55 \times 10^{-2})(3.00 \times 10^{-3})$
47. $(7.50 \times 10^4) \div (2.5 \times 10^3)$
48. $(6.80 \times 10^{-5}) \div (3.40 \times 10^{-8})$

Applying Skills

In 49–52, express each number in scientific notation.

49. A light-year, which is the distance light travels in 1 year, is approximately 9,500,000,000,000 kilometers.

50. A star that is about 12,000,000,000,000,000,000,000 miles away can be seen by the Palomar telescope.

51. The radius of an electron is about 0.0000000000005 centimeters.

52. The diameter of some white blood corpuscles is approximately 0.0008 inches.

In 53–57, express each number in ordinary decimal notation.

53. The diameter of the universe is 2×10^9 light-years.

54. The distance from the earth to the moon is 2.4×10^5 miles.

55. In a motion-picture film, the image of each picture remains on the screen approximately 6×10^{-2} seconds.

56. Light takes about 2×10^{-8} seconds to cross a room.

57. The mass of the earth is approximately 5.9×10^{24} kilograms.

5-8 DIVIDING BY A MONOMIAL

Dividing a Monomial by a Monomial

We know that

$$\frac{a}{b} \cdot \frac{c}{d} = \frac{ac}{bd}$$

We can rewrite this equality interchanging the left and right members.

$$\frac{ac}{bd} = \frac{a}{b} \cdot \frac{c}{d}$$

Using this relationship, we can write:

$$\frac{-30x^6}{2x^4} = \frac{-30}{2} \cdot \frac{x^6}{x^4} = -15x^2$$

$$\frac{-21a^5b^4}{-3a^4b} = \frac{-21}{-3} \cdot \frac{a^5}{a^4} \cdot \frac{b^4}{b} = 7a^1b^3 = 7ab^3$$

$$\frac{12y^2z^2}{4y^2z} = \frac{12}{4} \cdot \frac{y^2}{y^2} \cdot \frac{z^2}{z} = 3y^0z^1 = 3 \cdot 1 \cdot z = 3z$$

Procedure

To divide a monomial by a monomial:

1. Divide the numerical coefficients.
2. When variable factors are powers of the same base, divide by subtracting exponents.
3. Multiply the quotients from steps 1 and 2.

If the area of a rectangle is 42 and its length is 6, we can find its width by dividing the area, 42, by the length, 6. Thus, $42 \div 6 = 7$, which is the width.

Similarly, if the area of a rectangle is represented by $42x^2$ and its length by $6x$, we can find its width by dividing the area, $42x^2$, by the length, $6x$:

$$42x^2 \div 6x = 7x$$

Therefore, the width can be represented by $7x$.

EXAMPLE 1

Divide: *Answers*

a. $\frac{24a^5}{-3a^2}$ $= \frac{24}{-3} \cdot \frac{a^5}{a^2} = -8a^3$

b. $\frac{-18x^3y^2}{-6x^2y}$ $= \frac{-18}{-6} \cdot \frac{x^3}{x^2} \cdot \frac{y^2}{y} = 3xy$

c. $\frac{20a^3c^4d^2}{-5a^3c^3}$ $= \frac{20}{-5} \cdot \frac{a^3}{a^3} \cdot \frac{c^4}{c^3} \cdot d^2 = -4(1)cd^2 = -4cd^2$

EXAMPLE 2

The area of a rectangle is $24x^4y^3$. Express, in terms of x and y, the length of the rectangle if the width is $3xy^2$.

Solution The length of a rectangle can be found by dividing the area by the width.

$$\frac{24x^4y^3}{3xy^2} = 8x^3y \text{ Answer}$$

Dividing a Polynomial by a Monomial

We know that to divide by a number is the same as to multiply by its reciprocal. Therefore,

$$\frac{a + c}{b} = \frac{1}{b}(a + c) = \frac{a}{b} + \frac{c}{b}$$

Similarly,

$$\frac{2x + 2y}{2} = \frac{1}{2}(2x + 2y) = \frac{2x}{2} + \frac{2y}{2} = x + y$$

and

$$\frac{21a^2b - 3ab}{3ab} = \frac{1}{3ab}(21a^2b - 3ab) = \frac{21a^2b}{3ab} - \frac{3ab}{3ab} = 7a - 1$$

Usually, the two middle steps are done mentally.

Procedure

To divide a polynomial by a monomial, divide each term of the polynomial by the monomial.

EXAMPLE 3

Divide: *Answers*

a. $(8a^5 - 6a^4) \div 2a^2$ $= 4a^3 - 3a^2$

b. $\frac{24x^3y^4 - 18x^2y^2 - 6xy}{-6xy}$ $= -4x^2y^3 + 3xy + 1$

EXERCISES

Writing About Mathematics

1. Mikhail divided $(12ab^2 + 6ab)$ by $(6ab)$ and got $2b$ for his answer. Explain to Mikhail why his answer is incorrect.

2. Angelique divided $(15cd + 11c)$ by $5c$ and got $(3d + 2.2)$ as her answer. Do you agree with Angelique? Explain why or why not.

Developing Skills

In 3–26, divide in each case.

3. $14x^2y^2 \div -7$ **4.** $-36y^{10} \div 6y^2$ **5.** $\frac{18x^6}{2x^2}$ **6.** $\frac{5x^2y^3}{-5y^3}$

7. $\frac{-49c^4b^3}{7c^2b^2}$ **8.** $\frac{-24x^2y}{-3xy}$ **9.** $\frac{-56abc}{8abc}$ **10.** $\frac{-27xyz}{9xz}$

11. $(14x + 7) \div 7$ **12.** $\frac{cm + cn}{c}$ **13.** $\frac{tr - r}{r}$ **14.** $\frac{8c^2 - 12d^2}{-4}$

15. $\frac{p + prt}{p}$ **16.** $\frac{y^2 - 5y}{-y}$ **17.** $\frac{18d^3 + 12d^2}{6d}$ **18.** $\frac{18r^5 + 12r^3}{6r^2}$

19. $\frac{9y^9 - 6y^6}{-3y^3}$ **20.** $\frac{8a^3 - 4a^2}{-4a^2}$ **21.** $\frac{3ab^2 - 4a^2b}{ab}$ **22.** $\frac{4c^2d - 12cd^2}{4cd}$

23. $\frac{-2a^2 - 3a + 1}{-1}$ **24.** $\frac{2.4y^5 + 1.2y^4 - 0.6y^3}{-0.6y^3}$ **25.** $\frac{a^3 - 2a^2}{0.5a^2}$ **26.** $\frac{1.6cd - 4.0c^2d}{0.8cd}$

Applying Skills

27. If five oranges cost $15y$ cents, represent the average cost of one orange.

28. If the area of a triangle is $32ab$ and the base is $8a$, represent the height of the triangle.

29. If a train traveled $54r$ miles in 9 hours, represent the average distance traveled in 1 hour.

30. If $40ab$ chairs are arranged in $5a$ rows with equal numbers of chairs in each row, represent the number of chairs in one row.

5-9 DIVIDING BY A BINOMIAL

When we divide 736 by 32, we use repeated subtraction of multiples of 32 to determine how many times 32 is contained in 736. To divide a polynomial by a binomial, we will use a similar procedure to divide $x^2 + 6x + 8$ by $x + 2$.

How to Proceed

(1) Write the usual division form:

$$x + 2\overline{)x^2 + 6x + 8}$$

(2) Divide the first term of the dividend by the first term of the divisor to obtain the first term of the quotient:

$$\begin{array}{r} x \phantom{{}+6x+8} \\ x + 2\overline{)x^2 + 6x + 8} \end{array}$$

(3) Multiply the whole divisor by the first term of the quotient. Write each term of the product under the like term of the dividend:

$$\begin{array}{r} x \phantom{{}+6x+8} \\ x + 2\overline{)x^2 + 6x + 8} \\ \underline{x^2 + 2x} \phantom{{}+8} \end{array}$$

(4) Subtract and bring down the next term of the dividend to obtain a new dividend:

$$\begin{array}{r} x \phantom{{}+6x+8} \\ x + 2\overline{)x^2 + 6x + 8} \\ \underline{x^2 + 2x \phantom{{}+8}} \\ 4x + 8 \end{array}$$

(5) Divide the first term of the new dividend by the first term of the divisor to obtain the next term of the quotient:

$$\begin{array}{r} x + 4 \phantom{{}+8} \\ x + 2\overline{)x^2 + 6x + 8} \\ \underline{x^2 + 2x \phantom{{}+8}} \\ 4x + 8 \end{array}$$

(6) Repeat steps (3) and (4), multiplying the whole divisor by the new term of the quotient. Subtract this product from the new dividend. Here the remainder is zero and the division is complete:

$$\begin{array}{r} x + 4 \phantom{{}+8} \\ x + 2\overline{)x^2 + 6x + 8} \\ \underline{x^2 + 2x \phantom{{}+8}} \\ 4x + 8 \\ \underline{4x + 8} \\ 0 \end{array}$$

The division can be checked by multiplying the quotient by the divisor to obtain the dividend:

$$\begin{aligned} (x + 4)(x + 2) &= x(x + 2) + 4(x + 2) \\ &= x^2 + 2x + 4x + 8 = x^2 + 6x + 8 \end{aligned}$$

EXAMPLE 1

Divide $5s + 6s^2 - 6$ by $2s + 3$ and check.

Solution First arrange the terms of the dividend in descending order: $6s^2 + 5s - 6$

$$\begin{array}{r} 3s - 2 \\ 2s + 3\overline{)6s^2 + 5s - 6} \\ \underline{6s^2 \underset{(-)}{+} 9s} \\ -4s - 6 \\ \underline{-4s - 6} \\ 0 \end{array}$$

Check

$(3s - 2)(2s + 3)$

$= 3s(2s + 3) - 2(2s + 3)$

$= 6s^2 + 9s - 4s - 6$

$= 6s^2 + 5s - 6$ ✔

Note that we subtracted $9s$ from $5s$ by adding $-9s$ to $5s$.

Answer $3s - 2$

EXERCISES

Writing about Mathematics

1. Nate said that $\frac{x^3 - 1}{x + 1} = \frac{x^3}{x} + \frac{-1}{1} = x^2 - 1$. Is Nate correct? Explain why or why not.

2. Mason wrote $x^3 - 1$ as $x^3 + 0x^2 + 0x - 1$ before dividing by $x + 1$.

a. Does $x^3 - 1 = x^3 + 0x^2 + 0x - 1$?

b. Divide $x^3 - 1$ by $x - 1$ by writing $x^3 + 0x^2 + 0x - 1$ as the dividend. Check your answer to show that your computation is correct.

Developing Skills

In 3–14, divide and check.

3. $(b^2 + 5b + 6) \div (b + 3)$

4. $(y^2 + 3y + 2) \div (y + 2)$

5. $(m^2 - 8m + 7) \div (m - 1)$

6. $\frac{w^2 + 2w - 15}{w + 5}$

7. $\frac{y^2 + 21y + 68}{y + 17}$

8. $\frac{x^2 + 7x + 10}{x + 5}$

9. $(3a^2 - 8a + 4) \div (3a - 2)$

10. $(15t^2 - 19t - 56) \div (5t + 7)$

11. $\frac{10y^2 - y - 24}{2y + 3}$

12. $\frac{8 - 22c + 12c^2}{4c - 2}$

13. $(17x + 66 + x^2) \div (x + 6)$

14. $\frac{x^2 - 64}{x - 8}$

15. One factor of $x^2 - 4x - 21$ is $x - 7$. Find the other factor.

Applying Skills

16. The area of a rectangle is represented by $x^2 - 8x - 9$. If its length is represented by $x + 1$, how can the width be represented?

17. The area of a rectangle is represented by $3y^2 + 8y + 4$. If its length is represented by $3y + 2$, how can the width be represented?

CHAPTER SUMMARY

Two or more terms that contain the same variable, with corresponding variables having the same exponents, are called **like terms**. The sum of like terms is the sum of the coefficients of the terms times the common variable factor of the terms.

A term that has no variable in the denominator is called a **monomial**. A **polynomial** is the sum of monomials.

To subtract one polynomial from another, add the opposite of the polynomial to be subtracted (the subtrahend) to the polynomial from which it is to be subtracted (the minuend).

When x is a nonzero real number and a and b are integers:

$$x^a \cdot x^b = x^{a+b} \qquad (x^a)^b = x^{ab} \qquad x^a \div x^b = x^{a-b} \qquad x^0 = 1 \qquad x^{-a} = \frac{1}{x^a}$$

A number is in **scientific notation** when it is written as $a \times 10^n$, where $1 \leq a < 10$ and n is an integer.

If $x = a \times 10^n$. Then:

- When $x \geq 10$, n is positive.
- When $1 \leq x < 10$, n is zero.
- When $0 < x < 1$, n is negative.

To multiply a polynomial by a polynomial, multiply each term of one polynomial by each term of the other polynomial and write the product as the sum of these results in simplest form.

To divide a polynomial by a monomial, divide each term of the polynomial by the monomial and write the quotient as the sum of these results.

To divide a polynomial by a binomial, subtract multiples of the divisor from the dividend until the remainder is 0 or of degree less than the degree of the divisor.

VOCABULARY

5-1 Term • Like terms (similar terms) • Unlike terms • Monomial • Polynomial • Binomial • Trinomial • Simplest form • Descending order • Ascending order

5-4 FOIL

5-5 Zero exponent ($x^0 = 1$) • Negative integral exponent ($x^{-n} = \frac{1}{x^n}$)

5-7 Scientific notation

REVIEW EXERCISES

1. Explain why scientific notation is useful.

2. Is it possible to write a general rule for simplifying an expression such as $a^n + b^n$?

In 3–17, simplify each expression.

3. $5bc - bc$ **4.** $3y^2 - 2y + y^2 - 8y - 2$ **5.** $5t - (4 - 8t)$

6. $8mg(-3g)$ **7.** $3x^2(4x^2 + 2x - 1)$ **8.** $(4x + 3)(2x - 1)$

9. $(-6ab^3)^2$ **10.** $(-6a + b)^2$ **11.** $(2a + 5)(2a - 5)$

12. $(2a - 5)^2$ **13.** $2x - x(2x - 5)$ **14.** $\frac{40b^3c^6}{-8b^2c}$

15. $5y + \frac{6y^4}{-2y^3}$ **16.** $\frac{6w^3 - 8w^2 + 2w}{2w}$ **17.** $\frac{x^2 + x - 30}{x - 5}$

In 18–21, use the laws of exponents to perform the operations, and simplify.

18. $3^5 \cdot 3^4$ **19.** $(7^3)^2$ **20.** $[2(10^2)]^3$ **21.** $12^0 + 12^{-2} \cdot 12$

In 22–25, express each number in scientific notation.

22. 5,800 **23.** 14,200,000 **24.** 0.00006 **25.** 0.00000277

In 26–29, find the decimal number that is expressed by each given numeral.

26. 4×10^4 **27.** 3.06×10^{-3} **28.** 9.7×10^8 **29.** 1.03×10^{-4}

30. Express the area of each of the gardens and the total area of the two gardens described in the chapter opener on page 167.

31. If the length of one side of a square is $2h + 3$, express in terms of h:

a. the perimeter of the square.

b. the area of the square.

32. The perimeter of a triangle is $41px$. If the lengths of two sides are $18px$ and $7px$, represent the length of the third side.

33. If the length of a rectangle can be represented by $x + 5$, and the area of the rectangle by $x^2 + 7x + 10$, find the polynomial that represents:

a. the width of the rectangle.

b. the perimeter of the rectangle.

34. The cost of a pizza is 20 cents less than 9 times the cost of a soft drink. If x represents the cost, in cents, of a soft drink, express in simplest form the cost of two pizzas and six soft drinks.

Exploration

Study the squares of two-digit numbers that end in 5. From what you observe, can you devise a method for finding the square of such a number mentally? Can this method be applied to the square of a three-digit number that ends in 5?

Study the squares of the integers from 1 to 12. From what you observe, can you devise a method that uses the square of an integer to find the square of the next larger integer?

CUMULATIVE REVIEW — CHAPTERS 1–5

Part I

Answer all questions in this part. Each correct answer will receive 2 credits. No partial credit will be allowed.

1. Which of the numbers listed below has the largest value?

(1) $1\frac{2}{3}$ (2) 1.67 (3) $1.6\overline{7}$ (4) $\frac{12}{7}$

2. For which of the following values of x is $x^2 > x > \frac{1}{x}$?

(1) 1 (2) 0 (3) 3 (4) $\frac{2}{3}$

3. Which of the numbers given below is not a rational number?

(1) $\sqrt{2}$ (2) $1\frac{1}{2}$ (3) $1.\overline{3}$ (4) $\frac{7}{3}$

4. Which of the following inequalities is false?

(1) $1.5 < 1\frac{1}{2}$ (2) $1.5 \leq 1\frac{1}{2}$ (3) $-1.5 < 1.5$ (4) $-1.5 < -1$

5. Which of the following identities is an illustration of the associative property?

(1) $x + 7 = 7 + x$
(2) $3(x + 7) = 3x + 3(7)$
(3) $(x + 7) + 3 = 3 + (7 + x)$
(4) $(x + 7) + 3 = x + (7 + 3)$

6. The formula $C = \frac{5}{9}(F - 32)$ can be used to find the Celsius temperature, C, for a given Fahrenheit temperature, F. What Celsius temperature is equal to a Fahrenheit temperature of 68°?

(1) 3° (2) 20° (3) 35° (4) 180°

7. If the universe is the set of whole numbers, the solution set of $x \leq 3$ is

(1) {0, 1, 2} (2) {0, 1, 2, 3} (3) {1, 2} (4) {1, 2, 3}

8. The perimeter of a square whose area is 81 square centimeters is

(1) 9 cm (2) 18 cm (3) 20.25 cm (4) 36 cm

9. In simplest form, $(2x - 4)^2 + 3(x + 1)$ is equal to

(1) $4x^2 - 13x - 13$
(2) $4x^2 - 13x + 19$
(3) $4x^2 + 3x + 19$
(4) $4x^2 + 3x - 13$

10. To the nearest tenth of a meter, the circumference of a circle whose radius is 12.0 meters is

(1) 37.6 m (2) 37.7 m (3) 75.3 m (4) 75.4 m

Part II

Answer all questions in this part. Each correct answer will receive 2 credits. Clearly indicate the necessary steps, including appropriate formula substitutions, diagrams, graphs, charts, etc. For all questions in this part, a correct numerical answer with no work shown will receive only 1 credit.

11. The formula for the volume V of a cone is $V = \frac{1}{3}Bh$ where B is the area of the base and h is the height. Solve the formula for h in terms of V and B.

12. Each of the numbers given below is different from the others. Explain in what way each is different.

2 7 77 84

Part III

Answer all questions in this part. Each correct answer will receive 3 credits. Clearly indicate the necessary steps, including appropriate formula substitutions, diagrams, graphs, charts, etc. For all questions in this part, a correct numerical answer with no work shown will receive only 1 credit.

13. Solve the given equation for x. Show each step of the solution and name the property that is used in each step.

$$3(x - 4) = 5x + 8$$

14. Simplify the following expression. Show each step of the simplification and name the property that you used in each step.

$$4a - 7 + (7 - 3a)$$

Part IV

Answer all questions in this part. Each correct answer will receive 4 credits. Clearly indicate the necessary steps, including appropriate formula substitutions, diagrams, graphs, charts, etc. For all questions in this part, a correct numerical answer with no work shown will receive only 1 credit.

15. A small park is in the shape of a rectangle that measures 525 feet by 468 feet.

a. Find the number of feet of fencing that would be needed to enclose the park. Express your answer to the nearest foot.

b. If the entire park is to be planted with grass seed, find the number of square feet to be seeded. Express your answer to the correct number of significant digits based on the given dimensions.

c. The grass seed to be purchased is packaged in sacks, each of which holds enough seed to cover 25,000 square feet of ground. How many sacks of seed are needed to seed the park?

16. An ice cream stand sells single-dip cones for $1.75 and double-dip cones for $2.25. Yesterday, 500 cones were sold for $930. How many single-dip and how many double-dip cones were sold?

CHAPTER

6

RATIO AND PROPORTION

Everyone likes to save money by purchasing something at a reduced price. Because merchants realize that a reduced price may entice a prospective buyer to buy on impulse or to buy at one store rather than another, they offer discounts and other price reductions. These discounts are often expressed as a percent off of the regular price.

When the Acme Grocery offers a 25% discount on frozen vegetables and the Shop Rite Grocery advertises "Buy four, get one free," the price-conscious shopper must decide which is the better offer if she intends to buy five packages of frozen vegetables.

In this chapter, you will learn how ratios, and percents which are a special type of ratio, are used in many everyday problems.

CHAPTER TABLE OF CONTENTS

6-1 RATIO

A **ratio**, which is a comparison of two numbers by division, is the quotient obtained when the first number is divided by the second, nonzero number.

Since a ratio is the quotient of two numbers divided in a definite order, care must be taken to write each ratio in its intended order. For example, the ratio of 3 to 1 is written

$\frac{3}{1}$ (as a fraction) or $3:1$ (using a colon)

while the ratio of 1 to 3 is written

$\frac{1}{3}$ (as a fraction) or $1:3$ (using a colon)

In general, the ratio of a to b can be expressed as

$$\frac{a}{b} \quad \text{or} \quad a \div b \quad \text{or} \quad a:b$$

To find the ratio of two quantities, both quantities must be expressed in the same unit of measure before their quotient is determined. For example, to compare the value of a nickel and a penny, we first convert the nickel to 5 pennies and then find the ratio, which is $\frac{5}{1}$ or $5:1$. Therefore, a nickel is worth 5 times as much as a penny. The ratio has no unit of measure.

Equivalent Ratios

Since the ratio $\frac{5}{1}$ is a fraction, we can use the multiplication property of 1 to find many **equivalent ratios**. For example:

$$\frac{5}{1} = \frac{5}{1} \times \frac{2}{2} = \frac{10}{2} \qquad \frac{5}{1} = \frac{5}{1} \times \frac{3}{3} = \frac{15}{3} \qquad \frac{5}{1} = \frac{5}{1} \times \frac{x}{x} = \frac{5x}{1x} \quad (x \neq 0)$$

From the last example, we see that $5x$ and $1x$ represent two numbers whose ratio is $5:1$.

In general, if a, b, and x are numbers ($b \neq 0, x \neq 0$), ax and bx represent two numbers whose ratio is $a:b$ because

$$\frac{a}{b} = \frac{a}{b} \times 1 = \frac{a}{b} \times \frac{x}{x} = \frac{ax}{bx}$$

Also, since a ratio such as $\frac{24}{16}$ is a fraction, we can divide the numerator and the denominator of the fraction by the same nonzero number to find equivalent ratios. For example:

$$\frac{24}{16} = \frac{24 \div 2}{16 \div 2} = \frac{12}{8} \qquad \frac{24}{16} = \frac{24 \div 4}{16 \div 4} = \frac{6}{4} \qquad \frac{24}{16} = \frac{24 \div 8}{16 \div 8} = \frac{3}{2}$$

A ratio is expressed in **simplest form** when both terms of the ratio are whole numbers and when there is no whole number other than 1 that is a factor of both of these terms. Therefore, to express the ratio $\frac{24}{16}$ in simplest form, we divide both terms by 8, the largest integer that will divide both 24 and 16. Therefore, $\frac{24}{16}$ in simplest form is $\frac{3}{2}$.

Continued Ratio

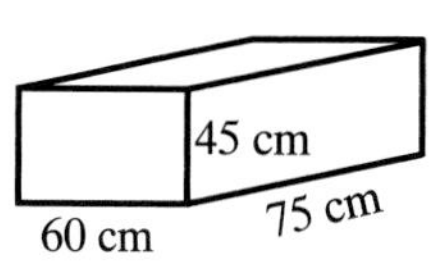

Comparisons can also be made for three or more quantities. For example, the length of a rectangular solid is 75 centimeters, the width is 60 centimeters, and the height is 45 centimeters. The ratio of the length to the width is 75 : 60, and the ratio of the width to the height is 60 : 45. We can write these two ratios in an abbreviated form as the continued ratio 75 : 60 : 45.

A **continued ratio** is a comparison of three or more quantities in a definite order. Here, the ratio of the measures of the length, width, and height (in that order) of the rectangular solid is 75 : 60 : 45 or, in simplest form, 5 : 4 : 3.

▶ **In general, the ratio of the numbers a, b, and c ($b \neq 0$, $c \neq 0$) is $a : b : c$.**

EXAMPLE 1

An oil tank with a capacity of 200 gallons contains 50 gallons of oil.

a. Find the ratio of the number of gallons of oil in the tank to the capacity of the tank.

b. What part of the tank is full?

Solution

a. Ratio $= \frac{\text{number of gallons of oil in tank}}{\text{capacity of tank}} = \frac{50}{200} = \frac{1}{4}$.

b. The tank is $\frac{1}{4}$ full.

Answers **a.** $\frac{1}{4}$ **b.** $\frac{1}{4}$ full

EXAMPLE 2

Compute the ratio of 6.4 ounces to 1 pound.

Solution First, express both quantities in the same unit of measure. Use the fact that 1 pound = 16 ounces.

$$\frac{6.4\text{ ounces}}{1\text{ pound}} = \frac{6.4\text{ ounces}}{16\text{ ounces}} = \frac{6.4}{16} = \frac{6.4}{16} \times \frac{10}{10} = \frac{64}{160} = \frac{64 \div 32}{160 \div 32} = \frac{2}{5}$$

Calculator Solution On a calculator, divide 6.4 ounces by 16 ounces.

ENTER: 6.4 [÷] 16 [ENTER]

DISPLAY:

Change the decimal in the display to a fraction.

ENTER: [2nd] [ANS] [MATH] [ENTER] [ENTER]

DISPLAY:

Ans▶Frac
2/5

Answer The ratio is 2 : 5.

EXAMPLE 3

Express the ratio $1\frac{3}{4}$ to $1\frac{1}{2}$ in simplest form.

Solution Since a ratio is the quotient obtained when the first number is divided by the second, divide $1\frac{3}{4}$ by $1\frac{1}{2}$.

$$1\frac{3}{4} \div 1\frac{1}{2} = \frac{7}{4} \div \frac{3}{2} = \frac{7}{4} \cdot \frac{2}{3} = \frac{14}{12} = \frac{7}{6}$$

Answer The ratio in simplest form is $\frac{7}{6}$ or 7 : 6.

EXERCISES

Writing About Mathematics

1. Last week, Melanie answered 24 out of 30 questions correctly on a test. This week she answered 20 out of 24 questions correctly. On which test did Melanie have better results? Explain your answer.

2. Explain why the ratio 1.5 : 4.5 is not in simplest form.

Developing Skills

In 3–12, express each ratio in simplest form: **a.** as a fraction **b.** using a colon

3. 36 to 12 **4.** 48 to 24 **5.** 40 to 25 **6.** 12 to 3 **7.** 5 to 4

8. 8 to 32 **9.** 40 to 5 **10.** 0.2 to 8 **11.** 72 to 1.2 **12.** $3c$ to $5c$

13. If the ratio of two numbers is 10 : 1, the larger number is how many times the smaller number?

14. If the ratio of two numbers is 8 : 1, the smaller number is what fractional part of the larger number?

In 15–19, express each ratio in simplest form.

15. $\frac{3}{4}$ to $\frac{1}{4}$ **16.** $1\frac{1}{8}$ to $\frac{3}{8}$ **17.** 1.2 to 2.4 **18.** 0.75 to 0.25 **19.** 6 to 0.25

In 20–31, express each ratio in simplest form.

20. 80 m to 16 m **21.** 75 g to 100 g **22.** 36 cm to 72 cm

23. 54 g to 90 g **24.** 75 cm to 350 cm **25.** 8 ounces to 1 pound

26. $1\frac{1}{2}$ hr to $\frac{1}{2}$ hr **27.** 3 in. to $\frac{1}{2}$ in. **28.** 1 ft to 1 in.

29. 1 yd to 1 ft **30.** 1 hr to 15 min **31.** 6 dollars to 50 cents

Applying Skills

32. A baseball team played 162 games and won 90.

a. What is the ratio of the number of games won to the number of games played?

b. For every nine games played, how many games were won?

33. A student did six of ten problems correctly.

a. What is the ratio of the number right to the number wrong?

b. For every two answers that were wrong, how many answers were right?

34. A cake recipe calls for $1\frac{1}{4}$ cups of milk to $1\frac{3}{4}$ cups of flour. Write, in simplest form, the ratio of the number of cups of milk to the number of cups of flour in this recipe.

35. The perimeter of a rectangular garden is 30 feet, and the width is 5 feet. Find the ratio of the length of the rectangle to its width in simplest form.

36. In a freshman class, there are b boys and g girls. Express the ratio of the number of boys to the total number of pupils.

37. The length of a rectangular classroom is represented by $3x$ and its width by $2x$. Find the ratio of the width of the classroom to its perimeter.

38. The ages of three teachers are 48, 28, and 24 years. Find, in simplest form, the continued ratio of these ages from oldest to youngest.

39. A woodworker is fashioning a base for a trophy. He starts with a block of wood whose length is twice its width and whose height is one-half its width. Write, in simplest form, the continued ratio of length to width to height.

40. Taya and Jed collect coins. The ratio of the number of coins in their collections, in some order, is 4 to 3. If Taya has 60 coins in her collection, how many coins could Jed have?

6-2 USING A RATIO TO EXPRESS A RATE

When two quantities have the same unit of measure, their ratio has no unit of measure. A **rate**, like a ratio, is a comparison of two quantities, but the quantities may have *different units of measures* and their ratio has a unit of measure.

For example, if a plane flies 1,920 kilometers in 3 hours, its *rate* of speed is a ratio that compares the distance traveled to the time that the plane was in flight.

$$\text{Rate} = \frac{\text{distance}}{\text{time}} = \frac{1{,}920\,\text{kilometers}}{3\,\text{hours}} = \frac{640\,\text{kilometers}}{1\,\text{hour}} = 640\text{ km/h}$$

The abbreviation km/h is read "kilometers per hour."

A rate may be expressed in **lowest terms** when the numbers in its ratio are whole numbers with no common factor other than 1. However, a rate is most frequently written as a ratio with 1 as its second term. As shown in the example above, the second term may be omitted when it is 1. A rate that has a denominator of 1 is called a **unit rate**. A rate that identifies the cost of an item per unit is called the **unit price**. For example, $0.15 per ounce or $3.79 per pound are unit prices.

EXAMPLE 1

Kareem scored 175 points in seven basketball games. Express, in lowest terms, the average rate of the number of points Kareem scored per game.

Solution

$$\text{Rate} = \frac{175\,\text{points}}{7\,\text{games}} = \frac{25\,\text{points}}{1\,\text{game}} = 25\text{ points per game}$$

Answer Kareem scored points at an average rate of 25 points per game.

EXAMPLE 2

There are 5 grams of salt in 100 cubic centimeters of a solution of salt and water. Express, in lowest terms, the ratio of the number of grams of salt per cubic centimeters in the solution.

Solution

$$\frac{5\,\text{g}}{100\,\text{cm}^3} = \frac{1\,\text{g}}{20\,\text{cm}^3} = \frac{1}{20}\text{ g/cm}^3$$

Answer The solution contains $\frac{1}{20}$ grams or 0.05 grams of salt per cubic centimeter of solution.

EXERCISES

Writing About Mathematics

1. How is a rate a special kind of ratio?
2. How does the way in which a rate is usually expressed differ from a ratio in simplest form?

Developing Skills

In 3–8, express each rate in lowest terms.

3. The ratio of 36 apples to 18 people.
4. The ratio of 48 patients to 6 nurses.
5. The ratio of $1.50 to 3 liters.
6. The ratio of 96 cents to 16 grams.
7. The ratio of $2.25 to 6.75 ounces.
8. The ratio of 62 miles to 100 kilometers.

Applying Skills

In 9–12, in each case, find the average rate of speed, expressed in miles per hour.

9. A vacationer traveled 230 miles in 4 hours.
10. A post office truck delivered mail on a 9-mile route in 2 hours.
11. A commuter drove 48 miles to work in $1\frac{1}{2}$ hours.
12. A race-car driver traveled 31 miles in 15 minutes. (Use 15 minutes $= \frac{1}{4}$ hour.)
13. If there are 240 tennis balls in 80 cans, how many tennis balls are in each can?
14. If an 11-ounce can of shaving cream costs 88 cents, what is the unit cost of the shaving cream in the can?
15. In a supermarket, the regular size of CleanRight cleanser contains 14 ounces and costs 49 cents. The giant size of CleanRight cleanser, which contains 20 ounces, costs 66 cents.

    a. Find, correct to the *nearest tenth* of a cent, the cost per ounce for the regular can.

    b. Find, correct to the *nearest tenth* of a cent, the cost per ounce for the giant can.

    c. Which is the better buy?
16. Johanna and Al use computers for word processing. Johanna can keyboard 920 words in 20 minutes, and Al can keyboard 1,290 words in 30 minutes. Who is faster at entering words on a keyboard?
17. Ronald runs 300 meters in 40 seconds. Carlos runs 200 meters in 30 seconds. Who is the faster runner for short races?

6-3 VERBAL PROBLEMS INVOLVING RATIO

Any pair of numbers in the ratio 3 : 5 can be found by multiplying 3 and 5 by the same nonzero number.

$3(3) = 9$	$3(7) = 21$	$3(0.3) = 0.9$	$3(x) = 3x$
$5(3) = 15$	$5(7) = 35$	$5(0.3) = 1.5$	$5(x) = 5x$
$3 : 5 = 9 : 15$	$3 : 5 = 21 : 35$	$3 : 5 = 0.9 : 1.5$	$3 : 5 = 3x : 5x$

Thus, for any nonzero number x, $3x : 5x = 3 : 5$.

In general, when we know the ratio of two or more numbers, we can use the terms of the ratio and a nonzero variable, x, to express the numbers. Any two numbers in the ratio $a : b$ can be written as ax and bx where x is a nonzero real number.

EXAMPLE 1

The perimeter of a triangle is 60 feet. If the sides are in the ratio 3 : 4 : 5, find the length of each side of the triangle.

Solution Let $3x$ = the length of the first side,

$4x$ = the length of the second side,

$5x$ = the length of the third side.

The perimeter of the triangle is 60 feet.

$$3x + 4x + 5x = 60$$
$$12x = 60$$
$$x = 5$$
$$3x = 3(5) = 15$$
$$4x = 4(5) = 20$$
$$5x = 5(5) = 25$$

Check

$15 : 20 : 25 = 3 : 4 : 5$ ✔

$15 + 20 + 25 = 60$ ✔

Answer The lengths of the sides are 15 feet, 20 feet, and 25 feet.

EXAMPLE 2

Two numbers have the ratio 2 : 3. The larger is 30 more than $\frac{1}{2}$ of the smaller. Find the numbers.

Solution Let $2x$ = the smaller number,

$3x$ = the larger number.

The larger number is 30 more than $\frac{1}{2}$ of the smaller number.

$$3x = \tfrac{1}{2}(2x) + 30$$
$$3x = x + 30$$
$$2x = 30$$
$$x = 15$$
$$2x = 2(15) = 30$$
$$3x = 3(15) = 45$$

Check
The ratio 30 : 45 in lowest terms is 2 : 3. ✔

One-half of the smaller number, 30, is 15. The larger number, 45, is 30 more than 15. ✔

Answer The numbers are 30 and 45.

EXERCISES

Writing About Mathematics

1. Two numbers in the ratio 2 : 3 can be written as $2x$ and $3x$. Explain why x cannot equal zero.
2. The ratio of the length of a rectangle to its width is 7 : 4. Pete said that the ratio of the length to the perimeter is 7 : 11. Do you agree with Pete? Explain why or why not.

Developing Skills

3. Two numbers are in the ratio 4 : 3. Their sum is 70. Find the numbers.
4. Find two numbers whose sum is 160 and that have the ratio 5 : 3.
5. Two numbers have the ratio 7 : 5. Their difference is 12. Find the numbers.
6. Find two numbers whose ratio is 4 : 1 and whose difference is 36.
7. The lengths of the sides of a triangle are in the ratio of 6 : 6 : 5. The perimeter of the triangle is 34 centimeters. Find the length of each side of the triangle.
8. The perimeter of a triangle is 48 centimeters. The lengths of the sides are in the ratio 3 : 4 : 5. Find the length of each side.
9. The perimeter of a rectangle is 360 centimeters. If the ratio of its length to its width is 11 : 4, find the dimensions of the rectangle.
10. The sum of the measures of two angles is 90°. The ratio of the measures of the angles is 2 : 3. Find the measure of each angle.
11. The sum of the measures of two angles is 180°. The ratio of the measures of the angles is 4 : 5. Find the measure of each angle.

12. The ratio of the measures of the three angles of a triangle is $2:2:5$. Find the measures of each angle.

13. In a triangle, two sides have the same length. The ratio of each of these sides to the third side is $5:3$. If the perimeter of the triangle is 65 inches, find the length of each side of the triangle.

14. Two positive numbers are in the ratio $3:7$. The larger exceeds the smaller by 12. Find the numbers.

15. Two numbers are in the ratio $3:5$. If 9 is added to their sum, the result is 41. Find the numbers.

Applying Skills

16. A piece of wire 32 centimeters in length is divided into two parts that are in the ratio $3:5$. Find the length of each part.

17. The ratio of the number of boys in a school to the number of girls is 11 to 10. If there are 525 pupils in the school, how many of them are boys?

18. The ratio of Carl's money to Donald's money is $7:3$. If Carl gives Donald $20, the two then have equal amounts. Find the original amount that each one had.

19. In a basketball free-throw shooting contest, the points made by Sam and Wilbur were in the ratio $7:9$. Wilbur made 6 more points than Sam. Find the number of points made by each.

20. A chemist wishes to make $12\frac{1}{2}$ liters of an acid solution by using water and acid in the ratio $3:2$. How many liters of each should she use?

6-4 PROPORTION

A **proportion** is an equation that states that two ratios are equal. Since the ratio $4:20$ or $\frac{4}{20}$ is equal to the ratio $1:5$ or $\frac{1}{5}$, we may write the proportion

$$4:20 = 1:5 \qquad \text{or} \qquad \frac{4}{20} = \frac{1}{5}$$

Each of these proportions is read as "4 is to 20 as 1 is to 5." The general form of a proportion may be written as:

$$a:b = c:d \qquad \text{or} \qquad \frac{a}{b} = \frac{c}{d} \; (b \neq 0, d \neq 0)$$

Each of these proportions is read as "a is to b as c is to d." There are four terms in this proportion, namely, a, b, c, and d. The outer terms, a and d, are called the **extremes** of the proportion. The inner terms, b and c, are the **means**.

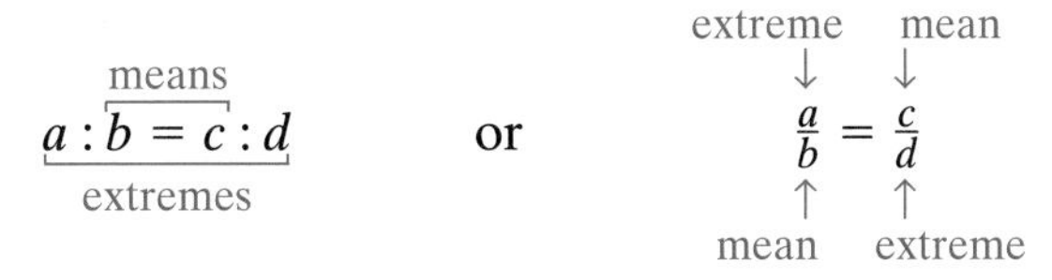

In the proportion, $4 : 20 = 1 : 5$, the product of the means, 20(1), is equal to the product of the extremes, 4(5).

In the proportion, $\frac{5}{15} = \frac{10}{30}$, the product of the means, 15(10), is equal to the product of the extremes, 5(30).

In any proportion $\frac{a}{b} = \frac{c}{d}$, we can show that the product of the means is equal to the product of the extremes, $ad = bc$. Since $\frac{a}{b} = \frac{c}{d}$ is an equation, we can multiply both members by bd, the least common denominator of the fractions in the equation.

$$\frac{a}{b} = \frac{c}{d}$$
$$bd\left(\frac{a}{b}\right) = bd\left(\frac{c}{d}\right)$$
$$\not{b}d\left(\frac{a}{\not{b}}\right) = b\not{d}\left(\frac{c}{\not{d}}\right)$$
$$d \cdot a = b \cdot c$$
$$ad = bc$$

Therefore, we have shown that the following statement is always true:

▶ **In a proportion, the product of the means is equal to the product of the extremes.**

Notice that the end result, $ad = bc$, is the result of multiplying the terms that are cross-wise from each other:

$$\frac{a}{b} \times \frac{c}{d}$$

This is called **cross-multiplying**, which we have just shown to be valid.

If the product of two cross-wise terms is called a **cross product**, then the following is also true:

▶ **In a proportion, the cross products are equal.**

If a, b, c, and d are nonzero numbers and $\frac{a}{b} = \frac{c}{d}$, then $ad = bc$. There are three other proportions using a, b, c and d for which $ad = bc$.

$$\frac{a}{c} = \frac{b}{d} \qquad \frac{d}{b} = \frac{c}{a} \qquad \frac{d}{c} = \frac{b}{a}$$

For example, we know that $\frac{6}{4} = \frac{15}{10}$ is a proportion because $6(10) = 4(15)$. Therefore, each of the following is also a proportion.

$$\frac{6}{15} = \frac{4}{10} \qquad \frac{10}{4} = \frac{15}{6} \qquad \frac{10}{15} = \frac{4}{6}$$

EXAMPLE 1

Show that $\frac{4}{16} = \frac{5}{20}$ is a proportion.

Solution Three methods are shown here. The first two use paper and pencil; the last makes use of a calculator.

METHOD 1 Reduce each ratio to simplest form.

$$\frac{4}{16} = \frac{4 \div 4}{16 \div 4} = \frac{1}{4} \quad \text{and} \quad \frac{5}{20} = \frac{5 \div 5}{20 \div 5} = \frac{1}{4}$$

Since each ratio equals $\frac{1}{4}$, the ratios are equal and $\frac{4}{16} = \frac{5}{20}$ is a proportion.

METHOD 2 Show that the cross products are equal.

$$\frac{4}{16} \times \frac{5}{20}$$

$$16 \times 5 = 4 \times 20$$

$$80 = 80$$

Therefore, $\frac{4}{16} = \frac{5}{20}$ is a proportion.

METHOD 3 Use a calculator. Enter the proportion. If the ratios are equal, then the calculator will display 1. If the ratios are not equal, the calculator will display 0.

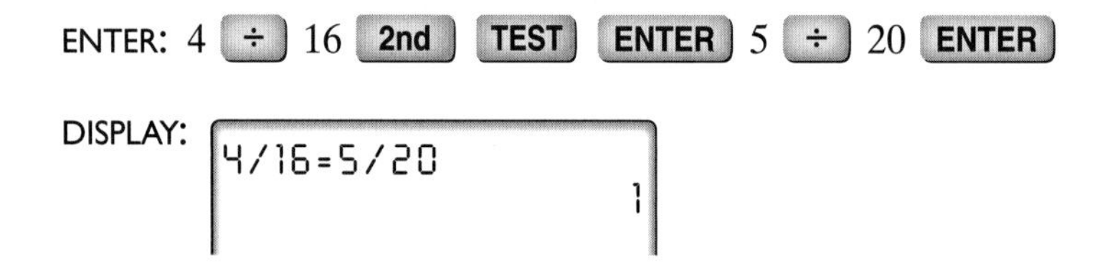

Since the calculator displays 1, the statement is true. The ratios are equal and $\frac{4}{16} = \frac{5}{20}$ is a proportion.

Answer Any one of the three methods shows that $\frac{4}{16} = \frac{5}{20}$ is a proportion.

EXAMPLE 2

Solve the proportion $25 : q = 5 : 2$ for q.

Solution Since $25 : q = 5 : 2$ is a proportion, the product of the means is equal to the product of the extremes. Therefore:

$$\underbrace{25 : \overbrace{q = 5}^{\text{means}} : 2}_{\text{extremes}}$$

$$\underbrace{5q}_{\text{means}} = \underbrace{25(2)}_{\text{extremes}}$$

$$5q = 50$$

$$q = 10$$

Check
Reduce each ratio to simplest form.

$$25 : q = 5 : 2$$

$$25 : \mathbf{10} \stackrel{?}{=} 5 : 2$$

$$5 : 2 = 5 : 2 \checkmark$$

Answer $q = 10$

Note: Example 2 could also have been solved by setting up the proportion $\frac{25}{q} = \frac{5}{2}$ and then using cross-multiplication to solve for the variable.

EXAMPLE 3

Solve for x: $\frac{12}{x-2} = \frac{32}{x+8}$

Solution Use the fact that the product of the means equals the product of the extremes (the cross products are equal).

$$\frac{12}{x-2} \times \frac{32}{x+8}$$

$$32(x - 2) = 12(x + 8)$$

$$\begin{array}{rcr} 32x - 64 & = & 12x + 96 \\ -12x + 64 & & -12x + 64 \\ \hline 20x & = & 160 \end{array}$$

$$\frac{20x}{20} = \frac{160}{20}$$

$$x = 8$$

Check

$$\frac{12}{x-2} = \frac{32}{x+8}$$

$$\frac{12}{\mathbf{8}-2} \stackrel{?}{=} \frac{32}{\mathbf{8}+8}$$

$$\frac{12}{6} \stackrel{?}{=} \frac{32}{16}$$

$$2 = 2 \checkmark$$

Answer $x = 8$

EXAMPLE 4

The denominator of a fraction exceeds the numerator by 7. If 3 is subtracted from the numerator of the fraction and the denominator is unchanged, the value of the resulting fraction becomes $\frac{1}{3}$. Find the original fraction.

Solution Let x = the numerator of original fraction,

$x + 7$ = the denominator of the original fraction.

$\frac{x}{x+7}$ = the original fraction.

$\frac{x-3}{x+7}$ = the new fraction.

The value of the new fraction is $\frac{1}{3}$.

$$\frac{x-3}{x+7} = \frac{1}{3}$$

$$1(x+7) = 3(x-3)$$

$$\begin{aligned} x + 7 &= 3x - 9 \\ -x + 9 &\quad -x + 9 \\ \hline 16 &= 2x \end{aligned}$$

$$x = 8$$

$$x + 7 = 15$$

Check

The original fraction was $\frac{8}{15}$.

The new fraction is

$\frac{8-3}{15} = \frac{5}{15} = \frac{1}{3}$ ✔

Answer The original fraction was $\frac{8}{15}$.

EXERCISES

Writing About Mathematics

1. Jeremy said that if the means and the extremes of a proportion are interchanged, the resulting ratios form a proportion. Do you agree with Jeremy? Explain why or why not.

2. Mike said that if the same number is added to each term of a proportion, the resulting ratios form a proportion. Do you agree with Mike? Explain why or why not.

Developing Skills

In 3–8, state, in each case, whether the given ratios may form a proportion.

3. $\frac{3}{4}, \frac{30}{40}$ **4.** $\frac{2}{3}, \frac{10}{5}$ **5.** $\frac{4}{5}, \frac{16}{25}$ **6.** $\frac{2}{5}, \frac{5}{2}$ **7.** $\frac{14}{18}, \frac{28}{36}$ **8.** $\frac{36}{30}, \frac{18}{15}$

In 9–16, find the missing term in each proportion.

9. $\frac{1}{2} = \frac{?}{8}$ **10.** $\frac{3}{5} = \frac{18}{?}$ **11.** $\frac{1}{4} = \frac{6}{?}$ **12.** $\frac{4}{6} = \frac{?}{42}$

13. $4 : ? = 12 : 60$ **14.** $? : 9 = 35 : 63$ **15.** $? : 60 = 6 : 10$ **16.** $16 : ? = 12 : 9$

In 17–25, solve each equation and check the solution.

17. $\frac{x}{60} = \frac{3}{20}$ **18.** $\frac{5}{4} = \frac{x}{12}$ **19.** $\frac{30}{4x} = \frac{10}{24}$

20. $\frac{5}{15} = \frac{x}{x + 8}$ **21.** $\frac{x}{12 - x} = \frac{10}{30}$ **22.** $\frac{16}{8} = \frac{21 - x}{x}$

23. $\frac{3x + 3}{3} = \frac{7x - 1}{5}$ **24.** $12 : 15 = x : 45$ **25.** $5 : x + 2 = 4 : x$

In 26–28, in each case solve for x in terms of the other variables.

26. $a : b = c : x$ **27.** $2r : s = x : 3s$ **28.** $2x : m = 4r : s$

Applying Skills

In 29–36, use a proportion to solve each problem.

29. The numerator of a fraction is 8 less than the denominator of the fraction. The value of the fraction is $\frac{3}{5}$. Find the fraction.

30. The denominator of a fraction exceeds twice the numerator of the fraction by 10. The value of the fraction is $\frac{5}{12}$. Find the fraction.

31. The denominator of a fraction is 30 more than the numerator of the fraction. If 10 is added to the numerator of the fraction and the denominator is unchanged, the value of the resulting fraction becomes $\frac{3}{5}$. Find the original fraction.

32. The numerator of a certain fraction is 3 times the denominator. If the numerator is decreased by 1 and the denominator is increased by 2, the value of the resulting fraction is $\frac{5}{2}$. Find the original fraction.

33. What number must be added to both the numerator and denominator of the fraction $\frac{7}{19}$ to make the resulting fraction equal to $\frac{3}{4}$?

34. The numerator of a fraction exceeds the denominator by 3. If 3 is added to the numerator and 3 is subtracted from the denominator, the resulting fraction is equal to $\frac{5}{2}$. Find the original fraction.

35. The numerator of a fraction is 7 less than the denominator. If 3 is added to the numerator and 9 is subtracted from the denominator, the new fraction is equal to $\frac{3}{2}$. Find the original fraction.

36. Slim Johnson was usually the best free-throw shooter on his basketball team. Early in the season, however, he had made only 9 of 20 shots. By the end of the season, he had made all the additional shots he had taken, thereby ending with a season record of 3 : 4. How many additional shots had he taken?

6-5 DIRECT VARIATION

s	1	2	3
P	4	8	12

If the length of a side, s, of a square is 1 inch, then the perimeter, P, of the square is 4 inches. Also, if s is 2 inches, P is 8 inches; if s is 3 inches, P is 12 inches. These pairs of values are shown in the table at the right.

From the table, we observe that, as s varies, P also varies. Comparing each value of P to its corresponding value of s, we notice that all three sets of values result in the same ratio when reduced to lowest terms:

$$\frac{P}{s} = \frac{4}{1} = \frac{8}{2} = \frac{12}{3}$$

If a relationship exists between two variables so that their ratio is a constant, that relationship between the variables is called a **direct variation.**

In every direct variation, we say that one variable **varies directly** as the other, or that one variable is **directly proportional** to the other. The constant ratio is called a **constant of variation.**

It is important to indicate the *order* in which the variables are being compared before stating the constant of variation. For example:

- In comparing P to s, $\frac{P}{s} = \frac{4}{1}$. The constant of variation is 4.
- In comparing s to P, $\frac{s}{P} = \frac{1}{4}$. The constant of variation is $\frac{1}{4}$.

Note that each proportion, $\frac{P}{s} = \frac{4}{1}$ and $\frac{s}{P} = \frac{1}{4}$, becomes $P = 4s$, the formula for the perimeter of a square.

In a direct variation, the value of each term of the ratio increases when we multiply each variable by a factor greater than 1; the value of each term of the ratio decreases when we divide each variable by a factor greater than 1, as shown below.

$$\frac{s}{P} = \frac{1}{4} = \frac{1 \times 2}{4 \times 2} = \frac{1 \times 3}{4 \times 3} \qquad \frac{s}{P} = \frac{1}{4} = \frac{2 \div 2}{8 \div 2} = \frac{3 \div 3}{12 \div 3}$$

EXAMPLE 1

If x varies directly as y, and $x = 1.2$ when $y = 7.2$, find the constant of variation by comparing x to y.

Solution

$$\text{Constant of variation} = \frac{x}{y} = \frac{1.2}{7.2} = \frac{1.2 \div 1.2}{7.2 \div 1.2} = \frac{1}{6}$$

Answer $\frac{1}{6}$

EXAMPLE 2

The table gives pairs of values for the variables x and y.

a. Show that one variable varies directly as the other.

b. Find the constant of variation by comparing y to x.

c. Express the relationship between the variables as a formula.

d. Find the values missing in the table.

x	1	2	3	10	?
y	8	16	24	?	1,600

Solution

a. $\frac{x}{y} = \frac{1}{8}$ $\quad \frac{x}{y} = \frac{2}{16} = \frac{2 \div 2}{16 \div 2} = \frac{1}{8}$ $\quad \frac{x}{y} = \frac{3}{24} = \frac{3 \div 3}{24 \div 3} = \frac{1}{8}$

Since all the given pairs of values have the same ratio, x and y vary directly.

b. Constant of variation $= \frac{8}{1} = 8$.

c. Write a proportion that includes both variables and the constant of variation:

$$\frac{x}{y} = \frac{1}{8}$$

$$y = 8x$$ Cross multiply to obtain the formula.

d. Substitute the known value in the equation written in **b**. Solve the equation.

When $x = 10$, find y.	When $y = 1{,}600$, find x.
$y = 8x$	$y = 8x$
$y = 8(10)$	$1{,}600 = 8x$
$y = 80$	$200 = x$

Answers **a.** The variables vary directly because the ratio of each pair is the same constant.

b. 8

c. $y = 8x$

d. When $x = 10, y = 80$; when $y = 1{,}600, x = 200$.

EXAMPLE 3

There are about 90 calories in 20 grams of a cheese. Reggie ate 70 grams of this cheese. About how many calories were there in the cheese she ate if the number of calories varies directly as the weight of the cheese?

Solution Let x = number of calories in 70 grams of cheese.

$$\frac{\text{number of calories}}{\text{number of grams of cheese}} = \frac{90}{20}$$

$$\frac{x}{70} = \frac{90}{20}$$

$$20x = 90(70)$$

$$20x = 6{,}300$$

$$x = 315$$

Answer There were about 315 calories in 70 grams of the cheese.

EXERCISES

Writing About Mathematics

1. On a cross-country trip, Natasha drives at an average speed of 65 miles per hour. She says that each day, her driving time and the distance that she travels are directly proportional. Do you agree with Natasha? Explain why or why not.

2. The cost of parking at the Center City Parking Garage is \$5.50 for the first hour or part of an hour and \$2.75 for each additional half hour or part of a half hour. The maximum cost for 24 hours is \$50. Does the cost of parking vary directly as the number of hours? Explain your answer.

Developing Skills

In 3–11, in each case one value is given for each of two variables that *vary directly*. Find the constant of variation.

3. $x = 12, y = 3$
4. $d = 120, t = 3$
5. $y = 2, z = 18$
6. $P = 12.8, s = 3.2$
7. $t = 12, n = 8$
8. $I = 51, t = 6$
9. $s = 88, t = 110$
10. $A = 212, P = 200$
11. $r = 87, s = 58$

In 12–17, tell, in each case, whether one variable varies directly as the other. If it does, express the relation between the variables by means of a formula.

12.

P	3	6	9
s	1	2	3

13.

n	3	4	5
c	6	8	10

14.

x	4	5	6
y	6	8	10

15.

t	1	2	3
d	20	40	60

16.

x	2	3	4
y	−6	−9	−12

17.

x	1	2	3
y	1	4	9

In 18–20, in each case one variable varies directly as the other. Write the formula that relates the variables and find the missing numbers.

18.

h	1	2	?
A	5	?	25

19.

h	4	8	?
S	6	?	15

20.

l	2	8	?
w	1	?	7

In 21–24, state whether the relation between the variables in each equation is a direct variation. In each case, give a reason for your answer.

21. $R + T = 80$ **22.** $15T = D$ **23.** $\frac{e}{i} = 20$ **24.** $bh = 36$

25. $C = 7n$ is a formula for the cost of n articles that sell for $7 each.

a. How do C and n vary?

b. How will the cost of nine articles compare with the cost of three articles?

c. If n is doubled, what change takes place in C?

26. $A = 12l$ is a formula for the area of any rectangle whose width is 12.

a. Describe how A and l vary.

b. How will the area of a rectangle whose length is 8 inches compare with the area of a rectangle whose length is 4 inches?

c. If l is tripled, what change takes place in A?

27. The variable d varies directly as t. If $d = 520$ when $t = 13$, find d when $t = 9$.

28. Y varies directly as x. If $Y = 35$ when $x = -5$, find Y when $x = -20$.

29. A varies directly as h. $A = 48$ when $h = 4$. Find h when $A = 36$.

30. N varies directly as d. $N = 10$ when $d = 8$. Find N when $d = 12$.

Applying Skills

In 31–48, the quantities vary directly. Solve algebraically.

31. If 3 pounds of apples cost $0.89, what is the cost of 15 pounds of apples at the same rate?

32. If four tickets to a show cost $17.60, what is the cost of seven such tickets?

33. If $\frac{1}{2}$ pound of meat sells for $3.50, how much meat can be bought for $8.75?

34. Willis scores an average of 7 foul shots in every 10 attempts. At the same rate, how many shots would he score in 200 attempts?

35. There are about 60 calories in 30 grams of canned salmon. About how many calories are there in a 210-gram can?

36. There are 81 calories in a slice of bread that weighs 30 grams. How many calories are there in a loaf of this bread that weighs 600 grams?

37. There are about 17 calories in three medium-size shelled peanuts. Joan ate 30 such peanuts. How many calories were there in the peanuts she ate?

38. A train traveled 90 miles in $1\frac{1}{2}$ hours. At the same rate, how long will the train take to travel 330 miles?

39. The weight of 20 meters of copper wire is 0.9 kilograms. Find the weight of 170 meters of the same wire.

40. A recipe calls for $1\frac{1}{2}$ cups of sugar for a 3-pound cake. How many cups of sugar should be used for a 5-pound cake?

41. In a certain concrete mixture, the ratio of cement to sand is 1 : 4. How many bags of cement would be used with 100 bags of sand?

42. The owner of a house that is assessed for $12,000 pays $960 in realty taxes. At the same rate, what should be the realty tax on a house assessed for $16,500?

43. The scale on a map is given as 5 centimeters to 3.5 kilometers. How far apart are two towns if the distance between these two towns on the map is 8 centimeters?

44. David received $8.75 in dividends on 25 shares of a stock. How much should Marie receive in dividends on 60 shares of the same stock?

45. A picture $3\frac{1}{4}$ inches long and $2\frac{1}{8}$ inches wide is to be enlarged so that its length will become $6\frac{1}{2}$ inches. What will be the width of the enlarged picture?

46. An 11-pound turkey costs $9.79. At this rate, find:

a. the cost of a 14.4-pound turkey, rounded to the nearest cent.

b. the cost of a 17.5-pound turkey, rounded to the nearest cent.

c. the price per pound at which the turkeys are sold.

d. the largest size turkey, to the *nearest tenth* of a pound, that can be bought for $20 or less.

47. If a man can buy p kilograms of candy for d dollars, represent the cost of n kilograms of this candy.

48. If a family consumes q liters of milk in d days, represent the amount of milk consumed in h days.

6-6 PERCENT AND PERCENTAGE PROBLEMS

Base, Rate, and Percent

Problems dealing with discounts, commissions, and taxes involve percents. A **percent**, which is a *ratio* of a number to 100, is also called a *rate*. Here, the word *rate* is treated as a comparison of a quantity to the whole. For example, 8% (read as 8 percent) is the ratio of 8 to 100, or $\frac{8}{100}$. A percent can be expressed as a fraction or as a decimal:

$$8\% = \frac{8}{100} = 0.08$$

If an item is taxed at a rate of 8%, then a $50 pair of jeans will cost an additional $4 for tax. Here, three quantities are involved.

1. The **base**, or the sum of money being taxed, is $50.
2. The **rate**, or the rate of tax, is 8% or 0.08 or $\frac{8}{100}$.
3. The **percentage**, or the amount of tax being charged, is $4.

These three related terms may be written as a proportion or as a formula:

As a proportion	*As a formula*
$\frac{\text{percentage}}{\text{base}} = \text{rate}$	$\text{base} \times \text{rate} = \text{percentage}$
For example:	For example:
$\frac{4}{50} = \frac{8}{100}$ or $\frac{4}{50} = 0.08$	$50 \times \frac{8}{100} = 4$ or $50 \times 0.08 = 4$

Just as we have seen two ways to look at this problem involving sales tax, we will see more than one approach to every percentage problem. Note that when we calculate using percent, we always use the fraction or decimal form of the percent.

Percent of Error

When we use a measuring device such as a ruler to obtain a measurement, the accuracy and precision of the measure is dependent on the type of instrument used and the care with which it is used. **Error** is the absolute value of the difference between a value found experimentally and the true theoretical value. For example, when the length and width of a rectangle are 13 inches and 84 inches, the true length of the diagonal, found by using the Pythagorean Theorem, is 85 inches. A student drew this rectangle and, using a ruler, found the measure of the diagonal to be $84\frac{7}{8}$ inches. The error of measurement would be

$85 - 84\frac{7}{8}$ or $\frac{1}{8}$ inches. The **percent of error** or is the ratio of the error to the true value, written as a percent.

$$\textbf{Percent of error} = \frac{|\text{measured value} - \text{true value}|}{\text{true value}} \times 100\%$$

In the example above, the percent of error is

$$\frac{\frac{1}{8}}{85} = \frac{1}{8} \div 85 = \frac{1}{8} \times \frac{1}{85} = \frac{1}{680} \approx 0.001470588 \approx 0.15\%$$

Note: The **relative error** is simply the percent of error written as a *decimal*.

EXAMPLE 1

Find the amount of tax on a \$60 radio when the tax rate is 8%.

Solution

METHOD 1 Use the proportion: $\frac{\text{percentage}}{\text{base}} = \text{rate}$.

Let t = the percentage or amount of tax.

$$\begin{aligned} \frac{\text{amount of tax}}{\text{base}} &= \frac{8}{100} \\ \frac{t}{60} &= \frac{8}{100} \\ 100t &= 480 \\ t &= 4.80 \end{aligned}$$

The tax is \$4.80.

METHOD 2 Use the formula: base × rate = percentage.

Let t = percentage or amount of tax.

Change 8% to a fraction.	Change 8% to a decimal.
base × rate = percentage	base × rate = percentage
$60 \times 8\% = t$	$60 \times 8\% = t$
$60 \times \frac{8}{100} = t$	$60 \times 0.08 = t$
$\frac{480}{100} = t$	$4.8 = t$
$4.8 = t$	

Whether the fraction or the decimal form of 8% is used, the tax is \$4.80.

Answer The tax is \$4.80.

EXAMPLE 2

During a sale, a store offers a discount of 25% off any purchase. What is the regular price of a dress that a customer purchased for \$73.50?

Solution The rate of the discount is 25%.

Therefore the customer paid (100 − 25)% or 75% of the regular price.

The percentage is given as \$73.50, and the base is not known.

Let n = the regular price, or base.

METHOD 1 Use the proportion.

$$\frac{\text{percentage}}{\text{base}} = \text{rate}$$
$$\frac{73.50}{n} = \frac{75}{100}$$
$$75n = 7{,}350$$
$$n = 98$$

Check

If 25% of 98 is subtracted from 98 does the difference equal 73.50?

$0.25 \times 98 = 24.50$

$98 - 24.50 = 73.50$ ✔

METHOD 2 Use the formula.

$$\text{base} \times \text{rate} = \text{percentage}$$
$$n \times 75\% = 73.50$$

Use fractions

$$n \times \frac{75}{100} = 73.50$$
$$n \times \frac{75}{100} \times \frac{100}{75} = 73.50 \times \frac{100}{75}$$
$$n = 98$$

Use decimals

$$n \times 0.75 = 73.50$$
$$\frac{0.75n}{0.75} = \frac{73.50}{0.75}$$
$$n = 98$$

Answer The regular price of the dress was \$98.

Alternative Solution Let n = the regular price of the dress.

Then, $0.25n$ = the discount.

The price of the dress minus the discount is the amount the customer paid.

$$n \quad - \quad 0.25n \quad = \quad 73.50$$

$$n - 0.25n = 73.50$$
$$1.00n - 0.25n = 73.50$$
$$0.75n = 73.50$$
$$\frac{0.75n}{0.75} = \frac{73.50}{0.75}$$
$$n = 98$$

The check is the same as that shown for Method 1.

Answer The regular price of the dress was \$98.

Percent of Increase or Decrease

A **percent of increase or decrease** gives the ratio of the amount of increase or decrease to the original amount. A sales tax is a percent of increase on the cost of a purchase. A discount is a percent of decrease on the regular price of a purchase. To find the percent of increase or decrease, find the difference between the original amount and the new amount. The original amount is the base, the absolute value of the difference is the percentage, and the percent of increase or decrease is the rate.

$$\text{Percent of increase or decrease} = \frac{|\text{original amount} - \text{new amount}|}{\text{original amount}} \times 100\%$$

EXAMPLE 3

Last year Marisa's rent was \$600 per month. This year, her rent increased to \$630 per month. What was the percent of increase in her rent?

Solution The original rent was \$600.

The new rent was \$630.

The amount of increase was $|\$600 - \$630| = \$30$.

$$\text{Percent of increase} = \frac{30}{600} = \frac{1}{20} = 0.05$$

Change 0.05 to a percent: $0.05 = 5\%$

Answer The percent of increase is 5%.

EXAMPLE 4

A store reduced the price of a television from \$840 to \$504. What was the percent of decrease in the price of the television?

Solution

$$\text{Original price} = \$840$$

$$\text{New price} = \$504$$

$$\text{Amount of decrease} = |\$840 - \$504| = \$336$$

$$\text{Percent of decrease} = \frac{336}{840} = 0.4$$

Change 0.4 to a percent: $0.4 = 40\%$

Answer The percent of decrease was 40%.

EXERCISES

Writing About Mathematics

1. Callie said that two decimal places can be used in place of the percent sign. Therefore, 3.6% can be written as 0.36. Do you agree with Callie. Explain why or why not.

2. If Ms. Edwards salary was increased by 4%, her current salary is what percent of her salary before the increase? Explain your answer.

Developing Skills

In 3–11, find each indicated percentage.

3. 2% of 36
4. 6% of 150
5. 15% of 48
6. 2.5% of 400
7. 60% of 56
8. 100% of 7.5
9. $12\frac{1}{2}$% of 128
10. $33\frac{1}{3}$% of 72
11. 150% of 18

In 12–19, find each number or base.

12. 20 is 10% of what number?
13. 64 is 80% of what number?
14. 8% of what number is 16?
15. 72 is 100% of what number?
16. 125% of what number is 45?
17. $37\frac{1}{2}$% of what number is 60?
18. $66\frac{2}{3}$% of what number is 54?
19. 3% of what number is 1.86?

In 20–27, find each percent.

20. 6 is what percent of 12?
21. 9 is what percent of 30?
22. What percent of 10 is 6?
23. What percent of 35 is 28?
24. 5 is what percent of 15?
25. 22 is what percent of 22?
26. 18 is what percent of 12?
27. 2 is what percent of 400?

Applying Skills

28. A newspaper has 80 pages. If 20 of the 80 pages are devoted to advertising, what percent of the newspaper consists of advertising?

29. A test was passed by 90% of a class. If 27 students passed the test, how many students are in the class?

30. Marie bought a dress that was marked \$24. The sales tax is 8%.

a. Find the sales tax.

b. Find the total amount Marie had to pay.

31. There were 120 planes on an airfield. If 75% of the planes took off for a flight, how many planes took off?

32. One year, the Ace Manufacturing Company made a profit of \$480,000. This represented 6% of the volume of business for the year. What was the volume of business for the year?

33. The price of a new motorcycle that Mr. Klein bought was \$5,430. Mr. Klein made a down payment of 15% of the price of the motorcycle and arranged to pay the rest in installments. How much was his down payment?

34. How much silver is in 75 kilograms of an alloy that is 8% silver?

35. In a factory, 54,650 parts were made. When they were tested, 4% were found to be defective. How many parts were good?

36. A baseball team won 9 games, which was 60% of the total number of games the team played. How many games did the team play?

37. The regular price of a sweater is \$40. The sale price of the sweater is \$34. What is the percent of decrease in the price?

38. A businessman is required to collect an 8% sales tax. One day, he collected \$280 in taxes. Find the total amount of sales he made that day.

39. A merchant sold a stereo speaker for \$150, which was 25% above its cost to her. Find the cost of the stereo speaker to the merchant.

40. Bill bought a wooden chess set at a sale. The original price was \$120; the sale price was \$90. What was the percent of decrease in the price?

41. If the sales tax on \$150 is \$7.50, what is the percent of the sales tax?

42. Mr. Taylor took a 2% discount on a bill. He paid the balance with a check for \$76.44. What was the original amount of the bill?

43. Mrs. Sims bought some stock for \$2,250 and sold the stock for \$2,520. What was the percent increase in the value of the stock?

44. When Sharon sold a vacuum cleaner for \$220, she received a commission of \$17.60. What was the rate of commission?

45. On the first day of a sale, a camera was reduced by \$8. This represented 10% of the original price. On the last day of the sale, the camera was sold for 75% of the original price. What was the final selling price of the camera?

46. The regular ticketed prices of four items at Grumbell's Clothier are as follows: coat, \$139.99; blouse, \$43.99; shoes, \$89.99; jeans, \$32.99.

a. These four items were placed on sale at 20% off the regular price. Find, correct to the nearest cent, the sale price of each of these four items.

b. Describe two different ways to find the sale prices.

47. At Relli's Natural Goods, all items are being sold today at 30% off their regular prices. However, customers must still pay an 8% tax on these items. Edie, a good-natured owner, allows each customer to choose one of two plans at this sale:

Plan 1. Deduct 30% of the cost of all items, then add 8% tax to the bill.

Plan 2. Add 8% tax to the cost of all items, then deduct 30% of this total.

Which plan if either, saves the customer more money? Explain why.

48. In early March, Phil Kalb bought shares of stocks in two different companies.

Stock ABC rose 10% in value in March, then decreased 10% in April.

Stock XYZ fell 10% in value in March, then rose 10% in April.

What percent of its original price is each of these stocks now worth?

49. A dairy sells milk in gallon containers. The containers are filled by machine and the amount of milk may vary slightly. A quality control employee selects a container at random and makes an accurate measure of the amount of milk as 16.25 cups. Find the percent of error to the nearest tenth of a percent.

50. A carpenter measures the length of a board as 50.5 centimeters. The exact measure of the length was 50.1 centimeters. Find the percent of error in the carpenter's measure to the nearest tenth of a percent.

51. A 5-pound weight is placed on a gymnasium scale. The scale dial displayed $5\frac{1}{2}$ pounds. If the scale is consistently off by the same percentage, how much does an athlete weigh, to the nearest tenth of a pound, if his weight displayed on this scale is 144 pounds?

52. Isaiah answered 80% of the questions correctly on the math midterm, and 90% of the questions correctly on the math final. Can you conclude that he answered 85% of *all* the questions correctly (the average of 80% and 90%)? Justify your answer or give a counter-example.

53. In January, Amy bought shares of stocks in two different companies. By the end of the year, shares of the first company had gone up by 12% while shares of the second company had gone up by 8%. Did Amy gain a total of $12\% + 8\% = 20\%$ in her investments? Explain why or why not.

6-7 CHANGING UNITS OF MEASURE

The weight and dimensions of a physical object are expressed in terms of units of measure. In applications, it is often necessary to change from one unit of measure to another by a process called **dimensional analysis**. To do this, we multiply by a fraction whose numerator and denominator are equal measures in two different units so that, in effect, we are multiplying by the identity element, 1. For example, since 100 centimeters and 1 meter are equal measures:

$$\frac{100\text{ cm}}{1\text{ m}} = \frac{100\text{ cm}}{100\text{ cm}} = 1$$

To change 4.25 meters to centimeters, multiply by $\frac{100\text{ cm}}{1\text{ m}}$.

$$4.25\text{ m} = 4.25\text{ m} \times \frac{100\text{ cm}}{1\text{ m}}$$
$$= 425\text{ cm}$$

$$\frac{1\text{ m}}{100\text{ cm}} = \frac{1\text{ m}}{1\text{ m}} = 1$$

To change 75 centimeters to meters, multiply by $\frac{1\text{ m}}{100\text{ cm}}$.

$$75\text{ cm} = 75\text{ cm} \times \frac{1\text{ m}}{100\text{ cm}}$$
$$= \frac{75}{100}\text{ m}$$
$$= 0.75\text{ m}$$

Note that in each case, the fraction was chosen so that the given unit of measure occurred in the denominator and could be "cancelled" leaving just the unit of measure that we wanted in the result.

Sometimes it is necessary to use more than one fraction to change to the required unit. For example, if we want to change 3.26 feet to centimeters and know that 1 foot = 12 inches and that 1 inch = 2.54 centimeters, it will be necessary to first use the fraction $\frac{12\text{ in.}}{1\text{ ft}}$ to change feet to inches.

$$3.26\text{ ft} = 3.26\text{ ft} \times \frac{12\text{ in.}}{1\text{ ft}} = \frac{39.12}{1}\text{ in.} = 39.12\text{ in.}$$

Then use the fraction $\frac{2.54\text{ cm}}{1\text{ in.}}$ to change inches to centimeters

$$39.12\text{ in.} = 39.12\text{ in.} \times \frac{2.54\text{ cm}}{1\text{ in.}} = \frac{99.3648}{1}\text{ cm} = 99.3648\text{ cm}$$

This answer, rounded to the nearest tenth, can be expressed as 99.4 centimeters.

EXAMPLE 1

If there are 5,280 feet in a mile, find, to the nearest hundredth, the number of miles in 1,200 feet.

Solution

How to Proceed

(1) Write a fraction equal to 1 with the required unit in the numerator and the given unit in the denominator: $\frac{1\text{ mi}}{5{,}280\text{ ft}}$

(2) Multiply the given measure by the fraction written in step 1:

$$1{,}200 \text{ ft} = 1{,}200 \cancel{\text{ft}} \times \frac{1 \text{ mi}}{5{,}280 \cancel{\text{ft}}}$$

$$= \frac{1{,}200}{5{,}280} \text{ mi}$$

$$= 0.2\overline{27} \text{ mi}$$

(3) Round the answer to the nearest hundredth:

$$1{,}200 \text{ ft} = 0.23 \text{ mi}$$

Answer 0.23 mi

EXAMPLE 2

In France, apples cost 4.25 euros per kilogram. In the United States, apples cost $1.29 per pound. If the currency exchange rate is 0.95 euros for 1 dollar, in which country are apples more expensive?

Solution Recall that "per" indicates division, that is, 4.25 euros per kilogram can be written as $\frac{4.25 \text{ euros}}{1 \text{ kilogram}}$ and $1.29 per pound as $\frac{1.29 \text{ dollars}}{1 \text{ pound}}$.

(1) Change euros in euros per kilogram to dollars. Use $\frac{1 \text{ dollar}}{0.95 \text{ euros}}$, a fraction equal to 1.

$$\frac{4.25 \cancel{\text{euros}}}{1 \text{ kilogram}} \times \frac{1 \text{ dollar}}{0.95 \cancel{\text{euros}}} = \frac{4.25 \text{ dollars}}{0.95 \text{ kilograms}}$$

(2) Now change kilograms in dollars per kilogram to pounds. One pound equals 0.454 kilograms. Since kilograms is in the denominator, use the fraction with kilograms in the numerator.

$$\frac{4.25 \text{ dollars}}{0.95 \cancel{\text{kilograms}}} \times \frac{0.454 \cancel{\text{kilogram}}}{1 \text{ pound}} = \frac{4.25(0.454) \text{ dollars}}{0.95(1) \text{ pounds}}$$

(3) Use a calculator for the computation.

(4) The number in the display, 2.031052632, is the cost of apples in France in dollars per pound. Round the number in the display to the nearest cent: $2.03

(5) Compare the cost of apples in France ($2.03 per pound) to the cost of apples in the United States ($1.29 per pound).

Answer Apples are more expensive in France.

EXAMPLE 3

Change 60 miles per hour to feet per second.

Solution Use dimensional analysis to change the unit of measure in the given rate to the required unit of measure.

(1) Write 60 miles per hour as a fraction: $\frac{60\text{ mi}}{1\text{ hr}}$

(2) Change miles to feet. Multiply by a ratio with miles in the denominator to cancel miles in the numerator: $= \frac{60\text{ mi}}{1\text{ hr}} \times \frac{5{,}280\text{ ft}}{1\text{ mi}}$

(3) Change hours to minutes. Multiply by a ratio with hours in the numerator to cancel hours in the denominator: $= \frac{60(5{,}280)\text{ ft}}{1\text{ hr}} \times \frac{1\text{ hr}}{60\text{ min}}$

(4) Change minutes to seconds. Multiply by a ratio with minutes in the numerator to cancel minutes in the denominator. $= \frac{60(5{,}280)\text{ ft}}{60\text{ min}} \times \frac{1\text{ min}}{60\text{ sec}}$

(5) Compute and simplify: $= \frac{60(5{,}280)\text{ ft}}{60(60)\text{ sec}} = \frac{88\text{ ft}}{1\text{ sec}}$

Alternative Solution Write the ratios in one expression and compute on a calculator.

$$\frac{60\text{ mi}}{1\text{ hr}} \times \frac{5{,}280\text{ ft}}{1\text{ mi}} \times \frac{1\text{ hr}}{60\text{ min}} \times \frac{1\text{ min}}{60\text{ sec}} = \frac{60(5{,}280)\text{ ft}}{60(60)\text{ sec}}$$

Answer 60 miles per hour = 88 feet per second.

EXERCISES

Writing About Mathematics

1. Sid cannot remember how many yards there are in a mile but knows that there are 5,280 feet in a mile and 3 feet in a yard. Explain how Sid can find the number of yards in a mile.

2. A recipe uses $\frac{3}{8}$ of a cup of butter. Abigail wants to use tablespoons to measure the butter and knows that 4 tablespoons equals $\frac{1}{4}$ cup. Explain how Abigail can find the number of tablespoons of butter needed for her recipe.

Developing Skills

In 3–16: **a.** write, in each case, the fraction that can be used to change the given units of measure, **b.** find the indicated unit of measure.

3. Change 27 inches to feet.

4. Change 175 centimeters to meters.

5. Change 40 ounces to pounds.

6. Change 7,920 feet to miles.

7. Change 850 millimeters to centimeters.

8. Change 12 pints to gallons.

9. Change 10.5 yards to inches

10. Change $3\frac{1}{2}$ feet to inches.

11. Change $\frac{4}{3}$ yard to feet.

12. Change 1.5 meters to centimeters.

13. Change 1.2 pounds to ounces.

14. Change 2.5 miles to feet.

15. Change 44 centimeters to millimeters.

16. Change $2\frac{1}{2}$ gallons to quarts.

Applying Skills

17. Miranda needs boards 0.8 meters long for a building project. The boards available at the local lumberyard are 2 feet, 3 feet, and 4 feet long.

a. Express the length, to the nearest hundredth of a foot, of the boards that Miranda needs to buy.

b. Which size board should Miranda buy? Explain your answer.

18. Carlos needs 24 inches of fabric for a pillow that he is making. The fabric store has a piece of material $\frac{3}{4}$ of a yard long that is already cut that he can buy for $5.50. If he has the exact size piece he needs cut from a bolt of fabric, it will cost $8.98 a yard.

a. Is the piece of material that is already cut large enough for his pillow?

b. What would be the cost of having exactly 24 inches of fabric cut?

c. Which is the better buy for Carlos?

19. A highway sign in Canada gives the speed limit as 100 kilometers per hour. Tracy is driving at 62 miles per hour. One mile is approximately equal to 1.6 kilometers.

a. Is Tracy exceeding the speed limit?

b. What is the difference between the speed limit and Tracy's speed in miles per hour?

20. Taylor has a painting for which she paid 1 million yen when she was traveling in Japan. At that time, the exchange rate was 1 dollar for 126 yen. A friend has offered her $2,000 for the painting.

a. Is the price offered larger or smaller than the purchase price?

b. If she sells the painting, what will be her profit or loss, in dollars?

c. Express the profit or loss as a percent of increase or decrease in the price of the painting.

CHAPTER SUMMARY

A **ratio**, which is a comparison of two numbers by division, is the quotient obtained when the first number is divided by a second, nonzero number. Quantities in a ratio are expressed in the same unit of measure before the quotient is found.

▶ **Ratio of a to b:** $\frac{a}{b}$ or $a : b$

A **rate** is a comparison of two quantities that may have *different* units of measure, such as a rate of speed in miles per hour. A rate that has a denominator of 1 is called a **unit rate.**

A **proportion** is an equation stating that two ratios are equal. Standard ways to write a proportion are shown below. In a proportion $a : b = c : d$, the outer terms are called the **extremes**, and the inner terms are the **means**.

▶ **Proportion:** $a : b = c : d$ (means: b, c; extremes: a, d) or $\frac{a}{b} = \frac{c}{d}$ (a extreme, b mean, c mean, d extreme)

In a proportion, the product of the means is equal to the product of the extremes, or alternatively, the **cross products** are equal. This process is also called **cross-multiplication**.

A **direct variation** is a relation between two variables such that their ratio is always the same value, called the **constant of variation.** For example, the diameter of a circle is always twice the radius, so $\frac{d}{r} = 2$ shows a direct variation between d and r with a constant of variation 2.

A **percent (%)**, which is a ratio of a number to 100, is also called a **rate.** Here, the word *rate* is treated as a comparison of a quantity to a whole. In basic formulas, such as those used with discounts and taxes, the **base** and **percentage** are numbers, and the rate is a percent.

$$\frac{\text{percentage}}{\text{base}} = \text{rate} \qquad \textit{or} \qquad \text{base} \times \text{rate} = \text{percentage}$$

The **percent of error** is the ratio of the absolute value of the difference between a measured value and a true value to the true value, expressed as a percent.

$$\text{Percent of error} = \frac{|\text{measured value} - \text{true value}|}{\text{true value}} \times 100\%$$

The **relative error** is the percent of error expressed as a decimal.

The **percent of increase or decrease** is the ratio of the absolute value of the difference between the original value and the new value to the original value.

$$\text{Percent of increase or decrease} = \frac{|\text{original value} - \text{new value}|}{\text{original value}} \times 100\%$$

VOCABULARY

6-1 Ratio • Equivalent ratios • Simplest form • Continued ratio

6-2 Rate • Lowest terms • Unit rate • Unit price

6-4 Proportion • Extremes • Means • Cross-multiplying • Cross product

6-5 Direct variation • Directly proportional • Constant of variation

6-6 Percent • Base • Rate • Percentage • Error • Percent of error • Relative error • Percent increase • Percent decrease

6-7 Dimensional analysis

REVIEW EXERCISES

1. Can an 8 inch by 12 inch photograph be reduced to a 3 inch by 5 inch photograph? Explain why or why not?

2. Karen has a coupon for an additional 20% off the sale price of any dress. She wants to buy a dress that is on sale for 15% off of the original price. Will the original price of the dress be reduced by 35%? Explain why or why not.

In 3–6, express each ratio in simplest form.

3. 30 : 35 **4.** $8w$ to $12w$ **5.** $\frac{3}{8}$ to $\frac{5}{8}$

6. 75 millimeters : 15 centimeters

In 7–9, in each case solve for x and check.

7. $\frac{8}{2x} = \frac{12}{9}$ **8.** $\frac{x}{x + 5} = \frac{1}{2}$ **9.** $\frac{4}{x} = \frac{6}{x + 3}$

10. The ratio of two numbers is 1 : 4, and the sum of these numbers is 40. Find the numbers.

In 11–13, in each case, select the numeral preceding the choice that makes the statement true.

11. In a class of 9 boys and 12 girls, the ratio of the number of girls to the number of students in the class is

(1) 3 : 4 (2) 4 : 3 (3) 4 : 7 (4) 7 : 4

12. The perimeter of a triangle is 45 centimeters, and the lengths of its sides are in the ratio 2 : 3 : 4. The length of the longest side is

(1) 5 cm (2) 10 cm (3) 20 cm (4) 30 cm

13. If $a : x = b : c$, then x equals

(1) $\frac{ac}{b}$ (2) $\frac{bc}{a}$ (3) $ac - b$ (4) $bc - a$

14. Seven percent of what number is 21?

15. What percent of 36 is 45?

16. The sales tax collected on each sale varies directly as the amount of the sale. What is the constant of variation if a sales tax of $0.63 is collected on a sale of $9.00?

17. If 10 paper clips weigh 11 grams, what is the weight in grams of 150 paper clips?

18. Thelma can type 150 words in 3 minutes. At this rate, how many words can she type in 10 minutes?

19. What is the ratio of six nickels to four dimes?

20. On a stormy February day, 28% of the students enrolled at Southside High School were absent. How many students are enrolled at Southside High School if 476 students were absent?

21. After a 5-inch-by-7-inch photograph is enlarged, the shorter side of the enlargement measures 15 inches. Find the length in inches of its longer side.

22. A student who is 5 feet tall casts an 8-foot shadow. At the same time, a tree casts a 40-foot shadow. How many feet tall is the tree?

23. If four carpenters can build four tables in 4 days, how long will it take one carpenter to build one table?

24. How many girls would have to leave a room in which there are 99 girls and 1 boy in order that 98% of the remaining persons would be girls?

25. On an Australian highway, the speed limit was 110 kilometers per hour. A motorist was going 70 miles per hour. (Use 1.6 kilometers = 1 mile)

a. Should the motorist be stopped for speeding?

b. How far over or under the speed limit was the motorist traveling?

26. The speed of light is 3.00×10^5 kilometers per second. Find the speed of light in miles per hour. Use 1.61 kilometers = 1 mile. Write your answer in scientific notation with three significant digits.

27. Which offer, described in the chapter opener on page 207, is the better buy? Does the answer depend on the price of a package of frozen vegetables? Explain.

28. A proposal was made in the state senate to raise the minimum wage from \$6.75 to \$7.15 an hour. What is the proposed percent of increase to the nearest tenth of a percent?

29. In a chemistry lab, a student measured 1.0 cubic centimeters of acid to use in an experiment. The actual amount of acid that the student used was 0.95 cubic centimeters. What was the percent of error in the student's measurement? Give your answer to the nearest tenth of a percent.

Exploration

a. Mark has two saving accounts at two different banks. In the first bank with a yearly interest rate of 2%, he invests \$585. In the second bank with a yearly interest rate of 1.5%, he invests \$360. Mark claims that at the end of one year, he will make a total of 2% + 1.5% = 3.5% on his investments. Is Mark correct?

b. Two different clothing items, each costing \$30, were on sale for 10% off the ticketed price. The manager of the store claims that if you buy both items, you will save a total of 20%. Is the manager correct?

c. In a class of 505 graduating seniors, 59% were involved in some kind of after-school club and 41% played in a sport. The principal of the school claims that 59% + 41% = 100% of the graduating seniors were involved in some kind of after-school activity. Is the principal correct?

d. A certain movie is shown in two versions, the original and the director's cut. However, movie theatres can play only one of the versions. A journalist for XYZ News, reports that since 30% of theatres are showing the director's cut and 60% are showing the original, the movie is playing in 90% of all movie theatres. Is the reporter correct?

e. Based on your answers for parts **a** through **d**, write a rule stating when it makes sense to add percents.

CUMULATIVE REVIEW CHAPTERS 1–6

Part I

Answer all questions in this part. Each correct answer will receive 2 credits. No partial credit will be allowed.

1. When $-4ab$ is subtracted from ab, the difference is
(1) $-3ab$ (2) -4 (3) $5ab$ (4) $-5ab$

2. If $n - 4$ represents an odd integer, the next larger odd integer is
(1) $n - 2$ (2) $n - 3$ (3) $n - 5$ (4) $n - 6$

3. The expression 2.3×10^{-3} is equal to
(1) 230 (2) 2,300 (3) 0.0023 (4) 0.023

4. The product $-2x^3y(-3xy^4)$ is equal to
(1) $-6x^3y^4$ (2) $6x^4y^5$ (3) $-6x^4y^5$ (4) $6x^3y^4$

5. What is the multiplicative inverse of $\frac{3}{2}$?
(1) $-\frac{3}{2}$ (2) $\frac{2}{3}$ (3) $-\frac{2}{3}$ (4) 1

6. If $0.2x + 4 = x - 0.6$, then x equals
(1) -46 (2) -460 (3) 5.75 (4) 0.575

7. The product $-3^4 \times 3^3 =$
(1) 3^7 (2) -3^7 (3) 9^7 (4) -9^7

8. The diameter of a circle whose area is 144π square centimeters is
(1) 24π cm (2) 6 cm (3) 12 cm (4) 24 cm

9. Jeannine paid \$88 for a jacket that was on sale for 20% off the original price. The original price of the jacket was
(1) \$105.60 (2) \$110.00 (3) \$440.00 (4) \$70.40

10. When $a = -2$ and $b = -5$, $a^2 - ab$ equals
(1) -14 (2) 14 (3) -6 (4) 6

Part II

Answer all questions in this part. Each correct answer will receive 2 credits. Clearly indicate the necessary steps, including appropriate formula substitutions, diagrams, graphs, charts, etc. For all questions in this part, a correct numerical answer with no work shown will receive only 1 credit.

11. Find two integers, x and $x + 1$, whose squares differ by 25.

12. Solve and check: $4(2x - 1) = 5x + 5$

Part III

Answer all questions in this part. Each correct answer will receive 3 credits. Clearly indicate the necessary steps, including appropriate formula substitutions, diagrams, graphs, charts, etc. For all questions in this part, a correct numerical answer with no work shown will receive only 1 credit.

13. Rectangle $ABCD$ is separated into three squares as shown in the diagram at the right. The length of DA is 20 centimeters.

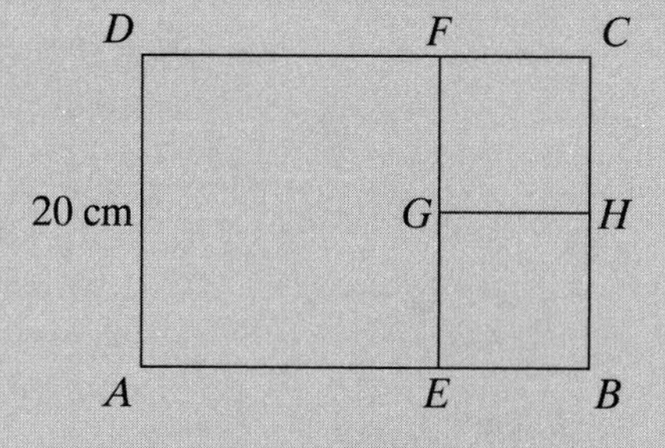

a. Find the measure of $\overline{AB}$.

b. Find the ratio of the area of square $AEFD$ to the area of rectangle $ABCD$.

14. Sam drove a distance of 410 miles in 7 hours. For the first part of the trip his average speed was 40 miles per hour and for the remainder of the trip his average speed was 60 miles per hour. How long did he travel at each speed?

Part IV

Answer all questions in this part. Each correct answer will receive 4 credits. Clearly indicate the necessary steps, including appropriate formula substitutions, diagrams, graphs, charts, etc. For all questions in this part, a correct numerical answer with no work shown will receive only 1 credit.

15. Three friends started a part-time business. They plan, each month, to share the profits in the ratio of the number of hours that each worked. During the first month, Rita worked 18 hours, Fred worked 30 hours, and Glen worked 12 hours.

a. Express, in lowest terms, the ratio of the times that they worked during the first month.

b. Find the amount each should receive if the profit for this first month was $540.

c. Find the amount each should receive if the profit for this first month was $1,270.

16. A skating rink is in the form of a rectangle with a semicircle at each end as shown in the diagram. The rectangle is 150 feet long and 64 feet wide. Scott skates around the rink 2.5 feet from the edge.

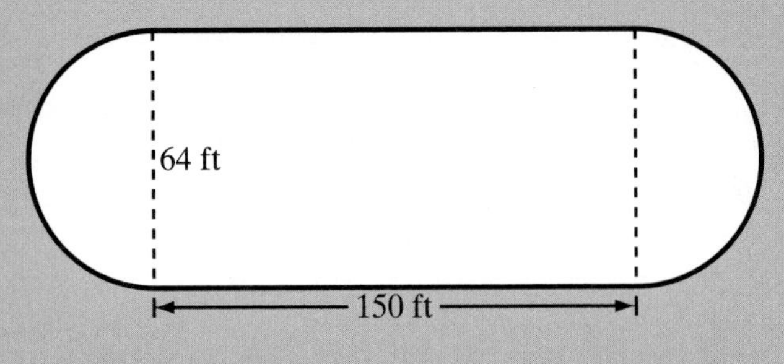

a. Scott skates once around the rink. Find, to the nearest ten feet, the distance that he skated.

b. Scott wants to skate at least 5 miles. What is the smallest number of complete trips around the rink that he must make?

CHAPTER

7

GEOMETRIC FIGURES, AREAS, AND VOLUMES

A carpenter is building a deck on the back of a house. As he works, he follows a plan that he made in the form of a drawing or blueprint. His blueprint is a model of the deck that he is building.

He begins by driving stakes into the ground to locate corners of the deck. Between each pair of stakes, he stretches a cord to indicate the edges of the deck. On the blueprint, the stakes are shown as points and the cords as segments of straight lines as shown in the sketch.

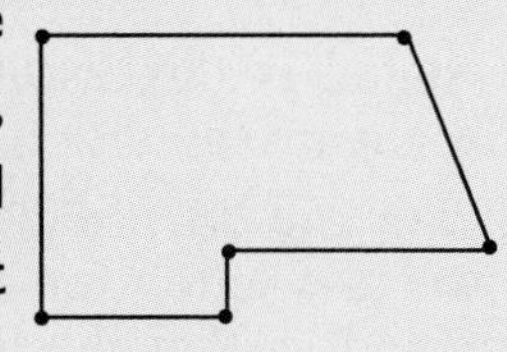

At each corner, the edges of the deck meet at an angle that can be classified according to its size.

Geometry combines points, lines, and planes to model the world in which we live. An understanding of geometry enables us to understand relationships involving the sizes of physical objects and the magnitude and direction of the forces that interact in daily life.

In this chapter, you will review some of the information that you already know to describe angles and apply this information to learn more about geometry.

CHAPTER TABLE OF CONTENTS

7-1 POINTS, LINES, AND PLANES

Undefined Terms

We ordinarily define a word by using a simpler term. The simpler term can be defined by using one or more still simpler terms. But this process cannot go on endlessly; there comes a time when the definition must use a term whose meaning is *assumed* to be clear to all people. Because the meaning is accepted without definition, such a term is called an **undefined term**.

In geometry, we use such ideas as *point*, *line*, and *plane*. Since we cannot give satisfactory definitions of these words by using simpler defined words, we will consider them to be undefined terms.

Although *point*, *line*, and *plane* are undefined words, we must have a clear understanding of what they mean. Knowing the properties and characteristics they possess helps us to achieve this understanding.

• A

A **point** indicates a place or position. It has no length, width, or thickness. A point is usually indicated by a dot and named with a capital letter. For example, point A is shown on the left.

A **line** is a set of points. The set of points may form a curved line, a broken line, or a straight line. A **straight line** is a line that is suggested by a stretched string but that extends without end in both directions.

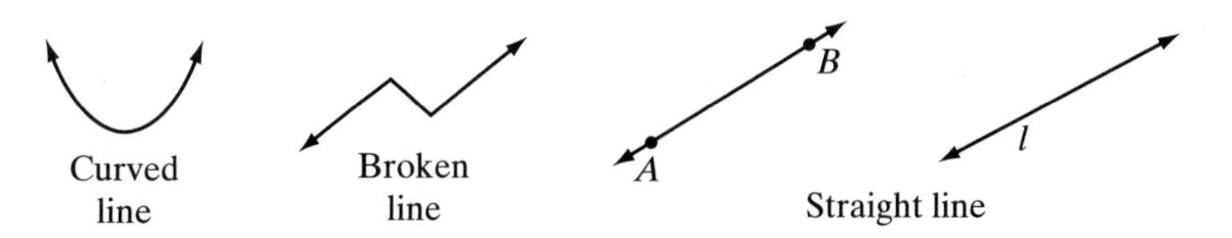

Unless otherwise stated, in this discussion the term *line* will mean *straight* line. A line is named by any two points on the line. For example, the straight line shown above is line AB or line BA, usually written as $\overleftrightarrow{AB}$ or $\overleftrightarrow{BA}$. A line can also be named by one lowercase letter, for example, line l shown above. The arrows remind us that the line continues beyond what is drawn in the diagram.

A **plane** is a set of points suggested by a flat surface. A plane extends endlessly in all directions. A plane may be named by a single letter, as plane P shown on the right.

A plane can also be named by three points of the plane, as plane ABC in the diagram on the right.

Facts About Straight Lines

A statement that is accepted as true without proof is called an **axiom** or a **postulate**. If we examine the three accompanying figures pictured on the next page, we see that it is reasonable to accept the following three statements as postulates:

1. In a plane, an infinite number of straight lines can be drawn through a given point.
2. One and only one straight line can be drawn that contains two given points. (Two points determine a straight line.)
3. In a plane, two different nonparallel straight lines will intersect at only one point.

Line Segments

The undefined terms, point, line, and plane, are used to define other geometric terms. A **line segment** or **segment** is a part of a line consisting of two points called **endpoints** and all points on the line between these endpoints.

At the left is pictured a line segment whose endpoints are points R and S. We use these endpoints to name this segment as segment RS, which may be written as $\overline{RS}$ or $\overline{SR}$.

Recall that the **measure of a line segment** or the **length of a line segment** is the distance between its endpoints. We use a number line to associate a number with each endpoint. Since the coordinate of A is 0 and the coordinate of B is 5, the length of $\overline{AB}$ is $5 - 0$ or $AB = 5$.

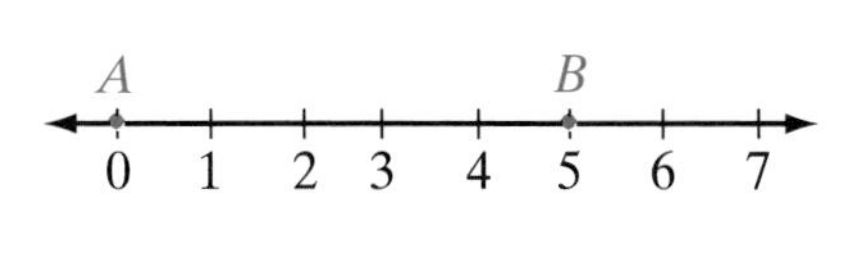

Note: The segment is written as $\overline{AB}$, with a bar over the letters that name the endpoints. The length of the segment is written as AB, with no bar over the letters that name the endpoints.

Half-Lines and Rays

When we choose any point A on a line, the two sets of points that lie on opposite sides of A are called **half-lines**. Note that point A is not part of the half-line. In the diagram below, point A separates $\overleftrightarrow{CD}$ into two half-lines.

The two points B and D belong to the same half-line since A is not a point of $\overline{BD}$. Points B and C, however, do not belong to the same half-line since A is a point of $\overline{CB}$. All points in the same half-line are said to be on the same side of A. We often talk about rays of sunlight, that is, the sun and the path that the sunlight travels to the earth. The ray of sunlight can be thought of as a point and a half-line.

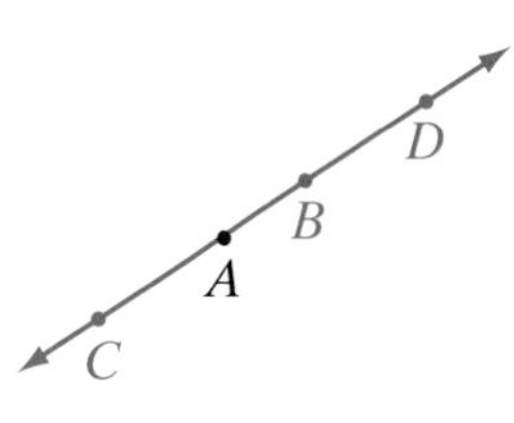

In geometry, a **ray** is a part of a line that consists of a point on the line, called an **endpoint**, and all the points on one side of the endpoint. To name a ray we use two capital letters and an arrow with one arrowhead. The first letter must be the letter that names the endpoint. The second letter is the name of any other point on the ray.

The figure on the right shows ray AB, which is written as $\overrightarrow{AB}$. This ray could also be called ray AC, written as $\overrightarrow{AC}$.

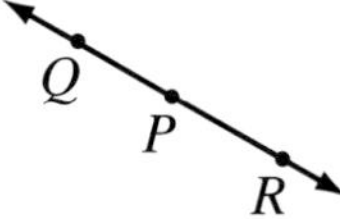

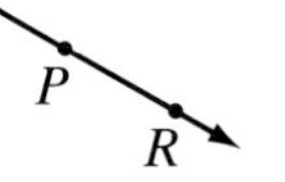
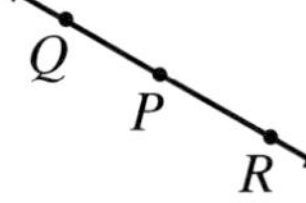

Two rays are called **opposite rays** if they are rays of the same line that have a common endpoint but no other points in common. In the diagram on the left, $\overrightarrow{PQ}$ and $\overrightarrow{PR}$ are opposite rays.

Angles

An **angle** is a set of points that is the union of two rays having the same endpoint. The common endpoint of the two rays is the **vertex** of the angle. The two rays forming the angle are also called the **sides** of the angle.

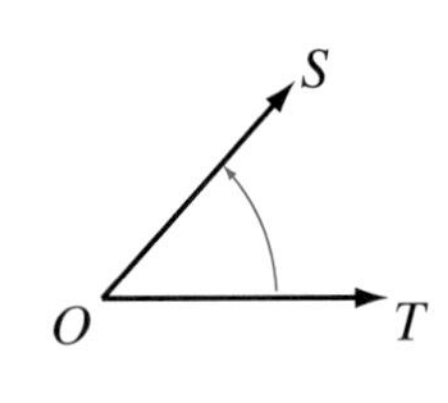

In the diagram on the left, we can think of $\angle TOS$ as having been formed by rotating $\overrightarrow{OT}$ in a counterclockwise direction about O to the position $\overrightarrow{OS}$. The union of the two rays, $\overrightarrow{OT}$ and $\overrightarrow{OS}$, that have the common endpoint O, is $\angle TOS$. Note that when three letters are used to name an angle, the letter that names the vertex is always in the middle. Since $\overrightarrow{OT}$ and $\overrightarrow{OS}$ are the only rays in the diagram that have the common endpoint O, the angle could also have been called $\angle O$.

Measuring Angles

To measure an angle means to determine the number of units of measure it contains. A common standard unit of measure of an angle is the **degree**; 1 degree is written as 1°. A degree is $\frac{1}{360}$ of the sum of all of the distinct angles about a point. In other words, if we think of an angle as having been formed by rotating a ray around its endpoint, then rotating a ray consecutively 360 times in 1-degree increments will result in one complete rotation.

Types of Angles

Angles are classified according to their measures.

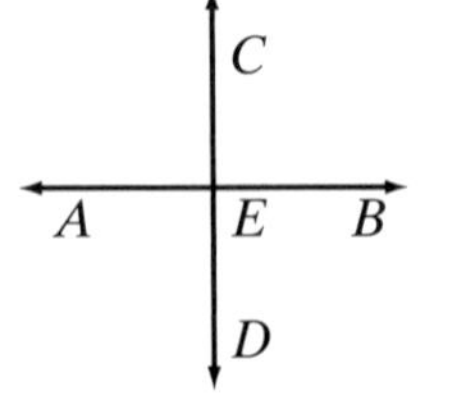

A **right angle** is an angle whose measure is 90°. In the diagram on the left, $\overleftrightarrow{AB}$ and $\overleftrightarrow{CD}$ intersect at E so that the four angles formed are right angles. The measure of each angle is 90° and the sum of the angles about point E is 360°. We can say that

$$m\angle AEC = m\angle CEB = m\angle BED = m\angle DEA = 90.$$

Note that the symbol $m\angle AEC$ is read "the measure of angle AEC." In this book, the angle measure will always be given in degrees and the degree symbol will be omitted when using the symbol "m" to designate measure.

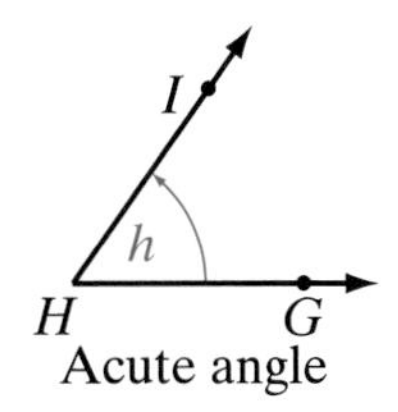

Acute angle

An **acute angle** is an angle whose measure is greater than 0° and less than 90°, that is, the measure of an acute angle is between 0° and 90°. Angle GHI, which can also be called $\angle h$, is an acute angle $(0 < m\angle h < 90)$.

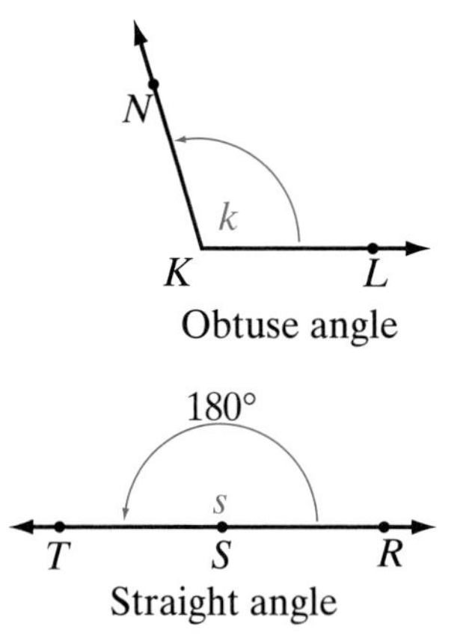

Obtuse angle

Straight angle

An **obtuse angle** is an angle whose measure is greater than 90° but less than 180°; that is, its measure is between 90° and 180°. Angle LKN is an obtuse angle where $90 < m\angle LKN < 180$.

A **straight angle** is an angle whose measure is 180°. Angle RST is a straight angle where $m\angle RST = 180$. A straight angle is the union of two opposite rays and forms a straight line.

Here are three important facts about angles:

1. The measure of an angle depends only on the amount of rotation, not on the pictured lengths of the rays forming the angle.
2. Since every right angle measures 90°, all right angles are equal in measure.
3. Since every straight angle measures 180°, all straight angles are equal in measure.

Perpendicularity

Two lines are **perpendicular** if and only if the two lines intersect to form right angles. The symbol for perpendicularity is $\perp$.

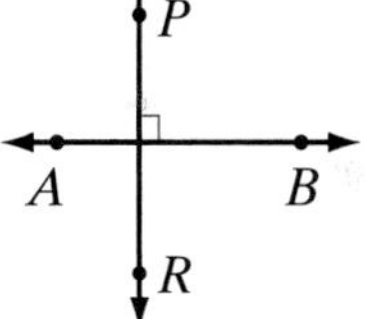

In the diagram, $\overleftrightarrow{PR}$ is perpendicular to $\overleftrightarrow{AB}$, symbolized as $\overleftrightarrow{PR} \perp \overleftrightarrow{AB}$. The symbol ⏋ is used in a diagram to indicate that a right angle exists where the perpendicular lines intersect.

Segments of perpendicular lines that contain the point of intersection of the lines are also perpendicular. In the diagram on the left, $\overline{ST} \perp \overleftrightarrow{CD}$.

EXERCISES

Writing About Mathematics

1. Explain how the symbols AB, $\overline{AB}$, and $\overleftrightarrow{AB}$ differ in meaning.

2. Explain the difference between a half-line and a ray.

Developing Skills

In 3–8, tell whether each angle appears to be acute, right obtuse, or straight.

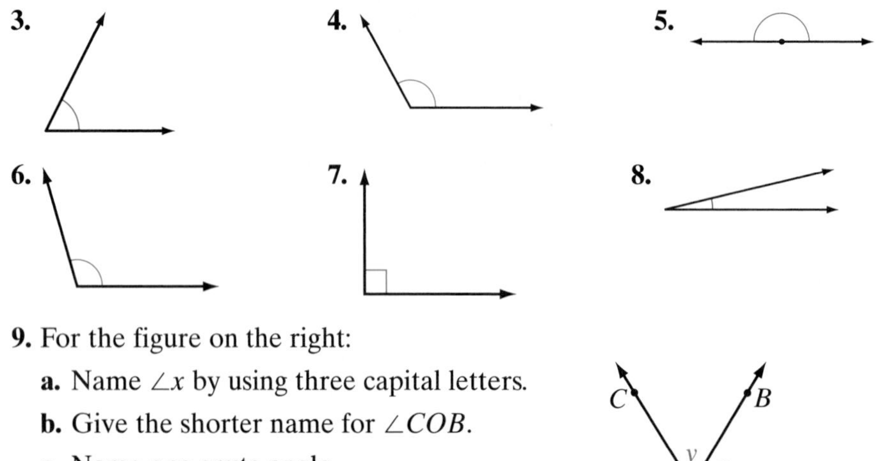

9. For the figure on the right:

a. Name $\angle x$ by using three capital letters.

b. Give the shorter name for $\angle COB$.

c. Name one acute angle.

d. Name one obtuse angle.

Applying Skills

In 10–13, find the number of degrees in the angle formed by the hands of a clock at each given time.

10. 1 P.M. **11.** 4 P.M. **12.** 6 P.M. **13.** 5:30 P.M.

14. At what time do the hands of the clock form an angle of 0°?

15. At what times do the hands of a clock form a right angle?

7-2 PAIRS OF ANGLES

Adjacent Angles

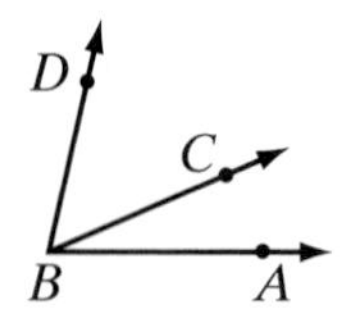

Adjacent angles are two angles in the same plane that have a common vertex and a common side but do not have any interior points in common. In the figure on the left, $\angle ABC$ and $\angle CBD$ are adjacent angles.

Complementary Angles

Two angles are **complementary angles** if and only if the sum of their measures is 90°. Each angle is the *complement* of the other.

In the figures shown below, because

$$m\angle CAB + m\angle FDE = 25 + 65 = 90,$$

$\angle CAB$ and $\angle FDE$ are complementary angles. Also, because

$$m\angle HGI + m\angle IGJ = 53 + 37 = 90,$$

$\angle HGI$ and $\angle IGJ$ are complementary angles.

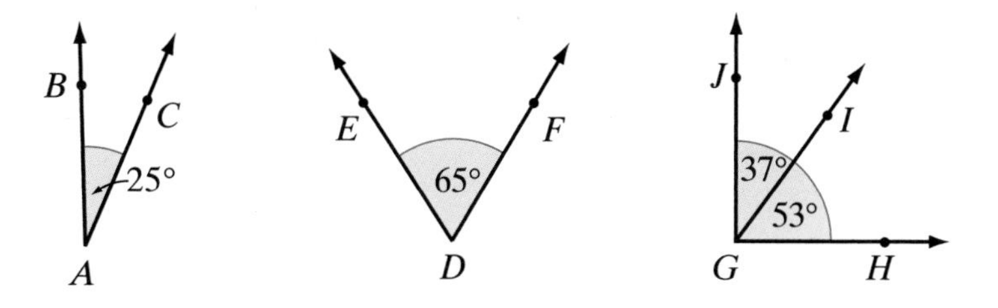

If the measure of an angle is 50°, the measure of its complement is (90 − 50)°, or 40°. In general,

▶ **If the measure of an angle is $x°$, the measure of its complement is $(90 - x)°$.**

Supplementary Angles

Two angles are **supplementary angles** if and only if the sum of their measures is 180°. Each angle is the *supplement* of the other. As shown in the figures below, because

$$m\angle LKM + m\angle ONP = 50 + 130 = 180,$$

$\angle LKM$ and $\angle ONP$ are supplementary angles. Also, $\angle RQS$ and $\angle SQT$ are supplementary angles because

$$m\angle RQS + m\angle SQT = 115 + 65 = 180.$$

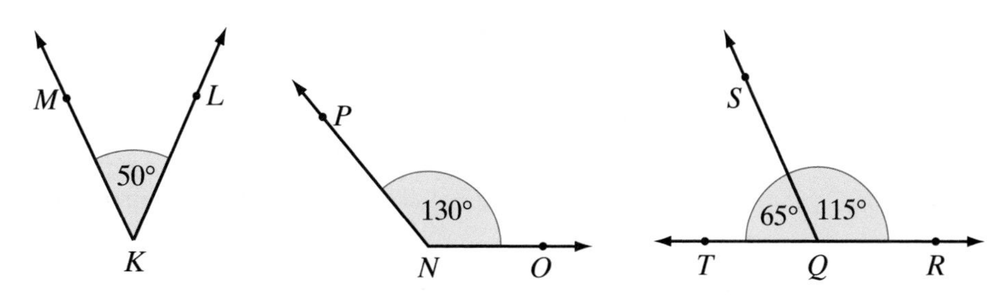

If the measure of an angle is 70°, the measure of its supplement is (180 − 70)°, or 110°. In general,

▶ **If the measure of an angle is $x°$, the measure of its supplement is $(180 - x)°$.**

Linear Pair

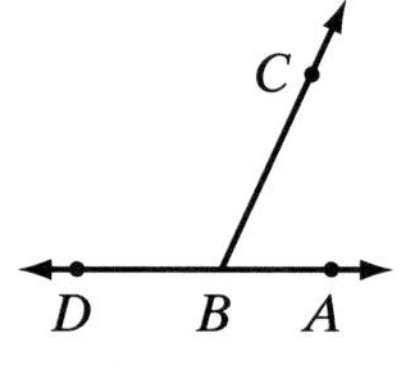

Through two points, D and A, draw a line. Choose any point B on $\overline{DA}$ and any point C not on $\overleftrightarrow{DA}$. Draw $\overrightarrow{BC}$. The adjacent angles formed, $\angle DBC$ and $\angle CBA$, are called a linear pair. A **linear pair of angles** are adjacent angles that are supplementary. The two sides that they do not share in common are opposite rays. The term *linear* tells us that a *line* is used to form this pair. Since the angles are supplementary, if $m\angle DBC = x$, then $m\angle CBA = (180 - x)$.

Vertical Angles

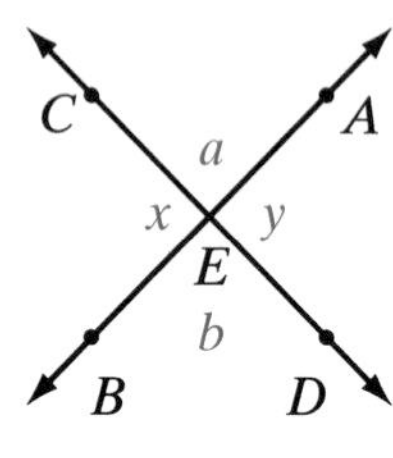

If two straight lines such as $\overleftrightarrow{AB}$ and $\overleftrightarrow{CD}$ intersect at E, then $\angle x$ and $\angle y$ share a common vertex at E but do not share a common side. Angles x and y are a pair of vertical angles. **Vertical angles** are two nonadjacent angles formed by two intersecting lines. In the diagram on the left, angles a and b are another pair of vertical angles.

If two lines intersect, four angles are formed that have no common interior point. In the diagram, $\overleftrightarrow{AB}$ and $\overleftrightarrow{CD}$ intersect at E. There are four linear pairs of angles:

$\angle AED$ and $\angle DEB$ $\quad$ $\angle BEC$ and $\angle CEA$
$\angle DEB$ and $\angle BEC$ $\quad$ $\angle CEA$ and $\angle AED$

The angles of each linear pair are supplementary.

- If $m\angle AED = 130$, then $m\angle DEB = 180 - 130 = 50$.
- If $m\angle DEB = 50$, then $m\angle BEC = 180 - 50 = 130$.

Therefore, $m\angle AED = m\angle BEC$.

DEFINITION

When two angles have equal measures, they are **congruent**.

We use the symbol $\cong$ to represent the phrase "is congruent to." Since $m\angle AED = m\angle BEC$, we can write $\angle AED \cong \angle BEC$, read as "angle AED is congruent to angle BEC." There are different correct ways to indicate angles with equal measures:

1. The angle measures are equal: $m\angle BEC = m\angle AED$

2. The angles are congruent: $\angle BEC \cong \angle AED$

It would not be correct to say that the angles are equal, or that the angle measures are congruent.

Notice that we have just shown that the two vertical angles $\angle BEC$ and $\angle AED$ are congruent. If we were to draw and measure additional pairs of vertical angles, we would find in each case that the vertical angles would be equal in measure. No matter how many examples of a given situation we consider, however, we cannot *assume* that a conclusion that we draw from these examples will always be true. We must *prove* the conclusion. Statements that we prove are called **theorems**. We will use algebraic expressions and properties to write an informal proof of the following statement.

► **If two lines intersect, the vertical angles formed are equal in measure; that is, they are congruent.**

(1) If $\overleftrightarrow{AB}$ and $\overleftrightarrow{CD}$ intersect at E, then $\angle AEB$ is a straight angle whose measure is 180°. Therefore, $m\angle AEC + m\angle CEB = 180$.

(2) Let $m\angle AEC = x$. Then $m\angle CEB = 180 - x$.

(3) Likewise, $\angle CED$ is a straight angle whose measure is 180°. Therefore, $m\angle CEB + m\angle BED = 180$.

(4) Since $m\angle CEB = 180 - x$, then $m\angle BED = 180 - (180 - x) = x$.

(5) Since both $m\angle AEC = x$ and $m\angle BED = x$, then $m\angle AEC = m\angle BED$; that is, $\angle AEC \cong \angle BED$.

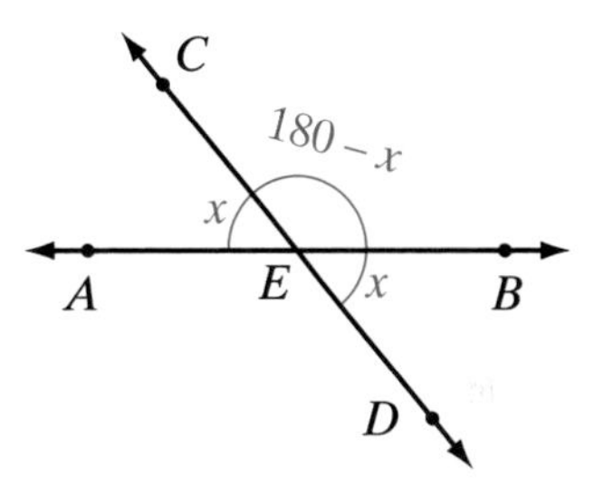

EXAMPLE 1

The measure of the complement of an angle is 4 times the measure of the angle. Find the measure of the angle.

Solution Let x = measure of angle.

Then $4x$ = measure of complement of angle.

The sum of the measures of an angle and its complement is 90°.

$$x + 4x = 90$$
$$5x = 90$$
$$x = 18$$

Check The measure of the first angle is 18°.

The measure of the second angle is $4(18°) = 72°$.

The sum of the measures of the angles is $18° + 72° = 90°$.

Thus, the angles are complementary. ✔

Answer The measure of the angle is 18°.

Note: The unit of measure is very important in the solution of a problem. While it is not necessary to include the unit of measure in each step of the solution, each term in an equation must represent the same unit and the unit of measure must be included in the answer.

EXAMPLE 2

The measure of an angle is 40° more than the measure of its supplement. Find the measure of the angle.

Solution Let x = the measure of the supplement of the angle.

Then $x + 40$ = the measure of the angle.

The sum of the measures of an angle and its supplement is 180°.

$$x + (x + 40) = 180$$
$$2x + 40 = 180$$
$$2x = 140$$
$$x = 70$$
$$x + 40 = 110$$

Check The sum of the measures is 110° + 70° = 180° and 110° is 40° more than 70°. ✔

Answer The measure of the angle is 110°.

EXAMPLE 3

The algebraic expressions $5w - 20$ and $2w + 16$ represent the measures in degrees of a pair of vertical angles.

a. Find the value of w.

b. Find the measure of each angle.

Solution **a.** Vertical angles are equal in measure.

$$5w - 20 = 2w + 16$$
$$3w - 20 = 16$$
$$3w = 36$$
$$w = 12$$

b. $5w - 20 = 5(12) - 20 = 60 - 20 = 40$

$2w + 16 = 2(12) + 16 = 24 + 16 = 40$

Check Since each angle has a measure of 40°, the vertical angles are equal in measure. ✔

Answers **a.** $w = 12$ **b.** The measure of each angle is 40°.

EXERCISES

Writing About Mathematics

1. Show that supplementary angles are always two right angles or an acute angle and an obtuse angle.

2. The measures of three angles are 15°, 26°, and 49°. Are these angles complementary? Explain why or why not.

Developing Skills

In 3–6, answer each of the following questions for an angle with the given measure.

a. What is the measure of the complement of the angle?

b. What is the measure of the supplement of the angle?

c. The measure of the supplement of the angle is how much larger than the measure of its complement?

3. 15° **4.** 37° **5.** 67° **6.** $x°$

In 7–10, $\angle A$ and $\angle B$ are complementary. Find the measure of each angle if the measures of the two angles are represented by the given expressions. Solve the problem algebraically using an equation.

7. $m\angle A = x, m\angle B = 5x$

8. $m\angle A = x, m\angle B = x + 20$

9. $m\angle A = x, m\angle B = x - 40$

10. $m\angle A = y, m\angle B = 2y + 30$

In 11–14, $\angle ABD$ and $\angle DBC$ are supplementary. Find the measure of each angle if the measures of the two angles are represented by the given expressions. Solve the problem algebraically using an equation.

11. $m\angle ABD = x, m\angle DBC = 3x$

12. $m\angle ABD = x, m\angle DBC = x + 80$

13. $m\angle DBC = x, m\angle ABD = x - 30$

14. $m\angle DBC = y, m\angle ABD = \frac{1}{4}y$

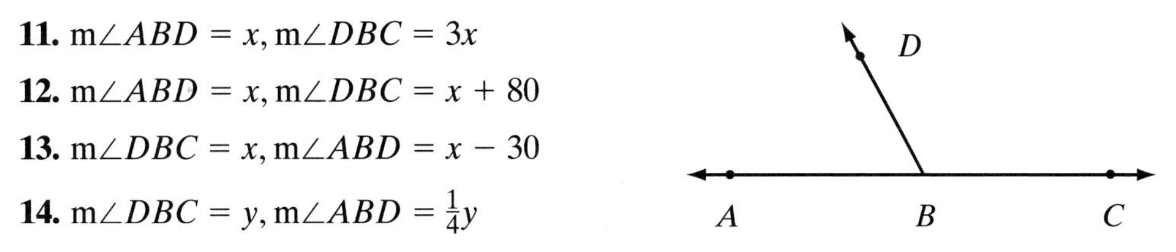

In 15–24, solve each problem algebraically using an equation.

15. Two angles are supplementary. The measure of one angle is twice as large as the measure of the other. Find the number of degrees in each angle.

16. The complement of an angle is 14 times as large as the angle. Find the measure of the complement.

17. The measure of the supplement of an angle is 40° more than the measure of the angle. Find the number of degrees in the supplement.

18. Two angles are complementary. One angle is twice as large as the other. Find the number of degrees in each angle.

19. The measure of the complement of an angle is one-ninth the measure of the angle. Find the measure of the angle.

20. Find the number of degrees in the measure of an angle that is 20° less than 4 times the measure of its supplement.

21. The difference between the measures of two supplementary angles is 80°. Find the measure of the larger of the two angles.

22. The complement of an angle measures 20° more than the angle. Find the number of degrees in the angle.

23. Find the number of degrees in an angle that measures 10° more than its supplement.

24. Find the number of degrees in an angle that measures 8° less than its complement.

25. The supplement of the complement of an acute angle is always:

(1) an acute angle (2) a right angle (3) an obtuse angle (4) a straight angle

In 26–28, $\overleftrightarrow{MN}$ and $\overleftrightarrow{RS}$ intersect at T.

26. If $m\angle RTM = 5x$ and $m\angle NTS = 3x + 10$, find $m\angle RTM$.

27. If $m\angle MTS = 4x - 60$ and $m\angle NTR = 2x$, find $m\angle MTS$.

28. If $m\angle RTM = 7x + 16$ and $m\angle NTS = 3x + 48$, find $m\angle NTS$.

In 29–34, find the measure of each angle named, based on the given information.

29. *Given*: $\overleftrightarrow{EF} \perp \overrightarrow{GH}$; $m\angle EGI = 62$.

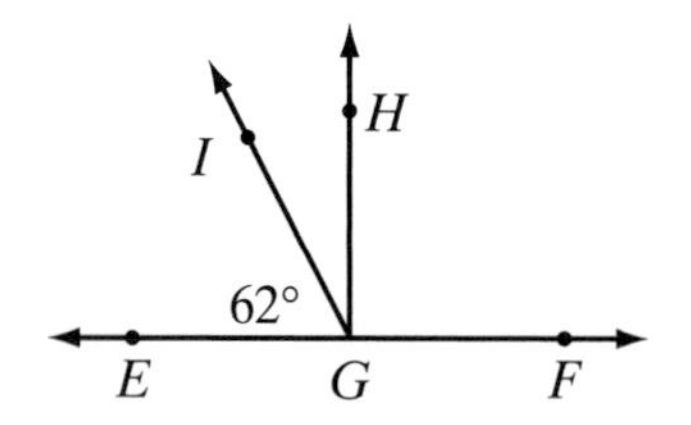

a. Find $m\angle FGH$.

b. Find $m\angle HGI$.

30. *Given*: $\overleftrightarrow{JK} \perp \overrightarrow{LM}$; $\overleftrightarrow{NLO}$ is a line; $m\angle NLM = 48$.

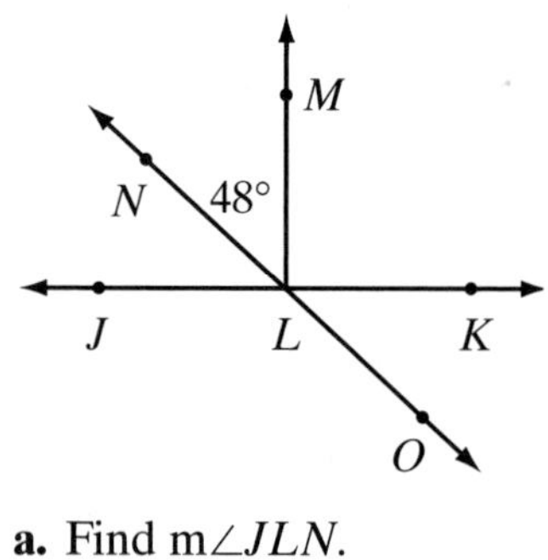

a. Find $m\angle JLN$.

b. Find $m\angle MLK$.

c. Find $m\angle KLO$.

d. Find $m\angle JLO$.

31. *Given*: $\angle GKH$ and $\angle HKI$ are a linear pair; $\overrightarrow{KH} \perp \overrightarrow{KJ}$; $m\angle IKJ = 34$.

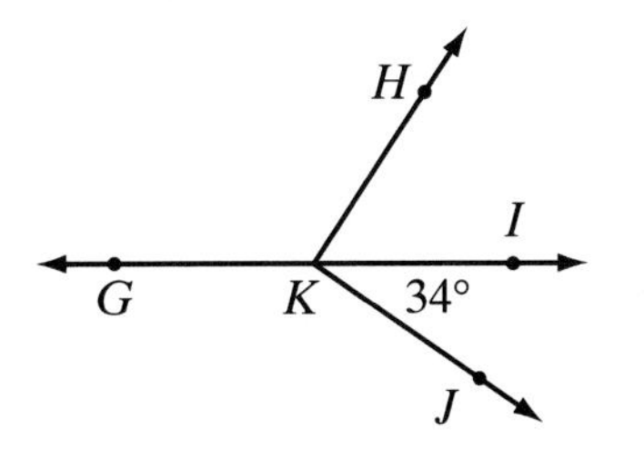

a. Find $m\angle HKI$.

b. Find $m\angle HKG$.

c. Find $m\angle GKJ$.

32. *Given*: $\overrightarrow{MO} \perp \overrightarrow{MP}$; $\overleftrightarrow{LMN}$ is a line; $m\angle PMN = 40$.

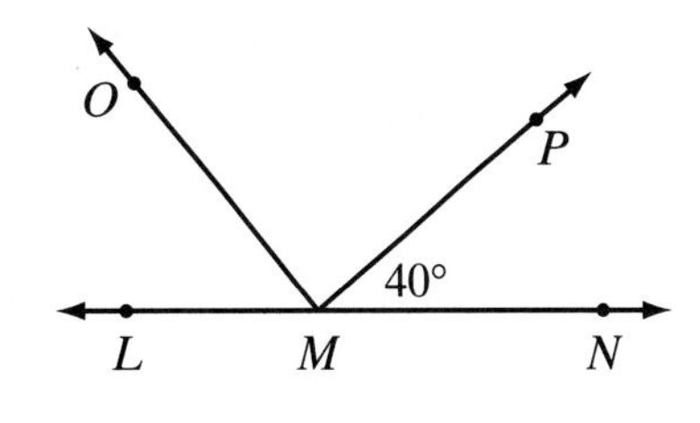

a. Find $m\angle PMO$.

b. Find $m\angle OML$.

33. *Given*: $\overleftrightarrow{ABC}$ is a line; $m\angle EBC = 40$; $\angle ABD \cong \angle DBE$.

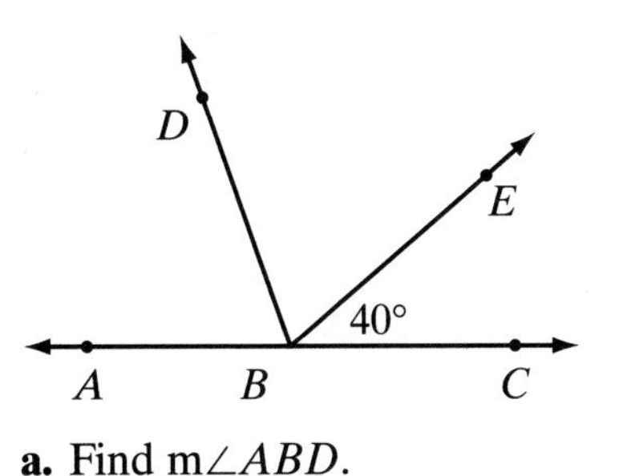

a. Find $m\angle ABD$.

b. Find $m\angle DBC$.

34. *Given*: $\overleftrightarrow{FI}$ intersects $\overleftrightarrow{JH}$ at K; $m\angle HKI = 40$; $m\angle FKG = m\angle FKJ$.

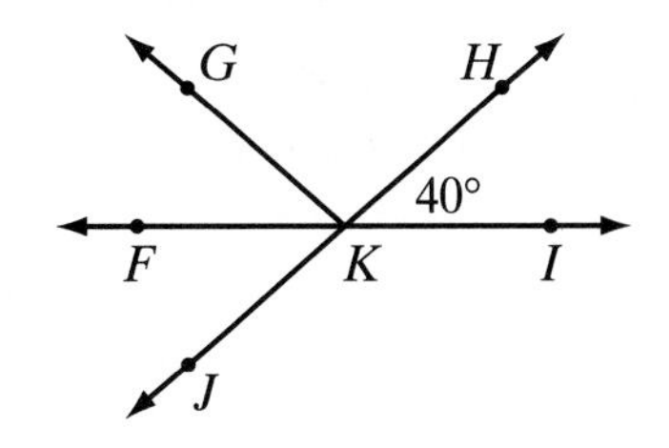

a. Find $m\angle FKJ$.

b. Find $m\angle FKG$.

c. Find $m\angle GKH$.

d. Find $m\angle JKI$.

In 35–38, sketch and label a diagram in each case and find the measure of each angle named.

35. $\overleftrightarrow{AB}$ intersects $\overleftrightarrow{CD}$ at E; $m\angle AED = 20$. Find:

a. $m\angle CEB$ **b.** $m\angle BED$ **c.** $m\angle CEA$

36. $\angle PQR$ and $\angle RQS$ are complementary; $m\angle PQR = 30$; $\overleftrightarrow{RQT}$ is a line. Find:

a. $m\angle RQS$ **b.** $m\angle SQT$ **c.** $m\angle PQT$

37. $\overleftrightarrow{LM}$ intersects $\overleftrightarrow{PQ}$ at R. The measure of $\angle LRQ$ is 80 more than m$\angle LRP$. Find:

a. m$\angle LRP$ **b.** m$\angle LRQ$ **c.** m$\angle PRM$

38. $\overleftrightarrow{CD} \perp \overleftrightarrow{AB}$ at E. Point F is in the interior of $\angle CEB$. The measure of $\angle CEF$ is 8 times the measure of $\angle FEB$. Find:

a. m$\angle FEB$ **b.** m$\angle CEF$ **c.** m$\angle AEF$

39. The angles, $\angle ABD$ and $\angle DBC$, form a linear pair and are congruent. What must be true about $\overleftrightarrow{ABC}$ and $\overleftrightarrow{BD}$?

7-3 ANGLES AND PARALLEL LINES

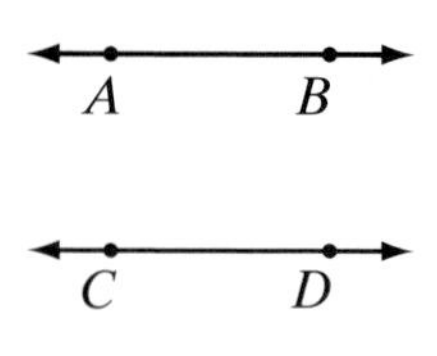

Not all lines in the same plane intersect. Two or more lines are called **parallel lines** if and only if the lines lie in the same plane but do not intersect.

In the figure on the left, $\overleftrightarrow{AB}$ and $\overleftrightarrow{CD}$ lie in the same plane but do not intersect. Hence, we say that $\overleftrightarrow{AB}$ is parallel to $\overleftrightarrow{CD}$. Using the symbol || for *is parallel to*, we write $\overleftrightarrow{AB} \parallel \overleftrightarrow{CD}$. When we speak of two parallel lines, we will mean two *distinct* lines. (In more advanced courses, you will see that a line is parallel to itself.) Line segments and rays are parallel if the lines that contain them are parallel. If two lines such as $\overleftrightarrow{AB}$ and $\overleftrightarrow{CD}$ lie in the same plane, they must be either intersecting lines or parallel lines, as shown in the following figures.

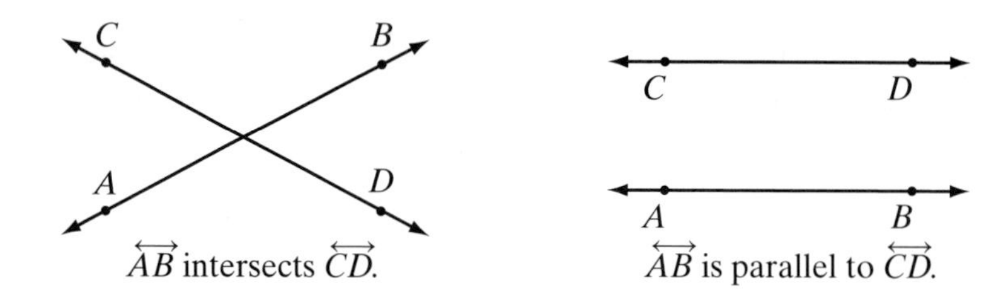

$\overleftrightarrow{AB}$ intersects $\overleftrightarrow{CD}$. $\overleftrightarrow{AB}$ is parallel to $\overleftrightarrow{CD}$.

When two lines such as $\overleftrightarrow{AB}$ and $\overleftrightarrow{CD}$ are parallel, they have no points in common. We can think of each line as a set of points. Hence, the intersection set of $\overleftrightarrow{AB}$ and $\overleftrightarrow{CD}$ is the empty set symbolized as $\overleftrightarrow{AB} \cap \overleftrightarrow{CD} = \varnothing$.

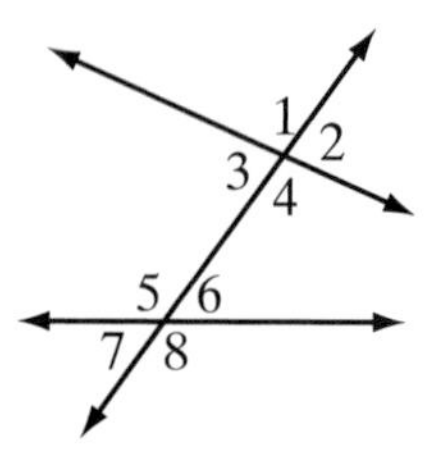

When two lines are cut by a third line, called a **transversal**, two sets of angles, each containing four angles, are formed. In the figure on the left:

- Angles 3, 4, 5, 6 are called **interior angles**.
- Angles 1, 2, 7, 8 are called **exterior angles**.
- Angles 4 and 5 are interior angles on opposite sides of the transversal and do not have the same vertex. They are called **alternate interior angles**. Angles 3 and 6 are another pair of alternate interior angles.

- Angles 1 and 8 are exterior angles on opposite sides of the transversal and do not have the same vertex. They are called **alternate exterior angles**. Angles 2 and 7 are another pair of alternate exterior angles.
- Angles 4 and 6 are **interior angles on the same side of the transversal**. Angles 3 and 5 are another pair of interior angles on the same side of the transversal.
- Angles 1 and 5 are on the same side of the transversal, one interior and one exterior, and at different vertices. They are called **corresponding angles**. Other pairs of corresponding angles are 2 and 6, 3 and 7, 4 and 8.

Alternate Interior Angles and Parallel Lines

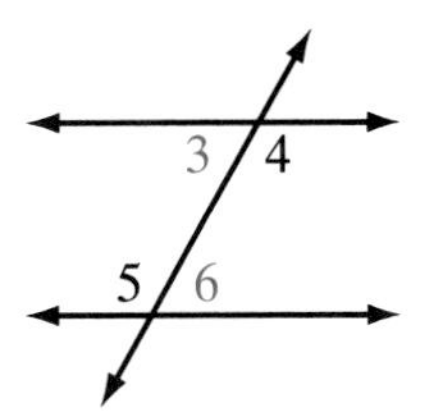

In the figure, a transversal intersects two parallel lines, forming a pair of alternate interior angles, $\angle 3$ and $\angle 6$. If we measure $\angle 3$ and $\angle 6$ with a protractor, we will find that each angle measures 60°. Here, alternate interior angles 3 and 6 have equal measures, and $\angle 3 \cong \angle 6$. If we draw other pairs of parallel lines intersected by transversals, we will find again that pairs of alternate interior angles have equal measures. Yet, we would be hard-pressed to prove that this is *always* true. Therefore, we accept, without proof, the following statement:

▶ **If two parallel lines are cut by a transversal, then the alternate interior angles that are formed have equal measures, that is, they are congruent.**

Note that $\angle 4$ and $\angle 5$ are another pair of alternate interior angles formed by a transversal that intersects the parallel lines. Therefore, $\angle 4 \cong \angle 5$.

Corresponding Angles and Parallel Lines

If two lines are cut by a transversal, four pairs of corresponding angles are formed.

One such pair of corresponding angles is $\angle 2$ and $\angle 6$, as shown in the figure on the left. If the original two lines are parallel, do these corresponding angles have equal measures? We are ready to prove in an informal manner that they do.

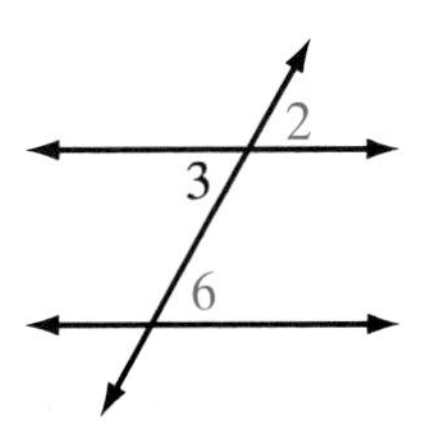

(1) Let $m\angle 2 = x$.

(2) If $m\angle 2 = x$, then $m\angle 3 = x$ (because $\angle 2$ and $\angle 3$ are vertical angles, and vertical angles are congruent).

(3) If $m\angle 3 = x$, then $m\angle 6 = x$ (because $\angle 3$ and $\angle 6$ are alternate interior angles of parallel lines, and alternate interior angles of parallel lines are congruent).

(4) Therefore $m\angle 2 = m\angle 6$ (because the measure of each angle is x).

The four steps on page 259 serve as a proof of the following theorem:

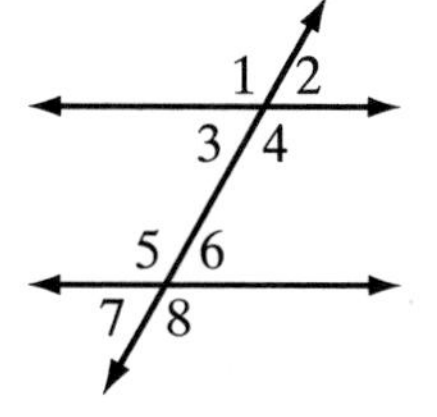

Theorem. If two parallel lines are cut by a transversal, then the corresponding angles formed have equal measures, that is, they are congruent.

Note that this theorem is true for each pair of corresponding angles:

$$\angle 1 \cong \angle 5; \angle 2 \cong \angle 6; \angle 3 \cong \angle 7; \angle 4 \cong \angle 8$$

Alternate Exterior Angles and Parallel Lines

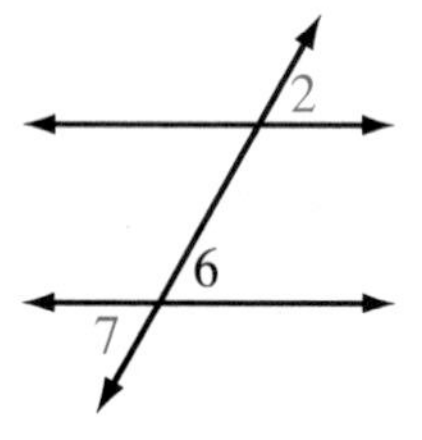

If two parallel lines are cut by a transversal, we can prove informally that the alternate exterior angles formed have equal measures. One such pair of alternate exterior angles consists of $\angle 2$ and $\angle 7$, as shown in the figure on the left.

(1) Let $m\angle 2 = x$.

(2) If $m\angle 2 = x$, then $m\angle 6 = x$ (because $\angle 2$ and $\angle 6$ are corresponding angles of parallel lines, proved to have the same measure).

(3) If $m\angle 6 = x$, then $m\angle 7 = x$ (because $\angle 6$ and $\angle 7$ are vertical angles, previously proved to have the same measure).

(4) Therefore $m\angle 2 = m\angle 7$ (because the measure of each angle is x).

These four steps serve as a proof of the following theorem:

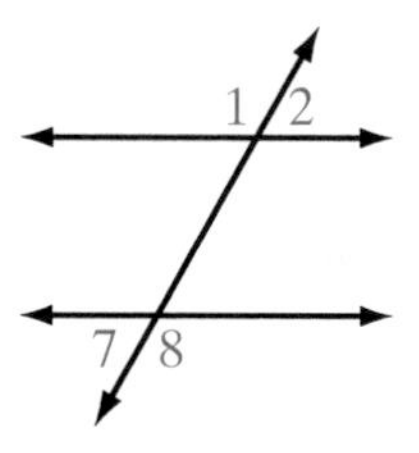

Theorem. If two parallel lines are cut by a transversal, then the alternate exterior angles formed have equal measures, that is, they are congruent.

Note that this theorem is true for each pair of alternate exterior angles:

$$\angle 1 \cong \angle 8; \angle 2 \cong \angle 7.$$

Interior Angles on the Same Side of the Transversal

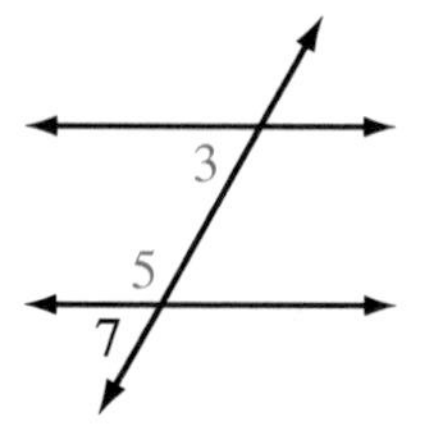

When two parallel lines are cut by a transversal, we can prove informally that the sum of the measures of the interior angles on the same side of the transversal is 180°. One such pair of interior angles on the same side of the transversal consists of $\angle 3$ and $\angle 5$, as shown in the figure on the left.

(1) $m\angle 5 + m\angle 7 = 180$ ($\angle 5$ and $\angle 7$ are supplementary angles).

(2) $m\angle 7 = m\angle 3$ ($\angle 7$ and $\angle 3$ are corresponding angles).

(3) $m\angle 5 + m\angle 3 = 180$ (by substituting $m\angle 3$ for its equal, $m\angle 7$).

These three steps serve as a proof of the following theorem:

> **Theorem.** If two parallel lines are cut by a transversal, then the sum of the measures of the interior angles on the same side of the transversal is 180°.

EXAMPLE 1

In the figure, the parallel lines are cut by a transversal. If $m\angle 1 = (5x - 10)$ and $m\angle 2 = (3x + 60)$, find the measures of $\angle 1$ and $\angle 2$.

Solution Since the lines are parallel, the alternate interior angles, $\angle 1$ and $\angle 2$, have equal measures. Write and solve an equation using the algebraic expressions for the measures of these angles:

$$5x - 10 = 3x + 60$$
$$5x = 3x + 70$$
$$2x = 70$$
$$x = 35$$

$$5x - 10 = 5(35) - 10 = 175 - 10 = 165$$
$$3x + 60 = 3(35) + 60 = 105 + 60 = 165$$

Answer $m\angle 1 = 165$ and $m\angle 2 = 165$.

EXERCISES

Writing About Mathematics

1. Give an example of a pair of lines that are neither parallel nor intersecting.

2. If a transversal is perpendicular to one of two parallel lines, is it perpendicular to the other? Prove your answer using definitions and theorems given in this chapter.

Developing Skills

In 3–10, the figure below shows two parallel lines cut by a transversal. For each given measure, find the measures of the other seven angles.

3. $m\angle 3 = 80$ **4.** $m\angle 6 = 150$

5. $m\angle 5 = 60$ **6.** $m\angle 1 = 75$

7. $m\angle 2 = 55$ **8.** $m\angle 4 = 10$

9. $m\angle 7 = 2$ **10.** $m\angle 8 = 179$

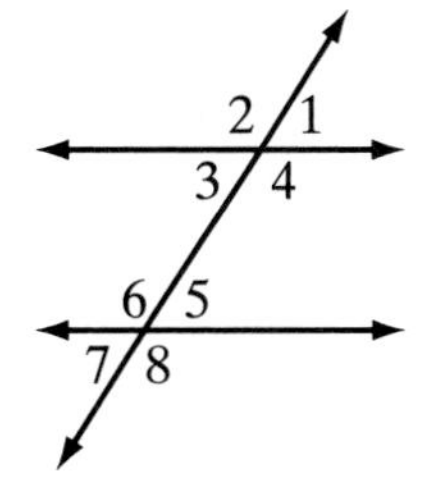

In 11–15, the figure below shows two parallel lines cut by a transversal. In each exercise, find the measures of all eight angles under the given conditions.

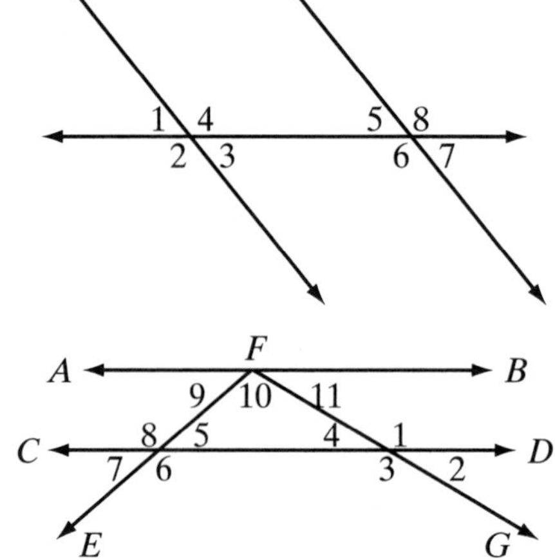

11. $m\angle 3 = 2x + 40$ and $m\angle 7 = 3x + 27$

12. $m\angle 4 = 4x - 10$ and $m\angle 6 = x + 80$

13. $m\angle 4 = 3x + 40$ and $m\angle 5 = 2x$

14. $m\angle 4 = 2x - 10$ and $m\angle 2 = x + 60$

15. $\angle 8 \cong \angle 1$

16. If $\overleftrightarrow{AB} \parallel \overleftrightarrow{CD}$, $m\angle 5 = 40$, and $m\angle 4 = 30$, find the measures of the other angles in the figure.

In 17–22, tell whether each statement is always, sometimes, or never true.

17. If two distinct lines intersect, then they are parallel.

18. If two distinct lines do not intersect, then they are parallel.

19. If two angles are alternate interior angles, then they are on opposite sides of the transversal.

20. If two parallel lines are cut by a transversal, then the alternate interior angles are congruent.

21. If two parallel lines are cut by a transversal, then the alternate interior angles are complementary.

22. If two parallel lines are cut by a transversal, then the corresponding angles are supplementary.

23. In the figure on the right, two parallel lines are cut by a transversal. Write an informal proof that $\angle 1$ and $\angle 2$ have equal measures.

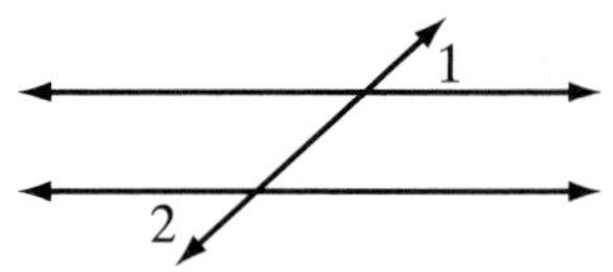

7-4 TRIANGLES

A **polygon** is a plane figure that consists of line segments joining three or more points. Each line segment is a **side** of the polygon and the endpoints of each side are called **vertices**. Each vertex is the endpoint of exactly two sides and no two sides have any other points in common.

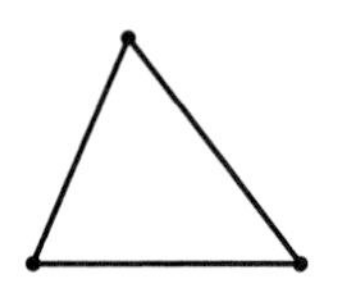

When three points that are not all on the same line are joined in pairs by line segments, the figure formed is a **triangle**. Each of the given points is the vertex of an angle of the triangle, and each line segment is a side of the triangle. There are many practical uses of the triangle, especially in construction work such as the building of bridges, radio towers, and airplane wings, because the triangle is a rigid figure. The shape of a triangle cannot be changed without changing the length of at least one of its sides.

We begin our discussion of the triangle by considering triangle ABC, shown below, which is symbolized as $\triangle ABC$. In $\triangle ABC$, points A, B, and C are the vertices, and $\overline{AB}$, $\overline{BC}$, and $\overline{CA}$ are the sides. Angles A, B, and C of the triangle are symbolized as $\angle A$, $\angle B$, and $\angle C$.

We make the following observations:

1. Side $\overline{AB}$ is included between $\angle A$ and $\angle B$.
2. Side $\overline{BC}$ is included between $\angle B$ and $\angle C$.
3. Side $\overline{CA}$ is included between $\angle C$ and $\angle A$.
4. Angle A is included between sides $\overline{AB}$ and $\overline{CA}$.
5. Angle B is included between sides $\overline{AB}$ and $\overline{BC}$.
6. Angle C is included between sides $\overline{CA}$ and $\overline{BC}$.

Classifying Triangles According to Angles

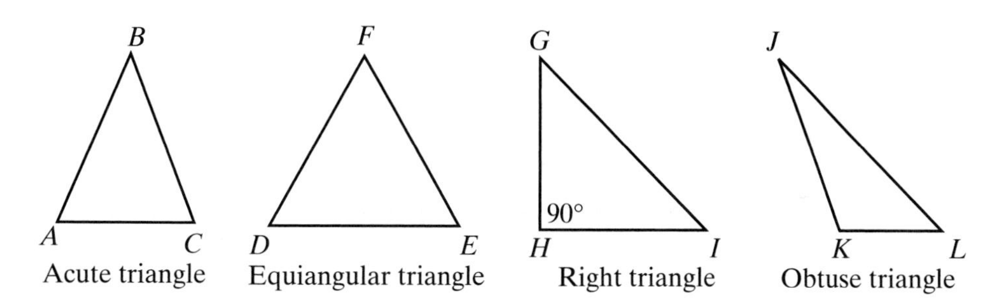

Acute triangle | Equiangular triangle | Right triangle | Obtuse triangle

- An **acute triangle** has three acute angles.
- An **equiangular triangle** has three angles equal in measure.
- A **right triangle** has one right angle.
- An **obtuse triangle** has one obtuse angle.

In right triangle GHI above, the two sides of the triangle that form the right angle, $\overline{GH}$ and $\overline{HI}$, are called the **legs** of the right triangle. The side opposite the right angle, $\overline{GI}$, is called the **hypotenuse**.

Sum of the Measures of the Angles of a Triangle

When we change the shape of a triangle, changes take place also in the measures of its angles. Is there any relationship among the measures of the angles of a triangle that does not change? Let us see.

Draw several triangles of different shapes. Measure the three angles of each triangle and find the sum of these measures. For example, in $\triangle ABC$ on the right,

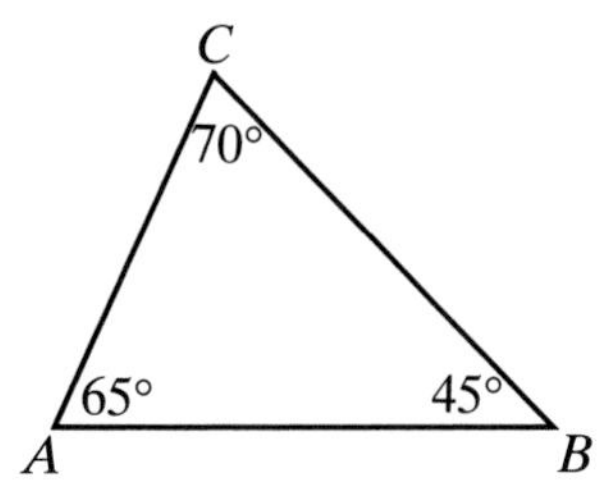

$$m\angle A + m\angle B + m\angle C = 65 + 45 + 70 = 180.$$

If you measured accurately, you should have found that in each triangle that you drew and measured, the sum of the measures of the three angles is 180°, regardless of its size or shape.

We can write an informal algebraic proof of the following statement:

► **The sum of the measures of the angles of a triangle is 180°.**

(1) In $\triangle ABC$, let $m\angle A = x$, $m\angle ACB = y$, and $m\angle B = z$.

(2) Let $\overleftrightarrow{DCE}$ be a line parallel to $\overleftrightarrow{AB}$.

(3) Since $\angle DCE$ is a straight angle, $m\angle DCE = 180$.

(4) $m\angle DCE = m\angle DCA + m\angle ACB + m\angle BCE$
$= 180$.

(5) $m\angle DCA = m\angle A = x$.

(6) $m\angle ACB = y$.

(7) $m\angle BCE = m\angle B = z$.

(8) Substitute from statements (5), (6), and (7) in statement (4):
$x + y + z = 180$.

EXAMPLE 1

Find the measure of the third angle of a triangle if the measures of two of the angles are 72.6° and 84.2°.

Solution Subtract the sum of the known measures from 180:

$$180 - (72.6 + 84.2) = 23.2$$

Answer 23.2°

EXAMPLE 2

In $\triangle ABC$, the measure of $\angle B$ is twice the measure of $\angle A$, and the measure of $\angle C$ is 3 times the measure of $\angle A$. Find the number of degrees in each angle of the triangle.

Solution Let x = the number of degrees in $\angle A$.

Then $2x$ = the number of degrees in $\angle B$.

Then $3x$ = the number of degrees in $\angle C$.

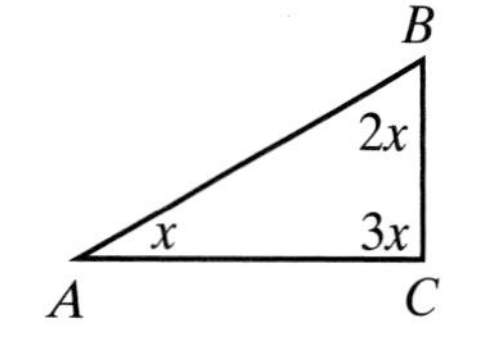

The sum of the measures of the angles of a triangle is 180°.

$$x + 2x + 3x = 180$$
$$6x = 180$$
$$x = 30$$
$$2x = 60$$
$$3x = 90$$

Check

$60 = 2(30)$

$90 = 3(30)$

$30 + 60 + 90 = 180$ ✔

Answer $m\angle A = 30, m\angle B = 60, m\angle C = 90$

Classifying Triangles According to Sides

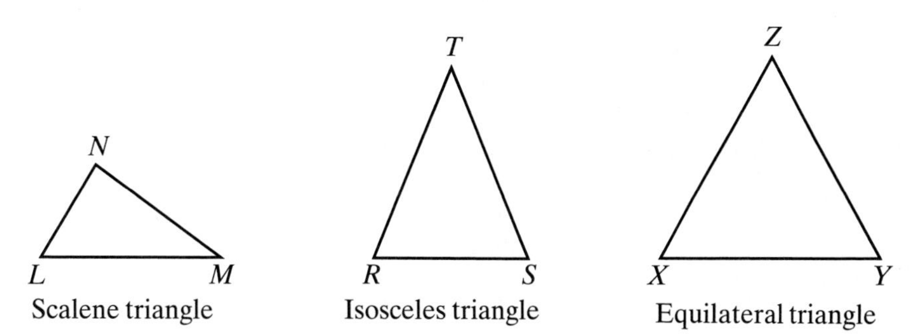

Scalene triangle

Isosceles triangle

Equilateral triangle

- A **scalene triangle** has no sides equal in length.
- An **isosceles triangle** has two sides equal in length.
- An **equilateral triangle** has three sides equal in length.

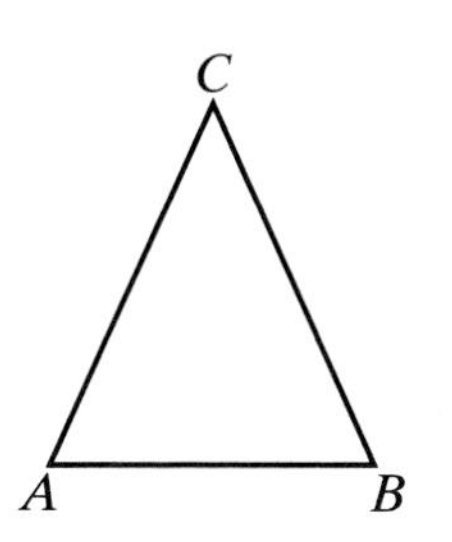

Isosceles Triangles

In isosceles triangle ABC, shown on the left, the two sides that are equal in measure, $\overline{AC}$ and $\overline{BC}$, are called the **legs**. The third side, $\overline{AB}$, is the **base**.

Two line segments that are equal in measure are said to be **congruent**. The angle formed by the two congruent sides, $\angle C$, is called the **vertex angle**. The two angles at the endpoints of the base, $\angle A$ and $\angle B$, are the **base angles**. In isosceles triangle ABC, if we measure the base angles, $\angle A$ and $\angle B$, we find that each angle contains 65°. Therefore, $m\angle A = m\angle B$. Similarly, if we measure the base angles in any other isosceles triangle, we find that they are equal in measure. Thus, we will accept the truth of the following statement:

▶ **The base angles of an isosceles triangle are equal in measure; that is, they are congruent.**

This statement may be rephrased in a variety of ways. For example:

1. If a triangle is isosceles, then its base angles are congruent.

2. If two sides of a triangle are equal in measure, then the angles opposite these sides are equal in measure.

The following statement is also true:

▶ **If two angles of a triangle are equal in measure, then the sides opposite these angles are equal in measure and the triangle is an isosceles triangle.**

This statement may be rephrased as follows:

1. If two angles of a triangle are congruent, then the sides opposite these angles are congruent.

2. If two angles of a triangle have equal measures, then the sides opposite these angles have equal measures.

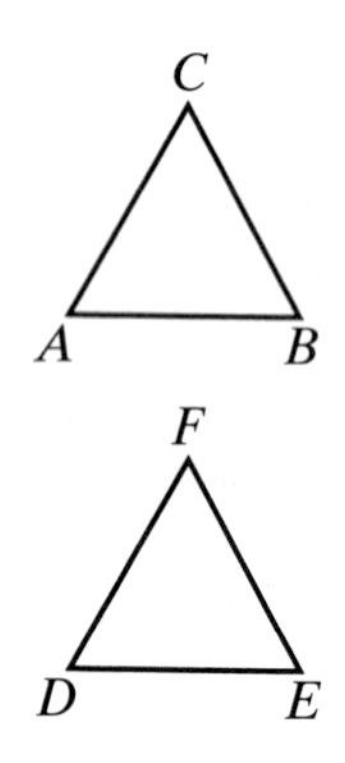

Equilateral Triangles

Triangle ABC is an equilateral triangle. Since $AB = BC$, $m\angle C = m\angle A$; also, since $AC = BC$, $m\angle B = m\angle A$. Therefore, $m\angle A = m\angle B = m\angle C$. In an equilateral triangle, the measures of all of the angles are equal.

In $\triangle DEF$, all of the angles are equal in measure. Since $m\angle D = m\angle E$, $EF = DF$; also, since $m\angle D = m\angle F$, $EF = DE$. Therefore, $DE = EF = DF$, and $\triangle DEF$ is equilateral.

▶ **If a triangle is equilateral, then it is equiangular.**

Properties of Special Triangles

1. If two sides of a triangle are equal in measure, the angles opposite these sides are also equal in measure. (The base angles of an isosceles triangle are equal in measure.)

2. If two angles of a triangle are equal in measure, the sides opposite these angles are also equal in measure.

3. All of the angles of an equilateral triangle are equal in measure. (An equilateral triangle is equiangular.)

4. If three angles of a triangle are equal in measure, the triangle is equilateral. (An equiangular triangle is equilateral.)

EXAMPLE 3

In isosceles triangle ABC, the measure of vertex angle C is 30° more than the measure of each base angle. Find the number of degrees in each angle of the triangle.

Solution Let x = number of degrees in one base angle, A.

Then x = number of degrees in the other base angle, B, and $x + 30$ = number of degrees in the vertex angle, C.

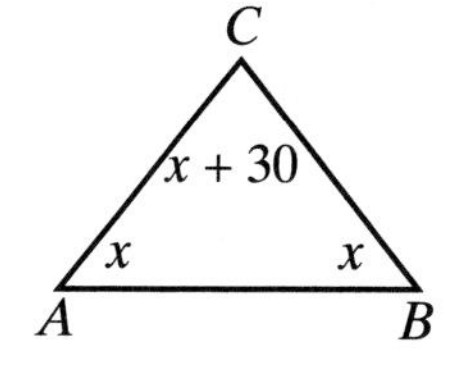

The sum of the measures of the angles of a triangle is 180°.

$$x + x + x + 30 = 180$$
$$3x + 30 = 180$$
$$3x = 150$$
$$x = 50$$
$$x + 30 = 80$$

Check $50 + 50 + 80 = 180$ ✔

Answer $m\angle A = 50, m\angle B = 50, m\angle C = 80$

EXERCISES

Writing About Mathematics

1. Ayyam said that if the sum of the measures of two angles of a triangle is equal to the measure of the third angle, the triangle is a right triangle. Prove or disprove Ayyam's statement.
2. Janice said that if two angles of a triangle each measure 60°, then the triangle is equilateral. Prove or disprove Janice's statement.

Developing Skills

In 3–5, state, in each case, whether the angles with the given measures can be the three angles of the same triangle.

3. 30°, 70°, 80° **4.** 70°, 80°, 90° **5.** 30°, 110°, 40°

In 6–9, find, in each case, the measure of the third angle of the triangle if the measures of two angles are:

6. 60°, 40° **7.** 100°, 20° **8.** 54.5°, 82.3° **9.** $24\frac{1}{4}°, 81\frac{1}{2}°$

10. What is the measure of each angle of an equiangular triangle?

11. Can a triangle have: **a.** two right angles? **b.** two obtuse angles? **c.** one right and one obtuse angle? Explain why or why not.

12. What is the sum of the measures of the two acute angles of a right triangle?

13. In $\triangle ABC$, $AC = 4$, $CB = 6$, and $AB = 6$.

a. What type of triangle is $\triangle ABC$?

b. Name two angles in $\triangle ABC$ whose measures are equal.

c. Why are these angles equal in measure?

d. Name the legs, base, base angles and vertex angle of this triangle.

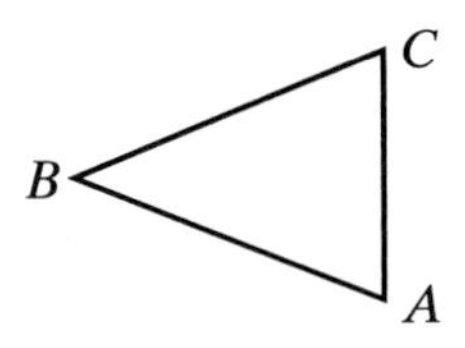

14. In $\triangle RST$, $m\angle R = 70$ and $m\angle T = 40$.

a. Find the measure of $\angle S$.

b. Name two sides in $\triangle RST$ that are congruent.

c. Why are the two sides congruent?

d. What type of triangle is $\triangle RST$?

e. Name the legs, base, base angles, and vertex angle of this triangle.

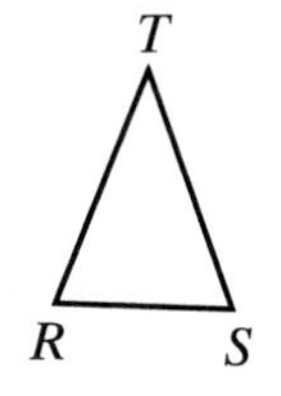

15. Find the measure of the vertex angle of an isosceles triangle if the measure of each base angle is:

a. 80° **b.** 55° **c.** 42° **d.** $22\frac{1}{2}°$ **e.** 51.5°

16. Find the measure of each base angle of an isosceles triangle if the measure of the vertex angle is:

a. 20° **b.** 50° **c.** 76° **d.** 100° **e.** 65°

17. What is the number of degrees in each acute angle of an isosceles right triangle?

18. If a triangle is equilateral, what is the measure of each angle?

Applying Skills

19. The measure of each base angle of an isosceles triangle is 7 times the measure of the vertex angle. Find the measure of each angle of the triangle.

20. The measure of each of the congruent angles of an isosceles triangle is one-half of the measure of the vertex angle. Find the measure of each angle of the triangle.

21. The measure of the vertex angle of an isosceles triangle is 3 times the measure of each base angle. Find the number of degrees in each angle of the triangle.

22. The measure of the vertex angle of an isosceles triangle is 15° more than the measure of each base angle. Find the number of degrees in each angle of the triangle.

23. The measure of each of the congruent angles of an isosceles triangle is 6° less than the measure of the vertex angle. Find the measure of each angle of the triangle.

24. The measure of each of the congruent angles of an isosceles triangle is 9° less than 4 times the vertex angle. Find the measure of each angle of the triangle.

25. In $\triangle ABC$, $m\angle A = x$, $m\angle B = x + 45$, and $m\angle C = 3x - 15$.

a. Find the measures of the three angles.

b. What kind of triangle is $\triangle ABC$?

26. The measures of the three angles of $\triangle DEF$ can be represented by $(x + 30)°$, $2x°$, and $(4x - 60)°$.

a. What is the measure of each angle?

b. What kind of triangle is $\triangle DEF$?

27. In a triangle, the measure of the second angle is 3 times the measure of the first angle, and the measure of the third angle is 5 times the measure of the first angle. Find the number of degrees in each angle of the triangle.

28. In a triangle, the measure of the second angle is 4 times the measure of the first angle. The measure of the third angle is equal to the sum of the measures of the first two angles. Find the number of degrees in each angle of the triangle. What kind of triangle is it?

29. In a triangle, the measure of the second angle is 30° more than the measure of the first angle, and the measure of the third angle is 45° more than the measure of the first angle. Find the number of degrees in each angle of the triangle.

30. In a triangle, the measure of the second angle is 5° more than twice the measure of the first angle. The measure of the third angle is 35° less than 3 times the measure of the first angle. Find the number of degrees in each angle of the triangle.

31. $\overleftrightarrow{AEFB}$ is a straight line, $m\angle AEG = 130$, and $m\angle BFG = 140$.

a. Find $m\angle x$, $m\angle y$, and $m\angle z$.

b. What kind of a triangle is $\triangle EFG$?

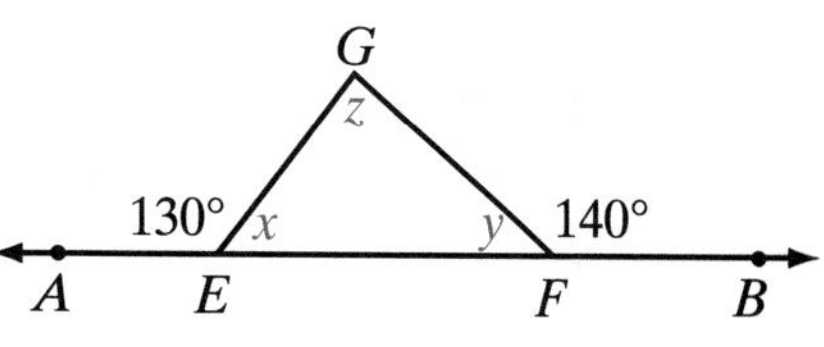

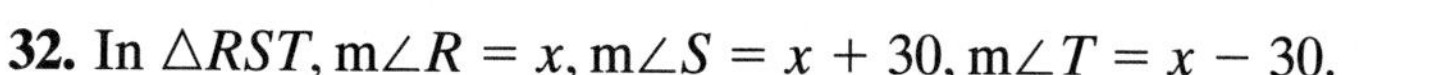

32. In $\triangle RST$, $m\angle R = x$, $m\angle S = x + 30$, $m\angle T = x - 30$.

a. Find the measures of the three angles of the triangle.

b. What kind of a triangle is $\triangle RST$?

33. In $\triangle KLM$, $m\angle K = 2x$, $m\angle L = x + 30$, $m\angle M = 3x - 30$.

a. Find the measures of the three angles of the triangle.

b. What kind of a triangle is $\triangle KLM$?

Hands-On Activity 1: Constructing a Line Segment Congruent to a Given Segment

To **construct** a geometric figure means that a specific design is accurately made by using only two instruments: a **compass** used to draw a complete circle or part of a circle and a **straightedge** used to draw a straight line.

In this activity, you will learn how to construct a line segment congruent to a given segment, that is, construct a *copy* of a line segment.

STEP 1. Use the straightedge to draw a line segment. Label the endpoints A and B.

STEP 2. Use the straightedge to draw a ray and label the endpoint C.

STEP 3. Place the compass so that the point of the compass is at A and the point of the pencil is at B.

STEP 4. Keeping the opening of the compass unchanged, place the point at C and draw an arc that intersects the ray. Label this intersection D.

Result: $\overline{AB} \cong \overline{CD}$

a. Now that you know how to construct congruent line segments, explain how you can construct a line segment that is three times the length of a given segment.

b. Explain how to construct a line segment with length equal to the *difference* of two given segments.

c. Explain how to construct a line segment whose length is the sum of the lengths of two given line segments.

Hands-On Activity 2: Constructing an Angle Congruent to a Given Angle

In this activity, you will learn how to construct an angle congruent to a given angle, that is, construct a copy of an angle.

STEP 1. Use the straightedge to draw an acute angle. Label the vertex S.

STEP 2. Use the straightedge to draw a ray and label the endpoint M.

STEP 3. With the point of the compass at S, draw an arc that intersects each ray of $\angle S$. Label the point of intersection on one ray R and the point of intersection on the other ray T.

STEP 4. Using the same opening of the compass as was used in step 3, place the point of the compass at M and draw an arc that intersects the ray and extends beyond the ray. (Draw at least half of a circle.) Label the point of intersection L.

STEP 5. Place the point of the compass at R and the point of the pencil at T.

STEP 6. Without changing the opening of the compass, place the point at L and draw an arc that intersects the arc drawn in step 4. Label this point of intersection N.

STEP 7. Draw $\overrightarrow{MN}$.

Result: $\angle RST \cong \angle LMN$

a. Now that you know how to construct congruent angles, explain how you can construct an angle that is three times the measure of a given angle.

b. Explain how to construct an angle with a measure equal to the *difference* of two given angles.

c. Explain how to construct a triangle congruent to a given triangle using two sides and the included angle.

Hands-On Activity 3: Constructing a Perpendicular Bisector

In this activity, you will learn how to construct a **perpendicular bisector** of a line segment. A perpendicular bisector of a line segment is the line that divides a segment into two equal parts and is perpendicular to the segment.

STEP 1. Use the straightedge to draw a line segment. Label one endpoint A and the other C.

STEP 2. Open the compass so that the distance between the point and the pencil point is more than half of the length of $\overline{AC}$.

STEP 3. With the point of the compass at A, draw an arc above $\overline{AC}$ and an arc below $\overline{AC}$.

STEP 4. With the same opening of the compass and the point of the compass at C, draw an arc above $\overline{AC}$ and an arc below $\overline{AC}$ that intersect the arcs drawn in step 3. Call one of these intersections E and the other F.

STEP 5. Use the straightedge to draw $\overleftrightarrow{EF}$, intersecting $\overline{AC}$ at B.

Result: $\overleftrightarrow{EF}$ is perpendicular to $\overline{AC}$, and B is the midpoint of $\overline{AC}$.

a. What is true about $\angle EAB$ and $\angle FAB$? What is true about $\angle EAB$ and $\angle ECB$?

b. Explain how to construct an isosceles triangle with a vertex angle that is twice the measure of a given angle.

c. Explain how to construct an isosceles right triangle.

Hands-On Activity 4: Constructing an Angle Bisector

In this activity, you will learn how to construct an **angle bisector**. An angle bisector is the line that divides an angle into two congruent angles.

STEP 1. Use the straightedge to draw an acute angle. Label the vertex S.

STEP 2. With any convenient opening of the compass, place the point at S and draw an arc that intersects both rays of $\angle S$. Call one of the intersections R and the other T.

STEP 3. Place the point of the compass at R and draw an arc in the interior of $\angle S$.

STEP 4. With the same opening of the compass, place the point of the compass at T and draw an arc that intersects the arc drawn in step 3. Label the intersection of the arcs P.

STEP 5. Draw $\overrightarrow{SP}$.

Result: $\angle RSP \cong \angle PST$; $\overrightarrow{SP}$ bisects angle S.

a. Now that you know how to construct an angle bisector, explain how you can construct an angle that is one and a half times the measure of a given angle.

b. Is it possible to use an angle bisector to construct a 90° angle? Explain.

c. Explain how to construct an isosceles triangle with a vertex angle that is congruent to a given angle.

7-5 QUADRILATERALS

In your study of mathematics you have learned many facts about polygons that have more than three sides. In this text you have frequently solved problems using the formulas for the perimeter and area of a rectangle. In this section we will review what you already know and use that knowledge to demonstrate the truth of many of the properties of polygons.

A **quadrilateral** is a polygon that has four sides. A point at which any two sides of the quadrilateral meet is a *vertex* of the quadrilateral. At each vertex, the two sides that meet form an angle of the quadrilateral. Thus, $ABCD$ on the left is a quadrilateral whose sides are $\overline{AB}$, $\overline{BC}$, $\overline{CD}$, and $\overline{DA}$. Its vertices are A, B, C, and D. Its angles are $\angle ABC$, $\angle BCD$, $\angle CDA$, and $\angle DAB$.

In a quadrilateral, two angles whose vertices are the endpoints of a side are called **consecutive angles**. For example, in quadrilateral $ABCD$, $\angle A$ and $\angle B$ are consecutive angles because their vertices are the endpoints of a side, $\overline{AB}$. Other pairs of consecutive angles are $\angle B$ and $\angle C$, $\angle C$ and $\angle D$, and $\angle D$ and $\angle A$. Two angles that are not consecutive angles are called **opposite angles**; $\angle A$ and $\angle C$ are opposite angles, and $\angle B$ and $\angle D$ are opposite angles.

Special Quadrilaterals

When we vary the shape of the quadrilateral by making some of its sides parallel, by making some of its sides equal in length, or by making its angles right angles, we get different members of the family of quadrilaterals, as shown and named below:

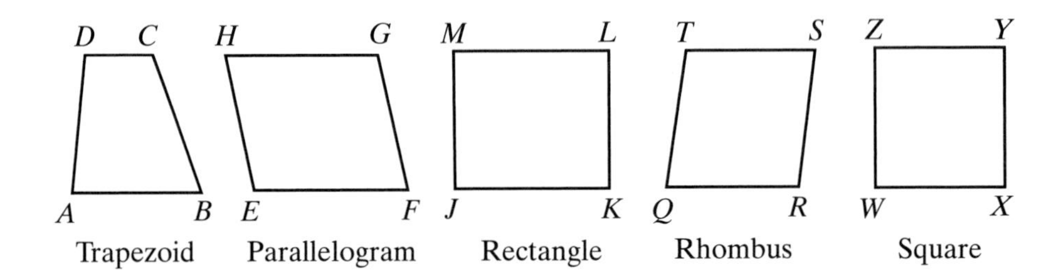

A **trapezoid** is a quadrilateral in which two and only two opposite sides are parallel. In trapezoid $ABCD$, $\overline{AB} \parallel \overline{CD}$. Parallel sides $\overline{AB}$ and $\overline{CD}$ are called the **bases** of the trapezoid.

A **parallelogram** is a quadrilateral in which both pairs of opposite sides are parallel. In parallelogram $EFGH$, $\overline{EF} \parallel \overline{GH}$ and $\overline{EH} \parallel \overline{FG}$. The symbol for a parallelogram is ▱.

A **rectangle** is a parallelogram in which all four angles are right angles. Rectangle $JKLM$ is a parallelogram in which $\angle J$, $\angle K$, $\angle L$, and $\angle M$ are right angles.

A **rhombus** is a parallelogram in which all sides are of equal length. Rhombus $QRST$ is a parallelogram in which $QR = RS = ST = TQ$.

A **square** is a rectangle in which all sides are of equal length. It is also correct to say that a square is a rhombus in which all angles are right angles. Therefore, square $WXYZ$ is also a parallelogram in which $\angle W$, $\angle X$, $\angle Y$, and $\angle Z$ are right angles, and $WX = XY = YZ = ZW$.

The Angles of a Quadrilateral

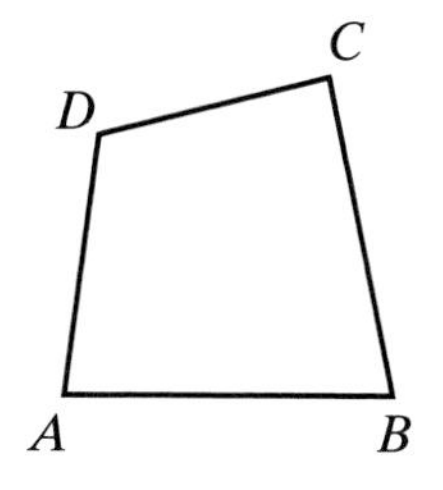

Draw a large quadrilateral like the one shown at the left and measure each of its four angles. Is the sum 360°? It should be. If we do the same with several other quadrilaterals of different shapes and sizes, is the sum of the four measures 360° in each case? It should be. Relying on what you have just verified by experimentation, it seems reasonable to make the following statement:

▶ **The sum of the measures of the angles of a quadrilateral is 360°.**

To prove informally that this statement is true, we draw diagonal $\overline{AC}$, whose endpoints are the vertices of the two opposite angles, $\angle A$ and $\angle C$. Then:

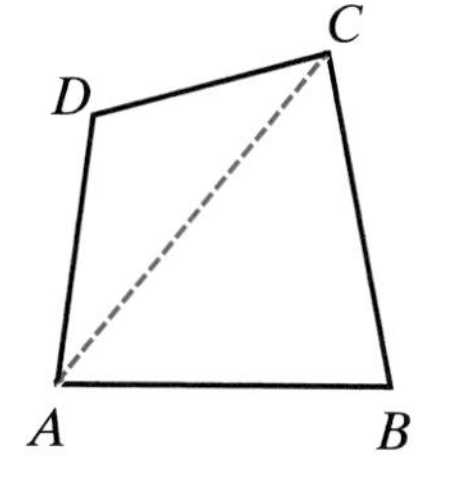

(1) Diagonal $\overline{AC}$ divides quadrilateral $ABCD$ into two triangles, $\triangle ABC$ and $\triangle ADC$.

(2) The sum of the measures of the angles of $\triangle ABC$ is 180°, and the sum of the measures of the angles of $\triangle ADC$ is 180°.

(3) The sum of the measures of all the angles of $\triangle ABC$ and $\triangle ADC$ together is 360°.

(4) Therefore, $m\angle A + m\angle B + m\angle C + m\angle D = 360$.

The Family of Parallelograms

Listed below are some relationships that are true for the family of parallelograms that includes rectangles, rhombuses, and squares.

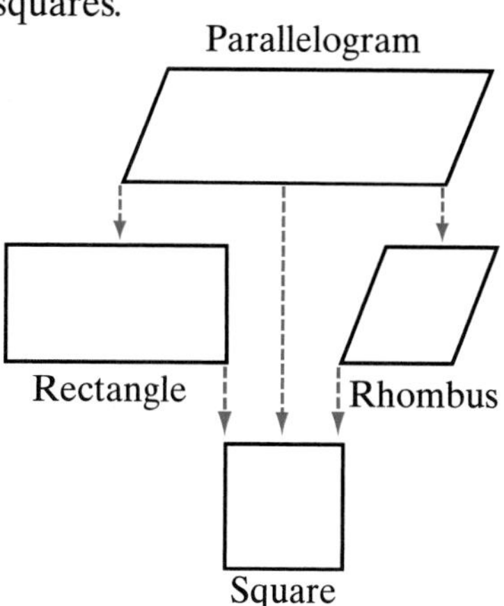

1. All rectangles, rhombuses, and squares are parallelograms. Therefore any property of a parallelogram must also be a property of a rectangle, a rhombus, or a square.
2. A square is also a rectangle. Therefore, any property of a rectangle must also be a property of a square.
3. A square is also a rhombus. Therefore, any property of a rhombus must also be a property of a square.

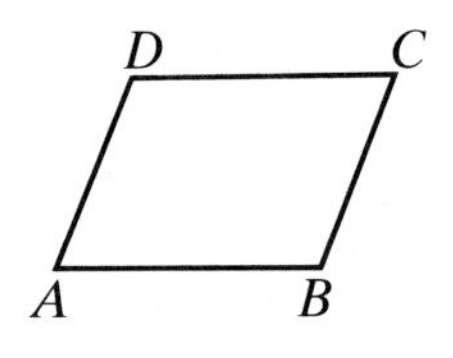

In parallelogram $ABCD$ at the left, opposite sides are parallel. Thus, $\overline{AB} \parallel \overline{DC}$ and $\overline{AD} \parallel \overline{BC}$. The following statements are true for any parallelograms.

▶ **Opposite sides of a parallelogram are congruent.**

Here, $\overline{AB} \cong \overline{DC}$ and $\overline{AD} \cong \overline{BC}$. Therefore $AB = DC$ and $AD = BC$.

▶ **Opposite angles of a parallelogram are congruent.**

Here, $\angle A \cong \angle C$ and $\angle B \cong \angle D$. Therefore $m\angle A = m\angle C$ and $m\angle B = m\angle D$.

▶ **Consecutive angles of a parallelogram are supplementary.**

Here, $m\angle A + m\angle B = 180$, $m\angle B + m\angle C = 180$, and so forth.

Since rhombuses, rectangles, and squares are also parallelograms, these statements will be true for any rhombus, any rectangle, and any square.

Informal Proofs for Statements About Angles in a Parallelogram

The statements about angles formed by parallel lines cut by a transversal that were shown to be true in Section 3 of this chapter can now be used to establish the relationships among the measures of the angles of a parallelogram.

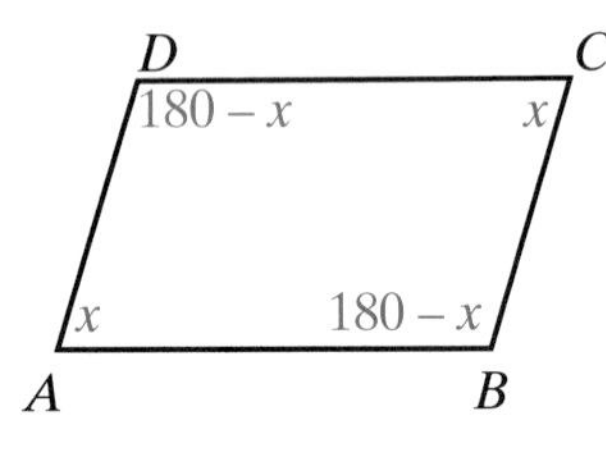

(1) In parallelogram $ABCD$, $\overline{AB}$ and $\overline{DC}$ are segments of parallel lines cut by transversal $\overline{AD}$.

(2) Let $m\angle A = x$. Then $m\angle D = 180 - x$ because $\angle A$ and $\angle D$ are interior angles on the same side of a transversal, and these angles have been shown to be supplementary.

(3) In parallelogram $ABCD$, $\overline{AD}$ and $\overline{BC}$ are segments of parallel lines cut by transversal $\overline{AB}$.

(4) Since $m\angle A = x$, $m\angle B = 180 - x$ because $\angle A$ and $\angle B$ are interior angles on the same side of a transversal, and these angles have been shown to be supplementary.

(5) Similarly, $\overline{AD}$ and $\overline{BC}$ are segments of parallel lines cut by transversal $\overline{DC}$. Since $m\angle D = 180 - x$, $m\angle C = 180 - (180 - x) = x$ because $\angle D$ and $\angle C$ are interior angles on the same side of the transversal.

Therefore, the consecutive angles of a parallelogram are supplementary:

$$m\angle A + m\angle D = x + (180 - x) = 180$$
$$m\angle A + m\angle B = x + (180 - x) = 180$$
$$m\angle C + m\angle B = x + (180 - x) = 180$$
$$m\angle C + m\angle D = x + (180 - x) = 180$$

Also, the opposite angles of a parallelogram have equal measures:

$$m\angle A = x \text{ and } m\angle C = x \qquad \text{or} \qquad m\angle A = m\angle C$$

$$m\angle B = 180 - x \text{ and } m\angle D = 180 - x \qquad \text{or} \qquad m\angle B = m\angle D$$

Polygons and Angles

We can use the sum of the measures of the interior angles of a triangle to find the sum of the measures of the interior angles of any polygon.

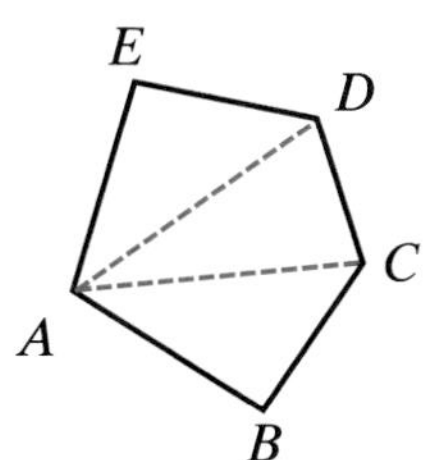

Polygon $ABCDE$ is a pentagon, a polygon with five sides. From vertex A, we draw diagonals to vertices C and D, the vertices that are not adjacent to A. These diagonals divide the pentagon into three triangles. The sum of the measures of the interior angles of $ABCDE$ is the sum of the measures of $\triangle ABC$, $\triangle ACD$, and $\triangle ADE$.

The sum of the measures of the interior angles of $ABCDE = 3(180°) = 540°$.

We can use this same method to find the sum of the interior angles of any polygon of more than three sides.

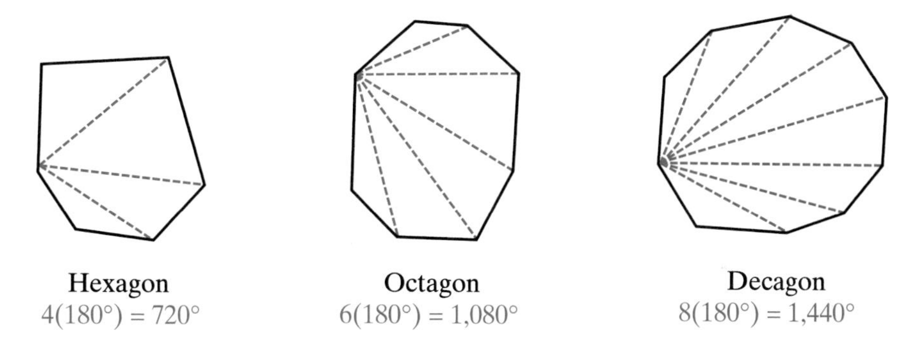

Hexagon $4(180°) = 720°$ Octagon $6(180°) = 1{,}080°$ Decagon $8(180°) = 1{,}440°$

In each case, the number of triangles into which the polygon can be divided is 2 fewer than the number of sides of the polygon. In general:

▶ **The sum of the measures of the interior angles of an n-sided polygon is $180(n - 2)$.**

Trapezoids

A trapezoid has one and only one pair of parallel lines. Each of the two parallel sides is called a *base* of the trapezoid. Therefore, we can use what we know about angles formed when parallel lines are cut by a transversal to demonstrate some facts about the angles of a trapezoid.

Quadrilateral $ABCD$ is a trapezoid with $\overline{AB} \parallel \overline{CD}$. Since $\angle DAB$ and $\angle CDA$ are interior angles on the same side of transversal $\overline{DA}$, they are supplementary. Also, $\angle CBA$ and $\angle DCB$ are interior angles on the same side of transversal $\overline{CB}$, and they are supplementary.

In a trapezoid, the parallel sides can never be congruent. But the nonparallel sides can be congruent. A trapezoid in which the nonparallel sides are congruent is called an **isosceles trapezoid**. In an isosceles trapezoid, the base angles, the two angles whose vertices are the endpoints of the same base, are congruent.

To write a formula for the area of trapezoid $ABCD$, draw diagonal $\overline{AC}$ and altitudes $\overline{AE}$ and $\overline{CF}$. Since $AFCE$ is a rectangle $AE = CF$.

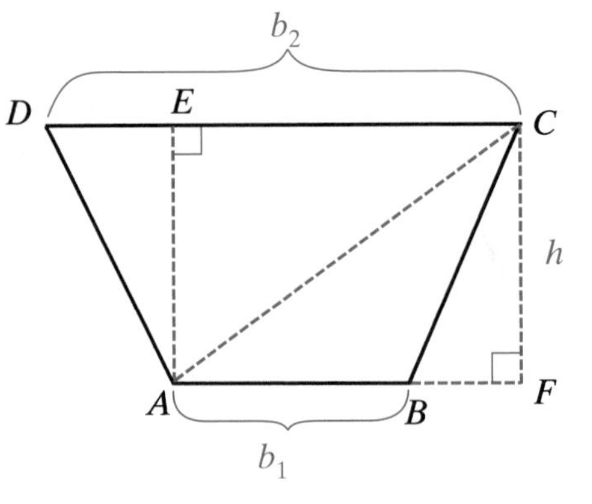

Let $AB = b_1$, $CD = b_2$, and $AE = CF = h$.

$$\text{Area of triangle } ABC = \tfrac{1}{2}b_1h$$

$$\text{Area of triangle } ACD = \tfrac{1}{2}b_2h$$

$$\begin{aligned}\text{Area of trapezoid } ABCD &= \tfrac{1}{2}b_1h + \tfrac{1}{2}b_2h \\ &= \tfrac{1}{2}h(b_1 + b_2)\end{aligned}$$

EXAMPLE 1

$ABCD$ is a parallelogram where $m\angle A = 2x + 50$ and $m\angle C = 3x + 40$.

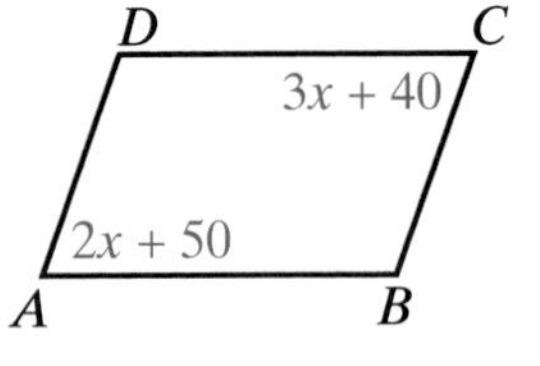

a. Find the value of x.

b. Find the measure of each angle.

Solution **a.** In $\square ABCD$, $m\angle C = m\angle A$ because the opposite angles of a parallelogram have equal measure. Thus:

$$3x + 40 = 2x + 50$$
$$x = 10$$

b. By substitution:

$$m\angle A = 2x + 50 = 2(10) + 50 = 70$$
$$m\angle B = 180 - m\angle A = 180 - 70 = 110$$
$$m\angle C = 3x + 40 = 3(10) + 40 = 70$$
$$m\angle D = 180 - m\angle C = 180 - 70 = 110$$

Answer **a.** $x = 10$

b. $m\angle A = 70$

$m\angle B = 110$

$m\angle C = 70$

$m\angle D = 110$

EXERCISES

Writing About Mathematics

1. Adam said that if a quadrilateral has four equal angles then the parallelogram is a rectangle. Do you agree with Adam? Explain why or why not.

2. Emmanuel said that if a parallelogram has one right angle then the parallelogram is a rectangle. Do you agree with Emmanuel? Explain why or why not.

Developing Skills

In 3–8, in each case, is the statement true or false?

3. If a polygon is a trapezoid, it is a quadrilateral.

4. If a polygon is a rectangle, it is a parallelogram.

5. If a polygon is a rhombus, it is a parallelogram.

6. If a polygon is a rhombus, it is a square.

7. If a polygon is a square, then it is a rhombus.

8. If two angles are opposite angles of a parallelogram, they are congruent.

In 9–12, the angle measures are represented in each quadrilateral. In each case:

a. Find the value of x.

b. Find the measure of each angle of the quadrilateral.

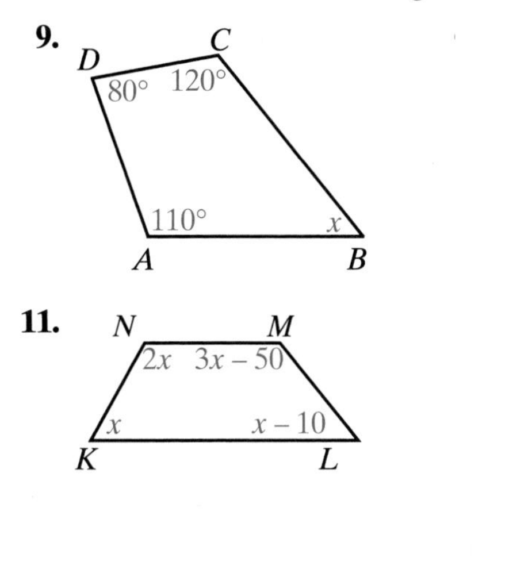

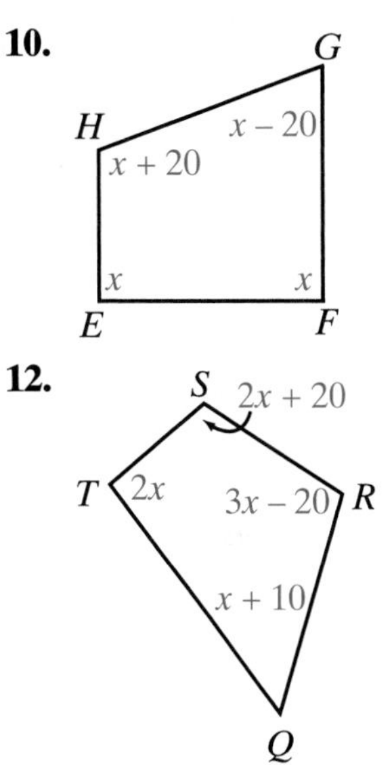

In 13 and 14, polygon $ABCD$ is a parallelogram.

13. $AB = 3x + 8$ and $DC = x + 12$. Find AB and DC.

14. $m\angle A = 5x - 40$ and $m\angle C = 3x + 20$. Find $m\angle A$, $m\angle B$, $m\angle C$, and $m\angle D$.

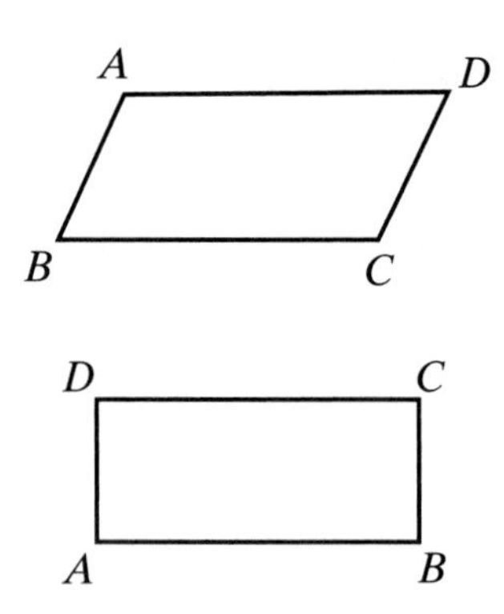

In 15 and 16, polygon $ABCD$ is a rectangle.

15. $BC = 4x - 5$, $AD = 2x + 3$. Find BC and AD.

16. $m\angle A = 5x - 10$. Find the value of x.

17. $ABCD$ is a square. If $AB = 8x - 6$ and $BC = 5x + 12$, find the length of each side of the square.

18. In rhombus $KLMN$, $KL = 3x$ and $LM = 2(x + 3)$. Find the length of each side of the rhombus.

Applying Skills

19. One side of a barn is in the shape of a seven-sided polygon called a heptagon. The heptagon, $ABCDEFG$, can be divided into rectangle $ABFG$, isosceles trapezoid $BCEF$, and isosceles triangle CDE, as shown in the diagram.

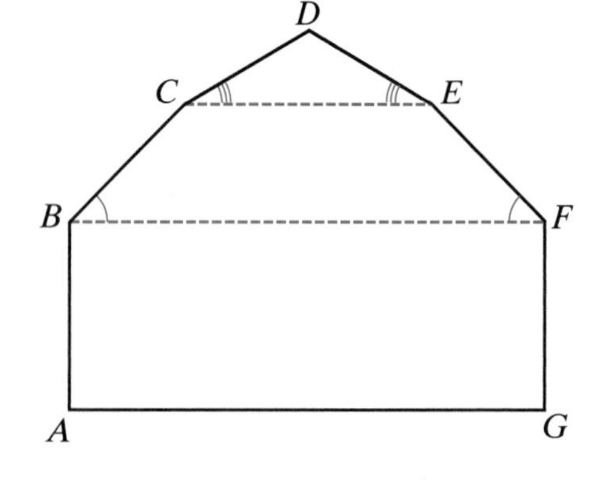

a. Find the sum of the measures of the interior angles of $ABCDEFG$ by using the sum of the measures of the angles of the two quadrilaterals and the triangle.

b. Sketch the heptagon on your answer paper and show how it can be divided into triangles by drawing diagonals from A. Use these triangles to find the sum of the measures of the interior angles of $ABCDEFG$.

c. If the measure of each lower base angle of trapezoid $BCEF$ is 45° and the measure of each base angle of isosceles triangle CDE is 30°, find the measure of each angle of heptagon $ABCDEFG$.

d. Find the sum of the measures of the interior angles of $ABCDEFG$ by adding the angle measures found in **c.**

20. Cassandra had a piece of cardboard that was in the shape of an equilateral triangle. She cut an isosceles triangle from each vertex of the cardboard. The length of each leg of the triangles that she cut was one third of the length of a side of the original triangle.

a. Show that each of the triangles that Cassandra cut off is an equilateral triangle.

b. What is the measure of each angle of the remaining piece of cardboard?

c. What is the shape of the remaining piece of cardboard?

21. A piece of land is bounded by two parallel roads and two roads that are not parallel forming a trapezoid. Along the parallel roads the land measures 1.3 miles and 1.7 miles. The distance between the parallel roads, measured perpendicular to the roads, is 2.82 miles.

a. Find the area of the land. Express the area to the nearest tenth of a mile.

b. An acre is a unit of area often used to measure land. There are 640 acres in a square mile. Express, to the nearest hundred acres, the area of the land.

22. If possible, draw the following quadrilaterals. If it is not possible, state why.

a. A quadrilateral with four acute angles.

b. A quadrilateral with four obtuse angles.

c. A quadrilateral with one acute angle and three obtuse angles.

d. A quadrilateral with three acute angles and one obtuse angle.

e. A quadrilateral with exactly three right angles.

7-6 AREAS OF IRREGULAR POLYGONS

Many polygons are irregular figures for which there is no formula for the area. However, the area of such figures can often be found by separating the figure into regions with known area formulas and adding or subtracting these areas to find the required area.

EXAMPLE 1

$ABCD$ is a square. E is a point on $\overline{DC}$ and F is a point on $\overline{BC}$. If $AB = 12$, $FC = 3$, and $DE = 8$, find the area of $ABFE$.

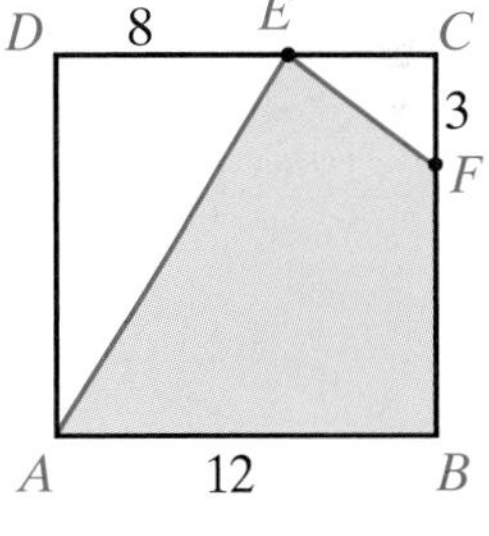

Solution Since $ABCD$ is a square and $AB = 12$, then $BC = 12$, $CD = 12$, and $DA = 12$.

Therefore, $CE = CD - DE = 12 - 8 = 4$.

$$\begin{aligned} \text{Area of } ABFE &= \text{Area of } ABCD - \text{Area of } \triangle FCE - \text{Area of } \triangle EDA \\ &= 12(12) - \tfrac{1}{2}(3)(4) - \tfrac{1}{2}(8)(12) \\ &= 144 - 6 - 48 \\ &= 90 \end{aligned}$$

Answer 90 square units

EXAMPLE 2

$ABCD$ is a quadrilateral. $AB = 5.30$ centimeters, $BC = 5.5$ centimeters, $CD = 3.2$ centimeters, and $DA = 3.52$ centimeters. Angle A and angle C are right angles. Find the area of $ABCD$ to the correct number of significant digits.

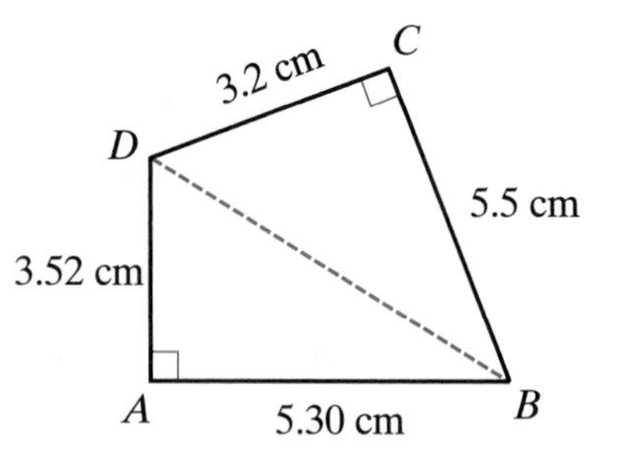

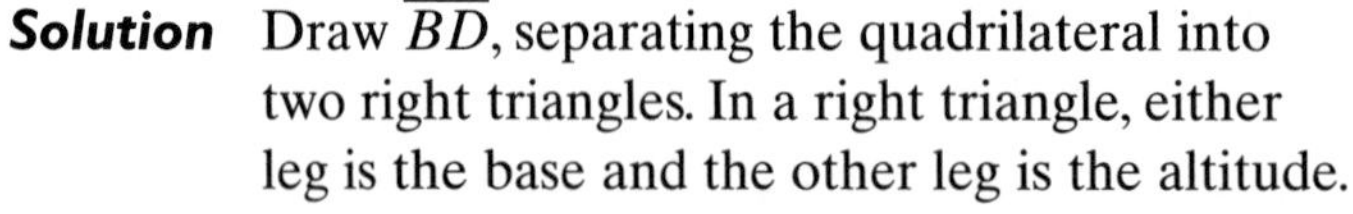

Solution Draw $\overline{BD}$, separating the quadrilateral into two right triangles. In a right triangle, either leg is the base and the other leg is the altitude.

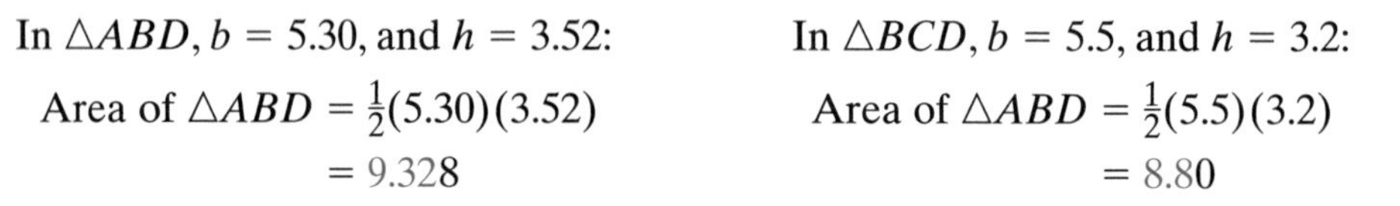

In $\triangle ABD$, $b = 5.30$, and $h = 3.52$:

$$\text{Area of } \triangle ABD = \tfrac{1}{2}(5.30)(3.52) = 9.328$$

In $\triangle BCD$, $b = 5.5$, and $h = 3.2$:

$$\text{Area of } \triangle ABD = \tfrac{1}{2}(5.5)(3.2) = 8.80$$

The first area should be given to three significant digits, the least number of significant digits involved in the calculation. Similarly, the second area should be given to two significant digits. However, the area of $ABCD$ should be no more precise than the least precise measurement of the areas that are added. Since the least precise measurement is 8.8, given to the nearest tenth, the area of $ABCD$ should also be written to the nearest tenth:

$$\begin{aligned}\text{Area of } ABCD &= \text{Area of } \triangle ABD + \text{Area of } \triangle BCD \\ &= 9.328 + 8.80 \\ &= 18.128 \\ &= 18.1\end{aligned}$$

Answer 18.1 square centimeters

Note: When working with significant digits involving multiple operations, make sure to keep at least one extra digit for intermediate calculations to avoid *round-off error*.

Alternatively, when working with a calculator, you can round at the end of the entire calculation.

EXERCISES

Writing About Mathematics

1. $ABCD$ is a trapezoid with $\overline{AB} \parallel \overline{CD}$. The area of $\triangle ABD$ is 35 square units. What is the area of $\triangle ABC$? Explain your answer.

2. $ABCD$ is a quadrilateral. The area of $\triangle ABD$ is 57 square inches, of $\triangle BDC$ is 62 square inches, and of $\triangle ABC$ is 43 square inches. What is the area of $\triangle ADC$? Explain your answer.

Developing Skills

In 3–10, find each measure to the correct number of significant digits.

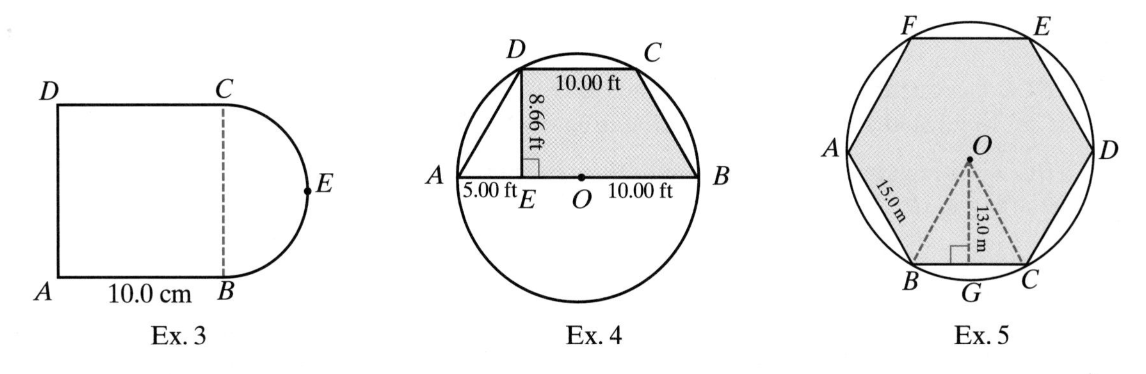

Ex. 3 Ex. 4 Ex. 5

3. $ABCD$ is a square and arc $\widehat{BEC}$ is a semicircle. If $AB = 10.0$ cm, find the area of the figure.

4. $ABCD$ is a trapezoid. The vertices of the trapezoid are on a circle whose center is at O. $\overline{DE} \perp \overline{AB}$, $OB = DC = 10.00$ ft, $DE = 8.66$ ft, and $AE = 5.00$ ft. Find the area of $EBCD$.

5. $ABCDEF$ is a regular hexagon. The vertices of the hexagon are on a circle whose center is at O. $\overline{OG} \perp \overline{BC}$, $AB = 15.0$ m and $OG = 13.0$ m. Find the area of $ABCDEF$.

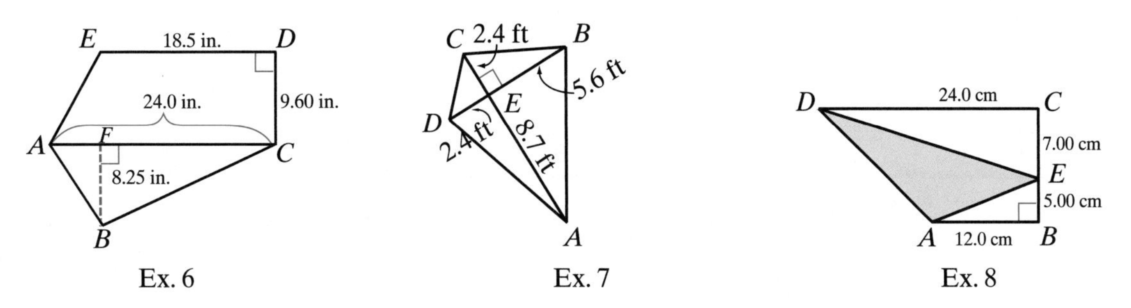

Ex. 6 Ex. 7 Ex. 8

6. $ABCDE$ is a pentagon. $\overline{DC} \perp \overline{ED}$, $\overline{ED} \parallel \overline{AC}$, and $\overline{BF} \perp \overline{AC}$. If $ED = 18.5$ in., $DC = 9.60$ in., $AC = 24.0$ in., and $BF = 8.25$ in., find the area of $ABCDE$.

7. $ABCD$ is a quadrilateral. The diagonals of the quadrilateral are perpendicular to each other at E. If $AE = 8.7$ ft, $BE = 5.6$ ft, $CE = 2.4$ ft, and $DE = 2.4$ ft, find the area of $ABCD$.

8. $ABCD$ is a trapezoid with $\overline{BC} \perp \overline{AB}$ and E a point on $\overline{BC}$. $AB = 12.0$ cm, $DC = 24.0$ cm, $BE = 5.00$ cm, and $EC = 7.00$ cm. Find the area of triangle AED.

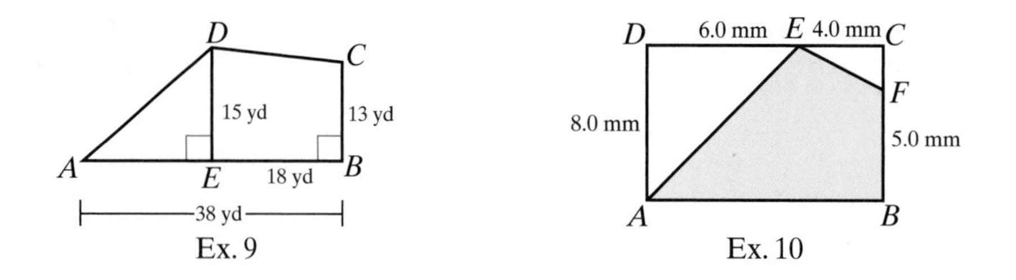

Ex. 9
Ex. 10

9. $ABCD$ is a quadrilateral with $\overline{BC} \perp \overline{AB}$ and $\overline{DE} \perp \overline{AB}$. If $AB = 38$ yd, $EB = 18$ yd, $BC = 13$ yd, and $DE = 15$ yd. Find the area of $ABCD$.

10. $ABCD$ is a rectangle. $AD = 8.0$ mm, $DE = 6.0$ mm, $EC = 4.0$ mm, and $BF = 5.0$ mm. Find the area of $ABFE$.

Applying Skills

11. Mason has a square of fabric that measures one yard on each side. He makes a straight cut from the center of one edge to the center of an adjacent edge. He now has one piece that is an isosceles triangle and another piece that is a pentagon. Find the area of each of these pieces in square inches.

12. A park is in the shape of isosceles trapezoid $ABCD$. The bases of the trapezoid, $\overline{AB}$ and $\overline{CD}$ measure 20.0 meters and 5.00 meters respectively. The measure of each base angle at $\overline{AB}$ is 60°, and the height of the trapezoid is 13.0 meters. A straight path from A to E, a point on $\overline{BC}$, makes an angle of 30° with $\overline{AB}$ and $BE = 10.0$ meters. The path separates the trapezoid into two regions; a quadrilateral planted with grass and a triangle planted with shrubs and trees. Find the area of the region planted with grass to the nearest meter.

7-7 SURFACE AREAS OF SOLIDS

A **right prism** is a solid with congruent bases and with a height that is perpendicular to these bases. Some examples of right prisms, as seen in the diagrams, include solids whose bases are triangles, trapezoids, and rectangles. The prism is named for the shape of its base. The two bases may be any polygons, as long as they have the same size and shape. The remaining sides, or **faces**, are rectangles.

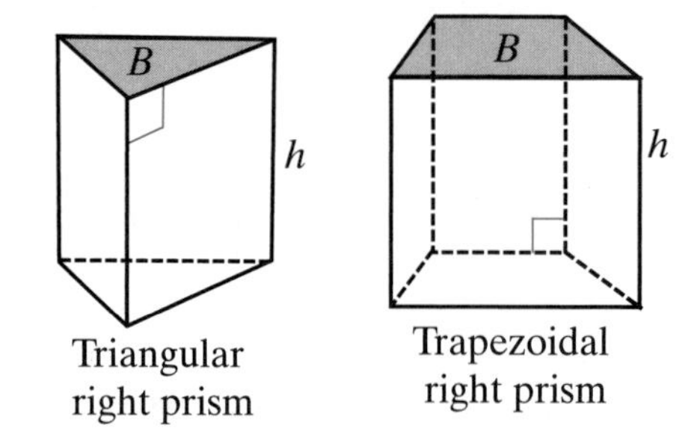

Triangular right prism
Trapezoidal right prism

The **surface area** of a solid is the sum of the areas of the surfaces that make up the solid. The surface area of a right prism is the sum of the areas of the bases and the faces of the solid. The number of faces of a prism is equal to the number of sides of a base.

- If the base of the prism is a triangle, it has 2 bases + 3 faces or 5 surfaces.
- If the base of the prism is a quadrilateral, it has 2 bases + 4 faces or 6 surfaces.
- If the base of the prism has n sides, it has 2 bases + n faces or $2 + n$ surfaces.

Two or more of the faces are congruent if and only if two or more of the sides of a base are equal in length. For example, if the bases are isosceles triangles, two of the faces will be congruent rectangles and if the bases are equilateral triangles, all three of the faces will be congruent rectangles. If the bases are squares, then the four faces will be congruent rectangles.

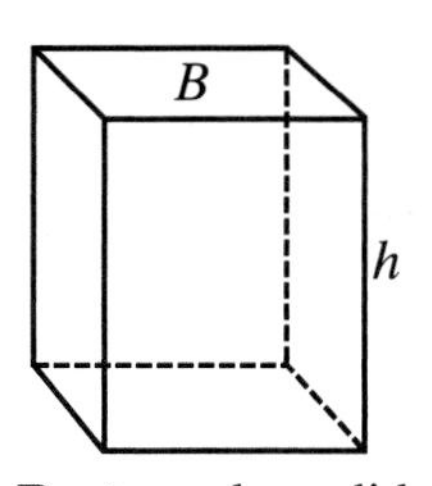

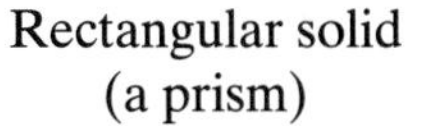
Rectangular solid (a prism)

The most common right prism is a **rectangular solid** that has rectangles as bases and as faces. Any two surfaces that have no edge in common can be the bases and the other four surfaces are the faces. If the dimensions of the rectangular solid are 3 by 5 by 4, then there are two rectangles that are 3 by 5, two that are 5 by 4 and two that are 3 by 4. The surface area of the rectangular solid is:

$$2(3)(5) + 2(5)(4) + 2(3)(4) = 30 + 40 + 24 = 94 \text{ square units}$$

In general, when the dimensions of a rectangular solid are represented by l, w, and h, then the formula for the surface area is:

Surface Area of a Rectangular Solid = $2lw + 2lh + 2wh$

If the rectangular solid has six squares as bases and sides, the solid is a cube. If s represents the length of each side, then $l = s$, $w = s$, and $h = s$. If we substitute in the formula for the area of a rectangular solid, then:

$$S = 2lw + 2lh + 2wh = 2s(s) + 2s(s) + 2s(s) = 2s^2 + 2s^2 + 2s^2 = 6s^2$$

Surface Area of a Cube = $6s^2$

A **right circular cylinder** is a solid with two bases that are circles of the same size, and with a height that is perpendicular to these bases.

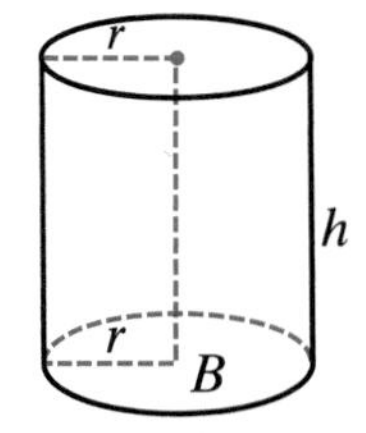

Right circular cylinder

Fruits and vegetables are often purchased in "cans" that are in the shape of a right circular cylinder. The label on the can corresponds to the surface area of the curved portion of the cylinder. That label is a rectangle whose length is the circumference of the can, $2\pi r$, and whose width is the height of the can, h. Therefore, the area of the curved portion of the cylinder is $2\pi rh$ and the area of each base is πr^2.

Surface Area of a Cylinder = 2 × the area of a base + the area of the curved portion

Surface Area of a Cylinder = $2\pi r^2 + 2\pi rh$

EXAMPLE 1

A rectangular solid has a square base. The height of the solid is 2 less than twice the length of one side of the square. The height of the solid is 22.2 centimeters. Find the surface area of the figure to the correct number of significant digits.

Solution

Let x = the length of a side of the base.

$2x - 2$ = the height of the figure.

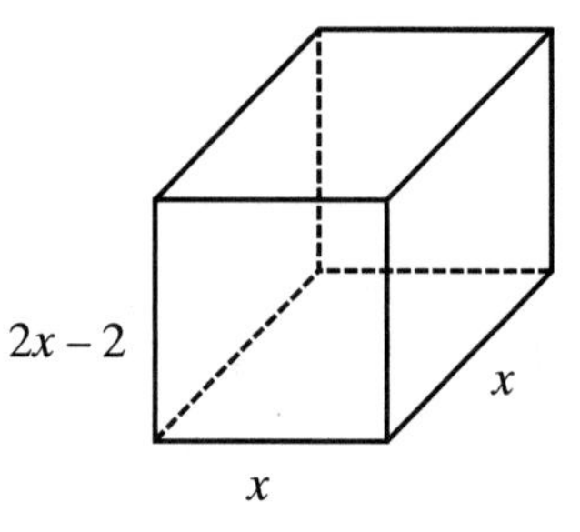

$$2x - 2 = 22.2$$
$$2x = 24.2$$
$$x = 12.1$$

The surface area consists of two square bases with sides that measure 12.1 centimeters and four rectangular faces with width 12.1 and length 22.2.

Surface area of the solid = surface area of the bases + surface area of the faces

$$= 2(12.1)^2 + 4(12.1)(22.2)$$
$$= 292.82 + 1{,}074.48$$
$$= 1{,}367.3$$
$$= 1{,}370$$

The areas of the bases should be given to three significant digits, the least number of significant digits involved in the calculation. Similarly, the areas of the faces should be given to three significant digits. However, the surface area of the solid should be no more precise than the least precise measurement of the areas that are added. Since the least precise measurement is 1,074.48, given to the nearest ten, the surface area of the solid should also be written to the nearest ten: 1,370 square centimeters.

Answer Surface area = 1,370 cm^2

Note: Recall that when working with significant digits involving multiple operations, you should keep at least one extra digit for intermediate calculations *or* wait until the end of the entire calculation to round properly.

EXERCISES

Writing About Mathematics

1. A right prism has bases that are regular hexagons. The measure of each of the six sides of the hexagon is represented by a and the height of the solid by $2a$.

a. How many surfaces make up the solid?

b. Describe the shape of each face

c. Express the dimensions and the area of each face in terms of a.

2. A regular hexagon can be divided into six equilateral triangles. If the length of a side of an equilateral triangle is a, the height is $\frac{\sqrt{3}}{2}a$. For the rectangular solid described in exercise 1:

a. Express the area of each base in terms of a.

b. Express the surface area in terms of a.

Developing Skills

In 3–9, find the surface area of each rectangular prism or cylinder to the nearest *tenth of a square unit.*

3. Bases are circles with a radius of 18 inches. The height of the cylinder is 48 inches.

4. Bases are squares with sides that measure 27 inches. The height is 12 inches.

5. Bases are rectangles with dimensions of 8 feet by 12 feet. The height is 3 feet.

6. Bases are isosceles trapezoids with parallel sides that measure 15 centimeters and 25 centimeters. The distance between the parallel sides is 12 centimeters and the length of each of the equal sides is 13 centimeters. The height of the prism is 20 centimeters.

7. Bases are isosceles right triangles with legs that measure 5 centimeters. The height is 7 centimeters.

8. Bases are circles with a diameter of 42 millimeters. The height is 3.4 centimeters.

9. Bases are circles each with an area of 314.16 square feet. The height is 15 feet.

Applying Skills

10. Agatha is using scraps of wallpaper to cover a box that is a rectangular solid whose base measures 8 inches by 5 inches and whose height is 3 inches. The box is open at the top. How many square inches of wallpaper does she need to cover the outside of the box?

11. Agatha wants to make a cardboard lid for the box described in exercise 10. Her lid will be a rectangular solid that is open at the top, with a base that is slightly larger than that of the box. She makes the base of the lid 8.1 inches by 5.1 inches with a height of 2.0 inches. To the nearest tenth of a square inch, how much wallpaper does she need to cover the outside of the lid?

12. Sandi wants to make a pillow in the shape of a right circular cylinder. The diameter of the circular ends is 10.0 inches and the length of the pillow (the height of the cylinder) is 16.0 inches. Find the number of square inches of fabric Sandi needs to make the pillow to the correct number of significant digits.

13. Mr. Breiner made a tree house for his son. The front and back walls of the house are trapezoids to allow for a slanted roof. The floor, roof, and remaining two sides are rectangles. The tree house is a rectangular solid. The front and back walls are the bases of this solid. The dimensions of the floor are 8 feet by 10 feet and the roof is 10 feet by 10 feet. The height of one side wall is 7 feet and the height of the other is 13 feet. The two side walls each contain a window measuring 3 feet by 3 feet. Mr. Breiner is going to buy paint for the floor and the exterior of the house, including the roof, walls, and the door. How many square feet must the paint cover, excluding the windows?

7-8 VOLUMES OF SOLIDS

The **volume** of a solid is the number of unit cubes (or cubic units) that it contains. Both the volume V of a right prism and the volume V of a right circular cylinder can be found by multiplying the area B of the base by the height h. This formula is written as

$$V = Bh$$

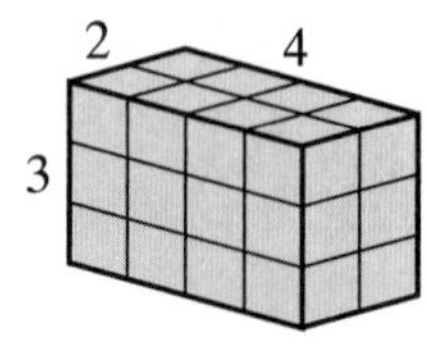

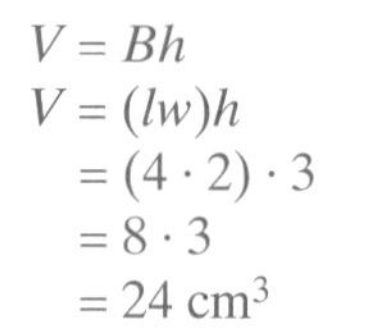

To find the volume of a solid, all lengths *must* be expressed in the same unit of measure. The volume is then expressed in cubic units of this length.

For example, in the rectangular solid or right prism shown at left, the base is a rectangle, 4 centimeters by 2 centimeters. The area, B, of this base is lw. Therefore, $B = 4(2) = 8 \text{ cm}^2$. Then, using a height h of 3 centimeters, the volume V of the rectangular solid $= Bh = (8)(3) = 24 \text{ cm}^3$.

For a rectangular solid, note that the volume formula can be written in two ways:

$$V = Bh \qquad \text{or} \qquad V = lwh$$

To understand volume, count the cubes in the diagram on the previous page. There are 3 layers, each containing 8 cubes, for a total of 24 cubes. Note that 3 corresponds to the height, h, that 8 corresponds to the area of the base, B, and that 24 corresponds to the volume V in cubic units.

A cube that measures 1 foot on each side represents 1 cubic foot. Each face of the cube is 1 square foot. Since each foot can be divided into 12 inches, the area of each face is 12×12 or 144 square inches and the volume of the cube is $12 \times 12 \times 12$ or 1,728 cubic inches.

1 square foot = 144 square inches

1 cubic foot = 1,728 cubic inches

A cube that measures 1 meter on each side represents 1 cubic meter. Each face of the cube is 1 square meter. Since each meter can be divided into 100 centimeters, the area of each face is 100×100 or 10,000 square centimeters and the volume of the cube is $100 \times 100 \times 100$ or 1,000,000 cubic centimeters.

1 square meter = 10,000 square centimeters

1 cubic meter = 1,000,000 cubic centimeters

EXAMPLE 1

A cylindrical can of soup has a radius of 1.5 inches and a height of 5 inches. Find the volume of this can:

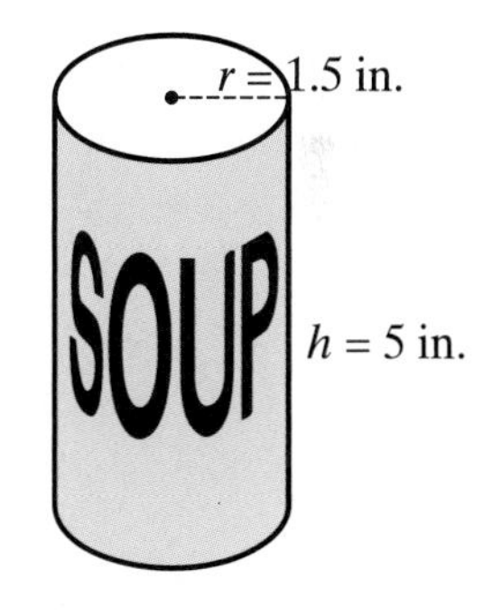

a. in terms of π.

b. to the nearest cubic inch.

Solution This can is a right circular cylinder. Use the formula $V = Bh$. Since the base is a circle whose area equals πr^2, the area B of the base can be replaced by πr^2.

a. $V = Bh = (\pi r^2)h = \pi(1.5)^2(5) = 11.25\pi$

b. When we use a calculator to evaluate 11.25π, the calculator gives 35.34291735 as a rational approximation. This answer rounded to the nearest integer is 35.

Answers **a.** 11.25π cu in. **b.** 35 cu in.

A **pyramid** is a solid figure with a base that is a polygon and triangular faces or sides that meet at a common point. The formula for the volume V of a pyramid is:

$$V = \frac{1}{3}Bh$$

In the pyramid and prism shown here, the bases are the same in size and shape, and the heights are equal in measure. If the pyramid could be filled with water and that water poured into the prism, exactly three pyramids of water would be needed to fill the prism. In other words, the volume of a pyramid is one-third the volume of a right prism with the same base and same height as those of the pyramid.

Base (B) Height (h) Base (B)

Pyramid $= \frac{1}{3}$ prism

The volume of a **cone** is one-third of the volume of a right circular cylinder, when both the cone and the cylinder have circular bases and heights that are equal in measure. The formula for the volume of the cone is

Height (h) r Base (B) r Base (B)

$$V = \frac{1}{3}Bh \qquad \text{or} \qquad V = \frac{1}{3}\pi r^2 h$$

As in the case of the pyramid and prism, if the cone could be filled with water and that water poured into the right circular cylinder, exactly three cones of water would be needed to fill the cylinder.

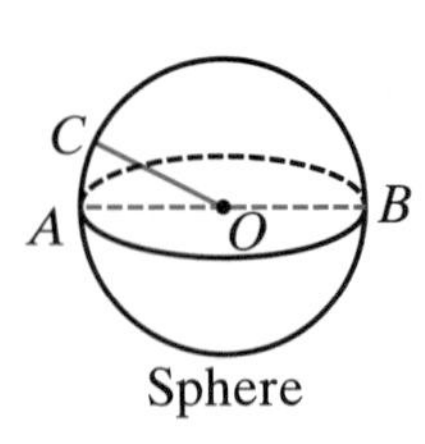

Sphere

A **sphere** is a *solid* figure whose points are all equally distant from one fixed point in space called its **center.** A sphere is *not* a circle, which is drawn on a flat surface or plane but similar terminology is used for a sphere. A line segment, such as $\overline{OC}$, that joins the center O to any point on the sphere is called a **radius** of the sphere. A line segment, such as $\overline{AB}$, that joins two points of the sphere and passes through its center is called a **diameter** of the sphere.

The volume of a sphere with radius r is found by using the formula:

$$V = \frac{4}{3}\pi r^3$$

EXAMPLE 2

An ice cream cone that has a diameter of 6.4 centimeters at its top and a height of 12.2 centimeters is filled with ice cream and topped with a scoop of ice cream that is approximately in the shape of half of a sphere. Using the correct number of significant digits, how many cubic centimeters of ice cream are needed?

Solution The amount of ice cream needed is approximately the volume of the cone plus one-half the volume of a sphere with a radius equal to the radius of the top of the cone.

$$\begin{aligned}\text{Volume of ice cream} &= \tfrac{1}{3}\pi r^2 h + \tfrac{1}{2}\left(\tfrac{4}{3}\pi r^3\right)\\ &= \tfrac{1}{3}\pi(3.2)^2(12.2) + \tfrac{1}{2}\left(\tfrac{4}{3}\right)\pi(3.2)^3\\ &= 130.824\ldots + 68.629\ldots\\ &= 199.453\ldots\end{aligned}$$

The least precise volume is the volume of the cone, given to the nearest ten. Therefore, the answer should be rounded to the nearest ten. Note that in the answer, the zero in the tens place is significant. The zero in the ones place is not significant.

Answer The volume of ice cream is 200 cubic centimeters.

Error in Geometric Calculations

When a linear measure is used to find area or volume, any error in the linear value will be increased in the higher-dimension calculations.

EXAMPLE 3

The length of a side of a cube that is actually 10.0 centimeters is measured to be 10.5 centimeters. Find the percent error in:

a. the linear measure.

b. the surface area.

c. the volume.

d. compare the results in **a**–**c** above.

Solution **a.** The true length is 10.0 centimeters. The measured value is 10.5 centimeters.

$$\begin{aligned}\text{Percent error} &= \frac{|\text{measured value} - \text{true value}|}{\text{true value}} \times 100\%\\ &= \frac{|10.5 - 10.0|}{10.0} \times 100\% = \frac{0.5}{10.0} \times 100\%\\ &= 0.05 \times 100\% = 5\% \quad \textit{Answer}\end{aligned}$$

b. For a cube, $S = 6s^2$.

Using the measured value: $S = 6(10.5)^2 = 661.5$

Using the true value: $S = 6(10.0)^2 = 600.0$

$$\text{Percent error} = \frac{|661.5 - 600.0|}{600.0} \times 100\% = \frac{61.5}{600.0} \times 100\%$$
$$= 0.1025 \times 100\% = 10.25\% \quad \textit{Answer}$$

c. For a cube, $V = s^3$.

Using the measured value: $V = (10.5)^3 = 1{,}157.625$

Using the true value: $V = (10.0)^3 = 1{,}000$

$$\text{Percent error} = \frac{|1{,}157.625 - 1{,}000|}{1{,}000} \times 100\% = \frac{157.625}{1{,}000} \times 100\%$$
$$= 0.157625 \times 100\% \approx 15.76\% \quad \textit{Answer}$$

d. The error in the linear measure is 5%. The error increases to 10.25% when the surface area is calculated. The error increased even more to 15.76% when the volume is calculated. *Answer*

EXERCISES

Writing About Mathematics

1. A chef purchases thin squares of dough that she uses for the top crust of pies. The squares measure 8 inches on each side. From each square of dough she cuts either one circle with a diameter of 8 inches or four circles, each with a diameter of 4 inches.

a. Compare the amount of dough left over when she makes one 8-inch circle with that left over when she makes four 4-inch circles.

b. The chef uses the dough to form the top crust of pot pies that are 1.5 inches deep. Each pie is approximately a right circular cylinder. Compare the volume of one 8-inch pie to that of one 4-inch pie.

2. Tennis balls can be purchased in a cylindrical can in which three balls are stacked one above the other. How does the radius of each ball compare with the height of the can in which they are packaged?

Developing Skills

In 3–6, use the formula $V = lwh$ and the given dimensions to find the volume of each rectangular solid using the correct number of significant digits.

3. $l = 5.0$ ft, $w = 4.0$ ft, $h = 7.0$ ft

4. $l = 8.5$ cm, $w = 4.2$ cm, $h = 6.0$ cm

5. $l = 2\frac{1}{2}$ m, $w = 3\frac{1}{4}$ cm, $h = 85$ cm

6. $l = 7.25$ in., $w = 6.40$ in., $h = 0.25$ ft

7. Find the volume of a cube if each edge measures $8\frac{3}{5}$ centimeters.

8. The measure of each edge of a cube is represented by $(2y - 3)$ centimeters. Find the volume of the cube when $y = 7.25$.

9. The base of a right prism is a triangle. One side of the triangular base measures 8 centimeters and the altitude to that side measures 6 centimeters. The height h of the prism measures 35 millimeters.

a. Find the area of the triangular base of the prism.

b. Find the volume of the prism.

10. The base of a right prism is a trapezoid. This trapezoid has bases that measure 6 feet and 10 feet and an altitude that measures 4 feet. The height h of the prism is 2 yards.

a. Find the area of the trapezoidal base of the prism.

b. Find the volume of the prism.

11. The base of a pyramid has an area B of 48 square millimeters and a height h of 13 millimeters. What is the volume of the pyramid?

12. The height of a pyramid is 4 inches, and the base is a rectangle 6 inches long and $3\frac{1}{2}$ inches wide. What is the volume of the pyramid?

13. A right circular cylinder has a base with a radius of 24.1 centimeters and a height of 17.3 centimeters.

a. Express the volume of the cylinder in terms of π.

b. Find the volume of the cylinder to the nearest hundred cubic centimeters.

14. A right circular cylinder has a base with a diameter of 25 meters and a height of 15 meters.

a. Express the volume of the cylinder in terms of π.

b. Find the volume of the cylinder to the nearest ten cubic meters.

15. The base of a cone has a radius of 7 inches. The height of the cone is 5 inches.

a. Find the volume of the cone in terms of π.

b. Find the volume of the cone to the nearest cubic inch.

16. The base of a cone has a radius of 7 millimeters. The height of the cone is 2 centimeters.

a. Find the volume of the cone in terms of π.

b. Express the volume of the cone to the nearest cubic centimeter.

17. A sphere has a radius of 12.5 centimeters.

a. Find the volume of the sphere in terms of π.

b. Express the volume of the sphere to the nearest cubic centimeter.

18. A sphere has a diameter of 3 feet.

a. Find the volume of the sphere in terms of π.

b. Express the volume of the sphere to the nearest cubic foot.

19. The side of a cube that is actually 8 inches is measured to be 7.6 inches. Find, to the nearest tenth of a percent, the percent error in:

a. the length of the side.

b. the surface area.

c. the volume.

Applying Skills

20. An official handball has a diameter of 4.8 centimeters. Find its volume:

a. to the nearest cubic centimeter **b.** to the nearest cubic inch.

21. A tank in the form of a right circular cylinder is used for storing water. It has a diameter of 12 feet and a height of 14 feet. How many gallons of water will it hold? (1 cubic foot contains 7.5 gallons.)

22. Four pieces of cardboard that are 8.0 inches by 12 inches and two pieces that are 12 inches by 12 inches are used to form a rectangular solid.

a. What is the surface area of the rectangular solid formed by the six pieces of cardboard?

b. What is the volume of the rectangular solid in cubic inches?

c. What is the volume of the rectangular solid in cubic feet?

23. A can of soda is almost in the shape of a cylinder with a diameter of 6.4 centimeters and a height of 12.3 centimeters.

a. What is the volume of the can?

b. If there are 1,000 cubic centimeters in a liter, find how many liters of soda the can holds.

In 24–26, express each answer to the correct number of significant digits.

24. Cynthia used a shipping carton that is a rectangular solid measuring 12.0 inches by 15.0 inches by 3.20 inches. What is the volume of the carton?

25. The highway department stores sand in piles that are approximately the shape of a cone. What is the volume of a pile of sand if the diameter of the base is 7.0 yards and the height of the pile is 8.0 yards?

26. The largest pyramid in the world was built around 2500 B.C. by Khufu, or Cheops, a king of ancient Egypt. The pyramid had a square base $23\dot{0}$ meters (756 feet) on each side, and a height of 147 meters (482 feet). (The length of a side of the base is given to the nearest meter. The zero in the ones place is significant.) Find the volume of Cheops' pyramid using:

a. cubic meters **b.** cubic feet

CHAPTER SUMMARY

Point, **line**, and **plane** are undefined terms that are used to define other terms. A **line segment** is a part of a line consisting of two points on the line, called **endpoints**, and all of the points on the line between the endpoints.

A **ray** is a part of a line that consists of a point on the line and all of the points on one side of that point. An **angle** is the union of two rays with a common endpoint.

Two angles are **complementary** if the sum of their measures is 90°. If the measure of an angle is $x°$, the measure of its complement is $(90 - x)°$.

Two angles are **supplementary** if the sum of their measures is 180°. If the measure of an angle is $x°$, the measure of its supplement is $(180 - x)°$.

A **linear pair of angles** are adjacent angles that are supplementary. Two angles are **vertical angles** if the sides of one are opposite rays of the sides of the other. Vertical angles are **congruent**.

If two parallel lines are cut by a transversal, then:

- The alternate interior angles are congruent.
- The alternate exterior angles are congruent.
- The corresponding angles are congruent.
- Interior angles on the same side of the transversal are supplementary.

The sum of the measures of the angles of a triangle is 180°.
The base angles of an isosceles triangle are congruent.
An equilateral triangle is equiangular.
The sum of the measures of the angles of a quadrilateral is 360°. The sum of the measures of the angles of any polygon with n sides is $180(n - 2)$.

If the measures of the bases of a trapezoid are b_1 and b_2 and the measure of the altitude is h, then the formula for the area of the trapezoid is $A = \frac{1}{2}h(b_1 + b_2)$.

Formulas for Surface Area
Rectangular solid: $S = 2lw + 2lh + 2wh$
Cube: $S = 6s^2$
Cylinder: $S = 2\pi r^2 + 2\pi rh$

Formulas for Volume
Any right prism: $V = Bh$
Rectangular solid: $V = lwh$
Right circular cylinder: $V = \pi r^2 h$
Pyramid: $V = \frac{1}{3}Bh$
Cone: $V = \frac{1}{3}\pi r^2 h$
Cube: $V = s^3$
Sphere: $V = \frac{4}{3}\pi r^3$

VOCABULARY

7-1 Undefined term • Point • Line • Straight line • Plane • Axiom (postulate) • Line segment (segment) • Endpoints of a segment • Measure of a line segment (length of a line segment) • Half-line • Ray • Endpoint of a ray • Opposite rays • Angle • Vertex of an angle • Sides an angle • Degree • Right angle • Acute angle • Obtuse angle • Straight angle • Perpendicular

7-2 Adjacent angles • Complementary angles • Complement • Supplementary angles • Linear pair of angles • Vertical angles • Congruent angles • Theorem

7-3 Parallel lines • Transversal • Interior angles • Alternate interior angles • Exterior angles • Alternate exterior angles • Interior angles on the same side of the transversal • Corresponding angles

7-4 Polygon • Sides of a polygon • Vertices • Triangle • Acute triangle • Equiangular triangle • Right triangle • Obtuse triangle • Legs of a right triangle • Hypotenuse • Scalene triangle • Isosceles triangle • Equilateral triangle • Legs of an isosceles triangle • Base of an isosceles triangle • Congruent line segments • Vertex angle of an isosceles triangle • Base angles of an isosceles triangle • Construction • Compass • Straightedge • Perpendicular bisector • Angle bisector

7-5 Quadrilateral • Consecutive angles • Opposite angles • Trapezoid • Bases of a trapezoid • Parallelogram • Rectangle • Rhombus • Square • Isosceles trapezoid

7-7 Right prism • Face • Surface area • Rectangular solid • Right circular cylinder

7-8 Volume • Pyramid • Cone • Sphere • Center of a sphere • Radius of a sphere • Diameter of a sphere

REVIEW EXERCISES

1. If $\overleftrightarrow{AB}$ and $\overleftrightarrow{CD}$ intersect at E, $m\angle AEC = x + 10$, and $m\angle DEB = 2x - 30$, find $m\angle AEC$.

2. If two angles of a triangle are complementary, what is the measure of the third angle?

3. The measure of the complement of an angle is 20° less than the measure of the angle. Find the number of degrees in the angle.

4. If each base angle of an isosceles triangle measures 55°, find the measure of the vertex angle of the triangle.

5. The measures of the angles of a triangle are consecutive even integers. What are the measures of the angles?

In 6–8, $\overleftrightarrow{AB}$ is parallel to $\overleftrightarrow{CD}$, and these lines are cut by transversal $\overleftrightarrow{EF}$ at points G and H, respectively.

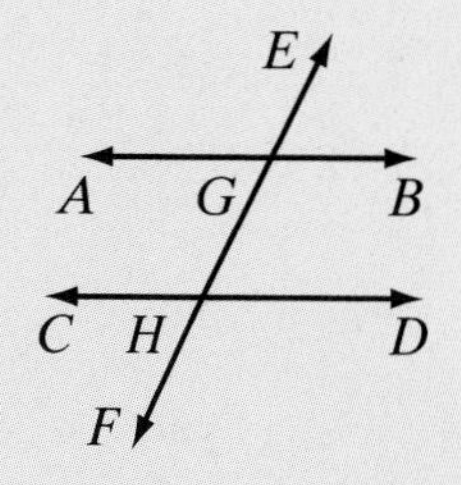

6. If $m\angle AGH = 73$, find $m\angle GHD$.

7. If $m\angle EGB = 70$ and $m\angle GHD = 3x - 2$, find x.

8. If $m\angle HGB = 2x + 10$ and $m\angle GHD = x + 20$, find $m\angle GHD$.

9. In $\triangle ART$, $m\angle A = y + 10$, $m\angle R = 2y$, and $m\angle T = 2y - 30$.

a. Find the measure of each of the three angles.

b. Is $\triangle ART$ acute, right, or obtuse?

c. Is $\triangle ART$ scalene, isosceles but not equilateral or equiangular?

10. The measure of each base angle of an isosceles triangle is 15° more than the measure of the vertex angle. Find the measure of each angle.

11. The measure of an angle is 20° less than 3 times the measure of its supplement. What is the measure of the angle and its supplement?

12. In $\triangle ABC$, the measure of $\angle B$ is $\frac{3}{2}$ the measure of $\angle A$, and the measure of $\angle C$ is $\frac{5}{2}$ the measure of $\angle A$. What are the measures of the three angles?

13. The measure of one angle is 3 times that of another angle, and the sum of these measures is 120°. What are the measures of the angles?

14. The measure of the smaller of two supplementary angles is $\frac{4}{5}$ of the measure of the larger. What are the measures of the angles?

15. The vertices of a trapezoid are $A(-3, -1)$, $B(7, -1)$, $C(5, 5)$, and $D(-7, 5)$.

a. Draw $ABCD$ on graph paper.

b. E is the point on $\overline{CD}$ such that $\overline{CD} \perp \overline{AE}$. What are the coordinates of E?

c. Find AB, CD, and AE.

d. Find the area of trapezoid $ABCD$.

16. In parallelogram $ABCD$, $AB = 3x + 4$, $BC = 2x + 5$, and $CD = x + 18$. Find the measure of each side of the parallelogram.

17. A flowerpot in the shape of a right circular cylinder has a height of 4.5 inches. The diameter of the base of the pot is 4.1 inches. Find the volume of the pot to the nearest tenth.

18. Natali makes a mat in the shape of an octagon (an eight-sided polygon) by cutting four isosceles right triangles of equal size from the corners of a 9 by 15 inch rectangle. If the measure of each of the equal sides of the triangles is 2 inches, what is the area of the octagonal mat?

19. Marvin measures a block of wood and records the dimensions as 5.0 centimeters by 3.4 centimeters by 4.25 centimeters. He places the block of wood in a beaker that contains 245 milliliters of water. With the block completely submerged, the water level rises to 317 milliliters.

a. Use the dimensions of the block to find the volume of the block of wood.

b. What is the volume of the block of wood based on the change in the water level?

c. Marvin knows that 1 ml = 1 cm^3. Can the answers to parts **a** and **b** both be correct? Explain your answer.

20. A watering trough for cattle is in the shape of a prism whose ends are the bases of the prism and whose length is the height of the prism. The ends of the trough are trapezoids with parallel edges 31 centimeters and 48 centimeters long and a height of 25 centimeters. The length of the trough is 496 centimeters.

a. Find the surface area of one end of the trough.

b. If the trough is filled to capacity, how many cubic centimeters of water does it hold?

c. How many liters of water does the trough hold? (1 liter = 1,000 cm^3)

Exploration

A **sector** of a circle is a fractional part of the interior of a circle, determined by an angle whose vertex is at the center of the circle (a **central angle**). The area of a sector depends on the measure of its central angle. For example,

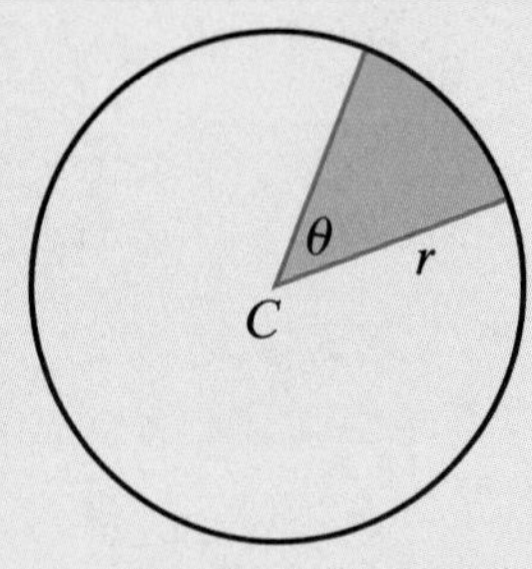

The shaded region represents a sector with central angle θ.

- If the central angle equals 90°, then the area of the sector is one-fourth the area of the circle, or $\frac{90}{360}\pi r^2$.
- If the central angle equals 180°, then the area of the sector is one-half the area of the circle, or $\frac{180}{360}\pi r^2$.
- If the central angle equals 270°, then the area of the sector is three-fourths the area of the circle, or $\frac{270}{360}\pi r^2$.

In general, if the measure of the central angle is θ (theta), then the area of the sector is

$$\textbf{Area of a sector} = \frac{\theta}{360}\pi r^2$$

For this Exploration, use your knowledge of Geometry and the formula for the area of a sector to find the areas of the shaded regions. Express your answers in terms of π. Assume that all the arcs that are drawn are circular.

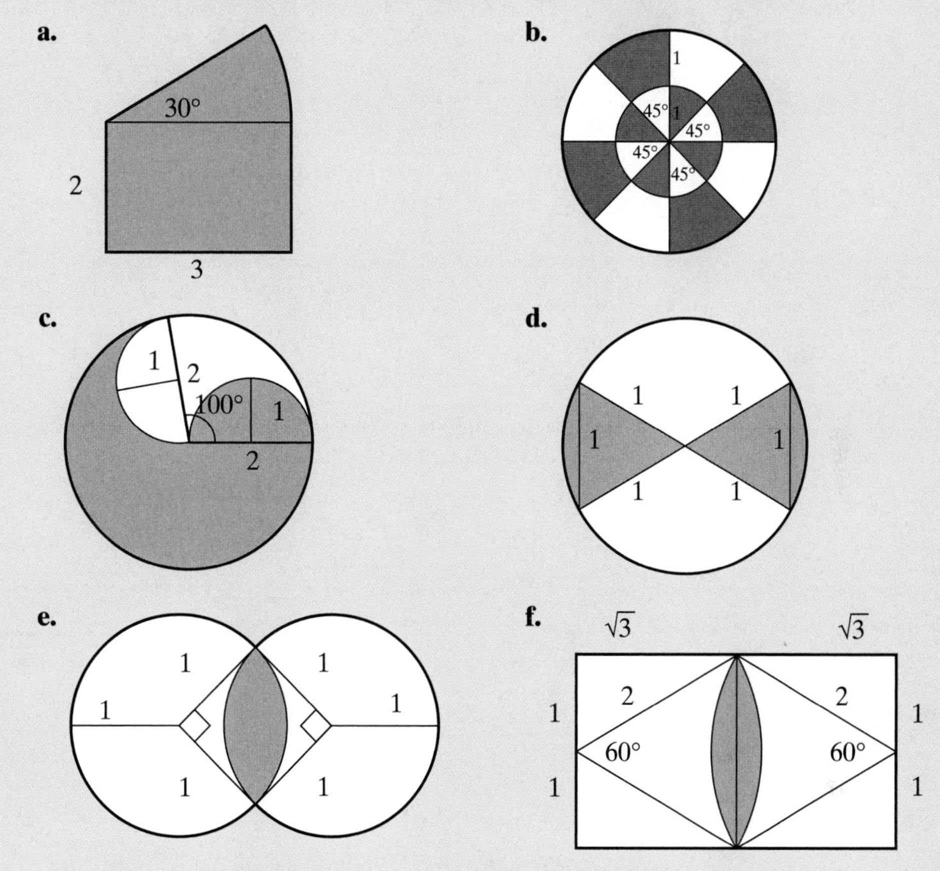

CUMULATIVE REVIEW CHAPTERS 1–7

Part I

Answer all questions in this part. Each correct answer will receive 2 credits. No partial credit will be allowed.

1. Which of the following inequalities is true?

(1) $-|-4| < -(-3) < |-5|$
(2) $-|-4| < |-5| < -(-3)$
(3) $-(-3) < -|-4| < |-5|$
(4) $|-5| < -|-4| < -(-3)$

2. Which of the following is an example of the use of the associative property?
(1) $2(x + 5) = 2(5 + x)$
(2) $2(x + 5) = 2x + 2(5)$
(3) $2 + (x + 5) = 2 + (5 + x)$
(4) $2 + (5 + x) = (2 + 5) + x$

3. Which of the following is not a rational number?
(1) $\sqrt{0.09}$ (2) $\sqrt{0.9}$ (3) 2^{-3} (4) $0.\overline{15}$

4. The product $(a^2b)(a^3b)$ is equivalent to
(1) a^6b (2) a^5b (3) a^5b^2 (4) a^6b^2

5. The solution set of $\frac{2}{3}x + 7 = \frac{1}{6}x - 5$ is
(1) $\{24\}$ (2) $\{-24\}$ (3) $\{6\}$ (4) $\{-6\}$

6. The sum of $b^2 - 7$ and $b^2 + 3b$ is
(1) $b^4 - 4b$ (2) $b^4 + 3b - 7$ (3) $2b^2 - 4b$ (4) $2b^2 + 3b - 7$

7. Two angles are supplementary. If the measure of one angle is 85°, the measure of the other is
(1) 5° (2) 85° (3) 95° (4) 180°

8. In trapezoid $ABCD$, $\overline{AB} \parallel \overline{CD}$. If $m\angle A$ is 75°, then $m\angle D$ is
(1) 15° (2) 75° (3) 105° (4) 165°

9. The graph at the right is the solution set of
(1) $-2 \leq x < 3$
(2) $-2 < x < 3$
(3) $-2 < x \leq 3$
(4) $(-2 > x)$ or $(x \geq 3)$

–4 –3 –2 –1 0 1 2 3 4

10. When $a = -1.5$, $-3a^2 - a$ equals
(1) -5.25 (2) 5.25 (3) 18.75 (4) 21.75

Part II

Answer all questions in this part. Each correct answer will receive 2 credits. Clearly indicate the necessary steps, including appropriate formula substitutions, diagrams, graphs, charts, etc. For all questions in this part, a correct numerical answer with no work shown will receive only 1 credit.

11. The area of a rectangle is $(x^2 + 6x + 8)$ square inches and its length is $(x + 4)$ inches. Express the width of the rectangle in terms of x.

12. Solve the following equation for x: $\frac{7}{x - 2} = \frac{3}{x + 2}$.

Part III

Answer all questions in this part. Each correct answer will receive 3 credits. Clearly indicate the necessary steps, including appropriate formula substitutions, diagrams, graphs, charts, etc. For all questions in this part, a correct numerical answer with no work shown will receive only 1 credit.

13. Mr. Popowich mailed two packages. The larger package weighed 12 ounces more than the smaller. If the total weight of the packages was 17 pounds, how much did each package weigh?

14. a. Solve for x in terms of a and b: $ax + 3b = 7$.

b. Find, to the nearest hundredth, the value of x when $a = \sqrt{3}$ and $b = \sqrt{5}$.

Part IV

Answer all questions in this part. Each correct answer will receive 4 credits. Clearly indicate the necessary steps, including appropriate formula substitutions, diagrams, graphs, charts, etc. For all questions in this part, a correct numerical answer with no work shown will receive only 1 credit.

15. Calvin traveled 600 miles, averaging 40 miles per hour for part of the trip and 60 miles per hour for the remainder of the trip. The entire trip took 11 hours. How long did Calvin travel at each rate?

16. A box used for shipping is in the shape of a rectangular prism. The bases are right triangles. The lengths of the sides of the bases are 9.0, 12, and 15 feet. The height of the prism is 4.5 feet.

a. Find the surface area of the prism. Express the answer using the correct number of significant digits.

b. Find the volume of the prism. Express the answer using the correct number of significant digits.

CHAPTER

8

TRIGONOMETRY OF THE RIGHT TRIANGLE

CHAPTER TABLE OF CONTENTS

The accurate measurement of land has been a critical challenge throughout the history of civilization. Today's land measurement problems are not unlike those George Washington might have solved by using measurements made with a transit, but the modern surveyor has available a total workstation including EDM (electronic distance measuring) and a theodolite for angle measurement. Although modern instruments can perform many measurements and calculations, the surveyor needs to understand the principles of indirect measurement and trigonometry to correctly interpret and apply these results.

In this chapter, we will begin the study of a branch of mathematics called **trigonometry**. The word *trigonometry* is Greek in origin and means "measurement of triangles." Although the trigonometric functions have applications beyond the study of triangles, in this chapter we will limit the applications to the study of right triangles.

8-1 THE PYTHAGOREAN THEOREM

The solutions of many problems require the measurement of line segments and angles. When we use a ruler or tape measure to determine the length of a segment, or a protractor to find the measure of an angle, we are taking a **direct measurement** of the segment or the angle. In many situations, however, it is inconvenient or impossible to make a measurement directly. For example, it is difficult to make the direct measurements needed to answer the following questions:

What is the height of a 100-year-old oak tree?
What is the width of a river?
What is the distance to the sun?

We can answer these questions by using methods that involve **indirect measurement**. Starting with some known lengths of segments or angle measures, we apply a formula or a mathematical relationship to indirectly find the measurement in question.

Engineers, surveyors, physicists, and astronomers frequently use these trigonometric methods in their work.

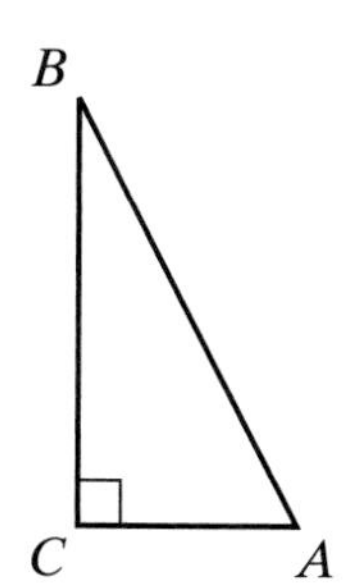

The figure at the left represents a *right triangle*. Recall that such a triangle contains one and only one right angle. In right triangle ABC, side $\overline{AB}$, which is opposite the right angle, is called the *hypotenuse*.

The hypotenuse is the longest side of the triangle. The other two sides of the triangle, $\overline{BC}$ and $\overline{AC}$, form the right angle. They are called the *legs* of the right triangle.

More than 2,000 years ago, the Greek mathematician Pythagoras demonstrated the following property of the right triangle, which is called the **Pythagorean Theorem**:

> **The Pythagorean Theorem** In a right triangle, the square of the length of the hypotenuse is equal to the sum of the squares of the lengths of the other two sides.

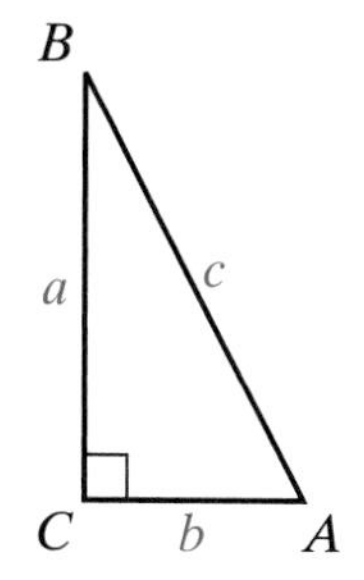

If we represent the length of the hypotenuse of right triangle ABC by c and the lengths of the other two sides by a and b, the Theorem of Pythagoras may be written as the following formula:

$$c^2 = a^2 + b^2$$

To show that this relationship is true for any right triangle ABC with the length of the hypotenuse represented by c and the lengths of the legs represented by a and b, consider a square with sides $(a + b)$.

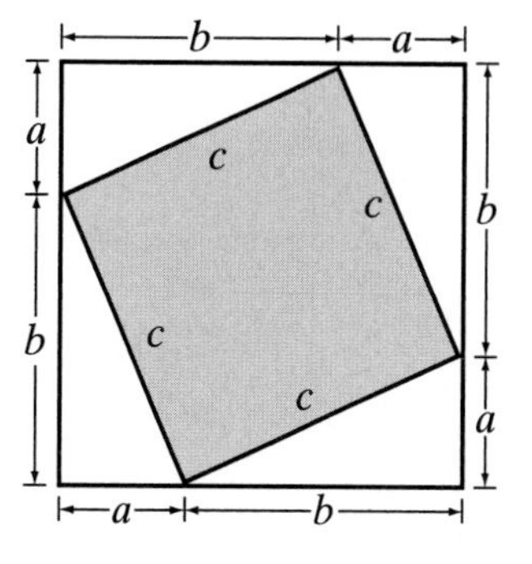

The area of the square is $(a + b)^2$. However, since it is divided into four triangles and one smaller square, its area can also be expressed as:

Area of the square = area of the four triangles + area of the smaller square

$= 4\left(\frac{1}{2}\right)ab + c^2$

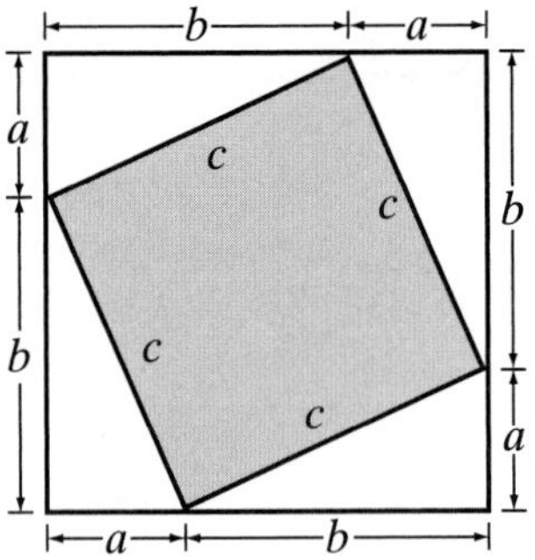

Although the area is written in two different ways, both expressions are equal.

Thus,

$$(a + b)^2 = 4\left(\tfrac{1}{2}\right)ab + c^2.$$

If we simplify, we obtain the relationship of the Pythagorean Theorem:

$$(a + b)^2 = 4\left(\tfrac{1}{2}\right)ab + c^2$$

$$a^2 + 2ab + b^2 = 2ab + c^2$$ Expand the binomial term $(a + b)^2$.

$$a^2 + 2ab - 2ab + b^2 = 2ab - 2ab + c^2$$ Subtract $2ab$ from both sides of the equality.

$$a^2 + b^2 = c^2$$

Statements of the Pythagorean Theorem

Two statements can be made for any right triangle where c represents the length of the hypotenuse (the longest side) and a and b represent the lengths of the other two sides.

1. If a triangle is a right triangle, then the square of the length of the hypotenuse is equal to the sum of the squares of the lengths of the other two sides. If a triangle is a right triangle, then $c^2 = a^2 + b^2$.
2. If the square of the length of the longest side of a triangle is equal to the sum of the squares of the lengths of the other two sides, the triangle is a right triangle. If $c^2 = a^2 + b^2$ in a triangle, then the triangle is a right triangle.

If we know the lengths of any two sides of a right triangle, we can find the length of the third side. For example, if the measures of the legs of a right triangle are 7 and 9, we can write:

$$c^2 = a^2 + b^2$$
$$c^2 = 7^2 + 9^2$$
$$c^2 = 49 + 81$$
$$c^2 = 130$$

To solve this equation for c, we must do the opposite of squaring, that is, we must find the square root of 130. There are two square roots of 130, $+\sqrt{130}$ and $-\sqrt{130}$ which we write as $\pm\sqrt{130}$.

There are two things that we must consider here when finding the value of c.

1. Since c represents the length of a line segment, only the positive number is an acceptable value. Therefore, $c = +\sqrt{130}$.
2. There is no rational number that has a square of 130. The value of c is an irrational number. However, we usually use a calculator to find a rational approximation for the irrational number.

ENTER: **2nd** √ 130 **ENTER**

DISPLAY:

```
√(130
          11.40175425
```

Therefore, to the nearest tenth, the length of the hypotenuse is 11.4. Note that the calculator gives only the positive rational approximation of the square root of 130.

EXAMPLE 1

A ladder is placed 5 feet from the foot of a wall. The top of the ladder reaches a point 12 feet above the ground. Find the length of the ladder.

Solution The ladder, the wall, and the ground form a right triangle. The length of the ladder is c, the length of the hypotenuse of the right triangle. The distance from the foot of the ladder to the wall is $a = 5$, and the distance from the ground to the top of the ladder is $b = 12$. Use the Theorem of Pythagoras.

$$c^2 = a^2 + b^2$$
$$c^2 = 5^2 + 12^2$$
$$c^2 = 25 + 144$$
$$c^2 = 169$$
$$c = \pm\sqrt{169} = \pm 13$$

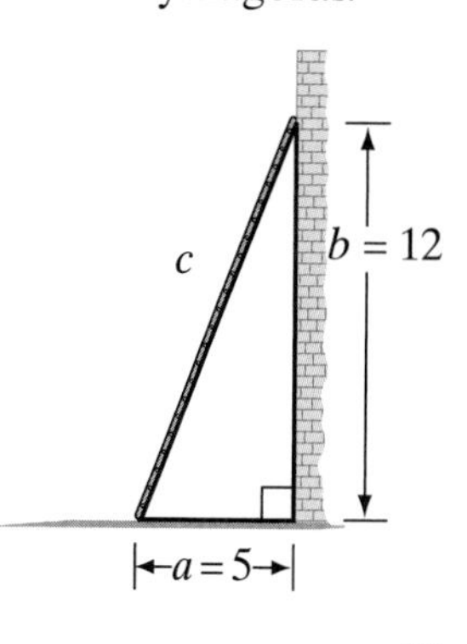

Reject the negative value. Note that in this case the exact value of c is a rational number because 169 is a perfect square.

Answer The length of the ladder is 13 feet.

EXAMPLE 2

The hypotenuse of a right triangle is 36.0 centimeters long and one leg is 28.5 centimeters long.

a. Find the length of the other leg to the nearest tenth of a centimeter.

b. Find the area of the triangle using the correct number of significant digits.

Solution **a.** The length of the hypotenuse is $c = 36.0$ and the length of one leg is $a = 28.5$. The length of the other leg is b. Substitute the known values in the Pythagorean Theorem.

$$\begin{aligned} c^2 &= a^2 + b^2 \\ 36.0^2 &= 28.5^2 + b^2 \\ 1{,}296 &= 812.25 + b^2 \\ 483.75 &= b^2 \\ \pm\sqrt{483.75} &= b \end{aligned}$$

Reject the negative value. Use a calculator to find a rational approximation of the value of b. A calculator displays 21.99431745. Round the answer to the nearest tenth.

Answer The length of the other leg is 22.0 centimeters.

b. Area of $\triangle ABC = \frac{1}{2}bh = \frac{1}{2}(28.5)(22.0) = 313.5$

Since the lengths are given to three significant digits, we will round the area to three significant digits.

Answer The area of $\triangle ABC$ is 314 square centimeters.

EXAMPLE 3

Is a triangle whose sides measure 8 centimeters, 7 centimeters, and 4 centimeters a right triangle?

Solution If the triangle is a right triangle, the longest side, whose measure is 8, must be the hypotenuse. Then:

$$\begin{aligned} c^2 &= a^2 + b^2 \\ 8^2 &\stackrel{?}{=} 7^2 + 4^2 \\ 64 &\stackrel{?}{=} 49 + 16 \\ 64 &\neq 65 \ ✗ \end{aligned}$$

Answer The triangle is not a right triangle.

EXERCISES

Writing About Mathematics

1. A **Pythagorean triple** is a set of three positive integers that make the equation $c^2 = a^2 + b^2$ true. Luz said that 3, 4, and 5 is a Pythagorean triple, and, for any positive integer k, $3k$, $4k$, and $5k$ is also a Pythagorean triple. Do you agree with Luz? Explain why or why not.

2. Regina said that if n is a positive integer, $2n + 1, 2n^2 + 2n$, and $2n^2 + 2n + 1$ is a Pythagorean triple. Do you agree with Regina? Explain why or why not.

Developing Skills

In 3–11, c represents the length of the hypotenuse of a right triangle and a and b represent the lengths of the legs. For each right triangle, find the length of the side whose measure is not given.

3. $a = 3, b = 4$ **4.** $a = 8, b = 15$ **5.** $c = 10, a = 6$

6. $c = 13, a = 12$ **7.** $c = 17, b = 15$ **8.** $c = 25, b = 20$

9. $a = \sqrt{2}, b = \sqrt{2}$ **10.** $a = 1, b = \sqrt{3}$ **11.** $a = \sqrt{8}, c = 3$

In 12–17, c represents the length of the hypotenuse of a right triangle and a and b represent the lengths of the legs. For each right triangle:

a. Express the length of the third side in radical form.

b. Express the length of the third side to the *nearest hundredth.*

12. $a = 2, b = 3$ **13.** $a = 3, b = 3$ **14.** $a = 4, c = 8$

15. $a = 7, b = 2$ **16.** $b = \sqrt{3}, c = \sqrt{14}$ **17.** $a = \sqrt{7}, c = 6$

In 18–21, find x in each case and express irrational results in radical form.

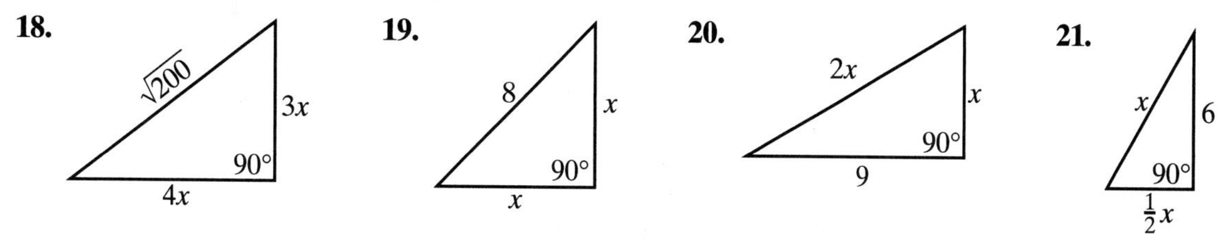

In 22–27, find, in each case, the length of the diagonal of a rectangle whose sides have the given measurements.

22. 7 inches by 24 inches **23.** 9 centimeters by 40 centimeters

24. 28 feet by 45 feet **25.** 17 meters by 144 meters

26. 15 yards by 20 yards **27.** 18 millimeters by 24 millimeters

28. The diagonal of a rectangle measures 65 centimeters. The length of the rectangle is 33 centimeters. Consider the measurements to be exact.

a. Find the width of the rectangle.

b. Find the area of the rectangle.

29. Approximate, to the nearest inch, the length of a rectangle whose diagonal measures 25.0 inches and whose width is 18.0 inches.

30. The altitude to the base of a triangle measures 17.6 meters. The altitude divides the base into two parts that are 12.3 meters and 15.6 meters long. What is the perimeter of the triangle to the nearest tenth of a meter?

Applying Skills

31. A ladder 39 feet long leans against a building and reaches the ledge of a window. If the foot of the ladder is 15 feet from the foot of the building, how high is the window ledge above the ground to the nearest foot?

32. Mr. Rizzo placed a ladder so that it reached a window 15.0 feet above the ground when the foot of the ladder was 5.0 feet from the wall. Find the length of the ladder to the nearest tenth of a foot.

33. Mrs. Culkowski traveled 24.0 kilometers north and then 10.0 kilometers east. How far was she from her starting point?

34. One day, Ronnie left his home at A and reached his school at C by walking along $\overline{AB}$ and $\overline{BC}$, the sides of a rectangular open field that was muddy. The dimensions of the field are 1,212 feet by 885 feet. When he was ready to return home, the field was dry and Ronnie decided to take a shortcut by walking diagonally across the field, along $\overline{AC}$. To the nearest whole foot, how much shorter was the trip home than the trip to school?

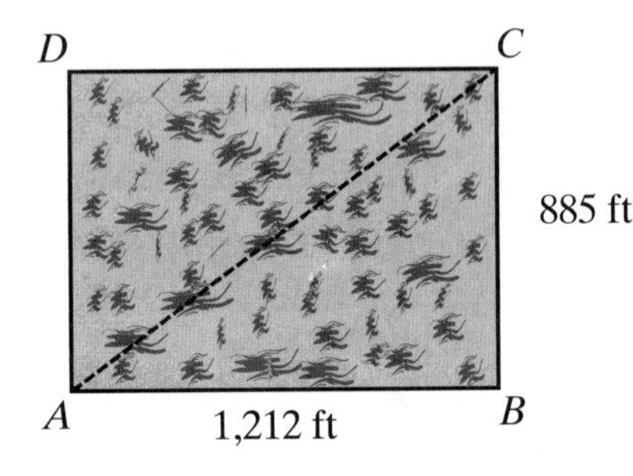

35. Corry and Torry have a two-way communication device that has a range of one-half mile (2,640 feet). Torry lives 3 blocks west and 2 blocks north of Corry. If the length of each block is 600 feet, can Corry and Torry communicate using this device when each is home? Explain your answer.

36. A baseball diamond has the shape of a square with the bases at the vertices of the square. If the distance from home plate to first base is 90.0 feet, approximate, to the nearest tenth of a foot, the distance from home plate to second base.

8-2 THE TANGENT RATIO

Naming Sides

In a right triangle, the hypotenuse, which is the longest side, is opposite the right angle. The other two sides in a right triangle are called the legs. However, in trigonometry of the right triangle, we call these legs the **opposite side** and the **adjacent side** to describe their relationship to one of the acute angles in the triangle.

Notice that $\triangle ABC$ is the same right triangle in both figures below, but the position names we apply to the legs change with respect to the angles.

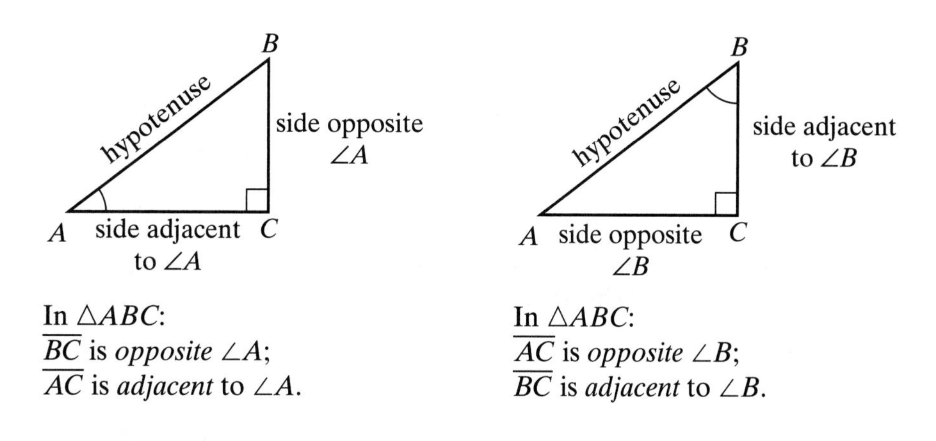

In $\triangle ABC$:
$\overline{BC}$ is *opposite* $\angle A$;
$\overline{AC}$ is *adjacent* to $\angle A$.

In $\triangle ABC$:
$\overline{AC}$ is *opposite* $\angle B$;
$\overline{BC}$ is *adjacent* to $\angle B$.

Similar Triangles

Three right triangles are drawn to coincide at vertex A. Since each triangle contains a right angle as well as $\angle A$, we know that the third angles of each triangle are congruent. When three angles of one triangle are congruent to the three angles of another, the triangles are **similar**.

The corresponding sides of similar triangles are in proportion. Therefore:

$$\frac{CB}{BA} = \frac{ED}{DA} = \frac{GF}{FA}$$

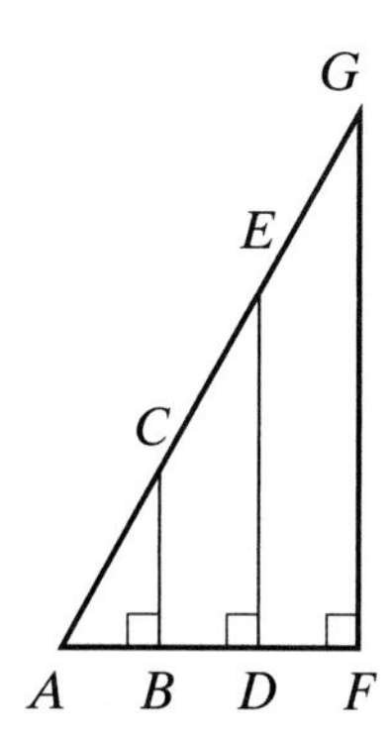

The similar triangles shown in the previous page, $\triangle ABC$, $\triangle ADE$, and $\triangle AFG$, are separated and shown below.

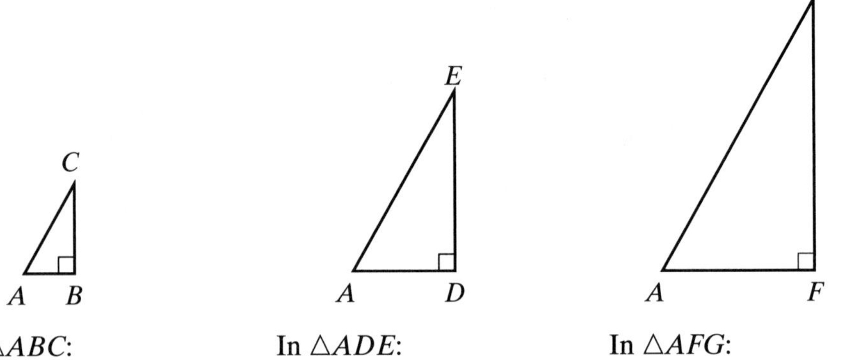

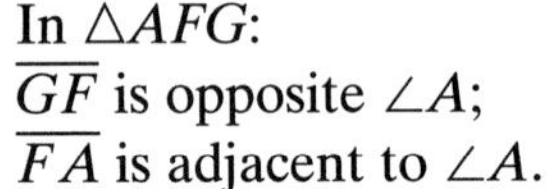

In $\triangle ABC$:
$\overline{CB}$ is opposite $\angle A$;
$\overline{BA}$ is adjacent to $\angle A$.

In $\triangle ADE$:
$\overline{ED}$ is opposite $\angle A$;
$\overline{DA}$ is adjacent to $\angle A$.

In $\triangle AFG$:
$\overline{GF}$ is opposite $\angle A$;
$\overline{FA}$ is adjacent to $\angle A$.

Therefore, $\frac{CB}{BA} = \frac{ED}{DA} = \frac{GF}{FA} = \frac{\text{length of side } \textit{opposite } \angle A}{\text{length of side } \textit{adjacent} \text{ to } \angle A}$ = a constant for $\triangle ABC$, $\triangle ADE$, $\triangle AFG$ and for any right triangle similar to these triangles, that is, for any right triangle with an acute angle congruent to $\angle A$. This ratio is called the **tangent of the angle.**

DEFINITION

The **tangent of an acute angle of a right triangle** is the ratio of the length of the side opposite the acute angle to the length of the side adjacent to the acute angle.

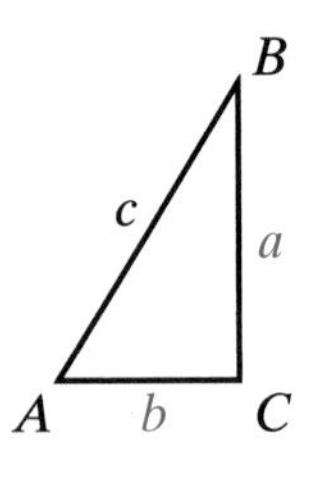

For right triangle ABC, with $m\angle C = 90$, the definition of the *tangent of* $\angle A$ is as follows:

$$\textbf{tangent } A = \frac{\text{length of side } \textit{opposite } \angle A}{\text{length of side } \textit{adjacent} \text{ to } \angle A} = \frac{BC}{AC} = \frac{a}{b}$$

By using "tan A" as an abbreviation for tangent A, "opp" as an abbreviation for the length of the leg opposite $\angle A$, and "adj" as an abbreviation for the length of the leg adjacent to $\angle A$, we can shorten the way we write the relationship given above as follows:

$$\tan A = \frac{\text{opp}}{\text{adj}} = \frac{BC}{AC} = \frac{a}{b}$$

Finding Tangent Ratios on a Calculator

The length of each side of equilateral triangle ABD is 2. The altitude $\overline{BC}$ from B to $\overline{AD}$ forms two congruent right triangles with $AC = 1$. We can use the hypotenuse rule to find BC.

$$(AC)^2 + (BC)^2 = (AB)^2$$
$$1^2 + (BC)^2 = 2^2$$
$$1 + (BC)^2 = 4$$
$$(BC)^2 = 3$$
$$BC = \sqrt{3}$$

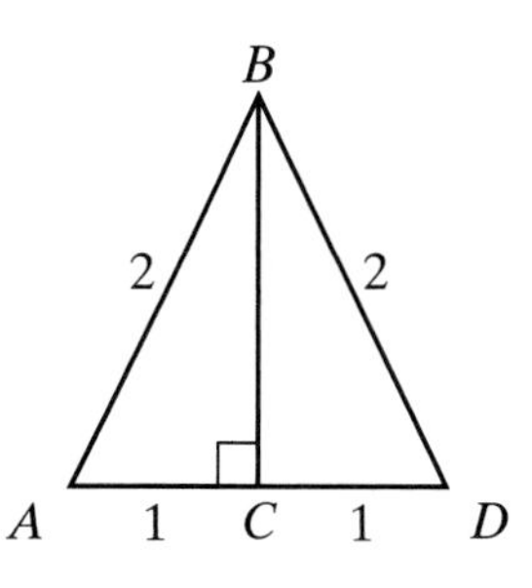

The measure of each angle of an equilateral triangle is 60°. Therefore we can use the lengths of $\overline{AC}$ and $\overline{BC}$ to find the exact value of the tangent of a 60° angle.

$$\tan 60^\circ = \frac{\text{opp}}{\text{adj}} = \frac{BC}{AC} = \frac{\sqrt{3}}{1} = \sqrt{3}$$

But how can we find the constant value of the tangent ratio when the right triangle has an angle of 40° or 76°? Since we want to work with the value of this ratio for any right triangle, no matter what the measures of the acute angles may be, mathematicians have compiled tables of the tangent values for angles with measures from 0° to 90°. Also, a calculator has the ability to display the value of this ratio for any angle. We will use a calculator to determine these values.

The measure of an angle can be given in degrees or in radians. In this book, we will always express the measure of an angle in degrees. A graphing calculator can use either radians or degrees. To place the calculator in degree mode, press **MODE**, then use the down arrow and the right arrow keys to highlight Deg. Press **ENTER** and **2nd** **QUIT**. Your calculator will be in degree mode each time you turn it on.

CASE I *Given an angle measure, find the tangent ratio.*

We saw that tan 60° is equal to $\sqrt{3}$. The calculator will display this value as an approximate decimal. To find tan 60°, enter the sequence of keys shown below.

ENTER: **TAN** 60 **)** **ENTER**

DISPLAY:
```
TAN (60)
           1.732050808
```

The value given in the calculator display is the rational approximation of $\sqrt{3}$, the value of tan 60° that we found using the ratio of the lengths of the legs of a right triangle with a 60° angle. Therefore, to the *nearest ten-thousandth*,

$$\tan 60^\circ = 1.7321.$$

CASE 2 *Given a tangent ratio, find the angle measure.*

The value of the tangent ratio is different for each different angle measure from 0° to 90°. Therefore, if we know the value of the tangent ratio, we can find the measure of the acute angle that has this tangent ratio. The calculator key used to do this is labeled **TAN⁻¹** and is accessed by first pressing **2nd**. We can think of $\tan^{-1}$ as "the angle whose tangent is." Therefore, $\tan^{-1}(0.9004)$ can be read as "the angle whose tangent is 0.9004."

To find the measure of $\angle A$ from the calculator, we use the following sequences of keys.

ENTER: **2nd** **TAN⁻¹** 0.9004 **)** **ENTER**

DISPLAY:

```
TAN⁻¹(0.9004)
         41.99987203
```

The measure of $\angle A$ to the *nearest degree* is 42°.

EXAMPLE 1

In $\triangle ABC$, $\angle C$ is a right angle, $BC = 3$, $AC = 4$, and $AB = 5$.

a. Find:

(1) $\tan A$

(2) $\tan B$

(3) $m\angle A$ to the *nearest degree*

(4) $m\angle B$ to the *nearest degree*

b. Show that the acute angles of the triangle are complementary.

Solution **a.** (1) $\tan A = \frac{\text{opp}}{\text{adj}} = \frac{BC}{AC} = \frac{3}{4}$ Answer

(2) $\tan B = \frac{\text{opp}}{\text{adj}} = \frac{AC}{BC} = \frac{4}{3}$ Answer

Use a calculator to find the measures of $\angle A$ and $\angle B$.

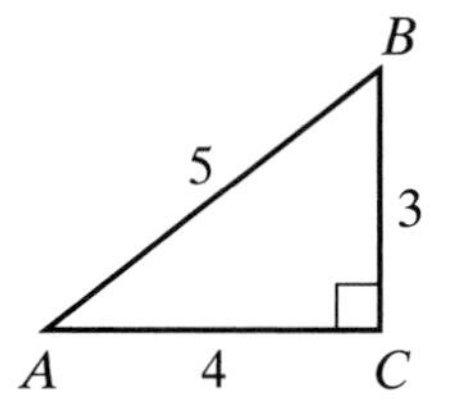

(3) ENTER: **2nd** **TAN⁻¹** 3 **÷** 4 **)** **ENTER**

DISPLAY:

```
TAN⁻¹(3/4)
         36.86989765
```

To the nearest degree, $m\angle A = 37$. Answer

(4) ENTER: 2nd TAN⁻¹ 4 ÷ 3) ENTER

DISPLAY:
```
TAN-1(4/3)
         53.13010235
```

To the nearest degree, m∠B = 53. *Answer*

b. m∠A + m∠B = 36.869889765 + 53.13010235 = 90.000000. Therefore, the acute angles of △ABC are complementary. *Answer*

Note: In a right triangle, the tangents of the two acute angles are reciprocals.

EXERCISES

Writing About Mathematics

1. Explain why the tangent of a 45° angle is 1.

2. Use one of the right triangles formed by drawing an altitude of an equilateral triangle to find tan 30°. Express the answer that you find to the nearest hundred-thousandth and compare this result to the valued obtained from a calculator.

Developing Skills

In 3–6, find: **a.** tan A **b.** tan B

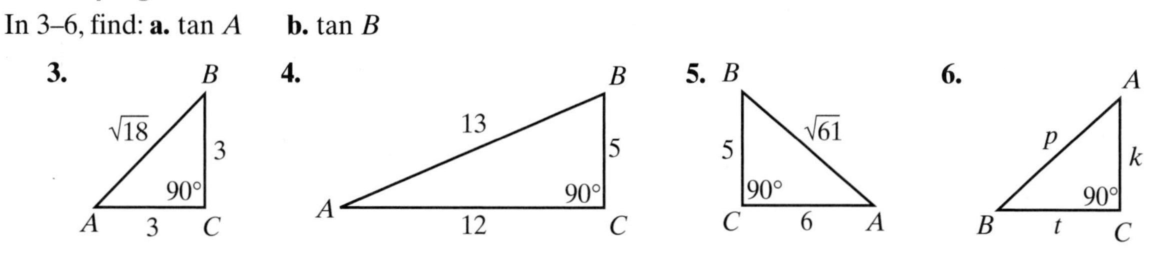

7. In △ABC, m∠C = 90, AC = 6, and AB = 10. Find tan A.

8. In △RST, m∠T = 90, RS = 13, and ST = 12. Find tan S.

In 9–16, use a calculator to find each of the following to the *nearest ten-thousandth*:

9. tan 10° **10.** tan 25° **11.** tan 70° **12.** tan 55°

13. tan 1° **14.** tan 89° **15.** tan 36° **16.** tan 67°

In 17–28, in each of the following, use a calculator to find the measure of ∠A to the *nearest degree*.

17. tan A = 0.0875 **18.** tan A = 0.3640 **19.** tan A = 0.5543

20. tan A = 1.0000 **21.** tan A = 2.0503 **22.** tan A = 3.0777

23. $\tan A = 0.3754$ **24.** $\tan A = 0.7654$ **25.** $\tan A = 1.8000$

26. $\tan A = 0.3500$ **27.** $\tan A = 0.1450$ **28.** $\tan A = 2.9850$

29. Does the tangent of an angle increase or decrease as the degree measure of the angle increases from 1° to 89°?

30. a. Use a calculator to find tan 20° and tan 40° to the *nearest ten-thousandth.*

b. Is the tangent of the angle doubled when the measure of the angle is doubled?

Applying Skills

31. In $\triangle ABC$, $m\angle C = 90$, $AC = 6$, and $BC = 6$.

a. Find $\tan A$.

b. Find the measure of $\angle A$.

32. In $\triangle ABC$, $m\angle C = 90$, $BC = 4$, and $AC = 9$.

a. Find $\tan A$.

b. Find the measure of $\angle A$ to the *nearest degree.*

c. Find $\tan B$.

d. Find the measure of $\angle B$ to the *nearest degree.*

33. In rectangle $ABCD$, $AB = 10$ and $BC = 5$.

a. Find $\tan \angle CAB$.

b. Find the measure of $\angle CAB$ to the *nearest degree.*

c. Find $\tan \angle CAD$.

d. Find the measure of $\angle CAD$ to the *nearest degree.*

34. In $\triangle ABC$, $\angle C$ is a right angle, $m\angle A = 45$, $AC = 4$, $BC = 4$, and $AB = 4\sqrt{2}$.

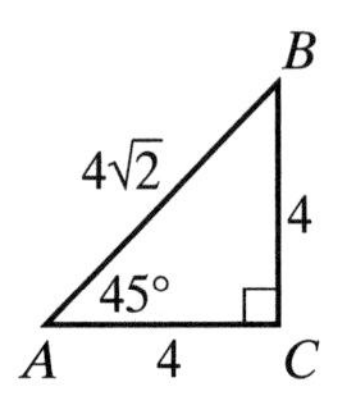

a. Using the given lengths, write the ratio for $\tan A$.

b. Use a calculator to find tan 45°.

35. In $\triangle RST$, $\angle T$ is a right angle and r, s, and t are lengths of sides. Using these lengths:

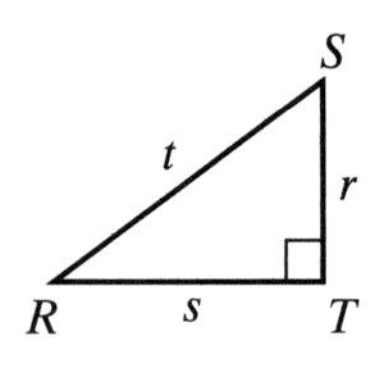

a. Write the ratio for $\tan R$.

b. Write the ratio for $\tan S$.

c. Use parts **a** and **b** to find the numerical value of the product $(\tan R)(\tan S)$.

8-3 APPLICATIONS OF THE TANGENT RATIO

The tangent ratio is often used to make indirect measurements when the measures of a leg and an acute angle of a right triangle are known.

Angle of Elevation and Angle of Depression

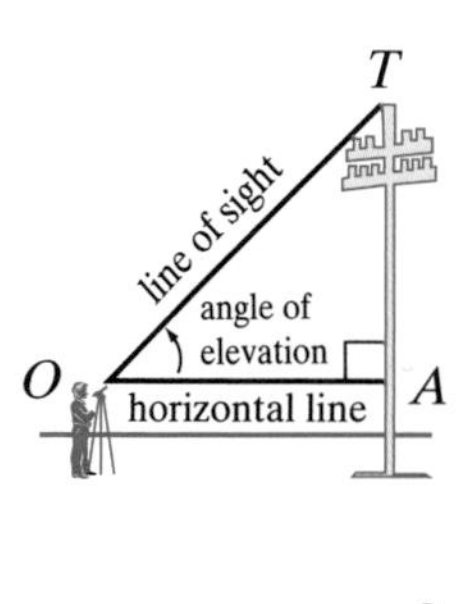

When a telescope or some similar instrument is used to sight the top of a telephone pole, the instrument is elevated (tilted upward) from a horizontal position. Here, $\overleftrightarrow{OT}$ is the line of sight and $\overleftrightarrow{OA}$ is the horizontal line. The **angle of elevation** is the angle determined by the rays that are parts of the horizontal line and the line of sight when looking upward. Here, $\angle TOA$ is the angle of elevation.

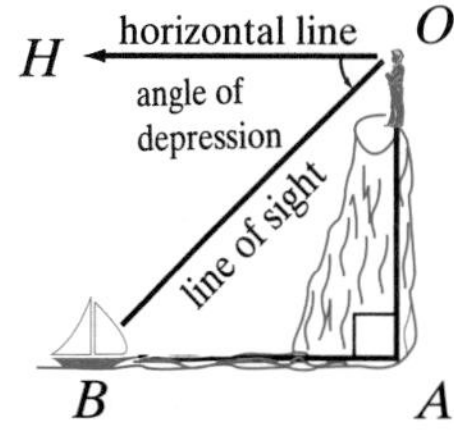

When an instrument is used to sight a boat from a cliff, the instrument is depressed (tilted downward) from a horizontal position. Here, $\overleftrightarrow{OB}$ is the line of sight and $\overleftrightarrow{OH}$ is the horizontal line. The **angle of depression** is the angle determined by the rays that are parts of the horizontal line and of the line of sight when looking downward. Here, $\angle HOB$ is the angle of depression.

Note that, if $\overleftrightarrow{BA}$ is a horizontal line and $\overleftrightarrow{BO}$ is the line of sight from the boat to the top of the cliff, $\angle ABO$ is called the angle of elevation from the boat to the top of the cliff. Since $\overleftrightarrow{HO} \parallel \overleftrightarrow{BA}$ and $\overleftrightarrow{OB}$ is a transversal, alternate interior angles are congruent, namely, $\angle HOB \cong \angle ABO$. Thus, the angle of elevation measured from B to O is congruent to the angle of depression measured from O to B.

Using the Tangent Ratio to Solve Problems

Procedure

To solve a problem by using the tangent ratio:

1. For the given problem, make a diagram that includes a right triangle. Label the known measures of the sides and angles. Identify the unknown quantity by a variable.
2. If for the right triangle either (1) the lengths of two legs or (2) the length of one leg and the measure of one acute angle are known, write a formula for the tangent of an acute angle.
3. Substitute known values in the formula and solve the resulting equation for the unknown value.

EXAMPLE 1

Find to the *nearest degree* the measure of the angle of elevation of the sun when a vertical pole 6.5 meters high casts a shadow 8.3 meters long.

Solution The angle of elevation of the sun is the same as $\angle A$, the angle of elevation to the top of the pole from A, the tip of the shadow. Since the vertical pole and the shadow are the legs of a right triangle opposite and adjacent to $\angle A$, use the tangent ratio.

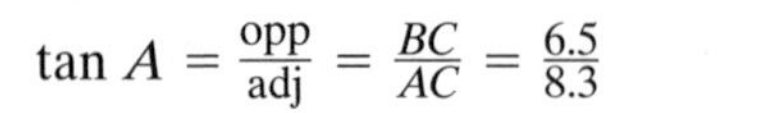

$$\tan A = \frac{\text{opp}}{\text{adj}} = \frac{BC}{AC} = \frac{6.5}{8.3}$$

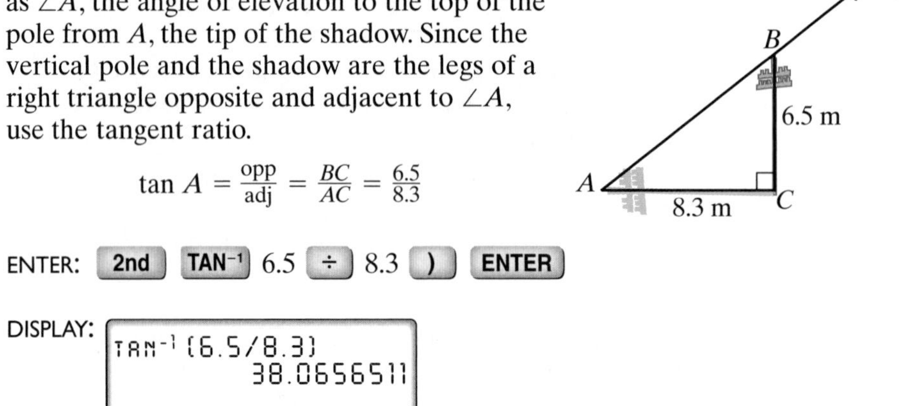

ENTER: 2nd TAN⁻¹ 6.5 ÷ 8.3) ENTER

DISPLAY:
```
TAN⁻¹(6.5/8.3)
          38.0656511
```

Answer To the *nearest degree*, the measure of the angle of elevation of the sun is 38°.

EXAMPLE 2

At a point on the ground 39 meters from the foot of a tree, the measure of the angle of elevation of the top of the tree is 42°. Find the height of the tree to the *nearest meter*.

Solution Let T be the top of the tree, A be the foot of the tree, and B be the point on the ground 39 meters from A. Draw $\triangle ABT$, and label the diagram: $m\angle B = 42$, $AB = 39$.

Let x = height of tree (AT). The height of the tree is the length of the perpendicular from the top of the tree to the ground. Since the problem involves the measure of an acute angle and the measures of the legs of a right triangle, use the tangent ratio:

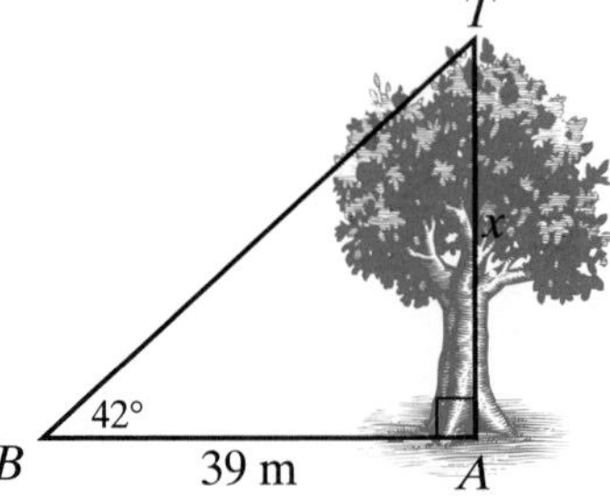

$$\tan B = \frac{\text{opp}}{\text{adj}}$$

$$\tan B = \frac{AT}{BA}$$

$$\tan 42° = \frac{x}{39} \qquad \text{Substitute the given values.}$$

$$x = 39 \tan 42° \qquad \text{Solve for } x.$$

Use a calculator for the computation: $x = 35.11575773$

Answer To the *nearest meter*, the height of the tree is 35 meters.

EXAMPLE 3

From the top of a lighthouse 165 feet above sea level, the measure of the angle of depression of a boat at sea is 35.0°. Find to the *nearest foot* the distance from the boat to the foot of the lighthouse.

Solution Let L be the top of the lighthouse, LA be the length of the perpendicular from L to sea level, and B be the position of the boat. Draw right triangle ABL, and draw $\overrightarrow{LH}$, the horizontal line through L.

Since $\angle HLB$ is the angle of depression, m$\angle HLB$ = 35.0, m$\angle LBA$ = 35.0, and m$\angle BLA$ = 90 − 35.0 = 55.0.

Let x = distance from the boat to the foot of the lighthouse (BA).

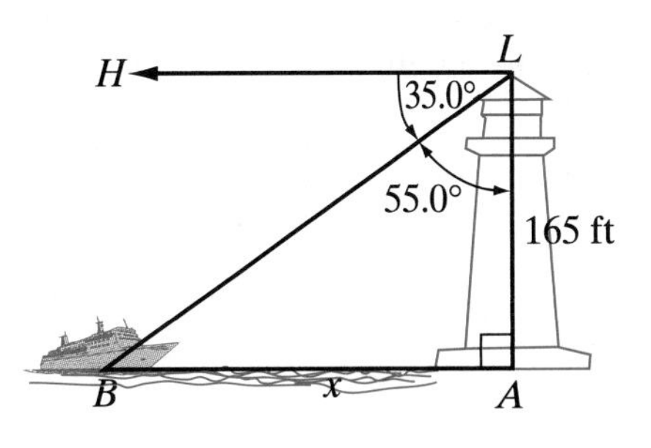

METHOD 1

Using $\angle BLA$, $\overline{BA}$ is the opposite side and $\overline{LA}$ is the adjacent side. Form the tangent ratio:

$$\tan \angle BLA = \frac{BA}{LA}$$
$$\tan 55.0° = \frac{x}{165}$$
$$x = 165 \tan 55.0°$$

Use a calculator to perform the computation. The display will read 235.644421.

METHOD 2

Using $\angle LBA$, $\overline{LA}$ is the opposite side and $\overline{BA}$ is the adjacent side. Form the tangent ratio:

$$\tan \angle LBA = \frac{LA}{BA}$$
$$\tan 35.0° = \frac{165}{x}$$
$$x \tan 35.0° = 165$$
$$x = \frac{165}{\tan 35.0°}$$

Use a calculator to perform the computation. The display will read 235.644421.

Answer To the *nearest foot*, the boat is 236 feet from the foot of the lighthouse.

EXERCISES

Writing About Mathematics

1. Zack is solving a problem in which the measure of the angle of depression from the top of a building to a point 85 feet from the foot of the building is 64°. To find the height of the building, Zack draws the diagram shown at the right. Explain why Zack's diagram is incorrect.

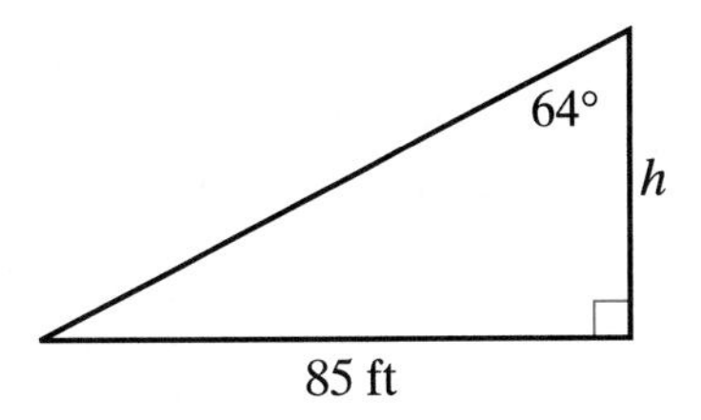

2. Explain why the angle of elevation from point A to point B is always congruent to the angle of depression from point B to point A.

Developing Skills

In 3–11, in each given triangle, find the length of the side marked x to the *nearest foot* or the measure of the angle marked x to the *nearest degree*.

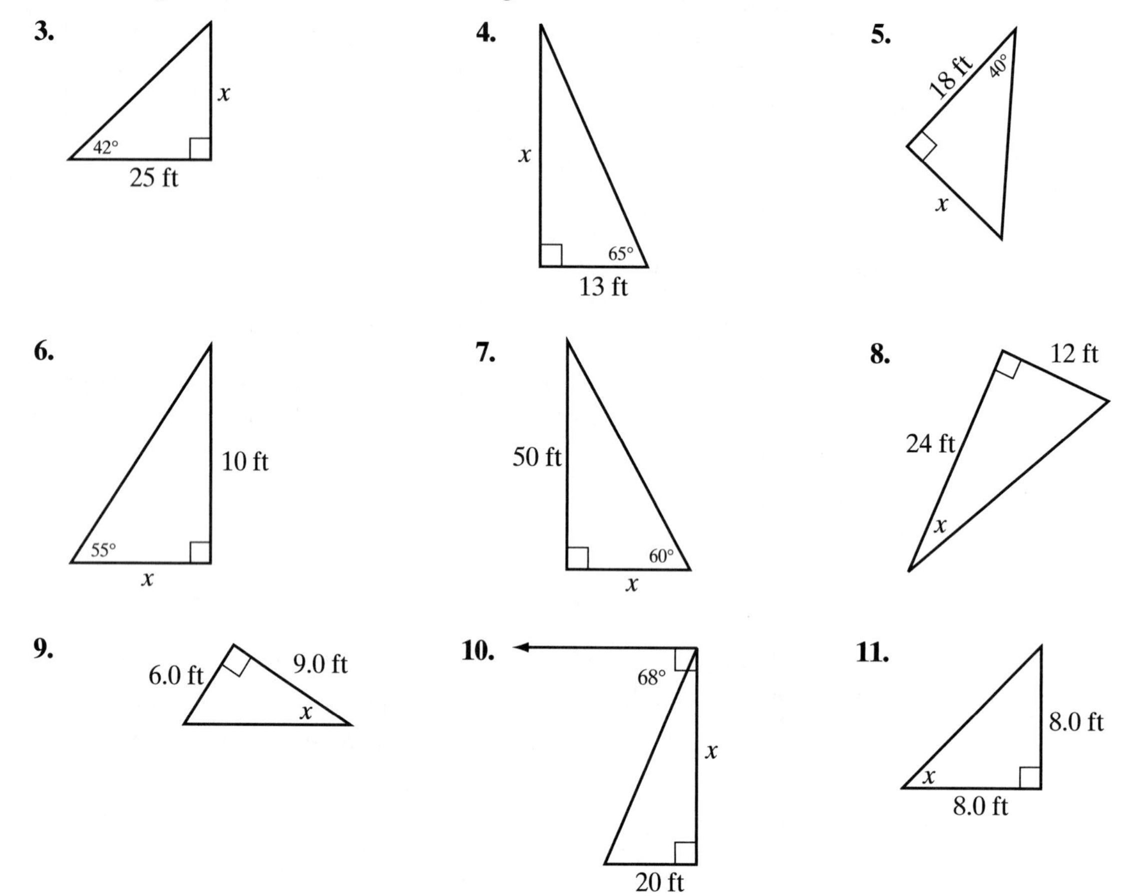

Applying Skills

12. At a point on the ground 52 meters from the foot of a tree, the measure of the angle of elevation of the top of the tree is 48°. Find the height of the tree to the *nearest meter*.

13. A ladder is leaning against a wall. The foot of the ladder is 6.25 feet from the wall. The ladder makes an angle of 74.5° with the level ground. How high on the wall does the ladder reach? Round the answer to the *nearest tenth of a foot*.

14. From a point, A, on the ground that is 938 feet from the foot, C, of the Empire State Building, the angle of elevation of the top, B, of the building has a measure of 57.5°. Find the height of the building to the *nearest ten feet.*

15. Find to the *nearest meter* the height of a building if its shadow is 18 meters long when the angle of elevation of the sun has a measure of 38°.

16. From the top of a lighthouse 50.0 meters high, the angle of depression of a boat out at sea has a measure of 15.0°. Find, to the *nearest meter*, the distance from the boat to the foot of the lighthouse, which is at sea level.

17. From the top of a school 61 feet high, the measure of the angle of depression to the road in front of the school is 38°. Find to the *nearest foot* the distance from the road to the school.

18. Find to the *nearest degree* the measure of the angle of elevation of the sun when a student 170 centimeters tall casts a shadow 170 centimeters long.

19. Find to the *nearest degree* the measure of the angle of elevation of the sun when a woman 150 centimeters tall casts a shadow 43 centimeters long.

20. A ladder leans against a building. The top of the ladder reaches a point on the building that is 18 feet above the ground. The foot of the ladder is 7.0 feet from the building. Find to the *nearest degree* the measure of the angle that the ladder makes with the level ground.

21. In any rhombus, the diagonals are perpendicular to each other and bisect each other. In rhombus $ABCD$, diagonals $\overline{AC}$ and $\overline{BD}$ meet at M. If $BD = 14$ and $AC = 20$, find the measure of each angle to the *nearest degree.*

a. m$\angle BCM$ **b.** m$\angle MBC$ **c.** m$\angle ABC$ **d.** m$\angle BCD$

8-4 THE SINE AND COSINE RATIOS

Since the tangent is the ratio of the lengths of the two legs of a right triangle, it is not directly useful in solving problems in which the hypotenuse is involved. In trigonometry of the right triangle, two ratios that involve the hypotenuse are called the **sine of an angle** and the **cosine of an angle**.

As in our discussion of the tangent ratio, we recognize that the figure at the right shows three similar triangles. Therefore, the ratios of corresponding sides are equal.

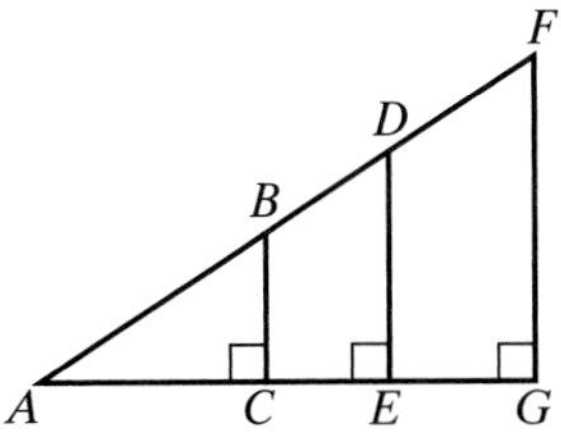

The Sine Ratio

From the figure, we see that

$$\frac{BC}{AB} = \frac{DE}{AD} = \frac{FG}{AF} = \text{a constant}$$

This ratio is called the **sine of** $\angle A$.

DEFINITION

The **sine of an acute angle of a right triangle** is the ratio of the length of the side opposite the acute angle to the length of the hypotenuse.

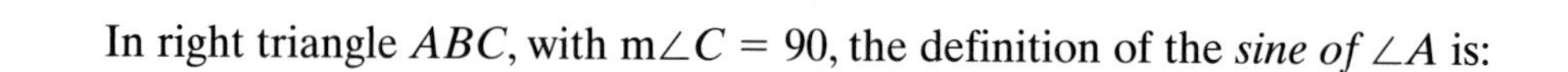

In right triangle ABC, with $m\angle C = 90$, the definition of the *sine of* $\angle A$ is:

$$\text{sine } A = \frac{\text{length of side } \textit{opposite } \angle A}{\text{length of } \textit{hypotenuse}} = \frac{BC}{AB} = \frac{a}{c}$$

By using "sin A" as an abbreviation for sine A, "opp" as an abbreviation for the length of the leg opposite $\angle A$, and "hyp" as an abbreviation for the length of the hypotenuse, we can shorten the way we write the definition of sine A as follows:

$$\sin A = \frac{\text{opp}}{\text{hyp}} = \frac{BC}{AB} = \frac{a}{c}$$

The Cosine Ratio

From the preceding figure on page 317, which shows similar triangles, $\triangle ABC$, $\triangle ADE$, and $\triangle AFG$, we see that

$$\frac{AC}{AB} = \frac{AE}{AD} = \frac{AG}{AF} = \text{a constant.}$$

This ratio is called the **cosine of $\angle A$**.

DEFINITION

The **cosine of an acute angle of a right triangle** is the ratio of the length of the side adjacent to the acute angle to the length of the hypotenuse.

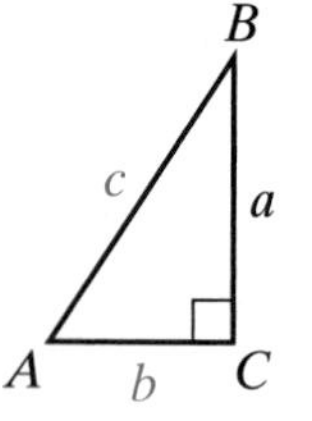

In right triangle ABC, with $m\angle C = 90$, the definition of the *cosine of* $\angle A$ is:

$$\text{cosine } A = \frac{\text{length of side } \textit{adjacent} \text{ to } \angle A}{\text{length of } \textit{hypotenuse}} = \frac{AC}{AB} = \frac{b}{c}$$

By using "cos A" as an abbreviation for cosine A, "adj" as an abbreviation for the length of the leg adjacent to $\angle A$, and "hyp" as an abbreviation for the length of the hypotenuse, we can shorten the way we write the definition of cosine A as follows:

$$\cos A = \frac{\text{adj}}{\text{hyp}} = \frac{AC}{AB} = \frac{b}{c}$$

Finding Sine and Cosine Ratios on a Calculator

CASE 1 *Given an angle measure, find the sine or cosine ratio.*

On a calculator we use the keys labeled SIN and COS to display the values of the sine and cosine of an angle. The sequence of keys that a calculator requires for tangent will be the same as the sequence for sine or cosine.

For example, to find sin 50° and cos 50°, we use the following:

ENTER: SIN 50) ENTER

DISPLAY:

```
sin (50)
        .7660444431
```

ENTER: COS 50) ENTER

DISPLAY:

```
cos (50)
        .6427876097
```

CASE 2 *Given a sine or cosine ratio, find the angle measure.*

A calculator will also find the measure of $\angle A$ when sin A or cos A is given. To do this we use the keys labeled SIN^{-1} and COS^{-1}. These are the second functions of SIN and COS and are accessed by first pressing 2nd. We can think of the meaning of $\sin^{-1}$ as "the angle whose sine is." Therefore, if sin $A = 0.2588$, then $\sin^{-1}(0.2588)$ can be read as "the angle whose sine is 0.2588."

To find the measure of $\angle A$ from the calculator, we use the following sequences of keys:

ENTER: 2nd SIN^{-1} 0.2588) ENTER

DISPLAY:

```
sin⁻¹(0.2588)
        14.99887031
```

The measure of $\angle A$ to the *nearest degree* is 15°.

EXAMPLE 1

In $\triangle ABC$, $\angle C$ is a right angle, $BC = 7$, $AC = 24$, and $AB = 25$. Find:

a. $\sin A$

b. $\cos A$

c. $\sin B$

d. $\cos B$

e. $m\angle B$, to the *nearest degree*

Answers

Solution **a.** $\sin A = \frac{\text{opp}}{\text{hyp}} = \frac{BC}{AB} = \frac{7}{25}$

b. $\cos A = \frac{\text{adj}}{\text{hyp}} = \frac{AC}{AB} = \frac{24}{25}$

c. $\sin B = \frac{\text{opp}}{\text{hyp}} = \frac{AC}{AB} = \frac{24}{25}$

d. $\cos B = \frac{\text{adj}}{\text{hyp}} = \frac{BC}{AB} = \frac{7}{25}$

e. Use a calculator. Start with the ratio in part **c** and use SIN⁻¹ or start with the ratio in part **d** and use COS⁻¹.

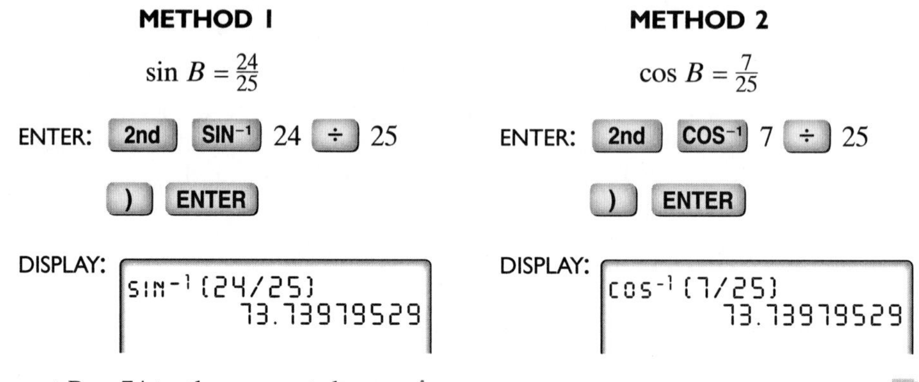

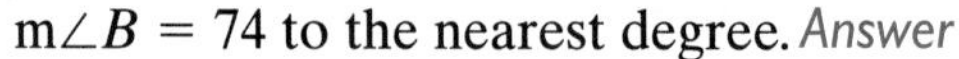

$m\angle B = 74$ to the nearest degree. *Answer*

EXERCISES

Writing About Mathematics

1. If $\angle A$ and $\angle B$ are the acute angles of right triangle ABC, show that $\sin A = \cos B$.

2. If $\angle A$ is an acute angle of right triangle ABC, explain why it is always true that $\sin A < 1$ and $\cos A < 1$.

Developing Skills

In 3–6, find: **a.** sin A **b.** cos A **c.** sin B **d.** cos B

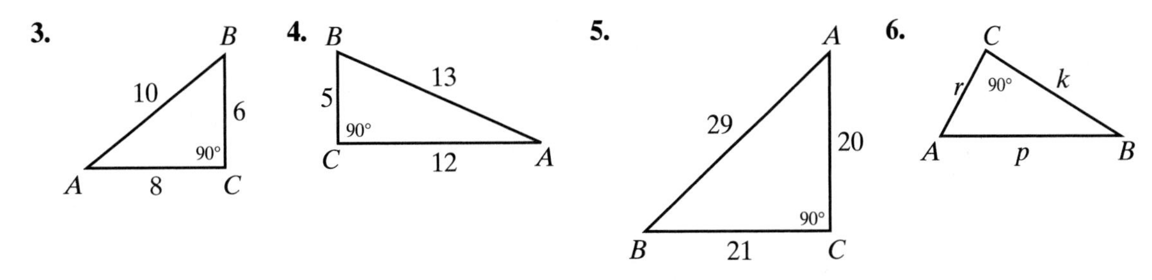

7. In $\triangle ABC$, $m\angle C = 90$, $AC = 4$, and $BC = 3$. Find sin A.

8. In $\triangle RST$, $m\angle S = 90$, $RS = 5$, and $ST = 12$. Find cos T.

In 9–20, for each of the following, use a calculator to find the trigonometric function value to the *nearest ten-thousandth*.

9. sin 18° **10.** sin 42° **11.** sin 58° **12.** sin 76°

13. sin 1° **14.** sin 89° **15.** cos 21° **16.** cos 35°

17. cos 40° **18.** cos 59° **19.** cos 74° **20.** cos 88°

In 21–38, for each of the following, use a calculator to find the measure of $\angle A$ to the *nearest degree*.

21. sin A = 0.1908 **22.** sin A = 0.8387 **23.** sin A = 0.3420

24. cos A = 0.9397 **25.** cos A = 0.0698 **26.** cos A = 0.8910

27. sin A = 0.8910 **28.** sin A = 0.9986 **29.** cos A = 0.9986

30. sin A = 0.1900 **31.** cos A = 0.9750 **32.** sin A = 0.8740

33. cos A = 0.8545 **34.** sin A = 0.5800 **35.** cos A = 0.5934

36. cos A = 0.2968 **37.** sin A = 0.1275 **38.** cos A = 0.8695

39. **a.** Use a calculator to find sin 25° and sin 50°.

b. If the measure of an angle is doubled, is the sine of the angle also doubled?

40. **a.** Use a calculator to find cos 25° and cos 50°.

b. If the measure of an angle is doubled, is the cosine of the angle also doubled?

41. As an angle increases in measure from 1° to 89°:

a. Does the sine of the angle increase or decrease?

b. Does the cosine of the angle increase or decrease?

In 42 and 43, complete each sentence by replacing ? with a degree measure that makes the sentence true.

42. **a.** $\sin 70° = \cos$?
b. $\sin 23° = \cos$?
c. $\sin 38° = \cos$?
d. $\sin x° = \cos$?

43. **a.** $\cos 50° = \sin$?
b. $\cos 17° = \sin$?
c. $\cos 82° = \sin$?
d. $\cos x° = \sin$?

Applying Skills

44. In $\triangle ABC$, $m\angle C = 90$, $BC = 20$, and $BA = 40$.
a. Find $\sin A$.
b. Find the measure of $\angle A$.

45. In $\triangle ABC$, $m\angle C = 90$, $AC = 40$, and $AB = 80$.
a. Find $\cos A$.
b. Find the measure of $\angle A$.

46. In $\triangle ABC$, $\angle C$ is a right angle, $AC = 8$, $BC = 15$, and $AB = 17$. Find:
a. $\sin A$
b. $\cos A$
c. $\sin B$
d. $\cos B$
e. the measure of $\angle A$ to the *nearest degree*
f. the measure of $\angle B$ to the *nearest degree*

47. In $\triangle RST$, $m\angle T = 90$, $ST = 11$, $RT = 60$, and $RS = 61$. Find:
a. $\sin R$
b. $\cos R$
c. $\sin S$
d. $\cos S$
e. the measure of $\angle R$ to the *nearest degree*
f. the measure of $\angle S$ to the *nearest degree*

48. In $\triangle ABC$, $\angle C$ is a right angle, $AC = 1.0$, $BC = 2.4$, and $AB = 2.6$. Find:
a. $\sin A$
b. $\cos A$
c. $\sin B$
d. $\cos B$
e. the measure of $\angle A$ to the *nearest degree.*
f. the measure of $\angle B$ to the *nearest degree.*

49. In rectangle $ABCD$, $AB = 3.5$ and $CB = 1.2$. Find:
a. $\sin \angle ABD$
b. $\cos \angle ABD$
c. $\sin \angle CBD$
d. $\cos \angle CBD$
e. the measure of $\angle ABD$ to the *nearest degree.*
f. the measure of $\angle CBD$ to the *nearest degree.*

50. In right triangle ABC, $\angle C$ is the right angle, $BC = 1$, $AC = \sqrt{3}$ and $AB = 2$.
a. Using the given lengths, write the ratios for $\sin A$ and $\cos A$.
b. Use a calculator to find $\sin 30°$ and $\cos 30°$.
c. What differences, if any, exist between the answers to parts **a** and **b**?

51. In $\triangle ABC$, $m\angle C = 90$ and $\sin A = \cos A$. Find $m\angle A$.

8-5 APPLICATIONS OF THE SINE AND COSINE RATIOS

Since the sine and cosine ratios each have the length of the hypotenuse of a right triangle as the second term of the ratio, we can use these ratios to solve problems in the following cases:

1. We know the length of one leg and the measure of one acute angle and want to find the length of the hypotenuse.
2. We know the length of the hypotenuse and the measure of one acute angle and want to find the length of a leg.
3. We know the lengths of the hypotenuse and one leg and want to find the measure of an acute angle.

EXAMPLE 1

While flying a kite, Betty lets out 322 feet of string. When the string is secured to the ground, it makes an angle of 38.0° with the ground. To the *nearest foot*, what is the height of the kite above the ground? (Assume that the string is stretched so that it is straight.)

Solution Let K be the position of the kite in the air, B be the point on the ground at which the end of the string is secured, and G be the point on the ground directly below the kite, as shown in the diagram. The height of the kite is the length of the perpendicular from the ground to the kite. Therefore, $m\angle G = 90$, $m\angle B = 38.0$, and the length of the string $BK = 322$ feet.

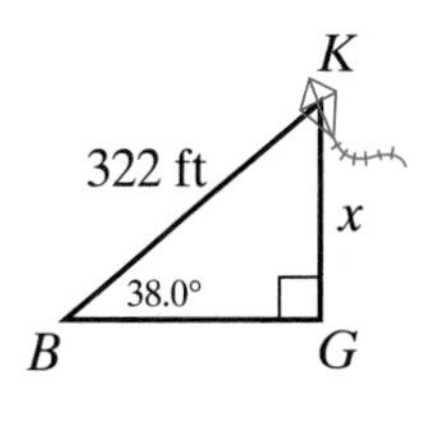

Let $x = KG$, the height of the kite. We know the length of the hypotenuse and the measure of one acute angle and want to find the length of a leg, KG.

In **Method 1** below, since leg KG is opposite $\angle B$, we can use $\sin B = \frac{\text{opp}}{\text{hyp}}$.

In **Method 2** below, since leg KG is adjacent to $\angle K$, we can use $\cos K = \frac{\text{adj}}{\text{hyp}}$ with $m\angle K = 90 - 38.0 = 52.0$.

	METHOD 1	**METHOD 2**
Write the ratio:	$\sin B = \frac{KG}{BK}$	$\cos K = \frac{KG}{BK}$
Substitute the given values:	$\sin 38.0° = \frac{x}{322}$	$\cos 52.0° = \frac{x}{322}$
Solve for x:	$x = 322 \sin 38.0°$	$x = 322 \cos 52.0°$
Compute using a calculator:	$x = 198.242995$	$x = 198.242995$

Answer The height of the kite to the *nearest foot* is 198 feet.

EXAMPLE 2

A wire reaches from the top of a pole to a stake in the ground 3.5 meters from the foot of the pole. The wire makes an angle of 65° with the ground. Find to the *nearest tenth of a meter* the length of the wire.

Solution In $\triangle BTS$, $\angle B$ is a right angle, $BS = 3.5$, $m\angle S = 65$, and $m\angle T = 90 - 65 = 25$.

Let $x = ST$, the length of the wire.

Since we know the length of one leg and the measure of one acute angle and want to find the length of the hypotenuse, we can use either the sine or the cosine ratio.

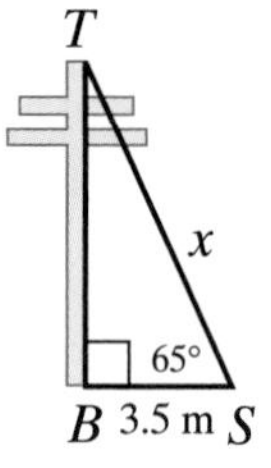

METHOD 1

$$\cos S = \frac{\text{adj}}{\text{hyp}} = \frac{BS}{ST}$$
$$\cos 65^\circ = \frac{3.5}{x}$$
$$x \cos 65^\circ = 3.5$$
$$x = \frac{3.5}{\cos 65^\circ}$$
$$x = 8.281705541$$

METHOD 2

$$\sin T = \frac{\text{opp}}{\text{hyp}} = \frac{BS}{ST}$$
$$\sin 25^\circ = \frac{3.5}{x}$$
$$x \sin 25^\circ = 3.5$$
$$x = \frac{3.5}{\sin 25^\circ}$$
$$x = 8.281705541$$

Answer The wire is 8.3 meters long to the *nearest tenth of a meter*.

EXAMPLE 3

A ladder 25 feet long leans against a building and reaches a point 23.5 feet above the ground. Find to the *nearest degree* the angle that the ladder makes with the ground.

Solution In right triangle ABC, AB, the length of the hypotenuse is 25 feet and BC, the side opposite $\angle A$, is 23.5 feet. Since the problem involves $\angle A$, $\overline{BC}$ (its opposite side), and $\overline{AB}$ (the hypotenuse), the sine ratio is used.

$$\sin A = \frac{\text{opp}}{\text{hyp}} = \frac{23.5}{25}$$

25.0 ft
23.5 ft
B
A C

ENTER: **2nd** **SIN⁻¹** 23.5 **÷** 25 **)** **ENTER**

DISPLAY:

```
SIN⁻¹(23.5/25.0)
         70.05155641
```

Answer To the nearest degree, the measure of the angle is 70°.

EXERCISES

Writing About Mathematics

1. Brittany said that for all acute angles, A, $(\tan A)(\cos A) = \sin A$. Do you agree with Brittany? Explain why or why not.

2. Pearl said that as the measure of an acute angle increases from 1° to 89°, the sine of the angle increases and the cosine of the angle decreases. Therefore, $\cos A$ is the reciprocal of $\sin A$. Do you agree with Pearl? Explain why or why not.

In 3–11, find to the *nearest centimeter* the length of the side marked x.

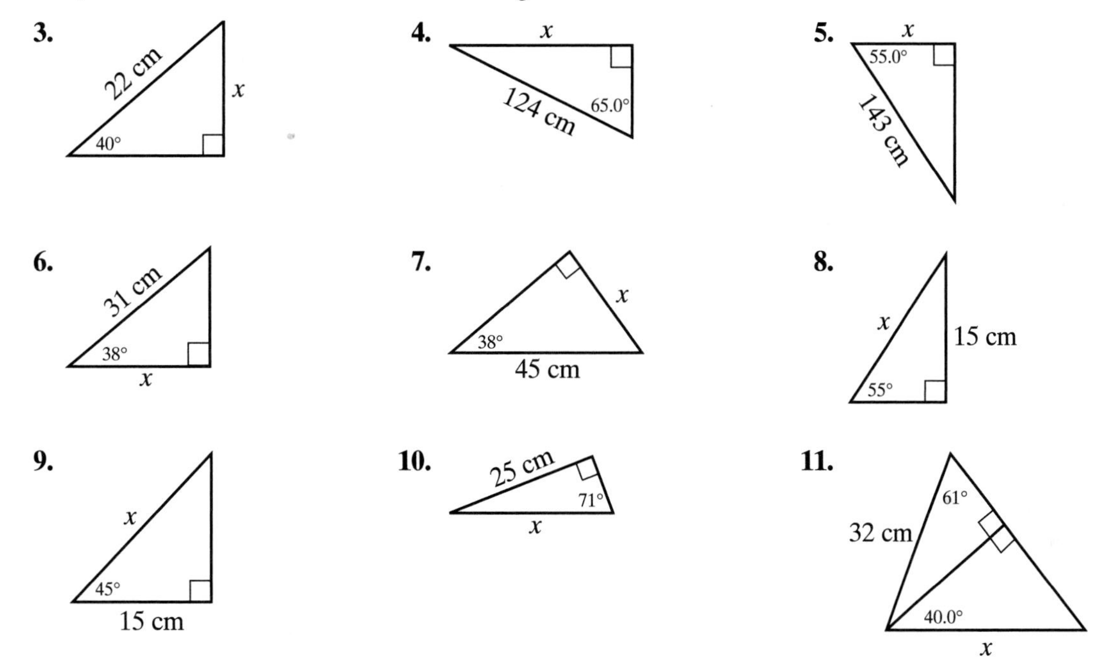

In 12–15, find to the *nearest degree* the measure of the angle marked x.

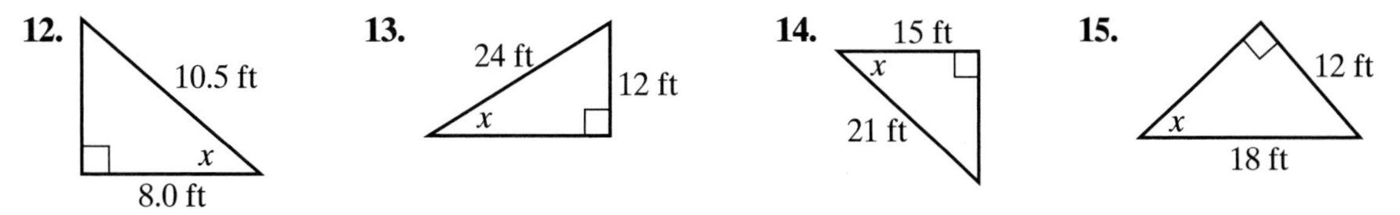

Applying Skills

16. A wooden beam 6.0 meters long leans against a wall and makes an angle of 71° with the ground. Find to the *nearest tenth of a meter* how high up the wall the beam reaches.

17. A boy flying a kite lets out 392 feet of string, which makes an angle of 52° with the ground. Assuming that the string is tied to the ground, find to the *nearest foot* how high the kite is above the ground.

18. A ladder that leans against a building makes an angle of 75° with the ground and reaches a point on the building 9.7 meters above the ground. Find to the *nearest meter* the length of the ladder.

19. From an airplane that is flying at an altitude of 3,500 feet, the angle of depression of an airport ground signal measures 27°. Find to the *nearest hundred feet* the distance between the airplane and the airport signal.

20. A 22-foot pole that is leaning against a wall reaches a point that is 18 feet above the ground. Find to the *nearest degree* the measure of the angle that the pole makes with the ground.

21. To reach the top of a hill that is 1.0 kilometer high, one must travel 8.0 kilometers up a straight road that leads to the top. Find to the *nearest degree* the measure of the angle that the road makes with the horizontal.

22. A 25-foot ladder leans against a building and makes an angle of 72° with the ground. Find to the *nearest foot* the distance between the foot of the ladder and the building.

23. A wire 2.4 meters in length is attached from the top of a post to a stake in the ground. The measure of the angle that the wire makes with the ground is 35°. Find to the *nearest tenth of a meter* the distance from the stake to the foot of the post.

24. An airplane rises at an angle of 14° with the ground. Find to the *nearest hundred feet* the distance the airplane has flown when it has covered a horizontal distance of 1,500 feet.

25. A kite string makes an angle of 43° with the ground. The string is staked to a point 104 meters from a point on the ground directly below the kite. Find to the *nearest meter* the length of the kite string, which is stretched taut.

26. The top of a 43-foot ladder touches a point on the wall that is 36 feet above the ground. Find to the *nearest degree* the measure of the angle that the ladder makes with the wall.

27. In a park, a slide 9.1 feet long is perpendicular to the ladder to the top of the slide. The distance from the foot of the ladder to the bottom of the slide is 10.1 feet. Find to the *nearest degree* the measure of the angle that the slide makes with the horizontal.

9.1 ft

10.1 ft

28. A playground has the shape of an isosceles trapezoid $ABCD$. The length of the shorter base, $\overline{CD}$, is 185 feet. The length of each of the equal sides is 115 feet and $m\angle A = 65.0$.

a. Find DE, the length of the altitude from D, to the *nearest foot.*

b. Find AE, to the *nearest tenth of a foot.*

c. Find AB, to the *nearest foot.*

d. What is the area of the playground to the *nearest hundred square feet*?

e. What is the perimeter of the playground?

29. What is the area of a rhombus, to the *nearest ten square feet*, if the measure of one side is 43.7 centimeters and the measure of one angle is 78.0°?

30. A roofer wants to reach the roof of a house that is 21 feet above the ground. The measure of the steepest angle that a ladder can make with the house when it is placed directly under the roof is 27°. Find the length of the shortest ladder that can be used to reach the roof, to the *nearest foot.*

8-6 SOLVING PROBLEMS USING TRIGONOMETRIC RATIOS

When the conditions of a problem can be modeled by a right triangle for which the measures of one side and an acute angle or of two sides are known, the trigonometric ratios can be used to find the measure of another side or of an acute angle.

Procedure

To solve a problem by using trigonometric ratios:

1. Draw the right triangle described in the problem.
2. Label the sides and angles with the given values.
3. Assign a variable to represent the measure to be determined.
4. Select the appropriate trigonometric ratio.
5. Substitute in the trigonometric ratio, and solve the resulting equation.

Given $\triangle ABC$ with $m\angle C = 90$:

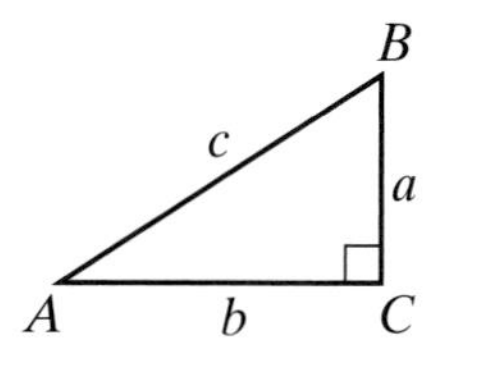

$$\sin A = \frac{\text{opp}}{\text{hyp}} = \frac{BC}{AB} = \frac{a}{c} \qquad \sin B = \frac{\text{opp}}{\text{hyp}} = \frac{AC}{AB} = \frac{b}{c}$$

$$\cos A = \frac{\text{adj}}{\text{hyp}} = \frac{AC}{AB} = \frac{b}{c} \qquad \cos B = \frac{\text{adj}}{\text{hyp}} = \frac{BC}{AB} = \frac{a}{c}$$

$$\tan A = \frac{\text{opp}}{\text{adj}} = \frac{BC}{AC} = \frac{a}{b} \qquad \tan B = \frac{\text{opp}}{\text{adj}} = \frac{AC}{BC} = \frac{b}{a}$$

EXAMPLE 1

Given: In isosceles triangle ABC, $AC = CB = 20$ and $m\angle A = m\angle B = 68$. $\overline{CD}$ is an altitude.

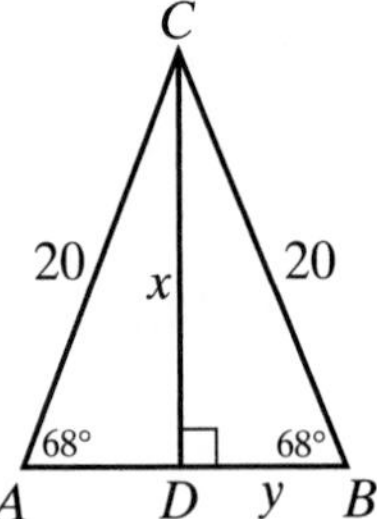

Find:

a. Length of altitude $\overline{CD}$ to the *nearest tenth.*

b. Length of $\overline{AB}$ to the *nearest tenth.*

Solution **a.** In right $\triangle BDC$, $\sin B = \frac{CD}{CB}$

Let $x = CD$.

$$\sin 68° = \frac{x}{20}$$
$$x = 20 \sin 68°$$
$$= 18.54367709$$
$$= 18.5$$

b. Since the altitude drawn to the base of an isosceles triangle bisects the base, $AB = 2DB$. Therefore, find DB in $\triangle BDC$ and double it to find AB.

In right $\triangle BDC$, $\cos B = \frac{DB}{CB}$.

Let $y = DB$.

$$\cos 68° = \frac{y}{20}$$
$$y = 20 \cos 68° = 7.492131868$$
$$AB = 2y = 2(7.492131868) = 14.9843736$$

Answers **a.** $CD = 18.5$, to the *nearest tenth.* **b.** $AB = 15.0$, to the *nearest tenth.*

EXERCISES

Writing About Mathematics

1. If the measures of two sides of a right triangle are given, it is possible to find the measures of the third side and of the acute angles. Explain how you would find these measures.

2. If the measures of the acute angles of a right triangle are given, is it possible to find the measures of the sides? Explain why or why not.

Developing Skills

In 3–10: In each given right triangle, find to the *nearest foot* the length of the side marked x; or find to the *nearest degree* the measure of the angle marked x. Assume that each measure is given to the nearest foot or to the nearest degree.

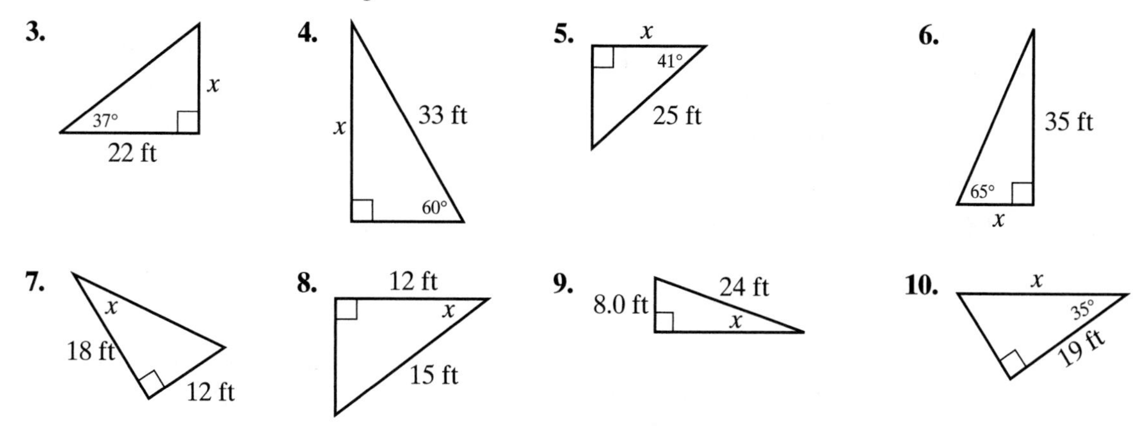

11. In $\triangle ABC$, $m\angle A = 42$, $AB = 14$, and $\overline{BD}$ is the altitude to $\overline{AC}$. Find BD to the *nearest tenth.*

12. In $\triangle ABC$, $\overline{AC} \cong \overline{BC}$, $m\angle A = 50$, and $AB = 30$. Find to the *nearest tenth* the length of the altitude from vertex C.

13. The legs of a right triangle measure 84 and 13. Find to the *nearest degree* the measure of the smallest angle of this triangle.

14. The length of hypotenuse $\overline{AB}$ of right triangle ABC is twice the length of leg $\overline{BC}$. Find the number of degrees in $\angle B$.

15. The longer side of a rectangle measures 10, and a diagonal makes an angle of 27° with this side. Find to the *nearest integer* the length of the shorter side.

16. In rectangle $ABCD$, diagonal $\overline{AC}$ measures 11 and side $\overline{AB}$ measures 7. Find to the *nearest degree* the measure of $\angle CAB$.

17. In right triangle ABC, $\overline{CD}$ is the altitude to hypotenuse $\overline{AB}$, $AB = 25$, and $AC = 20$. Find lengths AD, DB, and CD to the *nearest integer* and the measure of $\angle B$ to the *nearest degree.*

18. The lengths of the diagonals of a rhombus are 10 and 24.

a. Find the perimeter of the rhombus.

b. Find to the *nearest degree* the measure of the angle that the longer diagonal makes with a side of the rhombus.

19. The altitude to the hypotenuse of a right triangle ABC divides the hypotenuse into segments whose measures are 9 and 4. The measure of the altitude is 6. Find to the *nearest degree* the measure of the smaller acute angle of $\triangle ABC$.

20. In $\triangle ABC$, $AB = 30$, $m\angle B = 42$, $m\angle C = 36$, and $\overline{AD}$ is an altitude.

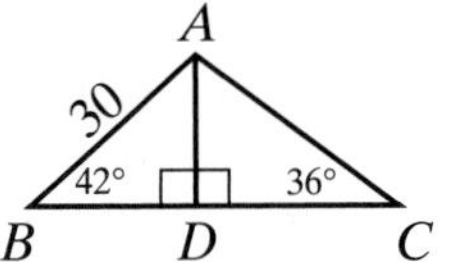

a. Find to the *nearest integer* the length of $\overline{AD}$.

b. Using the result of part **a**, find to the *nearest integer* the length of $\overline{DC}$.

21. Angle D in quadrilateral $ABCD$ is a right angle, and diagonal $\overline{AC}$ is perpendicular to $\overline{BC}$, $BC = 20$, $m\angle B = 35$, and $m\angle DAC = 65$.

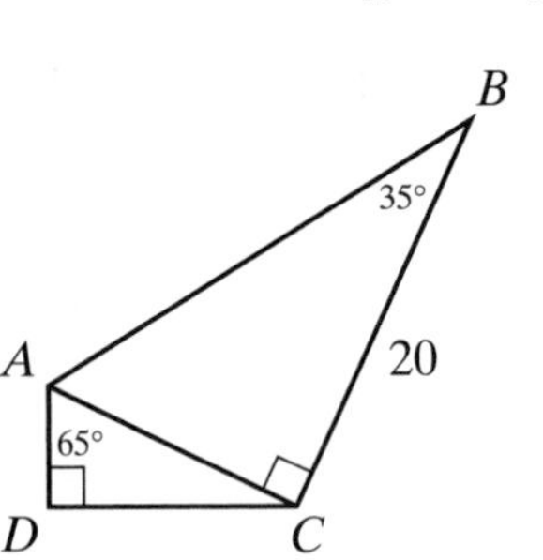

a. Find AC to the *nearest integer*.

b. Using the result of part **a**, find DC to the *nearest integer*.

22. The diagonals of a rectangle each measure 198 and intersect at an angle whose measure is 110°. Find to the *nearest integer* the length and width of the rectangle. *Hint:* The diagonals of a rectangle bisect each other.

23. In rhombus $ABCD$, the measure of diagonal $\overline{AC}$ is 80 and $m\angle BAC = 42$.

a. Find to the *nearest integer* the length of diagonal $\overline{BD}$.

b. Find to the *nearest integer* the length of a side of the rhombus.

24. In right triangle ABC, the length of hypotenuse $\overline{AB}$ is 100 and $m\angle A = 18$.

a. Find AC and BC to the *nearest integer*.

b. Show that the results of part **a** are approximately correct by using the relationship $(AB)^2 = (AC)^2 + (BC)^2$.

Applying Skills

25. Find to the *nearest meter* the height of a church spire that casts a shadow of 53.0 meters when the angle of elevation of the sun measures 68.0°.

26. From the top of a lighthouse 194 feet high, the angle of depression of a boat out at sea measures 34.0°. Find to the *nearest foot* the distance from the boat to the foot of the lighthouse.

27. A straight road to the top of a hill is 2,500 meters long and makes an angle of 12° with the horizontal. Find to the *nearest ten meters* the height of the hill.

28. A wire attached to the top of a pole reaches a stake in the ground 21 feet from the foot of the pole and makes an angle of 58° with the ground. Find to the *nearest foot* the length of the wire.

29. An airplane climbs at an angle of 11° with the ground. Find to the *nearest hundred feet* the distance the airplane has traveled when it has attained an altitude of 450 feet.

30. Find to the *nearest degree* the measure of the angle of elevation of the sun if a child 88 centimeters tall casts a shadow 180 centimeters long.

31. $\overline{AB}$ and $\overline{CD}$ represent cliffs on opposite sides of a river 125 meters wide. From B, the angle of elevation of D measures 20° and the angle of depression of C measures 25°. Find to the *nearest meter*:

a. the height of the cliff represented by $\overline{AB}$.

b. the height of the cliff represented by $\overline{CD}$.

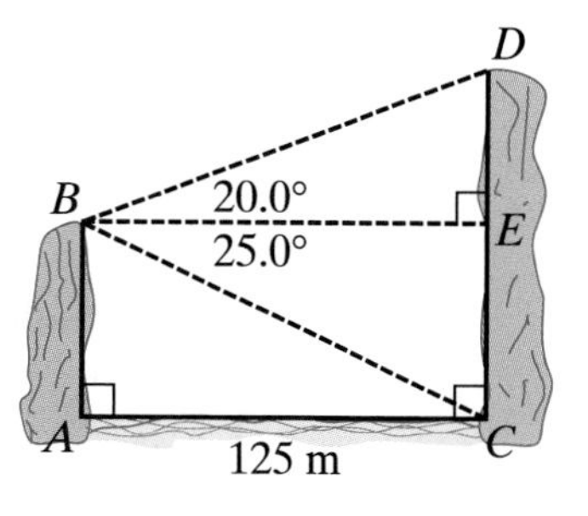

32. Points A, B, and D are on level ground. $\overline{CD}$ represents the height of a building, $BD = 86$ feet, and $m\angle D = 90$. At B, the angle of elevation of the top of the building, $\angle CBD$, measures 49°. At A, the angle of elevation of the top of the building, $\angle CAD$, measures 26°.

a. Find the height of the building, CD, to the *nearest foot*.

b. Find AD to the *nearest foot*.

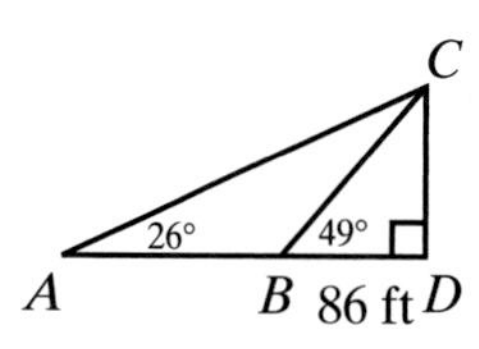

CHAPTER SUMMARY

The **Pythagorean Theorem** relates the lengths of the sides of a right triangle. If the lengths of the legs of a right triangle are a and b, and the length of the hypotenuse is c, then $c^2 = a^2 + b^2$.

The **trigonometric functions** associate the measure of each acute angle A with a number that is the ratio of the measures of two sides of a right triangle. The three most commonly used trigonometric functions are **sine**, **cosine**, and **tangent**. In the application of trigonometry to the right triangle, these ratios are defined as follows:

$$\sin A = \frac{\text{opp}}{\text{hyp}} \qquad \cos A = \frac{\text{adj}}{\text{hyp}} \qquad \tan A = \frac{\text{opp}}{\text{adj}}$$

In right triangle ABC, with hypotenuse $\overline{AB}$

- $\overline{BC}$ is opposite $\angle A$ and adjacent to $\angle B$;
- $\overline{AC}$ is opposite $\angle B$ and adjacent to $\angle A$.

$$\sin A = \frac{BC}{AB} \qquad \cos B = \frac{BC}{AB}$$

$$\cos A = \frac{AC}{AB} \qquad \sin B = \frac{AC}{AB}$$

$$\tan A = \frac{BC}{AC} \qquad \tan B = \frac{AC}{BC}$$

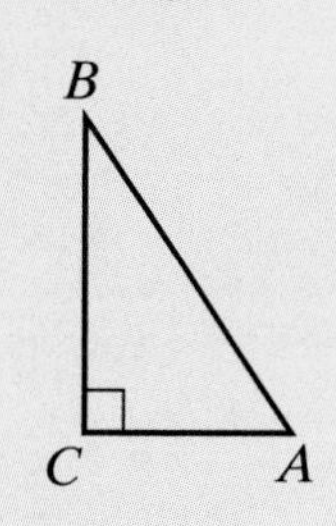

An **angle of elevation**, $\angle GDE$ in the diagram, is an angle between a horizontal line and a line of sight to an object at a higher elevation. An **angle of depression**, $\angle FED$ in the diagram, is an angle between a horizontal line and a line of sight to an object at a lower elevation.

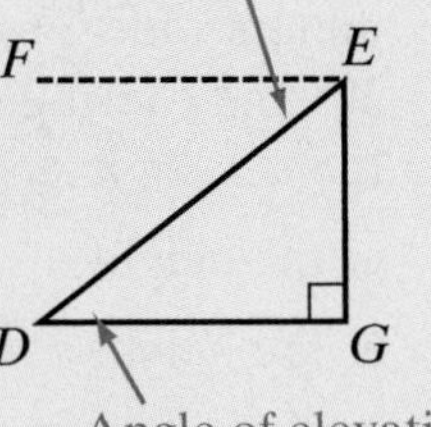

VOCABULARY

8-1 Trigonometry • Direct measurement • Indirect measurement • Pythagorean Theorem • Pythagorean triple

8-2 Opposite side • Adjacent side • Similar • Tangent of an acute angle of a right triangle

8-3 Angle of elevation • Angle of depression

8-4 Sine of an acute angle of a right triangle • Cosine of an acute angle of a right triangle

REVIEW EXERCISES

1. Talia's calculator is not functioning properly and does not give the correct value when she uses the TAN key. Assume that all other keys of the calculator are operating correctly.

a. Explain how Talia can find the measure of the leg $\overline{AC}$ of right triangle ABC when $BC = 4.5$ and $m\angle A = 43$.

b. Explain how Talia can use her calculator to find the tangent of any acute angle, given the measure of one side of a right triangle and the measure of an acute angle as in part **a**.

2. Jill made the following entry on her calculator:

ENTER: 2nd SIN⁻¹ 1.5) ENTER

Explain why the calculator displayed an error message.

In 3–8, refer to $\triangle RST$ and express the value of each ratio as a fraction.

3. $\sin R$ **4.** $\tan T$

5. $\sin T$ **6.** $\cos R$

7. $\cos T$ **8.** $\tan R$

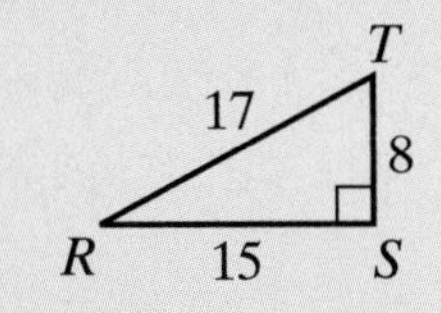

In 9–12: in each given triangle, find to the *nearest centimeter* the length of the side marked x. Assume that each given length is correct to the nearest centimeter.

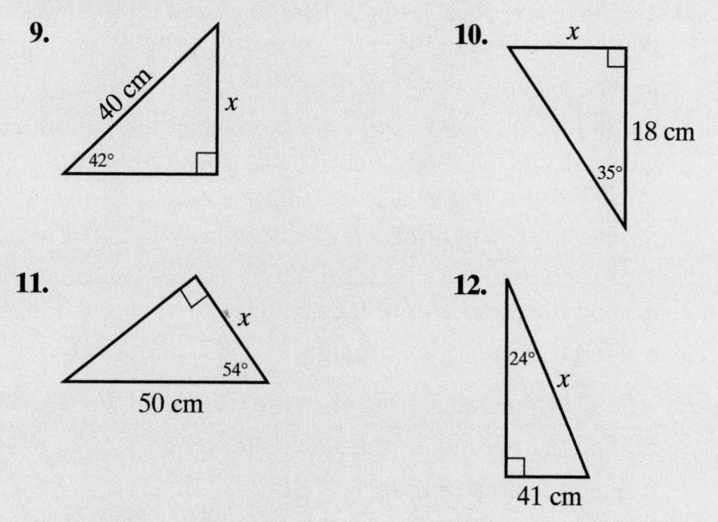

13. If $\cos A = \sin 30°$ and $0° \leq A \leq 90°$, what is the measure of $\angle A$?

14. In right triangle ACB, $m\angle C = 90$, $m\angle A = 66$, and $AC = 100$. Find BC to the *nearest integer*.

15. In right triangle ABC, $m\angle C = 90$, $m\angle B = 28$, and $BC = 30$. Find AB to the *nearest integer*.

16. In $\triangle ABC$, $m\angle C = 90$, $\tan A = 0.7$, and $AC = 40$. Find BC.

17. In $\triangle ABC$, $m\angle C = 90$, $AB = 30$, and $BC = 15$. What is the measure, in degrees, of $\angle A$?

18. In $\triangle ABC$, $m\angle C = 90$, $BC = 5$, and $AC = 9$. Find to the *nearest degree* the measure of $\angle A$.

19. Find to the *nearest meter* the height of a building if its shadow is 42 meters long when the angle of elevation of the sun measures 42°.

20. A 5-foot wire attached to the top of a tent pole reaches a stake in the ground 3 feet from the foot of the pole. Find to the *nearest degree* the measure of the angle made by the wire with the ground.

21. While flying a kite, Doris let out 425 feet of string. Assuming that the string is stretched taut and makes an angle of 48° with the ground, find to the *nearest ten feet* how high the kite is.

22. A rectangular field $ABCD$ is crossed by a path from A to C. If $m\angle BAC = 62$ and $BC = 84$ yards, find to the *nearest yard*:

a. the width of the field, AB.

b. the length of path, AC.

23. Find the length of a leg of an isosceles right triangle if the length of the hypotenuse is $\sqrt{72}$.

24. The measure of each of the base angles of an isosceles triangle is 15 degrees more than twice the measure of the vertex angle.

a. Find the measure of each angle of the triangle.

b. Find to the *nearest tenth of a centimeter* the measure of each of the equal sides of the triangle if the measure of the altitude to the base is 88.0 centimeters.

c. Find to the *nearest tenth of a centimeter* the measure of the base of the triangle.

d. Find the perimeter of the triangle.

e. Find the area of the triangle.

25. $ABCD$ is a rectangle with E a point on $\overline{BC}$. $AB = 12$, $BE = 5$, and $EC = 9$.

a. Find the perimeter of triangle AED.

b. Find the area of triangle AED.

c. Find the measure of $\angle CDE$.

d. Find the measure of $\angle BAE$.

e. Find the measure of $\angle AED$.

Exploration

A regular polygon with n sides can be divided into n congruent isosceles triangles.

a. Express, in terms of n, the measure of the vertex angle of one of the isosceles triangles.

b. Express, in terms of n, the measure of a base angle of one of the isosceles triangles.

c. Let s be the measure of a side of the regular polygon. Express, in terms of n and s, the measure of the altitude to the base of one of the isosceles triangles.

d. Express the area of one of the isosceles triangles in terms of n and s.

e. Write a formula for the area of a regular polygon in terms of the measure of a side, s, and the number of sides, n.

CUMULATIVE REVIEW — CHAPTERS 1–8

Part I

Answer all questions in this part. Each correct answer will receive 2 credits. No partial credit will be allowed.

1. Which of the following does not represent a real number when $x = 3$?

(1) $\frac{3}{x}$ (2) $\frac{x}{x}$ (3) $\frac{x-3}{x}$ (4) $\frac{x}{x-3}$

2. The coordinates of one point on the x-axis are

(1) $(1, 1)$ (2) $(-1, -1)$ (3) $(1, 0)$ (4) $(0, 1)$

3. The expression $-0.2a^2(10a^3 - 2a)$ is equivalent to

(1) $-2a^5 + 0.4a^3$
(2) $-20a^5 + 4a^3$
(3) $-2a^6 + 0.4a^2$
(4) $-20a^6 - 0.4a^3$

4. If $\frac{3}{4}x - 7 = \frac{1}{4}x + 3$, then x equals

(1) 20 (2) 10 (3) 5 (4) $\frac{5}{2}$

5. If $\sin A = 0.3751$, then, to the nearest degree, $m\angle A$ is

(1) 21 (2) 22 (3) 68 (4) 69

6. Which of the following statements is false?

(1) If a polygon is a square, then it is a parallelogram.
(2) If a polygon is a square, then it is a rhombus.
(3) If a polygon is a rectangle, then it is a parallelogram.
(4) If a polygon is a rectangle, then it is a rhombus.

7. If the measures of two legs of a right triangle are 7.0 feet and 8.0 feet, then, to the nearest tenth of a foot, the length of the hypotenuse is

(1) 10.6 (2) 15.0 (3) 41.2 (4) 48.9

8. The measure of the radius of a cylinder is 9.00 centimeters and the measure of its height is 24.00 centimeters. The surface area of the cylinder to the correct number of significant digits is

(1) 1,610 square centimeters
(2) 1,620 square centimeters
(3) 1,860 square centimeters
(4) 1,870 square centimeters

9. When $5b^2 + 2b$ is subtracted from $8b$ the difference is

(1) $6b - 5b^2$ (2) $5b^2 - 6b$ (3) $5b^2 - 10b$ (4) $-5b^2 - 6b$

10. When written in scientific notation, the fraction $\frac{(1.2 \times 10^{-4}) \times (3.5 \times 10^{8})}{8.4 \times 10^{-2}}$ equals

(1) 5.0×10^2 (2) 5.0×10^1 (3) 5.0×10^5 (4) 5.0×10^6

Part II

Answer all questions in this part. Each correct answer will receive 2 credits. Clearly indicate the necessary steps, including appropriate formula substitutions, diagrams, graphs, charts, etc. For all questions in this part, a correct numerical answer with no work shown will receive only 1 credit.

11. The vertices of pentagon $ABCDE$ are $A(2, -2), B(7, -2), C(7, 5), D(0, 5), E(-2, 0)$.

a. Draw pentagon $ABCDE$ on graph paper.

b. Find the area of the pentagon.

12. In $\triangle ABC$, $m\angle C = 90$, $m\angle B = 30$, $AC = 6x^2 - 4x$, and $AB = 2x$. Find the value of x.

Part III

Answer all questions in this part. Each correct answer will receive 3 credits. Clearly indicate the necessary steps, including appropriate formula substitutions, diagrams, graphs, charts, etc. For all questions in this part, a correct numerical answer with no work shown will receive only 1 credit.

13. Plank Road and Holt Road are perpendicular to each other. At a point 1.3 miles before the intersection of Plank and Holt, State Street crosses Plank at an angle of 57°. How far from the intersection of Plank and Holt will State Street intersect Holt? Write your answer to the nearest tenth of a mile.

14. Benny, Carlos, and Danny each play a different sport and have different career plans. Each of the four statements given below is true.

The boy who plays baseball plans to be an engineer.
Benny wants to be a lawyer.
Carlos plays soccer.
The boy who plans to be a doctor does not play basketball.

What are the career plans of each boy and what sport does he play?

Part IV

Answer all questions in this part. Each correct answer will receive 4 credits. Clearly indicate the necessary steps, including appropriate formula substitutions, diagrams, graphs, charts, etc. For all questions in this part, a correct numerical answer with no work shown will receive only 1 credit.

15. Bart wants to plant a garden around the base of a tree. To determine the amount of topsoil he will need to enrich his garden, he measured the circumference of the tree and found it to be 9.5 feet. His garden will be 2.0 feet wide, in the form of a ring around the tree. Find to the nearest square foot the surface area of the garden Bart intends to plant.

16. Samantha had a snapshot that is 3.75 inches wide and 6.5 inches high. She cut a strip off of the top of the snapshot so that an enlargement will fit into a frame that measures 5 inches by 8 inches. What were the dimensions of the strip that she cut off of the original snapshot?

CHAPTER

9

GRAPHING LINEAR FUNCTIONS AND RELATIONS

The Tiny Tot Day Care Center charges \$200 a week for children who stay at the center between 8:00 A.M. and 5:00 P.M. If a child is not picked up by 5:00 P.M., the center charges an additional \$4.00 per hour or any part of an hour.

Mr. Shubin often has to work late and is unable to pick up his daughter on time. For Mr. Shubin, the weekly cost of day care is a *function* of time; that is, his total cost depends on the time he arrives at the center.

This function for the daily cost of day care can be expressed in terms of an equation in two variables: $y = 30 + 4x$. The variable x represents the total time, in hours, after 5:00 P.M., that a child remains at the center, and the variable y represents the cost, in dollars, of day care for the day.

CHAPTER TABLE OF CONTENTS

9-1 SETS, RELATIONS, AND FUNCTIONS

Set-Builder Notation

In Chapters 1 and 2, we used **roster form** to describe sets. In roster form, the elements of a set are enclosed by braces and listed once. Repeated elements are not allowed. For example,

$$\{\ldots, -3, -2, -1, 0, 1, 2, 3, \ldots\}$$

is the set of integers, in roster form.

A second way to specify a set is to use **set-builder notation**. Set-builder notation is a mathematically concise way of describing a set without listing the elements of the set. For instance, using set-builder notation, the set of counting numbers from 1 to 100 is:

$$\{x \mid x \text{ is an integer and } 1 \leq x \leq 100\}$$

This reads as "the set of all x such that x is an integer and x is at least 1 and at most 100." The vertical bar "|" represents the phrase "such that," and the description to the right of the bar is the rule which defines the set.

Here are some other examples of set-builder notation:

1. $\{x \mid x \text{ is an integer and } x > 6\} = \{7, 8, 9, 10, \ldots\}$ = the set of integers greater than 6

2. $\{x \mid -1 \leq x \leq 3\}$ = any real number in the interval $[-1, 3]$

3. $\{2n + 1 \mid n \text{ is a whole number}\} = \{2(0) + 1, 2(1) + 1, 2(2) + 1, 2(3) + 1, \ldots\}$
$= \{1, 3, 5, 7, \ldots\}$
= the set of odd whole numbers

Frequently used with set-builder notation is the symbol $\in$, which means "is an element of." This symbol is used to indicate that an element is a member of a set. The symbol $\notin$ means "is not an element of," and is used to indicate that an element is not a member of a set. For instance,

$2 \in \{x \mid -1 \leq x \leq 3\}$ since 2 is between -1 and 3.

$-2 \notin \{x \mid -1 \leq x \leq 3\}$ since -2 is *not* between -1 and 3.

EXAMPLE 1

List the elements of each set or indicate that the set is the empty set.

a. $A = \{x \mid x \in \text{ of the set of natural numbers and } x < 0\}$

b. $B = \{2n \mid n \in \text{ of the set of whole numbers}\}$

c. $C = \{x \mid x + 3 = 5\}$

d. $D = \{m \mid m \text{ is a multiple of 5 and } m < -25\}$

Solution **a.** Since there are no natural numbers less than 0,

$$A = \varnothing$$

b. $B = \{2(0), 2(1), 2(2), 2(3), \ldots\}$
$= \{0, 2, 4, 6, \ldots\}$

c. C represents the solution set of the equation $x + 3 = 5$. Therefore,

$$C = \{2\}$$

d. D consists of the set of multiples of 5 that are less than -25.

$$D = \{\ldots, -45, -40, -35, -30\}$$

Answers **a.** $A = \varnothing$ **b.** $B = \{0, 2, 4, 6, \ldots\}$ **c.** $C = \{2\}$ **d.** $D = \{\ldots, -45, -40, -35, -30\}$

Relations That are Finite Sets

There are many instances in which one set of information is related to another. For example, we may identify the persons of a group who are 17, 18, or 19 to determine who is old enough to vote. This information can be shown in a diagram such as the one at the right, or as a set of ordered pairs.

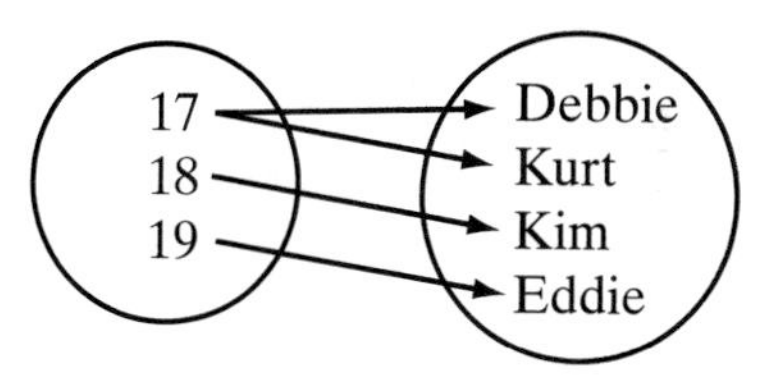

$$\{(17, \text{Debbie}), (17, \text{Kurt}), (18, \text{Kim}), (19, \text{Eddie})\}$$

DEFINITION

A **relation** is a set of ordered pairs.

The **domain of a relation** is the set of all first elements of the ordered pairs. For example, in the relation shown above, the domain is $\{17, 18, 19\}$, the set of ages.

The **range of a relation** is the set of all second elements of the ordered pairs. In the relation above, the range is {Debbie, Kurt, Kim, Eddie}, the set of people.

Let us consider the relation "is greater than," using the set $\{1, 2, 3, 4\}$ as both the domain and the range. Let (x, y) be an ordered pair of the relation. Then the relation can be shown in any of the following ways.

1. *A Rule*

   The relation can be described by the inequality $x > y$.

2. *Set of Ordered Pairs*

   The relation can be listed in set notation as shown below:

$$\{(2, 1), (3, 1), (3, 2), (4, 1), (4, 2), (4, 3)\}$$

3. *Table of Values*

The ordered pairs shown in 2 can be displayed in a table.

x	2	3	3	4	4	4
y	1	1	2	1	2	3

4. *Graph*

In the coordinate plane, the domain is a subset of the numbers on the x-axis and the range is a subset of the numbers on the y-axis. The points that correspond to the ordered pairs of numbers from the domain and range are shown. Of the 16 ordered pairs shown, only six are enclosed to indicate the relation $x > y$.

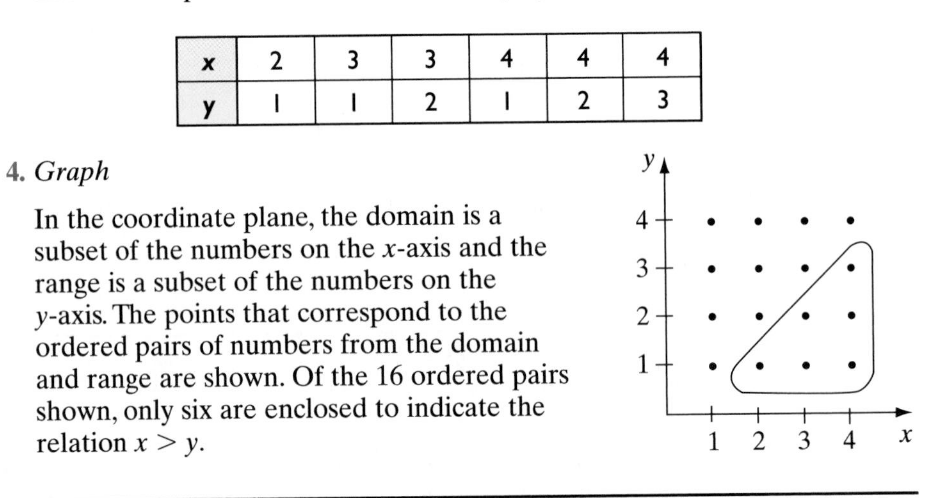

Relations That are Infinite Sets

Charita wants to enclose a rectangular garden. How much fencing will she need if the width of the garden is to be 5 feet? The answer to this question depends on the length of the garden. Therefore, we say that the length of the garden, x, and the amount of fencing, y, form a set of ordered pairs or a relation. The formula for the perimeter of a rectangle can be used to express y in terms of x:

$$P = 2l + 2w$$
$$y = 2x + 2(5)$$
$$y = 2x + 10$$

Some possible values for the amount of fencing can also be shown in a table. As the length of the garden changes, the amount of fencing changes, as the following table indicates.

x	2x + 10	y	(x, y)
1	2(1) + 10	12	(1, 12)
3	2(3) + 10	16	(3, 16)
4.5	2(4.5) + 10	19	(4.5, 19)
7.2	2(7.2) + 10	24.4	(7.2, 24.4)
9.1	2(9.1) + 10	28.2	(9.1, 28.2)

Each solution to $y = 2x + 10$ is a pair of numbers. In writing each pair, we place the value of x first and the value of y second. For example, when $x = 1$, $y = 12$, $(1, 12)$ is a solution. When $x = 3$, $y = 16$, $(3, 16)$ is another solution. It is not possible to list all ordered pairs in the solution set of the equation because the solution set is infinite. However, it is possible to determine whether a given ordered pair is a member of the solution set. Replace x with the first element of the pair (the x-coordinate) and y with the second element of the pair (the y-coordinate). If the result is a true statement, the ordered pair is a solution of the equation.

- $(1.5, 13) \in \{(x, y) \mid y = 2x + 10\}$ because $13 = 2(1.5) + 10$ is true.
- $(4, 14) \notin \{(x, y) \mid y = 2x + 10\}$ because $14 = 2(4) + 10$ is false.

Since in the equation $y = 2x + 10$, x represents the length of a garden, only positive numbers are acceptable replacements for x. Therefore, the domain of this relation is the set of positive real numbers. For every positive number x, $2x + 10$ will also be a positive number. Therefore, the range of this relation is the set of positive real numbers. Although a solution of $y = 2x + 10$ is $(-2, 6)$, it is not possible for -2 to be the length of a garden; $(-2, 6)$ is not a pair of the relation.

When we choose a positive real number to be the value of x, there is one and only one value of y that makes the equation true. Therefore, $y = 2x + 10$ is a special kind of relation called a function.

DEFINITION

A **function** is a relation in which no two ordered pairs have the same first element.

or

A **function** is a relation in which every element of the domain is paired with one and only one element of the range.

Since a function is a special kind of relation, the **domain of a function** is the set of all first elements of the ordered pairs of the function. Similarly, the **range of a function** is the set of all second elements of the ordered pairs of the function. The notation $f(a) = b$ or "f of a equals b," signifies that the value of the function, f, at a is equal to b.

The **independent variable** is the variable that represents the first element of an ordered pair. The domain of a function is the set of all values that the independent variable is allowed to take. The **dependent variable** is the variable that represents the second element of an ordered pair. The range of a function is the set of all values that the dependent variable is allowed to take.

EXAMPLE 2

Find five members of the solution set of the sentence $3x + y = 7$.

Solution *How to Proceed*

(1) Transform the equation into an equivalent equation with y alone as one member:

$$\begin{array}{rcl} 3x + y &=& 7 \\ -3x & & -3x \\ \hline y &=& -3x + 7 \end{array}$$

(2) Choose any five values for x. Since no replacement set is given, any real numbers can be used:

(3) For each selected value of x, determine y:

x	$-3x + 7$	y
-2	$-3(-2) + 7$	13
0	$-3(0) + 7$	7
$\frac{1}{3}$	$-3\left(\frac{1}{3}\right) + 7$	6
3	$-3(3) + 7$	-2
5	$-3(5) + 7$	-8

Answer $(-2, 13), (0, 7), \left(\frac{1}{3}, 6\right), (3, -2), (5, -8)$

Note that many other solutions are also possible, and for every real number x, one and only one real number y will make the equation $3x + y = 7$ true. Therefore, $3x + y = 7$ defines a function.

EXAMPLE 3

Determine whether each of the given ordered pairs is a solution of the inequality $y - 2x \geq 4$.

a. (4, 0) **b.** (–1, 2) **c.** (0, 6) **d.** (0, 7)

Solution

a.
$$\begin{aligned} y - 2x &\geq 4 \\ 0 - 2(4) &\stackrel{?}{\geq} 4 \\ 0 - 8 &\geq 4 \ ✗ \end{aligned}$$

b.
$$\begin{aligned} y - 2x &\geq 4 \\ 2 - 2(-1) &\stackrel{?}{\geq} 4 \\ 2 + 2 &\geq 4 \ ✓ \end{aligned}$$

c.
$$\begin{aligned} y - 2x &\geq 4 \\ 6 - 2(0) &\stackrel{?}{\geq} 4 \\ 6 - 0 &\geq 4 \ ✓ \end{aligned}$$

d.
$$\begin{aligned} y - 2x &\geq 4 \\ 7 - 2(0) &\stackrel{?}{\geq} 4 \\ 7 - 0 &\geq 4 \ ✓ \end{aligned}$$

Answers **a.** Not a solution **b.** A solution **c.** A solution **d.** A solution

Note that for the inequality $y - 2x \geq 4$, the pairs (0, 6) and (0, 7) have the same first element. Therefore $y - 2x \geq 4$ is a relation but not a function.

EXAMPLE 4

The cost of renting a car for 1 day is \$64.00 plus \$0.25 per mile. Let x represent the number of miles the car was driven, and let y represent the rental cost, in dollars, for a day.

a. Write an equation for the rental cost of the car in terms of the number of miles driven.

b. Find the missing member of each of the following ordered pairs, which are elements of the solution set of the equation written in part **a**, and explain the meaning of the pair.

(1) (155, ?) (2) (?, 69)

Solution **a.** Rental cost is \$64.00 plus \$0.25 times the number of miles.

$$y = 64.00 + 0.25x$$

b. (1) For the pair (155, ?), $x = 155$ and y is to be determined. Then:

$$\begin{aligned} y &= 64.00 + 0.25x \\ &= 64.00 + 0.25(155) \\ &= 64.00 + 38.75 \\ &= 102.75 \end{aligned}$$

When the car was driven 155 miles, the rental cost was \$102.75.

(2) For the pair (?, 69), $y = 69$ and x is to be determined. Then:

$$\begin{aligned} y &= 64 + 0.25x \\ 69 &= 64 + 0.25x \\ -64 &\quad -64 \\ 5 &= + 0.25x \\ \frac{5}{0.25} &= \frac{0.25x}{0.25} \\ 20 &= x \end{aligned}$$

When the rental cost was \$69, the car was driven 20 miles.

Answers **a.** $y = 64.00 + 0.25x$

b. (1) (155, 102.75) The car was driven 155 miles, and the cost was \$102.75.

(2) (20, 69) The car was driven 20 miles, and the cost was \$69.00.

EXAMPLE 5

Determine whether or not each set is a relation. If it is a relation, determine whether or not it is a function.

a. $\{x \mid 0 < x < 10\}$

b. $\{(x, y) \mid y = 4 - x\}$

c. $\{(a, b) \mid a = 2 \text{ and } b \text{ is a real number}\}$

d. $\{(x, y) \mid y = x^2\}$

Solution **a.** The set is not a set of ordered pairs. It is not a relation. *Answer*

b. The set is a relation and a function. It is a set of ordered pairs in which every first element is paired with one and only one second element. *Answer*

c. The set is a relation that is not a function. The same first element is paired with every second element. *Answer*

d. The set is a relation and a function. Every first element is paired with one and only one second element, its square. *Answer*

EXAMPLE 6

Jane would like to construct a rectangular pool having an area of 102 square feet. If l represents the length of the pool and w its height, express the set of all ordered pairs, (l, w), that represent the dimensions of the pool using set-builder notation.

Solution Since the area must be 102 square units, the variables l and w are related by the area formula: $A = lw$. Therefore, the set of ordered pairs representing the possible dimensions of the pool is:

$$\{(l, w) \mid 102 = lw, \text{ where } l \text{ and } w \text{ are both positive}\}$$ *Answer*

Note: Since the length and the width cannot be zero or negative, the rule must specify that the dimensions, l and w, of the pool are both positive.

EXERCISES

Writing About Mathematics

1. A function is a set of ordered pairs in which no two different pairs have the same first element. In Example 4, we can say that the cost of renting a car is a function of the number of miles driven. Explain how this example illustrates the definition of a function.

2. Usually when a car rental company charges for the number of miles driven, the number of miles is expressed as a whole number. For example, if the car was driven 145.3 miles, the driver would be charged for 146 miles. In Example 4, what would be the domain of the function? (Recall that the domain is the set of numbers that can replace the variable x.)

Developing Skills

In 3–7, find the missing member in each ordered pair if the second member of the pair is twice the first member.

3. $(3, ?)$ **4.** $(0, ?)$ **5.** $(-2, ?)$ **6.** $(?, 11)$ **7.** $(?, -8)$

In 8–12, find the missing member in each ordered pair if the first member of the pair is 4 more than the second member.

8. $(?, 5)$ **9.** $\left(?, \frac{1}{2}\right)$ **10.** $(?, 0)$ **11.** $\left(9\frac{1}{4}, ?\right)$ **12.** $(-8, ?)$

In 13–27, state whether each given ordered pair of numbers is a solution of the equation or inequality.

13. $y = 5x; (3, 15)$ **14.** $y = 4x; (16, 4)$ **15.** $y = 3x + 1; (7, 22)$

16. $3x - 2y = 0; (3, 2)$ **17.** $y > 4x; (2, 10)$ **18.** $y < 2x + 3; (0, 2)$

19. $3y > 2x + 1; (4, 3)$ **20.** $2x + 3y \leq 9; (0, 3)$ **21.** $x + y = 8; (4, 5)$

22. $4x + 3y = 2; \left(\frac{1}{4}, \frac{1}{3}\right)$ **23.** $3x = y + 4; (-7, -1)$ **24.** $x - 2y \leq 15; (1, -7)$

25. $y > 6x; (-1, -2)$ **26.** $3x < 4y; (5, 2)$ **27.** $5x - 2y \leq 19; (3, -2)$

In 28–31, state which sets of ordered pairs represent functions. If the set is not a function, explain why.

28. $\{(1, 2), (2, 3), (3, 4), (5, 6)\}$ **29.** $\{(1, 2), (2, 1), (3, 4), (4, 3)\}$

30. $\{(-5, -5), (-5, 5), (6, -6), (6, 6)\}$ **31.** $\{(81, 9), (81, -9), (25, 5), (25, -5)\}$

In 32–35, find the range of each function when the domain is:

a. $A = \{x \mid 10 < x < 13$ and $x \in$ of the set of integers$\}$ **b.** $B = \{6x \mid x \in$ of the set of whole numbers$\}$

32. $y = 2x + 3$ **33.** $y = \frac{x}{6}$ **34.** $y = -x - 2$ **35.** $y = x + 1$

Applying Skills

In 36–39, in each case: **a.** Write an equation or an inequality that expresses the relationship between x and y. **b.** Find two ordered pairs in the solution set of the equation that you wrote in part **a.** **c.** Is the relation determined by the equation or inequality a function?

36. The cost, y, of renting a bicycle is \$5.00 plus \$2.50 times the number of hours, x, that the bicycle is used. Assume that x can be a fractional part of an hour.

37. In an isosceles triangle whose perimeter is 54 centimeters, the length of the base in centimeters is x and the length of each leg in centimeters is y.

38. Jules has \$265 at the beginning of the month. What is the maximum amount, y, that he has left after 30 days if he spends at least x dollars a day?

39. At the beginning of the day, the water in the swimming pool was 2 feet deep. Throughout the day, water was added so that the depth increased by less than 0.4 foot per hour. What was the depth, y, of the water in the pool after being filled for x hours?

40. A fence to enclose a rectangular space along a river is to be constructed using 176 feet of fencing. The two sides perpendicular to the river have length x.

Length (x)	Width	Area (A)
1	$176 - 2(1)$	$1[176 - 2(1)] = 174$
2	$176 - 2(2)$	$2[176 - 2(2)] = 344$

a. Complete five more rows of the table shown on the right.

b. Is the area, A, a function of x? If so, write the function and determine its domain.

41. (1) At the Riverside Amusement Park, rides are paid for with tokens purchased at a central booth. Some rides require two tokens, and others one token. Tomas bought 10 tokens and spent them all on x two-token rides and y one-token rides. On how many two-token and how many one-token rides did Tomas go?

(2) Minnie used 10 feet of fence to enclose three sides of a rectangular pen whose fourth side was the side of a garage. She used x feet of fence for each side perpendicular to the garage and y feet of fence for the side parallel to the garage. What were the dimensions of the pen that Minnie built?

a. Write an equation that can be used to solve both problems.

b. Write the six solutions for problem (1).

c. Which of the six solutions for problem (1) are *not* solutions for problem (2)?

d. Write two solutions for problem (2) that are *not* solutions for problem (1).

e. There are only six solutions for problem (1) but infinitely many solutions for problem (2). Explain why.

42. A cylindrical wooden base for a trophy has radius r and height h. The volume of the base will be 102 cubic inches. Using set-builder notation, describe the set of ordered pairs, (r, h), that represent the possible dimensions of the base.

9-2 GRAPHING LINEAR FUNCTIONS USING THEIR SOLUTIONS

Think of two numbers that add up to 6. If x and y represent these numbers, then $x + y = 6$ is an equation showing that their sum is 6. If the replacement set for both x and y is the set of real numbers, we can find infinitely many solutions for the equation $x + y = 6$. Some of the solutions are shown in the following table.

x	8	7	6	5	4.5	4	3	2.2	2	0	-2
y	-2	-1	0	1	1.5	2	3	3.8	4	6	8

Each ordered pair of numbers, such as (8, –2), locates a point in the coordinate plane. When all the points that have the coordinates associated with the number pairs in the table are located, the points appear to lie on a straight line, as shown at the right.

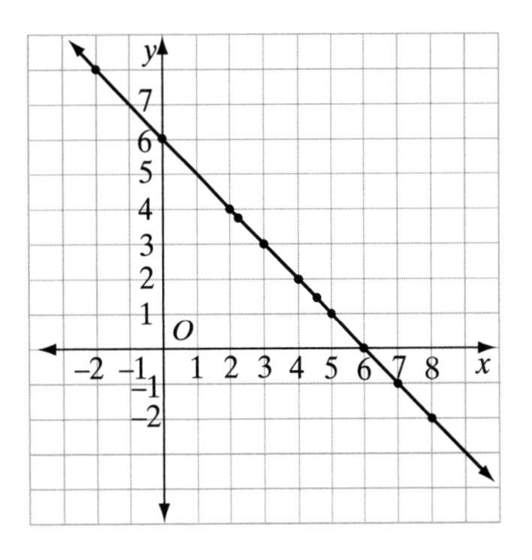

In fact, if the replacement set for both x and y is the set of real numbers, the following is true:

> All of the points whose coordinates are solutions of $x + y = 6$ lie on this same straight line, and all of the points whose coordinates are not solutions of $x + y = 6$ do *not* lie on this line.

This line, which is the set of all the points and only the points whose coordinates make the equation $x + y = 6$ true, is called the **graph** of $x + y = 6$.

DEFINITION

A first-degree equation in **standard form** is written as

$$Ax + By = C$$

where A, B, and C are real numbers, with A and B not both 0.

We call an equation that can be written in the form $Ax + By = C$ a **linear equation** since its graph is a straight line that contains all points and only those points whose coordinates make the equation true. The replacement set for both variables is the set of real numbers unless otherwise indicated.

When we graph a linear equation, we can determine the line by plotting two points whose coordinates satisfy that equation. However, we usually plot a third point as a check on the first two. If the third point lies on the line determined by the first two points, we have probably made no error.

Procedure

To graph a linear equation by means of its solutions:

1. Transform the equation into an equivalent equation that is solved for y in terms of x.
2. Find three solutions of the equation by choosing values for x and finding corresponding values for y.
3. In the coordinate plane, graph the ordered pairs of numbers found in Step 2.
4. Draw the line that passes through the points graphed in Step 3.

EXAMPLE 1

Does the point $(2, -3)$ lie on the graph of $x - 2y = -4$?

Solution The point $(2, -3)$ lies on the graph of $x - 2y = -4$ if and only if it is a solution of $x - 2y = -4$.

$$x - 2y = -4$$
$$2 - 2(-3) \stackrel{?}{=} -4$$
$$2 + 6 \stackrel{?}{=} -4$$
$$8 \neq -4 \text{ ✗}$$

Answer Since $8 = -4$ is not true, the point $(2, -3)$ does not lie on the line $x - 2y = -4$.

EXAMPLE 2

What must be the value of d if $(d, 4)$ lies on line $3x + y = 10$?

Solution The coordinates $(d, 4)$ must satisfy $3x + y = 10$.
Let $x = d$ and $y = 4$. Then:

$$3d + 4 = 10$$
$$3d = 6$$
$$d = 2$$

Answer The value of d is 2.

EXAMPLE 3

a. Write the following verbal sentence as an equation:

The sum of twice the x-coordinate of a point and the y-coordinate of that point is 4.

b. Graph the equation written in part **a**.

Solution **a.** Let x = the x-coordinate of the point, and y = the y-coordinate of the point.

Then:

$$2x + y = 4 \quad \textit{Answer}$$

b. *How to Proceed*

(1) Transform the equation into an equivalent equation that has y alone as one member:

$$2x + y = 4$$
$$y = -2x + 4$$

(2) Choose three values of x and find the corresponding values of y:

x	$-2x + 4$	y
0	$-2(0) + 4$	4
1	$-2(1) + 4$	2
2	$-2(2) + 4$	0

(3) Plot the points that are associated with the three solutions:

(4) Draw a line through the points that were plotted. Label the line with its equation:

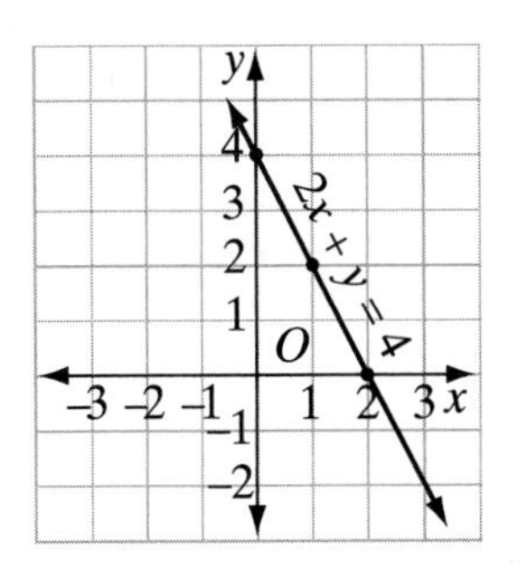

To display the graph of a function on a calculator, follow the steps below.

1. Solve the equation for y and enter it as Y_1.

ENTER: Y= (-) 2 X,T,Θ,n + 4 ENTER

DISPLAY:

Plot1 Plot2 Plot3
\Y1 = -2X+4

2. Use the standard viewing window to view the graph. The standard viewing window is a 20 by 20 graph centered at the origin.

ENTER: ZOOM 6 ENTER

DISPLAY:

ZOOM MEMORY
1: ZBox
2: Zoom In
3: Zoom Out
4: ZDecimal
5: ZSquare
6: ZStandard
7↓ZTrig

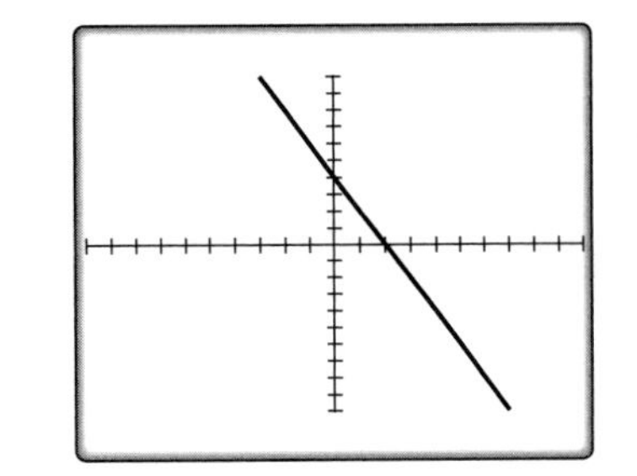

3. Display the coordinates of points on the graph.

ENTER: **TRACE**

In the viewing window, the equation appears at the top left of the screen. The point at which the graph intersects the y-axis is marked with a star and the coordinates of that point appear at the bottom of the screen. Press the right and left arrow keys to move the star along the line to display other coordinates.

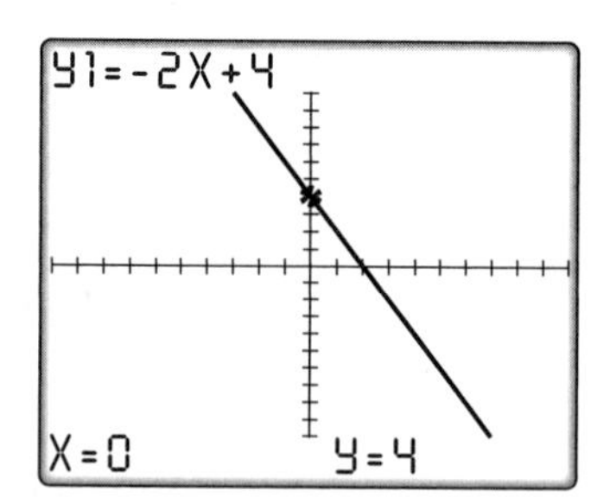

EXERCISES

Writing About Mathematics

1. The points whose coordinates are $(3, 1)$, $(5, -1)$ and $(7, -3)$ all lie on the same line. What could be the coordinates of another point that lies on that line? Explain how you found your answer.

2. Of the points $(0, 5)$, $(2, 4)$, $(3, 3)$, and $(6, 2)$, which one does not lie on the same line as the other three? Explain how you found your answer.

Developing Skills

In 3–5, state, in each case, whether the pair of values for x and y is a member of the solution set of the equation $2x - y = 6$.

3. $x = 4, y = 2$ **4.** $x = 0, y = 6$ **5.** $x = 4, y = -2$

For 6–9, state in each case whether the point whose coordinates are given is on the graph of the given equation.

6. $x + y = 7; (4, 3)$ **7.** $2y + x = 7; (1, 3)$

8. $3x - 2y = 8; (2, 1)$ **9.** $2y = 3x - 5; (-1, -4)$

In 10–13, find in each case the number or numbers that can replace k so that the resulting ordered number pair will be on the graph of the given equation.

10. $x + 2y = 5; (k, 2)$ **11.** $3x + 2y = 22; (k, 5)$

12. $x + 3y = 10; (13, k)$ **13.** $x - y = 0; (k, k)$

In 14–17, find in each case a value that can replace k so that the graph of the resulting equation will pass through the point whose coordinates are given.

14. $x + y = k; (2, 5)$

15. $x - y = k; (5, -3)$

16. $y - 2x = k; (-2, -1)$

17. $y = x + k; (5, 0)$

In 18–23, solve each equation for y in terms of x.

18. $3x + y = -1$

19. $4x - y = 6$

20. $2y = 6x$

21. $12x = \frac{3}{2}y$

22. $4x + 2y = 8$

23. $6x - 3y = 5$

In 24–26: **a.** Find the missing values of the variable needed to complete each table. **b.** Plot the points described by the pairs of values in each completed table; then, draw a line through the points.

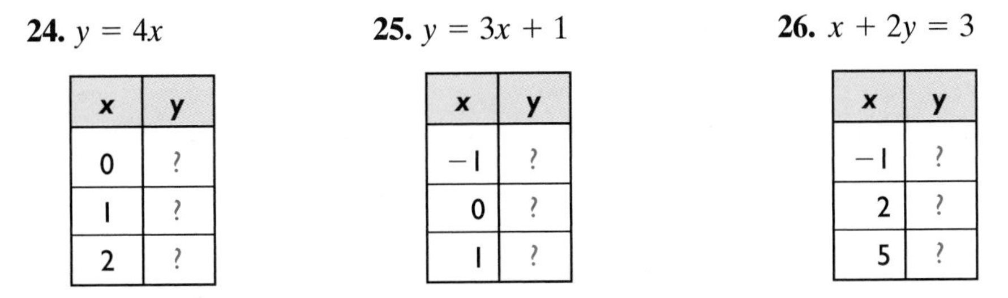

24. $y = 4x$

x	y
0	?
1	?
2	?

25. $y = 3x + 1$

x	y
−1	?
0	?
1	?

26. $x + 2y = 3$

x	y
−1	?
2	?
5	?

In 27–50, graph each equation.

27. $y = x$

28. $y = 5x$

29. $y = -3x$

30. $y = -x$

31. $x = 2y - 3$

32. $x = \frac{1}{2}y + 1$

33. $y = x + 3$

34. $y = 2x - 1$

35. $y = 3x + 1$

36. $y = -2x + 4$

37. $x + y = 8$

38. $x - y = 5$

39. $y - x = 0$

40. $3x + y = 12$

41. $x - 2y = 0$

42. $y - 3x = -5$

43. $2x - y = 6$

44. $3x - y = -6$

45. $x + 3y = 12$

46. $2x + 3y = 6$

47. $3x - 2y = -4$

48. $x - 3y = 9$

49. $2x = y - 4$

50. $4x = 3y$

51. **a.** Through points $(0, -2)$ and $(4, 0)$, draw a straight line.

b. Write the coordinates of two other points on the line drawn in part **a**.

52. **a.** Through points $(-2, 3)$ and $(1, -3)$, draw a straight line.

b. Does point $(0, -1)$ lie on the line drawn in part **a**?

Applying Skills

In 53–60: **a.** Write an equation that can be used to represent each sentence. **b.** Graph the equation.

53. The length of a rectangle, y, is twice the width, x.

54. The distance to school, y, is 2 miles more than the distance to the library, x.

55. The cost of a loaf of bread, x, plus the cost of a pound of meat, y, is 6 dollars.

56. The difference between Tim's height, x, and Sarah's height, y, is 1 foot.

57. The measure of each of three sides of the trapezoid is x, the measure of the remaining side is y, and the perimeter is 6.

58. The measure of each of the legs of an isosceles triangle is x, the measure of the base is y, and the perimeter is 9.

59. Bob's age, y, is equal to five more than twice Alice's present age, x.

60. The sum of the number of miles that Paul ran, x, and the number of miles that Sue ran, y, is 30 miles.

9-3 GRAPHING A LINE PARALLEL TO AN AXIS

Lines Parallel to the x-Axis

An equation such as $y = 2$ can be graphed in the coordinate plane. Any pair of values whose y-coordinate is 2, no matter what the x-coordinate is, makes the equation $y = 2$ true. Therefore,

$$(-3, 2), (-2, 2), (-1, 2), (0, 2), (1, 2),$$

or any other pair $(a, 2)$ for all values of a, are points on the graph of $y = 2$. As shown at the right, the graph of $y = 2$ is a horizontal line parallel to the x-axis and 2 units above it.

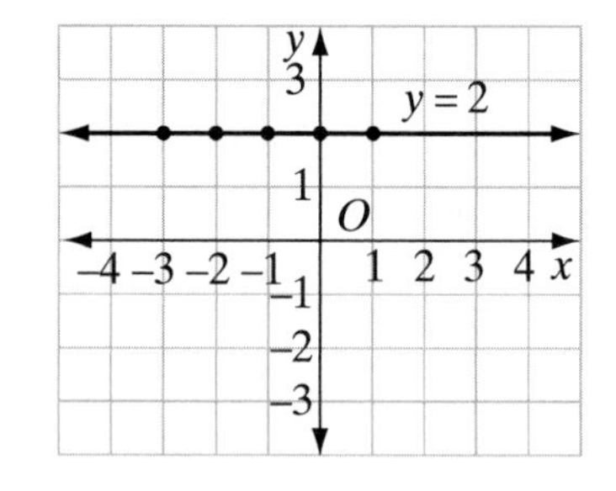

The equation $y = 2$ defines a function and can be displayed on a graphing calculator. If there are other equations entered in the Y= menu, press **CLEAR** before entering the new equation.

ENTER: **Y=** 2 **GRAPH**

DISPLAY:

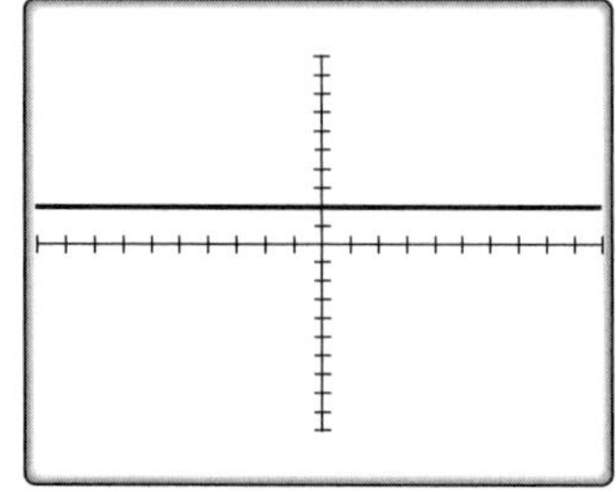

► **The equation of a line parallel to the x-axis and b units from the x-axis is $y = b$. If b is positive, the line is above the x-axis, and if b is negative, the line is below the x-axis.**

Lines Parallel to the y-Axis

An equation such as $x = -3$ can be graphed in the coordinate plane. Any pair of values whose x-coordinate is -3, no matter what the y-coordinate is, makes the equation $x = -3$ true. Therefore,

$$(-3, -2), (-3, -1), (-3, 0), (-3, 1), (-3, 2),$$

or any other pair $(-3, b)$ for all values of b, are points on the graph of $x = -3$. As shown at the right, the graph of $x = -3$ is a vertical line parallel to the y-axis and 3 units to the left of it.

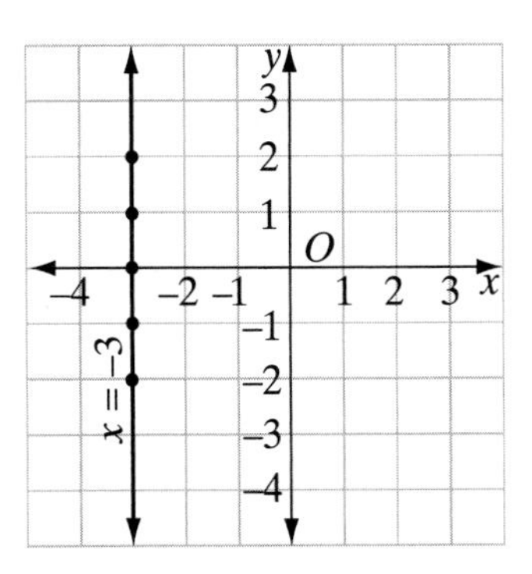

The equation $x = -3$ defines a relation that is not a function because it defines a set of ordered pairs but every ordered pair has the same first element. The method used to draw the graph of a function on a graphing calculator cannot be used to display this graph.

▶ **The equation of a line parallel to the y-axis and a units from the y-axis is $x = a$. If a is positive, the line is to the right of the y-axis, and if a is negative, the line is to the left of the x-axis.**

EXERCISES

Writing About Mathematics

1. Mike said that the equation of the x-axis is $x = 0$. Do you agree with Mike? Explain why or why not.

2. Does the line whose equation is $y = 4$ intersect the x-axis? Determine the coordinates of the point of intersection if one exists or explain why a point of intersection does not exist.

Developing Skills

In 3–17, draw the graph of each equation.

3. $x = 6$ **4.** $x = 4$ **5.** $x = 0$ **6.** $x = -3$ **7.** $x = -5$

8. $y = 4$ **9.** $y = 5$ **10.** $y = 0$ **11.** $y = -4$ **12.** $y = -7$

13. $x = \frac{1}{2}$ **14.** $y = 1\frac{1}{2}$ **15.** $y = -2.5$ **16.** $x = -\frac{3}{2}$ **17.** $y = 3.5$

18. Write an equation of the line that is parallel to the x-axis

a. 1 unit above the axis. **b.** 5 units above the axis. **c.** 4 units below the axis.

d. 8 units below the axis. **e.** 2.5 units above the axis. **f.** 3.5 units below the axis.

19. Write an equation of the line that is parallel to the y-axis

a. 3 units to the right of the axis.

b. 10 units to the right of the axis.

c. $4\frac{1}{2}$ units to the left of the axis.

d. 6 units to the left of the axis.

e. 2.5 units to the right of the axis.

f. 5.2 units to the left of the axis.

20. Which statement is true about the graph of the equation $y = 6$?

(1) It is parallel to the y-axis.

(2) It is parallel to the x-axis.

(3) It is *not* parallel to either axis.

(4) It goes through the origin.

21. Which statement is *not* true about the graph of the equation $x = 5$?

(1) It goes through $(5, 0)$.

(2) It is a vertical line.

(3) It is a function.

(4) It is parallel to the y-axis.

22. Which statement is true about the graph of the equation $y = x$?

(1) It is parallel to the y-axis.

(2) It is parallel to the x-axis.

(3) It goes through $(2, -2)$.

(4) It goes through the origin.

Applying Skills

23. The cost of admission to an amusement park on Family Night is $25.00 for a family of any size.

a. What is the cost of admission for the Gauger family of six persons?

b. Write an equation for the cost of admission, y, for a family of x persons.

24. The coordinates of the vertices of rectangle $ABCD$ are $A(-2, -1)$, $B(4, -1)$, $C(4, 5)$, and $D(-2, 5)$.

a. Draw $ABCD$ on graph paper.

b. Write the equations of the lines $\overleftrightarrow{AB}$, $\overleftrightarrow{BC}$, $\overleftrightarrow{CD}$, and $\overleftrightarrow{DA}$.

c. Draw a horizontal line and a vertical line that separate $ABCD$ into four congruent rectangles.

d. Write the equations of the lines drawn in part **c**.

25. During the years 2004, 2005, and 2006, there were 5,432 AnyClothes stores.

a. Write a linear equation that gives the number of stores, y.

b. Predict the number of stores for the years 2007 and 2008.

26. During the years 1999 through 2006, SmallTown News' circulation was 74,000 each year.

a. Write a linear equation that gives the circulation, y.

b. Predict the circulation of SmallTown News in the years 2007 and 2008.

9-4 THE SLOPE OF A LINE

Meaning of the Slope of a Line

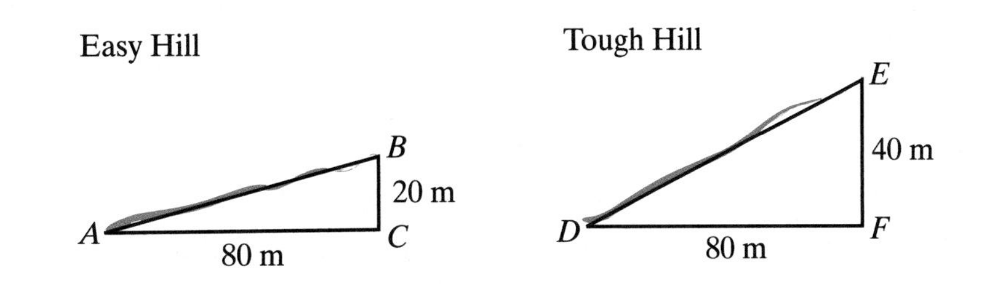

It is more difficult to hike up Tough Hill, shown above, than to hike up Easy Hill. Tough Hill rises 40 meters vertically over a horizontal distance of 80 meters, whereas Easy Hill rises only 20 meters vertically over the same horizontal distance of 80 meters. Therefore, Tough Hill is steeper than Easy Hill. To compare the steepness of roads $\overline{AB}$ and $\overline{DE}$, which lead up the two hills, we compare their **slopes**.

The slope of road $\overline{AB}$ is the ratio of the change in vertical distance, CB, to the change in horizontal distance, AC:

$$\text{slope of road } \overline{AB} = \frac{\text{change in vertical distance, } CB}{\text{change in horizontal distance, } AC} = \frac{20\text{ m}}{80\text{ m}} = \frac{1}{4}$$

Also:

$$\text{slope of road } \overline{DE} = \frac{\text{change in vertical distance, } FE}{\text{change in horizontal distance, } DF} = \frac{40\text{ m}}{80\text{ m}} = \frac{1}{2}$$

Note that the steeper hill has the larger slope.

Finding the Slope of a Line

To find the slope of the line determined by the two points (2, 3) and (5, 8), as shown at the right, we write the ratio of the difference in y-values to the difference in x-values as follows:

$$\text{slope} = \frac{\text{difference in } y\text{-values}}{\text{difference in } x\text{-values}} = \frac{8-3}{5-2} = \frac{5}{3}$$

Suppose we change the order of the points (2, 3) and (5, 8) in performing the computation. We then have:

$$\text{slope} = \frac{\text{difference in } y\text{-values}}{\text{difference in } x\text{-values}} = \frac{3-8}{2-5} = \frac{-5}{-3} = \frac{5}{3}$$

The result of both computations is the same.

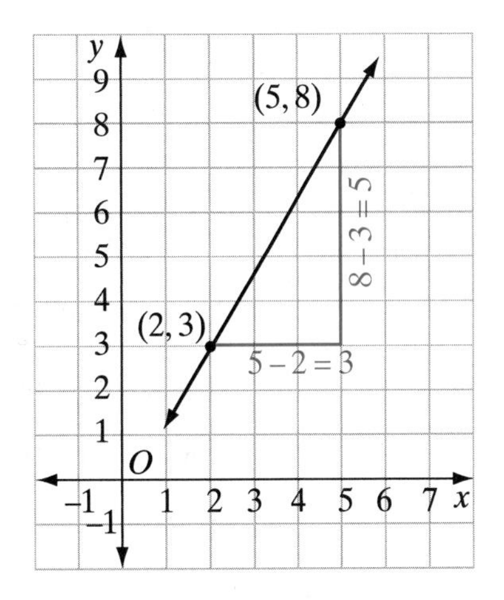

When we compute the slope of a line that is determined by two points, it does not matter which point is considered as the first point and which the second.

Also, when we find the slope of a line using two points on the line, it does not matter which two points on the line we use because all segments of a line have the same slope as the line.

Procedure

To find the slope of a line:

1. Select any two points on the line.
2. Find the vertical change, that is, the change in *y*-values by subtracting the *y*-coordinates in any order.
3. Find the horizontal change, that is, the change in *x*-values, by subtracting the *x*-coordinates in the same order as the *y*-coordinates.
4. Write the ratio of the vertical change to the horizontal change.

We can use this procedure to find the slopes of $\overleftrightarrow{LM}$ and $\overleftrightarrow{RS}$, the lines shown in the graph at the right.

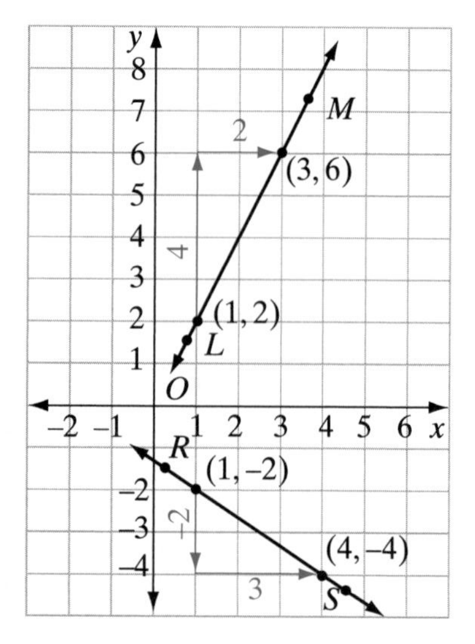

$$\text{slope of } \overleftrightarrow{LM} = \frac{\text{vertical change}}{\text{horizontal change}}$$
$$= \frac{6-2}{3-1} = \frac{4}{2} = \frac{2}{1} \text{ or } 2$$

$$\text{slope of } \overleftrightarrow{RS} = \frac{\text{vertical change}}{\text{horizontal change}}$$
$$= \frac{-4-(-2)}{4-1} = \frac{-2}{3} = -\frac{2}{3}$$

Slope is a rate of change. It is a measure of the rate at which y changes compared to x. For $\overleftrightarrow{LM}$ whose slope is 2 or $\frac{2}{1}$, y increases 2 units when x increases 1 unit. When the second element of a rate is 1, the rate is a **unit rate of change**. If x is the independent variable and y is the dependent variable, then the slope is the rate of change in the dependent variable compared to the independent variable.

For $\overleftrightarrow{RS}$ whose slope is $\frac{-2}{3}$, y decreases 2 units when x increases 3 units. We could also write this rate as $\frac{-\frac{2}{3}}{1}$ to write the slope as a unit rate of change, that is, y decreases $\frac{2}{3}$ of a unit when x increases 1 unit.

In general:

► **The slope, m, of a line that passes through any two points $P_1(x_1, y_1)$ and $P_2(x_2, y_2)$, where $x_1 \neq x_2$, is the ratio of the change or difference of the y-values of these points to the change or difference of the corresponding x-values.**

Thus:

$$\text{slope of a line} = \frac{\text{difference in } y\text{-values}}{\text{difference in } x\text{-values}}$$

Therefore:

$$\textbf{slope of } \overleftrightarrow{P_1P_2} = m = \frac{y_2 - y_1}{x_2 - x_1}$$

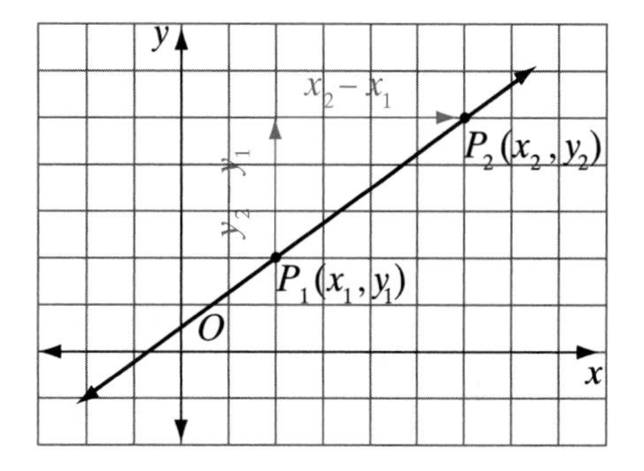

The difference in x-values, $x_2 - x_1$, can be represented by Δx, read as "delta x." Similarly, the difference in y-values, $y_2 - y_1$, can be represented by Δy, read as "delta y." Therefore, we write:

$$\textbf{slope of a line} = m = \frac{\Delta y}{\Delta x}$$

Positive Slopes

Examining $\overleftrightarrow{AB}$ from left to right and observing the path of a point, from C to D for example, we see that the line is rising. As the x-values increase, the y-values also increase. Between point C and point D, the change in y is 1, and the change in x is 2. Since both Δy and Δx are positive, the slope of $\overleftrightarrow{AB}$ must be positive. Thus:

$$\text{slope of } \overleftrightarrow{AB} = m = \frac{\Delta y}{\Delta x} = \frac{1}{2}$$

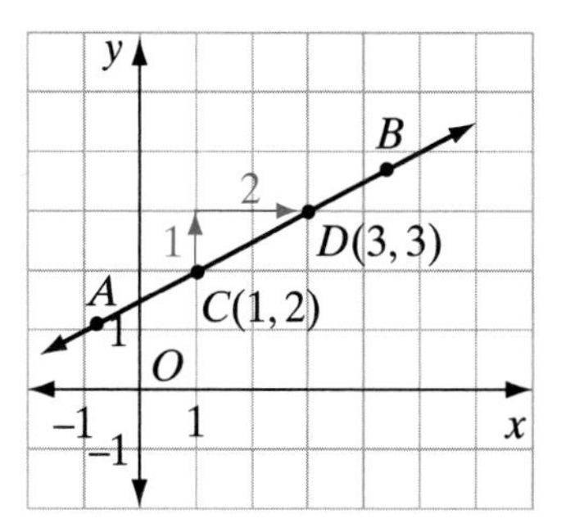

Principle 1. As a point moves from left to right along a line that is *rising*, y increases as x increases and the slope of the line is *positive*.

Negative Slopes

Now, examining $\overleftrightarrow{EF}$ from left to right and observing the path of a point from C to D, we see that the line is falling. As the x-values increase, the y-values decrease. Between point C and point D, the change in y is -2, and the change in x is 3. Since Δy is negative and Δx is positive, the slope of $\overleftrightarrow{EF}$ must be negative.

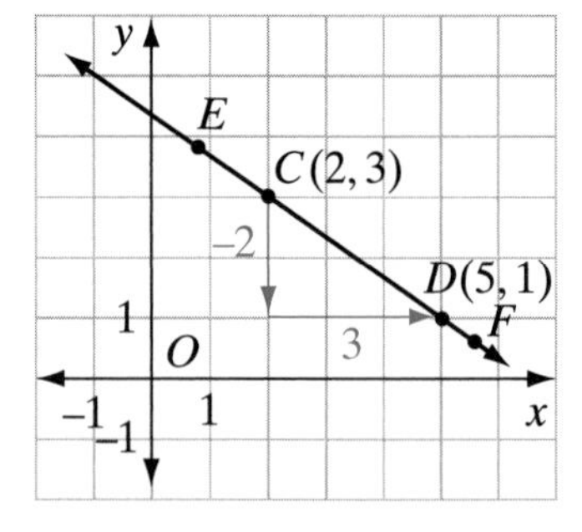

$$\text{slope of } \overleftrightarrow{EF} = m = \frac{\Delta y}{\Delta x} = \frac{-2}{3} = -\frac{2}{3}$$

Principle 2. As a point moves from left to right along a line that is *falling*, y decreases as x increases and the slope of the line is *negative*.

Zero Slope

On the graph, $\overleftrightarrow{GH}$ is parallel to the x-axis. We consider a point moving along $\overleftrightarrow{GH}$ from left to right, for example from C to D. As the x-values increase, the y-values are unchanged. Between point C and point D, the change in y is 0 and the change in x is 3. Since Δy is 0 and Δx is 3, the slope of $\overleftrightarrow{GH}$ must be 0. Thus:

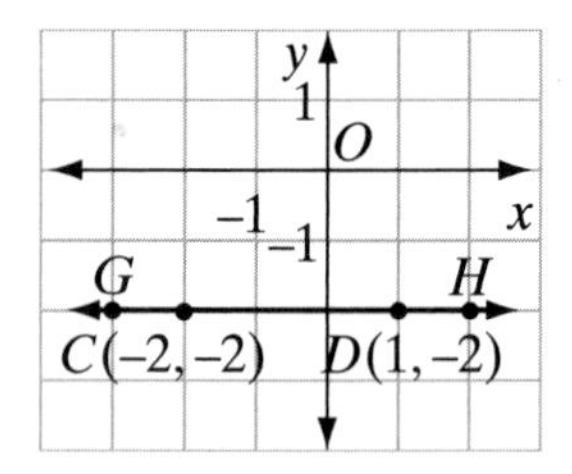

$$\text{slope of } \overleftrightarrow{GH} = m = \frac{\Delta y}{\Delta x} = \frac{0}{3} = 0$$

Principle 3. If a line is parallel to the x-axis, its slope is 0.

Note: The slope of the x-axis is also 0.

No Slope

On the graph, $\overleftrightarrow{LM}$ is parallel to the y-axis. We consider a point moving upward along $\overleftrightarrow{LM}$, for example from C to D. The x-values are unchanged, but the y-values increase. Between point C and point D, the change in y is 3, and the change in x is 0. Since the slope of $\overleftrightarrow{LM}$ is the change in y divided by the change in x and a number cannot be divided by 0, $\overleftrightarrow{LM}$ has no defined slope.

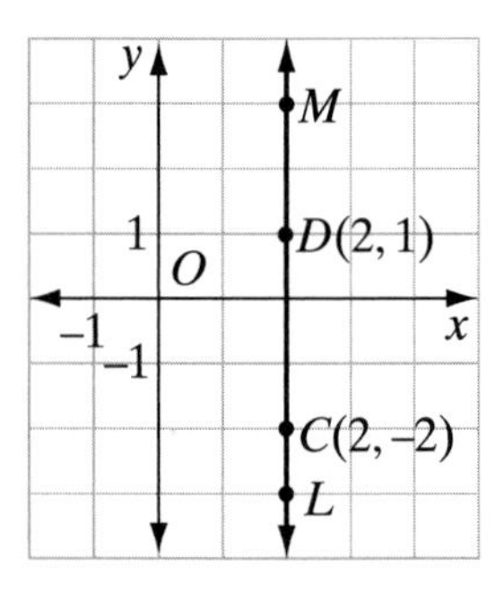

$$\text{slope of } \overleftrightarrow{MC} = m = \frac{\Delta y}{\Delta x} = \frac{1 - (-2)}{0} = \text{undefined}$$

Principle 4. If a line is parallel to the y-axis, it has no defined slope.

Note: The y-axis itself has no defined slope.

EXAMPLE 1

Find the slope of the line that is determined by points $(-2, 4)$ and $(4, 2)$.

Solution Plot points $(-2, 4)$ and $(4, 2)$. Let point $(-2, 4)$ be $P_1(x_1, y_1)$, and let point $(4, 2)$ be $P_2(x_2, y_2)$. Then, $x_1 = -2, y_1 = 4, x_2 = 4$, and $y_2 = 2$.

$$\begin{aligned} \text{slope of } \overleftrightarrow{P_1P_2} &= \frac{\Delta y}{\Delta x} = \frac{y_2 - y_1}{x_2 - x_1} \\ &= \frac{2 - 4}{4 - (-2)} \\ &= \frac{2 - 4}{4 + 2} \\ &= \frac{-2}{6} \\ &= -\tfrac{1}{3} \quad \textit{Answer} \end{aligned}$$

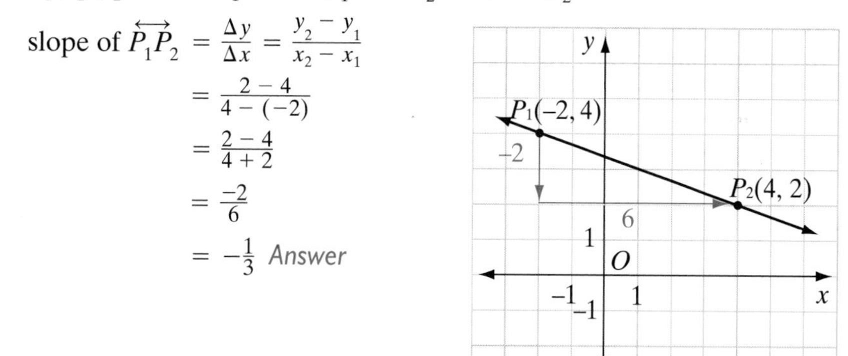

EXAMPLE 2

Through point $(2, -1)$, draw the line whose slope is $\frac{3}{2}$.

Solution *How to Proceed*

(1) Graph point $A(2, -1)$:

(2) Note that since slope $= \frac{\Delta y}{\Delta x} = \frac{3}{2}$, when y changes by 3, then x changes by 2. Start at point $A(2, -1)$ and move 3 units upward and 2 units to the right to locate point B:

(3) Start at B and repeat these movements to locate point C:

(4) Draw a line that passes through points A, B, and C:

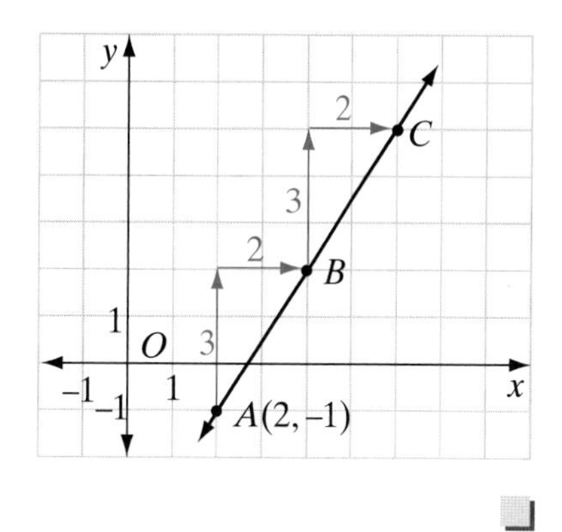

KEEP IN MIND A fundamental property of a straight line is that its slope is constant. Therefore, any two points on a line may be used to compute the slope of the line.

Applications of Slope

In real-world contexts, the slope is either a ratio or a rate. When both the x and y variables have the same unit of measure, slope represents a *ratio* since it will have no unit of measure. However, when the x and y variables have different units of measure, slope represents a *rate* since it will have a unit of measure. In both cases, the slope represents a *constant* ratio or rate of change.

EXAMPLE 3

The following are slopes of lines representing the daily sales, y, over time, x, for various sales representatives during the course of a year:

(1) $m = 20$ (2) $m = 45$

(3) $m = 39$ (4) $m = -7$

a. Interpret the meaning of the slopes if sales are given in thousands of dollars and time is given in months.

b. Which sales representative had the greatest monthly increase in sales?

c. Is it possible to determine which sales representative had the greatest total sales at the end of the year?

Solution **a.** The slopes represent the increase in sales *per month*:

(1) \$20,000 increase each month (2) \$45,000 increase each month

(3) \$39,000 increase each month (4) \$7,000 decrease each month

b. At \$45,000, (2) has the greatest monthly increase in sales.

c. The slopes give the *increase* in sales each month, not the total monthly sales. With the given information, it is not possible to determine which sales representative had the greatest total sales at the end of the year.

EXERCISES

Writing About Mathematics

1. Regina said that in Example 1, the answer would be the same if the formula for slope had been written as $\frac{y_1 - y_2}{x_1 - x_2}$. Do you agree with Regina? Explain why or why not. Use $\frac{y_1 - y_2}{x_1 - x_2}$ to find the answer to Example 1 to justify your response.

2. Explain why any two points that have the same x-coordinate lie on a line that has no slope.

Developing Skills

In 3–8: **a.** Tell whether each line has a positive slope, a negative slope, a slope of zero, or no slope. **b.** Find the slope of each line that has a slope.

3.

5.

7.

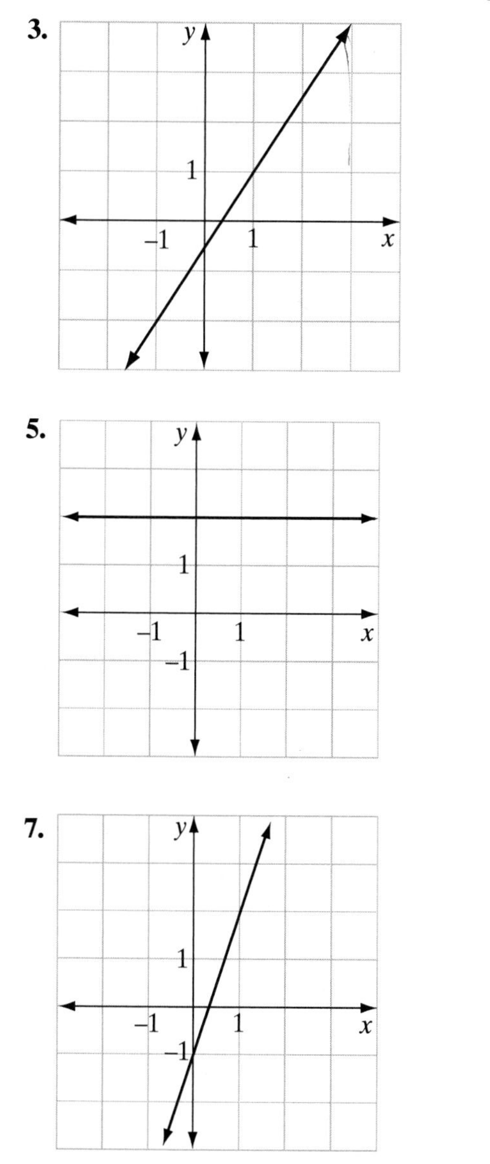

4.

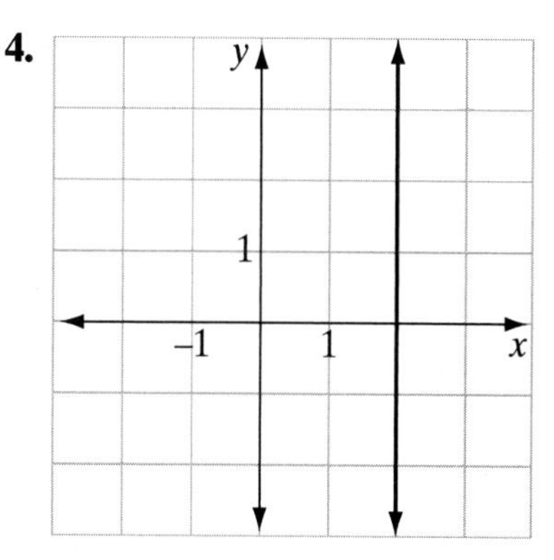

6.

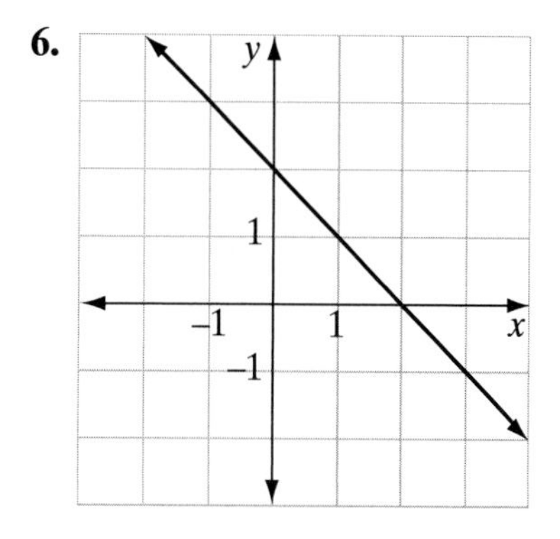

8.

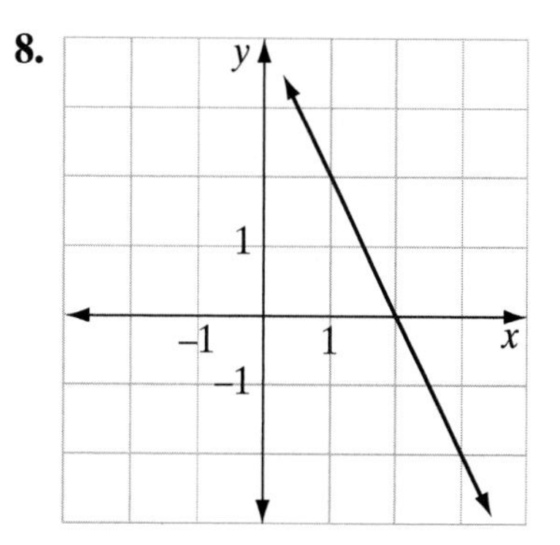

In 9–17, in each case: **a.** Plot both points, and draw the line that they determine. **b.** Find the slope of this line.

9. $(0, 0)$ and $(4, 4)$

10. $(0, 0)$ and $(4, 8)$

11. $(0, 0)$ and $(3, -6)$

12. $(1, 5)$ and $(3, 9)$

13. $(7, 3)$ and $(1, -1)$

14. $(-2, 4)$ and $(0, 2)$

15. $(5, -2)$ and $(7, -8)$

16. $(4, 2)$ and $(8, 2)$

17. $(-1, 3)$ and $(2, 3)$

In 18–29, in each case, draw a line with the given slope, m, through the given point.

18. $(0, 0)$; $m = 2$

19. $(1, 3)$; $m = 3$

20. $(2, -5)$; $m = 4$

21. $(-4, 5)$; $m = \frac{2}{3}$

22. $(3, 1)$; $m = 0$

23. $(-3, -4)$; $m = -2$

24. $(1, -5)$; $m = -1$

25. $(2, 4)$; $m = -\frac{3}{2}$

26. $(-2, 3)$; $m = -\frac{1}{3}$

27. $(-1, 0)$; $m = -\frac{5}{4}$

28. $(0, 2)$; $m = -\frac{2}{3}$

29. $(-2, 0)$; $m = \frac{1}{2}$

Applying Skills

30. Points $A(2, 4)$, $B(8, 4)$, and $C(5, 1)$ are the vertices of $\triangle ABC$. Find the slope of each side of $\triangle ABC$.

31. Points $A(3, -2)$, $B(9, -2)$, $C(7, 4)$, and $D(1, 4)$ are the vertices of a quadrilateral.

a. Graph the points and draw quadrilateral $ABCD$.

b. What type of quadrilateral does $ABCD$ appear to be?

c. Compute the slope of $\overleftrightarrow{BC}$ and the slope of $\overleftrightarrow{AD}$.

d. What is true of the slope of $\overleftrightarrow{BC}$ and the slope of $\overleftrightarrow{AD}$?

e. If two segments such as $\overline{AD}$ and $\overline{BC}$, or two lines such as $\overleftrightarrow{AD}$ and $\overleftrightarrow{BC}$, are parallel, what appears to be true of their slopes?

f. Since $\overline{AB}$ and $\overline{CD}$ are parallel, what might be true of their slopes?

g. Compute the slope of $\overline{AB}$ and the slope of $\overline{CD}$.

h. Is the slope of $\overline{AB}$ equal to the slope of $\overline{CD}$?

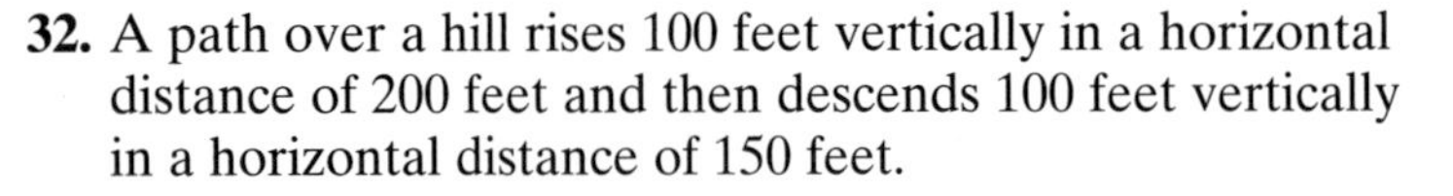

32. A path over a hill rises 100 feet vertically in a horizontal distance of 200 feet and then descends 100 feet vertically in a horizontal distance of 150 feet.

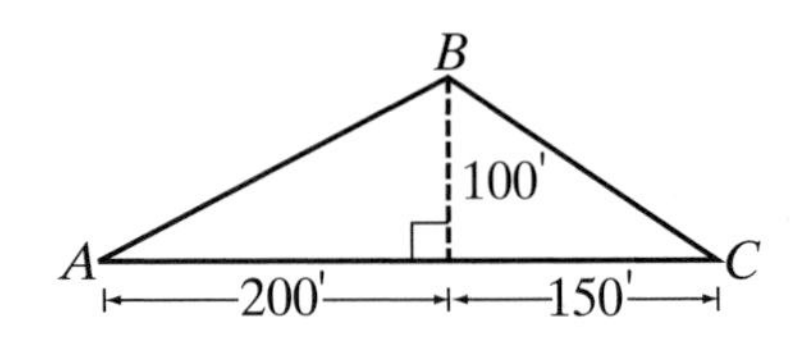

a. Find the slope of the path up the hill when walking from point A to point B.

b. Find the slope of the path down the hill when walking from point B to point C.

c. What is the unit rate of change from point A to point B?

d. What is the unit rate of change from point B to point C?

33. The amount that a certain internet phone company charges for international phone calls is represented by the equation $C = \$0.02t + \$1.00$, where C represents the total cost of the call and t represents time in minutes. Explain the meaning of the slope in terms of the information provided.

34. The monthly utility bill for a certain school can be represented by the equation $C = \$0.075g + \$100$, where C is the monthly bill and g is the amount of gas used by the hundred cubic feet (CCF). Explain the meaning of the slope in terms of the information provided.

35. A road that has an 8% grade rises 8 feet vertically for every 100 feet horizontally.

a. Find the slope of the road.

b. Explain the meaning of the slope in terms of the information provided.

c. If you travel a horizontal distance of 200 feet on this road, what will be the amount of vertical change in your position?

9-5 THE SLOPES OF PARALLEL AND PERPENDICULAR LINES

Parallel Lines

In the diagram, lines ℓ_1 and ℓ_2 are parallel. Right triangle 1 has been drawn with one leg coinciding with the x-axis and with its hypotenuse coinciding with ℓ_1. Similarly, right triangle 2 has been drawn with one leg coinciding with the x-axis and with its hypotenuse coinciding with ℓ_2.

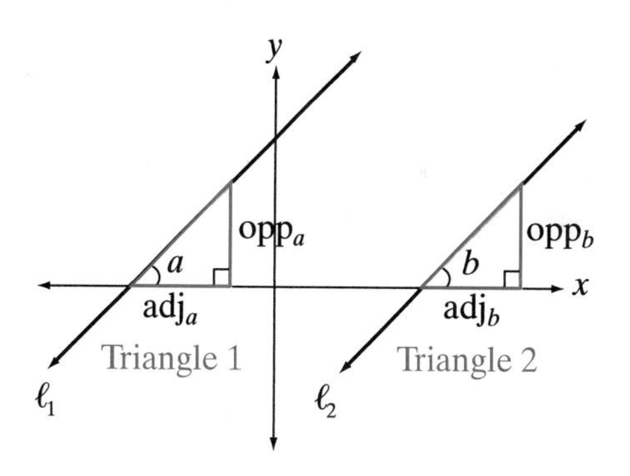

Since triangles 1 and 2 are right triangles,

$$\tan \angle a = \frac{\text{opp}_a}{\text{adj}_a} \qquad \tan \angle b = \frac{\text{opp}_b}{\text{adj}_b}$$

The x-axis is a transversal that intersects parallel lines ℓ_1 and ℓ_2. The corresponding angles, $\angle a$ and $\angle b$, are congruent. Therefore,

$$\frac{\text{opp}_a}{\text{adj}_a} = \frac{\text{opp}_b}{\text{adj}_b}$$

Since $\frac{\text{opp}_a}{\text{adj}_a}$ is the slope of ℓ_1 and $\frac{\text{opp}_b}{\text{adj}_b}$ is the slope of ℓ_2, we have shown that two parallel lines have the same slope.

► **If the slope of ℓ_1 is m_1, the slope of ℓ_2 is m_2, and $\ell_1 \parallel \ell_2$, then $m_1 = m_2$.**

The following statement is also true:

► **If the slope of ℓ_1 is m_1, the slope of ℓ_2 is m_2, and $m_1 = m_2$, then $\ell_1 \parallel \ell_2$.**

Perpendicular Lines

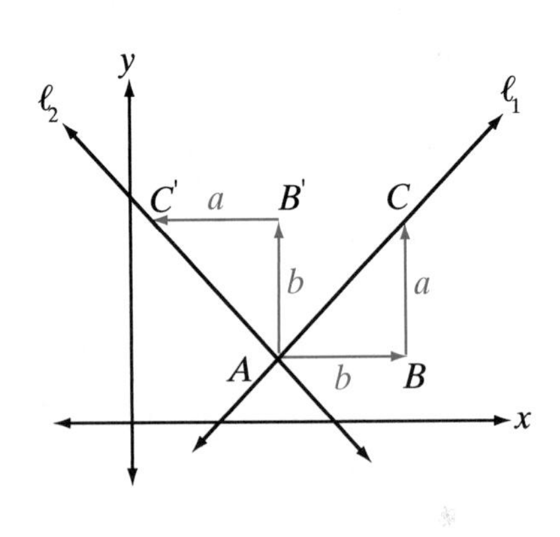

In the diagram, lines ℓ_1 and ℓ_2 are perpendicular. Right triangle ABC has been drawn with one leg parallel to the x-axis, the other leg parallel to the y-axis, and the hypotenuse coinciding with ℓ_1. The slope of ℓ_1 is $\frac{a}{b}$ since y increases by a units as x increases by b units. Triangle ABC is then rotated 90° counterclockwise about A so that the hypotenuse of $AB'C'$ coincides with ℓ_2.

Notice that since each leg has been rotated by 90°, the two legs are still parallel to the axes. However, they are now parallel to different axes. Also, in the rotated triangle, the change in y is an increase of b units and the change in x is a decrease of a units, that is, $-a$. Therefore, the slope of ℓ_2 is $\frac{b}{-a}$ or $-\frac{b}{a}$. In other words, the slope of ℓ_2 is the negative reciprocal of the slope of ℓ_1. We have thus shown that perpendicular lines have slopes that are negative reciprocals of each other.

In general:

▶ **If the slope of ℓ_1 is m_1, the slope of ℓ_2 is m_2, and $\ell_1 \perp \ell_2$, then $m_1 = -\frac{1}{m_2}$ or $m_1 \cdot m_2 = -1$.**

The following statement is also true:

▶ **If the slope of ℓ_1 is m_1, the slope of ℓ_2 is m_2, and $m_1 = -\frac{1}{m_2}$ or $m_1 \cdot m_2 = -1$, then $\ell_1 \perp \ell_2$.**

EXAMPLE 1

The equation of $\overleftrightarrow{AB}$ is $y = -2x + 1$.

a. Find the coordinates of any two points on $\overleftrightarrow{AB}$.

b. What is the slope of $\overleftrightarrow{AB}$?

c. What is the slope of a line that is parallel to $\overleftrightarrow{AB}$?

d. What is the slope of a line that is perpendicular to $\overleftrightarrow{AB}$?

Solution **a.** If $x = 0, y = -2(0) + 1 = 1$. One point is $(0, 1)$.

If $x = 1, y = -2(1) + 1 = -1$. Another point is $(1, -1)$.

b. Slope of $\overleftrightarrow{AB} = \frac{1 - (-1)}{0 - 1} = \frac{2}{-1} = -2$.

c. The slope of a line parallel to $\overleftrightarrow{AB}$ is equal to the slope of $\overleftrightarrow{AB}$, –2.

d. The slope of a line perpendicular to $\overleftrightarrow{AB}$ is $-\frac{1}{-2} = \frac{1}{2}$.

Answers **a.** (0, 1) and (1, −1) but many other answers are possible.

b. −2 **c.** −2 **d.** $\frac{1}{2}$

EXERCISES

Writing About Mathematics

1. a. What is the slope of a line that is perpendicular to a line whose slope is 0? Explain your answer.

b. What is the slope of a line that is perpendicular to a line that has no slope? Explain your answer.

2. What is the unit rate of change of y with respect to x if the slope of the graph of the equation for y in terms of x is $\frac{1}{4}$?

Developing Skills

In 3–11: **a.** Write the coordinates of two points on each line whose equation is given. **b.** Use the coordinates of the points found in part **a** to find the slope of the line. **c.** What is the slope of a line that is parallel to each line whose equation is given? **d.** What is the slope of a line that is perpendicular to the line whose equation is given?

3. $y = 2x + 6$ **4.** $y = x - 2$ **5.** $y = -3x + 7$

6. $3y = x$ **7.** $x - y = 4$ **8.** $2x - 3y = 6$

9. $x = 2y - 1$ **10.** $x = 4$ **11.** $y = -5$

Applying Skills

12. A taxi driver charges $3.00 plus $0.75 per mile.

a. What is the cost of a trip of 4 miles?

b. What is the cost of a trip of 8 miles?

c. Use the information from parts **a** and **b** to draw the graph of a linear function described by the taxi fares.

d. Write an equation for y, the cost of a taxi ride, in terms of x, the length of the ride in miles.

e. What is the change in the cost of a taxi ride when the length of the ride changes by 4 miles?

f. What is the unit rate of change of the cost with respect to the length of the ride in dollars per mile?

13. Mrs. Boyko is waiting for her oven to heat to the required temperature. When she first turns the oven on, the temperature registers 100°. She knows that the temperature will increase by 10° every 5 seconds.

a. What is the temperature of the oven after 10 seconds?

b. What is the temperature of the oven after 15 seconds?

c. Use the information from parts **a** and **b** to draw the graph of the linear function described by the oven temperatures.

d. Write an equation for y, the temperature of the oven, in terms of x, the number of seconds that it has been heating.

e. What is the unit rate in degrees per second at which the oven heats?

9-6 THE INTERCEPTS OF A LINE

The graph of any linear equation that is not parallel to an axis intersects both axes. The point at which the graph intersects the y-axis is a point whose x-coordinate is 0. The y-coordinate of this point is called the **y-intercept** of the equation. For example, for the equation $2x - y = 6$, let $x = 0$:

$$2x - y = 6$$
$$2(0) - y = 6$$
$$0 - y = 6$$
$$-y = 6$$
$$y = -6$$

The graph of $2x - y = 6$ intersects the *y-axis* at $(0, -6)$ and the y-intercept is -6.

The point at which the graph intersects the x-axis is a point whose y-coordinate is 0. The x-coordinate of this point is called the **x-intercept** of the equation. For example, for the equation $2x - y = 6$, let $y = 0$:

$$2x - y = 6$$
$$2x - 0 = 6$$
$$2x = 6$$
$$x = 3$$

The graph of $2x - y = 6$ intersects the *x-axis* at $(3, 0)$ and the x-intercept is 3.

Divide each term of the equation $2x - y = 6$ by the constant term, 6

$$\frac{2x}{6} - \frac{y}{6} = \frac{6}{6}$$
$$\frac{x}{3} - \frac{y}{6} = 1$$

Note that x is divided by the x-intercept, 3, and y is divided by the y-intercept, –6.

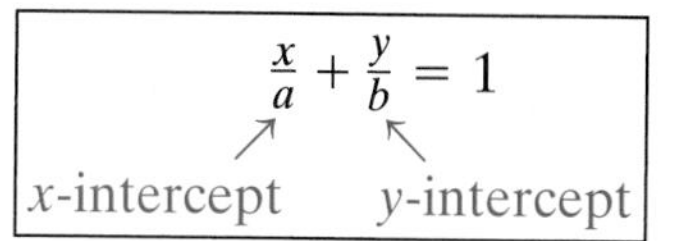

In general, if a linear equation is written in the form $\frac{x}{a} + \frac{y}{b} = 1$, the x-intercept is a and the y-intercept is b.

An equation of the form $\frac{x}{a} + \frac{y}{b} = 1$ is called **the intercept form of a linear equation**. Two points on the graph of $\frac{x}{a} + \frac{y}{b} = 1$ are $(a, 0)$ and $(0, b)$. We can use these two points to find the slope of the line.

$$\text{slope} = \frac{0 - b}{a - 0} = \frac{-b}{a} = -\frac{b}{a}$$

Slope and y-Intercept

Consider the equation $2x - y = 6$ from another point of view. Solve the equation for y:

$$\begin{aligned} 2x - y &= 6 \\ -y &= -2x + 6 \\ y &= 2x - 6 \end{aligned}$$

There are two numbers in the equation that has been solved for y: the coefficient of x, 2, and the constant term, -6. Each of these numbers gives us important information about the graph.

The y-intercept of a graph is the y-coordinate of the point at which the graph intersects the y-axis. For every point on the y-axis, $x = 0$. When $x = 0$,

$$\begin{aligned} y &= 2(0) - 6 \\ &= 0 - 6 \\ &= -6 \end{aligned}$$

Therefore, -6 is the y-intercept of the graph of the equation $y = 2x - 6$ and $(0, -6)$ is a point on its graph. The y-intercept is the constant term of the equation when the equation is solved for y.

Another point on the graph of $y = 2x - 6$ is $(1, -4)$. Use the two points, $(0, -6)$ and $(1, -4)$, to find the slope of the line.

$$\begin{aligned} \text{slope of } y = 2x - 6 &= \frac{-6 - (-4)}{0 - 1} \\ &= \frac{-2}{-1} \\ &= 2 \end{aligned}$$

The slope is the coefficient of x when the equation is solved for y.

For any equation in the form $y = mx + b$:

1. The coefficient of x is the slope of the line.
2. The constant term is the y-intercept.

The equation we have been studying can be compared to a general equation of the same type:

$$y = 2x - 3 \qquad\qquad y = mx + b$$

(In $y = 2x - 3$, 2 is the slope and -3 is the y-intercept; in $y = mx + b$, m is the slope and b is the y-intercept.)

The following statement is true for the general equation:

▶ **If the equation of a line is written in the form $y = mx + b$, then the slope of the line is m and the y-intercept is b.**

An equation of the form $y = mx + b$ is called **the slope-intercept form of a linear equation.**

EXAMPLE 1

Find the intercepts of the graph of the equation $5x + 2y = 6$.

Solution To find the x intercept, let $y = 0$:

$$5x + 2y = 6$$
$$5x + 2(0) = 6$$
$$5x + 0 = 6$$
$$5x = 6$$
$$x = \frac{6}{5}$$

To find the y intercept, let $x = 0$:

$$5x + 2y = 6$$
$$5(0) + 2y = 6$$
$$0 + 2y = 6$$
$$2y = 6$$
$$y = 3$$

Answer The x intercept is $\frac{6}{5}$ and the y intercept is 3.

EXAMPLE 2

Find the slope and the y-intercept of the line that is the graph of the equation $4x + 2y = 10$.

Solution

How to Proceed

(1) Solve the given equation for y to obtain an equivalent equation of the form $y = mx + b$:

$$4x + 2y = 10$$
$$2y = -4x + 10$$
$$y = -2x + 5$$

(2) The coefficient of x is the slope: slope $= -2$

(3) The constant term is the y-intercept: y-intercept $= 5$

Answer slope $= -2$, y-intercept $= 5$

EXERCISES

Writing About Mathematics

1. Amanda said that the y-intercept of the line whose equation is $2y + 3x = 8$ is 8 because the y-intercept is the constant term. Do you agree with Amanda? Explain why or why not.

2. Roberto said that the line whose equation is $y = x$ has no slope because x has no coefficient and no y-intercept because there is no constant term. Do you agree with Roberto? Explain why or why not.

Developing Skills

In 3–18, in each case, find the slope, y-intercept, and x-intercept of the line that is the graph of the equation.

3. $y = 3x + 1$
4. $y = 3 - x$
5. $y = 2x$
6. $y = x$
7. $x = -3$
8. $y = -2$
9. $y = -\frac{2}{3}x + 4$
10. $y - 3x = 7$
11. $2x + y = 5$
12. $3y = 6x + 9$
13. $2y = 5x - 4$
14. $\frac{1}{2}x + \frac{3}{4} = \frac{1}{3}y$
15. $4x - 3y = 0$
16. $2x = 5y - 10$
17. $\frac{x}{2} + \frac{y}{3} = 1$
18. $\frac{4x}{5} - 2y = 1$

19. What do the graphs of the equations $y = 4x$, $y = 4x + 2$, and $y = 4x - 2$ all have in common?

20. What do the lines that are the graphs of the equations $y = 2x + 1$, $y = 3x + 1$, and $y = -4x + 1$ all have in common?

21. If two lines are parallel, how are their slopes related?

22. What is true of two lines whose slopes are equal?

In 23–26, state in each case whether the lines are parallel, perpendicular, or neither.

23. $y = 3x + 2, y = 3x - 5$
24. $y = -2x - 6, y = 2x + 6$
25. $y = 4x - 8, 4y + x = 3$
26. $y = 2x, x = -2y$

27. Which of the following statements is true of the graph of $y = -3x$?
(1) It is parallel to the x-axis.
(2) It is parallel to the y-axis.
(3) Its slope is -3.
(4) It does not have a y-intercept.

28. Which of the following statements is true of the graph of $y = 8$?
(1) It is parallel to the x-axis.
(2) It is parallel to the y-axis.
(3) It has no slope.
(4) It goes through the origin.

9-7 GRAPHING LINEAR FUNCTIONS USING THEIR SLOPES

The slope and any one point can be used to draw the graph of a linear function. One convenient point is the point of intersection with the y-axis, that is, the point whose x-coordinate is 0 and whose y-coordinate is the y-intercept.

EXAMPLE 1

Draw the graph of $2x + 3y = 9$ using the slope and the y-intercept.

Solution *How to Proceed*

(1) Transform the equation into the form $y = mx + b$:

$$2x + 3y = 9$$
$$3y = -2x + 9$$
$$y = \frac{-2}{3}x + 3$$

(2) Find the slope of the line (the coefficient of x):

slope $= \frac{-2}{3}$

(3) Find the y-intercept of the line (the constant):

y-intercept $= 3$

(4) On the y-axis, graph point A, whose y-coordinate is the y-intercept:

(5) Use the slope to find two more points on the line. Since slope $= \frac{\Delta y}{\Delta x} = \frac{-2}{3}$, when y changes by -2, x changes by 3. Therefore, start at point A and move 2 units down and 3 units to the right to locate point B. Then start at point B and repeat the procedure to locate point C:

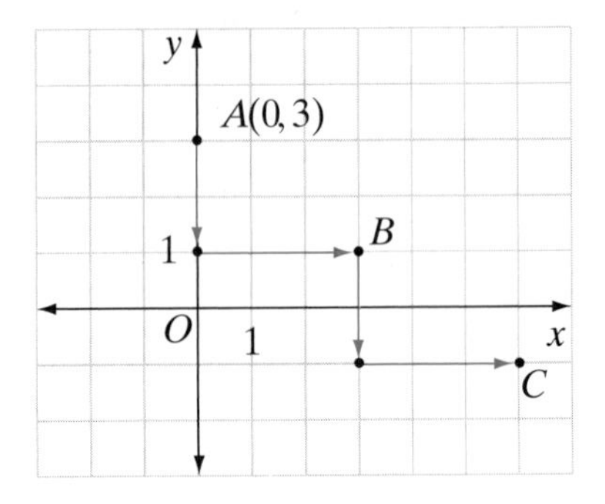

(6) Draw the line that passes through the three points:

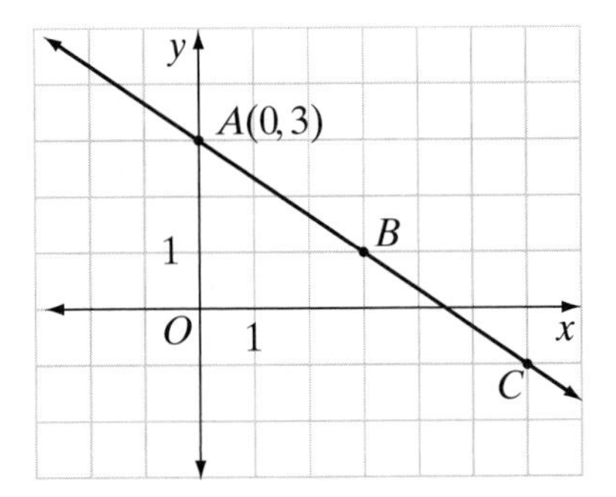

This line is the graph of $2x + 3y = 9$.

Check To check a graph, select two or more points on the line drawn and substitute their coordinates in the original equation. For example, check this graph using points (4.5, 0), (3, 1), and (6, -1).

$2x + 3y = 9$	$2x + 3y = 9$	$2x + 3y = 9$
$2(4.5) + 3(0) \stackrel{?}{=} 9$	$2(3) + 3(1) \stackrel{?}{=} 9$	$2(6) + 3(-1) \stackrel{?}{=} 9$
$9 + 0 \stackrel{?}{=} 9$	$6 + 3 \stackrel{?}{=} 9$	$12 - 3 \stackrel{?}{=} 9$
$9 = 9$ ✔	$9 = 9$ ✔	$9 = 9$ ✔

In this example, we used the slope and the y-intercept to draw the graph. We could use any point on the graph of the equation as a starting point.

EXAMPLE 2

Draw the graph of $3x - 2y = 9$ using the slope and any point.

Solution

How to Proceed

(1) Solve the equation for y. The slope is $\frac{3}{2}$ and the y-intercept is $-\frac{9}{2}$:

$$3x - 2y = 9$$
$$-2y = -3x + 9$$
$$y = \frac{-3x}{-2} + \frac{9}{-2}$$
$$y = \frac{3}{2}x - \frac{9}{2}$$

(2) Since $\left(0, -\frac{9}{2}\right)$ is not a convenient point to graph, choose another point. Let $x = 1$. A convenient point on the graph is $A(1, -3)$:

$$y = \frac{3}{2}(1) - \frac{9}{2}$$
$$y = \frac{3}{2} - \frac{9}{2}$$
$$y = -\frac{6}{2}$$
$$y = -3$$

(3) Graph point A whose coordinates are (1,–3):

(4) Use the slope to find two more points on the line. Since slope $= \frac{\Delta y}{\Delta x} = \frac{3}{2}$, when y changes by 3, x changes by 2. Therefore, start at point A and move 3 units up and 2 units to the right to locate point B. Then start at point B and repeat the procedure to locate point C:

(5) Draw the line the passes through the three points. This is the graph of $3x - 2y = 9$:

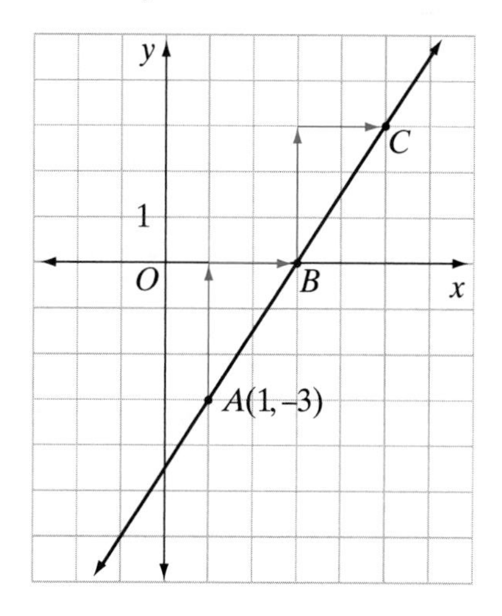

Note that the graph intersects the x-axis at $B(3, 0)$. The x-intercept is 3.

Translating, Reflecting, and Scaling Graphs of Linear Functions

An alternative way of looking at the effects of changing the values of the y-intercept and slope is that of translating, reflecting, or scaling the graph of the linear function $y = x$.

For instance, the graph of $y = x + 3$ can be thought of as the graph of $y = x$ shifted 3 units up. Similarly, the graph of $y = -x$ can be thought of the graph of $y = x$ reflected in the x-axis. On the other hand, the graph of $y = 4x$ can be thought of as the graph of $y = x$ stretched vertically by a factor of 4, while the graph of $y = \frac{1}{4}x$ is the graph of $y = x$ *compressed* vertically by a factor of $\frac{1}{4}$.

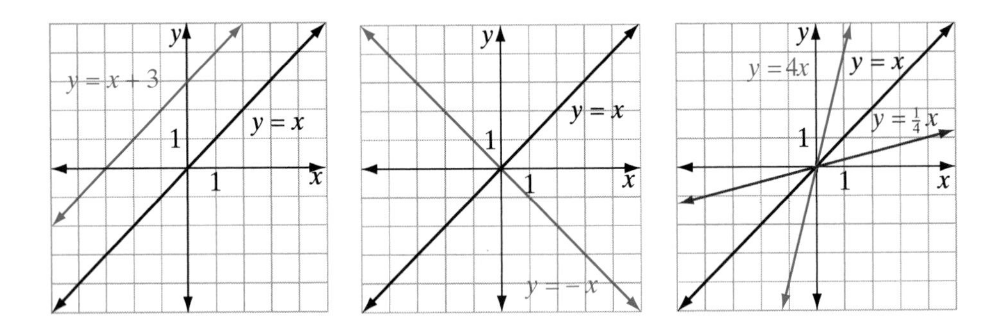

Translation Rules for Linear Functions

If c is *positive*:

- **The graph of $y = x + c$ is the graph of $y = x$ shifted c units up.**
- **The graph of $y = x - c$ is the graph of $y = x$ shifted c units down.**

Reflection Rule for Linear Functions

- **The graph of $y = -x$ is the graph of $y = x$ reflected in the x-axis.**

Scaling Rules for Linear Functions

- **When $c > 1$, the graph of $y = cx$ is the graph of $y = x$ stretched vertically by a factor of c.**
- **When $0 < c < 1$, the graph of $y = cx$ is the graph of $y = x$ compressed vertically by a factor of c.**

EXAMPLE 3

In **a–d**, write an equation for the resulting function if the graph of $y = x$ is:

a. shifted 6 units down

b. reflected in the x-axis and shifted 3 units up

c. compressed vertically by a factor of $\frac{1}{10}$ and shifted 1 unit up

d. stretched vertically by a factor of 4 and shifted 1 unit down

Solution **a.** $y = x - 6$ *Answer*

b. First, reflect in the x-axis: $y = -x$

Then, translate the resulting function 3 units up: $y = -x + 3$ *Answer*

c. First, compress vertically by a factor of $\frac{1}{10}$: $y = \frac{1}{10}x$

Then, translate the resulting function 1 unit up: $y = \frac{1}{10}x + 1$ *Answer*

d. First, stretch vertically by a factor of 4: $y = 4x$

Then, translate the resulting function 1 unit down: $y = 4x - 1$ *Answer*

EXERCISES

Writing About Mathematics

1. Explain why $(0, b)$ is always a point on the graph of $y = mx + b$.

2. Gunther said that in the first example in this section, since the slope of the line, $-\frac{2}{3}$, could have been written as $\frac{2}{-3}$, points on the line could have been located by moving up 2 units and to the left 3 units from the point $(0, 3)$. Do you agree with Gunther? Explain why or why not.

3. Hypatia said that, for linear functions, vertical translations have the same effect as *horizontal* translations and that reflecting across the x-axis has the same effect as reflecting across the y-axis. Do you agree with Hypatia? Explain why or why not.

Developing Skills

In 4–15, graph each equation using the slope and the y-intercept.

4. $y = 2x + 3$ **5.** $y = 2x$ **6.** $y = -2x$ **7.** $y = -3x - 2$

8. $y = \frac{2}{3}x + 2$ **9.** $y = \frac{1}{2}x - 1$ **10.** $y = -\frac{1}{3}x$ **11.** $y = -\frac{3}{4}x + 6$

12. $y - 2x = 8$ **13.** $3x + y = 4$ **14.** $\frac{x}{2} + \frac{y}{3} = 1$ **15.** $4x = 2y$

In 16–21, graph the equation using the slope and any point.

16. $3y = 4x + 9$ **17.** $5x - 2y = 3$ **18.** $3x + 4y = 7$

19. $2x = 3y - 1$ **20.** $3y - x = 5$ **21.** $2x - 3y - 6 = 0$

In 22–25, describe the translation, reflection, and/or scaling that must be applied to $y = x$ to obtain the graph of each given function.

22. $y = -x + 2$ **23.** $y = -\frac{1}{3}x - 2$ **24.** $y = 2x + 1.5$ **25.** $y = (x + 2) + 2$

Applying Skills

26. a. Draw the line through $(-2, -3)$ whose slope is 2.

b. What appears to be the y-intercept of this line?

c. Use the slope of the line and the answer to part **b** to write an equation of the line.

d. Do the coordinates of point $(-2, -3)$ satisfy the equation written in part **c**?

27. a. Draw the line through $(3, 5)$ whose slope is $\frac{2}{3}$.

b. What appears to be the y-intercept of this line?

c. Use the slope of the line and the answer to part **b** to write an equation of the line.

d. Do the coordinates of point $(3, 5)$ satisfy the equation written in part **c**?

28. a. Is $(1, 1)$ a point on the graph of $3x - 2y = 1$?

b. What is the slope of $3x - 2y = 1$?

c. Draw the graph of $3x - 2y = 1$ using point $(1, 1)$ and the slope of the line.

d. Why is it easier to use point $(1, 1)$ rather than the y-intercept to draw this graph?

9-8 GRAPHING DIRECT VARIATION

Recall that, when two variables represent quantities that are directly proportional or that vary directly, the ratio of the two variables is a constant. For example, when lemonade is made from a frozen concentrate, the amount of lemonade, y, that is obtained varies directly as the amount of concentrate, x, that is used. If we can make 3 cups of lemonade by using 1 cup of concentrate, the relationship can be expressed as

$$\frac{y}{x} = \frac{3}{1} \qquad \textit{or} \qquad y = 3x$$

The *constant of variation* is 3.

The equation $y = 3x$ can be represented by a line in the coordinate plane, as shown in the graph below.

x	3x	y
0	3(0)	0
1	3(1)	3
2	3(2)	6
3	3(3)	9

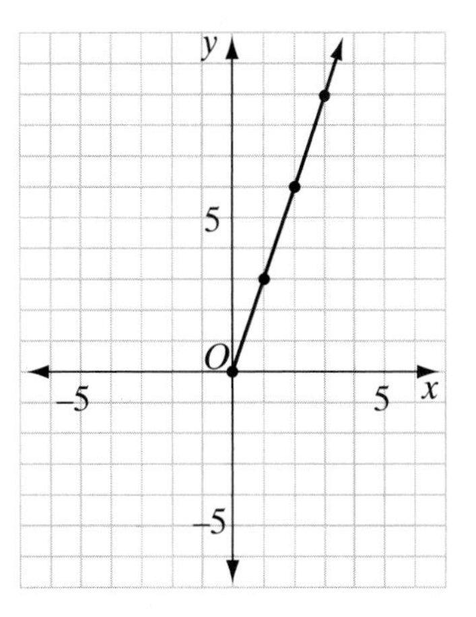

Here, the replacement set for x and y is the set of positive numbers and 0. Thus, the graph includes only points in the first quadrant.

In our example, if 0 cups of frozen concentrate are used, 0 cups of lemonade can be made. Thus, the ordered pair (0, 0) is a member of the solution set of $y = 3x$. Using any two points from the table above, for example. (0, 0) and (1, 3), we can write:

$$\frac{\Delta y}{\Delta x} = \frac{3 - 0}{1 - 0} = \frac{3}{1} = 3$$

Thus, we see that the slope of the line is also the constant of variation. An **equation of a direct variation** is a special case of an equation written in slope-intercept form; that is, $b = 0$ and $m =$ the constant of variation, k. A direct variation indicates that y is a multiple of x. Note that the graph of a direct variation is always a line through the origin.

The unit of measure for the lemonade and for the frozen concentrate is the same, namely, cups. Therefore, no unit of measure is associated with the ratio which, in this case, is $\frac{3}{1}$ or 3.

$$\frac{\text{lemonade}}{\text{concentrate}} = \frac{3\text{ cups}}{1\text{ cup}} = \frac{3}{1} \quad or \quad \frac{\text{lemonade}}{\text{concentrate}} = \frac{6\text{ cups}}{2\text{ cups}} = \frac{3}{1} \quad or \quad \frac{\text{lemonade}}{\text{concentrate}} = \frac{9\text{ cups}}{3\text{ cups}} = \frac{3}{1}$$

There are many applications of direct variation in business and science, and it is important to recognize how the choice of unit can affect the constant of variation. For example, if a machine is used to pack boxes of cereal in cartons, the rate at which the machine works can be expressed in cartons per minute or in cartons per second. If the machine fills a carton every 6 seconds, it will fill 10 cartons in 1 minute. The rate can be expressed as:

$$\frac{1\text{ carton}}{6\text{ seconds}} = \frac{1}{6}\text{ carton/second} \quad or \quad \frac{10\text{ cartons}}{1\text{ minute}} = 10\text{ cartons/minute}$$

Here each rate has been written as a unit rate.

As shown at right, each of these rates can be represented by a graph, where the rate, or constant of variation, is the slope of the line.

The legend of the graph, that is, the units in which the rate is expressed, must be clearly stated if the graph is to be meaningful.

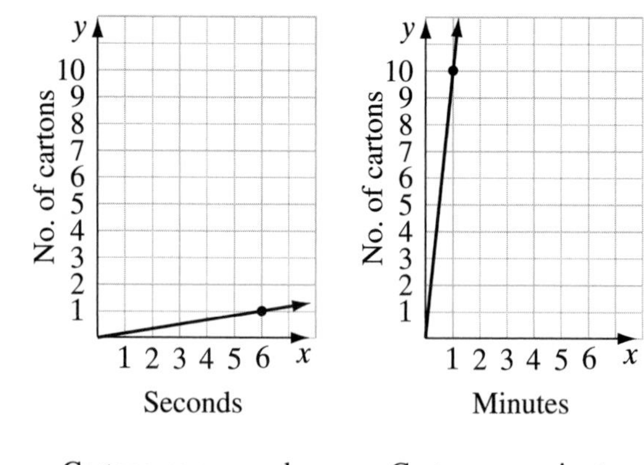

Cartons per second Cartons per minute

EXAMPLE 1

The amount of flour needed to make a white sauce varies directly as the amount of milk used. To make a white sauce, a chef used 2 cups of flour and 8 cups of milk. Write an equation and draw the graph of the relationship.

Solution Let x = the number of cups of milk, and y = the number of cups of flour. Then:

$$\frac{y}{x} = \frac{2}{8}$$
$$8y = 2x$$
$$y = \frac{2}{8}x$$
$$y = \frac{1}{4}x$$

x	$\frac{1}{4}$x	y
2	$\frac{1}{4}(2)$	$\frac{1}{2}$
4	$\frac{1}{4}(4)$	1
6	$\frac{1}{4}(6)$	$\frac{3}{2}$

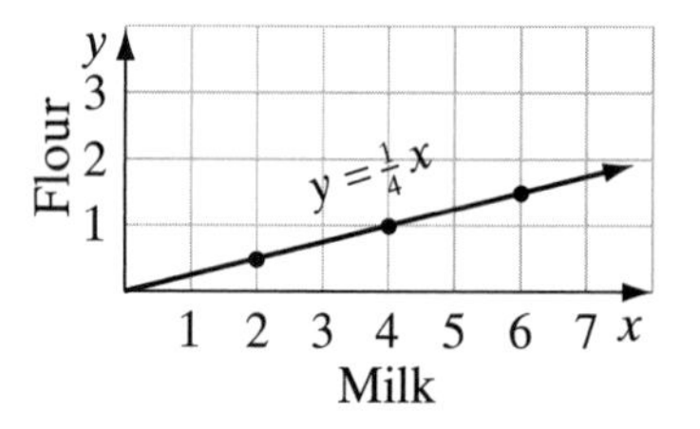

EXERCISES

Writing About Mathematics

1. A chef made the white sauce described in Example 1, measuring the flour and milk in ounces. Explain why the ratio of flour to milk should still be 1 : 4.

2. When Eduardo made white sauce, he used 1 cup of flour and 1 quart of milk. Is the ratio of flour to milk in Eduardo's recipe the same as in the recipe used by the chef in the example? Explain why or why not.

Developing Skills

In 3–12, y varies directly as x. In each case: **a.** What is the constant of variation? **b.** Write an equation for y in terms of x. **c.** Using an appropriate scale, draw the graph of the equation written in part **b**. **d.** What is the slope of the line drawn in part **c**?

3. The perimeter of a square (y) is 12 centimeters when the length of a side of the square (x) is 3 centimeters.

4. Jeanne can type 90 words (y) in 2 minutes (x).

5. A printer can type 160 characters (y) in 10 seconds (x).

6. A cake recipe uses 2 cups of flour (y) to $1\frac{1}{2}$ cups of sugar (x).

7. The length of a photograph (y) is 12 centimeters when the length of the negative from which it is developed (x) is 1.2 centimeters.

8. There are 20 slices (y) in 12 ounces of bread (x).

9. Three pounds of meat (y) will serve 15 people (x).

10. Twelve slices of cheese (y) weigh 8 ounces (x).

11. Willie averages 3 hits (y) for every 12 times at bat (x).

12. There are about 20 calories (y) in three crackers (x).

Applying Skills

13. If a car travels at a constant rate of speed, the distance that it travels varies directly as time. If a car travels 75 miles in 2.5 hours, it will travel 110 feet in 2.5 seconds.

    a. Find the constant of variation in miles per hour.

    b. Draw a graph that compares the distance that the car travels in miles to the number of hours traveled. Let the horizontal axis represent hours and the vertical axis represent distance.

    c. Find the constant of variation in feet per second.

    d. Draw a graph that compares the distance the car travels in feet to the number of seconds traveled.

14. In typing class, Russ completed a speed test in which he typed 420 characters in 2 minutes. His teacher told him to let 5 characters equal 1 word.

    a. Find Russ's rate on the test in characters per second.

    b. Draw a graph that compares the number of characters Russ typed to the number of minutes that he typed. (Let the horizontal axis represent minutes and the vertical axis represent characters.)

    c. Find Russ's rate on the test in words per minute.

    d. Draw a graph that compares the number of words Russ typed to the number of minutes that he typed. (Let the horizontal axis represent minutes and the vertical axis represent words.)

In 15–20, determine if the two variables are directly proportional. If so, write the equation of variation.

15. Perimeter of a square (P) and the length of a side (s).

16. Volume of a sphere (V) and radius (r).

17. The total amount tucked away in a piggy bank (s) and weekly savings (w), starting with an initial balance of $50.

18. Simple interest on an investment (I) in one year and the amount invested (P), at a rate of 2.5%.

19. Length measured in centimeters (c) and length measured in inches (i).

20. Total distance traveled (d) in one hour and average speed (s).

9-9 GRAPHING FIRST-DEGREE INEQUALITIES IN TWO VARIABLES

When a line is graphed in the coordinate plane, the line is a **plane divider** because it separates the plane into two regions called **half-planes.** One of these regions is a half-plane on one side of the line; the other is a half-plane on the other side of the line.

Let us consider, for example, the horizontal line $y = 3$ as a plane divider. As shown in the graph below, the line $y = 3$ and the two half-planes that it forms determine three sets of points:

1. The half-plane above the line $y = 3$ is the set of all points whose y-coordinates are greater than 3, that is, $y > 3$. For example, at point A, $y = 5$; at point B, $y = 4$.

2. The line $y = 3$ is the set of all points whose y-coordinates are equal to 3. For example, $y = 3$ at each point $C(-4, 3)$, $D(0, 3)$, and $G(6, 3)$.

3. The half-plane below the line $y = 3$ is the set of all points whose y-coordinates are less than 3, that is, $y < 3$. For example, at point E, $y = 1$; at point F, $y = -2$.

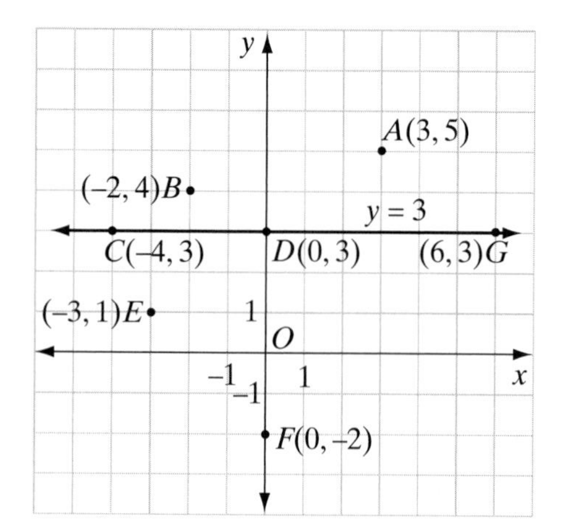

Together, the three sets of points form the entire plane.

To graph an inequality in the coordinate plane, we proceed as follows:

1. On the plane, represent the plane divider, for example, $y = 3$, by a *dashed line* to show that this divider does not belong to the graph of the half-plane.

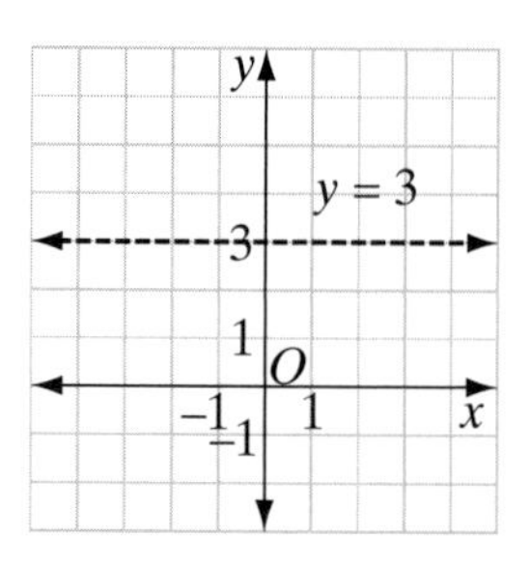

2. Shade the region of the half-plane whose points satisfy the inequality. To graph $y > 3$, shade the region above the plane divider.

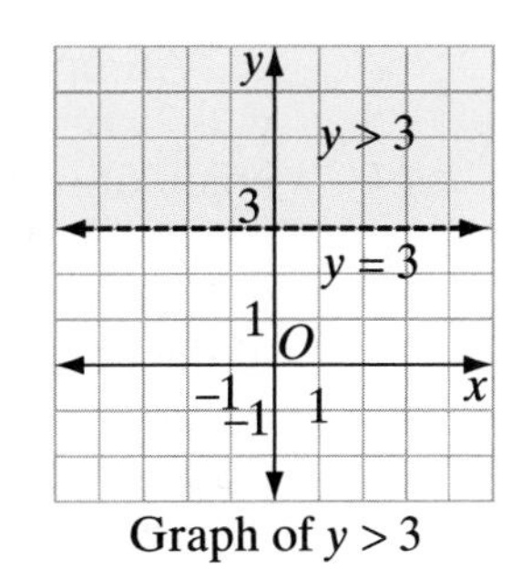

Graph of $y > 3$

3. To graph $y < 3$, shade the region below the plane divider.

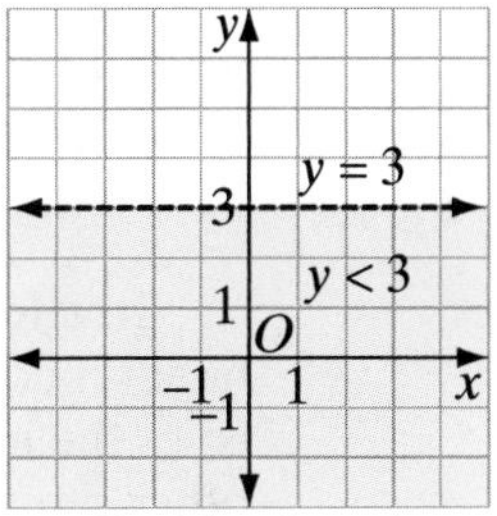

Graph of $y < 3$

Let us consider another example, where the plane divider is not a horizontal line. To graph the inequality $y > 2x$ or the inequality $y < 2x$, we use a dashed line to indicate that the line $y = 2x$ is not a part of the graph. This dashed line acts as a boundary line for the half-plane being graphed.

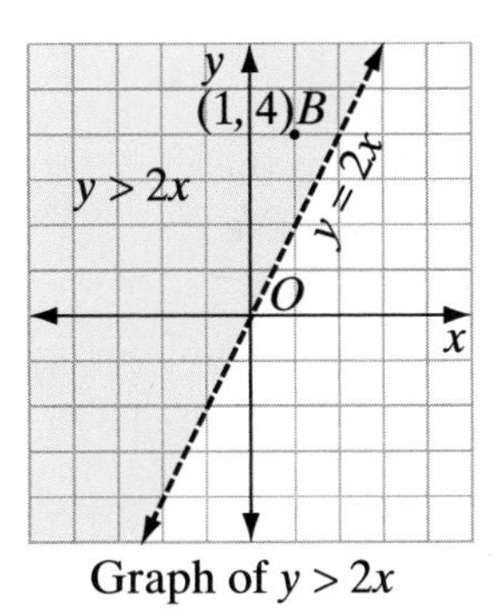

Graph of $y > 2x$

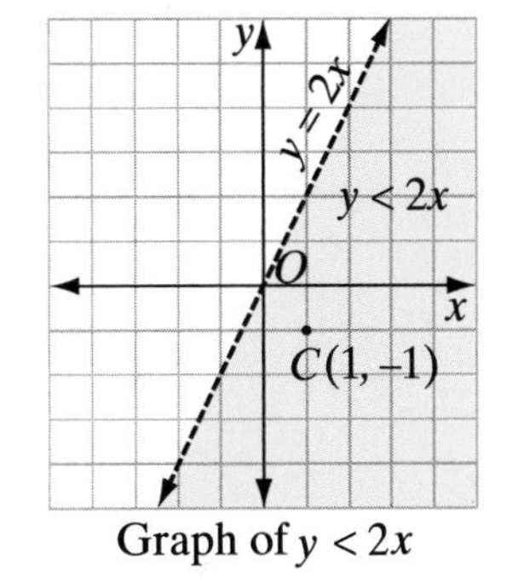

Graph of $y < 2x$

The graph of $y > 2x$ is the shaded half-plane above the line $y = 2x$. It is the set of all points in which the y-coordinate is greater than twice the x-coordinate.

The graph of $y < 2x$ is the shaded half-plane below the line $y = 2x$. It is the set of all points in which the y-coordinate is less than twice the x-coordinate.

An open sentence such as $y \geq 2x$ means $y > 2x$ or $y = 2x$. The graph of $y \geq 2x$ is the union of two disjoint sets. It includes all of the points in the solution set of the inequality $y > 2x$ and all of the points in the solution set of the equality $y = 2x$. To indicate that $y = 2x$ is part of the graph of $y \geq 2x$, we draw the graph of $y = 2x$ as a solid line. Then we shade the region above the line to include the points for which $y > 2x$.

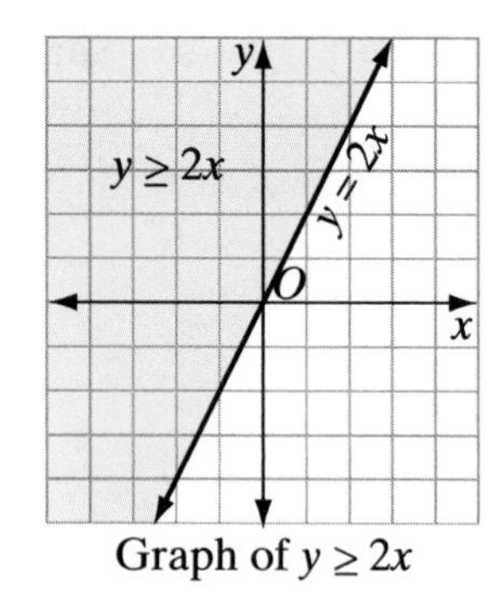

Graph of $y \geq 2x$

► **When the equation of a line is written in the form $y = mx + b$, the half-plane above the line is the graph of $y > mx + b$ and the half-plane below the line is the graph of $y < mx + b$.**

To check whether the correct half-plane has been chosen as the graph of a linear inequality, we select any point in that half-plane. If the selected point satisfies the inequality, every point in that half-plane satisfies the inequality. On the other hand, if the point chosen does not satisfy the inequality, then the other half-plane is the graph of the inequality.

EXAMPLE 1

Graph the inequality $y - 2x \geq 2$.

Solution *How to Proceed*

(1) Transform the inequality into one having y as the left member:

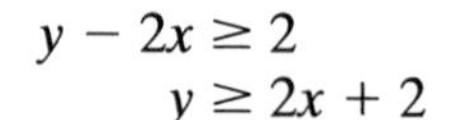

$$y - 2x \geq 2$$
$$y \geq 2x + 2$$

(2) Graph the plane divider, $y = 2x + 2$, by using the y-intercept, 2, to locate the first point $(0, 2)$ on the y-axis. Then use the slope, 2 or $\frac{2}{1}$, to find other points by moving up 2 and to the right 1:

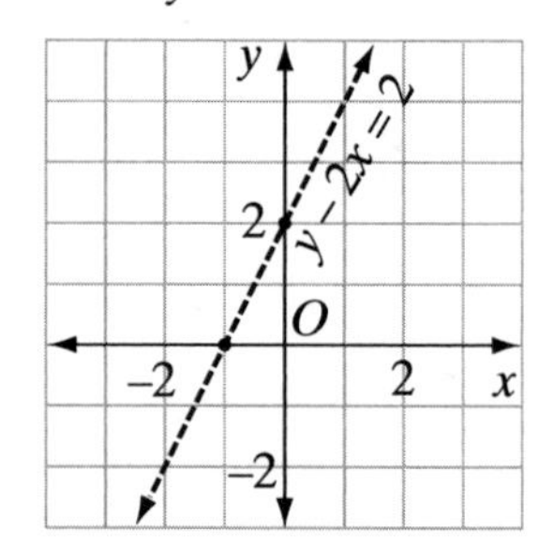

(3) Shade the half-plane above the line:

This region and the line are the required graph. The half-plane is the graph of $y - 2x > 2$, and the line is the graph of $y - 2x = 2$. Note that the line is now drawn solid to show that it is part of the graph.

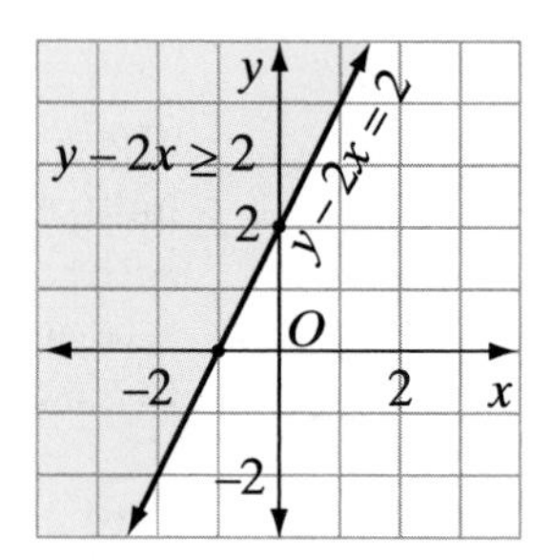

(4) Check the solution. Choose any point in the half-plane selected as the solution to see whether it satisfies the original inequality, $y - 2x \geq 2$:

Select point (0, 5) which is in the shaded region.

$$y - 2x \geq 2$$
$$5 - 2(0) \overset{?}{\geq} 2$$
$$5 \geq 2 \checkmark$$

The above graph is the graph of $y - 2x \geq 2$.

EXAMPLE 2

Graph each of the following inequalities in the coordinate plane.

a. $x > 1$ **b.** $x \leq 1$ **c.** $y \geq 1$ **d.** $y < 1$

Answers

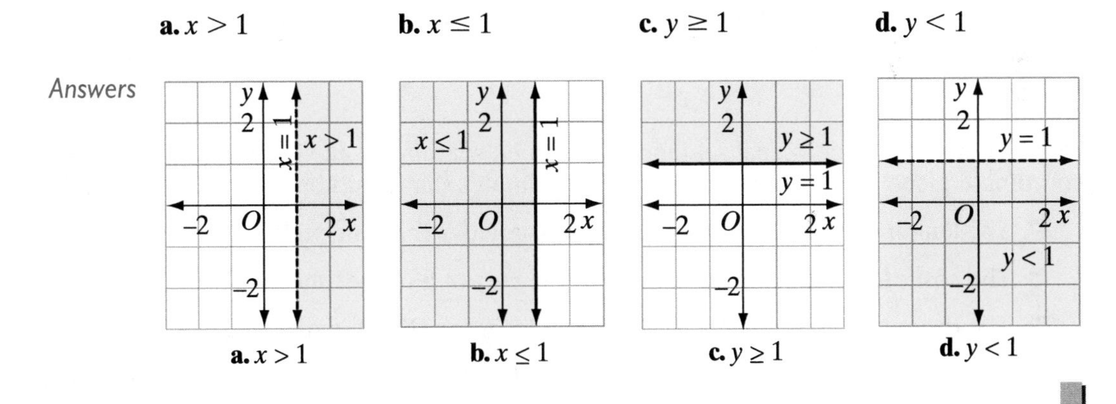

EXERCISES

Writing About Mathematics

1. Brittany said that the graph of $2x - y > 5$ is the region above the line that is the graph of $2x - y = 5$. Do you agree with Brittany? Explain why or why not.

2. Brian said that the union of the graph of $x > 2$ and the graph of $x < 2$ consists of every point in the coordinate plane. Do you agree with Brian? Explain why or why not.

Developing Skills

In 3–8, transform each sentence into one whose left member is y.

3. $y - 2x > 0$ **4.** $5x > 2y$ **5.** $y - x \geq 3$

6. $2x + y \leq 0$ **7.** $3x - y \geq 4$ **8.** $4y - 3x \leq 12$

In 9–23, graph each sentence in the coordinate plane.

9. $x > 4$
10. $x \leq -2$
11. $y > 5$
12. $y \leq -3$
13. $y < x - 2$
14. $y \geq \frac{1}{2}x + 3$
15. $x + y \leq -3$
16. $x - y \leq -1$
17. $x - 2y \leq 4$
18. $2y - 6x > 0$
19. $2x - 3y \geq 6$
20. $\frac{1}{2}y > \frac{1}{3}x + 1$
21. $9 - x \geq 3y$
22. $\frac{x}{2} + \frac{y}{5} \leq 1$
23. $2x + 2y - 6 \leq 0$

In 24–26: **a.** Write each verbal sentence as an open sentence. **b.** Graph each open sentence in the coordinate plane.

24. The y-coordinate of a point is equal to or greater than 3 more than the x-coordinate.

25. The sum of the x-coordinate and the y-coordinate of a point is less than or equal to 5.

26. The y-coordinate of a point decreased by 3 times the x-coordinate is greater than or equal to 2.

Applying Skills

In 27–31: **a.** Write each verbal sentence as an open sentence. **b.** Graph each open sentence in the coordinate plane. **c.** Choose one pair of coordinates that could be reasonable values for x and y.

27. The length of Mrs. Gauger's garden (y) is greater than the width (x).

28. The cost of a shirt (y) is less than half the cost of a pair of shoes (x).

29. The distance to school (y) is at least 2 miles more than the distance to the library (x).

30. The height of the flagpole (y) is at most 4 feet more than the height of the oak tree (x) nearby.

31. At the water park, the cost of a hamburger (x) plus the cost of a can of soda (y) is greater than 5 dollars.

9-10 GRAPHS INVOLVING ABSOLUTE VALUE

To draw the graph of the equation $y = |x|$, we can choose values of x and then find the corresponding values of y.

Let us consider the possible choices for x and the resulting y-values:

1. Choose $x = 0$. Since the absolute value of 0 is 0, y will be 0.
2. Choose x as any positive number. Since the absolute value of any positive number is that positive number, y will have the same value as x. For example, if $x = 5$, then $y = |5| = 5$.

3. Choose x as any negative number. Since the absolute value of any negative number is positive, y will be the opposite of x. For example, if $x = -3$, then $y = |-3| = 3$.

Thus, we conclude that x can be 0, positive, or negative, but y will be only 0 or positive.

Here is a table of values and the corresponding graph:

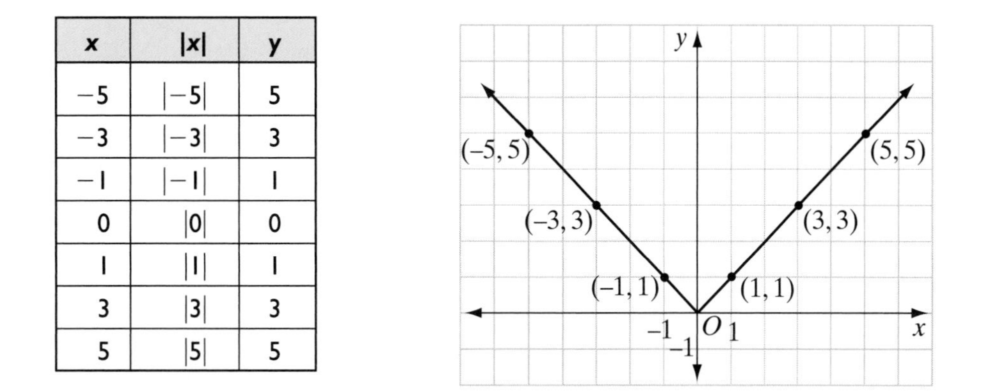

x	|x|	y
−5	|−5|	5
−3	|−3|	3
−1	|−1|	1
0	|0|	0
1	|1|	1
3	|3|	3
5	|5|	5

Notice that for positive values of x, the graph of $y = |x|$ is the same as the graph of $y = x$. For negative values of x, the graph of $y = |x|$ is the same as the graph of $y = -x$.

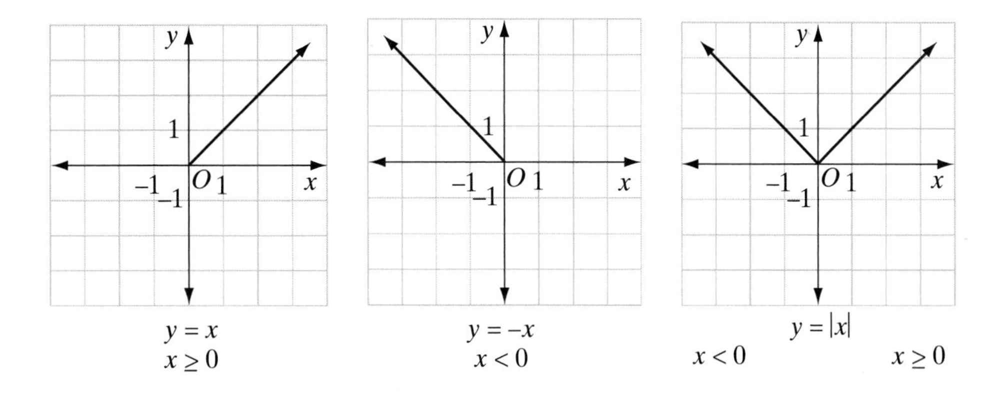

It should be noted that for all values of x, $y = -|x|$ results in a negative value for y. Therefore, for positive values of x, the graph of $y = -|x|$ is the same as the graph of $y = -x$. For negative values of x, the graph of $y = -|x|$ is the same as the graph of $y = x$.

EXAMPLE 1

Draw the graph of $y = |x| + 2$.

Solution (1) Make a table of values:

x	$\lvert x \rvert + 2$	y
−4	$\lvert -4 \rvert + 2$	6
−2	$\lvert -2 \rvert + 2$	4
−1	$\lvert -1 \rvert + 2$	3
0	$\lvert 0 \rvert + 2$	2
1	$\lvert 1 \rvert + 2$	3
3	$\lvert 3 \rvert + 2$	5
5	$\lvert 5 \rvert + 2$	7

(2) Plot the points and draw rays to connect the points that were graphed:

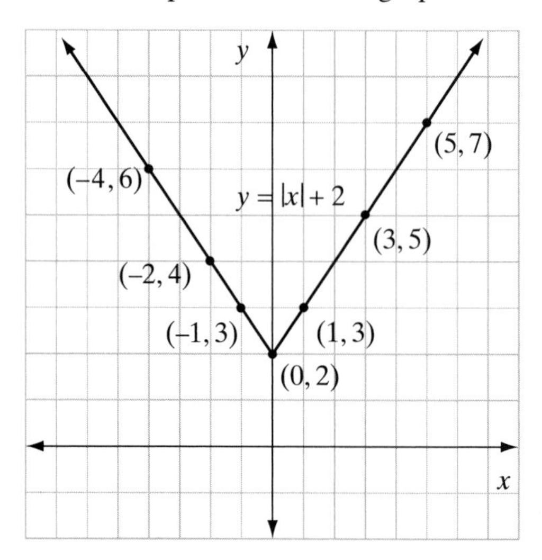

Calculator Solution (1) Enter the equation into Y_1:

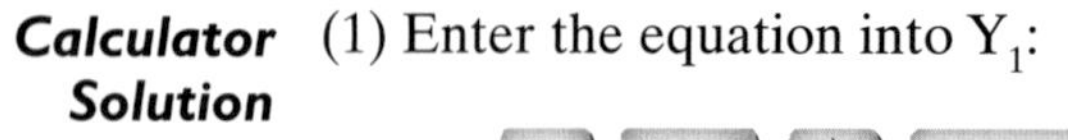

X,T,Θ,*n*) + 2

DISPLAY:

```
Plot1 Plot2 Plot3
\Y1=abs(X)+2
```

(2) Graph to the standard window:

ENTER: ZOOM 6

DISPLAY:

EXAMPLE 2

Draw the graph of $|x| + |y| = 3$.

Solution By the definition of absolute value, $|x| = |-x|$ and $|y| = |-y|$.

- Since $(1, 2)$ is a solution, $(-1, 2)$, $(-1, -2)$ and $(1, -2)$ are solutions.
- Since $(2, 1)$ is a solution, $(-2, 1)$, $(-2, -1)$ and $(2, -1)$ are solutions.
- Since $(0, 3)$ is a solution, $(0, -3)$ is a solution.
- Since $(3, 0)$ is a solution, $(-3, 0)$ is a solution.

Plot the points that are solutions, and draw the line segments joining them.

$|x| + |y| = 3$

Translating, Reflecting, and Scaling Graphs of Absolute Value Functions

Just as linear functions can be translated, reflected, or scaled, graphs of absolute value functions can also be manipulated by working with the graph of the absolute value function $y = |x|$.

For instance, the graph of $y = |x| - 5$ is the graph of $y = |x|$ shifted 5 units down. The graph of $y = -|x|$ is the graph of $y = |x|$ reflected in the x-axis. The graph of $y = 2|x|$ is the graph of $y = |x|$ stretched vertically by a factor of 2, while the graph of $y = \frac{1}{2}|x|$ is the graph of $y = |x|$ compressed vertically by a factor of $\frac{1}{2}$.

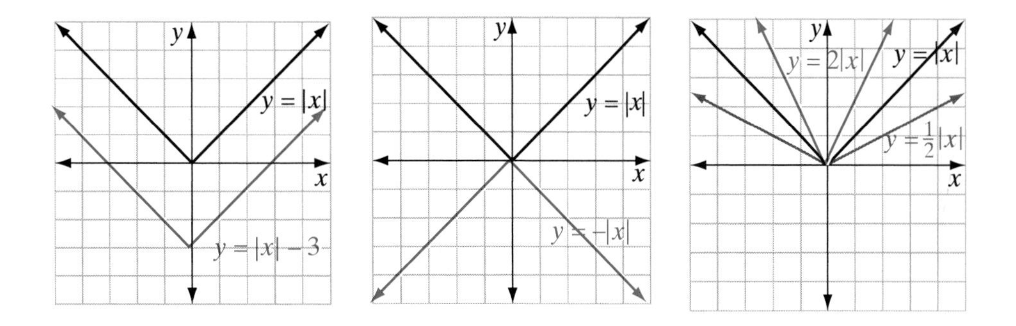

Translation Rules for Absolute Value Functions

If c is *positive*:

- ► **The graph of $y = |x| + c$ is the graph of $y = |x|$ shifted c units up.**
- ► **The graph of $y = |x| - c$ is the graph of $y = |x|$ shifted c units down.**

For absolute value functions, there are two additional translations that can be done to the graph of $y = |x|$, *horizontal* shifting to left or to the right.

If c is positive:

- **The graph of $y = |x + c|$ is the graph of $y = |x|$ shifted c units to the left.**
- **The graph of $y = |x - c|$ is the graph of $y = |x|$ shifted c units to the right.**

Reflection Rule for Absolute Value Functions

- **The graph of $y = -|x|$ is the graph of $y = |x|$ reflected across the x-axis.**

Scaling Rules for Absolute Value Functions

- **When $c > 1$, the graph of $y = c|x|$ is the graph of $y = |x|$ stretched vertically by a factor of c.**
- **When $0 < c < 1$, the graph of $y = c|x|$ is the graph of $y = |x|$ compressed vertically by a factor of c.**

EXAMPLE 3

In **a–e**, write an equation for the resulting function if the graph of $y = |x|$ is:

a. shifted 2.5 units down

b. shifted 6 units to the right

c. stretched vertically by a factor of 3 and shifted 5 units up

d. compressed vertically by a factor of $\frac{1}{3}$ and reflected in the x-axis

e. reflected in the x-axis, shifted 1 unit up, and shifted 1 unit to the left

Solution **a.** $y = |x| - 2.5$ Answer

b. $y = |x - 6|$ Answer

c. First, stretch vertically by a factor of 3: $y = 3|x|$

Then, translate the resulting function 5 units up: $y = 3|x| + 5$ Answer

d. First, compress vertically by a factor of $\frac{1}{3}$: $y = \frac{1}{3}|x|$

Then, reflect in the x-axis: $y = -\frac{1}{3}|x|$ Answer

e. First, reflect in the x-axis: $y = -|x|$

Then, translate the resulting function 1 unit up: $y = -|x| + 1$

Finally, translate the resulting function 1 unit to the left: $y = -|x + 1| + 1$ Answer

EXERCISES

Writing About Mathematics

1. Charity said that the graph of $y = |2x| + 1$ is the graph of $y = 2x + 1$ for $x \geq 0$ and the graph of $-2x + 1$ for $x < 0$. Do you agree with Charity? Explain why or why not.
2. April said that $|x| + |y| = 5$ is a function. Do you agree with April? Explain why or why not.
3. Euclid said that, for positive values of c, the graph of $y = |cx|$ is the same as the graph of $y = c|x|$. Do you agree with Euclid? If so, prove Euclid's statement.

Developing Skills

In 4–15, graph each equation.

4. $y = |x| - 1$ **5.** $y = |x| + 3$ **6.** $y = |x - 1|$ **7.** $y = |x + 3|$

8. $y = 2|x|$ **9.** $y = 2|x| + 1$ **10.** $|x| + |y| = 5$ **11.** $|x| + 2|y| = 7$

12. $|y| = |x|$ **13.** $|x| + |y| = 4$ **14.** $\frac{|x|}{2} + \frac{|y|}{4} = 1$ **15.** $2|x| + 4|y| - 6 = 0$

In 16–19, describe the translation, reflection, and/or scaling that must be applied to $y = |x|$ to obtain the graph of each given function.

16. $y = -|x| - 4$ **17.** $y = -2|x| + 2$ **18.** $y = |x + 2| - 3$ **19.** $y = -|x - 1.5| + 4$

9-11 GRAPHS INVOLVING EXPONENTIAL FUNCTIONS

A piece of paper is one layer thick. If we place another piece of paper on top of it, the stack is two layers thick. If we place another piece of paper on top, the stack is now three layers thick.

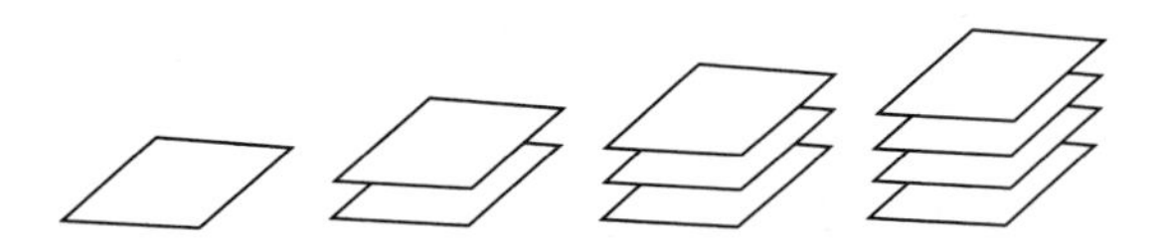

As the process continues, we can describe the number of layers, y, in terms of the number of sheets added, x, by the linear function $y = 1 + x$. For example, after we have added seven pieces of paper, the stack is eight layers thick. This is an example of **linear growth** because the change can be described by a linear function.

Now consider another experiment. A piece of paper is one layer thick. If the paper is folded in half, the stack is two layers thick. If the stack is folded again, the stack is four layers thick. If the stack is folded a third time, it is eight layers thick.

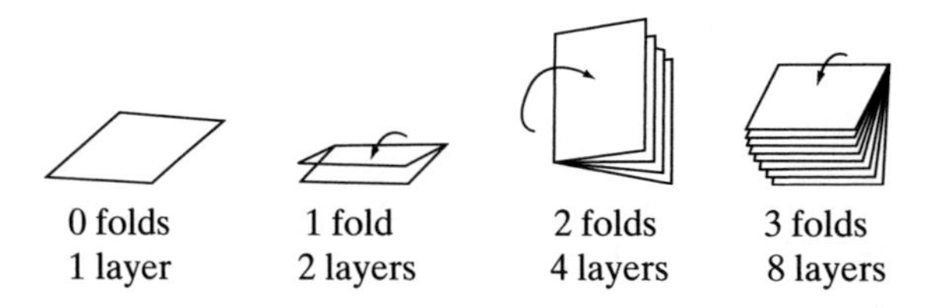

x	2^x	$y = 2^x$
0	2^0	1
1	2^1	2
2	2^2	4
3	2^3	8
4	2^4	16
5	2^5	32

Although we will reach a point at which it is impossible to fold the paper, imagine that this process can continue. We can describe the number of layers of paper, y, in terms of the number of folds, x, by an **exponential function**, $y = 2^x$. The table at the right shows some values for this function. Such functions are said to be **nonlinear**.

After five folds, the stack is 32 layers thick. After ten folds, the stack would be 2^{10} or 1,024 layers thick. In other words, it would be thinker than this book. This is an example of **exponential growth** because the growth is described by an exponential function $y = b^x$. When the base b is a positive number greater than 1 ($b > 1$), y increases as x increases.

We can draw the graph of the exponential function, adding zero and negative integral values to those given above. Locate the points on the coordinate plane and draw a smooth curve through them. When we draw the curve, we are assuming that the domain of the independent variable, x, is the set of real numbers. You will learn about powers that have exponents that are not integers in more advanced math courses.

x	2^x	$y = 2^x$
0	2^0	1
−1	2^{-1}	$\frac{1}{2}$
−2	2^{-2}	$\frac{1}{4}$
−3	2^{-3}	$\frac{1}{8}$

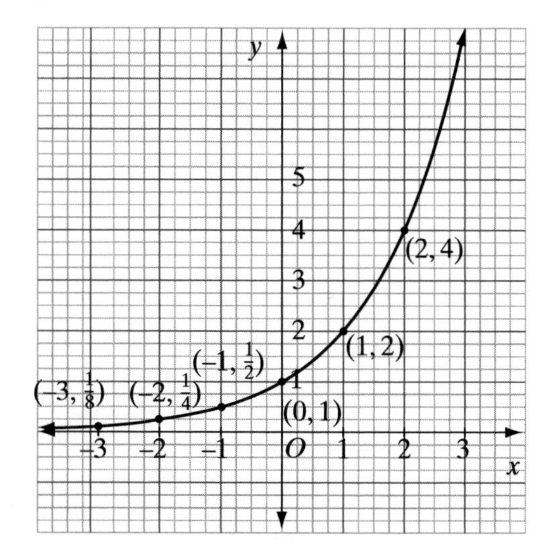

There are many examples of exponential growth in the world around us. Over an interval, certain populations—for example of bacteria, or rabbits, or people—grow exponentially.

Compound interest is also an example of exponential growth. If a sum of money, called the principal, P, is invested at 4% interest, then after one year the value of the investment, A, is the principal plus the interest. Recall that $I = Prt$. In this case, $r = 0.04$ and $t = 1$.

$$A = P + I = P + Prt = P(1 + rt) = P(1+ 0.04(1)) = P(1.04)$$

After 2 years, the new principal is $(1.04)^2P$. Replace P by $1.04P$ in the formula $A = P(1.04)$:

$$A = 1.04P(1.04) = P(1.04)^2$$

After 3 years, the new principal is $(1.04)^3P$. Replace P by $P(1.04)^2$ in the formula $A = P(1.04)$:

$$A = (1.04)^2P(1.04) = P(1.04)^3$$

In general, after n years, the value of an investment for which the rate of interest is r can be found by using the formula

$$A = P(1 + r)^n.$$

Note that in this formula for exponential growth, P is the amount present at the beginning; r, a positive number, is the rate of growth; and n is the number of intervals of time during which the growth has been taking place. Since r is a positive number, the base, $1 + r$, is a number greater than 1.

Compare the formula $A = P(1 + r)^n$ with the exponential function $y = 2^x$ used to determine the number of layers in a stack after x folds. $P = 1$ since we started with 1 layer. The base, $(1 + r) = 2$, so $r = 1$ or 100%. Doubling the number of layers in the stack is an increase equal to the size of the stack, that is, an increase of 100%.

An exponential change can be a decrease as well as an increase. Consider this example. Start with a large piece of paper. If we cut the paper into two equal parts, each part is one-half of the original piece. Now if we cut one of these pieces into two equal parts, each of the parts is one-fourth of the original piece. As this process continues, we can describe the part of the original piece that results from each cut in terms of the number of cuts. The function $y = \left(\frac{1}{2}\right)^x$ gives the part of the original piece, y, that is the result of x cuts. This is an example of **exponential decay**. For the exponential function $y = b^x$, when the base b is a positive number less than 1 $(0 < b < 1)$, y decreases as x increases.

x	$\left(\frac{1}{2}\right)^x$	y
0	$\left(\frac{1}{2}\right)^0$	1
1	$\left(\frac{1}{2}\right)^1$	$\frac{1}{2}$
2	$\left(\frac{1}{2}\right)^2$	$\frac{1}{4}$
3	$\left(\frac{1}{2}\right)^3$	$\frac{1}{8}$
4	$\left(\frac{1}{2}\right)^4$	$\frac{1}{16}$
5	$\left(\frac{1}{2}\right)^5$	$\frac{1}{32}$

Some examples of exponential decay include a decrease in population, a fund, or the value of a machine when that decrease can be represented by a constant rate at regular intervals. The radioactive decay of a chemical element such as carbon is also an example of exponential decay.

Problems of exponential decay can be solved using the same formula as that for exponential growth. In exponential decay, the rate of change is a negative number between -1 and 0 so that the base is a number between 0 and 1. This is illustrated in Example 3.

EXAMPLE 1

Draw the graph of $y = \left(\frac{2}{3}\right)^x$.

Solution Make a table of values from $x = -3$ to $x = 3$. Plot the points and draw a smooth curve through them.

x	$\left(\frac{2}{3}\right)^x$	y
-3	$\left(\frac{2}{3}\right)^{-3} = \left(\frac{3}{2}\right)^3$	$\frac{27}{8}$
-2	$\left(\frac{2}{3}\right)^{-2} = \left(\frac{3}{2}\right)^2$	$\frac{9}{4}$
-1	$\left(\frac{2}{3}\right)^{-1} = \left(\frac{3}{2}\right)^1$	$\frac{3}{2}$
0	$\left(\frac{2}{3}\right)^0$	1
1	$\left(\frac{2}{3}\right)^1$	$\frac{2}{3}$
2	$\left(\frac{2}{3}\right)^2$	$\frac{4}{9}$
3	$\left(\frac{2}{3}\right)^3$	$\frac{8}{27}$

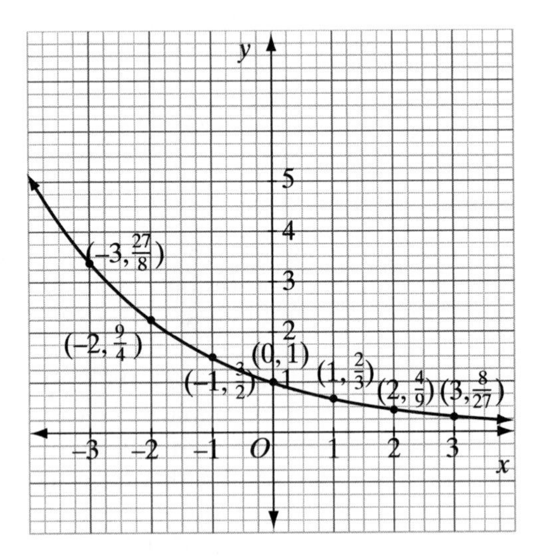

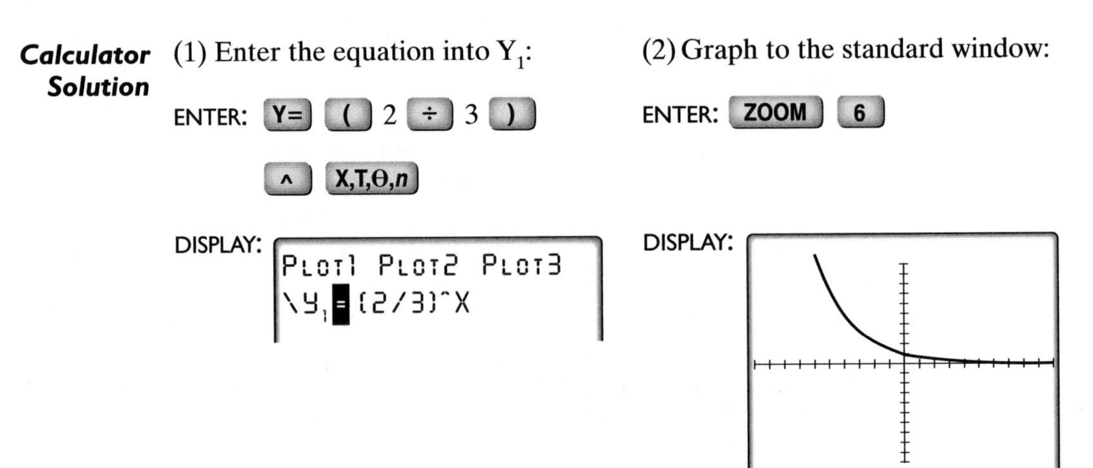

Calculator Solution (1) Enter the equation into Y_1:

ENTER: Y= (2 ÷ 3) ^ X,T,Θ,n

DISPLAY:

(2) Graph to the standard window:

ENTER: ZOOM 6

DISPLAY:

EXAMPLE 2

A bank is advertising a rate of 5% interest compounded annually. If $2,000 is invested in an account at that rate, find the amount of money in the account after 10 years.

Solution This is an example of exponential growth. Use the formula $A = P(1 + r)^n$ with $r = 0.05$, $P =$ the initial investment, and $n = 10$ years.

$$A = 2{,}000(1 + 0.05)^{10} = 2{,}000(1.05)^{10} = 3{,}257.78954$$

Answer The value of the investment is $3,257.79 after 10 years.

EXAMPLE 3

The population of a town is decreasing at the rate of 2.5% per year. If the population in the year 2000 was 28,000, what will be the expected population in 2015 if this rate of decrease continues? Give your answer to the nearest thousand.

Solution Use the formula for exponential growth or decay. The rate of decrease is entered as a negative number so that the base is a number between 0 and 1.

$r = -0.025$

$P =$ the initial population

$n = 15$ years ($x = 0$ corresponds to the year 2000)

$A = P(1 + r)^n = 28{,}000(1 + (-0.025))^{15} = 28{,}000(0.975)^{15} = 19{,}152.5792$

Answer The population will be about 19,000 in 2015.

EXERCISES

Writing About Mathematics

1. The equation $y = b^x$ is an exponential function. If the graph of the function is a smooth curve, explain why the value of b cannot be negative.

2. The equation $A = P(1 + r)^n$ can be used for both exponential growth and exponential decay when r represents the percent of increase or decrease.

a. How does the value of r that is used in this equation for exponential growth differ from that of exponential decay?

b. If the equation represents exponential growth, can the rate be greater than 100%? Explain why or why not.

c. If the equation represents exponential decay, can the rate be greater than 100%? Explain why or why not.

3. Euler says that the graph of $y = 3^{x+2}$ is the graph of $y = 3^x$ shifted 2 units to the left and that the graph of $y = 3^x + 2$ is the graph of $y = 3^x$ shifted 2 units up. Do you agree with Euler? Explain why or why not.

Developing Skills

In 4–9, sketch each graph using as values of x the integers in the given interval.

4. $y = 3^x, [-3, 3]$ **5.** $y = 4^x, [-2, 2]$ **6.** $y = 1.5^x, [-4, 4]$

7. $y = \left(\frac{2}{5}\right)^x, [-3, 3]$ **8.** $y = \left(\frac{1}{3}\right)^x, [-2, 2]$ **9.** $y = 1.25^x, [0, 10]$

10. Compare each of the exponential graphs drawn in exercises 4 through 9.

a. What point is common to all of the graphs?

b. If the value of the base is greater than 1, in what direction does the graph curve?

c. If the value of the base is between 0 and 1, in what direction does the graph curve?

11. a. Sketch the graph of $y = \left(\frac{1}{2}\right)^x$ in the interval $[-3, 3]$.

b. Sketch the graph of $y = 2^{-x}$ in the interval $[-3, 3]$.

c. What do you observe about the graphs drawn in **a** and **b**?

Applying Skills

12. In 2005, the population of a city was 25,000. The population increased by 20% in each of the next three years. If this rate of increase continues, what will be the population of the city in 2012?

13. Alberto invested $5,000 at 6% interest compounded annually. What will be the value of Alberto's investment after 8 years?

14. Mrs. Boyko has a trust fund from which she withdraws 5% each year. If the fund has a value of $50,000 this year, what will be the value of the fund after 10 years?

15. Hailey has begun a fitness program. The first week she ran 1 mile every day. Each week she increases the amount that she runs each day by 20%. In week 10, how many miles does she run each day? Give your answer to the nearest mile.

16. Alex received $75 for his birthday. In the first week after his birthday, he spent one-third of the money. In the second week and each of the following weeks, he spent one-third of the money he had left. How much money will Alex have left after 5 weeks?

17. During your summer vacation, you are offered a job at which you can work as many days as you choose. If you work 1 day, you will be paid $0.01. If your work 2 days you will be paid a total of $0.02. If your work 3 days you will be paid a total of $0.04. If you continue to work, your total pay will continue to double each day.

a. Would you accept this job if you planned to work 10 days? Explain why or why not.

b. Would you accept this job if you planned to work 25 days? Explain why or why not.

CHAPTER SUMMARY

The solutions of equations or inequalities in two variables are ordered pairs of numbers. The set of points whose coordinates make an equation or inequality true is the graph of that equation or inequality. The **graph of a linear function** of the form $Ax + By = C$ is a straight line.

The **domain of a function** is the set of all values the **independent variable**, the x-coordinate, is allowed to take. This determines the **range of a function**, that is, the set of all values the **dependent variable**, the y-coordinate, will take.

Set-builder notation provides a mathematically concise method of describing the elements of a set. **Roster form** lists every element of a set exactly once.

A linear function can be written in the form $y = mx + b$, where m is the slope and b is the y-intercept of the line that is the graph of the function. A line parallel to the x-axis has a slope of 0, and a line parallel to the y-axis has no slope. The **slope** of a line is the ratio of the change in the vertical direction to the change in the horizontal direction. If (x_1, y_1) and (x_2, y_2) are two points on a line, the slope of the line is

$$\textbf{slope} = m = \frac{y_2 - y_1}{x_2 - x_2}.$$

If y varies directly as x, the ratio of y to x is a constant. **Direct variation** can be represented by a line through the origin whose slope is the **constant of variation**.

The graph of $y = mx + b$ separates the plane into two **half-planes**. The half-plane above the graph of $y = mx + b$ is the graph of $y > mx + b$, and the half-plane below the graph of $y = mx + b$ is the graph of $y < mx + b$.

The graph of $y = |x|$ is the **union** of the graph of $y = -x$ for $x < 0$ and the graph of $y = x$ for $x \geq 0$.

The equation $y = b^x$, $b > 0$ and $b \neq 1$, is an example of an exponential function. The exponential equation $A = P(1 + r)^n$ is a formula for exponential growth or decay. A is the amount present after n intervals of time, P is the amount present at time 0, and r is the rate of increase or decrease. In exponential growth, r is a positive number. In exponential decay, r is a negative number in the interval $(-1, 0)$.

The function $y = x$ and $y = |x|$ can be translated, reflected, or scaled to graph other linear and absolute value functions.

	Linear Function	**Absolute Value**
Vertical translations ($c > 0$)	$y = x + c$ (up)	$y = \|x\| + c$ (up)
	$y = x - c$ (down)	$y = \|x\| - c$ (down)
Horizontal translations ($c > 0$)	$y = x + c$ (left)	$y = \|x + c\|$ (left)
	$y = x - c$ (right)	$y = \|x - c\|$ (right)
Reflection across the x-axis	$y = -x$	$y = -\|x\|$
Vertical stretching ($c > 1$)	$y = cx$	$y = c\|x\|$
Vertical compression ($0 < c < 1$)	$y = cx$	$y = cx$

VOCABULARY

9-1 Roster form • Set-builder notation • Member of a set (∈) • Not a member of a set (∉) • Relation • Domain of a relation • Range of a relation • Function • Domain of a function • Range of a function • Independent variable • Dependent variable

9-2 Graph • Standard form • Linear equation

9-4 Slope • Unit rate of change

9-6 y-intercept • x-intercept • Intercept form a linear equation • Slope-intercept form of a linear equation

9-8 Equation of a direct variation

9-9 Plane divider • Half-plane

9-11 Linear growth • Exponential function • Nonlinear functions • Exponential growth • Exponential decay

REVIEW EXERCISES

1. Determine **a.** the domain and **b.** the range of the relation shown in the graph to the right. **c.** Is the relation a function? Explain why or why not.

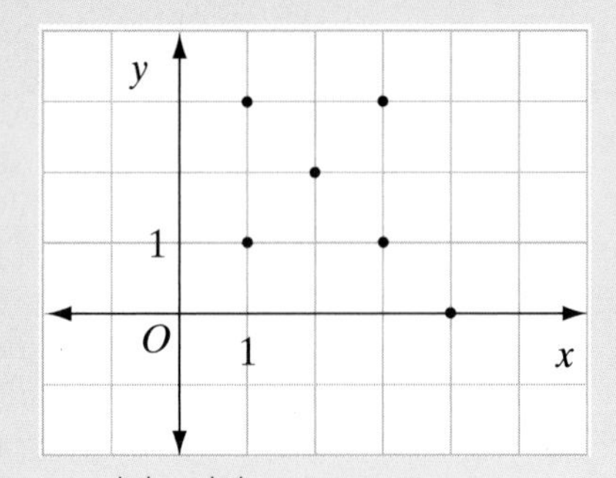

2. Draw the graph of $|x| + |y| = 5$. Is the graph of $|x| + |y| = 5$ a square? Prove your answer.

3. A function is a set of ordered pairs in which no two ordered pairs have the same first element. Explain why $\{(x, y) \mid y < 2x + 1\}$ is not a function.

4. What is the slope of the graph of $y = -2x + 5$?

5. What is the slope of the line whose equation is $3x - 2y = 12$?

6. What are the intercepts of the line whose equation is $3x - 2y = 12$?

7. What is the slope of the line that passes through points (4, 5) and (6, 1)?

In 8–13, in each case: **a.** Graph the equation or inequality. **b.** Find the domain and range of the equation or inequality.

8. $y = -x + 2$ **9.** $y = 3$ **10.** $y = \frac{2}{3}x$

11. $x + 2y = 8$ **12.** $y - x > 2$ **13.** $2x - y \geq 4$

In 14–22, refer to the coordinate graph to answer each question.

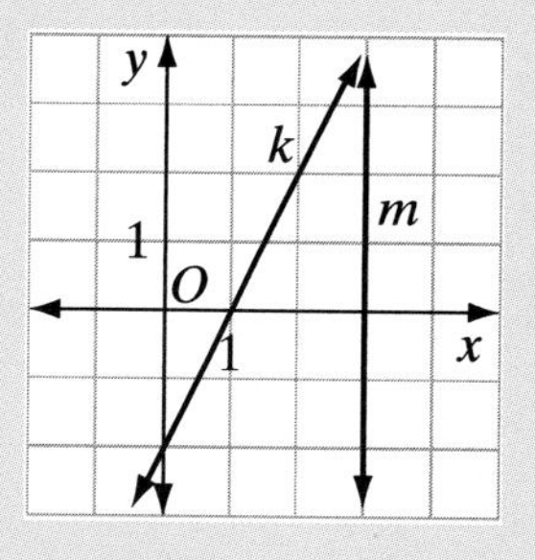

14. What is the slope of line k?

15. What is the x-intercept of line k?

16. What is the y-intercept of line k?

17. What is the slope of a line that is parallel to line k?

18. What is the slope of a line that is perpendicular to line k?

19. What is the slope of line m?

20. What is the x-intercept of line m?

21. What is the y-intercept of line m?

22. What is the slope of a line that is perpendicular to line m?

23. If point $(d, 3)$ lies on the graph of $3x - y = 9$, what is the value of d?

In 24–33, in each case select the numeral preceding the correct answer.

24. Which set of ordered pairs represents a function?
(1) $\{(-6, 1), (-7, 1), (1, -7), (1, -6)\}$
(2) $\{(0, 13), (1, 13), (6, 13), (112, 13)\}$
(3) $\{(3, -3), (3, -4), (-3, -4), (-3, -3)\}$
(4) $\{(1, 12), (11, 1), (112, 11), (11, 112)\}$

25. Which point does not lie on the graph of $3x - y = 9$?
(1) $(1, -6)$ (2) $(2, 3)$ (3) $(3, 0)$ (4) $(0, -9)$

26. Which ordered pair is in the solution set of $y < 2x - 4$?
(1) $(0, -5)$ (2) $(2, 0)$ (3) $(3, 3)$ (4) $(0, 2)$

27. Which equation has a graph parallel to the graph of $y = 5x - 2$?
(1) $y = -5x$ (2) $y = 5x + 3$ (3) $y = -2x$ (4) $y = 2x - 5$

28. The graph of $2x + y = 8$ intersects the x-axis at point
(1) $(0, 8)$ (2) $(8, 0)$ (3) $(0, 4)$ (4) $(4, 0)$

29. What is the slope of the graph of the equation $y = 4$?
(1) The line has no slope.
(2) 0 (3) -4 (4) 4

30. In which ordered pair is the x-coordinate 3 more than the y-coordinate?
(1) $(1, 4)$ (2) $(1, 3)$ (3) $(3, 1)$ (4) $(4, 1)$

31. Which of the following is *not* a graph of a function?

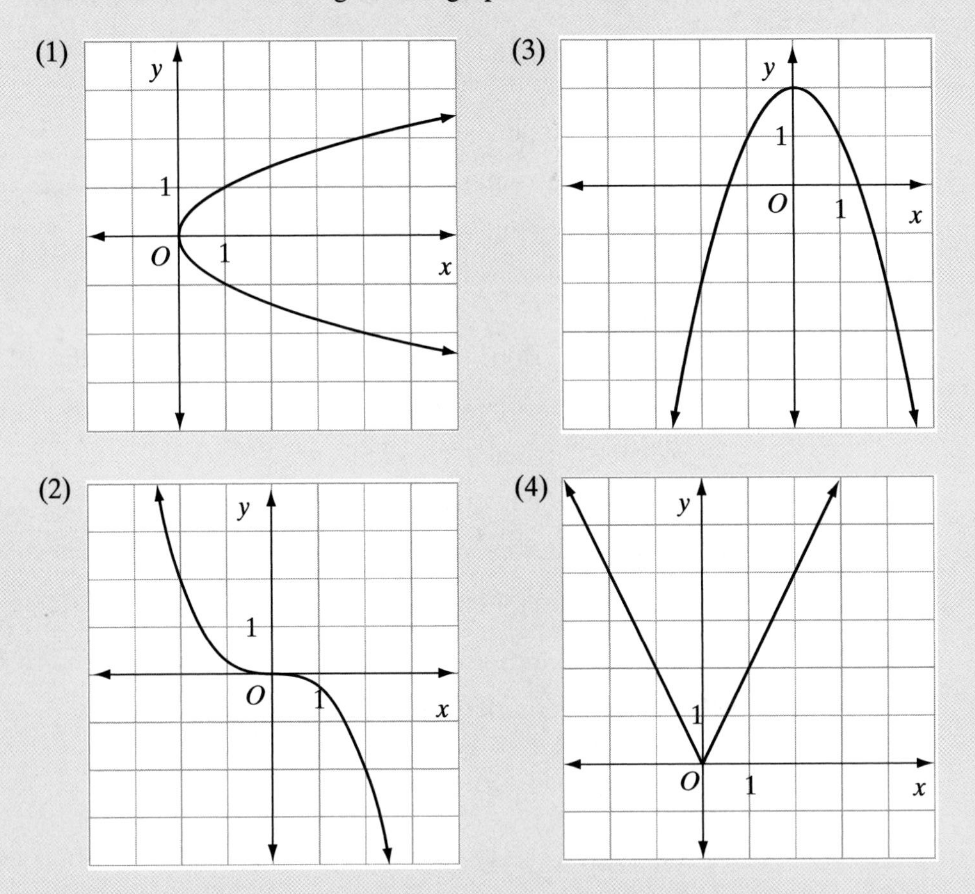

32. Using the domain $D = \{4x \mid x \in \text{the set of integers and } 10 \leq 4x \leq 20\}$, what is the range of the function $y = |2 - x| + 2$?

(1) $\{-10, -14, -18\}$ (3) $\{10, 14, 16, 20\}$
(2) $\{12, 16, 20\}$ (4) $\{4, 14, 24, 34\}$

33. Which equation best describes the graph of $y = |x|$ reflected across the x-axis, shifted 9 units up, and shifted 2 units to the left?

(1) $y = |-x + 2| + 9$ (3) $y = |x - 2| + 9$
(2) $y = -|x + 2| + 9$ (4) $y = -|x - 2| - 9$

34. **a.** Plot points $A(5, -2)$, $B(3, 3)$, $C(-3, 3)$, and $D(-5, -2)$.

b. Draw polygon $ABCD$.

c. What kind of polygon is $ABCD$?

d. Find DA and BC.

e. Find the length of the altitude from C to $\overline{DA}$.

f. Find the area of $ABCD$.

In 35–38, graph each equation.

35. $y = |x - 2|$ **36.** $y = |x| - 2$ **37.** $|x| + 2|y| = 6$ **38.** $y = 2.5^x$

39. Tiny Tot Day Care Center has changed its rates. It now charges $350 a week for children who stay at the center between 8:00 A.M. and 5:00 P.M. The center charges an additional $2.00 for each day that the child is not picked up by 5:00 P.M.

a. Write an equation for the cost of day care for the week, y, in terms of x, the number of days that the child stays beyond 5:00 P.M.

b. Define the slope of the equation found in part **a** in terms of the information provided.

c. What is the charge for 1 week for a child who was picked up at the following times: Monday at 5:00, Tuesday at 5:10, Wednesday at 6:00, Thursday at 5:00, and Friday at 5:25?

40. Each time Raphael put gasoline into his car, he recorded the number of gallons of gas he needed to fill the tank and the number of miles driven since the last fill-up. The chart below shows his record for one month.

Gallons of Gasoline	12	8.5	10	4.5	13
Number of Miles	370	260	310	140	420

a. Plot points on a graph to represent the information in the chart. Let the horizontal axis represent the number of gallons of gasoline and the vertical axis the number of miles driven.

b. Find the average number of gallons of gasoline per fill-up.

c. Find the average number of miles driven between fill-ups.

d. Locate a point on the graph whose x-coordinate represents the average number of gallons of gasoline per fill-up and whose y-coordinate is the average number of miles driven between fill-ups.

e. It is reasonable that $(0, 0)$ is a point on the graph? Draw, through the point that you plotted in **d** and the point $(0, 0)$, a line that could represent the information in the chart.

f. Raphael drove 200 miles since his last fill-up. How many gallons of gasoline should he expect to need to fill the tank based on the line drawn in **e**?

41. Mandy and Jim are standing 20 feet apart. Each second, they decrease the distance between each other by one-half.

a. Write an equation to show their distance apart, D, at s seconds.

b. How far apart will they be after 5 seconds?

c. According to your equation, will Mandy and Jim ever meet?

42. According to the curator of zoology at the Rochester Museum and Science Center, if you count the number of chirps of a tree cricket in 15 seconds and add 40, you will have a close approximation of the actual air temperature in degrees Fahrenheit. To test this statement, Alexa recorded, on several summer evenings, the temperature in degrees Fahrenheit and the number of chirps of a tree cricket in 1 minute. She obtained the following results:

Chirps/minute	150	170	140	125	108	118	145	68	95	110
Temperature	80	85	76	71	65	70	76	56	63	67

a. Write an equation that states the relationship between the number of chirps per minute of the tree cricket, c, to the temperature in degrees Fahrenheit, F. (To change chirps per minute to chirps in 15 seconds, divide c by 4.)

b. Draw a graph of the data given in the table. Record the number of chirps per minute on the horizontal axis, and the temperature on the vertical axis.

c. Draw the graph of the equation that you wrote in part **a** on the same set of axes that you used for part **b**.

d. Can the data that Alexa recorded be represented by the equation that you wrote for part **a**? Explain your answer.

43. In order to control the deer population in a local park, an environmental group plans to reduce the number of deer by 5% each year. If the deer population is now estimated to be 4,000, how many deer will there be after 8 years?

Exploration

STEP 1. Draw the graph of $y = x - 2$.

STEP 2. Let A be the point at which the graph intersects the x-axis, B be any point on the line that has an x-coordinate greater than A, and C be the point at which a vertical line from B intersects the x-axis.

STEP 3. Compare the slope of $y = x - 2$ with tan $\angle BAC$.

STEP 4. Use a calculator to find m$\angle BAC$.

STEP 5. Repeat Steps 2 through 4 for three other lines that have a positive slope.

STEP 6. Draw the graph of $y = -x + 2$ and repeat Step 2. Let the measure of an acute angle between the x-axis and a line that slants upward be positive and the measure of an acute angle between the x-axis and a line that slants downward be negative.

STEP 7. Repeat Steps 3 and 4 for $y = -x + 2$.

STEP 8. Repeat Steps 2 through 4 for three other lines that have a negative slope.

STEP 9. What conclusion can you draw?

CUMULATIVE REVIEW CHAPTERS 1–9

Part I

Answer all questions in this part. Each correct answer will receive 2 credits. No partial credit will be allowed.

1. The product of $-5a^3$ and $-3a^4$ is

(1) $15a^7$ (2) $15a^{12}$ (3) $-15a^7$ (4) $-15a^{12}$

2. A basketball team won b games and lost 4. The ratio of games won to games played is

(1) $\frac{b}{4}$ (2) $\frac{b}{b+4}$ (3) $\frac{4}{b}$ (4) $\frac{4}{b+4}$

3. In decimal notation, 8.72×10^{-2} is

(1) 87,200 (2) 872 (3) 0.0872 (4) 0.00872

4. Which equation is not an example of direct variation?

(1) $y = 2x$ (2) $\frac{y}{x} = 2$ (3) $y = \frac{2}{x}$ (4) $y = \frac{x}{2}$

5. The graph of $y - 2x = 4$ is parallel to the graph of

(1) $y = 2x + 5$ (2) $2x + y = 7$ (3) $y = -2x + 3$ (4) $2x + y = 0$

6. The area of a triangle is 48.6 square centimeters. The length of the base of the triangle is 3.00 centimeters. The length of the altitude of the triangle is

(1) 32.4 centimeters
(2) 16.2 centimeters
(3) 8.10 centimeters
(4) 3.24 centimeters

7. The measure of the larger acute angle of a right triangle is 15 more than twice the measure of the smaller. What is the measure of the larger acute angle?

(1) 25 (2) 37.5 (3) 55 (4) 65

8. Solve for x: $1 - 2(x - 4) = x$.

(1) $-\frac{1}{2}$ (2) 2 (3) 3 (4) 4.5

9. The measure of one leg of a right triangle is 9 and the measure of the hypotenuse is 41. The measure of the other leg is

(1) 32 (2) 40 (3) 41.98 (4) $\sqrt{1,762}$

10. Which of the following geometric figures does not always have a pair of congruent angles?

(1) a parallelogram
(2) an isosceles triangle
(3) a rhombus
(4) a trapezoid

Part II

Answer all questions in this part. Each correct answer will receive 2 credits. Clearly indicate the necessary steps, including appropriate formula substitutions, diagrams, graphs, charts, etc. For all questions in this part, a correct numerical answer with no work shown will receive only 1 credit.

11. Cory has planted a rectangular garden. The ratio of the length to the width of the garden is 5 : 7. Cory bought 100 feet of fencing. After enclosing the garden with the fence, he still had 4 feet of fence left. What are the dimensions of his garden?

12. A square, $ABCD$, has a vertex at $A(4, 2)$. Side $\overline{AB}$ is parallel to the x-axis and $AB = 7$. What could be the coordinates of the other three vertices? Explain how you know that for the coordinates you selected $ABCD$ is a square.

Part III

Answer all questions in this part. Each correct answer will receive 3 credits. Clearly indicate the necessary steps, including appropriate formula substitutions, diagrams, graphs, charts, etc. For all questions in this part, a correct numerical answer with no work shown will receive only 1 credit.

13. A straight line with a slope of -2 contains the point $(2, 4)$. Find the y-coordinate of a point on this line whose x-coordinate is 5.

14. Mrs. Gantrish paid $42 for 8 boxes of file folders. She bought some on sale for $4 a box and the rest at a later time for $6 a box. How many boxes of file folders did she buy on sale?

Part IV

Answer all questions in this part. Each correct answer will receive 4 credits. Clearly indicate the necessary steps, including appropriate formula substitutions, diagrams, graphs, charts, etc. For all questions in this part, a correct numerical answer with no work shown will receive only 1 credit.

15. Concentric circles (circles that have the same center) are used to divide the circular face of a dartboard into 4 regions of equal area. If the radius of the board is 12.0 inches, what is the radius of each of the concentric circles? Express your answers to the *nearest tenth of an inch.*

16. A ramp is to be built from the ground to a doorway that is 4.5 feet above the ground. What will be the length of the ramp if it makes an angle of 22° with the ground? Express your answer to the *nearest foot.*

CHAPTER

10

WRITING AND SOLVING SYSTEMS OF LINEAR FUNCTIONS

Architects often add an outdoor stairway to a building as a design feature or as an approach to an entrance above ground level. But stairways are an obstacle to persons with disabilities, and most buildings are now approached by means of ramps in addition to or in place of stairways.

In designing a ramp, the architect must keep the slant or slope gradual enough to easily accommodate a wheelchair. If the slant is too gradual, however, the ramp may become inconveniently long or may require turns to fit it into the available space. The architect will also want to include design features that harmonize with the rest of the building and its surroundings.

Solving a problem such as the design of a ramp often involves writing and determining the solution of several equations or inequalities at the same time.

CHAPTER TABLE OF CONTENTS

10-1 WRITING AN EQUATION GIVEN SLOPE AND ONE POINT

You have learned to draw a line using the slope and the coordinates of one point on the line. With this same information, we can also write an equation of a line.

EXAMPLE 1

Write an equation of the line that has a slope of 4 and that passes through point (3, 5).

Solution **METHOD 1.** Use the definition of slope.

How to Proceed

(1) Write the definition of slope: $\text{slope} = \frac{y_2 - y_1}{x_2 - x_1}$

(2) Let $(x_1, y_1) = (3, 5)$ and $(x_2, y_2) = (x, y)$. Substitute these values in the definition of slope: $4 = \frac{y - 5}{x - 3}$

(3) Solve the equation for y in terms of x:

$$y - 5 = 4(x - 3)$$
$$y - 5 = 4x - 12$$
$$y = 4x - 7$$

METHOD 2. Use the slope-intercept form of an equation.

How to Proceed

(1) In the equation of a line, $y = mx + b$, replace m by the given slope, 4:

$$y = mx + b$$
$$y = 4x + b$$

(2) Since the given point, (3, 5), is on the line, its coordinates satisfy the equation $y = 4x + b$. Replace x by 3 and y by 5:

$$5 = 4(3) + b$$

(3) Solve the resulting equation to find the value of b, the y-intercept:

$$5 = 12 + b$$
$$-7 = b$$

(4) In $y = 4x + b$, replace b by -7:

$$y = 4x - 7$$

Answer $y = 4x - 7$

Note: In standard form $(Ax + By = C)$, the equation of the line is $4x - y = 7$.

EXERCISES

Writing About Mathematics

1. Micha says that there is another set of information that can be used to find the equation of a line: the y-intercept and the slope of the line. Jen says that that is the same information as the coordinates of one point and the slope of the line. Do you agree with Jen? Explain why or why not.

2. In Method 1 used to find the equation of a line with a slope of 4 and that passes through point (3, 5):

   a. How can the slope, 4, be written as a ratio?

   b. What are the principles used in each step of the solution?

Developing Skills

In 3–6, in each case, write an equation of the line that has the given slope, m, and that passes through the given point.

3. $m = 2; (1, 4)$ 4. $m = 2; (-3, 4)$ 5. $m = -3; (-2, -1)$ 6. $m = -\frac{5}{3}; (-3, 0)$

7. Write an equation of the line that has a slope of $\frac{1}{2}$ and a y-intercept of 2.

8. Write an equation of the line that has a slope of $\frac{-3}{4}$ and a y-intercept of 0.

9. Write an equation of the line, in the form $y = mx + b$, that is:

   a. parallel to the line $y = 2x - 4$, and has a y-intercept of 1.

   b. perpendicular to the line $y = 2x - 4$, and has a y-intercept of 1.

   c. parallel to the line $y - 3x = 6$, and has a y-intercept of -2.

   d. perpendicular to the line $y - 3x = 6$, and has a y-intercept of -2.

   e. parallel to the line $2x + 3y = 12$, and passes through the origin.

   f. perpendicular to the line $2x + 3y = 12$, and passes through the origin.

10. Write an equation of the line, in the form $Ax + By = C$, that is:

   a. parallel to the line $y = 4x + 1$, and passes through point (2, 3).

   b. perpendicular to the line $y = 4x + 1$, and passes through point (2, 3).

   c. parallel to the line $2y - 6x = 9$, and passes through point $(-2, 1)$.

   d. perpendicular to the line $2y - 6x = 9$, and passes through point $(-2, 1)$.

   e. parallel to the line $y = 4x + 3$, and has the same y-intercept as the line $y = 5x - 3$.

   f. perpendicular to the line $y = 4x + 3$, and has the same y-intercept as the line $y = 5x - 3$.

11. For **a–c**, write an equation, in the form $y = mx + b$, that describes each graph.

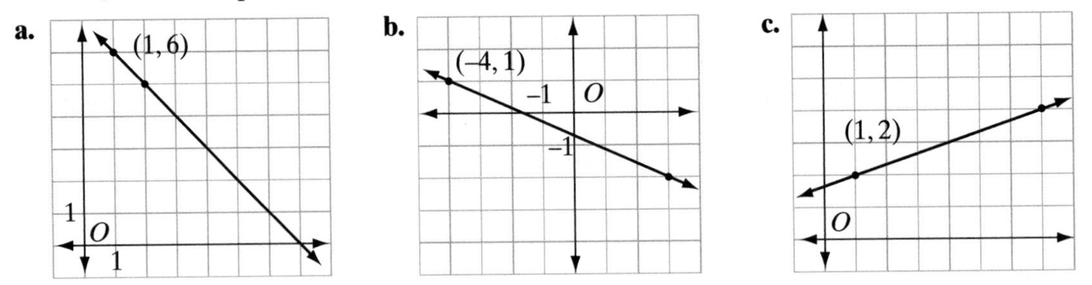

Applying Skills

12. Follow the directions below to draw a map in the coordinate plane.

a. The map begins at the point $A(0, 3)$ with a straight line segment that has a slope of 1. Write an equation for this segment.

b. Point B has an x-coordinate of 4 and is a point on the line whose equation you wrote in **a**. Find the coordinates of B and draw $\overline{AB}$.

c. Next, write an equation for a segment from point B that has a slope of $-\frac{1}{2}$.

d. Point C has an x-coordinate of 6 and is a point on the line whose equation you wrote in **c**. Find the coordinates of C and draw $\overline{BC}$.

e. Finally, write an equation for a line through C that has a slope of $\frac{1}{2}$.

f. Point D lies on the line whose equation you wrote in **e** and on the y-axis. Does point D coincide with point A?

13. If distance is represented by y and time by x, then the rate at which a car travels along a straight road can be represented as slope. Tom leaves his home and drives for 12 miles to the thruway entrance. On the thruway, he travels 65 miles per hour.

a. Write an equation for Tom's distance from his home in terms of the number of hours that he traveled *on the thruway*. (*Hint:* Tom enters the thruway at time 0 when he is at a distance of 12 miles from his home.)

b. How far from home is Tom after 3 hours?

c. How many hours has Tom driven on the thruway when he is 285 miles from home?

10-2 WRITING AN EQUATION GIVEN TWO POINTS

You have learned to draw a line using two points on the line. With this same information, we can also write an equation of a line.

EXAMPLE 1

Write an equation of the line that passes through points (2, 5) and (4, 11).

Solution **METHOD 1.** Use the definition of slope and the coordinates of the two points to find the slope of the line. Then find the equation of the line using the slope and the coordinates of one point.

How to Proceed

(1) Find the slope of the line that passes through the two given points, (2, 5) and (4, 11):

Let $P_1 = (2, 5)$. $[x_1 = 2 \text{ and } y_1 = 5]$
Let $P_2 = (4, 11)$. $[x_2 = 4 \text{ and } y_2 = 11]$

$$m = \frac{y_2 - y_1}{x_2 - x_1} = \frac{11 - 5}{4 - 2} = \frac{6}{2} = 3$$

(2) In $y = mx + b$, replace m by the slope, 3:

$y = mx + b$
$y = 3x + b$

(3) Select one point that is on the line, for example, (2, 5). Its coordinates must satisfy the equation $y = 3x + b$. Replace x by 2 and y by 5:

$5 = 3(2) + b$

(4) Solve the resulting equation to find the value of b, the y-intercept:

$5 = 6 + b$
$-1 = b$

(5) In $y = 3x + b$, replace b by -1:

$y = 3x - 1$

METHOD 2. Let $A(2, 5)$ and $B(4, 11)$ be the two given points on the line and $P(x, y)$ be any point on the line. Use the fact that the slope of $\overline{PA}$ equals the slope of $\overline{AB}$ to write an equation.

How to Proceed

(1) Set slope of $\overline{PA}$ equal to slope of $\overline{AB}$:

$$\frac{y - 5}{x - 2} = \frac{5 - 11}{2 - 4}$$

(2) Solve the resulting equation for y:

$$\frac{y - 5}{x - 2} = \frac{-6}{-2}$$
$$\frac{y - 5}{x - 2} = 3$$
$$y - 5 = 3(x - 2)$$
$$y - 5 = 3x - 6$$
$$y = 3x - 1$$

Check Do the coordinates of the second point, (4, 11), satisfy the equation $y = 3x - 1$?

$$11 \stackrel{?}{=} 3(4) - 1$$
$$11 = 11 \checkmark$$

Answer $y = 3x - 1$

EXERCISES

Writing About Mathematics

1. In step 3 of Method 1, the coordinates (2, 5) were substituted into the equation. Could the coordinates (4, 11), the coordinates of the other point on the line, be substituted instead? Explain your answer and show that you are correct.

2. Name the principle used in each step of the solution of the equation in Method 2.

Developing Skills

In 3–6, in each case write an equation of the line, in the form $y = mx + b$, that passes through the given points.

3. $(0, 5), (-2, 0)$ **4.** $(0, -3), (1, -1)$ **5.** $(1, 4), (3, 8)$ **6.** $(3, 1), (9, 7)$

In 7–10, in each case write an equation of the line, in the form $Ax + By = C$, that passes through the given points.

7. $(1, 2), (10, 14)$ **8.** $(0, -1), (6, 8)$ **9.** $(-2, -5), (-1, -2)$ **10.** $(0, 0), (-3, 5)$

11. A triangle is determined by the three points $A(3, 5)$, $B(6, 4)$, and $C(1, -1)$. Write the equation of each line in the form $y = mx + b$:

a. $\overleftrightarrow{AB}$ **b.** $\overleftrightarrow{BC}$ **c.** $\overleftrightarrow{CA}$

d. Is triangle ABC a right triangle? Explain your answer.

12. A quadrilateral is determined by the four points $W(1, 1)$, $X(-4, 6)$, $Y\left(\frac{1}{2}, 10\frac{1}{2}\right)$, $Z\left(5\frac{3}{5}, 5\frac{3}{5}\right)$. Is the quadrilateral a trapezoid? Explain your answer.

Applying Skills

13. Latonya uploads her digital photos to an internet service that archives them onto CDs for a fee per CD plus a fixed amount for postage and handling; that is, the amount for postage and handling is the same no matter how many archive CDs she purchases. Last month Latonya paid $7.00 for two archive CDs, and this month she paid $13.00 for five archive CDs.

a. Write two ordered pairs such that the x-coordinate is the number of archive CDs purchased and the y-coordinate is the total cost of archiving the photos.

b. Write an equation for the total cost, y, for x archive CDs.

c. What is the domain of the equation found in **b**? Write your answer in set-builder notation.

d. What is the range of the equation found in **b**? Write your answer in set-builder notation.

14. Every repair bill at Chickie's Service Spot includes a fixed charge for an estimate of the repairs plus an hourly fee for labor. Jack paid \$123 for a TV repair that required 3 hours of labor. Nina paid \$65 for a DVD player repair that required 1 hour of labor.

a. Write two ordered pairs (x, y), where x represents the number of hours of labor and y is the total cost for repairs.

b. Write an equation in the form $y = mx + b$ that expresses the value of y, the total cost of repairs, in terms of x, the number of hours of labor.

c. What is the fixed charge for an estimate at Chickie's Service Spot?

d. What is the hourly fee for labor at Chickie's?

15. A copying service charges a uniform rate for the first one hundred copies or less and a fee for each additional copy. Nancy Taylor paid \$7.00 to make 200 copies and Rosie Barbi paid \$9.20 to make 310 copies.

a. Write two ordered pairs (x, y), where x represents the number of copies over one hundred and y represents the cost of the copies.

b. Write an equation in the form $y = mx + b$ that expresses the value of y, the total cost of the copies, in terms of x, the number of copies over one hundred.

c. What is the cost of the first one hundred copies?

d. What is the cost of each additional copy?

10-3 WRITING AN EQUATION GIVEN THE INTERCEPTS

The x-intercept and the y-intercept are two points on the graph of a line. If we know these two points, we can graph the line and we can write the equation of the line. For example, to write the equation of the line whose x-intercept is 5 and whose y-intercept is -3, we can use two points.

- If the x-intercept is 5, one point is $(5, 0)$.
- If the y-intercept is -3, one point is $(0, -3)$.

Let (x, y) be any other point on the line. Set the slope of the line from (x, y) to $(5, 0)$ equal to the slope of the line from $(5, 0)$ to $(0, -3)$.

$$\frac{y - 0}{x - 5} = \frac{0 - (-3)}{5 - 0}$$

$$\frac{y}{x - 5} = \frac{3}{5}$$

Solve the resulting equation for y.

$$5y = 3(x - 5)$$

$$5y = 3x - 15$$

$$y = \tfrac{3}{5}x - 3$$

The slope-intercept form of the equation shows that the y-intercept is -3. Recall from Section 9-6 that we can write this equation in *intercept form* to show both the x-intercept and the y-intercept.

Start with the equation:	$5y = 3x - 15$
Add $-3x$ to each side:	$-3x + 5y = -15$
Divide each term by the constant term:	$\frac{-3x}{-15} + \frac{5y}{-15} = \frac{-15}{-15}$
	$\frac{x}{5} + \frac{y}{-3} = 1$

When the equation is in this form, x is divided by the x-intercept and y is divided by the y-intercept. In other words, when the equation of a line is written in the form $\frac{x}{a} + \frac{y}{b} = 1$, the x-intercept is a and the y-intercept is b.

EXAMPLE 1

Write an equation of the line whose x-intercept is -1 and whose y-intercept is $\frac{2}{3}$. Write the answer in standard form ($Ax + By = C$).

Solution *How to Proceed*

(1) In $\frac{x}{a} + \frac{y}{b} = 1$, replace a by -1 and b by $\frac{2}{3}$:

$$\frac{x}{a} + \frac{y}{b} = 1$$

$$\frac{x}{-1} + \frac{y}{\frac{2}{3}} = 1$$

(2) Write this equation in simpler form:

$$\frac{x}{-1} + \frac{y}{1} \cdot \frac{3}{2} = 1$$

$$\frac{x}{-1} + \frac{3y}{2} = 1$$

Note: In the original form of the equation, y is divided by $\frac{2}{3}$. To divide by $\frac{2}{3}$ is the same as to multiply by the reciprocal, $\frac{3}{2}$.

(3) Write the equation with integral coefficients by multiplying each term of the equation by the product of the denominators, -2:

$$-2\left(\frac{x}{-1}\right) - 2\left(\frac{3y}{2}\right) = -2(1)$$

$$2x - 3y = -2$$

Answer $2x - 3y = -2$

EXAMPLE 2

Find the x-intercept and y-intercept of the equation $x + 5y = 2$.

Solution Write the equation in the form $\frac{x}{a} + \frac{y}{b} = 1$. Divide each member of the equation by 2, the constant term:

$$x + 5y = 2$$
$$\frac{x}{2} + \frac{5y}{2} = \frac{2}{2}$$
$$\frac{x}{2} + \frac{5y}{2} = 1$$

The number that divides x, 2, is the x-intercept. The variable y is multiplied by $\frac{5}{2}$, which is equivalent to saying that it is divided by the reciprocal, $\frac{2}{5}$, so the y-intercept is $\frac{2}{5}$.

Answer The x-intercept is 2 and the y-intercept is $\frac{2}{5}$.

EXERCISES

Writing About Mathematics

1. a. Can the equation $y = 4$ be put into the form $\frac{x}{a} + \frac{y}{b} = 1$ to find the x-intercept and y-intercept of the line? Explain your answer.

b. Can the equation $x = 4$ be put into the form $\frac{x}{a} + \frac{y}{b} = 1$ to find the x-intercept and y-intercept of the line? Explain your answer.

2. Can the equation $x + y = 0$ be put into the form $\frac{x}{a} + \frac{y}{b} = 1$ to find the x-intercept and y-intercept of the line? Explain your answer.

Developing Skills

In 3–10, find the intercepts of each equation if they exist.

3. $x + y = 5$ **4.** $x - 3y = 6$ **5.** $5y = x - 10$ **6.** $2x + y - 2 = 0$

7. $3x + 4y = 8$ **8.** $7 - x = 2y$ **9.** $y = 3x - 1$ **10.** $y = -4$

11. Express the slope of $\frac{x}{a} + \frac{y}{b} = 1$ in terms of a and b.

12. For **a–c**, use the x- and y-intercepts to write an equation for each graph.

a.

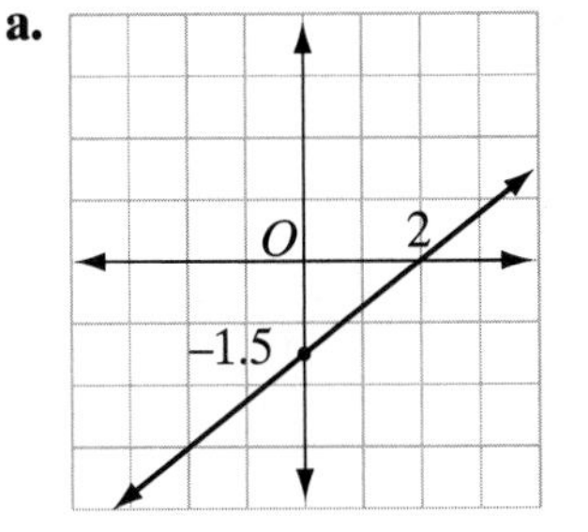

b.

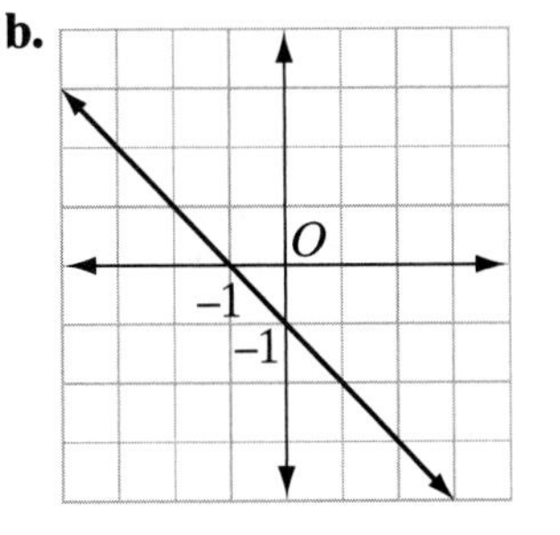

c.

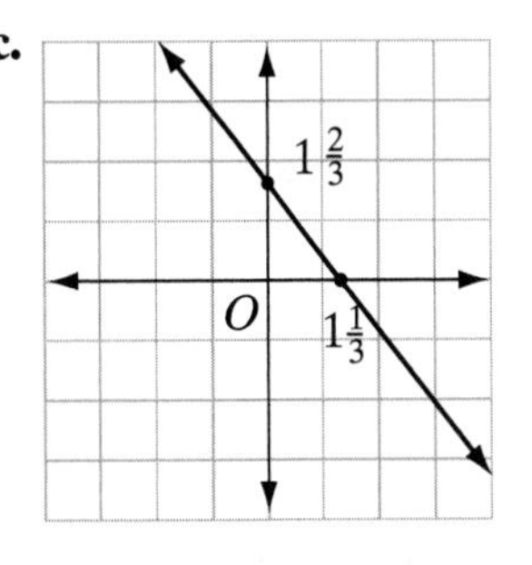

Applying Skills

13. Triangle ABC is drawn on the coordinate plane. Point A is at (4, 0), point B is at (0, 3), and point C is at the origin.

a. What is the equation of $\overleftrightarrow{AB}$ written in the form $\frac{x}{a} + \frac{y}{b} = 1$?

b. What is the length of $\overline{AB}$?

14. a. Isosceles right triangle ABC is drawn on the coordinate plane. Write the equation of each side of the triangle if $\angle C$ is the right angle at the origin and the length of each leg is 7.

b. Is more than one answer to **a** possible? Explain your answer.

10-4 USING A GRAPH TO SOLVE A SYSTEM OF LINEAR EQUATIONS

Consistent Equations

The perimeter of a rectangle is 10 feet. When we let x represent the width of the rectangle and y represent the length, the equation $2x + 2y = 10$ expresses the perimeter of the rectangle. This equation, which can be simplified to $x + y = 5$, has infinitely many solutions.

If we also know that the length of the rectangle is 1 foot more than the width, the dimensions of the rectangle can be represented by the equation $y = x + 1$. We want both of the equations, $x + y = 5$ and $y = x + 1$, to be true for the same pair of numbers. The two equations are called a **system of simultaneous equations** or a **linear system.**

> Let x = the width of the rectangle, and y = the length of the rectangle.
>
> Perimeter:
>
> $2x + 2y = 10$
>
> $x + y = 5$ (Simplified)
>
> Measures of the sides:
>
> $y = x + 1$
>
> System of simultaneous equations:
>
> $x + y = 5$
>
> $y = x + 1$

The solution of a system of simultaneous equations in two variables is an ordered pair of numbers that satisfies both equations.

The graphs of $x + y = 5$ and $y = x + 1$, drawn using the same set of axes in a coordinate plane, are shown at the right. The possible solutions of $x + y = 5$ are all ordered pairs that are coordinates of the points on the line $x + y = 5$. The possible solutions of $y = x + 1$ are all ordered pairs that are coordinates of the points on the line $y = x + 1$. The coordinates of the point of intersection, (2, 3), are a solution of both equations. The ordered pair (2, 3) is a solution of the system of simultaneous equations.

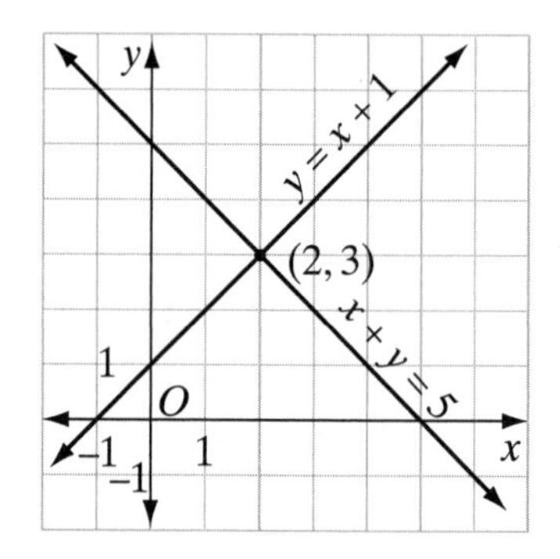

Check: Substitute (2, 3) in both equations:

$$x + y = 5 \qquad\qquad y = x + 1$$
$$2 + 3 \stackrel{?}{=} 5 \qquad\qquad 3 \stackrel{?}{=} 2 + 1$$
$$5 = 5 \checkmark \qquad\qquad 3 = 3 \checkmark$$

Since two straight lines can intersect in no more than one point, there is no other ordered pair that is a solution of this system. Therefore $x = 2$ and $y = 3$, or (2, 3), is the solution of the system of equations. The width of the rectangle is 2 feet, and the length of the rectangle is 3 feet.

When two lines are graphed in the same coordinate plane on the same set of axes, one and only one of the following three possibilities can occur. The pair of lines will:

1. intersect in one point and have one ordered number pair in common;

2. be parallel and have no ordered number pairs in common;

3. coincide, that is, be the same line with an infinite number of ordered number pairs in common.

If a system of linear equations such as $x + y = 5$ and $y = x + 1$ has at least one common solution, it is called a **system of consistent equations**. If a system has exactly one solution, it is a **system of independent equations**.

Inconsistent Equations

Sometimes, as shown at the right, when two linear equations are graphed in a coordinate plane using the same set of axes, the lines are parallel and fail to intersect, as in the case of $x + y = 2$ and $x + y = 4$. There is no common solution for the system of equations $x + y = 2$ and $x + y = 4$. It is obvious that there can be no ordered number pair (x, y) such that the sum of those numbers, $x + y$, is both 2 and 4. Since the solution set of the system has no members, it is the empty set.

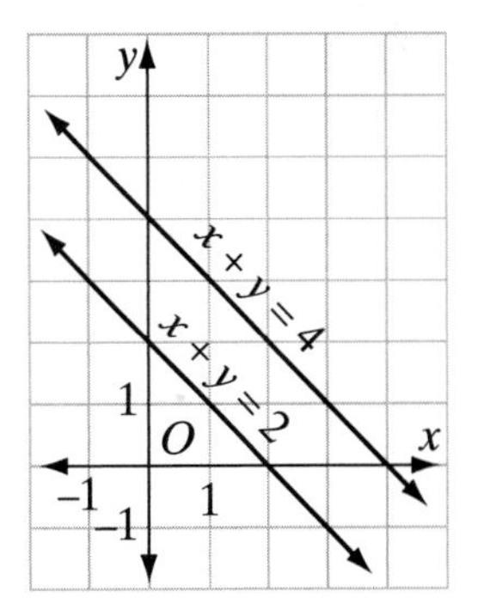

If a system of linear equations such as $x + y = 2$ and $x + y = 4$ has no common solution, it is called a **system of inconsistent equations**. The graphs of two inconsistent linear equations are lines that have equal slopes or lines that have no slopes. Such lines are parallel.

Dependent Equations

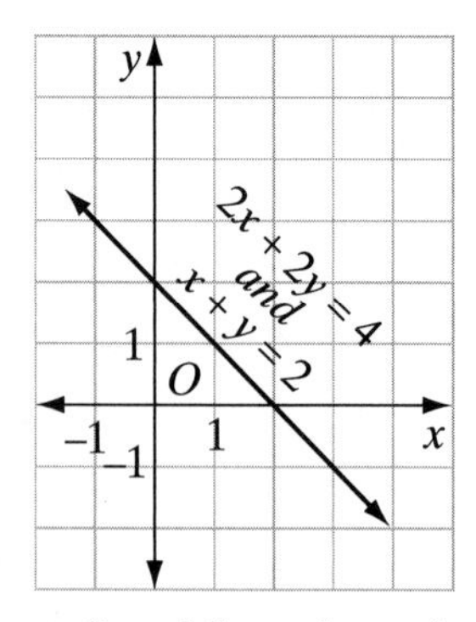

Sometimes, as shown at the right, when two linear equations are graphed in a coordinate plane using the same set of axes, the graphs turn out to be the same line; that is, they coincide. This happens in the case of the equations $x + y = 2$ and $2x + 2y = 4$. Every one of the infinite number of solutions of $x + y = 2$ is also a solution of $2x + 2y = 4$. Thus, we see that $2x + 2y = 4$ and $x + y = 2$ are equivalent equations with identical solutions. We note that, when both sides of the equation $2x + 2y = 4$ are divided by 2, the result is $x + y = 2$.

If a system of two linear equations, for example, $x + y = 2$ and $2x + 2y = 4$, is such that every solution of one of the equations is also a solution of the other, it is called a **system of dependent equations**. The graphs of two dependent linear equations are the same line. Note that a system of dependent equations is considered consistent because there is at least one solution.

Procedure

To solve a pair of linear equations graphically:

1. Graph one equation in a coordinate plane.
2. Graph the other equation using the same set of coordinate axes.
3. One of three relationships will apply:
   a. If the graphs intersect in one point, the common solution is the ordered pair of numbers that are the coordinates of the point of intersection of the two graphs. The equations are independent and consistent.
   b. If the graphs have no points in common, that is, the graphs are parallel, there is no solution. The equations are inconsistent.
   c. If the graphs have all points in common, that is, the graphs are the same line, every point on the line is a solution. The equations are consistent and dependent.
4. Check any solution by verifying that the ordered pair satisfies both equations.

EXAMPLE 1

Solve graphically and check:

$$2x + y = 8$$
$$y - x = 2$$

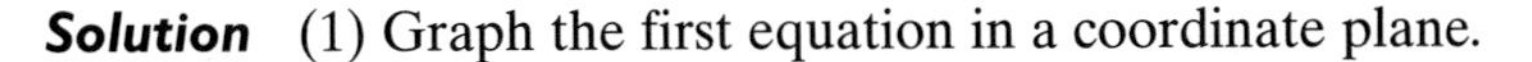

Solution (1) Graph the first equation in a coordinate plane.

Solve the first equation for y:

$$2x + y = 8$$
$$y = -2x + 8$$

The y-intercept is 8 and the slope is -2. Start at the point $(0, 8)$ and move down 2 units and to the right 1 unit to determine two other points. Or, choose three values of x and find the corresponding values of y to determine three points. Draw the line that is the graph of $2x + y = 8$:

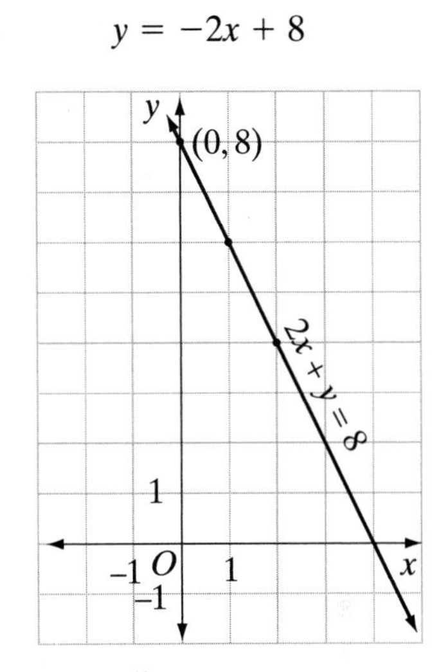

(2) Graph the second equation using the same set of coordinate axes.

Solve the second equation for y:

$$y - x = 2$$
$$y = x + 2$$

The y-intercept is 2 and the slope is 1. Start at the point $(0, 2)$ and move up 1 unit and to the right 1 unit to determine two other points. Or, choose three values of x and find the corresponding values of y to determine three points. Draw the line that is the graph of $y - x = 2$ on the same set of coordinate axes:

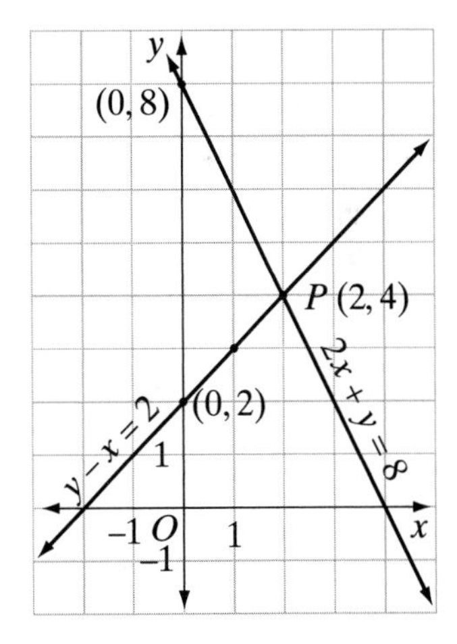

(3) The graphs intersect at one point, $P(2, 4)$.

(4) *Check:* Substitute $(2, 4)$ in each equation:

$2x + y = 8$	$y - x = 2$
$2(2) + 4 \stackrel{?}{=} 8$	$4 - 2 \stackrel{?}{=} 2$
$4 + 4 = 8$ ✔	$2 = 2$ ✔

Calculator Solution

(1) Solve the equations for y:

$2x + y = 8$	$y - x = 2$
$y = -2x + 8$	$y = x + 2$

(2) Enter the equations in the **Y=** menu.

ENTER: **Y=** **(-)** 2 **X,T,Θ,n** **+** 8 **ENTER**

X,T,Θ,n **+** 2

DISPLAY:

```
Plot1 Plot2 Plot3
\Y1=-2X+8
\Y2=X+2
```

(3) Use ZStandard to display the equations.

ENTER: **ZOOM** 6

DISPLAY:

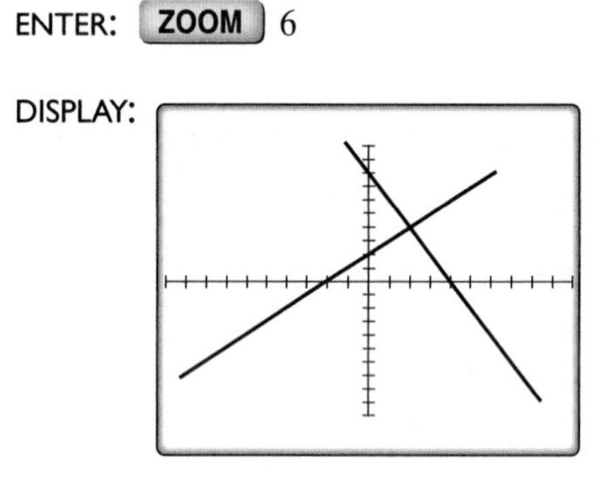

(4) Use **TRACE** to determine the coordinates of the point of intersection. Use the right and left arrow keys to move along the first equation to what appears to be the point of intersection and note the coordinates, (2, 4). Use the up arrow key to change to the second equation. Again note the coordinates of the point of intersection, (2, 4).

ENTER: **TRACE**

DISPLAY:

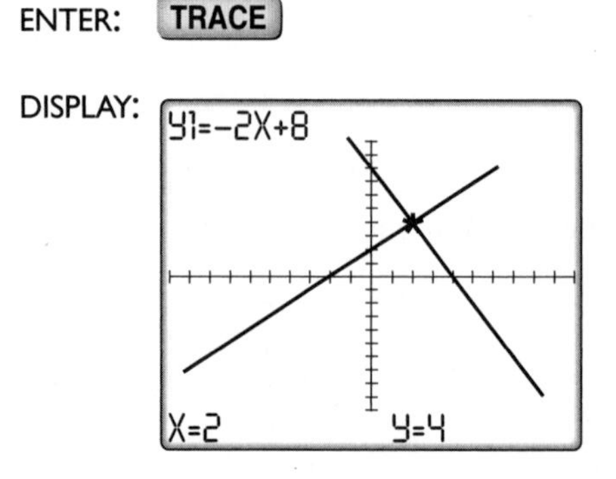

Answer (2, 4), or $x = 2$, $y = 4$, or the solution set $\{(2, 4)\}$

EXERCISES

Writing About Mathematics

1. Are there any ordered pairs that satisfy both the equations $2x + y = 7$ and $2x = 5 - y$? Explain your answer.

2. Are there any ordered pairs that satisfy the equation $y - x = 4$ but do *not* satisfy the equation $2y = 8 + 2x$? Explain your answer.

Developing Skills

In 3–22, solve each system of equations graphically, and check.

3. $y = 2x + 5$
$y = x + 4$

4. $y = -2x + 3$
$y = \frac{1}{2}x + 3$

5. $x + y = 1$
$x - y = 7$

6. $x + y = 5$
$x + 3y = 9$

7. $y = 2x + 1$
$x + 2y = 7$

8. $x + 2y = 12$
$y = 2x + 6$

9. $y = 3x$
$2x + y = 10$

10. $x + 3y = 9$
$x = 3$

11. $y - x = -2$
$x - 2y = 4$

12. $3x + y = 6$
$y = 3$

13. $y = 2x + 4$
$x = y - 5$

14. $2x - y = 1$
$x = y + 1$

15. $y = \frac{1}{3}x - 3$
$2x - y = 8$

16. $3x + y = 13$
$x + 6y = -7$

17. $2x = y + 9$
$6x + 3y = 15$

18. $5x - 3y = 9$
$5y = 13 - x$

19. $x = 0$
$y = 0$

20. $x + y + 2 = 0$
$x = y - 8$

21. $y + 2x + 6 = 0$
$y = x$

22. $7x - 4y + 7 = 0$
$3x - 5y + 3 = 0$

In 23–28, in each case: **a.** Graph both equations. **b.** State whether the system is consistent and independent, consistent and dependent, or inconsistent.

23. $x + y = 1$
$x + y = 3$

24. $x + y = 5$
$2x + 2y = 10$

25. $y = 2x + 1$
$y = 3x + 3$

26. $2x - y = 1$
$2y = 4x - 2$

27. $y - 3x = 2$
$y = 3x - 2$

28. $x + 4y = 6$
$x = 2$

Applying Skills

In 29–32, in each case: **a.** Write a system of two first-degree equations involving the variables x and y that represent the conditions stated in the problem. **b.** Solve the system graphically.

29. The sum of two numbers is 8. The difference of these numbers is 2. Find the numbers.

30. The sum of two numbers is 5. The larger number is 7 more than the smaller number. Find the numbers.

31. The perimeter of a rectangle is 12 meters. Its length is twice its width. Find the dimensions of the rectangle.

32. The perimeter of a rectangle is 14 centimeters. Its length is 3 centimeters more than its width. Find the length and the width.

33. a. The U-Drive-It car rental agency rents cars for \$50 a day with unlimited free mileage. Write an equation to show the cost of renting a car from U-Drive-It for one day, y, if the car is driven for x miles.

b. The Safe Travel car rental agency rents cars for \$30 a day plus \$0.20 a mile. Write an equation to show the cost of renting a car from Safe Travel for one day, y, if the car is driven for x miles.

c. Draw, on the same set of axes, the graphs of the equations written in parts **a** and **b**.

d. If Greg will drive the car he rents for 200 miles, which agency offers the less expensive car?

e. If Sarah will drive the car she rents for 50 miles, which agency offers the less expensive car?

f. If Philip finds that the price for both agencies will be the same, how far is he planning to drive the car?

10-5 USING ADDITION TO SOLVE A SYSTEM OF LINEAR EQUATIONS

In the preceding section, graphs were used to find solutions of systems of simultaneous equations. Since most of the solutions were integers, the values of x and y were easily read from the graphs. However, it is not always possible to read values accurately from a graph.

For example, the graphs of the system of equations $2x - y = 2$ and $x + y = 2$ are shown at the right. The solution of this system of equations is not a pair of integers. We could approximate the solution and then determine whether our approximation was correct by checking. However, there are other, more direct methods of solution.

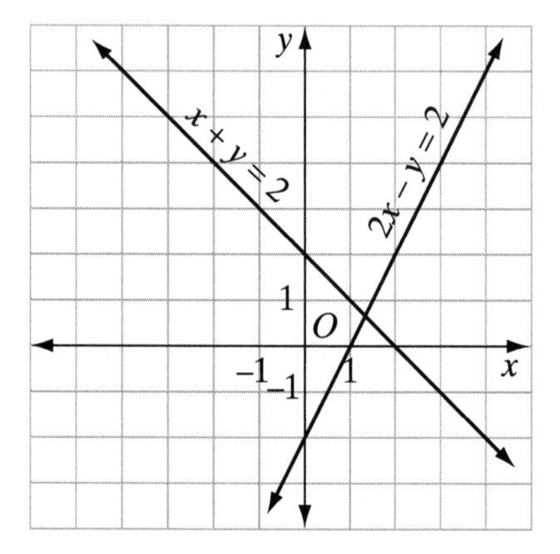

Algebraic methods can be used to solve a system of linear equations in two variables. Solutions by these methods often take less time and lead to more accurate results than the graphic method used in Section 4 of this chapter.

To solve a system of linear equations such as $2x - y = 2$ and $x + y = 2$, we make use of the properties of equality to obtain an equation in one variable. When the coefficient of one of the variables in the first equation is the additive inverse of the coefficient of the same variable in the second, that variable can be eliminated by adding corresponding members of the two equations.

The system $2x - y = 2$ and $x + y = 2$ can be solved by the **addition method** as follows:

Solve for x:

(1) Since the coefficients of y in the two equations are additive inverses, add the equations to obtain an equation that has only one variable:

$$\begin{array}{r} 2x - y = 2 \\ \underline{x + y = 2} \\ 3x \quad\;\; = 4 \end{array}$$

(2) Solve the resulting equation for x:

$$x = \tfrac{4}{3}$$

Solve for y:

(3) Replace x by its value in either of the given equations:

$$x + y = 2$$
$$\tfrac{4}{3} + y = 2$$

(4) Solve the resulting equation for y:

$$-\tfrac{4}{3} + \tfrac{4}{3} + y = 2 - \tfrac{4}{3}$$
$$y = \tfrac{2}{3}$$

Check: Substitute $\frac{4}{3}$ for x and $\frac{2}{3}$ for y in each of the given equations, and show that these values make the given equations true:

$$\begin{array}{rcl} 2x - y &=& 2 \\ 2\left(\tfrac{4}{3}\right) - \tfrac{2}{3} &\stackrel{?}{=}& 2 \\ \tfrac{8}{3} - \tfrac{2}{3} &\stackrel{?}{=}& 2 \\ \tfrac{6}{3} &\stackrel{?}{=}& 2 \\ 2 &=& 2 \checkmark \end{array} \qquad \begin{array}{rcl} x + y &=& 2 \\ \tfrac{4}{3} + \tfrac{2}{3} &\stackrel{?}{=}& 2 \\ \tfrac{6}{3} &\stackrel{?}{=}& 2 \\ 2 &=& 2 \checkmark \end{array}$$

We were able to add the equations in two variables to obtain an equation in one variable because the coefficients of one of the variables were additive inverses. If the coefficients of neither variable are additive inverses, we can multiply one or both equations by a convenient constant or constants. For clarity, it is often helpful to label the equations in a system. The procedure is shown in the following examples.

KEEP IN MIND When solving a system of linear equations for a variable using the addition method, if the result is:

- the equation $0 = 0$, then the system is dependent;
- a false statement (such as $0 = 3$), then the system is inconsistent.

EXAMPLE 1

Solve the system of equations and check:

$$x + 3y = 13 \quad [A]$$
$$x + y = 5 \quad [B]$$

Solution

How to Proceed

(1) Since the coefficients of the variable x are the same in both equations, write an equation equivalent to equation [B] by multiplying both sides of equation [B] by -1. Now, since the coefficients of x are additive inverses, add the two equations so that the resulting equation involves one variable, y:

$$\begin{array}{rl} x + 3y = 13 & [A] \\ -x - y = -5 & -1[B] \\ \hline 2y = 8 & \end{array}$$

(2) Solve the resulting equation for the variable y:

$$y = 4$$

(3) Replace y by its value in either of the given equations:

$$x + y = 5 \quad [B]$$
$$x + 4 = 5$$

(4) Solve the resulting equation for the remaining variable, x:

$$x = 1$$

Check Substitute 1 for x and 4 for y in each of the given equations to verify the solution:

$$x + 3y = 13 \qquad\qquad x + y = 5$$
$$1 + 3(4) \stackrel{?}{=} 13 \qquad\qquad 1 + 4 \stackrel{?}{=} 5$$
$$1 + 12 \stackrel{?}{=} 13 \qquad\qquad 5 = 5 \checkmark$$
$$13 = 13 \checkmark$$

Answer $x = 1, y = 4$ or $(1, 4)$

Note: If equation [A] in Example 1 was $x + y = 13$, then the system of equations would be inconsistent.

$$\begin{array}{rl} x + y = 13 & [A] \\ -x - y = -5 & -[B] \\ \hline 0 = 8 \ ✗ & \end{array}$$

EXAMPLE 2

Solve the system of equations and check:

$$5a + b = 13 \quad [A]$$
$$4a - 3b = 18 \quad [B]$$

Solution *How to Proceed*

(1) Multiply both sides of equation [A] by 3 to obtain an equivalent equation, 3[A]:

$$5a + b = 13 \quad [A]$$
$$15a + 3b = 39 \quad 3[A]$$

Note that the coefficient of b in 3[A] is the additive inverse of the coefficient of b in [B].

(2) Add the corresponding members of equations 3[A] and [B] to eliminate the variable b:

$$15a + 3b = 39 \quad 3[A]$$
$$\underline{4a - 3b = 18} \quad [B]$$
$$19a = 57$$

(3) Solve the resulting equation for a:

$$a = 3$$

(4) Replace a by its value in either of the given equations and solve for b:

$$5a + b = 13 \quad [A]$$
$$5(3) + b = 13$$
$$15 + b = 13$$
$$b = -2$$

Check Substitute 3 for a and -2 for b in the given equations:

$5a + b = 13$	$4a - 3b = 18$
$5(3) + (-2) \stackrel{?}{=} 13$	$4(3) - 3(-2) \stackrel{?}{=} 18$
$15 + (-2) \stackrel{?}{=} 13$	$12 + 6 \stackrel{?}{=} 18$
$13 = 13$ ✔	$18 = 18$ ✔

Answer $a = 3, b = -2$, or $(a, b) = (3, -2)$

Order is critical when expressing solutions. While $(a, b) = (3, -2)$ is the solution to the above system of equations, $(a, b) = (-2, 3)$ is *not*. Be sure that all variables correspond to their correct values in your answers.

EXAMPLE 3

Solve and check:

$$7x = 5 - 2y \quad [A]$$
$$3y = 16 - 2x \quad [B]$$

Solution

How to Proceed

(1) Transform each of the given equations into equivalent equations in which the terms containing the variables appear on one side and the constant appears on the other side:

$$7x = 5 - 2y \quad [A]$$
$$7x + 2y = 5$$
$$3y = 16 - 2x \quad [B]$$
$$2x + 3y = 16$$

(2) To eliminate y, find the least common multiple of the coefficients of y in equations [A] and [B]. That least common multiple is 6. We want to write one equation in which the coefficient of y is 6 and the other in which the coefficient of y is -6. Multiply both sides of equation [A] by 3, and multiply both sides of equation [B] by -2 so that the new coefficients of y will be additive inverses, 6 and -6:

$$3(7x + 2y = 5) \quad 3[A]$$
$$-2(2x + 3y = 16) \quad -2[B]$$

(3) Add the corresponding members of this last pair of equations to eliminate variable y:

$$21x + 6y = 15 \quad 3[A]$$
$$\underline{-4x - 6y = -32} \quad -2[B]$$
$$17x \quad = -17$$

(4) Solve the resulting equation for variable x:

$$x = -1$$

(5) Replace x by its value in any equation containing both variables:

$$3y = 16 - 2x \quad [B]$$
$$3y = 16 - 2(-1)$$
$$3y = 16 + 2$$

(6) Solve the resulting equation for the remaining variable, y:

$$3y = 18$$
$$y = 6$$

Check Substitute -1 for x and 6 for y in each of the original equations to verify the answer. The check is left to you.

Answer $x = -1, y = 6$ or $(-1, 6)$

EXERCISES

Writing About Mathematics

1. Raphael said that the solution to Example 3 could have been found by multiplying equation [A] by -2 and equation [B] by 7. Do you agree with Raphael? Explain why or why not.

2. Fernando said that the solution to Example 3 could have been found by multiplying equation [A] by -3 and equation [B] by 2. Do you agree with Fernando? Explain why or why not.

Developing Skills

In 3–26, solve each system of equations by using addition to eliminate one of the variables. Check your solution.

3. $x + y = 12$
$x - y = 4$

4. $a + b = 13$
$a - b = 5$

5. $3x + y = 16$
$2x + y = 11$

6. $c - 2d = 14$
$c + 3d = 9$

7. $a - 4b = -8$
$a - 2b = 0$

8. $8a + 5b = 9$
$2a - 5b = -4$

9. $-2m + 4n = 13$
$6m + 4n = 9$

10. $4x - y = 10$
$2x + 3y = 12$

11. $5x + 8y = 1$
$3x + 4y = -1$

12. $5x - 2y = 20$
$2x + 3y = 27$

13. $2x - y = 26$
$3x - 2y = 42$

14. $2x + 3y = 6$
$3x + 5y = 15$

15. $5r - 2s = 8$
$3r - 7s = -1$

16. $3x + 7y = -2$
$2x + 3y = -3$

17. $4x + 3y = -1$
$5x + 4y = 1$

18. $4a - 6b = 15$
$6a - 4b = 10$

19. $2x + y = 17$
$5x = 25 + y$

20. $5r + 3s = 30$
$2r = 12 - 3s$

21. $3x - 4y = 2$
$x = 2(7 - y)$

22. $3x + 5(y + 2) = 1$
$8y = -3x$

23. $\frac{1}{3}x + \frac{1}{4}y = 10$
$\frac{1}{3}x - \frac{1}{2}y = 4$

24. $\frac{1}{2}a + \frac{1}{3}b = 8$
$\frac{3}{2}a - \frac{4}{3}b = -4$

25. $c - 2d = 1$
$\frac{2}{3}c + 5d = 26$

26. $2a = 3b$
$\frac{2}{3}a - \frac{1}{2}b = 2$

Applying Skills

27. Pepe invested x dollars in a savings account and y dollars in a certificate of deposit. His total investment was $500. After 1 year he received 4% interest on the money in his savings account and 6% interest on the certificate of deposit. His total interest was $26. To find how much he invested at each rate, solve the following system of equations:

$$x + y = 500$$
$$0.04x + 0.06y = 26$$

28. Mrs. Briggs deposited a total of \$400 in two different banks. One bank paid 3% interest and the other paid 5%. In one year, the total interest was \$17. Let x be the amount invested at 3% and y be the amount invested at 5%. To find the amount invested in each bank, solve the following system of equations:

$$x + y = 400$$
$$0.03x + 0.05y = 17$$

29. Heather deposited a total of \$600 in two different banks. One bank paid 3% interest and the other 6%. The interest on the account that pays 3% was \$9 more than the interest on the account that pays 6%. Let x be the amount invested at 3% and y be the amount invested at 6%. To find the amount Heather invested in each account, solve the following system of equations:

$$x + y = 600$$
$$0.03x = 0.06y + 9$$

30. Greta is twice as old as Robin. In 3 years, Robin will be 4 years younger than Greta is now. Let g represent Greta's current age, and let r represent Robin's current age. To find their current ages, solve the following system of equations:

$$g = 2r$$
$$r + 3 = g - 4$$

31. At the grocery store, Keith buys 3 kiwis and 4 zucchinis for a total of \$4.95. A kiwi costs \$0.25 more than a zucchini. Let k represent the cost of a kiwi, and let z represent the cost of a zucchini. To find the cost of a kiwi and the cost of a zucchini, solve the following system of equations:

$$3k + 4z = 4.95$$
$$k = z + 0.25$$

32. A 4,000 gallon oil truck is loaded with gasoline and kerosene. The profit on one gallon of gasoline is \$0.10 and \$0.13 for a gallon of kerosene. Let g represent the number of gallons of gasoline loaded onto the truck and k the number of gallons of kerosene. To find the number of gallons of each fuel that were loaded into the truck when the profit is \$430, solve the following system of equations:

$$g + k = 4{,}000$$
$$0.10g + 0.13k = 430$$

10-6 USING SUBSTITUTION TO SOLVE A SYSTEM OF LINEAR EQUATIONS

Another algebraic method, called the **substitution method**, can be used to eliminate one of the variables when solving a system of equations. When we use this method, we apply the substitution principle to transform one of the equations of the system into an equivalent equation that involves only one variable.

Substitution Principle: In any statement of equality, a quantity may be substituted for its equal.

To use the substitution method, we must express one variable in terms of the other. Often one of the given equations already expresses one of the variables in terms of the other, as seen in Example 1.

EXAMPLE 1

Solve the system of equations and check:

$$4x + 3y = 27 \quad [A]$$
$$y = 2x - 1 \quad [B]$$

Solution *How to Proceed*

(1) Since in equation [B], both y and $2x - 1$ name the same number, eliminate y in equation [A] by replacing it with $2x - 1$:

$$4x + 3y = 27 \quad [A]$$
$$y = 2x - 1 \quad [B]$$
$$4x + 3(2x - 1) = 27 \quad [B \rightarrow A]$$

(2) Solve the resulting equation for x:

$$4x + 6x - 3 = 27$$
$$10x - 3 = 27$$
$$10x = 30$$
$$x = 3$$

(3) Replace x with its value in any equation involving both variables:

$$y = 2x - 1 \quad [B]$$
$$y = 2(3) - 1$$

(4) Solve the resulting equation for y:

$$y = 6 - 1$$
$$y = 5$$

Check Substitute 3 for x and 5 for y in each of the given equations to verify that the resulting sentences are true.

$4x + 3y = 27$	$y = 2x - 1$
$4(3) + 3(5) \stackrel{?}{=} 27$	$5 \stackrel{?}{=} 2(3) - 1$
$12 + 15 \stackrel{?}{=} 27$	$5 \stackrel{?}{=} 6 - 1$
$27 = 27$ ✔	$5 = 5$ ✔

Answer $x = 3, y = 5$ or $(3, 5)$

EXAMPLE 2

Solve the system of equations and check:

$$3x - 4y = 26 \quad [A]$$
$$x + 2y = 2 \quad [B]$$

Solution In neither equation is one of the variables expressed in terms of the other. We will use equation [B] in which the coefficient of x is 1 to solve for x in terms of y.

How to Proceed

(1) Transform one of the equations into an equivalent equation in which one of the variables is expressed in terms of the other. In equation [B], solve for x in terms of y:

$$x + 2y = 2$$
$$x = 2 - 2y \quad [B]$$

(2) Eliminate x from equation [A] by replacing it with its equal, $2 - 2y$, from step (1):

$$3x - 4y = 26 \quad [A]$$
$$3(2 - 2y) - 4y = 26 \quad [B{\rightarrow}A]$$

(3) Solve the resulting equation for y:

$$6 - 6y - 4y = 26$$
$$6 - 10y = 26$$
$$-10y = 20$$
$$y = -2$$

(4) Replace y by its value in any equation that has both variables:

$$x = 2 - 2y \quad [B]$$
$$x = 2 - 2(-2)$$

(5) Solve the resulting equation for x:

$$x = 2 + 4$$
$$x = 6$$

Check Substitute 6 for x and -2 for y in each of the given equations to verify that the resulting sentences are true. This check is left to you.

Answer $x = 6, y = -2$ or $(6, -2)$

EXERCISES

Writing About Mathematics

1. In Example 2, the system of equations could have been solved by first solving the equation $3x - 4y = 26$ for y. Explain why the method used in Example 2 was easier.

2. Try to solve the system of equations $x - 2y = 5$ and $y = \frac{1}{2}x + 1$ by substituting $\frac{1}{2}x + 1$ for y in the first equation. What conclusion can you draw?

Developing Skills

In 3–14, solve each system of equations by using substitution to eliminate one of the variables. Check.

3. $y = x$
$x + y = 14$

4. $y = 2x$
$x + y = 21$

5. $a = -2b$
$5a - 3b = 13$

6. $y = x + 1$
$x + y = 9$

7. $a = 3b + 1$
$5b - 2a = 1$

8. $a + b = 11$
$3a - 2b = 8$

9. $a - 2b = -2$
$2a - b = 5$

10. $7x - 3y = 23$
$x + 2y = 13$

11. $4d - 3h = 25$
$3d - 12h = 9$

12. $2x = 3y$
$4x - 3y = 12$

13. $4y = -3x$
$5x + 8y = 4$

14. $2x + 3y = 7$
$4x - 5y = 25$

In 15–26, solve each system of equations by using any convenient algebraic method. Check.

15. $s + r = 0$
$r - s = 6$

16. $3a - b = 13$
$2a + 3b = 16$

17. $y = x - 2$
$3x - y = 16$

18. $3x + 8y = 16$
$5x + 10y = 25$

19. $y = 3x$
$\frac{1}{3}x + \frac{1}{2}y = 11$

20. $a - \frac{2}{3}b = 4$
$\frac{3}{5}a + b = 15$

21. $3(y - 6) = 2x$
$3x + 5y = 11$

22. $x + y = 300$
$0.1x + 0.3y = 78$

23. $3d = 13 - 2c$
$\frac{3c + d}{2} = 8$

24. $3x = 4y$
$\frac{3x + 8}{5} = \frac{3y - 1}{2}$

25. $\frac{a}{3} = \frac{a + b}{6}$
$5a + 2b = 49$

26. $a + 3(b - 1) = 0$
$2(a - 1) + 2b = 16$

Applying Skills

27. The length of the base of an isosceles triangle is 6 centimeters less than the sum of the lengths of the two congruent sides. The perimeter of the triangle is 78 centimeters.

a. If x represents the length of each of the congruent sides and y represents the length of the base, express y in terms of x.

b. Write an equation that represents the perimeter in terms of x and y.

c. Solve the equations that you wrote in **a** and **b** to find the lengths of the sides of the triangle.

28. A package of batteries costs $1.16 more than a roll of film. Martina paid $11.48 for 3 rolls of film and a package of batteries.

a. Express the cost of a package of batteries, y, in terms of the cost of a roll of film, x.

b. Write an equation that expresses the amount that Martina paid for the film and batteries in terms of x and y.

c. Solve the equations that you wrote in **a** and **b** to find what Martina paid for a roll of film and of a package of batteries.

29. The cost of the hotdog was \$0.30 less than twice the cost of the cola. Jules paid \$3.90 for a hotdog and a cola.

a. Express the cost of a hotdog, y, in terms of the cost of a cola, x.

b. Express what Jules paid for a hotdog and a cola in terms of x and y.

c. Solve the equations that you wrote in **a** and **b** to find what Jules paid for a hotdog and for a cola.

30. Terri is 12 years older than Jessica. The sum of their ages is 54.

a. Express Terri's age, y, in terms of Jessica's age, x.

b. Express the sum of their ages in terms of x and y.

c. Solve the equations that you wrote in **a** and **b** to find Terri's age and Jessica's age.

10-7 USING SYSTEMS OF EQUATIONS TO SOLVE VERBAL PROBLEMS

You have previously learned how to solve word problems by using one variable. Frequently, however, a problem can be solved more easily by using two variables rather than one variable. For example, we can use two variables to solve the following problem:

The sum of two numbers is 8.6. Three times the larger number decreased by twice the smaller is 6.3. What are the numbers?

First, we will represent each number by a different variable:

Let x = the larger number and y = the smaller number.

Now use the conditions of the problem to write two equations:

The sum of the numbers is 8.6: $x + y = 8.6$

Three times the larger decreased by twice the smaller is 6.3: $3x - 2y = 6.3$

Solve the system of equations to find the numbers:

$$x + y = 8.6 \quad \rightarrow \quad 2(x + y) = 2(8.6)$$

$$\begin{aligned} 2x + 2y &= 17.2 \\ 3x - 2y &= 6.3 \\ \hline 5x &= 23.5 \\ x &= 4.7 \end{aligned}$$

$$\begin{aligned} x + y &= 8.6 \\ 4.7 + y &= 8.6 \\ y &= 3.9 \end{aligned}$$

The two numbers are 4.7 and 3.9.

Procedure

To solve a word problem by using a system of two equations involving two variables:

1. Use two different variables to represent the different unknown quantities in the problem.
2. Use formulas or information given in the problem to write a system of two equations in two variables.
3. Solve the system of equations.
4. Use the solution to determine the answer(s) to the problem.
5. Check the answer(s) in the original problem.

EXAMPLE 1

The owner of a men's clothing store bought six belts and eight hats for \$140. A week later, at the same prices, he bought nine belts and six hats for \$132. Find the price of a belt and the price of a hat.

Solution (1) Let b = the price, in dollars, of a belt,

and h = the price, in dollars, of a hat.

(2)

6 *belts*	and	8 *hats*	cost	\$140.
↓	↓	↓	↓	↓
$6b$	$+$	$8h$	$=$	140

9 *belts*	and	6 *hats*	cost	\$132.
↓	↓	↓	↓	↓
$9b$	$+$	$6h$	$=$	132

(3) The least common multiple of the coefficients of b is 18. To eliminate b, write equivalent equations with 18 and -18 as the coefficients of b. Obtain these equations by multiplying both sides of the first equation by 3 and both members of the second equation by -2. Then add the equations and solve for h.

$$\begin{array}{rcl} 3(6b + 8h) = 3(140) & \rightarrow & 18b + 24h = 420 \\ -2(9b + 6h) = -2(132) & \rightarrow & \underline{-18b - 12h = -264} \\ & & 12h = 156 \\ & & h = 13 \end{array}$$

Substitute 13 for h in any equation containing both variables.

$$\begin{aligned} 6b + 8h &= 140 \\ 6b + 8(13) &= 140 \\ 6b + 104 &= 140 \\ 6b &= 36 \\ b &= 6 \end{aligned}$$

(4) *Check:* 6 belts and 8 hats cost 6($6) + 8($13) = $36 + $104 = $140. ✔
9 belts and 6 hats cost 9($6) + 6($13) = $54 + $78 = $132. ✔

Answer A belt costs $6; a hat costs $13.

EXAMPLE 2

When Angelo cashed a check for $170, the bank teller gave him 12 bills, some $20 bills and the rest $10 bills. How many bills of each denomination did Angelo receive?

Solution (1) Represent the unknowns using two variables:

Let x = number of $10 bills,

and y = number of $20 bills.

In this problem, part of the information is in terms of the *number* of bills (the bank teller gave Angelo 12 bills) and part of the information is in terms of the *value* of the bills (the check was for $170).

(2) Write one equation using the *number* of bills: $x + y = 12$
Write a second equation using the *value* of the bills. The value of x $10 bills is $10x$, the value of y $20 bills is $20y$, and the total value is $170: $10x + 20y = 170$

(3) Solve the system of equations:

$$\begin{aligned} & & 10x + 20y &= 170 \\ -10(x + y) = -10(12) & \rightarrow & -10x - 10y &= -120 \\ & & 10y &= 50 \\ & & y &= 5 \end{aligned} \qquad \begin{aligned} x + y &= 12 \\ x + 5 &= 12 \\ x &= 7 \end{aligned}$$

(4) *Check* the number of bills:

7 ten-dollar bills and 5 twenty-dollar bills = 12 bills ✔

Check the value of the bills:

7 ten-dollar bills are worth $70 and 5 twenty-dollar bills are worth $100.

Total value = $70 + $100 = $170 ✔

Answer Angelo received 7 ten-dollar bills and 5 twenty-dollar bills.

EXERCISES

Writing About Mathematics

1. Midori solved the equations in Example 2 by using the substitution method. Which equation do you think she would have solved for one of the variables? Explain your answer.

2. The following system can be solved by drawing a graph, by using the addition method, or by using the substitution method.

$$x + y = 1$$
$$5x + 10y = 8$$

a. Which method do you think is the more efficient way of solving this system of equations? Explain why you chose this method.

b. Which method do you think is the less efficient way of solving this system of equations? Explain why you chose this method.

Developing Skills

In 3–9, solve each problem algebraically, using two variables.

3. The sum of two numbers is 36. Their difference is 24. Find the numbers.

4. The sum of two numbers is 74. The larger number is 3 more than the smaller number. Find the numbers.

5. The sum of two numbers is 104. The larger number is 1 less than twice the smaller number. Find the numbers.

6. The difference between two numbers is 25. The larger exceeds 3 times the smaller by 4. Find the numbers.

7. If 5 times the smaller of two numbers is subtracted from twice the larger, the result is 16. If the larger is increased by 3 times the smaller, the result is 63. Find the numbers.

8. One number is 15 more than another. The sum of twice the larger and 3 times the smaller is 182. Find the numbers.

9. The sum of two numbers is 900. When 4% of the larger is added to 7% of the smaller, the sum is 48. Find the numbers.

Applying Skills

In 10–31, solve each problem algebraically using two variables.

10. The perimeter of a rectangle is 50 centimeters. The length is 9 centimeters more than the width. Find the length and the width of the rectangle.

11. A rectangle has a perimeter of 38 feet. The length is 1 foot less than 3 times the width. Find the dimensions of the rectangle.

12. Two angles are supplementary. The larger angle measures 120° more than the smaller. Find the degree measure of each angle.

13. Two angles are supplementary. The larger angle measures 15° less than twice the smaller. Find the degree measure of each angle.

14. Two angles are complementary. The measure of the larger angle is 30° more than the measure of the smaller angle. Find the degree measure of each angle.

15. The measure of the larger of two complementary angles is 6° less than twice the measure of the smaller angle. Find the degree measure of each angle.

16. In an isosceles triangle, each base angle measures 30° more than the vertex angle. Find the degree measures of the three angles of the triangle.

17. At a snack bar, 3 pretzels and 1 can of soda cost $2.75. Two pretzels and 1 can of soda cost $2.00. Find the cost of a pretzel and the cost of a can of soda.

18. On one day, 4 gardeners and 4 helpers earned $360. On another day, working the same number of hours and at the same rate of pay, 5 gardeners and 6 helpers earned $480. Each gardener receives the same pay for a day's work and each helper receives the same pay for a day's work. How much does a gardener and how much does a helper earn each day?

19. A baseball manager bought 4 bats and 9 balls for $76.50. On another day, she bought 3 bats and 1 dozen balls at the same prices and paid $81.00. How much did she pay for each bat and each ball?

20. Mrs. Black bought 2 pounds of veal and 3 pounds of pork, for which she paid $20.00. Mr. Cook, paying the same prices, paid $11.25 for 1 pound of veal and 2 pounds of pork. Find the price of a pound of veal and the price of a pound of pork.

21. One day, Mrs. Rubero paid $18.70 for 4 kilograms of brown rice and 3 kilograms of basmati rice. Next day, Mrs. Leung paid $13.30 for 3 kilograms of brown rice and 2 kilograms of basmati rice. If the prices were the same on each day, find the price per kilogram for each type of rice.

22. Tickets for a high school dance cost $10 each if purchased in advance of the dance, but $15 each if bought at the door. If 100 tickets were sold and $1,200 was collected, how many tickets were sold in advance and how many were sold at the door?

23. A dealer sold 200 tennis racquets. Some were sold for $33 each, and the rest were sold on sale for $18 each. The total receipts from these sales were $4,800. How many racquets did the dealer sell at $18 each?

24. Mrs. Rinaldo changed a $100 bill in a bank. She received $20 bills and $10 bills. The number of $20 bills was 2 more than the number of $10 bills. How many bills of each kind did she receive?

25. Linda spent $4.50 for stamps to mail packages. Some were 39-cent stamps and the rest were 24-cent stamps. The number of 39-cent stamps was 3 less than the number of 24-cent stamps. How many stamps of each kind did Linda buy?

26. At the Savemore Supermarket, 3 pounds of squash and 2 pounds of eggplant cost $2.85. The cost of 4 pounds of squash and 5 pounds of eggplant is $5.41. What is the cost of one pound of squash, and what is the cost of one pound of eggplant?

27. One year, Roger Jackson and his wife Wilma together earned $67,000. If Roger earned $4,000 more than Wilma earned that year, how much did each earn?

28. Mrs. Moto invested \$1,400, part at 5% and part at 8%. Her total annual income from both investments was \$100. Find the amount she invested at each rate.

29. Mr. Stein invested a sum of money in certificates of deposit yielding 4% a year and another sum in bonds yielding 6% a year. In all, he invested \$4,000. If his total annual income from the two investments was \$188, how much did he invest at each rate?

30. Mr. May invested \$21,000, part at 8% and the rest at 6%. If the annual incomes from both investments were equal, find the amount invested at each rate.

31. A dealer has some hard candy worth \$2.00 a pound and some worth \$3.00 a pound. He makes a mixture of these candies worth \$45.00. If he used 10 pounds more of the less expensive candy than he used of the more expensive candy, how many pounds of each kind did he use?

10-8 GRAPHING THE SOLUTION SET OF A SYSTEM OF INEQUALITIES

To find the solution set of a **system of inequalities**, we must find the ordered pairs that satisfy the open sentences of the system. We do this by a graphic procedure that is similar to the method used in finding the solution set of a system of equations.

EXAMPLE 1

Graph the solution set of this system:

$$x > 2$$
$$y < -2$$

Solution (1) Graph $x > 2$ by first graphing the plane divider $x = 2$, which, in the figure at the right, is represented by the dashed line labeled l. The half-plane to the right of this line is the graph of the solution set of $x > 2$.

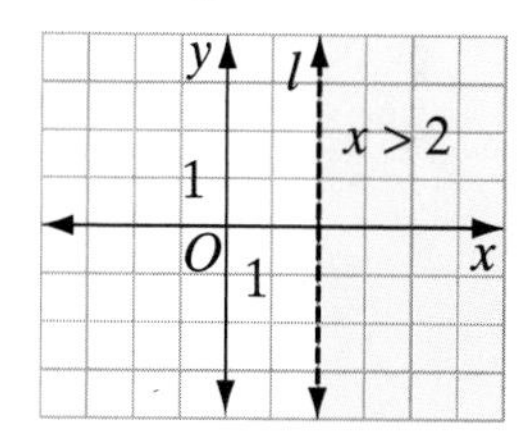

(2) Using the same set of axes, graph $y < -2$ by first graphing the plane divider $y = -2$, which, in the figure at the right, is represented by the dashed line labeled m. The half-plane below this line is the graph of the solution set of $y < -2$.

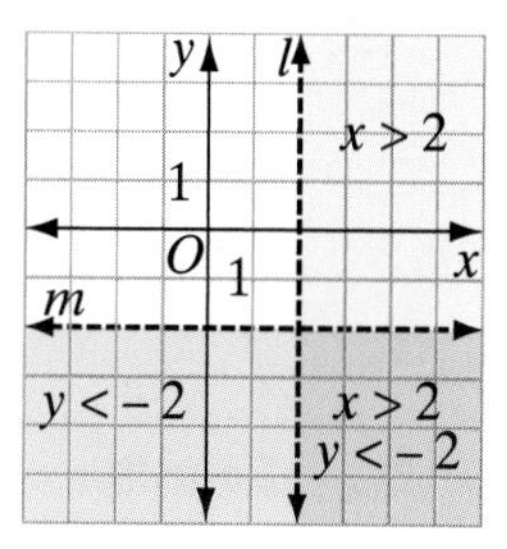

(3) The solution set of the system $x > 2$ and $y < -2$ consists of the intersection of the solution sets of $x > 2$ and $y < -2$. Therefore, the dark colored region in the lower figure, which is the intersection of the graphs made in steps (1) and (2), is the graph of the solution set of the system $x > 2$ and $y < -2$.

From the graph on page 431, all points in the solution region, and no others, satisfy both inequalities of the system. For example, point $(4, -3)$, which lies in the region, satisfies the system because its x-value satisfies one of the given inequalities, $4 > 2$, and its y-value satisfies the other inequality, $-3 < -2$.

EXAMPLE 2

Graph the solution set of $3 < x < 5$ in a coordinate plane.

Solution The inequality $3 < x < 5$ means $3 < x$ and $x < 5$. This may be written as $x > 3$ and $x < 5$.

(1) Graph $x > 3$ by first graphing the plane divider $x = 3$, which, in the figure at the right, is represented by the dashed line labeled l. The half-plane to the right of line $x = 3$ is the graph of the solution set of $x > 3$.

(2) Using the same set of axes, graph $x < 5$ by first graphing the plane divider $x = 5$, which, in the figure at the right, is represented by the dashed line labeled m. The half-plane to the left of line $x = 5$ is the graph of the solution set of $x < 5$.

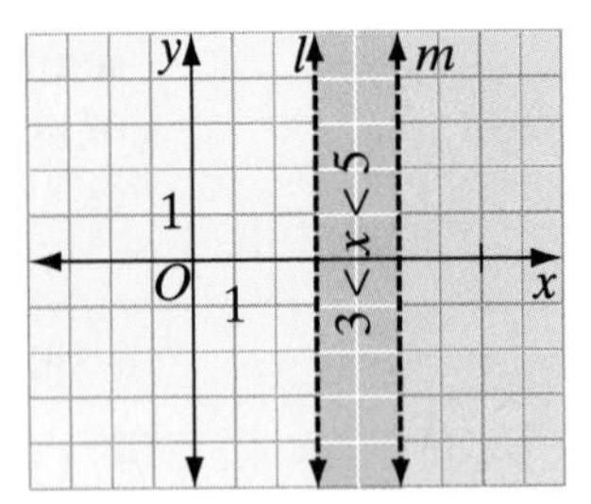

(3) The dark colored region, which is the intersection of the graphs made in steps (1) and (2), is the graph of the solution set of $3 < x < 5$, or $x > 3$ and $x < 5$. All points in this region, and no others, satisfy $3 < x < 5$. For example, point $(4, 3)$, which lies in the region and whose x-value is 4, satisfies $3 < x < 5$ because $3 < 4 < 5$ is a true statement.

EXAMPLE 3

Graph the following system of inequalities and label the solution set R:

$$x + y \geq 4$$
$$y \leq 2x - 3$$

Solution (1) Graph $x + y \geq 4$ by first graphing the plane divider $x + y = 4$, which, in the figure at the top of page 433, is represented by the solid line labeled l. The line $x + y = 4$ and the half-plane above this line together form the graph of the solution set of $x + y \geq 4$.

(2) Using the same set of axes, graph $y \leq 2x - 3$ by first graphing the plane divider $y = 2x - 3$. In the figure at the right, this is the solid line labeled m. The line $y = 2x - 3$ and the half-plane below this line together form the graph of the solution set of $y \leq 2x - 3$.

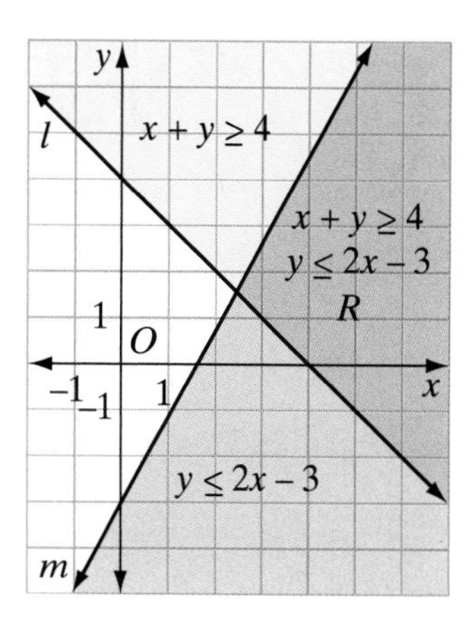

(3) The dark colored region, the intersection of the graphs made in steps (1) and (2), is the solution set of the system $x + y \geq 4$ and $y \leq 2x - 3$. Label it R. Any point in the region R, such as (5, 2), will satisfy $x + y \geq 4$ because $5 + 2 \geq 4$, or $7 \geq 4$, is true, and will satisfy $y \leq 2x - 3$ because $2 \leq 2(5) - 3$, or $2 \leq 7$, is true.

EXAMPLE 4

Sandi boards cats and dogs while their owners are away. Each week she can care for no more than 12 animals. For next week she already has reservations for 4 cats and 5 dogs, but she knows those numbers will probably increase. Draw a graph to show the possible numbers of cats and dogs that Sandi might board next week and list all possible numbers of cats and dogs.

Solution (1) Let x = the number of cats Sandi will board, and y = the number of dogs.

(2) Use the information in the problem to write inequalities.

Sandi can care for no more than 12 animals: $x + y \leq 12$

She expects at least 4 cats: $x \geq 4$

She expects at least 5 dogs: $y \geq 5$

(3) Draw the graphs of these three inequalities.

(4) The possible numbers of cats and dogs are represented by the ordered pairs of positive integers that are included in the solution set of the inequalities. These ordered pairs are shown as points on the graph.

The points (4, 5), (4, 6), (4, 7), (4, 8), (5, 5), (5, 6), (5, 7), (6, 5), (6, 6), (7, 5), all satisfy the given conditions.

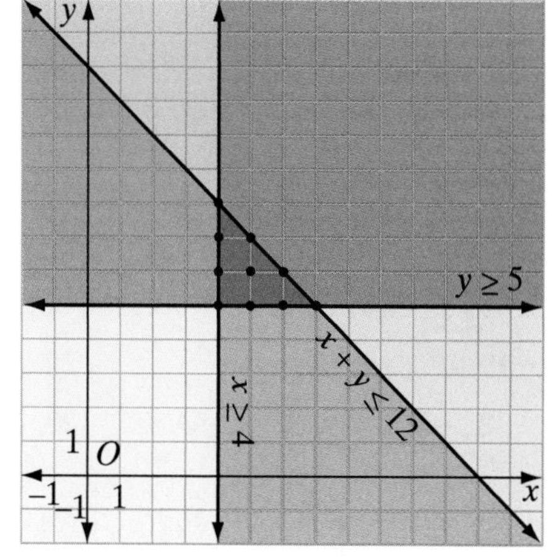

Answer If there are 4 cats, there can be 5, 6, 7, or 8 dogs.

If there are 5 cats, there can be 5, 6, or 7 dogs.

If there are 6 cats, there can be 5 or 6 dogs.

If there are 7 cats, there can be 5 dogs.

EXERCISES

Writing About Mathematics

1. Write a system of inequalities whose solution set is the unshaded region of the graph drawn in Example 3. Explain your answer.

2. What points on the graph drawn in Example 3 are in the solution set of the system of open sentences $x + y \geq 4$ and $y = 2x - 3$? Explain your answer.

3. Describe the solution set of the system of inequalities $y > 4x$ and $4x - y \geq 3$. Explain why this occurs.

Developing Skills

In 4–18, graph each system of inequalities and label the solution set S. Check one solution.

4. $x \geq 1$
$y > -2$

5. $x < 0$
$y > 0$

6. $y \geq 1$
$y < x - 1$

7. $y > x$
$y < 2x + 3$

8. $y \geq 2x$
$y > x + 3$

9. $y \leq 2x + 3$
$y \geq -x$

10. $y - x \geq 5$
$y - 2x \leq 7$

11. $y < x - 1$
$x + y \geq 2$

12. $y + 3x \geq 6$
$y < 2x - 4$

13. $x + y > 3$
$x - y < 6$

14. $x - y \leq -2$
$x + y \geq 2$

15. $2x + y \leq 6$
$x + y - 2 > 0$

16. $2x + 3y \geq 6$
$x + y - 4 \leq 0$

17. $y \geq \frac{1}{2}x$
$x > 0$

18. $x + \frac{3}{4}y \leq 3$
$\frac{1}{3}y - \frac{2}{3}x < 0$

In 19–23, in each case, graph the solution set in a coordinate plane.

19. $1 < x < 4$

20. $-5 \leq x \leq -1$

21. $2 < y \leq 6$

22. $-2 \leq y \leq 3$

23. $(y \geq -1)$ and $(y < 5)$

In 24 and 25, write the system of equation whose solution set is labeled S.

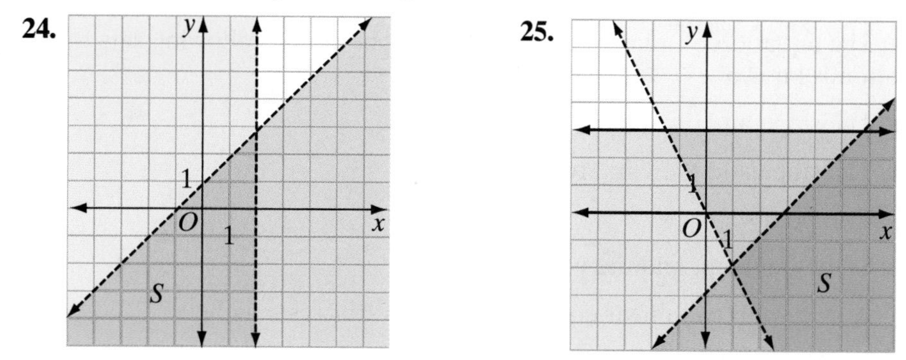

Applying Skills

26. In Ms. Dwyer's class, the number of boys is more than twice the number of girls. There are at least 2 girls. There are no more than 12 boys.

a. Write the three sentences given above as three inequalities, letting x equal the number of girls and y equal the number of boys.

b. On one set of axes, graph the three inequalities written in part **a**.

c. Label the solution set of the system of inequalities S.

d. Do the coordinates of every point in the region labeled S represent the possible number of girls and number of boys in Ms. Dwyer's class? Explain your answer.

e. Write one ordered pair that could represent the number of girls and boys in Ms. Dwyer's class.

27. When Mr. Ehmke drives to work, he drives on city streets for part of the trip and on the expressway for the rest of the trip. The total trip is less than 8 miles. He drives at least 1 mile on city streets and at least 2 miles on the expressway.

a. Write three inequalities to represent the information given above, letting x equal the number of miles driven on city streets and y equal the number of miles driven on the expressway.

b. On one set of axes, graph the three inequalities written in part **a**.

c. Label the solution set of the system of inequalities R.

d. Do the coordinates of every point in the region labeled R represent the possible number of miles driven on city streets and driven on the expressway? Explain your answer.

e. Write one ordered pair that could represent the number of miles Mr. Ehmke drives on city streets and on the expressway.

28. Mildred bakes cakes and pies for sale. She takes orders for the baked goods but also makes extras to sell in her shop. On any day, she can make a total of no more than 12 cakes and pies. For Monday, she had orders for 4 cakes and 6 pies. Draw a graph and list the possible number of cakes and pies she might make on Monday.

CHAPTER SUMMARY

The equation of a line can be written if one of the following sets of information is known:

1. The slope and one point
2. The slope and the y-intercept
3. Two points
4. The x-intercept and the y-intercept

A system of two linear equations in two variables may be:

1. **Consistent:** its solution is at least one ordered pair of numbers. (The graphs of the equations are intersecting lines or lines that coincide.)
2. **Inconsistent:** its solution is the empty set. (The graphs of the equations are parallel with no points in common.)
3. **Dependent:** its solution is an infinite set of number pairs. (The graphs of the equations are the same line.)
4. **Independent:** its solution set is exactly one ordered pair of numbers.

The solution of an independent linear system in two variables may be found **graphically** by determining the coordinates of the point of intersection of the graphs or **algebraically** by using **addition** or **substitution**. Systems of linear equations can be used to solve verbal problems.

The solution set of a system of inequalities can be shown on a graph as the **intersection** of the solution sets of the inequalities.

VOCABULARY

10-4 System of simultaneous equations • Linear system • System of consistent equations • System of independent equations • System of inconsistent equations • System of dependent equations

10-5 Addition method

10-6 Substitution method • Substitution principle

10-8 System of inequalities

REVIEW EXERCISES

1. Solve the following system of equations for x and y and check the solution:

$$5x - 2y = 22$$
$$x + 2y = 2$$

2. a. Graph the following system of inequalities and check one solution:

$$y \leq 2x + 3$$
$$y \leq 2x$$

b. Describe the solution set of the system in terms of the solution sets of the individual inequalities.

In 3–8, write the equation of the line, in the form $y = mx + b$, that satisfies each of the given conditions.

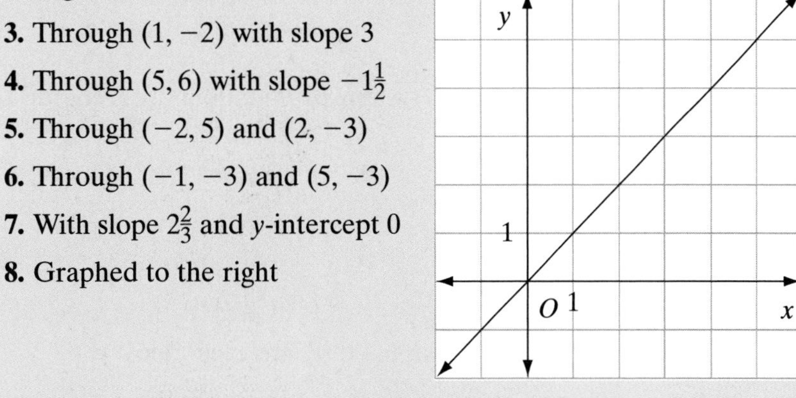

3. Through $(1, -2)$ with slope 3

4. Through $(5, 6)$ with slope $-1\frac{1}{2}$

5. Through $(-2, 5)$ and $(2, -3)$

6. Through $(-1, -3)$ and $(5, -3)$

7. With slope $2\frac{2}{3}$ and y-intercept 0

8. Graphed to the right

In 9–14, write the equation of the line, in the form $\frac{x}{a} + \frac{y}{b} = 1$, that satisfies each of the given conditions.

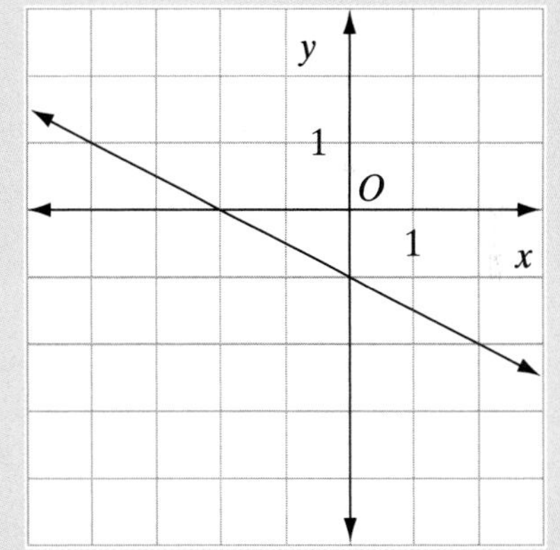

9. With slope 1 and y-intercept -1

10. With slope -3.5 and y-intercept -1.5

11. With x-intercept 5 and y-intercept -2

12. With x-intercept $\frac{3}{4}$ and y-intercept $-\frac{5}{7}$

13. Through $(1, 3)$ and $(5, 6)$

14. Graphed to the right

In 15–17, write the equation of the line, in the form $Ax + By = C$, that satisfies each of the given conditions.

15. Parallel to the y-axis and 3 units to the left of the y-axis

16. Through $(0, 4)$ and $(2, 0)$

17. Through $(1, 1)$ and $(4, 5)$

In 18–20, solve each system of equations *graphically* and check.

18. $x + y = 6$
$y = 2x - 6$

19. $y = -x$
$2x + y = 3$

20. $2y = x + 4$
$x - y + 4 = 0$

In 21–23, solve each system of equations by using *addition* to eliminate one of the variables. Check.

21. $2x + y = 10$
$x + y = 3$

22. $x + 4y = 1$
$5x - 6y = -8$

23. $3c + d = 0$
$c - 4d = 52$

In 24–26, solve each system of equations by using *substitution* to eliminate one of the variables. Check.

24 $x + 2y = 7$
$x = y - 8$

25. $3r + 2s = 20$
$r = -2s$

26. $x + y = 7$
$2x + 3y = 21$

In 27–32, solve each system of equations by using an appropriate algebraic method. Check.

27. $x + y = 0$
$3x + 2y = 5$

28. $5a + 3b = 17$
$4a - 5b = 21$

29. $t + u = 12$
$t = \frac{1}{3}u$

30. $3a = 4b$
$4a - 5b = 2$

31. $x + y = 1{,}000$
$0.06x = 0.04y$

32. $10t + u = 24$
$t + u = \frac{1}{7}(10u + t)$

33. **a.** Solve this system of equations algebraically:

$$x - y = 3$$
$$x + 3y = 9$$

b. On a set of coordinate axes, graph the system of equations given in part **a.**

In 34–36, graph each system of inequalities and label the solution set A.

34. $y > 2x - 3$
$y \leq 5 - x$

35. $y \leq \frac{1}{2}x$
$x \geq -4$

36. $2x + y < 4$
$x - y < -2$

In 37–41, for each problem, write two equations in two variables and solve algebraically.

37. The sum of two numbers is 7. Their difference is 18. Find the numbers.

38. At a store, 3 notebooks and 2 pencils cost \$2.80. At the same prices, 2 notebooks and 5 pencils cost \$2.60. Find the cost of a notebook and of a pencil.

39. Two angles are complementary. The larger angle measures 15° less than twice the smaller angle. Find the degree measure of each angle.

40. The measure of the vertex angle of an isosceles triangle is 3 times the measure of each of the base angles. Find the measure of each angle of the triangle.

41. At the beginning of the year, Lourdes's salary increased by 3%. If her weekly salary is now \$254.41, what was her weekly salary last year?

Exploration

Jason and his brother Robby live 1.25 miles from school. Jason rides his bicycle to school and Robby walks. The graph below shows their relative positions on their way to school one morning.

Use the information provided by the graph to describe Robby's and Jason's journeys to school. Who left first? Who arrived at school first? If each boy took the same route to school, when did Jason pass Robby? How fast did Robby walk? How fast did Jason ride his bicycle?

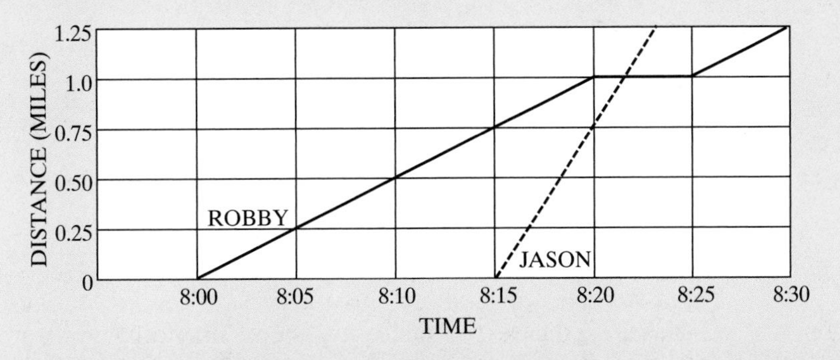

CUMULATIVE REVIEW CHAPTERS 1–10

Part I

Answer all questions in this part. Each correct answer will receive 2 credits. No partial credit will be allowed.

1. The multiplicative inverse of -0.4 is
(1) 0.4 (2) 1 (3) 2.5 (4) -2.5

2. Which of the following numbers has the greatest value?
(1) -1.5 (2) $-1.\overline{5}$ (3) $-\sqrt{2}$ (4) -0.5π

3. Which of the following is an example of the associative property of addition?
(1) $6 + (x + 3) = 6 + (3 + x)$
(2) $2(x + 2) = 2x + 4$
(3) $3(4a) = (3 \times 4)a$
(4) $3 + (4 + a) = (3 + 4) + a$

4. Which of the following is a point on the line whose equation is $x - y = -5$?
(1) $(-1, -4)$ (2) $(1, -4)$ (3) $(-1, 4)$ (4) $(1, 4)$

5. The expression $5 - 2(3x^2 + 8)$ is equivalent to
(1) $9x^2 + 24$ (2) $-6x^2 - 21$ (3) $9x^2 + 8$ (4) $-6x^2 - 11$

6. The solution of the equation $0.2x - 3 = x + 1$ is
(1) -5 (2) 5 (3) 0.5 (4) 50

7. The product of 3.40×10^{-3} and 8.50×10^{2} equals
(1) 2.89×10^{0} (2) 2.89×10^{1} (3) 2.89×10^{2} (4) 2.89×10^{-1}

8. The measure of $\angle A$ is 12° less than twice the measure of its complement. What is the measure of $\angle A$?
(1) 51° (2) 34° (3) 39° (4) 56°

9. The y-intercept of the graph of $2x + 3y = 6$ is
(1) 6 (2) 2 (3) 3 (4) −3

10. When $2a^2 - 5a$ is subtracted from $5a^2 + 1$, the difference is
(1) $-3a^2 - 5a - 1$
(2) $3a^2 - 5a + 1$
(3) $3a^2 + 6a$
(4) $3a^2 + 5a + 1$

Part II

Answer all questions in this part. Each correct answer will receive 2 credits. Clearly indicate the necessary steps, including appropriate formula substitutions, diagrams, graphs, charts, etc. For all questions in this part, a correct numerical answer with no work shown will receive only 1 credit.

11. A straight line has a slope of −3 and contains the point (0, 6). What are the coordinates of the point at which the line intersects the x-axis?

12. Last year the Edwards family spent $6,200 for food. This year, the cost of food for the family was $6,355. What was the percent of increase in the cost of food for the Edwards family?

Part III

Answer all questions in this part. Each correct answer will receive 3 credits. Clearly indicate the necessary steps, including appropriate formula substitutions, diagrams, graphs, charts, etc. For all questions in this part, a correct numerical answer with no work shown will receive only 1 credit.

13. Dana paid $3.30 for 2 muffins and a cup of coffee. At the same prices, Damion paid $5.35 for 3 muffins and 2 cups of coffee. What is the cost of a muffin and of a cup of coffee?

14. In trapezoid $ABCD$, $\overline{BC} \perp \overline{AB}$ and $\overline{BC} \perp \overline{DC}$. If $AB = 54.8$ feet, $DC = 37.2$ feet, and $BC = 15.8$ feet, find to the *nearest degree* the measure of $\angle A$.

Part IV

Answer all questions in this part. Each correct answer will receive 4 credits. Clearly indicate the necessary steps, including appropriate formula substitutions, diagrams, graphs, charts, etc. For all questions in this part, a correct numerical answer with no work shown will receive only 1 credit.

15. Tamara wants to develop her own pictures but does not own photographic equipment. The Community Darkroom makes the use of photographic equipment available for a fee. Membership costs $25 a month and members pay $1.50 an hour for the use of the equipment. Non-members may also use the equipment for $4.00 an hour.

a. Draw a graph that shows the cost of becoming a member and using the photographic equipment. Let the vertical axis represent cost in dollars and the horizontal axis represent hours of use.

b. On the same graph, show the cost of using the photographic equipment for non-members.

c. When Tamara became a member of the Community Darkroom in June, she found that her cost for that month would have been the same if she had not been a member. How many hours did Tamara use the facilities of the darkroom in June?

16. $ABCD$ is a rectangle, E is a point on $\overline{DC}$, $AD = 24$ centimeters, $DE = 32$ centimeters, and $EC = 10$ centimeters.

a. Find the area of $\triangle ABE$.

b. Find the perimeter of $\triangle ABE$.

CHAPTER

11

SPECIAL PRODUCTS AND FACTORS

Chapter Table of Contents

The owners of a fruit farm intend to extend their orchards by planting 100 new apple trees. The trees are to be planted in rows, with the same number of trees in each row. In order to decide how the trees are to be planted, the owners will use whole numbers that are factors of 100 to determine the possible arrangements:

1 row of 100 trees	20 rows of 5 trees each
2 rows of 50 trees each	25 rows of 4 trees each
4 rows of 25 trees each	50 rows of 2 trees each
5 rows of 20 trees each	100 rows of 1 tree each
10 rows of 10 trees each	

From this list of possibilities, the arrangement that best fits the dimensions of the land to be planted can be chosen.

If the owner had intended to plant 90 trees, how many possible arrangements would there be?

From earliest times, the study of factors and prime numbers has fascinated mathematicians, leading to the discovery of many important principles. In this chapter you will extend your knowledge of the factors of whole numbers, study the products of special binomials, and learn to write polynomials in factored form.

11-1 FACTORS AND FACTORING

When two numbers are multiplied, the result is called their **product**. The numbers that are multiplied are *factors* of the product. Since $3(5) = 15$, the numbers 3 and 5 are factors of 15.

Factoring a number is the process of finding those numbers whose product is the given number. Usually, when we factor, we are finding the factors of an integer and we find only those factors that are integers. We call this **factoring over the set of integers.**

Factors of a product can be found by using division. Over the set of integers, if the divisor and the quotient are both integers, then they are factors of the dividend. For example, $35 \div 5 = 7$. Thus, $35 = 5(7)$, and 5 and 7 are factors of 35.

Every positive integer that is the product of two positive integers is also the product of the opposites of those integers.

$$+21 = (+3)(+7) \qquad\qquad +21 = (-3)(-7)$$

Every negative integer that is the product of a positive integer and a negative integer is also the product of the opposites of those integers.

$$-21 = (+3)(-7) \qquad\qquad -21 = (-3)(+7)$$

Usually, when we factor a positive integer, we write only the positive integral factors.

Two factors of any number are 1 and the number itself. To find other integral factors, if they exist, we use division, as stated above. We let the number being factored be the dividend, and we divide this number in turn by the whole numbers 2, 3, 4, and so on. If the quotient is an integer, then both the divisor and the quotient are factors of the dividend.

For example, use a calculator to find the integral factors of 126. We will use integers as divisors and look for quotients that are integers. Pairs of factors of 126 are listed to the right of the quotients.

Quotients	*Pairs of Factors of 126*	*Quotients*	*Pairs of Factors of 126*
$126 \div 1 = 126$	$1 \cdot 126$	$126 \div 7 = 18$	$7 \cdot 18$
$126 \div 2 = 63$	$2 \cdot 63$	$126 \div 8 = 15.75$	—
$126 \div 3 = 42$	$3 \cdot 42$	$126 \div 9 = 14$	$9 \cdot 14$
$126 \div 4 = 31.5$	—	$126 \div 10 = 12.6$	—
$126 \div 5 = 25.2$	—	$126 \div 11 = 11.\overline{45}$	—
$126 \div 6 = 21$	$6 \cdot 21$	$126 \div 12 = 10.5$	—

When the quotient is smaller than the divisor (here, $10.5 < 12$), we have found all possible positive integral factors.

The factors of 126 are 1, 2, 3, 6, 7, 9, 14, 18, 21, 42, 63, and 126.

Recall that a *prime number* is an integer greater than 1 that has no positive integral factors other than itself and 1. The first seven prime numbers are 2, 3, 5, 7, 11, 13, and 17. Integers greater than 1 that are not prime are called *composite numbers*.

In general, a positive integer greater than 1 is a prime or can be expressed as the product of prime factors. Although the factors may be written in any order, there is one and only one combination of prime factors whose product is a given composite number. As shown below, a prime factor may occur in the product more than once.

$$21 = 3 \cdot 7$$
$$20 = 2 \cdot 2 \cdot 5 \text{ or } 2^2 \cdot 5$$

To express a positive integer, for example 280, as the product of primes, we start with any pair of positive integers, say 28 and 10, whose product is the given number. Then, we factor these factors and continue to factor the factors until all are primes. Finally, we rearrange these factors in numerical order, as shown at the right.

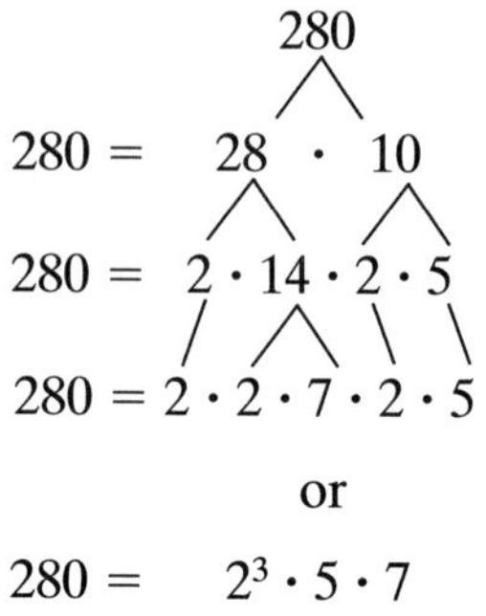

$$280 = 28 \cdot 10$$
$$280 = 2 \cdot 14 \cdot 2 \cdot 5$$
$$280 = 2 \cdot 2 \cdot 7 \cdot 2 \cdot 5$$
or
$$280 = 2^3 \cdot 5 \cdot 7$$

Expressing each of two integers as the product of prime factors makes it possible to discover the greatest integer that is a factor of both of them. We call this factor the **greatest common factor (GCF)** of these integers.

Let us find the greatest common factor of 180 and 54.

$180 =$	$2 \cdot 2 \cdot 3 \cdot 3 \cdot 5$	or	$2^2 \cdot 3^2 \cdot 5$	
	$\downarrow \quad \downarrow \; \downarrow$			
$54 =$	$2 \cdot 3 \cdot 3 \cdot 3$	or	$2 \cdot 3^3$	
	$\downarrow \quad \downarrow \; \downarrow$			
Greatest common factor $=$	$2 \cdot \quad 3 \cdot 3$	or	$2 \cdot 3^2$	or 18

Only the prime numbers 2 and 3 are factors of both 180 and 54. We see that the greatest number of times that 2 appears as a factor of both 180 and 54 is once; the greatest number of times that 3 appears as a factor of both 180 and 54 is twice. Therefore, the greatest common factor of 180 and 54 is $2 \cdot 3 \cdot 3$, or $2 \cdot 3^2$, or 18.

To find the greatest common factor of two or more monomials, find the product of the numerical and variable factors that are common to the monomials. For example, let us find the greatest common factor of $24a^3b^2$ and $18a^2b$.

$24a^3b^2 =$	$2 \cdot 2 \cdot 2 \cdot 3 \cdot$	$a \cdot a \cdot a \cdot b \cdot b$	
	$\downarrow \qquad \downarrow$	$\downarrow \; \downarrow \quad \downarrow$	
$18a^2b =$	$2 \cdot \qquad 3 \cdot 3 \cdot$	$a \cdot a \cdot \quad b$	
	$\downarrow \qquad \downarrow$	$\downarrow \; \downarrow \quad \downarrow$	
Greatest common factor $=$	$2 \cdot \qquad 3 \cdot$	$a \cdot a \cdot \quad b$	$= 6a^2b$

The greatest common factor of $24a^3b^2$ and $18a^2b$ is $6a^2b$.

When we are expressing an algebraic factor, such as $6a^2b$, we will agree that:

- Numerical coefficients need not be factored. (6 need not be written as $2 \cdot 3$.)
- Powers of variables need not be represented as the product of several equal factors. (a^2b need not be written as $a \cdot a \cdot b$.)

EXAMPLE 1

Write all positive integral factors of 72.

Solution

Quotients	*Pairs of Factors of 72*	Quotients	*Pairs of Factors of 72*
$72 \div 1 = 72$	$1 \cdot 72$	$72 \div 5 = 14.4$	—
$72 \div 2 = 36$	$2 \cdot 36$	$72 \div 6 = 12$	$6 \cdot 12$
$72 \div 3 = 24$	$3 \cdot 24$	$72 \div 7 = 10.\overline{285714}$	—
$72 \div 4 = 18$	$4 \cdot 18$	$72 \div 8 = 9$	$8 \cdot 9$

It is not necessary to divide by 9, since 9 has just been listed as a factor.

Answer The factors of 72 are 1, 2, 3, 4, 6, 8, 9, 12, 18, 24, 36, and 72.

EXAMPLE 2

Express 700 as a product of prime factors.

Solution

$$700 = 2 \cdot 350$$
$$700 = 2 \cdot 2 \cdot 175$$
$$700 = 2 \cdot 2 \cdot 5 \cdot 35$$
$$700 = 2 \cdot 2 \cdot 5 \cdot 5 \cdot 7 \text{ or } 2^2 \cdot 5^2 \cdot 7$$

Answer $2^2 \cdot 5^2 \cdot 7$

EXAMPLE 3

Find the greatest common factor of the monomials $60r^2s^4$ and $36rs^2t$.

Solution To find the greatest common factor of two monomials, write each as the product of primes and variables to the first power and choose all factors that occur in each product.

$$60r^2s^4 = 2 \cdot 2 \cdot 3 \cdot 5 \cdot r \cdot r \cdot s \cdot s \cdot s \cdot s$$
$$36rs^2t = 2 \cdot 2 \cdot 3 \cdot 3 \cdot r \cdot s \cdot s \cdot t$$
$$\text{Greatest common factor} = 2 \cdot 2 \cdot 3 \cdot r \cdot s \cdot s$$

Answer $12rs^2$

EXERCISES

Writing About Mathematics

1. Ross said that some pairs of number such as 5 and 12 have no greatest common factor. Do you agree with Ross? Explain why or why not.

2. To find the factors of 200, Chaz tried all of the integers from 2 to 14. Should Chaz have used more numbers? Explain why or why not.

Developing Skills

In 3–12, tell whether each integer is prime, composite, or neither.

3. 5 **4.** 8 **5.** 13 **6.** 18 **7.** 73

8. 36 **9.** 41 **10.** 49 **11.** 57 **12.** 1

In 13–22, express each integer as a product of prime numbers.

13. 35 **14.** 18 **15.** 144 **16.** 77 **17.** 128

18. 400 **19.** 202 **20.** 129 **21.** 590 **22.** 316

In 23–34, write all the positive integral factors of each given number.

23. 26 **24.** 50 **25.** 36 **26.** 88 **27.** 100 **28.** 242

29. 37 **30.** 62 **31.** 253 **32.** 102 **33.** 70 **34.** 169

35. The product of two monomials is $36x^3y^4$. Find the second factor if the first factor is:

a. $3x^2y^3$ **b.** $6x^3y^2$ **c.** $12xy^4$ **d.** $9x^3y$ **e.** $18x^3y^2$

36. The product of two monomials is $81c^9d^{12}e$. Find the second factor if the first factor is:

a. $27e$ **b.** c^2d^{11} **c.** $3cd^3e$ **d.** $9c^4d^4$ **e.** $81c^5d^9e$

In 37–42, find, in each case, the greatest common factor of the given integers.

37. 10; 15 **38.** 12; 28 **39.** 14; 35

40. 18; 24; 36 **41.** 75; 50 **42.** 72; 108

In 43–51, find, in each case, the greatest common factor of the given monomials.

43. $4x$; $4y$ **44.** $4r$; $6r^2$ **45.** $8xy$; $6xz$

46. $10x^2$; $15xy^2$ **47.** $36xy^2z$; $-27xy^2z^2$ **48.** $24ab^2c^3$; $18ac^2$

49. $14a^2b$; $13ab$ **50.** $36xyz$; $25xyz$ **51.** $2ab^2c$; $3x^2yz$

11-2 COMMON MONOMIAL FACTORS

To **factor a polynomial** over the set of integers means to express the given polynomial as the product of polynomials whose coefficients are integers. For example, since $2(x + y) = 2x + 2y$, the polynomial $2x + 2y$ can be written in factored form as $2(x + y)$. The monomial 2 is a factor of each term of the polynomial $2x + 2y$. Therefore, 2 is called a **common monomial factor** of the polynomial $2x + 2y$.

To factor a polynomial, we look first for the **greatest common monomial factor**, that is, the greatest monomial that is a factor of each term of the polynomial.

For example:

1. Factor $4rs + 8st$. There are many common factors of $4rs$ and $8st$ such as 2, 4, $2s$, and $4s$. The greatest common monomial factor is $4s$. We divide $4rs + 8st$ by $4s$ to obtain the quotient $r + 2t$, which is the second factor. Therefore, the polynomial $4rs + 8st = 4s(r + 2t)$.
2. Factor $3x + 4y$. We notice that 1 is the only common factor of $3x$ and $4y$, so the second factor is $3x + 4y$. We say that $3x + 4y$ is a **prime polynomial**. A polynomial with integers as coefficients is a prime polynomial if its only factors are 1 and the polynomial itself.

Procedure

To factor a polynomial whose terms have a common monomial factor:

1. Find the greatest monomial that is a factor of each term of the polynomial.
2. Divide the polynomial by the factor found in step 1. The quotient is the other factor.
3. Express the polynomial as the product of the two factors.

We can check by multiplying the factors to obtain the original polynomial.

EXAMPLE 1

Write in factored form: $6c^3d - 12c^2d^2 + 3cd$.

Solution (1) $3cd$ is the greatest common factor of $6c^3d$, $-12c^2d^2$, and $3cd$.

(2) To find the other factor, divide $6c^3d - 12c^2d^2 + 3cd$ by $3cd$.

$$\begin{aligned}(6c^3d - 12c^2d^2 + 3cd) \div 3cd &= \frac{6c^3d}{3cd} - \frac{12c^2d^2}{3cd} + \frac{3cd}{3cd}\\ &= 2c^2 - 4cd + 1\end{aligned}$$

Answer $3cd(2c^2 - 4cd + 1)$

EXERCISES

Writing About Mathematics

1. Ramón said that the factored form of $3a + 6a^2 + 15a^3$ is $3a(2a + 5a^2)$. Do you agree with Ramón? Explain why or why not.

2. a. The binomial $12a^2 + 20ab$ can be written as $4a(3a + 5b)$. Does that mean that for all positive integral values of a and b, the value of $12a^2 + 20ab$ has at least two factors other than itself and 1? Justify your answer.

   b. The binomial $3a + 5b$ is a prime polynomial. Does that mean that for all positive integral values of a and b, $3a + 5b$ is a prime number? Justify your answer.

Developing Skills

In 3–29, write each expression in factored form.

3. $2a + 2b$

4. $3x - 3y$

5. $bx + by$

6. $xc - xd$

7. $4x + 8y$

8. $3m - 6n$

9. $12x - 18y$

10. $18c + 27d$

11. $8x + 16$

12. $7y - 7$

13. $6 - 18c$

14. $y^2 - 3y$

15. $2x^2 + 5x$

16. $ax - 5ab$

17. $3y^4 + 3y^2$

18. $10x - 15x^3$

19. $2x - 4x^3$

20. $p + prt$

21. $\pi r^2 + \pi rl$

22. $\pi r^2 + 2\pi rh$

23. $3a^2 - 9$

24. $12y^2 - 4y$

25. $3ab^2 - 6a^2b$

26. $21r^3s^2 - 14r^2s$

27. $3x^2 - 6x - 30$

28. $c^3 - c^2 + 2c$

29. $9ab^2 - 6ab - 3a$

Applying Skills

30. The perimeter of a rectangle is represented by $2l + 2w$. Express the perimeter as the product of two factors.

31. The lengths of the parallel sides of a trapezoid are represented by a and b and its height by h. The area of the trapezoid can be written as $\frac{1}{2}ah + \frac{1}{2}bh$. Express this area as the product of two factors.

32. A cylinder is cut from a cube whose edge measures $2s$.

   a. Express the volume of the cube in terms of s.

   b. If the largest possible cylinder is made, express the volume of the cylinder in terms of s.

   c. Use the answers to parts **a** and **b** to express, in terms of s, the volume of the material cut away to form the cylinder.

   d. Express the volume of the waste material as the product of two factors.

11-3 THE SQUARE OF A MONOMIAL

To **square a monomial** means to multiply the monomial by itself. For example:

$$(3x)^2 = (3x)(3x) = (3)(3)(x)(x) = (3)^2(x)^2 \text{ or } 9x^2$$
$$(5y^2)^2 = (5y^2)(5y^2) = (5)(5)(y^2)(y^2) = (5)^2(y^2)^2 \text{ or } 25y^4$$
$$(-6b^4)^2 = (-6b^4)(-6b^4) = (-6)(-6)(b^4)(b^4) = (-6)^2(b^4)^2 \text{ or } 36b^8$$
$$(4c^2d^3)^2 = (4c^2d^3)(4c^2d^3) = (4)(4)(c^2)(c^2)(d^3)(d^3) = (4)^2(c^2)^2(d^3)^2 \text{ or } 16c^4d^6$$

When a monomial is a perfect square, its numerical coefficient is a perfect square and the exponent of each variable is an even number. This statement holds true for each of the results shown above.

Procedure

To square a monomial:

1. Find the square of the numerical coefficient.
2. Find the square of each literal factor by multiplying its exponent by 2.
3. The square of the monomial is the product of the expressions found in steps 1 and 2.

EXAMPLE 1

Square each monomial mentally.

	Think	*Write*
a. $(4s^3)^2$	$(4)^2(s^3)^2$	$16s^6$
b. $\left(\frac{2}{5}ab\right)^2$	$\left(\frac{2}{5}\right)^2(a)^2(b)^2$	$\frac{4}{25}a^2b^2$
c. $(-7xy^2)^2$	$(-7)^2(x)^2(y^2)^2$	$49x^2y^4$
d. $(0.3y^2)^2$	$(0.3)^2(y^2)^2$	$0.09y^4$

EXERCISES

Writing About Mathematics

1. Explain why the exponent of each variable in the square of a monomial is an even number.
2. When the square of a number is written in scientific notation, is the exponent of 10 always an even number? Explain why or why not and give examples to justify your answer.

Developing Skills

In 3–22, square each monomial.

3. $(a^2)^2$ **4.** $(b^3)^2$ **5.** $(-d^5)^2$ **6.** $(rs)^2$

7. $(m^2n^2)^2$ **8.** $(-x^3y^2)^2$ **9.** $(3x^2)^2$ **10.** $(-5y^4)^2$

11. $(9ab)^2$ **12.** $(10x^2y^2)^2$ **13.** $(-12cd^3)^2$ **14.** $\left(\frac{3}{4}a\right)^2$

15. $\left(\frac{5}{7}xy\right)^2$ **16.** $\left(-\frac{7}{8}a^2b^2\right)^2$ **17.** $\left(\frac{x}{6}\right)^2$ **18.** $\left(-\frac{4x^2}{5}\right)^2$

19. $(0.8x)^2$ **20.** $(0.5y^2)^2$ **21.** $(0.01xy)^2$ **22.** $(-0.06a^2b)^2$

Applying Skills

In 23–28, each monomial represents the length of a side of a square. Write the monomial that represents the area of the square.

23. $4x$ **24.** $10y$ **25.** $\frac{2}{3}x$ **26.** $1.5x$ **27.** $3x^2$ **28.** $4x^2y^3$

11-4 MULTIPLYING THE SUM AND THE DIFFERENCE OF TWO TERMS

Recall that when two binomials are multiplied, the product contains *four terms*. This fact is illustrated in the diagram below and is also shown by using the distributive property.

$$\begin{aligned}(a + b)(c + d) &= a(c + d) + b(c + d)\\ &= ac + ad + bc + bd\end{aligned}$$

	a	b
c	ac	bc
d	ad	bd

If two of the four terms of the product are similar, the similar terms can be combined so that the product is a trinomial. For example:

$$\begin{aligned}(x + 2)(x + 3) &= x(x + 3) + 2(x + 3)\\ &= x^2 + 3x + 2x + 6\\ &= x^2 + 5x + 6\end{aligned}$$

There is, however, a special case in which the sum of two terms is multiplied by the difference of the *same* two terms. In this case, the sum of the two middle terms of the product is 0, and the product is a binomial, as in the following examples:

$$\begin{aligned}(a + 4)(a - 4) &= a(a - 4) + 4(a - 4)\\ &= a^2 - 4a + 4a - 16\\ &= a^2 - 16\end{aligned}$$

$$\begin{aligned}(3x^2 - 5y)(3x^2 + 5y) &= 3x^2(3x^2 + 5y) - 5y(3x^2 + 5y)\\ &= 9x^4 + 15x^2y - 15x^2y - 25y^2\\ &= 9x^4 - 25y^2\end{aligned}$$

These two examples illustrate the following procedure, which enables us to find the product of the sum and difference of the same two terms mentally.

Procedure

To multiply the sum of two terms by the difference of the same two terms:

1. Square the first term.
2. From this result, subtract the square of the second term.

$$(a + b)(a - b) = a^2 - b^2$$

EXAMPLE 1

Find each product.

	Think	*Write*
a. $(y + 7)(y - 7)$	$(y)^2 - (7)^2$	$y^2 - 49$
b. $(3a - 4b)(3a + 4b)$	$(3a)^2 - (4b)^2$	$9a^2 - 16b^2$

EXERCISES

Writing About Mathematics

1. Rose said that the product of two binomials is a binomial only when the two binomials are the sum and the difference of the same two terms. Miranda said that that cannot be true because $(5a + 10)(a - 2) = 5a^2 - 20$, a binomial. Show that Rose is correct by writing the factors in the example that Miranda gave in another way.

2. Ali wrote the product $(x + 2)^2(x - 2)^2$ as $(x^2 - 4)^2$. Do you agree with Ali? Explain why or why not.

Developing Skills

In 3–17, find each product.

3. $(x + 8)(x - 8)$ **4.** $(y + 10)(y - 10)$ **5.** $(n - 9)(n + 9)$

6. $(12 + a)(12 - a)$ **7.** $(c + d)(c - d)$ **8.** $(3x + 1)(3x - 1)$

9. $(8x + 3y)(8x - 3y)$ **10.** $(x^2 + 8)(x^2 - 8)$ **11.** $(3 - 5y^3)(3 + 5y^3)$

12. $\left(a + \frac{1}{2}\right)\left(a - \frac{1}{2}\right)$ **13.** $(r + 0.5)(r - 0.5)$ **14.** $(0.3 + m)(0.3 - m)$

15. $(a + 5)(a - 5)(a^2 + 25)$ **16.** $(x - 3)(x + 3)(x^2 + 9)$ **17.** $(a + b)(a - b)(a^2 + b^2)$

Applying Skills

In 18–21, express the area of each rectangle whose length l and width w are given.

18. $l = x + 7$, $w = x - 7$ **19.** $l = 2x + 3$, $w = 2x - 3$ **20.** $l = c + d$, $w = c - d$ **21.** $l = 2a + 3b$, $w = 2a - 3b$

In 22–25, find each product mentally by thinking of the factors as the sum and difference of the same two numbers.

22. $(27)(33) = (30 - 3)(30 + 3)$ **23.** $(52)(48) = (50 + 2)(50 - 2)$

24. $(65)(75)$ **25.** $(19)(21)$

11-5 FACTORING THE DIFFERENCE OF TWO PERFECT SQUARES

An expression of the form $a^2 - b^2$ is called a **difference of two perfect squares.** Factoring the difference of two perfect squares is the reverse of multiplying the sum of two terms by the difference of the same two terms. Since the product $(a + b)(a - b)$ is $a^2 - b^2$, the factors of $a^2 - b^2$ are $(a + b)$ and $(a - b)$. Therefore:

$$a^2 - b^2 = (a + b)(a - b)$$

Procedure

To factor a binomial that is a difference of two perfect squares:

1. Express each of its terms as the square of a monomial.
2. Apply the rule $a^2 - b^2 = (a + b)(a - b)$.

Remember that a monomial is a square if and only if its numerical coefficient is a square and the exponent of each of its variables is an even number.

EXAMPLE 1

Factor each polynomial.

	Think	*Write*
a. $r^2 - 1$	$(r)^2 - (1)^2$	$(r + 1)(r - 1)$
b. $25x^2 - \frac{1}{49}y^2$	$(5x)^2 - \left(\frac{1}{7}y\right)^2$	$\left(5x + \frac{1}{7}y\right)\left(5x - \frac{1}{7}y\right)$
c. $0.04 - c^6d^4$	$(0.2)^2 - (c^3d^2)^2$	$(0.2 + c^3d^2)(0.2 - c^3d^2)$

EXAMPLE 2

Express $x^2 - 100$ as the product of two binomials.

Solution Since $x^2 - 100$ is a difference of two squares, the square of x and the square of 10, the factors of $x^2 - 100$ are $(x + 10)$ and $(x - 10)$.

Answer $(x + 10)(x - 10)$

EXERCISES

Writing About Mathematics

1. If 391 can be written as $400 - 9$, find two factors of 391 without using a calculator or pencil and paper. Explain how you found your answer.
2. Does $5a^2 - 45$ have two binomial factors? Explain your answer.

Developing Skills

In 3–18, factor each binomial.

3. $a^2 - 4$ **4.** $c^2 - 100$ **5.** $9 - x^2$ **6.** $144 - c^2$

7. $16a^2 - b^2$ **8.** $25m^2 - n^2$ **9.** $d^2 - 4c^2$ **10.** $r^4 - 9$

11. $25 - s^4$ **12.** $100x^2 - 81y^2$ **13.** $w^2 - \frac{1}{64}$ **14.** $x^2 - 0.64$

15. $0.04 - 49r^2$ **16.** $0.16y^2 - 9$ **17.** $0.81 - y^2$ **18.** $81m^4 - 49$

Applying Skills

In 19–23, each given polynomial represents the area of a rectangle. Express the area as the product of two binomials.

19. $x^2 - 4$ **20.** $y^2 - 9$ **21.** $t^2 - 49$ **22.** $t^4 - 64$ **23.** $4x^2 - y^2$

In 24–26, express the area of each shaded region as: **a.** the difference of the areas shown, and **b.** the product of two binomials.

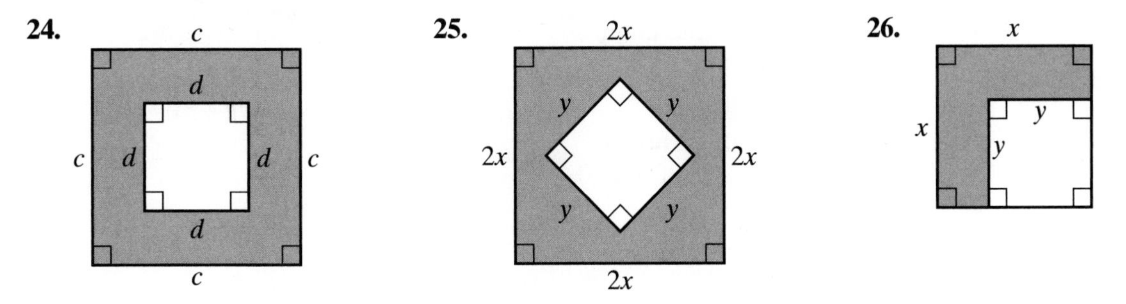

In 27 and 28, express the area of each shaded region as the product of two binomials.

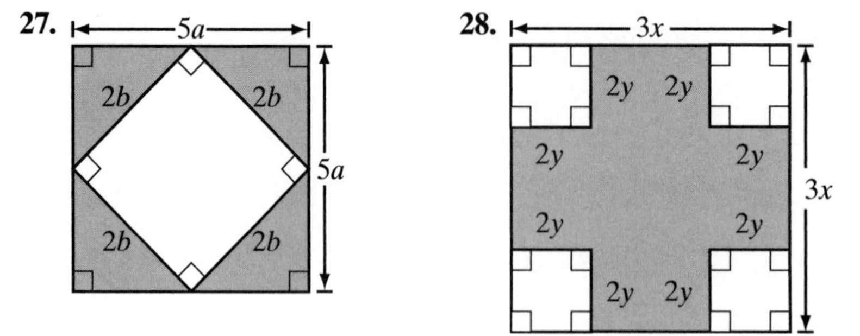

11-6 MULTIPLYING BINOMIALS

We have used the "FOIL" method from Section 5-4 to multiply two binomials of the form $ax + b$ and $cx + d$. The pattern of the multiplication in the following example shows us how to find the product of two binomials *mentally*.

$$\begin{aligned}(2x - 3)(4x + 5) &= 2x(4x) + 2x(5) - 3(4x) - 3(5)\\ &= 8x^2 + 10x - 12x - 15\\ &= 8x^2 - 2x - 15\end{aligned}$$

Examine the trinomial $8x^2 - 2x - 15$ and note the following:

1. The first term of the trinomial is the product of the first terms of the binomials:

$$\overbrace{(2x - 3)(4x}^{+8x^2} + 5) \rightarrow 2x(4x) = 8x^2$$

2. The middle term of the trinomial is the product of the outer terms *plus* the product of the inner terms of the binomials:

Think

$$\underbrace{(2x - \overbrace{3)(4x}^{-12x} + 5)}_{+10x} \rightarrow (-12x) + (+10x) = -2x$$

3. The last term of the trinomial is the product of the last terms of the binomials:

$$(2x - \underbrace{3)(4x + 5)}_{-15} \rightarrow (-3)(+5) = -15$$

Procedure

To find the product of two binomials ($ax + b$) and ($cx + d$) where a, b, c, and d have numerical values:

1. Multiply the first terms of the binomials.
2. Multiply the first term of each binomial by the last term of the other binomial (the outer and inner terms), and add these products.
3. Multiply the last terms of the binomials.
4. Write the results of steps 1, 2, and 3 as a sum.

EXAMPLE 1

Write the product $(x - 5)(x - 7)$ as a trinomial.

Solution

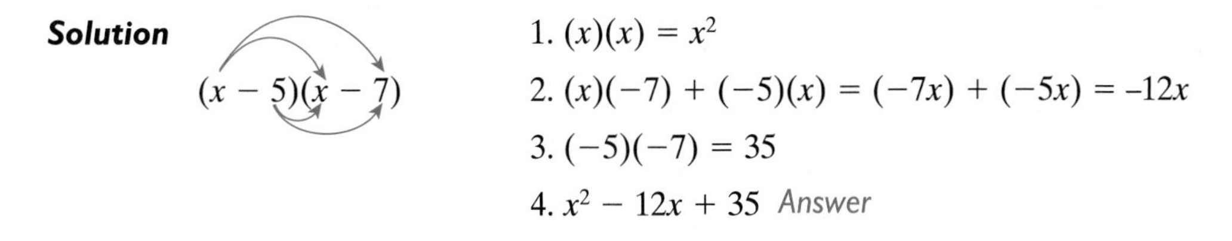

1. $(x)(x) = x^2$
2. $(x)(-7) + (-5)(x) = (-7x) + (-5x) = -12x$
3. $(-5)(-7) = 35$
4. $x^2 - 12x + 35$ *Answer*

EXAMPLE 2

Write the product $(3y - 8)(4y + 3)$ as a trinomial.

Solution

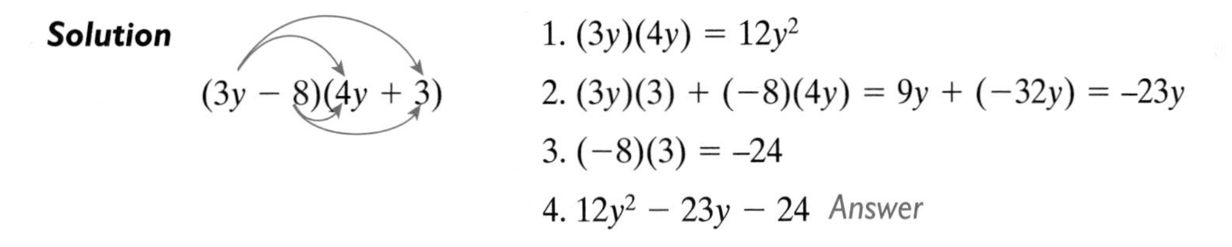

1. $(3y)(4y) = 12y^2$
2. $(3y)(3) + (-8)(4y) = 9y + (-32y) = -23y$
3. $(-8)(3) = -24$
4. $12y^2 - 23y - 24$ *Answer*

EXAMPLE 3

Write the product $(z + 4)^2$ as a trinomial.

Solution $(z + 4)^2 = (z + 4)(z + 4) = z^2 + 4z + 4z + 16 = z^2 + 8z + 16$ *Answer*

A trinomial, such as $z^2 + 8z + 16$, that is the square of a binomial is called a **perfect square trinomial**. Note that the first and last terms are the squares of the terms of the binomial, and the middle term is twice the product of the terms of the binomial.

EXERCISES

Writing About Mathematics

1. What is a shorter procedure for finding the product of binomials of the form $(ax + b)$ and $(ax - b)$?

2. Which part of the procedure for multiplying two binomials of the form $(ax + b)$ and $(cx + d)$ does not apply to the product of binomials such as $(3x + 5)$ and $(2y + 7)$? Explain your answer.

Developing Skills

In 3–29, perform each indicated operation mentally.

3. $(x + 5)(x + 3)$ **4.** $(6 + d)(3 + d)$ **5.** $(x - 10)(x - 5)$

6. $(8 - c)(3 - c)$ **7.** $(x + 7)(x - 2)$ **8.** $(n - 20)(n + 3)$

9. $(5 - t)(9 + t)$ **10.** $(3x + 2)(x + 5)$ **11.** $(c - 5)(3c - 1)$

12. $(y + 8)^2$ **13.** $(a - 4)^2$ **14.** $(2x + 1)^2$

15. $(3x - 2)^2$ **16.** $(7x + 3)(2x - 1)$ **17.** $(2y + 3)(3y + 2)$

18. $(3x + 4)^2$ **19.** $(2x - 5)^2$ **20.** $(3t - 2)(4t + 7)$

21. $(5y - 4)(5y - 4)$ **22.** $(2t + 3)(5t + 1)$ **23.** $(2a - b)(2a - 3b)$

24. $(5x + 7y)(3x - 4y)$ **25.** $(2c - 3d)(5c - 2d)$ **26.** $(a + b)^2$

27. $(a + b)(2a - 3)$ **28.** $(6t - 1)(4t + z)$ **29.** $(9y - w)(9y + 3w)$

Applying Skills

30. Write the trinomial that represents the area of a rectangle whose length and width are:

a. $(x + 5)$ and $(x + 4)$ **b.** $(2x + 3)$ and $(x - 1)$

31. Write the perfect square trinomial that represents the area of a square whose sides are:

a. $(x + 6)$ **b.** $(x - 2)$ **c.** $(2x + 1)$ **d.** $(3x - 2)$

32. The length of a garden is 3 feet longer than twice the width.

a. If the width of the garden is represented by x, represent the length of the garden in terms of x.

b. If the width of the garden is increased by 5 feet, represent the new width in terms of x.

c. If the length of the enlarged garden is also increased so that it is 3 feet longer than twice the new width, represent the length of the enlarged garden in terms of x.

d. Express the area of the enlarged garden in terms of x.

11-7 FACTORING TRINOMIALS

We have learned that $(x + 3)(x + 5) = x^2 + 8x + 15$. Therefore, factors of $x^2 + 8x + 15$ are $(x + 3)$ and $(x + 5)$. Factoring a trinomial of the form $ax^2 + bx + c$ is the reverse of multiplying binomials of the form $(dx + e)$ and $(fx + g)$. When we factor a trinomial of this form, we use combinations of factors of the first and last terms. We list the possible pairs of factors, then test the pairs, one by one, until we find the pair that gives the correct middle term.

For example, let us factor $x^2 + 7x + 10$:

1. The product of the first terms of the binomials must be x^2. Therefore, for each first term, we use x. We write:

$$x^2 + 7x + 10 = (x \qquad)(x \qquad)$$

2. Since the product of the last terms of the binomials must be +10, these last terms must be either both positive or both negative. The pairs of integers whose products is +10 are:

$$(+1)(+10) \qquad (+5)(+2) \qquad (-1)(-10) \qquad (-2)(-5)$$

3. From the products obtained in steps 1 and 2, we see that the possible pairs of factors are:

$$(x + 10)(x + 1) \qquad (x + 5)(x + 2)$$

$$(x - 10)(x - 1) \qquad (x - 5)(x - 2)$$

4. Now, we test each pair of factors:

$(x + 10)(x + 1)$ is not correct because the middle term is $+11x$, not $+7x$.

$$(x + 10)(x + 1) \rightarrow 10x + 1x = 11x \quad ✗$$

(inner: $+10x$; outer: $+1x$)

$(x + 5)(x + 2)$ is correct because the middle term is $+7x$.

$$(x + 5)(x + 2) \rightarrow 2x + 5x = 7x \quad ✓$$

(inner: $+5x$; outer: $+2x$)

Neither of the remaining pairs of factors is correct because each would result in a negative middle term.

5. The factors of $x^2 + 7x + 10$ are $(x + 5)$ and $(x + 2)$.

Observe that in this trinomial, the first term, x^2, is positive and must have factors with the same signs. We usually choose positive factors of the first terms. The last term, +10, is positive and must have factors with the same signs. Since the middle term of the trinomial is also *positive*, the last terms of both binomial factors must be *positive* (+5 and +2).

In this example, if we had chosen $-x$ times $-x$ as the factors of x^2, the factors of $x^2 + 7x + 10$ would have been written as $(-x - 5)$ and $(-x - 2)$. Every trinomial that has two binomial factors also has the opposites of these binomials as factors. Usually, however, we write only the pair of factors whose first terms have positive coefficients as the factors of a trinomial.

Procedure

To factor a trinomial of the form $ax^2 + bx + c$, find two binomials that have the following characteristics:

1. The product of the first terms of the binomials is equal to the first term of the trinomial (ax^2).
2. The product of the last terms of the binomials is equal to the last term of the trinomial (c).
3. When the first term of each binomial is multiplied by the last term of the other binomial and the sum of these products is found, the result is equal to the middle term of the trinomial.

EXAMPLE 1

Factor $y^2 - 8y + 12$.

Solution (1) The product of the first terms of the binomials must be y^2. Therefore, for each first term, we use y. We write:

$$y^2 - 8y + 12 = (y \qquad)(y \qquad)$$

(2) Since the product of the last terms of the binomials must be +12, these last terms must be either both positive or both negative. The pairs of integers whose product is +12 are:

$$(+1)(+12) \qquad (+2)(+6) \qquad (+3)(+4)$$
$$(-1)(-12) \qquad (-2)(-6) \qquad (-3)(-4)$$

(3) The possible factors are:

$$(y + 1)(y + 12) \qquad (y + 2)(y + 6) \qquad (y + 3)(y + 4)$$
$$(y - 1)(y - 12) \qquad (y - 2)(y - 6) \qquad (y - 3)(y - 4)$$

(4) When we find the middle term in each of the trinomial products, we see that only for the factors $(y - 6)(y - 2)$ is the middle term $-8y$:

$$\overset{-6y}{(y - 6)(y - 2)} \rightarrow -2y - 6y = -8y$$
$$-2y$$

Answer $y^2 - 8y + 12 = (y - 6)(y - 2)$

When the first and last terms are both positive (y^2 and $+12$ in this example) and the middle term of the trinomial is *negative*, the last terms of both binomial factors must be *negative* (-6 and -2 in this example).

EXAMPLE 2

Factor: **a.** $c^2 + 5c - 6$ **b.** $c^2 - 5c + 6$

Solution **a.** (1) The product of the first terms of the binomials must be c^2. Therefore, for each first term, we use c. We write:

$$c^2 + 5c - 6 = (c \quad)(c \quad)$$

(2) Since the product of the last terms of the binomials must be –6, one of these last terms must be positive and the other negative. The pairs of integers whose product is –6 are:

$$(+1)(-6) \qquad (+3)(-2)$$
$$(-1)(+6) \qquad (-3)(+2)$$

(3) The possible factors are:

$$(c + 1)(c - 6) \qquad (c + 3)(c - 2)$$
$$(c - 1)(c + 6) \qquad (c - 3)(c + 2)$$

(4) When we find the middle term in each of the trinomial products, we see that only for the factors $(c - 1)(c + 6)$ is the middle term $+5c$:

$$(c - 1)(c + 6) \rightarrow 6c - 1c = 5c$$

(inner product: $-1c$; outer product: $+6c$)

b. (1) As in the solution to part **a**, we write:

$$c^2 - 5c + 6 = (c \quad)(c \quad)$$

(2) Since the product of the last terms of the binomials must be +6, these last terms must be either both positive or both negative. The pairs of integers whose product is +6 are:

$$(+1)(+6) \qquad (+3)(+2)$$
$$(-1)(-6) \qquad (-3)(-2)$$

(3) The possible factors are:

$$(c + 1)(c + 6) \qquad (c + 3)(c + 2)$$
$$(c - 1)(c - 6) \qquad (c - 3)(c - 2)$$

(4) When we find the middle term in each of the trinomial products, we see that only for the factors $(c - 3)(c - 2)$ is the middle term $-5c$.

$$(c - 3)(c - 2) \rightarrow -2c - 3c = -5c$$

(inner product: $-3c$; outer product: $-2c$)

Answer **a.** $c^2 + 5c - 6 = (c - 1)(c + 6)$ **b.** $c^2 - 5c + 6 = (c - 3)(c - 2)$

EXAMPLE 3

Factor $2x^2 - 7x - 15$.

Solution (1) Since the product of the first terms of the binomials must be $2x^2$, we use $2x$ and x as the first terms. We write:

$$2x^2 - 7x - 15 = (2x \quad)(x \quad)$$

(2) Since the product of the last terms of the binomials must be -15, one of these last terms must be positive and the other negative. The pairs of integers whose product is -15 are:

$$(+1)(-15) \qquad (+3)(-5)$$
$$(-1)(+15) \qquad (-3)(+5)$$

(3) These four pairs of integers will form eight pairs of binomial factors since there are two ways in which the first terms can be arranged. Note how $(2x + 1)(x - 15)$ is not the same product as $(2x - 15)(x + 1)$. The possible pairs of factors are:

$$(2x + 1)(x - 15) \quad (2x + 3)(x - 5) \quad (2x - 1)(x + 15) \quad (2x - 3)(x + 5)$$
$$(2x + 15)(x - 1) \quad (2x + 5)(x - 3) \quad (2x - 15)(x + 1) \quad (2x - 5)(x + 3)$$

(4) When we find the middle term in each of the trinomial products, we see that only for the factors $(2x + 3)(x - 5)$ is the middle term $-7x$:

$$\overset{+3x}{}$$
$$(2x + 3)(x - 5) \rightarrow 3x - 10x = -7x$$
$$\underset{-10x}{}$$

Answer $2x^2 - 7x - 15 = (2x + 3)(x - 5)$

In factoring a trinomial of the form $ax^2 + bx + c$, when a is a positive integer $(a > 0)$:

1. The coefficients of the first terms of the binomial factors are usually written as positive integers.

2. If the last term, c, is positive, the last terms of the binomial factors must be either both positive (when the middle term, b, is positive), or both negative (when the middle term, b, is negative).

3. If the last term, c, is negative, one of the last terms of the binomial factors must be positive and the other negative.

EXERCISES

Writing About Mathematics

1. Alicia said that the factors of $x^2 + bx + c$ are $(x + d)(x + e)$ if $c = de$ and $b = d + e$. Do you agree with Alicia? Explain why or why not.
2. Explain why there is no positive integral value of c for which $x^2 + x + c$ has two binomial factors.

Developing Skills

In 3–33, factor each trinomial.

3. $a^2 + 3a + 2$
4. $c^2 + 6c + 5$
5. $x^2 + 8x + 7$
6. $x^2 - 11x + 10$
7. $y^2 - 6y + 8$
8. $y^2 + 6y + 8$
9. $y^2 - 9y + 8$
10. $y^2 + 9y + 8$
11. $y^2 - 2y - 8$
12. $y^2 + 2y - 8$
13. $y^2 - 7y - 8$
14. $y^2 + 7y - 8$
15. $x^2 + 11x + 24$
16. $a^2 - 11a + 18$
17. $z^2 + 10z + 25$
18. $x^2 - 5x + 6$
19. $x^2 - 10x + 24$
20. $x^2 - x - 2$
21. $x^2 - 6x - 7$
22. $y^2 + 4y - 5$
23. $c^2 + 2c - 35$
24. $x^2 - 7x - 18$
25. $z^2 + 9z - 36$
26. $2x^2 + 5x + 2$
27. $3x^2 + 10x + 8$
28. $16x^2 + 8x + 1$
29. $2x^2 + x - 3$
30. $4x^2 - 12x + 5$
31. $10a^2 - 9a + 2$
32. $3a^2 - 7ab + 2b^2$
33. $4x^2 - 5xy - 6y^2$

Applying Skills

In 34–36, each trinomial represents the area of a rectangle. In each case, find two binomials that could be expressions for the dimensions of the rectangle.

34. $x^2 + 10x + 9$
35. $x^2 + 9x + 20$
36. $3x^2 + 14x + 15$

In 37–39, each trinomial represents the area of a square. In each case, find a binomial that could be an expression for the measure of each side of the square.

37. $x^2 + 10x + 25$
38. $81x^2 + 18x + 1$
39. $4x^2 + 12x + 9$

11-8 FACTORING A POLYNOMIAL COMPLETELY

Some polynomials, such as $x^2 + 4$ and $x^2 + x + 1$, cannot be factored into other polynomials with integral coefficients. We say that these polynomials are **prime over the set of integers.**

Factoring a polynomial completely means finding the prime factors of the polynomial over a designated set of numbers. In this book, we will consider a polynomial factored when it is written as a product of monomials or prime polynomials over the set of integers.

Procedure

To factor a polynomial completely:

1. Look for the greatest common factor. If it is greater than 1, factor the given polynomial into a monomial times a polynomial.
2. Examine the polynomial factor or the given polynomial if it has no common factor (the greatest common factor is 1). Then:
   - Factor any trinomial into the product of two binomials if possible.
   - Factor any binomials that are the difference of two perfect squares as such.
3. Write the given polynomial as the product of all of the factors. Make certain that all factors except the monomial factor are prime polynomials.

EXAMPLE 1

Factor $by^2 - 4b$.

Solution *How to Proceed*

(1) Find the greatest common factor of the terms: $by^2 - 4b = b(y^2 - 4)$

(2) Factor the difference of two squares: $by^2 - 4b = b(y + 2)(y - 2)$ *Answer*

EXAMPLE 2

Factor $3x^2 - 6x - 24$.

Solution *How to Proceed*

(1) Find the greatest common factor of the terms: $3x^2 - 6x - 24 = 3(x^2 - 2x - 8)$

(2) Factor the trinomial: $3x^2 - 6x - 24 = 3(x - 4)(x + 2)$ *Answer*

EXAMPLE 3

Factor $4d^2 - 6d + 2$.

Solution *How to Proceed*

(1) Find the greatest common factor of the terms: $4d^2 - 6d + 2 = 2(2d^2 - 3d + 1)$

(2) Factor the trinomial: $4d^2 - 6d + 2 = 2(2d - 1)(d - 1)$ *Answer*

EXAMPLE 4

Factor $x^4 - 16$.

Solution *How to Proceed*

(1) Find the greatest common factor of the terms:	The greatest common factor is 1.
(2) Factor the binomial as the difference of two squares:	$x^4 - 16 = (x^2 + 4)(x^2 - 4)$
(3) Factor the difference of two squares:	$x^4 - 16 = (x^2 + 4)(x + 2)(x - 2)$ Answer

EXERCISES

Writing About Mathematics

1. Greta said that since $4a^2 - a^2b^2$ is the difference of two squares, the factors are $(2a + ab)(2a - ab)$. Has Greta factored $4a^2 - a^2b^2$ into prime polynomial factors? Explain why or why not.

2. Raul said that the factors of $x^3 - 1$ are $(x + 1)(x + 1)(x - 1)$. Do you agree with Raul? Explain why or why not.

Developing Skills

In 3–29, factor each polynomial completely.

3. $2a^2 - 2b^2$
4. $4x^2 - 4$
5. $ax^2 - ay^2$
6. $st^2 - 9s$
7. $2x^2 - 32$
8. $3x^2 - 27y^2$
9. $18m^2 - 8$
10. $63c^2 - 7$
11. $x^3 - 4x$
12. $z^3 - z$
13. $4a^2 - 36$
14. $x^4 - 1$
15. $y^4 - 81$
16. $\pi c^2 - \pi d^2$
17. $3x^2 + 6x + 3$
18. $4r^2 - 4r - 48$
19. $x^3 + 7x^2 + 10x$
20. $4x^2 - 6x - 4$
21. $d^3 - 8d^2 + 16d$
22. $2ax^2 - 2ax - 12a$
23. $16x^2 - x^2y^4$
24. $a^4 - 10a^2 + 9$
25. $y^4 - 13y^2 + 36$
26. $5x^4 + 10x^2 + 5$
27. $2a^2b + 7ab + 3b$
28. $16x^2 - 16x + 4$
29. $25x^2 + 100xy + 100y^2$

Applying Skills

30. The volume of a rectangular solid is represented by $12a^3 - 5a^2b - 2ab^2$. Find the algebraic expressions that could represent the dimensions of the solid.

31. A rectangular solid has a square base. The volume of the solid is represented by $3m^2 + 12m + 12$.

a. Find the algebraic expression that could represent the height of the solid.

b. Find the algebraic expression that could represent the length of each side of the base.

32. A rectangular solid has a square base. The volume of the solid is represented by $10a^3 + 20a^2 + 10a$.

a. Find the algebraic expression that could represent the height of the solid.

b. Find the algebraic expression that could represent the length of each side of the base

CHAPTER SUMMARY

Factoring a number or a polynomial is the process of finding those numbers or polynomials whose product is the given number or polynomial. A **perfect square trinomial** is the square of a binomial, whereas the **difference of two perfect squares** is the product of binomials that are the sum and difference of the same two. A **prime polynomial**, like a prime number, has only two factors, 1 and itself.

To factor a polynomial completely:

1. Factor out the greatest common monomial factor if it is greater than 1.
2. Write any factor of the form $a^2 - b^2$ as $(a + b)(a - b)$.
3. Write any factor of the form $ax^2 + bx + c$ as the product of two binomial factors if possible.
4. Write the given polynomial as the product of these factors.

VOCABULARY

11-1 Product • Factoring a number • Factoring over the set of integers • Greatest common factor

11-2 Factoring a polynomial • Common monomial factor • Greatest common monomial factor • Prime polynomial

11-3 Square of a monomial

11-5 Difference of two perfect squares

11-6 Perfect square trinomial

11-8 Prime over the set of integers • Factoring a polynomial completely

REVIEW EXERCISES

1. Leroy said that $4x^2 + 16x + 12 = (2x + 2)(2x + 6) = 2(x + 1)(x + 3)$. Do you agree with Leroy? Explain why or why not.

2. Express 250 as a product of prime numbers.

3. What is the greatest common factor of $8ax$ and $4ay$?

4. What is the greatest common factor of $16a^3bc^2$ and $24a^2bc^4$?

In 5–8, square each monomial.

5. $(3g^3)^2$ **6.** $(-4x^4)^2$ **7.** $(0.2c^2y)^2$ **8.** $\left(\frac{1}{2}a^3b^5\right)^2$

In 9–14, find each product.

9. $(x - 5)(x + 9)$ **10.** $(y - 8)(y - 6)$ **11.** $(ab + 4)(ab - 4)$

12. $(3d + 1)(d - 2)$ **13.** $(2w + 1)^2$ **14.** $(2x + 3c)(x + 4c)$

In 15–29, in each case factor completely.

15. $6x + 27b$ **16.** $3y^2 + 10y$ **17.** $m^2 - 81$

18. $x^2 - 16h^2$ **19.** $x^2 - 4x - 5$ **20.** $y^2 - 9y + 14$

21. $64b^2 - 9$ **22.** $121 - k^2$ **23.** $x^2 - 8x + 16$

24. $a^2 - 7a - 30$ **25.** $x^2 - 16x + 60$ **26.** $16y^2 - 16$

27. $2x^2 + 12bx - 32b^2$ **28.** $x^4 - 1$ **29.** $3x^3 - 6x^2 - 24x$

30. Express the product $(k + 15)(k - 15)$ as a binomial.

31. Express $4ez^2(4e - z)$ as a binomial.

32. Factor completely: $60a^2 + 37a - 6$.

33. Which of the following polynomials has a factor that is a perfect square trinomial and a factor that is a perfect square?

(1) $a^2y + 10ay + 25y$ (3) $18m^2 + 24m + 8$

(2) $2ax^2 - 2ax - 12a$ (4) $c^2z^2 - 18cz^2 + 81z^2$

34. Of the four polynomials given below, explain how each is different from the others.

$x^2 - 9$ $x^2 - 2x + 1$ $x^2 - 2x - 1$ $x^3 + 5x^2 + 6x$

35. If the length and width of a rectangle are represented by $2x - 3$ and $3x - 2$, respectively, express the area of the rectangle as a trinomial.

36. Find the trinomial that represents the area of a square if the measure of a side is $8m + 1$.

37. If $9x^2 + 30x + 25$ represents the area of a square, find the binomial that represents the length of a side of the square.

38. A group of people wants to form several committees, each of which will have the same number of persons. Everyone in the group is to serve on one and only one committee. When the group tries to make 2, 3, 4, 5, or 6 committees, there is always one extra person. However, they are able to make more than 6 but fewer than 12 committees of equal size.

a. What is the smallest possible number of persons in the group?

b. Using the group size found in **a**, how many persons are on each committee that is formed?

Exploration

a. Explain how the expression $(a + b)(a - b) = a^2 - b^2$ can be used to find the product of two consecutive even or two consecutive odd integers.

b. The following diagrams illustrate the formula:

$$1 + 3 + 5 + \ldots + (2n - 3) + (2n - 1) = n^2$$

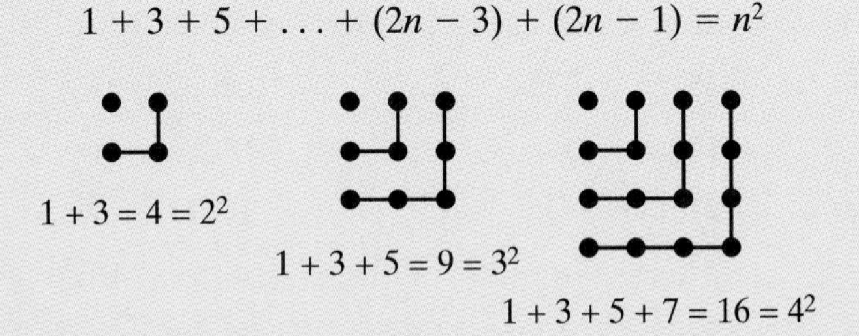

(1) Explain how 3, 5, and 7 are the difference of two squares using the diagrams.

(2) Use the formula to explain how any odd number, $2n - 1$, can be written as the difference of squares.
[*Hint:* $1 + 3 + 5 + \ldots + (2n - 3) = (n - 1)^2$]

CUMULATIVE REVIEW — CHAPTERS 1–11

Part I

Answer all questions in this part. Each correct answer will receive 2 credits. No partial credit will be allowed.

1. If apples cost \$3.75 for 3 pounds, what is the cost, at the same rate, of 7 pounds of apples?

(1) \$7.25 (2) \$8.25 (3) \$8.50 (4) \$8.75

2. When $3a^2 - 4$ is subtracted from $2a^2 - 4a$, the difference is
(1) $-a^2 - 4a + 4$ (2) $a^2 + 4a - 4$ (3) $a^2 + 8a$ (4) $-a^2 - a$

3. The solution of the equation $x - 3(2x + 4) = 7x$ is
(1) -6 (2) 6 (3) -1 (4) 1

4. The area of a circle is 16π square centimeters. The circumference of the circle is
(1) 16π centimeters (3) 4π centimeters
(2) 8π centimeters (4) 2π centimeters

5. The x-intercept of the line whose equation is $2x - y = 5$ is
(1) 5 (2) -5 (3) $\frac{2}{5}$ (4) $\frac{5}{2}$

6. The factors of $x^2 - 3x - 10$ are
(1) $(x + 5)(x - 2)$ (3) $(x - 5)(x + 2)$
(2) $(x - 5)(x - 2)$ (4) $(x + 5)(x + 2)$

7. The fraction $\frac{2.5 \times 10^{-5}}{5.0 \times 10^{-8}}$ is equal to
(1) 5×10^{-3} (2) 5×10^{-2} (3) 5×10^{2} (4) 5×10^{3}

8. When factored completely, $5a^3 - 45a$ is equal to
(1) $5a(a + 3)^2$ (3) $a(5a + 9)(a - 5)$
(2) $5(a^2 + 3)(a - 3)$ (4) $5a(a + 3)(a - 3)$

9. When $a = -3$, $-a^2 + 5a$ is equal to
(1) 6 (2) -6 (3) 24 (4) -24

10. The graph of the equation $2x + y = 7$ is parallel to the graph of
(1) $y = 2x + 3$ (2) $y = -2x + 5$ (3) $x + 2y = 4$ (4) $y - 2x = 2$

Part II

Answer all questions in this part. Each correct answer will receive 2 credits. Clearly indicate the necessary steps, including appropriate formula substitutions, diagrams, graphs, charts, etc. For all questions in this part, a correct numerical answer with no work shown will receive only 1 credit.

11. The measures of the sides of a right triangle are 32, 60, and 68. Find, to the nearest degree, the measure of the smallest angle of the triangle.

12. Solve and check: $\frac{x}{x + 8} = \frac{2}{3}$.

Part III

Answer all questions in this part. Each correct answer will receive 3 credits. Clearly indicate the necessary steps, including appropriate formula substitutions, diagrams, graphs, charts, etc. For all questions in this part, a correct numerical answer with no work shown will receive only 1 credit.

13. The width of Mattie's rectangular garden is 10 feet less than the length. She bought 25 yards of fencing and used all but 7 feet to enclose the garden.

a. Write an equation or a system of equations that can be used to find the dimensions, in feet, of the garden.

b. What are the dimensions, in feet, of the garden?

14. The population of a small town at the beginning of the year was 7,000. Records show that during the year there were 5 births, 7 deaths, 28 new people moved into town, and 12 residents moved out. What was the percent of increase or decrease in the town population?

Part IV

Answer all questions in this part. Each correct answer will receive 4 credits. Clearly indicate the necessary steps, including appropriate formula substitutions, diagrams, graphs, charts, etc. For all questions in this part, a correct numerical answer with no work shown will receive only 1 credit.

15. Admission to a museum is $5.00 for adults and $3.00 for children. In one day $2,400 was collected for 620 paid admissions. How many adults and how many children paid to visit the museum that day?

16. a. Write an equation of a line whose slope is -2 and whose y-intercept is 5.

b. Sketch the graph of the equation written in **a**.

c. Are $(-1, 3)$ the coordinates of a point on the graph of the equation written in **a**?

d. If $(k, 1)$ are the coordinates of a point on the graph of the equation written in **a**, what is the value of k?

CHAPTER

12

OPERATIONS WITH RADICALS

Whenever a satellite is sent into space, or astronauts are sent to the moon, technicians at earthbound space centers monitor activities closely. They continually make small corrections to help the spacecraft stay on course.

The distance from the earth to the moon varies, from 221,460 miles to 252,700 miles, and both the earth and the moon are constantly rotating in space. A tiny error can send the craft thousands of miles off course. Why do such errors occur?

Space centers rely heavily on sophisticated computers, but computers and calculators alike work with *approximations* of numbers, not necessarily with *exact* values.

We have learned that irrational numbers, such as $\sqrt{2}$ and $\sqrt{5}$, are shown on a calculator as decimal approximations of their true values. All irrational numbers, which include radicals, are nonrepeating decimals that never end. How can we work with them?

In this chapter, we will learn techniques to compute with radicals to find exact answers. We will also look at methods for working with radicals on a calculator to understand how to minimize errors when using these devices.

CHAPTER TABLE OF CONTENTS

12-1 RADICALS AND THE RATIONAL NUMBERS

Squares and Square Roots

Recall from Section 1-3 that to *square* a number means to multiply the number by itself. To square 8, we write:

$$8^2 = 8 \times 8 = 64$$

On a calculator:

ENTER: 8 **x^2** **ENTER**

DISPLAY:

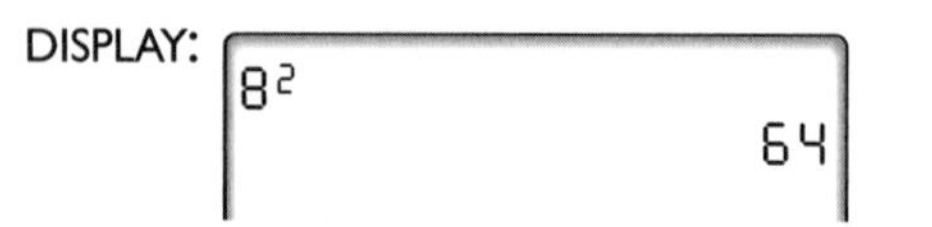

To find the *square root* of a number means to find the value that, when multiplied by itself, is equal to the given number. To express the square root of 64, we write:

$$\sqrt{64} = 8$$

On a calculator:

ENTER: **2nd** **√** 64 **ENTER**

DISPLAY:

√(64
8

The symbol $\sqrt{\ }$ is called the *radical sign*, and the quantity under the radical sign is the **radicand**. For example, in $\sqrt{64}$, which we read as "the square root of 64," the radicand is 64.

A **radical**, which is any term containing both a radical sign and a radicand, is a root of a quantity. For example, $\sqrt{64}$ is a radical.

▶ **In general, the square root of b is x (written as $\sqrt{b} = x$) if and only if $x \geq 0$ and $x^2 = b$.**

Some radicals, such as $\sqrt{4}$ and $\sqrt{9}$, are rational numbers; others, such as $\sqrt{2}$ and $\sqrt{3}$, are irrational numbers. We begin this study of radicals by examining radicals that are rational numbers.

Perfect Squares

Any number that is the square of a rational number is called a **perfect square**. For example,

$3 \times 3 = 9$ $\quad 0 \times 0 = 0$ $\quad 1.4 \times 1.4 = 1.96$ $\quad \frac{2}{7} \times \frac{2}{7} = \frac{4}{49}$

Therefore, perfect squares include 9, 0, 1.96, and $\frac{4}{49}$.

Then, by applying the inverse operation, we know that:

► **The square root of every perfect square is a *rational* number.**

$\sqrt{9} = 3$ $\quad \sqrt{0} = 0$ $\quad \sqrt{1.96} = 1.4$ $\quad \sqrt{\frac{4}{49}} = \frac{2}{7}$

Radicals That Are Square Roots

Certain generalizations can be made for *all* radicals that are square roots, whether they are rational numbers or irrational numbers:

1. Since the square root of 36 is a number whose square is 36, we can write the statement $\left(\sqrt{36}\right)^2 = 36$. We notice that $\left(\sqrt{36}\right)^2 = (6)^2 = 36$. It is also true that $\sqrt{(6)^2} = \sqrt{36} = 6$.

► **In general, for every nonnegative real number n: $\left(\sqrt{n}\right)^2 = n$ and $\sqrt{n^2} = n$.**

2. Since $(+6)(+6) = 36$ and $(-6)(-6) = 36$, both $+6$ and -6 are square roots of 36. This example illustrates the following statement:

► **Every positive number has two square roots: one root is a positive number called the *principal square root*, and the other root is a negative number. These two roots have the same absolute value.**

To indicate the positive or principal square root only, place a radical sign over the number:

$$\sqrt{25} = 5 \qquad \text{and} \qquad \sqrt{0.49} = 0.7$$

To indicate the negative square root only, place a negative sign before the radical:

$$-\sqrt{25} = -5 \qquad \text{and} \qquad -\sqrt{0.49} = -0.7$$

To indicate both square roots, place both a positive and a negative sign before the radical:

$$\pm\sqrt{25} = \pm 5 \qquad \text{and} \qquad \pm\sqrt{0.49} = \pm 0.7$$

3. Every real number, when squared, is either positive or 0. Therefore:

► **The square root of a negative number does not exist in the set of real numbers.**

For example, $\sqrt{-25}$ does not exist in the set of real numbers because there is no real number that, when multiplied by itself, equals -25.

Calculators and Square Roots

A calculator will return only the principal square root of a positive number. To display the negative square root of a number, the negative sign must be entered before the radical.

ENTER: (-) 2nd √ 25 ENTER

DISPLAY:

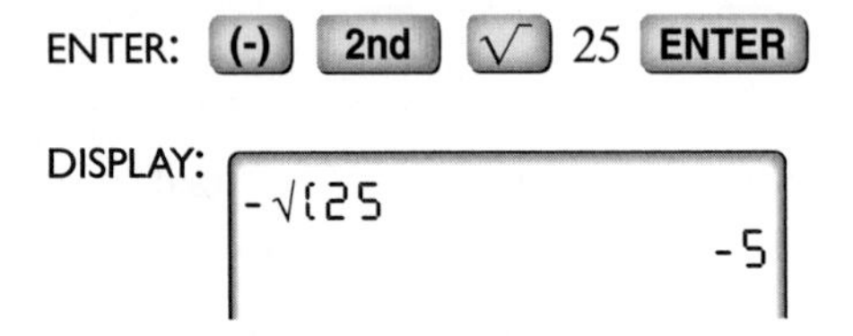

The calculator will display an error message if it is set in "real" mode and the square root of a negative number is entered.

ENTER: 2nd √ (-) 25 ENTER

DISPLAY:

ERR:NONREAL ANS
1:QUIT
2:GOTO

Cube Roots and Other Roots

A **cube root** of a number is one of the three equal factors of the number. For example, 2 is a cube root of 8 because $2 \times 2 \times 2 = 8$, or $2^3 = 8$. A cube root of 8 is written as $\sqrt[3]{8}$.

► **Finding a cube root of a number is the inverse operation of cubing a number. In general, the cube root of b is x (written as $\sqrt[3]{b} = x$) if and only if $x^3 = b$.**

We have said that $\sqrt{-25}$ does not exist in the set of real numbers. However, $\sqrt[3]{-8}$ does exist in the set of real numbers. Since $(-2)^3 = (-2)(-2)(-2) = -8$, then $\sqrt[3]{-8} = -2$.

In the set of real numbers, every number has one cube root. The cube root of a positive number is positive, and the cube root of a negative number is negative.

In the expression $\sqrt[n]{b}$, the integer n that indicates the root to be taken, is called the **index** of the radical. Here are two examples:

- In $\sqrt[3]{8}$, read as "the cube root of 8," the index is 3.
- In $\sqrt[4]{16}$, read as "the fourth root of 16," the index is 4. Since $2^4 = 16$, 2 is one of the four equal factors of 16, and $\sqrt[4]{16} = 2$.

When no index appears, the index is understood to be 2. Thus, $\sqrt{25} = \sqrt[2]{25} = 5$.

When the index of the root is even and the radicand is positive, the value is a real number. That real number is positive if a plus sign or no sign precedes the radical, and negative if a minus sign precedes the radical.

When the index of the root is even and the radicand is negative, the root has no real number value.

When the index of the root is odd and the radicand either positive or negative, the value is a real number. That real number is positive if the radicand is positive and negative if the radicand is negative.

Calculators and Roots

A radical that has an index of 3 or larger can be evaluated on most graphing calculators. To do so, first press the MATH key to display a list of choices. Cube root is choice 4. To show that $\sqrt[3]{64} = 4$:

ENTER: MATH 4 64 ENTER

DISPLAY:

```
3√(64
                4
```

Any root with an index greater than 3 can be found using choice 5 of the MATH menu. The index, indicated by x in the menu, must be entered first.

To show that $\sqrt[4]{625} = 5$:

ENTER: 4 MATH 5 625 ENTER

DISPLAY:

```
4ˣ√625
                5
```

EXAMPLE 1

Find the principal square root of 361.

Solution Since $19 \times 19 = 361$, then $\sqrt{361} = 19$.

Calculator Solution ENTER: **2nd** **√** 361 **ENTER**

DISPLAY:

```
√(361
                19
```

Answer 19

EXAMPLE 2

Find the value of $-\sqrt{0.0016}$.

Solution Since $(0.04)(0.04) = 0.0016$, then $-\sqrt{0.0016} = -0.04$.

Calculator Solution ENTER: **(-)** **2nd** **√** .0016 **ENTER**

DISPLAY:

```
-√(.0016
               -.04
```

Answer -0.04

EXAMPLE 3

Is $\left(\sqrt{13}\right)^2$ a rational or an irrational number? Explain your answer.

Solution Since $\left(\sqrt{n}\right)^2 = n$ for $n \geq 0$, then $\left(\sqrt{13}\right)^2 = 13$.

Answer The quantity $\left(\sqrt{13}\right)^2$ is a rational number since its value, 13, can be written as the quotient of two integers, $\frac{13}{1}$, where the denominator is not 0.

Note: $\left(\sqrt{-13}\right)^2$ does not exist in the set of real numbers since, by the order of operations, $\sqrt{-13}$ must be evaluated first. There is no real number that, when squared, equals -13.

EXAMPLE 4

Solve for x: $x^2 = 36$.

Solution If $x^2 = a$, then $x = \pm\sqrt{a}$ when a is a positive number.

	Check	
$x^2 = 36$	$x^2 = 36$	$x^2 = 36$
$x = \pm\sqrt{36}$	$(+6)^2 = 36$	$(-6)^2 = 36$
$x = \pm 6$	$36 = 36$ ✔	$36 = 36$ ✔

Answer $x = +6$ or $x = -6$; the solution set is $\{+6, -6\}$.

EXERCISES

Writing About Mathematics

1. Explain the difference between $-\sqrt{9}$ and $\sqrt{-9}$.

2. We know that $5^3 = 125$ and $5^4 = 625$. Explain why $\sqrt[3]{-125} = -5$ but $\sqrt[4]{-625} \neq -5$.

Developing Skills

In 3–22, express each radical as a rational number with the appropriate signs.

3. $\sqrt{16}$ **4.** $-\sqrt{64}$ **5.** $\pm\sqrt{100}$ **6.** $\pm\sqrt{169}$ **7.** $\sqrt{400}$

8. $-\sqrt{625}$ **9.** $\sqrt{\frac{1}{4}}$ **10.** $-\sqrt{\frac{9}{16}}$ **11.** $\pm\sqrt{\frac{25}{81}}$ **12.** $\sqrt{0.64}$

13. $-\sqrt{1.44}$ **14.** $\pm\sqrt{0.09}$ **15.** $\pm\sqrt{0.0004}$ **16.** $\sqrt[3]{1}$ **17.** $\sqrt[4]{81}$

18. $\sqrt[5]{32}$ **19.** $\sqrt[3]{-8}$ **20.** $-\sqrt[3]{-125}$ **21.** $\sqrt[4]{0.1296}$ **22.** $-\sqrt{\frac{36}{4}}$

In 23–32, evaluate each radical by using a calculator.

23. $\sqrt{10.24}$ **24.** $-\sqrt{46.24}$ **25.** $\sqrt[3]{2.197}$ **26.** $\sqrt[3]{-3,375}$ **27.** $\sqrt[4]{4,096}$

28. $\sqrt[5]{-1,024}$ **29.** $-\sqrt[3]{-1,000}$ **30.** $-\sqrt{32.49}$ **31.** $\sqrt[3]{-0.125}$ **32.** $\pm\sqrt{5.76}$

In 33–38, find the value of each expression in simplest form.

33. $\left(\sqrt{8}\right)^2$ **34.** $\sqrt{\left(\frac{1}{2}\right)^2}$ **35.** $\left(\sqrt{0.7}\right)^2$

36. $\sqrt{\left(\frac{9}{3}\right)^2}$ **37.** $\left(\sqrt{97}\right)\left(\sqrt{97}\right)$ **38.** $\sqrt{(-9)^2} + \left(\sqrt{83}\right)^2$

In 39–47, replace each ? with >, <, or = to make a true statement.

39. $\frac{3}{4} \ ? \ \left(\frac{3}{4}\right)^2$ **40.** $1 \ ? \ 1^2$ **41.** $\frac{3}{2} \ ? \ \left(\frac{3}{2}\right)^2$

42. $\frac{1}{9} \ ? \ \sqrt{\frac{1}{9}}$ **43.** $\frac{4}{25} \ ? \ \sqrt{\frac{4}{25}}$ **44.** $1 \ ? \ \sqrt{1}$

45. $m \ ? \ \sqrt{m}, 0 < m < 1$ **46.** $m \ ? \ \sqrt{m}, m = 1$ **47.** $m \ ? \ \sqrt{m}, m > 1$

In 48–55, solve each equation for the variable when the replacement set is the set of real numbers.

48. $x^2 = 4$ **49.** $y^2 = \frac{4}{81}$ **50.** $x^2 = 0.49$ **51.** $x^2 - 16 = 0$

52. $y^2 - 30 = 6$ **53.** $2x^2 = 50$ **54.** $3y^2 - 27 = 0$ **55.** $x^3 = 8$

Applying Skills

In 56–59, in each case, find the length of the hypotenuse of a right triangle when the lengths of the legs have the given values.

56. 6 inches and 8 inches **57.** 5 centimeters and 12 centimeters

58. 15 meters and 20 meters **59.** 15 feet and 36 feet

In 60–63, find, in each case: **a.** the length of each side of a square that has the given area **b.** the perimeter of the square.

60. 36 square feet **61.** 196 square yards

62. 121 square centimeters **63.** 225 square meters

64. Express in terms of x, $(x > 0)$, the perimeter of a square whose area is represented by x^2.

65. Write each of the integers from 101 to 110 as the sum of the smallest possible number of perfect squares. (For example, $99 = 7^2 + 7^2 + 1^2$.) Use positive integers only.

12-2 RADICALS AND THE IRRATIONAL NUMBERS

We have learned that $\sqrt{n}$ is a rational number when n is a perfect square. What type of number is $\sqrt{n}$ when n is *not* a perfect square? As an example, let us examine $\sqrt{5}$ using a calculator.

ENTER: 2nd √ 5 ENTER

DISPLAY:

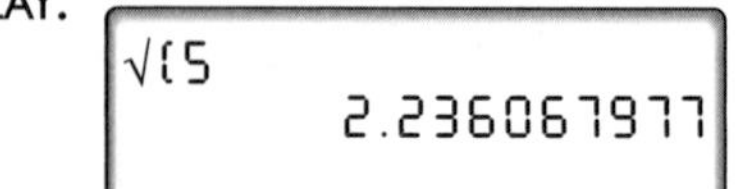

Can we state that $\sqrt{5} = 2.236067977$? Or is 2.236067977 a rational approximation of $\sqrt{5}$? To answer this question, we will find the square of 2.236067977.

ENTER: 2.236067977 **x²** **ENTER**

DISPLAY:

```
2.236067977²
      4.999999998
```

We know that if $\sqrt{b} = x$, then $x^2 = b$. The calculator displays shown above demonstrate that $\sqrt{5} \neq 2.236067977$ because $(2.236067977)^2 \neq 5$. The rational number 2.236067977 is an approximate value for $\sqrt{5}$.

Recall that a number such as $\sqrt{5}$ is called an *irrational number*. Irrational numbers cannot be expressed in the form $\frac{a}{b}$, where a and b are integers and $b \neq 0$. Furthermore, irrational numbers cannot be expressed as terminating or repeating decimals. The example above illustrates the truth of the following statement:

▶ **If n is any positive number that is *not* a perfect square, then $\sqrt{n}$ is an irrational number.**

Radicals and Estimation

The radical $\sqrt{5}$ represents the *exact value* of the irrational number whose square is 5. The calculator display for $\sqrt{5}$ is a *rational approximation* of the irrational number. It is a number that is close to, but not equal to $\sqrt{5}$. There are other values correctly rounded from the calculator display that are also approximations of $\sqrt{5}$:

2.236067977	calculator display, to nine decimal places
2.236068	rounded to six decimal places (nearest millionth)
2.23607	rounded to five decimal places (nearest hundred-thousandth)
2.2361	rounded to four decimal places (nearest ten-thousandth)
2.236	rounded to three decimal places (nearest thousandth)
2.24	rounded to two decimal places (nearest hundredth)

Each rational approximation of $\sqrt{5}$, as seen above, indicates that $\sqrt{5}$ is greater than 2 but less than 3. This fact can be further demonstrated by placing the square of $\sqrt{5}$, which is 5, between the squares of two consecutive integers, one less than 5 and one greater than 5, and then finding the square root of each number.

$$\begin{array}{ll} \text{Since} & 4 < 5 < 9, \\ \text{then} & \sqrt{4} < \sqrt{5} < \sqrt{9}, \\ \text{or} & 2 < \sqrt{5} < 3. \end{array}$$

In the same way, to get a quick estimate of any square-root radical, we place its square between the squares of two consecutive integers. Then we take the square root of each term to show between which two consecutive integers the radical lies. For example, to estimate $\sqrt{73}$:

$$\begin{aligned} &\text{Since } & 64 &< 73 < 81, \\ &\text{then } & \sqrt{64} &< \sqrt{73} < \sqrt{81}, \\ &\text{or } & 8 &< \sqrt{73} < 9. \end{aligned}$$

Basic Rules for Radicals That Are Irrational Numbers

There are general rules to follow when working with radicals, especially those that are irrational numbers:

1. If the degree of accuracy is not specified in a question, it is best to give the *exact* answer in radical form. In other words, if the answer involves a radical, leave the answer in radical form.

   For example, the sum of 2 and $\sqrt{5}$ is written as $2 + \sqrt{5}$, an *exact* value.

2. If the degree of accuracy is not specified in a question and a rational approximation is to be given, the approximation should be correct to *two or more* decimal places.

   For example, an exact answer is $2 + \sqrt{5}$. By using a calculator, a student discovers that $2 + \sqrt{5}$ is approximately $2 + 2.236067977 = 4.236067977$. *Acceptable answers* would include the calculator display and correctly rounded approximations of the display to two or more decimal places:

4.2360680	(seven places)	4.2361	(four places)
4.236068	(six places)	4.236	(three places)
4.23607	(five places)	4.24	(two places)

Unacceptable answers would include values that are not rounded correctly, as well as values with fewer than two decimal places such as 4.2 (one decimal place) and 4 (no decimal place).

3. When the solution to a problem involving radicals has two or more steps, no radical should be rounded until the very last step.

   For example, to find the value of $3 \times \sqrt{5}$, rounded to the *nearest hundredth*, first multiply the calculator approximation for $\sqrt{5}$ by 3 and then round the product to two decimal places.

Correct Solution: $3(\sqrt{5}) = 3(2.236067977) = 6.708203931 \approx 6.71$

Incorrect Solution: $3(\sqrt{5}) = 3(2.236067977) \approx 3(2.24) = 6.72$

The solution, 6.72, is incorrect because the rational approximation of $\sqrt{5}$ was rounded too soon. To the nearest hundredth, the correct answer is 6.71.

EXAMPLE 1

Between which two consecutive integers is $-\sqrt{42}$?

Solution *How to Proceed*

(1) Place 42 between the squares of consecutive integers: $36 < 42 < 49$

(2) Take the square root of each number: $\sqrt{36} < \sqrt{42} < \sqrt{49}$

(3) Simplify terms: $6 < \sqrt{42} < 7$

(4) Multiply each term of the inequality by -1: $-6 < -\sqrt{42} < -7$

Recall that when an inequality is multiplied by a negative number, the order of the inequality is reversed. $-7 > -\sqrt{42} > -6$

Answer $-\sqrt{42}$ is between -7 and -6.

EXAMPLE 2

Is $\sqrt{56}$ a rational or an irrational number?

Solution Since 56 is a positive integer that is not a perfect square, there is no rational number that, when squared, equals 56. Therefore, $\sqrt{56}$ is irrational.

Calculator Solution **STEP 1.** Evaluate $\sqrt{56}$.

ENTER: 2nd √ 56 ENTER

DISPLAY:

```
√(56
          7.483314774
```

STEP 2. To show that the calculator displays a rational approximation, not an exact value, show that the square of 7.483314774 does not equal 56.

ENTER: 7.483314774 x² ENTER

DISPLAY:

```
7.483314774²
          56.00000001
```

Since $(7.483314774)^2 \neq 56$, then 7.483314774 is a rational approximation for $\sqrt{56}$, not an exact value.

Answer $\sqrt{56}$ is an irrational number.

EXAMPLE 3

Is $\sqrt{8.0656}$ rational or irrational?

Calculator Solution **STEP 1.** Evaluate $\sqrt{8.0656}$.

ENTER: 2nd √ 8.0656 ENTER

DISPLAY:

```
√(8.0656
                2.84
```

STEP 2. It appears that the rational number in the calculator display is an exact value of $\sqrt{8.0656}$. To verify this, show that the square of 2.84 does equal 8.0656.

ENTER: 2.84 x^2 ENTER

DISPLAY:

```
2.84²
              8.0656
```

Since $(2.84)^2 = 8.0656$, then $\sqrt{8.0656} = 2.84$.

Answer $\sqrt{8.0656}$ is a rational number.

If n is a positive rational number written as a terminating decimal, then n^2 has twice as many decimal places as n. For example, the square of 2.84 has four decimal places. Also, since the last digit of 2.84 is 4, note that the last digit of 2.84^2 must be 6 because $4^2 = 16$.

EXAMPLE 4

Approximate the value of the expression $\sqrt{8^2 + 13} - 4$

a. to the *nearest thousandth* **b.** to the *nearest hundredth*

Calculator Solution ENTER: 2nd √ 8 x^2 + 13) − 4 ENTER

DISPLAY:

```
√(8²+13)-4
         4.774964387
```

a. To round to the *nearest thousandth*, look at the digit in the ten-thousandth (4th) decimal place. Since this digit (9) is greater than 5, add 1 to the digit in the thousandth (3rd) decimal place. When rounded to the nearest thousandth, 4.774964387 is approximately equal to 4.775

b. To round to the *nearest hundredth*, look at the digit in the thousandth (3rd) decimal place. Since this digit (4) is not greater than or equal to 5, drop this digit and all digits to the right. When rounded to the nearest hundredth, 4.774964387 is approximately equal to 4.77

Answers **a.** 4.775 **b.** 4.77

Note: It is not correct to round 4.774964387 to the nearest hundredth by rounding 4.775, the approximation to the nearest thousandth.

EXERCISES

Writing About Mathematics

1. a. Use a calculator to evaluate $\sqrt{999,999}$.

b. Enter your answer to part **a** and square the answer. Is the result 999,999?

c. Is $\sqrt{999,999}$ rational or irrational? Explain your answer.

2. Ursuline said that $\sqrt{\frac{18}{50}}$ is an irrational number because it is the ratio of $\sqrt{18}$ which is irrational and $\sqrt{50}$ which is irrational. Do you agree with Ursuline? Explain why or why not.

Developing Skills

In 3–12, between which consecutive integers is each given number?

3. $\sqrt{8}$ **4.** $\sqrt{13}$ **5.** $\sqrt{40}$ **6.** $-\sqrt{2}$ **7.** $-\sqrt{14}$

8. $\sqrt{52}$ **9.** $\sqrt{73}$ **10.** $-\sqrt{125}$ **11.** $\sqrt{143}$ **12.** $\sqrt{9 + 36}$

In 13–18, in each case, write the given numbers in order, starting with the smallest.

13. $2, \sqrt{3}, -1$ **14.** $4, \sqrt{17}, 3$ **15.** $-\sqrt{15}, -3, -4$

16. $0, \sqrt{7}, -\sqrt{7}$ **17.** $5, \sqrt{21}, \sqrt{30}$ **18.** $-\sqrt{11}, -\sqrt{23}, -\sqrt{19}$

In 19–33, state whether each number is rational or irrational.

19. $\sqrt{25}$ **20.** $\sqrt{40}$ **21.** $-\sqrt{36}$ **22.** $-\sqrt{54}$ **23.** $-\sqrt{150}$

24. $\sqrt{400}$ **25.** $\sqrt{\frac{1}{2}}$ **26.** $-\sqrt{\frac{4}{9}}$ **27.** $\sqrt{0.36}$ **28.** $\sqrt{0.1}$

29. $\sqrt{1,156}$ **30.** $\sqrt{951}$ **31.** $\sqrt{6.1504}$ **32.** $\sqrt{2,672.89}$ **33.** $\sqrt{5.8044}$

In 34–48, for each irrational number, write a rational approximation: **a.** as shown on a calculator display **b.** rounded to four decimal places

34. $\sqrt{2}$ **35.** $\sqrt{3}$ **36.** $\sqrt{21}$ **37.** $\sqrt{39}$ **38.** $\sqrt{80}$

39. $\sqrt{90}$ **40.** $\sqrt{108}$ **41.** $\sqrt{23.5}$ **42.** $\sqrt{88.2}$ **43.** $-\sqrt{115.2}$

44. $\sqrt{28.56}$ **45.** $\sqrt{67.25}$ **46.** $\sqrt{4{,}389}$ **47.** $\sqrt{123.7}$ **48.** $\sqrt{134.53}$

In 49–56, use a calculator to approximate each expression: **a.** to the *nearest hundredth* **b.** to the *nearest thousandth*.

49. $\sqrt{7} + 7$ **50.** $\sqrt{2} + \sqrt{6}$ **51.** $\sqrt{8} + \sqrt{8}$ **52.** $\sqrt{50} + 17$

53. $2 + \sqrt{3}$ **54.** $19 - \sqrt{3}$ **55.** $\sqrt{130} - 4$ **56.** $\sqrt{27} + 4.0038$

57. Both $\sqrt{58}$ and $\sqrt{58.01}$ are irrational numbers. Find a rational number n such that $\sqrt{58} < n < \sqrt{58.01}$.

Applying Skills

In 58–63, in each case, find to the *nearest tenth* of a centimeter the length of a side of a square whose area is the given measure.

58. 18 square centimeters **59.** 29 square centimeters **60.** 96 square centimeters

61. 140 square centimeters **62.** 202 square centimeters **63.** 288 square centimeters

In 64–67, find the perimeter of each figure, rounded to the *nearest hundredth*.

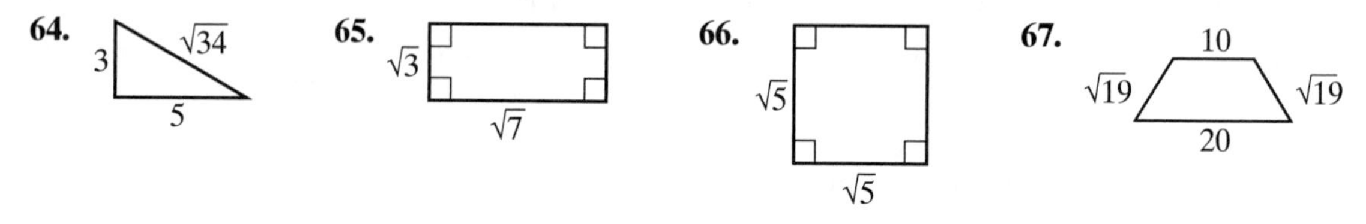

12-3 FINDING THE PRINCIPAL SQUARE ROOT OF A MONOMIAL

Just as we can find the principal square root of a number that is a perfect square, we can find the principal square roots of variables and monomials that represent perfect squares.

- Since $(6)(6) = 36$, then $\sqrt{36} = 6$.
- Since $(x)(x) = x^2$, then $\sqrt{x^2} = x\ (x \geq 0)$.
- Since $(a^2)(a^2) = a^4$, then $\sqrt{a^4} = a^2$.
- Since $(6a^2)(6a^2) = 36a^4$, then $\sqrt{36a^4} = 6a^2$.

In the last case, where the square root contains both numerical and variable factors, we can determine the square root by finding the square roots of its factors and multiplying:

$$\sqrt{36a^4} = \sqrt{36} \cdot \sqrt{a^4} = 6 \cdot a^2 = 6a^2$$

Procedure

To find the square root of a monomial that has two or more factors, write the indicated product of the square roots of its factors.

Note: In our work, we limit the domain of all variables under a radical sign to *nonnegative* numbers.

EXAMPLE 1

Find the principal square root of each monomial. Assume that all variables represent positive numbers.

a. $25y^2$ **b.** $1.44a^6b^2$ **c.** $1{,}369m^{10}$ **d.** $\frac{81}{4}g^6$

Solution **a.** $\sqrt{25y^2} = \left(\sqrt{25}\right)\left(\sqrt{y^2}\right) = (5)(y) = 5y$ *Answer*

b. $\sqrt{1.44a^6b^2} = \left(\sqrt{1.44}\right)\left(\sqrt{a^6}\right)\left(\sqrt{b^2}\right) = (1.2)(a^3)(b) = 1.2a^3b$ *Answer*

c. $\sqrt{1{,}369m^{10}} = \left(\sqrt{1{,}369}\right)\left(\sqrt{m^{10}}\right) = (37)(m^5) = 37m^5$ *Answer*

d. $\sqrt{\frac{81}{4}g^6} = \left(\sqrt{\frac{81}{4}}\right)(g^6) = \left(\frac{9}{2}\right)(g^3) = \frac{9}{2}g^3$ *Answer*

EXERCISES

Writing About Mathematics

1. Is it true that for $x < 0$, $\sqrt{x^2} = -x$? Explain your answer.

2. Melanie said that when a is an even integer and $x \geq 0$, $\sqrt{x^a} = x^{\frac{a}{2}}$. Do you agree with Melanie? Explain why or why not.

Developing Skills

In 3–14, find the principal square root of each monomial. Assume that all variables represent positive numbers.

3. $4a^2$ **4.** $49z^2$ **5.** $\frac{16}{25}r^2$ **6.** $0.81w^2$

7. $9c^2$ **8.** $36y^4$ **9.** c^2d^2 **10.** $4x^2y^2$

11. $144a^4b^2$ **12.** $0.36m^2$ **13.** $0.49a^2b^2$ **14.** $70.56b^2x^{10}$

Applying Skills

In 15–18, where all variables represent positive numbers:

a. Represent each side of the square whose area is given.

b. Represent the perimeter of that square.

15. $49c^2$ **16.** $64x^2$ **17.** $100x^2y^2$ **18.** $144a^2b^2$

19. The length of the legs of a right triangle are represented by $9x$ and $40x$. Represent the length of the hypotenuse of the right triangle.

12-4 SIMPLIFYING A SQUARE-ROOT RADICAL

A radical that is an irrational number, such as $\sqrt{12}$, can often be simplified. To understand this procedure, let us first consider some radicals that are rational numbers. We know that $\sqrt{36} = 6$ and that $\sqrt{400} = 20$.

Since $\sqrt{4 \cdot 9} = \sqrt{36} = 6$
and $\sqrt{4} \cdot \sqrt{9} = 2 \cdot 3 = 6$,
then $\sqrt{4 \cdot 9} = \sqrt{4} \cdot \sqrt{9}$.

Since $\sqrt{16 \cdot 25} = \sqrt{400} = 20$.
and $\sqrt{16} \cdot \sqrt{25} = 4 \cdot 5 = 20$,
then $\sqrt{16 \cdot 25} = \sqrt{16} \cdot \sqrt{25}$.

These examples illustrate the following property of square-root radicals:

► **The square root of a product of nonnegative numbers is equal to the product of the square roots of these numbers.**

In general, if a and b are nonnegative numbers:

$$\sqrt{a \cdot b} = \sqrt{a} \cdot \sqrt{b} \quad \text{and} \quad \sqrt{a} \cdot \sqrt{b} = \sqrt{a \cdot b}$$

Now we will apply this rule to a square-root radical with a radicand that is not a perfect square.

1. Express 50 as the product of 25 and 2, where 25 is the *greatest perfect-square* factor of 50: $\sqrt{50} = \sqrt{25 \cdot 2}$

2. Write the product of the two square roots: $= \sqrt{25} \cdot \sqrt{2}$

3. Replace $\sqrt{25}$ with 5 to obtain the simplified expression with the smallest possible radicand: $= 5\sqrt{2}$

The expression $5\sqrt{2}$ is called the *simplest form* of $\sqrt{50}$. **The simplest form of a square-root radical** is one in which the radicand is an integer that has no perfect-square factor other than 1.

If the radicand is a fraction, change it to an equivalent fraction that has a denominator that is a perfect square. Write the radicand as the product of a fraction that is a perfect square and an integer that has no perfect-square factor other than 1. For example:

$$\sqrt{\tfrac{8}{3}} = \sqrt{\tfrac{8}{3} \times \tfrac{3}{3}} = \sqrt{\tfrac{24}{9}} = \sqrt{\tfrac{4}{9} \times 6} = \sqrt{\tfrac{4}{9}} \times \sqrt{6} = \tfrac{2}{3}\sqrt{6}$$

Procedure

To simplify the square root of a product:

1. If the radicand is a fraction, write it as an equivalent fraction with a denominator that is a perfect square.
2. Find, if possible, two factors of the radicand such that one of the factors is a perfect square and the other is an integer that has no factor that is a perfect square.
3. Express the square root of the product as the product of the square roots of the factors.
4. Find the square root of the factor with the perfect-square radicand.

EXAMPLE 1

Simplify each expression. Assume that $y > 0$.

Answers

a. $\sqrt{18}$ $= \sqrt{9 \cdot 2} = \sqrt{9} \cdot \sqrt{2} = 3\sqrt{2}$

b. $4\sqrt{150}$ $= 4\sqrt{25 \cdot 6} = 4\sqrt{25} \cdot \sqrt{6} = 4 \cdot 5\sqrt{6} = 20\sqrt{6}$

c. $\sqrt{\tfrac{9}{2}}$ $= \sqrt{\tfrac{9}{2}} = \sqrt{\tfrac{9}{2} \times \tfrac{2}{2}} = \sqrt{\tfrac{18}{4}} = \sqrt{\tfrac{9}{4}} \times \sqrt{2} = \tfrac{3}{2}\sqrt{2}$

d. $\sqrt{4y^3}$ $= \sqrt{4y^3} = \sqrt{4y^2 \cdot y} = \sqrt{4y^2} \cdot \sqrt{y} = 2y\sqrt{y}$

EXERCISES

Writing About Mathematics

1. Does $\frac{1}{3}\sqrt{27} = \sqrt{9}$? Explain why or why not.

2. Abba simplified the expression $\sqrt{192}$ by writing $\sqrt{192} = \sqrt{16 \cdot 12} = 4\sqrt{12}$.

a. Explain why $4\sqrt{12}$ is *not* the simplest form of $\sqrt{192}$.

b. Show how it is possible to find the simplest form of $\sqrt{192}$ by starting with $4\sqrt{12}$.

c. What is the simplest form of $\sqrt{192}$?

Developing Skills

In 3–22, write each expression in simplest form. Assume that all variables represent positive numbers.

3. $\sqrt{12}$ **4.** $\sqrt{27}$ **5.** $\sqrt{63}$ **6.** $\sqrt{98}$ **7.** $4\sqrt{12}$

8. $2\sqrt{20}$ **9.** $5\sqrt{24}$ **10.** $\frac{1}{4}\sqrt{48}$ **11.** $\frac{3}{4}\sqrt{96}$ **12.** $\frac{3}{2}\sqrt{80}$

13. $\sqrt{\frac{1}{2}}$ **14.** $\sqrt{\frac{36}{5}}$ **15.** $\sqrt{\frac{32}{3}}$ **16.** $2\sqrt{\frac{1}{8}}$ **17.** $15\sqrt{\frac{2}{5}}$

18. $8\sqrt{9x}$ **19.** $\sqrt{3x^3}$ **20.** $\sqrt{49x^5}$ **21.** $\sqrt{36r^2s}$ **22.** $\sqrt{243xy^2}$

23. The expression $\sqrt{48}$ is equivalent to

(1) $2\sqrt{3}$ (2) $4\sqrt{12}$ (3) $4\sqrt{3}$ (4) $16\sqrt{3}$

24. The expression $4\sqrt{2}$ is equivalent to

(1) $\sqrt{8}$ (2) $\sqrt{42}$ (3) $\sqrt{32}$ (4) $\sqrt{64}$

25. The expression $3\sqrt{18}$ is equivalent to

(1) $\sqrt{54}$ (2) $3\sqrt{2}$ (3) $9\sqrt{2}$ (4) $3\sqrt{6}$

26. The expression $3\sqrt{3}$ is equivalent to

(1) $\sqrt{9}$ (2) $\sqrt{6}$ (3) $\sqrt{12}$ (4) $\sqrt{27}$

In 27–30, for each expression:

a. Use a calculator to find a rational approximation of the expression.

b. Write the original expression in simplest radical form.

c. Use a calculator to find a rational approximation of the answer to part **b**.

d. Are the approximations in parts **a** and **c** equal?

27. $\sqrt{300}$ **28.** $\sqrt{180}$ **29.** $2\sqrt{288}$ **30.** $\frac{1}{3}\sqrt{252}$

31. a. Does $\sqrt{9 + 16} = \sqrt{9} + \sqrt{16}$? Explain your answer.

b. Is finding a square root distributive over addition?

32. a. Does $\sqrt{169 - 25} = \sqrt{169} - \sqrt{25}$? Explain your answer.

b. Is finding a square root distributive over subtraction?

12-5 ADDITION AND SUBTRACTION OF RADICALS

Radicals are exact values. Computations sometimes involve many radicals, as in the following example:

$$\sqrt{12} + \sqrt{75} + 3\sqrt{3}$$

Rather than use approximations obtained with a calculator to find this sum, it may be important to express the answer as an *exact* value in radical form. To learn how to perform computations with radicals, we need to recall some algebraic concepts.

Adding and Subtracting Like Radicals

We have learned how to add like terms in algebra:

$$2x + 5x + 3x = 10x$$

If we replace the variable x by an irrational number, $\sqrt{3}$, the following statement must be true:

$$2\sqrt{3} + 5\sqrt{3} + 3\sqrt{3} = 10\sqrt{3}$$

Like radicals are radicals that have the same index and the same radicand. For example, $2\sqrt{3}$, $5\sqrt{3}$, and $3\sqrt{3}$ are like radicals and $6\sqrt[3]{7}$ and $\sqrt[3]{7}$ are like radicals.

To demonstrate that the sum of like radicals can be written as a single term, we can apply the concept used to add like terms, namely, the distributive property:

$$2\sqrt{3} + 5\sqrt{3} + 3\sqrt{3} = \sqrt{3}(2 + 5 + 3) = \sqrt{3}(10) = 10\sqrt{3}$$

Similarly,

$$6\sqrt[3]{7} - \sqrt[3]{7} = \sqrt[3]{7}(6 - 1) = \sqrt[3]{7}(5) = 5\sqrt[3]{7}$$

Procedure

To add or subtract like terms that contain like radicals:

1. Add or subtract the coefficients of the radicals.
2. Multiply the sum or difference obtained by the common radical.

Adding and Subtracting Unlike Radicals

Unlike radicals are radicals that have different radicands or different indexes, or both. For example:

- $3\sqrt{5}$ and $2\sqrt{2}$ are unlike radicals because their radicands are different.
- $\sqrt[3]{2}$ and $\sqrt{2}$ are unlike radicals because their indexes are different.
- $9\sqrt{10}$ and $\sqrt[3]{4}$ are unlike radicals because their radicands and indexes are different.

The sum or difference of unlike radicals cannot always be expressed as a single term. For instance:

- The sum of $3\sqrt{5}$ and $2\sqrt{2}$ is $3\sqrt{5} + 2\sqrt{2}$.
- The difference of $5\sqrt{7}$ and $\sqrt{11}$ is $5\sqrt{7} - \sqrt{11}$.

However, it is sometimes possible to transform unlike radicals into equivalent like radicals. These like radicals can then be combined into a single term. Let us return to the problem posed at the start of this section:

$$\begin{aligned}\sqrt{12} + \sqrt{75} + 3\sqrt{3} &= \sqrt{4 \cdot 3} + \sqrt{25 \cdot 3} + 3\sqrt{3} \\ &= 2\sqrt{3} + 5\sqrt{3} + 3\sqrt{3} \\ &= 10\sqrt{3}\end{aligned}$$

Procedure

To combine unlike radicals:

1. Simplify each radical if possible.
2. Combine like radicals by using the distributive property.
3. Write the indicated sum or difference of the unlike radicals.

EXAMPLE 1

Combine: **a.** $8\sqrt{5} + \sqrt{5} - 2\sqrt{5}$ **b.** $6\sqrt{n} - 2\sqrt{49n}$

Solution **a.** $8\sqrt{5} + \sqrt{5} - 2\sqrt{5} = \sqrt{5}(8 + 1 - 2) = \sqrt{5}(7) = 7\sqrt{5}$

b. $6\sqrt{n} - 2\sqrt{49n} = 6\sqrt{n} - 2(7)\sqrt{n} = \sqrt{n}(6 - 14) = -8\sqrt{n}$

Answers **a.** $7\sqrt{5}$ **b.** $-8\sqrt{n}$

EXAMPLE 2

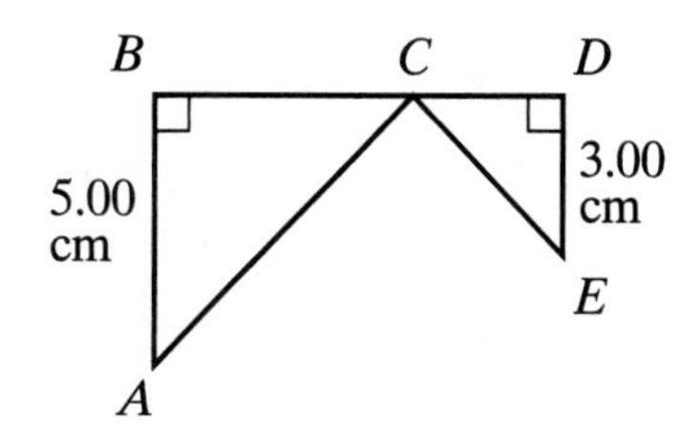

Alexis drew the figure at the right. Triangles ABC and CDE are isosceles right triangles. $AB = 5.00$ centimeters and $DE = 3.00$ centimeters.

a. Find AC and CE.

b. Find, in simplest radical form, $AC + CE$.

c. Alexis measured $AC + CE$ and found the measure to be 11.25 centimeters. Find the percent of error of the measurement that Alexis made. Express the answer to the nearest hundredth of a percent.

Solution **a.** Use the Pythagorean Theorem.

$$\begin{aligned} AC^2 &= AB^2 + BC^2 \\ &= 5^2 + 5^2 \\ &= 25 + 25 \\ &= 50 \\ AC &= \sqrt{50} \end{aligned} \qquad \begin{aligned} CE^2 &= CD^2 + DE^2 \\ &= 3^2 + 3^2 \\ &= 9 + 9 \\ &= 18 \\ CE &= \sqrt{18} \end{aligned}$$

b. $AC + CE = \sqrt{50} + \sqrt{18} = \sqrt{25} \cdot \sqrt{2} + \sqrt{9} \cdot \sqrt{2} = 5\sqrt{2} + 3\sqrt{2} = 8\sqrt{2}$

c. Error $= 8\sqrt{2} - 11.25$

Percent of error $= \frac{8\sqrt{2} - 11.25}{8\sqrt{2}}$

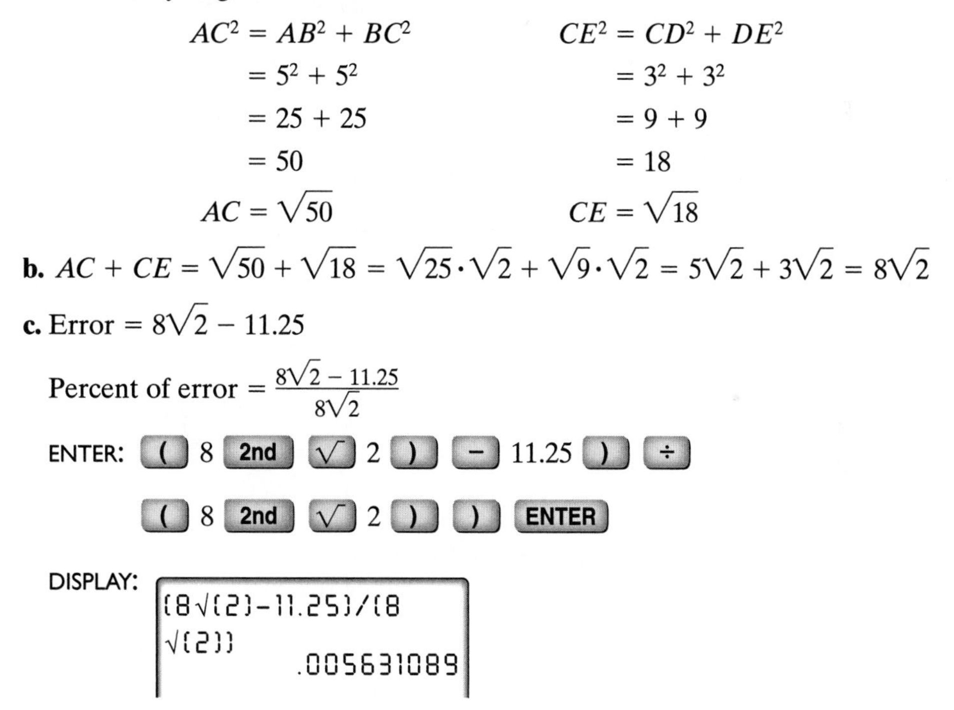

Multiply the number in the display by 100 to change to percent.

Percent of error $= 0.5631089\% = 0.56\%$

Answers **a.** $AC = \sqrt{50}$ cm, $CE = \sqrt{18}$ cm **b.** $8\sqrt{2}$ cm **c.** 0.56%

EXERCISES

Writing About Mathematics

1. Compare adding fractions with adding radicals. How are the two operations alike and how are they different?

2. Marc said that $3\sqrt{5} - \sqrt{5} = 3$. Do you agree with Marc? Explain why or why not.

Developing Skills

In 3–23, in each case, combine the radicals. Assume that all variables represent positive numbers.

3. $8\sqrt{5} + \sqrt{5}$

4. $5\sqrt{3} + 2\sqrt{3} + 8\sqrt{3}$

5. $7\sqrt{2} - \sqrt{2}$

6. $4\sqrt{3} + 2\sqrt{3} - 6\sqrt{3}$

7. $5\sqrt{3} + \sqrt{3} - 2\sqrt{3}$

8. $4\sqrt{7} - \sqrt{7} - 5\sqrt{7}$

9. $3\sqrt{5} + 6\sqrt{2} - 3\sqrt{2} + \sqrt{5}$

10. $9\sqrt{x} + 3\sqrt{x}$

11. $15\sqrt{y} - 7\sqrt{y}$

12. $\sqrt{2} + \sqrt{50}$

13. $\sqrt{27} + \sqrt{75}$

14. $\sqrt{80} - \sqrt{5}$

15. $\sqrt{12} - \sqrt{48} + \sqrt{3}$

16. $\sqrt{0.98} - 4\sqrt{0.08} + 3\sqrt{1.28}$

17. $\sqrt{0.2} + \sqrt{0.45}$

18. $\sqrt{\frac{8}{9}} - \sqrt{72}$

19. $\sqrt{\frac{3}{4}} + \sqrt{\frac{1}{3}}$

20. $\sqrt{7a} + \sqrt{28a}$

21. $\sqrt{100b} - \sqrt{64b} + \sqrt{9b}$

22. $3\sqrt{3x} - \sqrt{12x}$

23. $\sqrt{3a^2} + \sqrt{12a^2}$

24. $x\sqrt{a^2} + 6\sqrt{a} - 3\sqrt{a}$

In 25–27, in each case, select the numeral preceding the correct choice.

25. The difference $5\sqrt{2} - \sqrt{32}$ is equivalent to

(1) $\sqrt{2}$ (2) $9\sqrt{2}$ (3) $4\sqrt{30}$ (4) $5\sqrt{30}$

26. The sum $3\sqrt{8} + 6\sqrt{2}$ is equivalent to

(1) $9\sqrt{10}$ (2) $\sqrt{72}$ (3) $18\sqrt{10}$ (4) $12\sqrt{2}$

27. The sum of $\sqrt{12}$ and $\sqrt{27}$ is equivalent to

(1) $\sqrt{39}$ (2) $5\sqrt{6}$ (3) $13\sqrt{3}$ (4) $5\sqrt{3}$

Applying Skills

In 28 and 29: **a.** Express the perimeter of the figure in simplest radical form. **b.** Using a calculator, approximate the expression obtained in part **a** to the *nearest thousandth*.

30. On the way to softball practice, Maggie walks diagonally through a square field and a rectangular field. The square field has a length of 60 yards. The rectangular field has a length of 70 yards and a width of 10 yards. What is the total distance Maggie walks through the fields?

12-6 MULTIPLICATION OF SQUARE-ROOT RADICALS

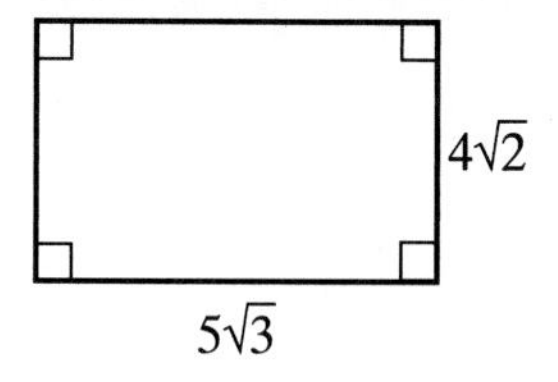

To find the area of the rectangle pictured at the right, we multiply $5\sqrt{3}$ by $4\sqrt{2}$. We have learned that $\sqrt{a} \cdot \sqrt{b} = \sqrt{ab}$ when a and b are nonnegative numbers. To multiply $4\sqrt{2}$ by $5\sqrt{3}$, we use the commutative and associative laws of multiplication as follows:

$$\left(4\sqrt{2}\right)\left(5\sqrt{3}\right) = (4)(5)\left(\sqrt{2}\right)\left(\sqrt{3}\right) = (4 \cdot 5)\left(\sqrt{2 \cdot 3}\right) = 20\sqrt{6}$$

In general, if a and b are nonnegative numbers:

$$\left(x\sqrt{a}\right)\left(y\sqrt{b}\right) = xy\sqrt{ab}$$

Procedure

To multiply two monomial square roots:

1. Multiply the coefficients to find the coefficient of the product.
2. Multiply the radicands to find the radicand of the product.
3. If possible, simplify the result.

EXAMPLE 1

a. Multiply $(3\sqrt{6})(5\sqrt{2})$ and write the product in simplest radical form.

b. Check the work performed in part **a** using a calculator.

Solution **a.**

$$\begin{aligned}(3\sqrt{6})(5\sqrt{2}) &= 3(5)\sqrt{6(2)}\\ &= 15\sqrt{12}\\ &= 15(\sqrt{4})(\sqrt{3})\\ &= 15(2)\sqrt{3}\\ &= 30\sqrt{3}\ \textit{Answer}\end{aligned}$$

b. To check, evaluate $(3\sqrt{6})(5\sqrt{2})$.

ENTER: 3 [2nd] [√] 6 [)] [×] 5 [2nd] [√] 2 [)] [ENTER]

DISPLAY:
```
3√(6)*5√(2)
         51.96152423
```

Then evaluate $30\sqrt{3}$.

ENTER: 30 [2nd] [√] 3 [ENTER]

DISPLAY:
```
30√(3
         51.96152423
```

Therefore, $(3\sqrt{6})(5\sqrt{2}) = 30\sqrt{3}$ appears to be true.

EXAMPLE 2

Find the value of $(2\sqrt{3})^2$.

Solution $(2\sqrt{3})^2 = (2\sqrt{3})(2\sqrt{3}) = 2(2)(\sqrt{3})(\sqrt{3}) = 4\sqrt{9} = 4(3) = 12$

Alternative Solution $(2\sqrt{3})^2 = (2)^2(\sqrt{3})^2 = 4(3) = 12$

Answer 12

EXAMPLE 3

Find the indicated product: $\sqrt{3x} \cdot \sqrt{6x}$, $(x > 0)$.

Solution $\sqrt{3x} \cdot \sqrt{6x} = \sqrt{3x \cdot 6x} = \sqrt{18x^2} = \sqrt{9x^2} \cdot \sqrt{2} = 3x\sqrt{2}$ *Answer*

EXERCISES

Writing About Mathematics

1. When a and b are unequal prime numbers, is $\sqrt{ab}$ rational or irrational? Explain your answer.

2. Is $\sqrt{4a^2}$ a rational number for all values of a? Explain your answer.

Developing Skills

In 3–26: in each case multiply, or raise to the power, as indicated. Then simplify the result. Assume that all variables represent positive numbers.

3. $\sqrt{3} \cdot \sqrt{3}$ **4.** $\sqrt{7} \cdot \sqrt{7}$ **5.** $\sqrt{a} \cdot \sqrt{a}$

6. $\sqrt{2x} \cdot \sqrt{2x}$ **7.** $\sqrt{12} \cdot \sqrt{12}$ **8.** $8\sqrt{18} \cdot 3\sqrt{8}$

9. $\sqrt{14} \cdot \sqrt{2}$ **10.** $\sqrt{60} \cdot \sqrt{5}$ **11.** $3\sqrt{6} \cdot \sqrt{3}$

12. $5\sqrt{8} \cdot 7\sqrt{3}$ **13.** $\frac{2}{3}\sqrt{24} \cdot 9\sqrt{3}$ **14.** $5\sqrt{6} \cdot \frac{2}{3}\sqrt{15}$

15. $\left(-4\sqrt{a}\right)\left(3\sqrt{a}\right)$ **16.** $\left(-\frac{1}{2}\sqrt{y}\right)\left(-6\sqrt{y}\right)$ **17.** $\left(\sqrt{2}\right)^2$

18. $\left(\sqrt{y}\right)^2$ **19.** $\left(\sqrt{t}\right)^2$ **20.** $\left(3\sqrt{6}\right)^2$

21. $\left(\sqrt{25x}\right)\left(\sqrt{4x}\right)$ **22.** $\left(\sqrt{27a}\right)\left(\sqrt{3a}\right)$ **23.** $\left(\sqrt{15x}\right)\left(\sqrt{3x}\right)$

24. $\left(\sqrt{9a}\right)\left(\sqrt{ab}\right)$ **25.** $\left(\sqrt{5x}\right)^2$ **26.** $\left(2\sqrt{t}\right)^2$

In 27–33: **a.** Perform each indicated operation. **b.** State whether the product is an irrational number or a rational number.

27. $\left(5\sqrt{12}\right)\left(4\sqrt{3}\right)$ **28.** $\left(3\sqrt{2}\right)\left(2\sqrt{32}\right)$ **29.** $\left(4\sqrt{6}\right)\left(9\sqrt{3}\right)$ **30.** $\left(8\sqrt{5}\right)\left(\frac{1}{2}\sqrt{10}\right)$

31. $\left(\frac{2}{3}\sqrt{5}\right)^2$ **32.** $\left(\frac{1}{6}\sqrt{8}\right)\left(\frac{1}{2}\sqrt{18}\right)$ **33.** $\left(\frac{5}{7}\sqrt{7}\right)^3$ **34.** $\left(11\sqrt{38}\right)\left(\frac{1}{11}\sqrt{45}\right)$

35. Two square-root radicals that are *irrational numbers* are multiplied.

a. Give two examples where the product of these radicals is also an irrational number.

b. Give two examples where the product of these radicals is a rational number.

Applying Skills

In 36–39, in each case, find the area of the square in which the length of each side is given.

36. $\sqrt{2}$ **37.** $2\sqrt{3}$ **38.** $6\sqrt{2}$ **39.** $5\sqrt{3}$

In 40 and 41: **a.** Express the area of the figure in simplest radical form. **b.** Check the work performed in part **a** by using a calculator.

40. **41.**

12-7 DIVISION OF SQUARE-ROOT RADICALS

Since	$\sqrt{\frac{4}{9}} = \frac{2}{3}$	Since	$\sqrt{\frac{16}{25}} = \frac{4}{5}$
and	$\frac{\sqrt{4}}{\sqrt{9}} = \frac{2}{3}$,	and	$\frac{\sqrt{16}}{\sqrt{25}} = \frac{4}{5}$,
then	$\sqrt{\frac{4}{9}} = \frac{\sqrt{4}}{\sqrt{9}}$.	then	$\sqrt{\frac{16}{25}} = \frac{\sqrt{16}}{\sqrt{25}}$.

These examples illustrate the following property of square-root radicals:

► **The square root of a fraction that is the quotient of two positive numbers is equal to the square root of the numerator divided by the square root of the denominator.**

In general, if a and b are positive numbers:

$$\sqrt{\frac{a}{b}} = \frac{\sqrt{a}}{\sqrt{b}} \quad \text{and} \quad \frac{\sqrt{a}}{\sqrt{b}} = \sqrt{\frac{a}{b}}$$

We use this principle to divide $\sqrt{72}$ by $\sqrt{8}$. In this example, notice that the quotient of two irrational numbers is a *rational* number.

$$\frac{\sqrt{72}}{\sqrt{8}} = \sqrt{\frac{72}{8}} = \sqrt{9} = 3$$

We can also divide radical terms by using the property of fractions:

$$\frac{ac}{bd} = \frac{a}{b} \cdot \frac{c}{d}$$

Note, in the following example, that the quotient of two irrational numbers is irrational:

$$\frac{6\sqrt{10}}{3\sqrt{2}} = \frac{6}{3} \cdot \frac{\sqrt{10}}{\sqrt{2}} = \frac{6}{3} \cdot \sqrt{\frac{10}{2}} = 2\sqrt{5}$$

In general, if a and b are positive, and $y \neq 0$:

$$\frac{x\sqrt{a}}{y\sqrt{b}} = \frac{x}{y}\sqrt{\frac{a}{b}}$$

Procedure

To divide two monomial square roots:

1. Divide the coefficients to find the coefficient of the quotient.
2. Divide the radicands to find the radicand of the quotient.
3. If possible, simplify the result.

EXAMPLE 1

Divide $8\sqrt{48}$ by $4\sqrt{2}$, and simplify the quotient.

Solution $8\sqrt{48} \div 4\sqrt{2} = \frac{8}{4}\sqrt{\frac{48}{2}} = 2\sqrt{24} = 2\left(\sqrt{4}\right)\left(\sqrt{6}\right) = 2(2)\left(\sqrt{6}\right) = 4\sqrt{6}$ *Answer*

EXAMPLE 2

Find the indicated division: $\frac{\sqrt{2x^2y^3z}}{\sqrt{6y}}$, $(x > 0, y > 0, z > 0)$.

Solution $\frac{\sqrt{2x^2y^3z}}{\sqrt{6y}} = \sqrt{\frac{1}{3}} \cdot \sqrt{x^2y^2z} = \sqrt{\frac{3}{9}} \cdot \sqrt{x^2y^2z} = \sqrt{\frac{1}{9}x^2y^2} \cdot \sqrt{3z} = \frac{1}{3}xy\sqrt{3z}$ *Answer*

EXERCISES

Writing About Mathematics

1. Ross simplified $\sqrt{\frac{16}{81}}$ by writing $\frac{\sqrt{16}}{\sqrt{81}} = \frac{4}{9} = \frac{2}{3}$. Do you agree with Ross? Explain why or why not.

2. Is $\sqrt{\frac{1}{16}}$ rational or irrational? Explain your answer.

Developing Skills

In 3–18, divide. Write the quotient in simplest form. Assume that all variables represent positive numbers.

3. $\sqrt{72} \div \sqrt{2}$ **4.** $\sqrt{75} \div \sqrt{3}$ **5.** $\sqrt{70} \div \sqrt{10}$ **6.** $\sqrt{14} \div \sqrt{2}$

7. $8\sqrt{48} \div 2\sqrt{3}$ **8.** $\sqrt{24} \div \sqrt{2}$ **9.** $\sqrt{150} \div \sqrt{3}$ **10.** $21\sqrt{40} \div \sqrt{5}$

11. $9\sqrt{6} \div 3\sqrt{6}$ **12.** $7\sqrt{3} \div 3\sqrt{3}$ **13.** $2\sqrt{2} \div 8\sqrt{2}$ **14.** $\sqrt{9y} \div \sqrt{y}$

15. $\frac{12\sqrt{20}}{3\sqrt{5}}$ **16.** $\frac{20\sqrt{50}}{4\sqrt{2}}$ **17.** $\frac{a\sqrt{b^3c^4}}{\sqrt{a^2}}$ **18.** $\frac{3\sqrt{x^3y}}{6\sqrt{z}}$

In 19–26, state whether each quotient is a rational number or an irrational number.

19. $\frac{\sqrt{5}}{7}$ **20.** $\frac{\sqrt{50}}{\sqrt{2}}$ **21.** $\frac{\sqrt{18}}{\sqrt{3}}$ **22.** $\frac{\sqrt{49}}{\sqrt{7}}$

23. $\frac{\sqrt{9}}{\sqrt{16}}$ **24.** $\frac{\sqrt{18}}{\sqrt{25}}$ **25.** $\frac{25\sqrt{24}}{5\sqrt{2}}$ **26.** $\frac{3\sqrt{54}}{6\sqrt{3}}$

In 27–34, simplify each expression. Assume that all variables represent positive numbers.

27. $\sqrt{\frac{36}{49}}$ **28.** $\sqrt{\frac{3}{4}}$ **29.** $4\sqrt{\frac{5}{16}}$ **30.** $\sqrt{\frac{8}{49}}$

31. $10\sqrt{\frac{8}{25}}$ **32.** $\frac{9}{18}\sqrt{\frac{xy^2}{y}}$ **33.** $\sqrt{\frac{a^6b^5c^4}{a^4b^3c^2}}$ **34.** $\frac{15}{3}\sqrt{\frac{a^5}{a^2b^2}}$

Applying Skills

In 35–38, in each case, the area A of a parallelogram and the measure of its base b are given. Find the height h of the parallelogram, expressed in simplest form.

35. $A = 7\sqrt{12}, b = 7\sqrt{3}$ **36.** $A = \sqrt{640}, b = \sqrt{32}$

37. $A = 8\sqrt{45}, b = 2\sqrt{15}$ **38.** $A = 2\sqrt{98}, b = \sqrt{32}$

CHAPTER SUMMARY

A **radical**, which is the root of a quantity, is written in its general form as $\sqrt[n]{b}$. A radical consists of a **radicand**, b, placed under a **radical sign**, $\sqrt{\ }$, with an **index**, n.

Finding the square root of a quantity reverses the result of the operation of squaring. A **square-root radical** has an index of 2, which is generally not written. Thus, $\sqrt[2]{49} = \sqrt{49} = 7$ because $7^2 = 49$. In general, for nonnegative numbers b and x, $\sqrt{b} = x$ if and only if $x^2 = b$.

If k is a positive number that is a perfect square, then $\sqrt{k}$ is a rational number. If k is positive but not a perfect square, then $\sqrt{k}$ is an irrational number.

Every positive number has two square roots: a positive root called the **principal square root**, and a negative root. These roots have the same absolute value:

Principal Square Root	*Negative Square Root*	*Both Square Roots*
$\sqrt{x^2} = \lvert x \rvert$	$-\sqrt{x^2} = -\lvert x \rvert$	$\pm\sqrt{x^2} = \pm x$

Finding the **cube root** of a number is the inverse of the operation of cubing. Thus, $\sqrt[3]{64} = 4$ because $4^3 = 4 \times 4 \times 4 = 64$. In general, $\sqrt[3]{b} = x$ if and only if $x^3 = b$.

Like radicals have the same radicand and the same index. For example, $2\sqrt{7}$ and $3\sqrt{7}$. **Unlike radicals** can differ in their radicands ($\sqrt{7}$ and $\sqrt{2}$), in their indexes ($\sqrt{7}$ and $\sqrt[3]{7}$), or in both ($\sqrt[4]{7}$ and $\sqrt[3]{6}$).

A square root of a positive integer is **simplified** by factoring out the square root of its greatest perfect square. The radicand of a simplified radical, then, has no perfect square factor other than 1. When a radical is irrational, the radical expresses its *exact* value. Most calculators display only **rational approximations** of radicals that are irrational numbers.

Operations with radicals include:

1. **Addition and subtraction:** Combine like radicals by adding or subtracting their coefficients and then multiplying this result by their common radical. The sum or difference of unlike radicals, unless transformed to equivalent like radicals, cannot be expressed as a single term.

2. **Multiplication and division:** For all radicals whose denominators are not equal to 0, multiply (or divide) coefficients, multiply (or divide) radicands, and simplify. The general rules for these operations are as follows:

$$\left(x\sqrt{a}\right)\left(y\sqrt{b}\right) = xy\sqrt{ab} \quad \text{and} \quad \frac{x\sqrt{a}}{y\sqrt{b}} = \frac{x}{y}\sqrt{\frac{a}{b}}$$

VOCABULARY

12-1 Radicand • Radical • Perfect square • Principal square root • Cube root • Index

12-4 Simplest form of a square-root radical

12-5 Like radicals • Unlike radicals

REVIEW EXERCISES

1. When $\sqrt{\frac{a}{b}}$ is an integer, what is the relationship between a and b?

2. When a is a positive perfect square and b is a positive number that is not a perfect square, is $\sqrt{ab}$ rational or irrational? Explain your answer.

3. Write the principal square root of 1,225.

4. Write the following numbers in order starting with the smallest: $\sqrt{18}, 2\sqrt{2}, 3$.

In 5–14, write each number in simplest form.

5. $\sqrt{\frac{9}{25}}$ **6.** $-\sqrt{49}$ **7.** $\sqrt[3]{-27}$ **8.** $\pm\sqrt{1.21}$

9. $\sqrt{400y^4}$ **10.** $\sqrt{180}$ **11.** $3\sqrt{18}$ **12.** $\frac{1}{2}\sqrt{28}$

13. $\sqrt{48b^3}, b > 0$ **14.** $\sqrt{0.01m^{16}}$ **15.** $\sqrt{\frac{9}{27}x^3y^5}, x > 0, y > 0$ **16.** $\sqrt{0.25a^8b^{10}}$

In 17–20, in each case, solve for the variable, using the set of real numbers as the replacement set.

17. $y^2 = 81$ **18.** $m^2 = 0.09$ **19.** $3x^2 = 600$ **20.** $2k^2 - 144 = 0$

21. a. Use a calculator to evaluate $\sqrt{315.4176}$.

b. Is $\sqrt{315.4176}$ a rational or an irrational number? Explain your answer.

In 22–33, in each case, perform the indicated operation and simplify the result.

22. $\sqrt{18} + \sqrt{8} - \sqrt{32}$

23. $3\sqrt{20} - 2\sqrt{45}$

24. $2\sqrt{50} - \sqrt{98} + \frac{1}{2}\sqrt{72}$

25. $\sqrt{75} - 3\sqrt{12}$

26. $8\sqrt{2}(2\sqrt{2})$

27. $(3\sqrt{5})^2$

28. $2\sqrt{7}(\sqrt{70})$

29. $\sqrt{98} \div \sqrt{2}$

30. $\frac{16\sqrt{21}}{2\sqrt{7}}$

31. $\frac{5\sqrt{162}}{9\sqrt{50}}$

32. $\sqrt{3}(\sqrt{24}) - \sqrt{5}(\sqrt{10})$

33. $\frac{\sqrt{80}}{\sqrt{2}} + \sqrt{6}(\sqrt{60})$

In 34 and 35, in each case, select the numeral preceding the correct choice.

34. The expression $\sqrt{108} - \sqrt{3}$ is equivalent to

(1) $\sqrt{105}$ (2) $35\sqrt{3}$ (3) $5\sqrt{3}$ (4) 6

35. The sum of $9\sqrt{2}$ and $\sqrt{32}$ is

(1) $9\sqrt{34}$ (2) $13\sqrt{2}$ (3) $10\sqrt{34}$ (4) 15

In 36–39, for each irrational number given, write a rational approximation: **a.** as shown on a calculator display **b.** rounded to four significant digits.

36. $\sqrt{194}$ **37.** $\sqrt[3]{16}$ **38.** $-\sqrt{0.7}$ **39.** $\sqrt[5]{-27}$

40. The area of a square is 28.00 square meters.

a. Find, to the *nearest thousandth* of a meter, the length of one side of the square.

b. Find, to the *nearest thousandth* of a meter, the perimeter of the square.

c. Explain why the answer to part **b** is *not* equal to 4 times the answer to part **a**.

41. Write as a polynomial in simplest form: $(2x + \sqrt{3})(x - \sqrt{3})$.

42. What is the product of $-3.5x^2$ and $-6.2x^3$?

Exploration

STEP 1. On a sheet of graph paper, draw the positive ray of the real number line. Draw a square, using the interval from 0 to 1 as one side of the square. Draw the diagonal of this square from 0 to its opposite vertex.

The length of the diagonal is $\sqrt{2}$. Why? Place the point of a compass at 0 and the pencil of a compass at the opposite vertex of the square so that the measure of the opening of the pair of compasses is $\sqrt{2}$. Keep the point of the compass at 0 and use the pencil to mark the endpoint of a segment of length $\sqrt{2}$ on the number line.

STEP 2. Using the same number line, draw a rectangle whose dimensions are $\sqrt{2}$ by 1, using the interval on the number line from 0 to $\sqrt{2}$ as one side. Draw the diagonal of this rectangle from 0 to the opposite vertex. The length of the diagonal is $\sqrt{3}$. Place the point of a pair of compasses at 0 and the pencil at the opposite vertex of the rectangle so that the measure of the opening of the pair of compasses is $\sqrt{3}$. Keep the point of the compasses at 0 and use the pencil to mark the endpoint of a segment of length $\sqrt{3}$ on the number line.

STEP 3. Repeat step 2, drawing a rectangle whose dimensions are $\sqrt{3}$ by 1 to locate $\sqrt{4}$ on the number line. This point should coincide with 2 on the number line.

STEP 4. Explain how these steps can be used to locate $\sqrt{n}$ for any positive integer n.

CUMULATIVE REVIEW — CHAPTERS 1–12

Part I

Answer all questions in this part. Each correct answer will receive 2 credits. No partial credit will be allowed.

1. A flagpole casts a shadow five feet long at the same time that a man who is six feet tall casts a shadow that is two feet long. How tall is the flagpole?
(1) 12 feet (2) 15 feet (3) 18 feet (4) 24 feet

2. Which of the following is an irrational number?
(1) $\sqrt{12}(\sqrt{3})$ (2) $\sqrt{24} \div \sqrt{6}$ (3) $\sqrt{5} - \sqrt{5}$ (4) $3\sqrt{5} - \sqrt{5}$

3. Parallelogram $ABCD$ is drawn on the coordinate plane with the vertices $A(0, 0)$, $B(8, 0)$, $C(10, 5)$, and $D(2, 5)$. The number of square units in the area of $ABCD$ is
(1) 40 (2) 20 (3) 16 (4) 10

4. The product $(3a - 2)(2a + 3)$ can be written as
(1) $6a^2 - 6$
(2) $6a^2 + a - 6$
(3) $6a^2 - 5a - 6$
(4) $6a^2 + 5a - 6$

5. If $0.2x - 8 = x + 4$, then x equals
(1) 120 (2) 12 (3) −15 (4) 15

6. If the height of a right circular cylinder is 12 centimeters and the measure of the diameter of a base is 8 centimeters, then the volume of the cylinder is
(1) 768π (2) 192π (3) 96π (4) 48π

7. The identity $3(a + 7) = 3a + 21$ is an example of
(1) the additive inverse property
(2) the associative property for addition
(3) the commutative property for addition
(4) the distributive property of multiplication over addition

8. The value of a share of stock decreased from \$24.50 to \$22.05. The percent of decrease was
(1) 1% (2) 10% (3) 11% (4) 90%

9. When written in scientific notation, 384.5 is equal to
(1) 38.45×10^1 (2) 3.845×10^2 (3) 3.845×10^3 (4) 3.845×10^{-2}

10. In right triangle ABC, $\angle C$ is the right angle. The cosine of $\angle B$ is
(1) $\frac{AC}{AB}$ (2) $\frac{BC}{AC}$ (3) $\frac{BC}{AB}$ (4) $\frac{AC}{BC}$

Part II

Answer all questions in this part. Each correct answer will receive 2 credits. Clearly indicate the necessary steps, including appropriate formula substitutions, diagrams, graphs, charts, etc. For all questions in this part, a correct numerical answer with no work shown will receive only 1 credit.

11. A car uses $\frac{3}{4}$ of a tank of gasoline to travel 600 kilometers. The tank holds 48 liters of gasoline. How far can the car go on one liter of gasoline?

12. A ramp that is 20.0 feet long makes an angle of 12.5° with the ground. What is the perpendicular distance from the top of the ramp to the ground?

Part III

Answer all questions in this part. Each correct answer will receive 3 credits. Clearly indicate the necessary steps, including appropriate formula substitutions, diagrams, graphs, charts, etc. For all questions in this part, a correct numerical answer with no work shown will receive only 1 credit.

13. $ABCD$ is a trapezoid with $\overline{BC} \perp \overline{AB}$ and $\overline{BC} \perp \overline{CD}$, $AB = 13$, $BC = 12$, and $CD = 8$. A line segment is drawn from A to E, the midpoint of $\overline{CD}$.

a. Find the area of $\triangle AED$.

b. Find the perimeter of $\triangle AED$.

14. Maria's garden is in the shape of a rectangle that is twice as long as it is wide. Maria increases the width by 2 feet, making the garden 1.5 times as long as it is wide. What are the dimensions of the original garden?

Part IV

Answer all questions in this part. Each correct answer will receive 4 credits. Clearly indicate the necessary steps, including appropriate formula substitutions, diagrams, graphs, charts, etc. For all questions in this part, a correct numerical answer with no work shown will receive only 1 credit.

15. The length of the base of an isosceles triangle is $10\sqrt{2}$ and the length of the altitude is $12\sqrt{2}$. Express the perimeter as an exact value in simplest form.

16. In the coordinate plane, O is the origin, A is a point on the y-axis, and B is a point on the x-axis. The slope of $\overleftrightarrow{AB}$ is -2 and its y-intercept is 8.

a. Write the equation of $\overleftrightarrow{AB}$.

b. Draw $\overleftrightarrow{AB}$ on graph paper.

c. What are the coordinates of B?

d. What is the x-intercept of $\overleftrightarrow{AB}$?

CHAPTER

13

QUADRATIC RELATIONS AND FUNCTIONS

CHAPTER TABLE OF CONTENTS

When a baseball is hit, its path is not a straight line. The baseball rises to a maximum height and then falls, following a curved path throughout its flight. The maximum height to which it rises is determined by the force with which the ball was hit and the angle at which it was hit. The height of the ball at any time can be found by using an equation, as can the maximum height to which the ball rises and the distance between the batter and the point where the ball hits the ground.

In this chapter we will study the quadratic equation that models the path of a baseball as well as functions and relations that are not linear.

13-1 SOLVING QUADRATIC EQUATIONS

The equation $x^2 - 3x - 10 = 0$ is a polynomial equation in one variable. This equation is of degree two, or second degree, because the greatest exponent of the variable x is 2. The equation is in **standard form** because all terms are collected in descending order of exponents in one side, and the other side is 0.

A **polynomial equation of degree two** is also called a **quadratic equation**. The standard form of a quadratic equation in one variable is

$$ax^2 + bx + c = 0$$

where a, b, and c are real numbers and $a \neq 0$.

To write an equation such as $x(x - 4) = 5$ in standard form, rewrite the left side without parentheses and add -5 to both sides to make the right side 0.

$$x(x - 4) = 5$$
$$x^2 - 4x = 5$$
$$x^2 - 4x - 5 = 0$$

Solving a Quadratic Equation by Factoring

When 0 is the product of two or more factors, at least one of the factors must be equal to 0. This is illustrated by the following examples:

$$5 \times 0 = 0 \qquad (-2) \times 0 = 0 \qquad \tfrac{1}{2} \times 0 = 0$$
$$0 \times 7 = 0 \qquad 0 \times (-3) = 0 \qquad 0 \times 0 = 0$$

In general:

▶ **When a and b are real numbers, $ab = 0$ if and only if $a = 0$ or $b = 0$.**

This principle is used to solve quadratic equations. For example, to solve the quadratic equation $x^2 - 3x + 2 = 0$, we can write the left side as $(x - 2)(x - 1)$. The factors $(x - 2)$ and $(x - 1)$ represent real numbers whose product is 0. The equation will be true if the first factor is 0, that is, if $(x - 2) = 0$ or if the second factor is 0, that is, if $(x - 1) = 0$.

$$x^2 - 3x + 2 = 0$$
$$(x - 2)(x - 1) = 0$$

$$\begin{array}{rcr} x - 2 &=& 0 \\ +2 && +2 \\ \hline x &=& 2 \end{array} \qquad \begin{array}{rcr} x - 1 &=& 0 \\ +1 && +1 \\ \hline x &=& 1 \end{array}$$

A check will show that both 2 and 1 are values of x for which the equation is true.

Check for $x = 2$:	*Check for* $x = 1$:
$x^2 - 3x + 2 = 0$	$x^2 - 3x + 2 = 0$
$(2)^2 - 3(2) + 2 \stackrel{?}{=} 0$	$(1)^2 - 3(1) + 2 \stackrel{?}{=} 0$
$4 - 6 + 2 \stackrel{?}{=} 0$	$1 - 3 + 2 \stackrel{?}{=} 0$
$0 = 0$ ✔	$0 = 0$ ✔

Since both 2 and 1 satisfy the equation $x^2 - 3x + 2 = 0$, the solution set of this equation is {2, 1}. The **roots of the equation**, that is, the values of the variable that make the equation true, are 2 and 1.

Note that the factors of the trinomial $x^2 - 3x + 2$ are $(x - 2)$ and $(x - 1)$. If the trinomial $x^2 - 3x + 2$ is set equal to zero, then an equation is formed and this equation has a solution set, {2, 1}.

Procedure

To solve a quadratic equation by factoring:

1. If necessary, transform the equation into standard form.
2. Factor the quadratic expression.
3. Set each factor containing the variable equal to 0.
4. Solve each of the resulting equations.
5. Check by substituting each value of the variable in the original equation.

► **The real number k is a root of $ax^2 + bx + c = 0$ if and only if $(x - k)$ is a factor of $ax^2 + bx + c$.**

EXAMPLE 1

Solve and check: $x^2 - 7x = -10$

Solution *How to Proceed*

(1) Write the equation in standard form:

$$x^2 - 7x = -10$$

$$x^2 - 7x + 10 = 0$$

(2) Factor the quadratic expression:

$$(x - 2)(x - 5) = 0$$

(3) Let each factor equal 0:

$$x - 2 = 0 \quad | \quad x - 5 = 0$$

(4) Solve each equation:

$$x = 2 \quad | \quad x = 5$$

(5) Check both values in the original equation:

Check for $x = 2$:	*Check for* $x = 5$:
$x^2 - 7x = -10$	$x^2 - 7x = -10$
$(2)^2 - 7(2) \stackrel{?}{=} -10$	$(5)^2 - 7(5) \stackrel{?}{=} -10$
$4 - 14 \stackrel{?}{=} -10$	$25 - 35 \stackrel{?}{=} -10$
$-10 = -10$ ✔	$-10 = -10$ ✔

Answer $x = 2$ or $x = 5$; the solution set is $\{2, 5\}$.

EXAMPLE 2

List the members of the solution set of $2x^2 = 3x$.

Solution

How to Proceed

(1) Write the equation in standard form:

$$2x^2 = 3x$$
$$2x^2 - 3x = 0$$

(2) Factor the quadratic expression:

$$x(2x - 3) = 0$$

(3) Let each factor equal 0:

$$x = 0 \quad \Big| \quad 2x - 3 = 0$$

(4) Solve each equation:

$$2x = 3$$
$$x = \tfrac{3}{2}$$

(5) Check both values in the original equation.

Check for $x = 0$:	*Check for* $x = \frac{3}{2}$:
$2x^2 = 3x$	$2x^2 = 3x$
$2(0)^2 \stackrel{?}{=} 3(0)$	$2\left(\frac{3}{2}\right)^2 \stackrel{?}{=} 3\left(\frac{3}{2}\right)$
$0 = 0$ ✔	$2\left(\frac{9}{4}\right) \stackrel{?}{=} 3\left(\frac{3}{2}\right)$
	$\frac{9}{2} = \frac{9}{2}$ ✔

Answer: The solution set is $\left\{0, \frac{3}{2}\right\}$.

Note: We never divide both sides of an equation by an expression containing a variable. If we had divided $2x^2 = 3x$ by x, we would have obtained the equation $2x = 3$, whose solution is $\frac{3}{2}$ but would have lost the solution $x = 0$.

EXAMPLE 3

Find the solution set of the equation $x(x - 8) = 2x - 25$.

Solution *How to Proceed*

(1) Use the distributive property on the left side of the equation:
$$x(x - 8) = 2x - 25$$
$$x^2 - 8x = 2x - 25$$

(2) Write the equation in standard form:
$$x^2 - 10x + 25 = 0$$

(3) Factor the quadratic expression:
$$(x - 5)(x - 5) = 0$$

(4) Let each factor equal 0:
$$x - 5 = 0 \quad \Big| \quad x - 5 = 0$$

(5) Solve each equation:
$$x = 5 \quad \Big| \quad x = 5$$

(6) Check the value in the original equation:
$$x(x - 8) = 2x - 25$$
$$5(5 - 8) \stackrel{?}{=} 2(5) - 25$$
$$-15 \stackrel{?}{=} 10 - 25$$
$$-15 = -15 \checkmark$$

Answer $x = 5$; the solution set is $\{5\}$.

Every quadratic equation has two roots, but as Example 3 shows, the two roots are sometimes the same number. Such a root, called a **double root**, is written only once in the solution set.

EXAMPLE 4

The height h of a ball thrown into the air with an initial vertical velocity of 24 feet per second from a height of 6 feet above the ground is given by the equation $h = -16t^2 + 24t + 6$ where t is the time, in seconds, that the ball has been in the air. After how many seconds is the ball at a height of 14 feet?

Solution (1) In the equation, let $h = 14$:
$$h = -16t^2 + 24t + 6$$
$$14 = -16t^2 + 24t + 6$$

(2) Write the equation in standard form:
$$0 = -16t^2 + 24t - 8$$

(3) Factor the quadratic expression:
$$0 = -8(2t^2 - 3t + 1)$$
$$0 = -8(2t - 1)(t - 1)$$

(4) Solve for t:
$$2t - 1 = 0 \quad \Big| \quad t - 1 = 0$$
$$2t = 1 \quad \Big| \quad t = 1$$
$$t = \tfrac{1}{2}$$

Answer The ball is at a height of 14 feet after $\frac{1}{2}$ second as it rises and after 1 second as it falls.

EXAMPLE 5

The area of a circle is equal to 3 times its circumference. What is the radius of the circle?

Solution *How to Proceed*

(1) Write an equation from the given information: $\pi r^2 = 3(2\pi r)$

(2) Set the equation in standard form: $\pi r^2 - 6\pi r = 0$

(3) Factor the quadratic expression: $\pi r(r - 6) = 0$

(4) Solve for r: $\pi r = 0$ | $r - 6 = 0$

(5) Reject the zero value. Use the positive value to write the answer. $r = 0$ | $r = 6$

Answer The radius of the circle is 6 units.

EXERCISES

Writing About Mathematics

1. Can the equation $x^2 = 9$ be solved by factoring? Explain your answer.

2. In Example 4, the trinomial was written as the product of three factors. Only two of these factors were set equal to 0. Explain why the third factor was not used to find a solution of the equation.

Developing Skills

In 3–38, solve each equation and check.

3. $x^2 - 3x + 2 = 0$ **4.** $z^2 - 5z + 4 = 0$ **5.** $x^2 - 8x + 16 = 0$

6. $r^2 - 12r + 35 = 0$ **7.** $c^2 + 6c + 5 = 0$ **8.** $m^2 + 10m + 9 = 0$

9. $x^2 + 2x + 1 = 0$ **10.** $y^2 + 11y + 24 = 0$ **11.** $x^2 - 4x - 5 = 0$

12. $x^2 + x - 6 = 0$ **13.** $x^2 + 2x - 15 = 0$ **14.** $r^2 - r - 72 = 0$

15. $x^2 - x - 12 = 0$ **16.** $x^2 - 49 = 0$ **17.** $z^2 - 4 = 0$

18. $m^2 - 64 = 0$ **19.** $3x^2 - 12 = 0$ **20.** $d^2 - 2d = 0$

21. $s^2 - s = 0$ **22.** $x^2 + 3x = 0$ **23.** $z^2 + 8z = 0$

24. $x^2 - x = 6$ **25.** $y^2 - 3y = 28$ **26.** $c^2 - 8c = -15$

27. $r^2 = 4$ **28.** $x^2 = 121$ **29.** $y^2 = 6y$

30. $s^2 = -4s$ **31.** $y^2 = 8y + 20$ **32.** $x^2 = 9x - 20$

33. $30 + x = x^2$ **34.** $x^2 + 3x - 4 = 50$ **35.** $2x^2 + 7 = 5 - 5x$

36. $x(x - 2) = 35$ **37.** $y(y - 3) = 4$ **38.** $x(x + 3) = 40$

In 39–44, solve each equation and check.

39. $\frac{x + 2}{2} = \frac{12}{x}$

40. $\frac{y + 3}{3} = \frac{6}{y}$

41. $\frac{x}{3} = \frac{12}{x}$

42. $\frac{x + 4}{-1} = \frac{4}{x}$

43. $\frac{-4x}{x - 3} = \frac{x - 1}{2}$

44. $\frac{2x - 2}{x + 3} = \frac{x - 1}{x - 2}$

Applying Skills

45. The height h of a ball thrown into the air with an initial vertical velocity of 48 feet per second from a height of 5 feet above the ground is given by the equation

$$h = -16t^2 + 48t + 5$$

where t is the time in seconds that the ball has been in the air. After how many seconds is the ball at a height of 37 feet?

46. A batter hit a baseball at a height of 4 feet with a force that gave the ball an initial vertical velocity of 64 feet per second. The equation

$$h = -16t^2 + 64t + 4$$

gives the height h of the baseball t seconds after the ball was hit. If the ball was caught at a height of 4 feet, how long after the batter hit the ball was the ball caught?

47. The length of a rectangle is 12 feet more than twice the width. The area of the rectangle is 320 square feet.

a. Write an equation that can be used to find the length and width of the rectangle.

b. What are the dimensions of the rectangle?

48. A small park is enclosed by four streets, two of which are parallel. The park is in the shape of a trapezoid. The perpendicular distance between the parallel streets is the height of the trapezoid. The portions of the parallel streets that border the park are the bases of the trapezoid. The height of the trapezoid is equal to the length of one of the bases and 20 feet longer than the other base. The area of the park is 9,000 square feet.

a. Write an equation that can be used to find the height of the trapezoid.

b. What is the perpendicular distance between the two parallel streets?

49. At a kennel, each dog run is a rectangle whose length is 4 feet more than twice the width. Each run encloses 240 square feet. What are the dimensions of the runs?

50. One leg of a right triangle is 14 centimeters longer than the other leg. The length of the hypotenuse is 26 centimeters. What are the lengths of the legs?

13-2 THE GRAPH OF A QUADRATIC FUNCTION

A batter hits a baseball at a height of 3 feet off the ground, with an initial vertical velocity of 72 feet per second. The height, y, of the baseball can be found using the equation $y = -16x^2 + 72x + 3$ when x represents the number of seconds from the time the ball was hit. The graph of the equation—and, incidentally, the actual path of the ball—is a curve called a **parabola**. The special properties of parabolas are discussed in this section.

An equation of the form $y = ax^2 + bx + c$ $(a \neq 0)$ is called a **second-degree polynomial function** or a **quadratic function**. It is a function because for every ordered pair in its solution set, each different value of x is paired with one and only one value of y. The graph of any quadratic function is a parabola.

Because the graph of a quadratic function is nonlinear, a larger number of points are needed to draw the graph than are needed to draw the graph of a linear function. The graphs of two equations of the form $y = ax^2 + bx + c$, one that has a positive coefficient of x^2 $(a > 0)$ and the other a negative coefficient of x^2 $(a < 0)$, are shown below.

CASE 1 *The graph of* $ax^2 + bx + c$ *where* $a > 0$.

Graph the quadratic function $y = x^2 - 4x + 1$ for integral values of x from -1 to 5 inclusive:

(1) Make a table using integral values of x from -1 to 5.

(2) Plot the points associated with each ordered pair (x, y).

(3) Draw a smooth curve through the points.

x	$x^2 - 4x + 1$	y
−1	1 + 4 + 1	6
0	0 − 0 + 1	1
1	1 − 4 + 1	−2
2	4 − 8 + 1	−3
3	9 − 12 + 1	−2
4	16 − 16 + 1	1
5	25 − 20 + 1	6

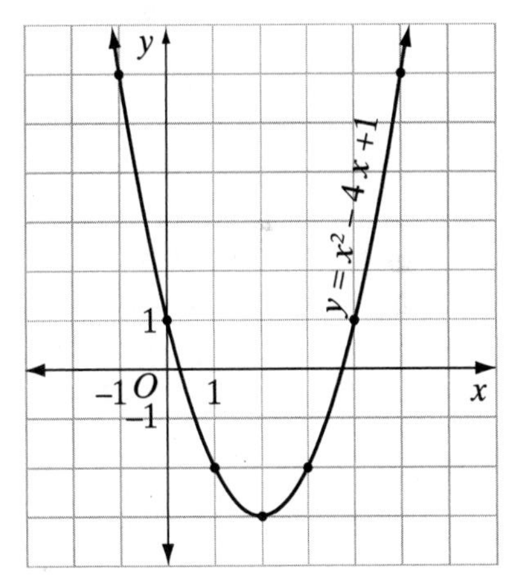

The values of x that were chosen to draw this graph are not a random set of numbers. These numbers were chosen to produce the pattern of y-values shown in the table. Notice that as x increases from -1 to 2, y decreases from 6 to -3. Then the graph reverses and as x continues to increase from 2 to 5, y increases from -3 to 6. The smallest value of y occurs at the point $(2, -3)$. The point is called the **minimum** because its y-value, -3, is the smallest value of y for the equation. The minimum point is also called the **turning point** or **vertex** of the parabola.

The graph is symmetric with respect to the vertical line, called the **axis of symmetry of the parabola.** The axis of symmetry of the parabola is determined by the formula

$$x = \frac{-b}{2a}$$

where a and b are the coefficients x^2 and x, respectively, from the standard form of the quadratic equation.

For the function $y = x^2 - 4x + 1$, the equation of the vertical line of symmetry is $x = \frac{-(-4)}{2(1)}$ or $x = 2$. Every point on the parabola to the left of $x = 2$ matches a point on the parabola to the right of $x = 2$, and vice versa.

This example illustrates the following properties of the graph of the quadratic equation $y = x^2 - 4x + 1$:

1. The graph of the equation is a parabola.
2. The parabola is symmetric with respect to the vertical line $x = 2$.
3. The parabola opens upward and has a minimum point at $(2, -3)$.
4. The equation defines a function. For every x-value there is one and only one y-value.
5. The constant term, 1, is the y-intercept. The y-intercept is the value of y when x is 0.

CASE 2 *The graph of $ax^2 + bx + c$ where $a < 0$.*

Graph the quadratic function $y = -x^2 - 2x + 5$ using integral values of x from -4 to 2 inclusive:

(1) Make a table using integral values of x from -4 to 2.

(2) Plot the points associated with each ordered pair (x, y).

(3) Draw a smooth curve through the points.

x	$-x^2 - 2x + 5$	y
−4	−16 + 8 + 5	−3
−3	−9 + 6 + 5	2
−2	−4 + 4 + 5	5
−1	−1 + 2 + 5	6
0	0 − 0 + 5	5
1	−1 − 2 + 5	2
2	−4 − 4 + 5	−3

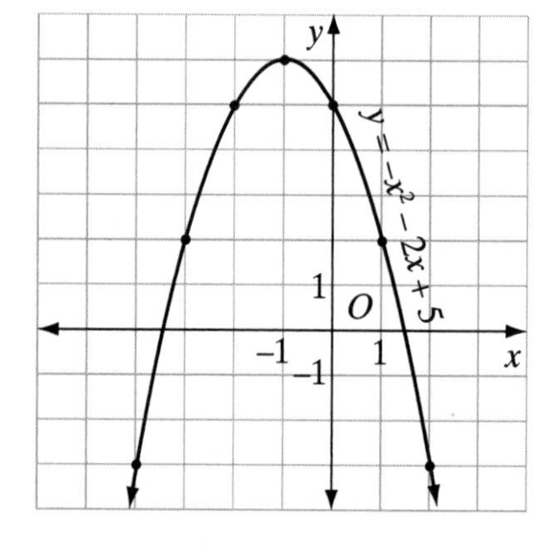

Again, the values of x that were chosen to produce the pattern of y-values shown in the chart. Notice that as x increases from -4 to -1, y increases from

-3 to 6. Then the graph reverses, and as x continues to increase from -1 to 2, y decreases from 6 to -3. The largest value of y occurs at the point $(-1, 6)$. This point is called the **maximum** because its y-value, 6, is the largest value of y for the equation. In this case, the maximum point is the turning point or vertex of the parabola.

The graph is symmetric with respect to the vertical line whose equation is $x = -1$. As shown in Case 1, this value of x is again given by the formula

$$x = \frac{-b}{2a}$$

where a and b are the coefficients of x^2 and x, respectively. For the function $y = -x^2 - 2x + 5$, the equation of the axis of symmetry is $x = \frac{-(-2)}{2(-1)}$ or $x = -1$.

This example illustrates the following properties of the graph of the quadratic equation $y = -x^2 - 2x + 5$:

1. The graph of the equation is a parabola.
2. The parabola is symmetric with respect to the vertical line $x = -1$.
3. The parabola opens downward and has a maximum point at $(-1, 6)$
4. The equation defines a function.
5. The constant term, 5, is the y-intercept.

When the equation of a parabola is written in the form $y = ax^2 + bx + c = 0$, the equation of the axis of symmetry is $x = \frac{-b}{2a}$ and the x-coordinate of the turning point is $\frac{-b}{2a}$. This can be used to find a convenient set of values of x to be used when drawing a parabola. Use $\frac{-b}{2a}$ as a middle value of x with three values that are smaller and three that are larger.

KEEP IN MIND

1. The graph of $y = ax^2 + bx + c$, with $a \neq 0$, is a parabola.
2. The axis of symmetry of the parabola is a vertical line whose equation is $x = \frac{-b}{2a}$.
3. A parabola has a turning point on the axis of symmetry. The x-coordinate of the turning point is $\frac{-b}{2a}$. The y-coordinate of the turning point is found by substituting $\frac{-b}{2a}$ into the equation of the parabola. The turning point is the vertex of the parabola.
4. If a is positive, the parabola opens upward and the turning point is a minimum. The minimum value of y for the parabola is the y-coordinate of the turning point.
5. If a is negative, the parabola opens downward and the turning point is a maximum. The maximum value of y for the parabola is the y-coordinate of the turning point.

EXAMPLE 1

a. Write the equation of the axis of symmetry of $y = x^2 - 3$.

b. Graph the function.

c. Does the function have a maximum or a minimum?

d. What is the maximum or minimum value of the function?

e. Write the coordinates of the vertex.

Solution **a.** In this equation, $a = 1$. Since there is no x term in the equation, the equation can be written as $y = x^2 + 0x - 3$ with $b = 0$. The equation of the axis of symmetry is

$$x = \frac{-b}{2a} = \frac{-(0)}{2(1)} = 0 \text{ or } x = 0.$$

b. (1) Since the vertex of the parabola is on the axis of symmetry, the x-coordinate of the vertex is 0. Use three values of x that are less than 0 and three values of x that are greater than 0. Make a table using integral values of x from -3 to 3.

(2) Plot the points associated with each ordered pair (x, y).

(3) Draw a smooth curve through the points to draw a parabola.

x	$x^2 - 3$	y
−3	$(-3)^2 - 3$	6
−2	$(-2)^2 - 3$	1
−1	$(-1)^2 - 3$	−2
0	$(0)^2 - 3$	−3
1	$(1)^2 - 3$	−2
2	$(2)^2 - 3$	1
3	$(3)^2 - 3$	6

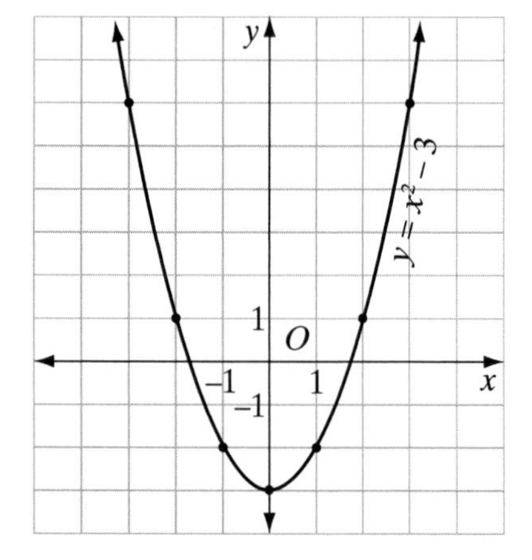

c. Since $a = 1 > 0$, the function has a minimum.

d. The minimum value of the function is the y-coordinate of the vertex, -3, which can be read from the table of values.

e. Since the vertex is the turning point of this parabola, the coordinates of the vertex are $(0, -3)$.

Calculator Solution

a. Determine the equation of the axis of symmetry as before.

b. Enter the equation in the Y= list of functions and graph the function. Clear any equations already in the list.

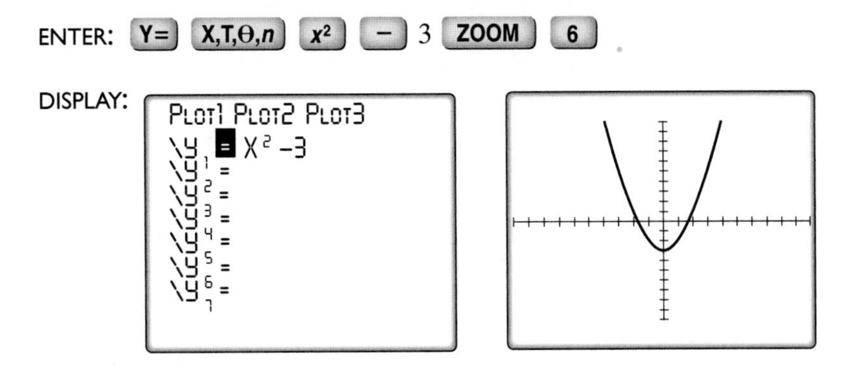

c. The graph shows that the function has a minimum.

d. Since the minimum of the function occurs at the vertex, use value (2nd CALC 1) from the CALC menu to evaluate the function at $x = 0$, the x-coordinate of the vertex:

ENTER: 2nd CALC 1 0 ENTER

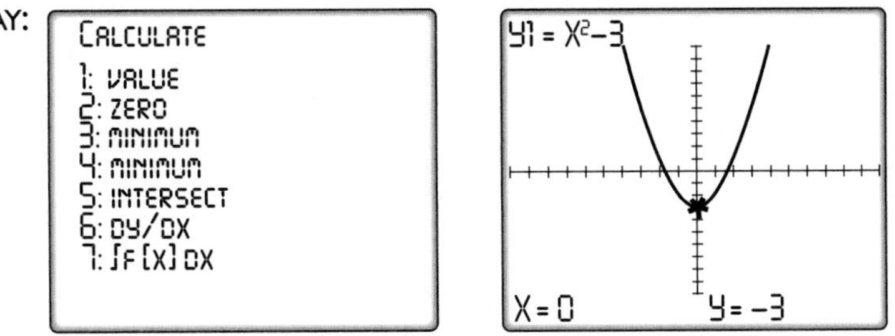

The calculator displays the minimum value, -3.

e. The coordinates of the vertex are $(0, -3)$.

Answers

a. The axis of symmetry is the y-axis. The equation is $x = 0$.

b. Graph

c. The function has a minimum.

d. The minimum value is -3.

e. The vertex is $(0, -3)$.

When the coordinates of the turning point are rational numbers, we can use minimum (2nd CALC 3) or maximum (2nd CALC 4) from the CALC menu to find the vertex. In Example 1, since the turning point is a minimum, use minimum:

ENTER: 

When the calculator asks "LeftBound?" move the cursor to any point to the left of the vertex using the left or right arrow keys, and then press enter. When the calculator asks "RightBound?" move the cursor to the right of the vertex, and then press enter. When the calculator asks "Guess?" move the cursor near the vertex and then press enter. The calculator displays the coordinates of the vertex at the bottom of the screen.

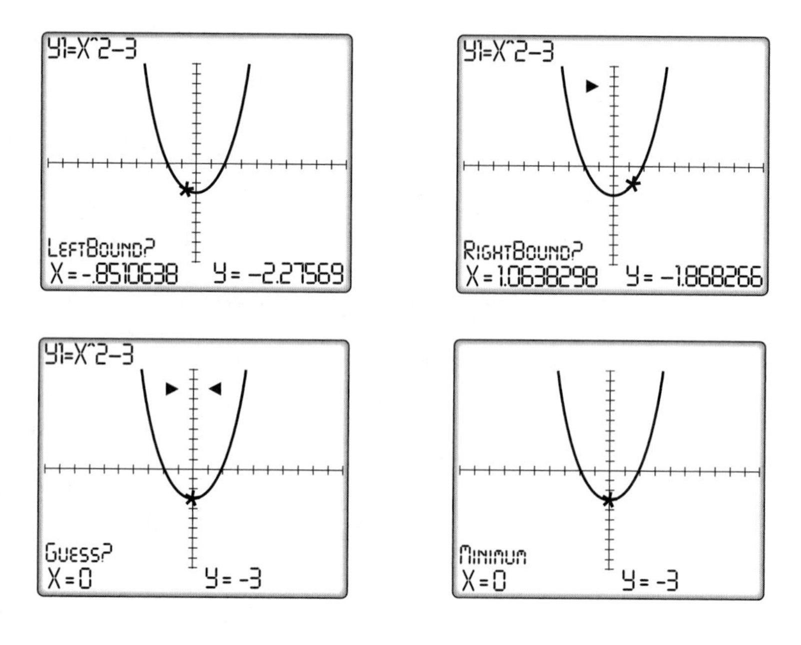

EXAMPLE 2

Sketch the graph of the function $y = x^2 - 7x + 10$.

Solution (1) Find the equation of the axis of symmetry and the x-coordinate of the vertex:

$$x = \frac{-b}{2a} = \frac{-(-7)}{2(1)} = 3.5$$

(2) Make a table of values using three integral values smaller than and three larger than 3.5.

(3) Plot the points whose coordinates are given in the table and draw a smooth curve through them.

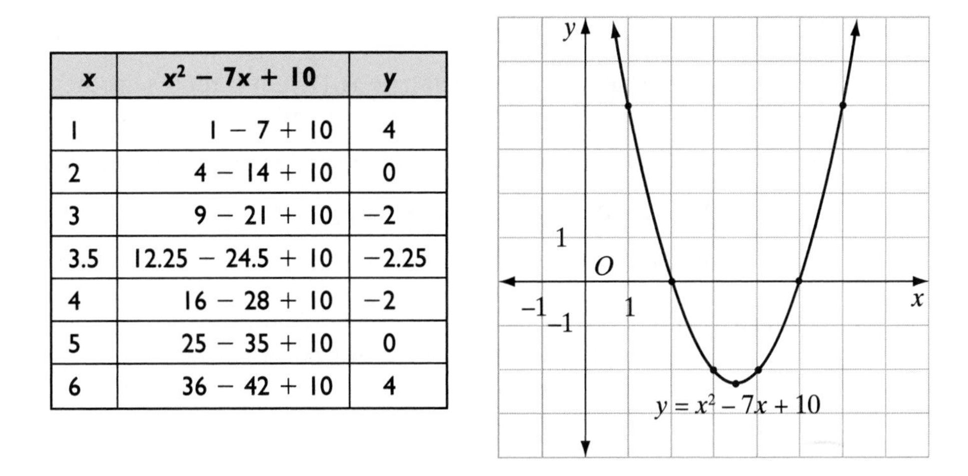

x	$x^2 - 7x + 10$	y
1	1 − 7 + 10	4
2	4 − 14 + 10	0
3	9 − 21 + 10	−2
3.5	12.25 − 24.5 + 10	−2.25
4	16 − 28 + 10	−2
5	25 − 35 + 10	0
6	36 − 42 + 10	4

Note: The table can also be displayed on the calculator. First enter the equation into Y_1.

ENTER: Y= X,T,Θ,*n* x^2 − 7 X,T,Θ,*n* + 10

Then enter the starting value and the interval between the *x* values. We will use 1 as the starting value and 0.5 as the interval in order to include 3.5, the *x*-value of the vertex.

ENTER: 2nd TBLSET 1 ENTER .5 ENTER

Before creating the table, make sure that "Indpnt:" and "Depend:" are set to "auto." If they are not, press ENTER ▼ ENTER. Finally, press 2nd TABLE to create the table. Scroll up and down to view the values of *x* and *y*.

X	Y1	
1	4	
1.5	1.75	
2	0	
2.5	-1.25	
3	-2	
3.5	-2.25	
4	-2	

X=1

Calculator Solution Enter the equation in the Y= menu and sketch the graph of the function in the standard window.

ENTER: Y= X,T,Θ,*n* x^2 − 7 X,T,Θ,*n* + 10 ZOOM 6

DISPLAY:

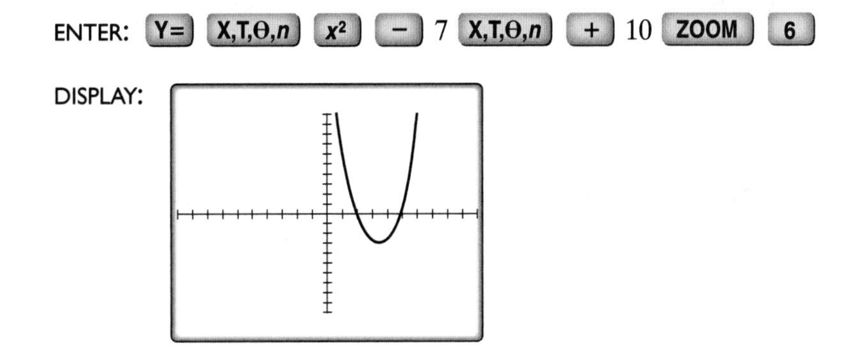

EXAMPLE 3

The perimeter of a rectangle is 12. Let x represent the measure of one side of the rectangle and y represent the area.

a. Write an equation for the area of the rectangle in terms of x.

b. Draw the graph of the equation written in **a.**

c. What is the maximum area of the rectangle?

Solution **a.** Let x be the measure of the length of the rectangle. Use the formula for perimeter to express the measure of the width in terms of x:

$$P = 2l + 2w$$
$$12 = 2x + 2w$$
$$6 = x + w$$
$$w = 6 - x$$

Write the formula for area in terms of x and y:

$$A = lw$$
$$y = x(6 - x)$$
$$y = 6x - x^2$$
$$y = -x^2 + 6x$$

b. The equation of the axis of symmetry is $x = \frac{-6}{2(-1)}$ or $x = 3$. Make a table of values using values of x on each side of 3.

x	$-x^2 + 6x$	y
0	$-0 + 0$	0
1	$-1 + 6$	5
2	$-4 + 12$	8
3	$-9 + 18$	9
4	$-16 + 24$	8
5	$-25 + 30$	5
6	$-36 + 36$	0

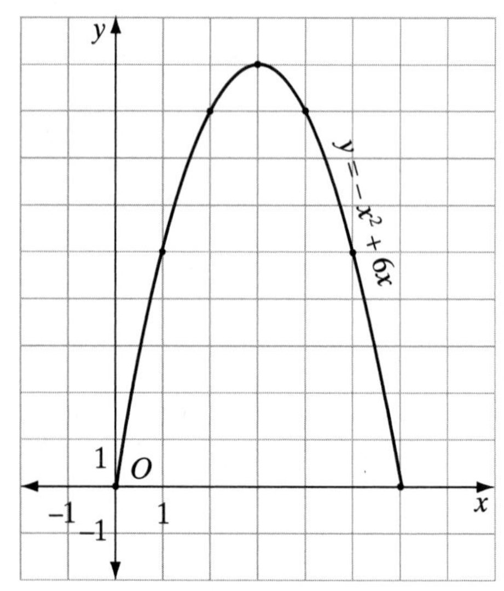

c. The maximum value of the area, y, is 9.

Note: The graph shows all possible values of x and y. Since both the measure of a side of the rectangle, x, and the area of the rectangle, y, must be positive, $0 < x < 6$ and $0 < y \leq 9$. Since (2, 8) is a point on the graph, one possible rectangle has dimensions 2 by $(6 - 2)$ or 2 by 4 and an area of 8. The rectangle with maximum area, 9, has dimensions 3 by $(6 - 3)$ or 3 by 3, a square.

Translating, Reflecting, and Scaling Graphs of Quadratic Functions

Just as linear and absolute value functions can be translated, reflected, or scaled, graphs of quadratic functions can also be manipulated by working with the graph of the quadratic function $y = x^2$.

For instance, the graph of $y = x^2 + 2.5$ is the graph of $y = x^2$ shifted 2.5 units up. The graph of $y = -x^2$ is the graph of $y = x^2$ reflected in the x-axis. The graph of $y = 3x^2$ is the graph of $y = x^2$ stretched vertically by a factor of 3, while the graph of $y = \frac{1}{3}x^2$ is the graph of $y = x^2$ compressed vertically by a factor of $\frac{1}{3}$.

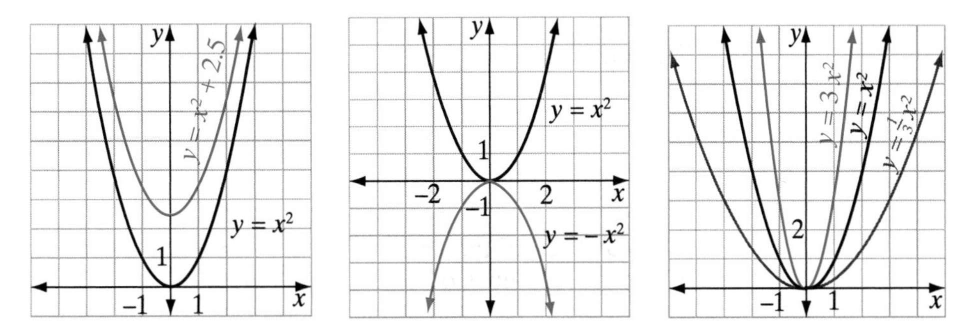

Translation Rules for Quadratic Functions

If c is *positive*:

- **The graph of $y = x^2 + c$ is the graph of $y = x^2$ shifted c units up.**
- **The graph of $y = x^2 - c$ is the graph of $y = x^2$ shifted c units down.**
- **The graph of $y = (x + c)^2$ is the graph of $y = x^2$ shifted c units to the left.**
- **The graph of $y = (x - c)^2$ is the graph of $y = x^2$ shifted c units to the right.**

Reflection Rule for Quadratic Functions

- **The graph of $y = -x^2$ is the graph of $y = x^2$ reflected in the x-axis.**

Scaling Rules for Quadratic Functions

- **When $c > 1$, the graph of $y = cx^2$ is the graph of $y = x^2$ stretched vertically by a factor of c.**
- **When $0 < c < 1$, the graph of $y = cx^2$ is the graph of $y = x^2$ compressed vertically by a factor of c.**

EXAMPLE 4

In **a–e**, write an equation for the resulting function if the graph of $y = x^2$ is:

a. shifted 5 units down and 1.5 units to the left

b. stretched vertically by a factor of 4 and shifted 2 units down

c. compressed vertically by a factor of $\frac{1}{6}$ and reflected in the x-axis

d. reflected in the x-axis, shifted 2 units up, and shifted 2 units to the right

Solution **a.** $y = (x + 1.5)^2 - 5$ *Answer*

b. First, stretch vertically by a factor of 4: $y = 4x^2$

Then, translate the resulting function 2 units down: $y = 4x^2 - 2$ *Answer*

c. First, compress vertically by a factor of $\frac{1}{6}$: $y = \frac{1}{6}x^2$

Then, reflect in the x-axis: $y = -\frac{1}{6}x^2$ *Answer*

d. First, reflect in the x-axis: $y = -x^2$

Then, translate the resulting function 2 units up: $y = -x^2 + 2$

Finally, translate the resulting function 2 units to the right: $y = -(x - 2)^2 + 2$ *Answer*

EXERCISES

Writing About Mathematics

1. Penny drew the graph of $h(x) = \frac{1}{4}x^2 - 2x$ from $x = -4$ to $x = 4$. Her graph is shown to the right. Explain why Penny's graph does not look like a parabola.

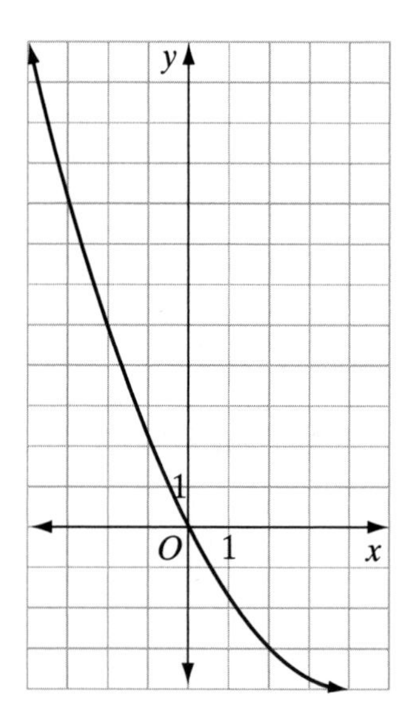

2. What values of x would you choose to draw the graph of $h(x) = \frac{1}{4}x^2 - 2x$ so that points on both sides of the turning point would be shown on the graph? Explain your answer.

Developing Skills

In 3–14: **a.** Graph each quadratic function on graph paper using the integral values for x indicated in parentheses to prepare the necessary table of values. **b.** Write the equation of the axis of symmetry of the graph. **c.** Write the coordinates of the turning point of the graph.

3. $y = x^2$ $(-3 \leq x \leq 3)$

4. $y = -x^2$ $(-3 \leq x \leq 3)$

5. $y = x^2 + 1$ $(-3 \leq x \leq 3)$

6. $y = x^2 - 1$ $(-3 \leq x \leq 3)$

7. $y = -x^2 + 4$ $(-3 \leq x \leq 3)$

8. $y = x^2 - 2x$ $(-2 \leq x \leq 4)$

9. $y = -x^2 + 2x$ $(-2 \leq x \leq 4)$

10. $y = x^2 - 6x + 8$ $(0 \leq x \leq 6)$

11. $y = x^2 - 4x + 3$ $(-1 \leq x \leq 5)$

12. $y = x^2 - 2x + 1$ $(-2 \leq x \leq 4)$

13. $y = -x^2 - 2x + 3$ $(-4 \leq x \leq 2)$

14. $y = -x^2 + 4x - 3$ $(-1 \leq x \leq 5)$

In 15–20: **a.** Write the equation of the axis of symmetry of the graph of the function. **b.** Find the coordinates of the vertex. **c.** Draw the graph on graph paper or on a calculator, showing at least three points with integral coefficients on each side of the vertex.

15. $y = x^2 - 6x - 1$

16. $y = x^2 - 2x + 8$

17. $y = x^2 + 8x + 12$

18. $y = x^2 + 4x + 3$

19. $y = x^2 - 3x + 7$

20. $y = x^2 + x + 5$

21. Write an equation for the resulting function if the graph of $y = x^2$ is:

a. reflected in the x-axis and shifted 3 units left.

b. compressed vertically by a factor of $\frac{2}{7}$ and shifted 9 units up.

c. reflected in the x-axis, stretched vertically by a factor of 6, shifted 1 unit down, and shifted 4 units to the right.

In 22–25, each graph is a translation and/or a reflection of the graph of $y = x^2$. For each graph, **a.** determine the vertex and the axis of symmetry, and **b.** write the equation of each graph.

22.

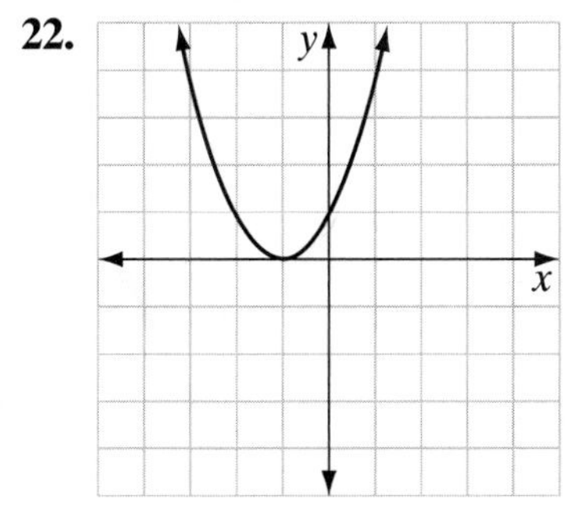

23.

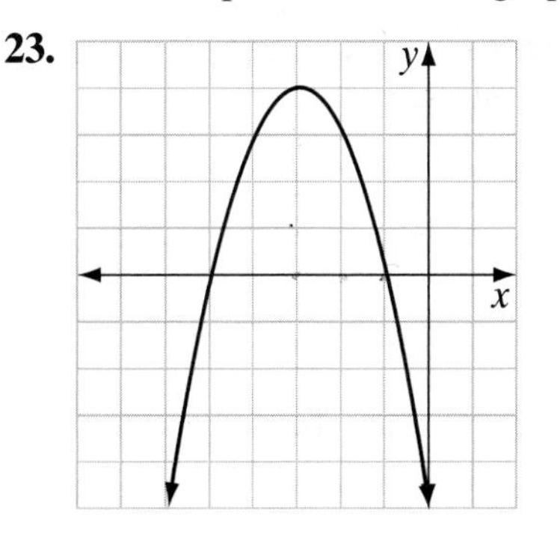

24. 25.

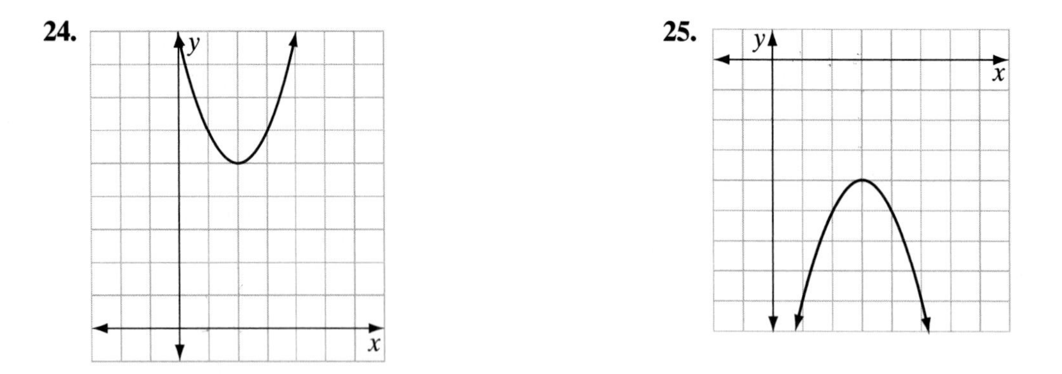

26. Of the graphs below, which is the graph of a quadratic function and the graph of an absolute value function?

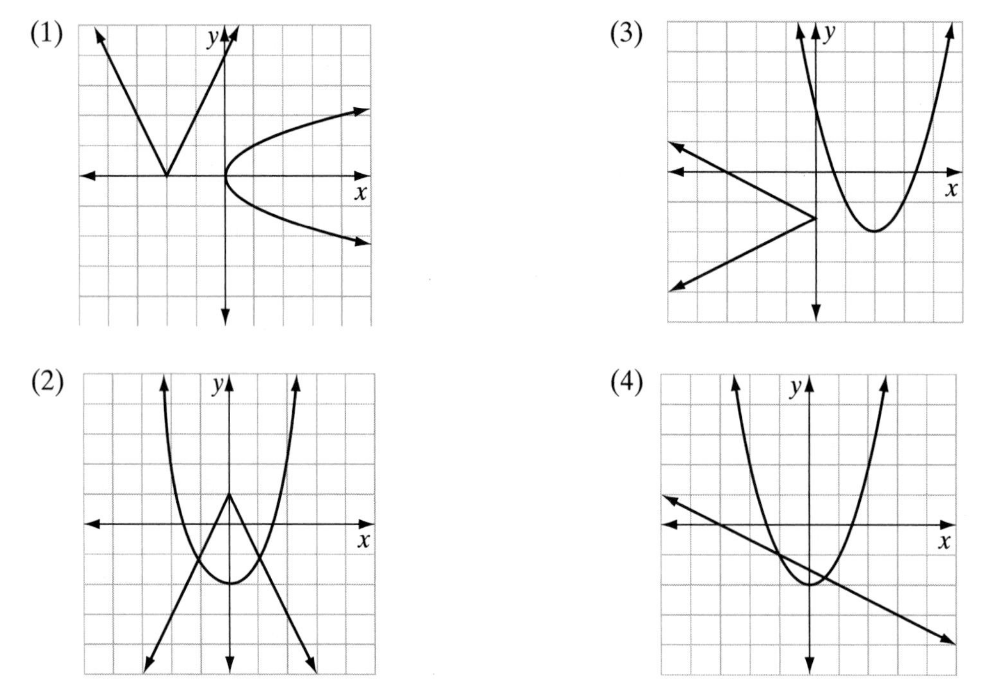

Applying Skills

27. The length of a rectangle is 4 more than its width.

   a. If x represents the width of the rectangle, represent the length of the rectangle in terms of x.

   b. If y represents the area of the rectangle, write an equation for y in terms of x.

   c. Draw the graph of the equation that you wrote in part **b**.

   d. Do all of the points on the graph that you drew represent pairs of values for the width and area of the rectangle? Explain your answer.

28. The height of a triangle is 6 less than twice the length of the base.

a. If x represents the length of the base of the triangle, represent the height in terms of x.

b. If y represents the area of the triangle, write an equation for y in terms of x.

c. Draw the graph of the equation that you wrote in part **b**.

d. Do all of the points on the graph that you drew represent pairs of values for the length of the base and area of the triangle? Explain your answer.

29. The perimeter of a rectangle is 20 centimeters. Let x represent the measure of one side of the rectangle and y represent the area of the rectangle.

a. Use the formula for perimeter to express the measure of a second side of the rectangle.

b. Write an equation for the area of the rectangle in terms of x.

c. Draw the graph of the equation written in **b**.

d. What are the dimensions of the rectangle with maximum area?

e. What is the maximum area of the rectangle?

f. List four other possible dimensions and areas for the rectangle.

30. A batter hit a baseball at a height 3 feet off the ground, with an initial vertical velocity of 64 feet per second. Let x represent the time in seconds, and y represent the height of the baseball. The height of the ball can be determined over a limited period of time by using the equation $y = -16x^2 + 64x + 3$.

a. Make a table using integral values of x from 0 to 4 to find values of y.

b. Graph the equation. Let one horizontal unit $= \frac{1}{4}$ second, and one vertical unit $= 10$ feet. (See suggested coordinate grid below.)

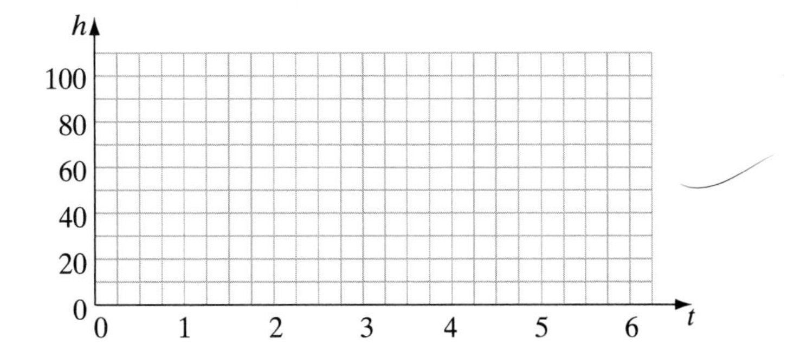

c. If the ball was caught after 4 seconds, what was its height when it was caught?

d. From the table and the graph, determine:

(1) the maximum height reached by the baseball;

(2) the time required for the ball to reach this height.

13-3 FINDING ROOTS FROM A GRAPH

In Section 1, you learned to find the solution of an equation of the form $ax^2 + bx + c = 0$ by factoring. In Section 2, you learned to draw the graph of a function of the form $y = ax^2 + bx + c$. How are these similar expressions related? To answer this question, we will consider three possible cases.

CASE 1 *A quadratic equation can have two distinct real roots.*

The equation $x^2 - 7x + 10 = 0$ has exactly two roots or solutions, 5 and 2, that make the equation true. The function, $y = x^2 - 7x + 10$ has infinitely many pairs of numbers that make the equation true. The graph of this function shown at the right intersects the x-axis in two points, (5, 0) and (2, 0). Since the y-coordinates of these points are 0, the x-coordinates of these points are the roots of the equation $x^2 - 7x + 10 = 0$.

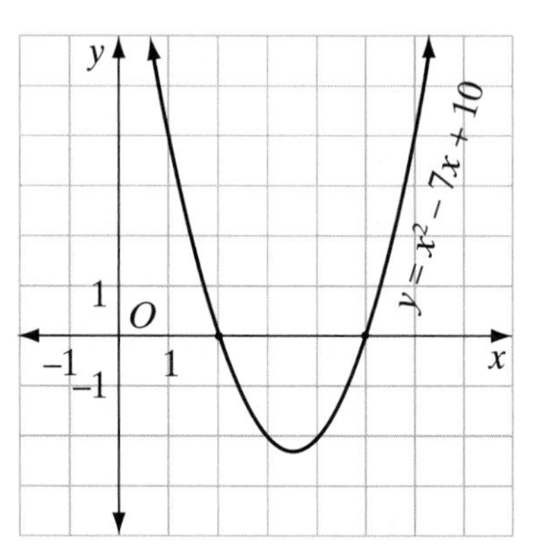

CASE 2 *A quadratic equation can have only one distinct real root.*

The two roots of the equation $x^2 - 6x + 9 = 0$ are equal. There is only one number, 3, that makes the equation true. The function $y = x^2 - 6x + 9$ has infinitely many pairs of numbers that make the equation true. The graph of this function shown at the right intersects the x-axis in only one point, (3, 0). Since the y-coordinate of this point is 0, the x-coordinate of this point is the root of the equation $x^2 - 6x + 9 = 0$.

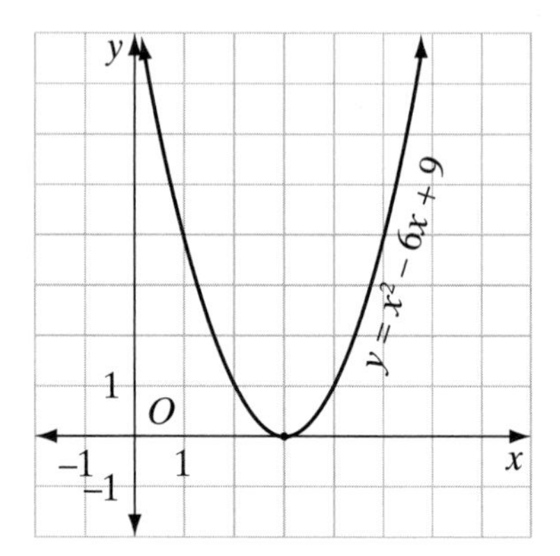

Recall from Section 1 that when a quadratic equation has only one distinct root, the root is said to be a *double root*. In other words, when the root of a quadratic equation is a double root, the graph of the corresponding quadratic function intersects the x-axis exactly once.

CASE 3 *A quadratic equation can have no real roots.*

The equation $x^2 - 2x + 3 = 0$ has no real roots. There is no real number that makes the equation true. The function, $y = x^2 - 2x + 3$ has infinitely many pairs of numbers that make the equation true. The graph of this function shown at the right does not intersect the x-axis. There is no point on the graph whose y-coordinate is 0. Since there is no point whose y-coordinate is 0, there are no real numbers that are roots of the equation $x^2 - 2x + 3 = 0$.

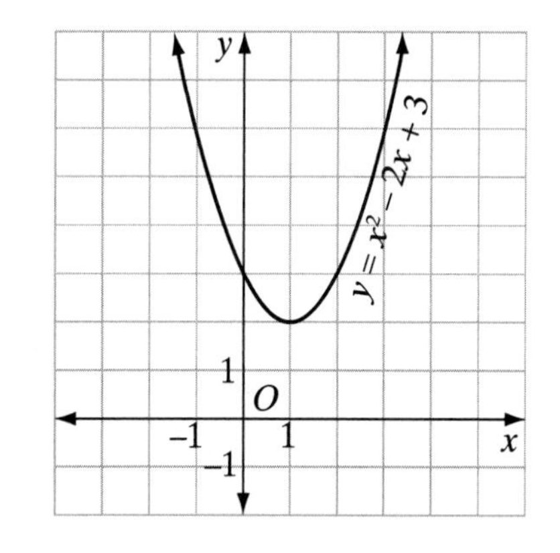

The equation of the x-axis is $y = 0$. The x-coordinates of the points at which the graph of $y = ax^2 + bx + c$ intersects the x-axis are the roots of the equation $ax^2 + bx + c = 0$. The graph of the function $y = ax^2 + bx + c$ can intersect the x-axis in 0, 1, or 2 points, and the equation $ax^2 + bx + c = 0$ can have 0, 1, or 2 real roots.

A real number k is a root of the quadratic equation $ax^2 + bx + c = 0$ if and only if the graph of $y = ax^2 + bx + c$ intersects the x-axis at $(k, 0)$.

EXAMPLE 1

Use the graph of $y = x^2 - 5x + 6$ to find the roots of $x^2 - 5x + 6 = 0$.

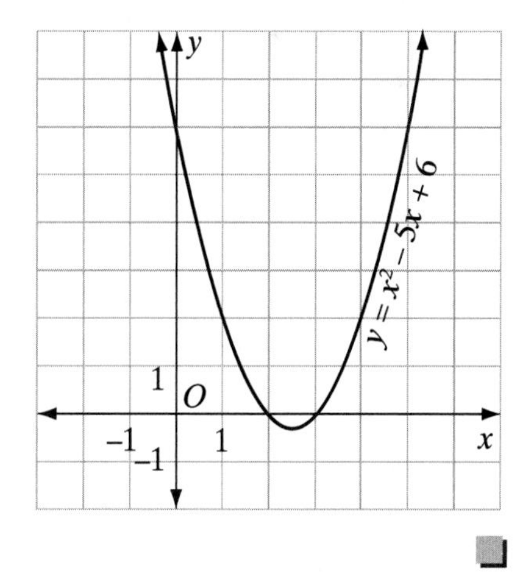

Solution The graph intersects the x-axis at $(2, 0)$ and $(3, 0)$.

The x-coordinates of these points are the roots of $x^2 - 5x + 6 = 0$.

Answer The roots are 2 and 3. The solution set is $\{2, 3\}$.

Note that in Example 1 the quadratic expression $x^2 - 5x + 6$ can be factored into $(x - 2)(x - 3)$, from which we can obtain the solution set $\{2, 3\}$.

EXAMPLE 2

Use the graph of $y = x^2 + 3x - 4$ to find the linear factors of $x^2 + 3x - 4$.

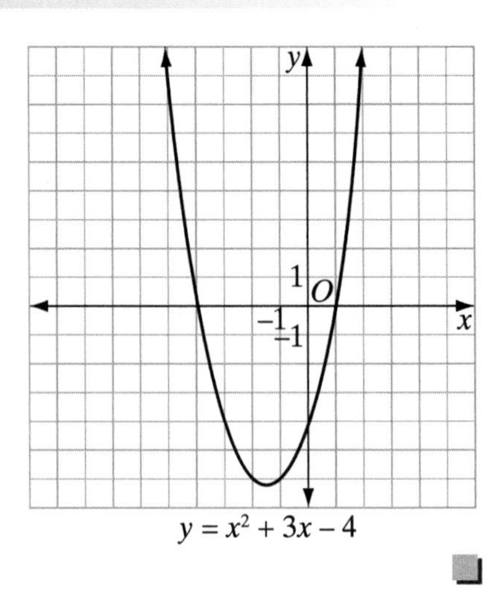

Solution The graph intersects the x-axis at $(-4, 0)$ and $(1, 0)$.

Therefore, the roots are -4 and 1.

If $x = -4$, then $x - (-4) = (x + 4)$ is a factor.

If $x = 1$, then $(x - 1)$ is a factor.

Answer The linear factors of $x^2 + 3x - 4$ are $(x + 4)$ and $(x - 1)$.

EXERCISES

Writing About Mathematics

1. The coordinates of the vertex of a parabola $y = x^2 + 2x + 5$ are $(-1, 4)$. Does the equation $x^2 + 2x + 5 = 0$ have real roots? Explain your answer.

2. The coordinates of the vertex of a parabola $y = -x^2 - 2x + 3$ are $(-1, 4)$. Does the equation $-x^2 - 2x + 3 = 0$ have real roots? Explain your answer.

Developing Skills

In 3–10: **a.** Draw the graph of the parabola. **b.** Using the graph, find the real numbers that are elements of the solution set of the equation. **c.** Using the graph, factor the corresponding quadratic expression if possible.

3. $y = x^2 + 6x + 5; 0 = x^2 + 6x + 5$

4. $y = x^2 + 2x + 1; 0 = x^2 + 2x + 1$

5. $y = x^2 - 2x - 3; 0 = x^2 - 2x - 3$

6. $y = -x^2 + x + 2; 0 = -x^2 + x + 2$

7. $y = x^2 - 2x + 1; 0 = x^2 - 2x + 1$

8. $y = -x^2 + 3x - 2; 0 = -x^2 + 3x - 2$

9. $y = x^2 + 4x + 5; 0 = x^2 + 4x + 5$

10. $y = -x^2 + 5x + 6; 0 = -x^2 + 5x + 6$

11. If the graph of a quadratic function, $f(x)$, crosses the x-axis at $x = 6$ and $x = 8$, what are two factors of $f(x)$?

12. If the factors of a quadratic function, $h(x)$, are $(x + 6)$ and $(x - 3)$, what is the solution set for the equation $h(x) = 0$?

In 13–15, refer to the graph of the parabola shown below.

13. Which of the following is the equation of the parabola?

(1) $y = (x + 2)(x - 3)$

(2) $y = -(x + 2)(x - 3)$

(3) $y = (x - 2)(x + 3)$

(4) $y = -(x - 2)(x + 3)$

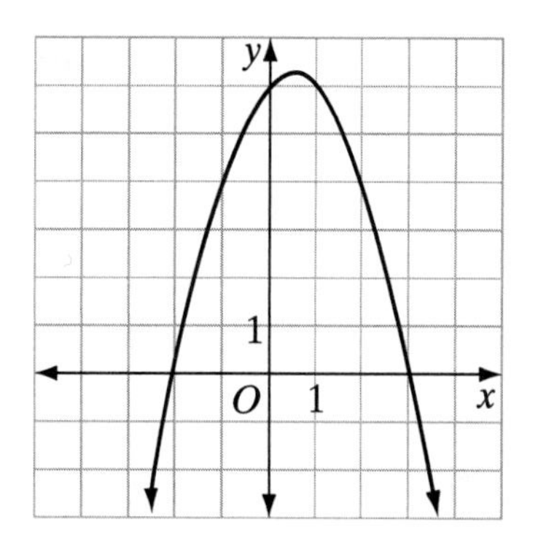

14. Set the equation of the parabola equal to 0. What are the roots of this quadratic equation?

15. If the graph of the parabola is reflected in the x-axis, how many roots will its corresponding quadratic equation have?

13-4 GRAPHIC SOLUTION OF A QUADRATIC-LINEAR SYSTEM

In Chapter 10 you learned how to solve a *system of linear equations* by graphing. For example, the graphic solution of the given system of linear equations is shown below.

$$y = -\tfrac{1}{2}x + 4$$
$$y = 2x - 1$$

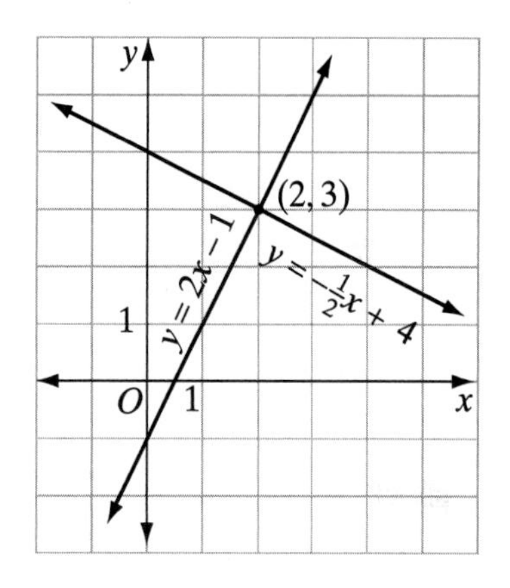

Since the point of intersection, (2, 3), is a solution of both equations, the common solution of the system is $x = 2$ and $y = 3$.

A **quadratic-linear system** consists of a quadratic equation and a linear equation. The solution of a quadratic-linear system is the set of ordered pairs of numbers that make both equations true. As shown below, the line may intersect the curve in two, one, or no points. Thus the solution set may contain two ordered pairs, one ordered pair, or no ordered pairs.

Two points of intersection

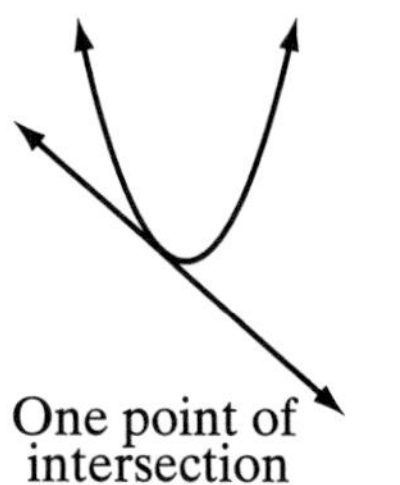
One point of intersection

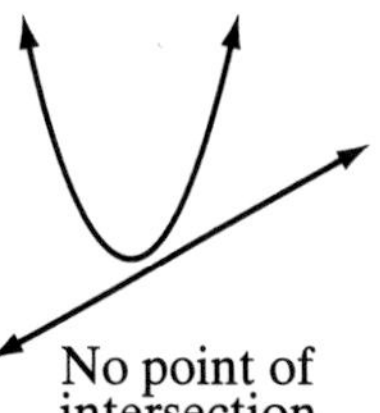
No point of intersection

EXAMPLE I

Solve the quadratic-linear system graphically: $y = x^2 - 6x + 6$
$y = x - 4$

Solution (1) Draw the graph of $y = x^2 - 6x + 6$.

The axis of symmetry of a parabola is $x = \frac{-b}{2a}$. Therefore, the axis of symmetry for the graph of $y = x^2 - 6x + 6$ is $x = \frac{-(-6)}{2(1)}$ or $x = 3$. Make a table of values using integral values of x less than 3 and greater than 3. Plot the points associated with each pair (x, y) and join them with a smooth curve.

x	$x^2 - 6x + 6$	y
0	0 − 0 + 6	6
1	1 − 6 + 6	1
2	4 − 12 + 6	−2
3	9 − 18 + 6	−3
4	16 − 24 + 6	−2
5	25 − 30 + 6	1
6	36 − 36 + 6	6

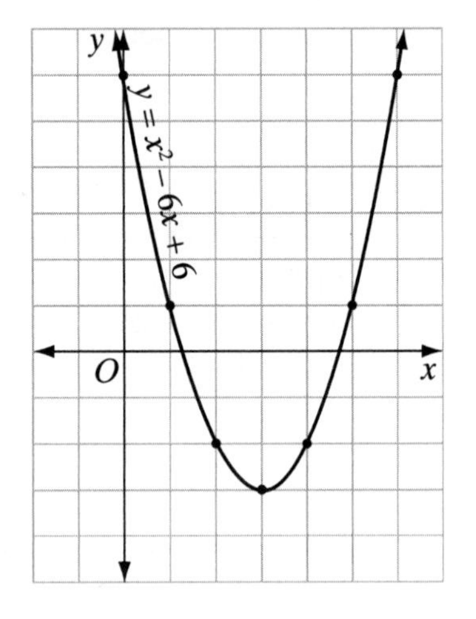

(2) On the same set of axes, draw the graph of $y = x - 4$. Make a table of values and plot the points.

x	x − 4	y
0	0 − 4	−4
2	2 − 4	−2
4	4 − 4	0

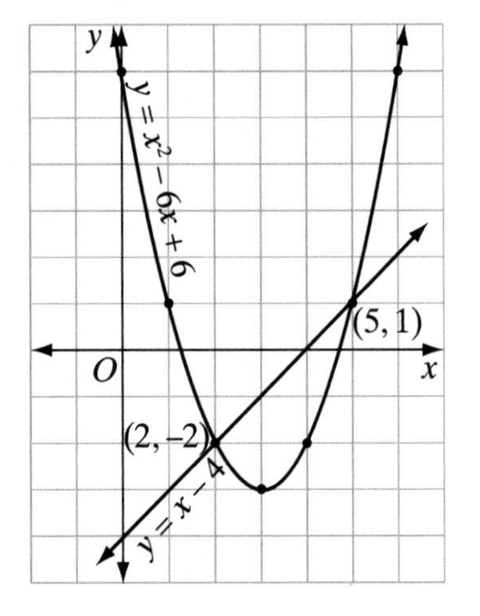

The line could also have been graphed by using the slope $= 1$ or $\frac{1}{1}$, and the y-intercept, -4. Starting at the point $(0, -4)$ move 1 unit up and 1 unit to the right to locate a second point. Then again, move 1 unit up and 1 unit to the right to locate a third point. Draw a line through these points.

(3) Find the coordinates of the points at which the graphs intersect. The graphs intersect at $(2, -2)$ and at $(5, 1)$. Check each solution in each equation. Four checks are required in all. The checks are left for you.

Calculator Solution (1) Enter the equations into the Y= menu.

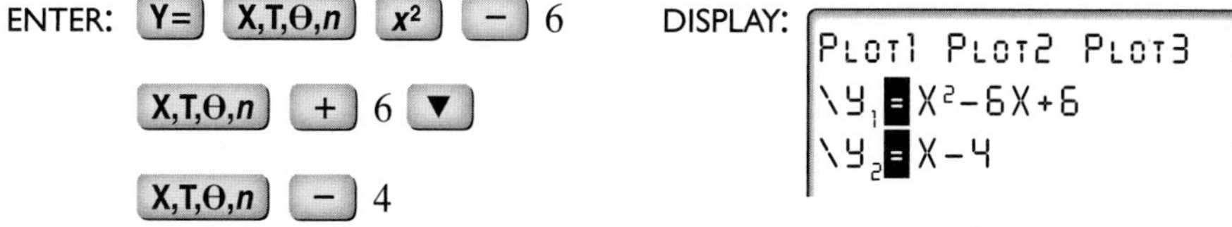

(2) Calculate the first intersection by choosing intersect from the CALC menu. Accept $Y^1 = X^2 - 6X + 6$ as the first curve and $Y^2 = X - 4$ as the second curve. Then press enter when the screen prompts "Guess?".

ENTER: 2nd CALC 5 ENTER ENTER ENTER

DISPLAY:

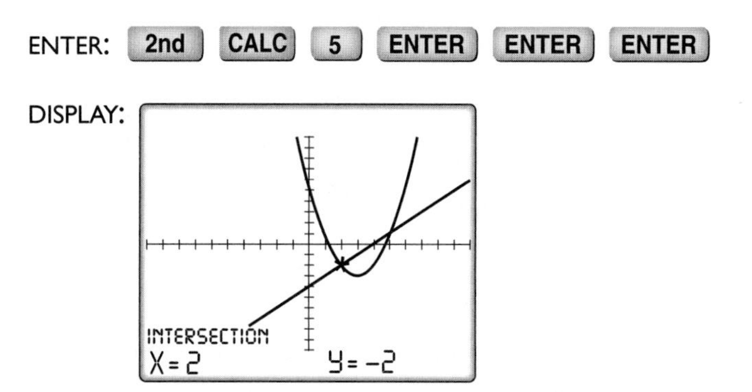

The calculator displays the coordinates of the intersection point at the bottom of the screen.

(3) To calculate the second intersection point, repeat the process from (2), but when the screen prompts "Guess?" move the cursor with the left and right arrow keys to the approximate position of the second intersection point.

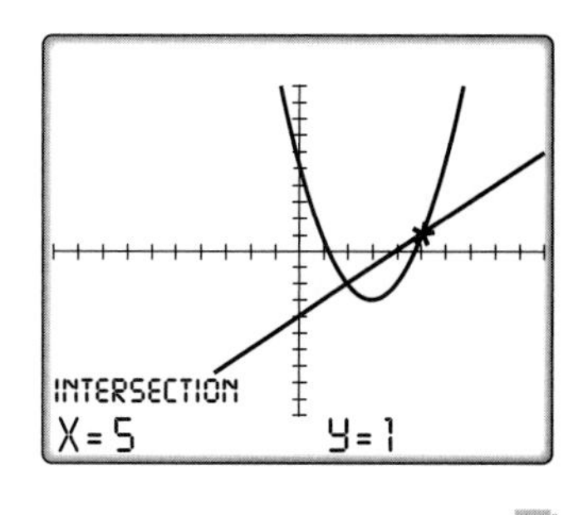

Answer The solution set is $\{(2, -2), (5, 1)\}$.

EXERCISES

Writing About Mathematics

1. What is the solution set of a system of equations when the graphs of the equations do not intersect? Explain your answer.

2. Melody said that the equations $y = x^2$ and $y = -2$ do not have a common solution even before she drew their graphs. Explain how Melody was able to justify her conclusion.

Developing Skills

In 3–6, use the graph on the right to find the common solution of the system.

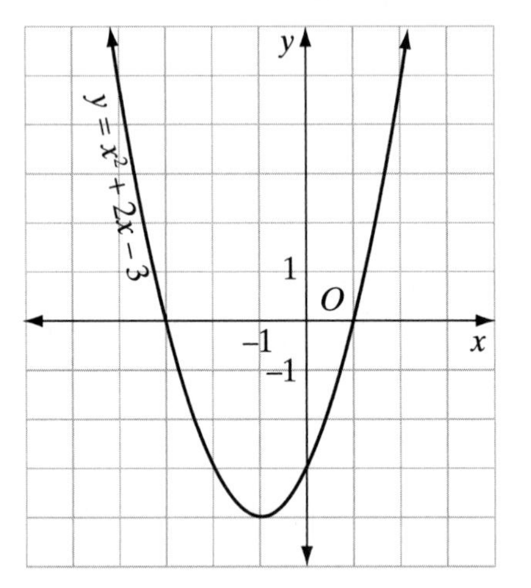

3. $y = x^2 + 2x - 3$
   $y = 0$

4. $y = x^2 + 2x - 3$
   $y = -3$

5. $y = x^2 + 2x - 3$
   $y = -4$

6. $y = x^2 + 2x - 3$
   $y = 5$

7. For what values of c do the equations $y = x^2 + 2x - 3$ and $y = c$ have no points in common?

8. **a.** Draw the graph of $y = x^2 - 4x - 2$, in the interval $-1 \leq x \leq 5$.
   b. On the same set of axes, draw the graph of $y = -x - 2$.
   c. Write the coordinates of the points of intersection of the graphs made in parts **a** and **b**.
   d. Check the common solutions found in part **c** in both equations.

In 9–16, find graphically and check the solution set of each system of equations.

9. $y = x^2$
   $y = x + 2$

10. $y = x^2 - 2x - 4$
    $y = x$

11. $y = x^2 + 2x + 1$
    $y = 2x + 5$

12. $y = 4x - x^2$
    $y = x - 4$

13. $y = x^2 - 8x + 15$
    $x + y = 5$

14. $y = -x^2 + 6x - 5$
    $y = 3$

15. $y = x^2 + 4x + 1$
    $y = 2x + 1$

16. $y = x^2 + x - 4$
    $2x - y = 2$

Applying Skills

17. When a stone is thrown upward from a height of 5 feet with an initial velocity of 48 feet per second, the height of the stone y after x seconds is given by the function $y = -16x^2 + 48x + 5$.

a. Draw a graph of the given function. Let each horizontal unit equal $\frac{1}{2}$ second and each vertical unit equal 5 feet. Plot points for $0, \frac{1}{2}, 1, \frac{3}{2}, 2, \frac{5}{2}$, and 3 seconds.

b. On the same set of axes, draw the graph of $y = 25$.

c. From the graphs drawn in parts **a** and **b**, determine when the stone is at a height of 25 feet.

18. If you drop a baseball on Mars, the gravity accelerates the baseball at 12 feet per second squared. Let's suppose you drop a baseball from a height of 100 feet. A formula for the height, y, of the baseball after x seconds is given by $y = -6x^2 + 100$.

a. On a calculator, graph the given function. Set Xmin to –10, Xmax to 10, Xscl to 1, Ymin to −5, Ymax to 110, and Yscl to 10.

b. Graph the functions $y = 46$ and $y = 0$ as Y_2 and Y_3.

c. Using a calculator method, determine, to the *nearest tenth of a second*, when the baseball has a height of 46 feet and when the baseball hits the ground (reaches a height of 0 feet).

13-5 ALGEBRAIC SOLUTION OF A QUADRATIC-LINEAR SYSTEM

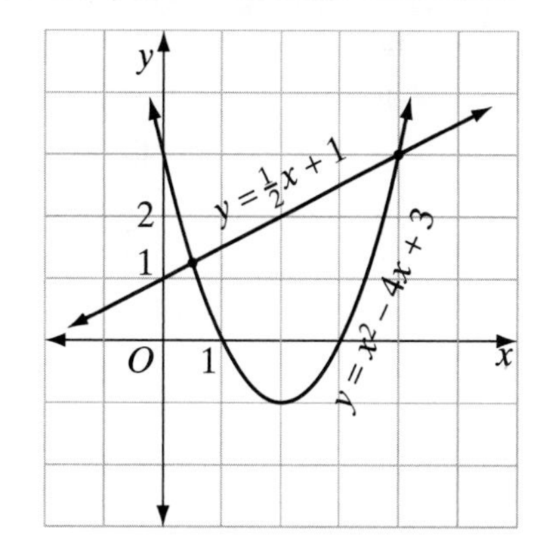

In the last section, we learned to solve a quadratic-linear system by finding the points of intersection of the graphs. The solutions of most of the systems that we solved were integers that were easy to read from the graphs. However, not all solutions are integers. For example, the graphs of $y = x^2 - 4x + 3$ and $y = \frac{1}{2}x + 1$ are shown at the right. They intersect in two points. One of those points, (4, 3), has integral coordinates and can be read easily from the graph. However, the coordinates of the other point are not integers, and we are not able to identify the exact values of x and y from the graph.

In Chapter 10 we learned that a system of linear equations can be solved by an algebraic method of substitution. This method can also be used for a quadratic-linear system. The algebraic solution of the system graphed on the previous page is shown in Example 1.

EXAMPLE 1

Solve algebraically and check: $y = x^2 - 4x + 3$
$y = \frac{1}{2}x + 1$

Solution (1) Since y is expressed in terms of x in the linear equation, substitute the expression $\frac{1}{2}x + 1$ for y in the quadratic equation to form an equation in one variable:

$$y = x^2 - 4x + 3$$
$$\tfrac{1}{2}x + 1 = x^2 - 4x + 3$$

(2) To eliminate fractions as coefficients, multiply both sides of the equation by 2:

$$2\left(\tfrac{1}{2}x + 1\right) = 2(x^2 - 4x + 3)$$
$$x + 2 = 2x^2 - 8x + 6$$

(3) Write the quadratic equation in standard form:

$$0 = 2x^2 - 9x + 4$$

(4) Solve the quadratic equation by factoring:

$$0 = (2x - 1)(x - 4)$$

(5) Set each factor equal to 0 and solve for x.

$2x - 1 = 0$	$x - 4 = 0$
$2x = 1$	$x = 4$
$x = \frac{1}{2}$	

(6) Substitute each value of x in the linear equation to find the corresponding value of y:

$y = \frac{1}{2}x + 1$	$y = \frac{1}{2}x + 1$
$y = \frac{1}{2}\left(\frac{1}{2}\right) + 1$	$y = \frac{1}{2}(4) + 1$
$y = \frac{1}{4} + 1$	$y = 2 + 1$
$y = \frac{5}{4}$	$y = 3$

(7) Write each solution as coordinates: $(x, y) = \left(\frac{1}{2}, \frac{5}{4}\right)$ | $(x, y) = (4, 3)$

(8) Check each ordered pair in each of the given equations:

Check for $x = \frac{1}{2}, y = \frac{5}{4}$

$y = x^2 - 4x + 3$	$y = \frac{1}{2}x + 1$
$\frac{5}{4} \stackrel{?}{=} \left(\frac{1}{2}\right)^2 - 4\left(\frac{1}{2}\right) + 3$	$\frac{5}{4} \stackrel{?}{=} \frac{1}{2}\left(\frac{1}{2}\right) + 1$
$\frac{5}{4} \stackrel{?}{=} \frac{1}{4} - 2 + 3$	$\frac{5}{4} \stackrel{?}{=} \frac{1}{4} + 1$
$\frac{5}{4} = \frac{5}{4}$ ✔	$\frac{5}{4} = \frac{5}{4}$ ✔

Check for $x = 4, y = 3$

$y = x^2 - 4x + 3$	$y = \frac{1}{2}x + 1$
$3 \stackrel{?}{=} (4)^2 - 4(4) + 3$	$3 \stackrel{?}{=} \frac{1}{2}(4) + 1$
$3 \stackrel{?}{=} 16 - 16 + 3$	$3 \stackrel{?}{=} 2 + 1$
$3 = 3$ ✔	$3 = 3$ ✔

Answer $\left\{\left(\frac{1}{2}, \frac{5}{4}\right), (4, 3)\right\}$

EXAMPLE 2

The length of the longer leg of a right triangle is 2 units more than twice the length of the shorter leg. The length of the hypotenuse is 13 units. Find the lengths of the legs of the triangle.

Solution Let a = the length of the longer leg and b = the length of the shorter leg.

(1) Use the Pythagorean Theorem to write an equation:

$$a^2 + b^2 = (13)^2$$
$$a^2 + b^2 = 169$$

(2) Use the information in the first sentence of the problem to write another equation:

$$a = 2b + 2$$

(3) Substitute the expression for a from step 2 in the equation in step 1:

$$a^2 + b^2 = 169$$
$$(2b + 2)^2 + b^2 = 169$$

(4) Square the binomial and write the equation in standard form:

$$4b^2 + 8b + 4 + b^2 = 169$$
$$5b^2 + 8b - 165 = 0$$

(5) Factor the left member of the equation:

$$(5b + 33)(b - 5) = 0$$

(6) Set each factor equal to 0 and solve for b:

$5b + 33 = 0$	$b - 5 = 0$
$5b = -33$	$b = 5$
$b = \frac{-33}{5}$	$b = 5$

(7) For each value of b, find the value of a. Use the linear equation in step 2:

(8) Reject the negative values. Use the pair of positive values to write the answer.

$a = 2b + 2$	$a = 2b + 2$
$a = 2\left(\frac{-33}{5}\right) + 2$	$a = 2(5) + 2$
$a = \frac{-56}{5}$	$a = 12$

Answer The lengths of the legs are 12 units and 5 units.

EXERCISES

Writing About Mathematics

1. Explain why the equations $y = x^2$ and $y = -4$ have no common solution in the set of real numbers.
2. Explain why the equations $x^2 + y^2 = 49$ and $x = 8$ have no common solution in the set of real numbers.

Developing Skills

In 3–20, solve each system of equations algebraically and check all solutions.

3. $y = x^2 - 2x$ $y = x$	**4.** $y = x^2 + 5$ $y = x + 5$	**5.** $y = x^2 - 4x + 3$ $y = x - 1$
6. $y = x^2 + 2x + 1$ $y = x + 3$	**7.** $x^2 + y = 9$ $y = x + 9$	**8.** $x^2 + y = 2$ $y = -x$
9. $x^2 + 2y = 5$ $y = x + 1$	**10.** $y = 2x^2 - 6x + 5$ $y = x + 2$	**11.** $y = 2x^2 + 2x + 3$ $y - x = 3$
12. $y = 3x^2 - 8x + 5$ $x + y = 3$	**13.** $y = x^2 + 3x + 1$ $y = \frac{1}{3}x + 2$	**14.** $y = x^2 - 6x + 8$ $y = -\frac{1}{2}x + 2$
15. $x^2 + y^2 = 25$ $x = 2y - 5$	**16.** $x^2 + y^2 = 100$ $y = x + 2$	**17.** $x^2 + y^2 = 50$ $x = y$
18. $x^2 + y^2 = 40$ $y = 2x + 2$	**19.** $x^2 + y^2 = 20$ $y = x + 2$	**20.** $x^2 + y^2 = 2$ $y = x + 2$

Applying Skills

21. A rectangular tile pattern consists of a large square, two small squares, and a rectangle arranged as shown in the diagram. The height of the small rectangle is 1 unit and the area of the tile is 70 square units.

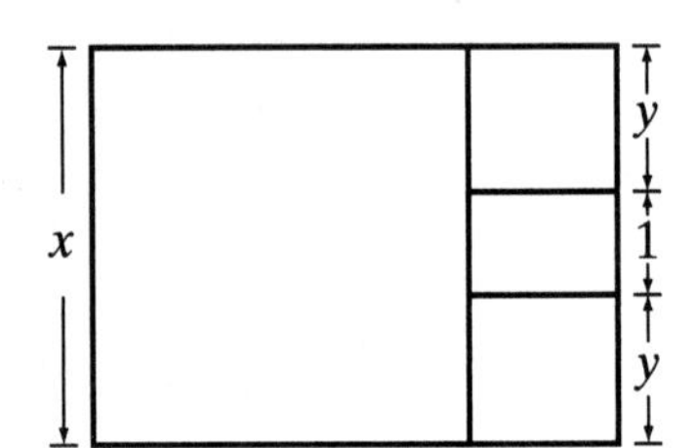

   a. Write a linear equation that expresses the relationship between the length of the sides of the quadrilaterals that make up the tile.

   b. Write a second-degree equation using the sum of the areas of the quadrilaterals that make up the tile.

   c. Solve algebraically the system of equations written in parts **a** and **b**.

   d. What are the dimensions of each quadrilateral in the tile?

22. A doorway to a store is in the shape of an arch whose equation is $h = -\frac{3}{4}x^2 + 6x$, where x represents the horizontal distance, in feet, from the left end of the base of the doorway and h is the height, in feet, of the doorway x feet from the left end of the base.

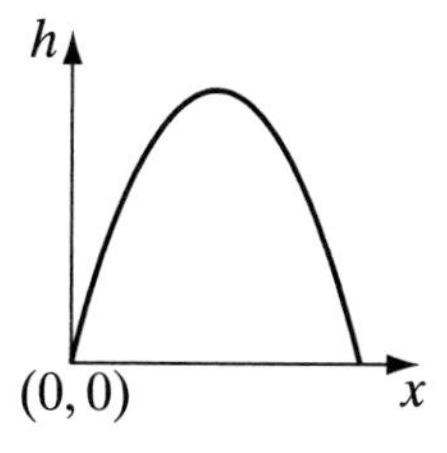

a. How wide is the doorway at its base?

b. What is the maximum height of the doorway?

c. Can a box that is 6 feet wide, 6 feet long, and 5 feet high be moved through the doorway? Explain your answer.

23. The length of the diagonal of a rectangle is $\sqrt{85}$ meters. The length of the rectangle is 1 meter longer than the width. Find the dimensions of the rectangle.

24. The length of one leg of an isosceles triangle is $\sqrt{29}$ feet. The length of the altitude to the base of the triangle is 1 foot more than the length of the base.

a. Let a = the length of the altitude to the base and b = the distance from the vertex of a base angle to the vertex of the right angle that the altitude makes with the base. Use the Pythagorean Theorem to write an equation in terms of a and b.

b. Represent the length of the base in terms of b.

c. Represent the length of the altitude, a, in terms of b.

d. Solve the system of equations from parts **a** and **c**.

e. Find the length of the base and the length of the altitude to the base.

f. Find the perimeter of the triangle.

g. Find the area of the triangle.

CHAPTER SUMMARY

The equation $y = ax^2 + bx + c$, where $a \neq 0$, is a **quadratic function** whose domain is the set of real numbers and whose graph is a **parabola**. The **axis of symmetry of the parabola** is the vertical line $x = \frac{-b}{2a}$. The **vertex** or **turning point** of the parabola is on the axis of symmetry. If $a > 0$, the parabola opens upward and the y-value of the vertex is a **minimum** value for the range of the function. If $a < 0$, the parabola opens downward and the y-value of the vertex is a **maximum** value for the range of the function.

A **quadratic-linear system** consists of two equations one of which is an equation of degree two and the other a linear equation (an equation of degree one). The common solution of the system may be found by graphing the equations on the same set of axes or by using the algebraic method of substitution. A quadratic-linear system of two equations may have two, one, or no common solutions.

The **roots of the equation** $ax^2 + bx + c = 0$ are the x-coordinates of the points at which the function $y = ax^2 + bx + c$ intersects the x-axis. The real number k is a root of $ax^2 + bx + c = 0$ if and only if $(x - k)$ is a factor of $ax^2 + bx + c$.

When the graph of $y = x^2$ is translated by k units in the vertical direction, the equation of the image is $y = x^2 + k$. When the graph of $y = x^2$ is translated k units in the horizontal direction, the equation of the image is $y = (x - k)^2$. When the graph of $y = x^2$ is reflected over the x-axis, the equation of the image is $y = -x^2$. The graph of $y = kx^2$ is the result of stretching the graph of $y = x^2$ in the vertical direction when $k > 1$ or of compressing the graph of $y = x^2$ when $0 < k < 1$.

VOCABULARY

13-1 Standard form • Polynomial equation of degree two • Quadratic equation • Roots of an equation • Double root

13-2 Parabola • Second-degree polynomial function • Quadratic function • Minimum • Turning point • Vertex • Axis of symmetry of a parabola • Maximum

13-4 Quadratic-linear system

REVIEW EXERCISES

1. Explain why $x = y^2$ is not a function when the domain and range are the set of real numbers.

2. Explain why $x = y^2$ is a function when the domain and range are the set of positive real numbers.

In 3–6, the set of ordered pairs of a relation is given. For each relation, **a.** list the elements of the domain, **b.** list the elements of the range, **c.** determine if the relation is a function.

3. $\{(1, 3), (2, 2), (3, 1), (4, 0), (5, -1)\}$

4. $\{(1, 3), (1, 2), (1, 1), (1, 0), (1, -1)\}$

5. $\{(-3, 9), (-2, 4), (-1, 1), (0, 0), (1, 1), (2, 4), (3, 9)\}$

6. $\{(0, 1), (1, 1), (2, 1), (3, 1), (4, 1)\}$

In 7–10, for each of the given functions:

a. Write the equation of the axis of symmetry.

b. Draw the graph.

c. Write the coordinates of the turning point.

d. Does the function have a maximum or a minimum?

e. What is the range of the function?

7. $y = x^2 - 6x + 6$ **8.** $y = x^2 - 4x - 1$

9. $f(x) = -x^2 - 2x + 6$ **10.** $f(x) = -x^2 + 6x - 1$

In 11–16, solve each system of equations graphically and check the solution(s) if they exist.

11. $y = x^2 - 6$
$x + y = 6$

12. $y = -x^2 + 2x + 1$
$y = x - 5$

13. $y = x^2 - x - 3$
$y = x$

14. $y = x^2 - 4x$
$x + y = 4$

15. $y = 5 - x^2$
$y = 4$

16. $y = 2x - x^2$
$x + y = 2$

In 17–22, solve each system of equations algebraically and check the solutions.

17. $x^2 - y = 5$
$y = 3x - 1$

18. $y = x^2 - 4x + 9$
$y - 1 = 2x$

19. $x^2 - 2y = 11$
$y = x - 4$

20. $y = x^2 - 6x + 5$
$y = x - 1$

21. $x^2 + y^2 = 40$
$y = 3x$

22. $x^2 + y^2 = 5$
$y = \frac{1}{2}x$

23. Write an equation for the resulting function if the graph of $y = x^2$ is shifted 3 units up, 2.5 units to the left, and is reflected over the x-axis.

24. The sum of the areas of two squares is 85. The length of a side of the larger square minus the length of a side of the smaller square is 1. Find the length of a side of each square.

25. Two square pieces are cut from a rectangular piece of carpet as shown in the diagram. The area of the original piece is 144 square feet, and the width of the small rectangle that is left is 2 feet. Find the dimensions of the original piece of carpet.

Exploration

Write the square of each integer from 2 to 20. Write the prime factorization of each of these squares. What do you observe about the prime factorization of each of these squares?

Let n be a positive integer and a, b, and c be prime numbers. If the prime factorization of $n = a^3 \times b^2 \times c^4$, is n a perfect square? Is $\sqrt{n}$ rational or irrational? Express $\sqrt{n}$ in terms of a, b, and c.

CUMULATIVE REVIEW — CHAPTERS 1–13

Part I

Answer all questions in this part. Each correct answer will receive 2 credits. No partial credit will be allowed.

1. When $a = -7$, $5 - 2a^2$ equals

(1) 147 (2) −93 (3) 103 (4) −147

2. Which expression is rational?

(1) π (2) $\sqrt{\frac{1}{3}}$ (3) $\sqrt{\frac{2}{8}}$ (4) $\sqrt{0.4}$

3. When factored completely, $3x^2 - 75$ can be expressed as

(1) $(3x + 15)(x - 5)$
(2) $(x + 5)(3x - 15)$
(3) $3x(x + 5)(x - 5)$
(4) $3(x + 5)(x - 5)$

4. When $b^2 + 4b$ is subtracted from $3b^2 - 3b$ the difference is

(1) $3 - 7b$ (2) $-2b^2 + 7b$ (3) $2b^2 - 7b$ (4) $2b^2 + b$

5. The solution set of the equation $0.5x + 4 = 2x - 0.5$ is

(1) {3} (2) {1.4} (3) {30} (4) {14}

6. Which of these represents the quadratic function $y = x^2 + 5$ shifted 2 units down and 4 units to the right?

(1) $y = (x - 4)^2 - 2$
(2) $y = (x - 4)^2 + 3$
(3) $y = (x + 4)^2 + 7$
(4) $y = (x - 2)^2 + 9$

7. The slope of the line whose equation is $x - 2y = 4$ is

(1) −2 (2) 2 (3) $\frac{1}{2}$ (4) $-\frac{1}{2}$

8. The solution set of the equation $x^2 - 7x + 10 = 0$ is

(1) {−2, −5}
(2) {2, 5}
(3) $(x - 2)(x - 5)$
(4) $(x + 2)(x + 5)$

9. Which of these shows the graph of a linear function intersecting the graph of an absolute value function?

(1)

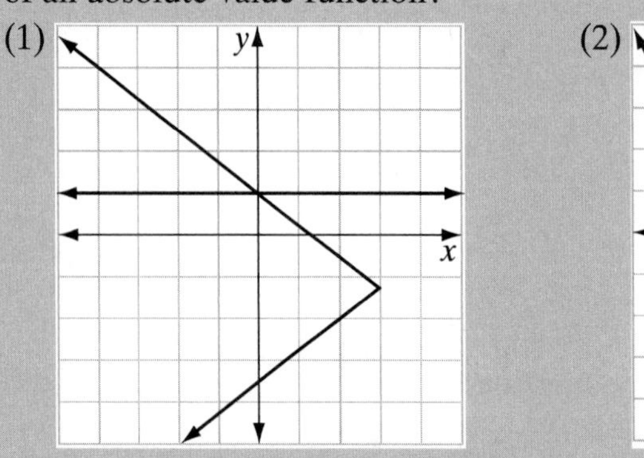

(2)

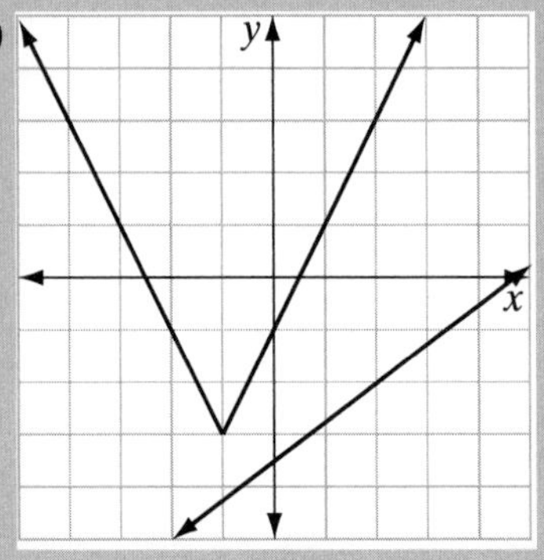

(3) (4)

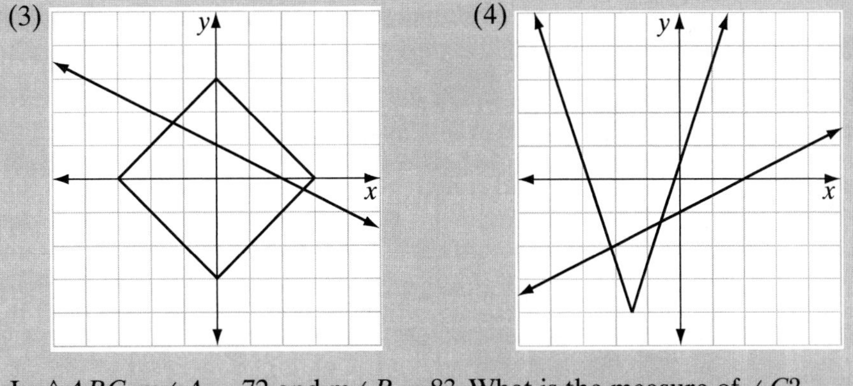

10. In $\triangle ABC$, $m\angle A = 72$ and $m\angle B = 83$. What is the measure of $\angle C$?
(1) 155° (2) 108° (3) 97° (4) 25°

Part II

Answer all questions in this part. Each correct answer will receive 2 credits. Clearly indicate the necessary steps, including appropriate formula substitutions, diagrams, graphs, charts, etc. For all questions in this part, a correct numerical answer with no work shown will receive only 1 credit.

11. In a class of 300 students, 242 take math, 208 take science, and 183 take both math and science. How many students take neither math nor science?

12. Each of the equal sides of an isosceles triangle is 3 centimeters longer than the base. The perimeter of the triangle is 54 centimeters, what is the measure of each side of the triangle?

Part III

Answer all questions in this part. Each correct answer will receive 3 credits. Clearly indicate the necessary steps, including appropriate formula substitutions, diagrams, graphs, charts, etc. For all questions in this part, a correct numerical answer with no work shown will receive only 1 credit.

13. The base of a right circular cylinder has a diameter of 5.00 inches. Sally measured the circumference of the base of the cylinder and recorded it to be 15.5 inches. What is the percent of error in her measurement? Express your answer to the nearest tenth of a percent.

14. Solve the following system of equations and check your solutions.

$$y = -x^2 + 3x + 1$$
$$y = x + 1$$

Part IV

Answer all questions in this part. Each correct answer will receive 4 credits. Clearly indicate the necessary steps, including appropriate formula substitutions, diagrams, graphs, charts, etc. For all questions in this part, a correct numerical answer with no work shown will receive only 1 credit.

15. Find to the nearest degree the measure of the acute angle that the graph of $y = 2x - 4$ makes with the x-axis.

16. Jean Forester has a small business making pies and cakes. Today, she must make at least 4 cakes to fill her orders and at least 3 pies. She has time to make a total of no more than 10 pies and cakes.

a. Let x represent the number of cakes that Jean makes and y represent the number of pies. Write three inequalities that can be used to represent the number of pies and cakes that she can make.

b. In the coordinate plane, graph the inequalities that you wrote and indicate the region that represents their common solution.

c. Write at least three ordered pairs that represent the number of pies and cakes that Jean can make.

CHAPTER

14

ALGEBRAIC FRACTIONS, AND EQUATIONS AND INEQUALITIES INVOLVING FRACTIONS

Although people today are making greater use of decimal fractions as they work with calculators, computers, and the metric system, common fractions still surround us.

We use common fractions in everyday measures: $\frac{1}{4}$-inch nail, $2\frac{1}{2}$-yard gain in football, $\frac{1}{2}$ pint of cream, $1\frac{1}{3}$ cups of flour. We buy $\frac{1}{2}$ dozen eggs, not 0.5 dozen eggs. We describe 15 minutes as $\frac{1}{4}$ hour, not 0.25 hour. Items are sold at a third $\left(\frac{1}{3}\right)$ off, or at a fraction of the original price.

Fractions are also used when sharing. For example, Andrea designed some beautiful Ukrainian eggs this year. She gave one-fifth of the eggs to her grandparents. Then she gave one-fourth of the eggs she had left to her parents. Next, she presented her aunt with one-third of the eggs that remained. Finally, she gave one-half of the eggs she had left to her brother, and she kept six eggs. Can you use some problem-solving skills to discover how many Ukrainian eggs Andrea designed?

In this chapter, you will learn operations with algebraic fractions and methods to solve equations and inequalities that involve fractions.

CHAPTER TABLE OF CONTENTS

14-1 THE MEANING OF AN ALGEBRAIC FRACTION

A *fraction* is a quotient of any number divided by any nonzero number. For example, the arithmetic fraction $\frac{3}{4}$ indicates the quotient of 3 divided by 4.

An **algebraic fraction** is a quotient of two algebraic expressions. An algebraic fraction that is the quotient of two polynomials is called a **fractional expression** or a **rational expression**. Here are some examples of algebraic fractions that are rational expressions:

$$\frac{x}{2} \qquad \frac{2}{x} \qquad \frac{4c}{3d} \qquad \frac{x+5}{x-2} \qquad \frac{x^2+4x+3}{x+1}$$

The fraction $\frac{a}{b}$ means that the number represented by a, the numerator, is to be divided by the number represented by b, the denominator. Since division by 0 is not possible, the value of the denominator, b, cannot be 0. An algebraic fraction is defined or has meaning only for values of the variables for which the denominator is not 0.

EXAMPLE I

Find the value of x for which $\frac{12}{x-9}$ is not defined.

Solution The fraction $\frac{12}{x-9}$ is not defined when the denominator, $x - 9$, is equal to 0.

$$x - 9 = 0$$

$$x = 9 \text{ Answer}$$

EXERCISES

Writing About Mathematics

1. Since any number divided by itself equals 1, the solution set for $\frac{x}{x} = 1$ is the set of all real numbers. Do you agree with this statement? Explain why or why not.

2. Aaron multiplied $\frac{b-1}{1+\frac{1}{b}}$ by $\frac{b}{b}$ (equal to 1) to obtain the fraction $\frac{b^2-b}{b+1}$. Is the fraction $\frac{b-1}{1+\frac{1}{b}}$ equal to the fraction $\frac{b^2-b}{b+1}$ for all values of b? Explain your answer.

Developing Skills

In 3–12, find, in each case, the value of the variable for which the fraction is not defined.

3. $\frac{2}{x}$
4. $\frac{-5}{6x}$
5. $\frac{12}{y^2}$
6. $\frac{1}{x-5}$
7. $\frac{7}{2-x}$
8. $\frac{y+5}{y+2}$
9. $\frac{10}{2x-1}$
10. $\frac{2y+3}{4y+2}$
11. $\frac{1}{x^2-4}$
12. $\frac{3}{x^2-5x-14}$

Applying Skills

In 13–17, represent the answer to each problem as a fraction.

13. What is the cost of one piece of candy if five pieces cost c cents?

14. What is the cost of 1 meter of lumber if p meters cost 980 cents?

15. If a piece of lumber $10x + 20$ centimeters in length is cut into y pieces of equal length, what is the length of each of the pieces?

16. What fractional part of an hour is m minutes?

17. If the perimeter of a square is $3x + 2y$ inches, what is the length of each side of the square?

14-2 REDUCING FRACTIONS TO LOWEST TERMS

A fraction is said to be **reduced to lowest terms** or is a **lowest terms fraction** when its numerator and denominator have no common factor other than 1 or −1.

Each of the fractions $\frac{5}{10}$ and $\frac{a}{2a}$ can be expressed in lowest terms as $\frac{1}{2}$.

The arithmetic fraction $\frac{5}{10}$ is reduced to lowest terms when both its numerator and denominator are divided by 5:

$$\frac{5}{10} = \frac{5 \div 5}{10 \div 5} = \frac{1}{2}$$

The algebraic fraction $\frac{a}{2a}$ is reduced to lowest terms when both its numerator and denominator are divided by a, where $a \neq 0$:

$$\frac{a}{2a} = \frac{a \div a}{2a \div a} = \frac{1}{2}$$

Fractions that are equal in value are called **equivalent fractions**. Thus, $\frac{5}{10}$ and $\frac{1}{2}$ are equivalent fractions, and both are equivalent to $\frac{a}{2a}$, when $a \neq 0$.

The examples shown above illustrate the **division property of a fraction**: if the numerator and the denominator of a fraction are divided by the same nonzero number, the resulting fraction is equal to the original fraction.

In general, for any numbers a, b, and x, where $b \neq 0$ and $x \neq 0$:

$$\frac{ax}{bx} = \frac{ax \div x}{bx \div x} = \frac{a}{b}$$

When a fraction is reduced to lowest terms, we list the values of the variables that must be excluded so that the original fraction is equivalent to the reduced form and also has meaning. For example:

$$\frac{4x}{5x} = \frac{4x \div x}{5x \div x} = \frac{4}{5} \text{ (where } x \neq 0\text{)}$$

$$\frac{cy}{dy} = \frac{cy \div y}{dy \div y} = \frac{c}{d} \text{ (where } y \neq 0, d \neq 0\text{)}$$

When reducing a fraction, the division of the numerator and the denominator by a common factor may be indicated by a **cancellation**.

Here, we use cancellation to divide the numerator and the denominator by 3:

$$\frac{3(x+5)}{18} = \frac{\overset{1}{\cancel{3}}(x+5)}{\underset{6}{\cancel{18}}} = \frac{x+5}{6}$$

Here, we use cancellation to divide the numerator and the denominator by $(a - 3)$:

$$\frac{a^2-9}{3a-9} = \frac{\overset{1}{\cancel{(a-3)}}(a+3)}{3\underset{1}{\cancel{(a-3)}}} = \frac{a+3}{3}$$

(where $a \neq 3$)

By re-examining one of the examples just seen, we can show that the *multiplication property of one* is used whenever a fraction is reduced:

$$\frac{3(x+5)}{18} = \frac{3 \cdot (x+5)}{3 \cdot 6} = \frac{3}{3} \cdot \frac{(x+5)}{6} = 1 \cdot \frac{(x+5)}{6} = \frac{x+5}{6}$$

However, when the multiplication property of one is applied to fractions, it is referred to as the **multiplication property of a fraction.** In general, for any numbers a, b, and x, where $b \neq 0$ and $x \neq 0$:

$$\frac{a}{b} = \frac{a}{b} \cdot \frac{x}{x} = \frac{a}{b} \cdot 1 = \frac{a}{b}$$

Procedure

To reduce a fraction to lowest terms:

METHOD 1

1. Factor completely both the numerator and the denominator.
2. Determine the greatest common factor of the numerator and the denominator.
3. Express the given fraction as the product of two fractions, one of which has as its numerator and its denominator the greatest common factor determined in step 2.
4. Use the multiplication property of a fraction.

METHOD 2

1. Factor both the numerator and the denominator.
2. Divide both the numerator and the denominator by their greatest common factor.

EXAMPLE 1

Reduce $\frac{15x^2}{35x^4}$ to lowest terms.

Solution **METHOD 1**

$$\frac{15x^2}{35x^4} = \frac{3}{7x^2} \cdot \frac{5x^2}{5x^2} = \frac{3}{7x^2} \cdot 1 = \frac{3}{7x^2}$$

METHOD 2

$$\frac{15x^2}{35x^4} = \frac{3 \cdot 5x^2}{7x^2 \cdot 5x^2} = \frac{3 \cdot \overset{1}{\cancel{5x^2}}}{7x^2 \cdot \underset{1}{\cancel{5x^2}}} = \frac{3}{7x^2}$$

Answer $\frac{3}{7x^2}$ $(x \neq 0)$ ■

EXAMPLE 2

Express $\frac{2x^2 - 6x}{10x}$ as a lowest terms fraction.

Solution **METHOD 1**

$$\frac{2x^2 - 6x}{10x} = \frac{2x(x-3)}{2x \cdot 5} = \frac{2x}{2x} \cdot \frac{(x-3)}{5} = 1 \cdot \frac{(x-3)}{5} = \frac{x-3}{5}$$

METHOD 2

$$\frac{2x^2 - 6x}{10x} = \frac{2x(x-3)}{10x} = \frac{\overset{1}{\cancel{2x}}(x-3)}{\underset{5}{\cancel{10x}}} = \frac{x-3}{5}$$

Answer $\frac{x-3}{5}$ $(x \neq 0)$ ■

EXAMPLE 3

Reduce each fraction to lowest terms.

a. $\frac{x^2 - 16}{x^2 - 5x + 4}$ **b.** $\frac{2 - x}{4x - 8}$

Solution **a.** Use Method 1:

$$\frac{x^2 - 16}{x^2 - 5x + 4} = \frac{(x+4)(x-4)}{(x-1)(x-4)} = \frac{x+4}{x-1} \cdot \frac{x-4}{x-4} = \frac{x+4}{x-1} \cdot 1 = \frac{x+4}{x-1}$$

b. Use Method 2:

$$\frac{2-x}{4x-8} = \frac{-1(x-2)}{4(x-2)} = \frac{-1\overset{1}{\cancel{(x-2)}}}{4\underset{1}{\cancel{(x-2)}}} = -\frac{1}{4}$$

Answers **a.** $\frac{x+4}{x-1}$ $(x \neq 1, x \neq 4)$ **b.** $-\frac{1}{4}$ $(x \neq 2)$ ■

EXERCISES

Writing About Mathematics

1. Kevin used cancellation to reduce $\frac{a+4}{a+8}$ to lowest terms as shown below. What is the error in Kevin's work?

$$\frac{a+4}{a+8} = \frac{\overset{1}{\cancel{a}} + \overset{1}{\cancel{4}}}{\underset{1}{\cancel{a}} + \underset{2}{\cancel{8}}} = \frac{1+1}{1+2} = \frac{2}{3}$$

2. Kevin let $a = 4$ to prove that when reduced to lowest terms, $\frac{a+4}{a+8} = \frac{2}{3}$. Explain to Kevin why his reasoning is incorrect.

Developing Skills

In 3–54, reduce each fraction to lowest terms. In each case, list the values of the variables for which the fractions are not defined.

3. $\frac{4x}{12x}$ **4.** $\frac{27y^2}{36y}$ **5.** $\frac{24c}{36d}$ **6.** $\frac{9r}{10r}$

7. $\frac{ab}{cb}$ **8.** $\frac{3ay^2}{6by^2}$ **9.** $\frac{5xy}{9xy}$ **10.** $\frac{2abc}{4abc}$

11. $\frac{15x^2}{5x}$ **12.** $\frac{5x^2}{25x^4}$ **13.** $\frac{27a}{36a^2}$ **14.** $\frac{8xy^2}{24x^2y}$

15. $\frac{+12a^2b}{-8ac}$ **16.** $\frac{-20x^2y^2}{-90xy^2}$ **17.** $\frac{-32a^3b^3}{+48a^3b^3}$ **18.** $\frac{5xy}{45x^2y^2}$

19. $\frac{3x+6}{4}$ **20.** $\frac{8y-12}{6}$ **21.** $\frac{5x-35}{5x}$ **22.** $\frac{8m^2+40m}{8m}$

23. $\frac{2ax+2bx}{6x^2}$ **24.** $\frac{5a^2-10a}{5a^2}$ **25.** $\frac{12ab-3b^2}{3ab}$ **26.** $\frac{6x^2y+9xy^2}{12xy}$

27. $\frac{18b^2+30b}{9b^3}$ **28.** $\frac{4x}{4x+8}$ **29.** $\frac{7d}{7d+14}$ **30.** $\frac{5y}{5y+5x}$

31. $\frac{2a^2}{6a^2-2ab}$ **32.** $\frac{14}{7r-21s}$ **33.** $\frac{12a+12b}{3a+3b}$ **34.** $\frac{x^2-9}{3x+9}$

35. $\frac{x^2-1}{5x-5}$ **36.** $\frac{1-x}{x-1}$ **37.** $\frac{3-b}{b^2-9}$ **38.** $\frac{2s-2r}{s^2-r^2}$

39. $\frac{16-a^2}{2a-8}$ **40.** $\frac{x^2-y^2}{3y-3x}$ **41.** $\frac{2b(3-b)}{b^2-9}$ **42.** $\frac{r^2-r-6}{3r-9}$

43. $\frac{x^2+7x+12}{x^2-16}$ **44.** $\frac{x^2+x-2}{x^2+4x+4}$ **45.** $\frac{3y-3}{y^2-2y+1}$ **46.** $\frac{x^2-3x}{x^2-4x+3}$

47. $\frac{x^2-25}{x^2-2x-15}$ **48.** $\frac{a^2-a-6}{a^2-9}$ **49.** $\frac{a^2-6a}{a^2-7a+6}$ **50.** $\frac{2x^2-50}{x^2+8x+15}$

51. $\frac{r^2-4r-5}{r^2-2r-15}$ **52.** $\frac{48+8x-x^2}{x^2+x-12}$ **53.** $\frac{2x^2-7x+3}{(x-3)^2}$ **54.** $\frac{x^2-7xy+12y^2}{x^2+xy-20y^2}$

55. a. Use substitution to find the *numerical* value of $\frac{x^2 - 5x}{x - 5}$, then reduce each *numerical fraction* to lowest terms when:

(1) $x = 7$ (2) $x = 10$ (3) $x = 20$
(4) $x = 2$ (5) $x = -4$ (6) $x = -10$

b. What pattern, if any, do you observe for the answers to part **a**?

c. Can substitution be used to evaluate $\frac{x^2 - 5x}{x - 5}$ when $x = 5$? Explain your answer.

d. Reduce the algebraic fraction $\frac{x^2 - 5x}{x - 5}$ to lowest terms.

e. Using the answer to part **d**, find the value of $\frac{x^2 - 5x}{x - 5}$, reduced to lowest terms, when $x = 38{,}756$.

f. If the fraction $\frac{x^2 - 5x}{x - 5}$ is multiplied by $\frac{x}{x}$ to become $\frac{x(x^2 - 5x)}{x(x - 5)}$, will it be equivalent to $\frac{x^2 - 5x}{x - 5}$? Explain your answer.

14-3 MULTIPLYING FRACTIONS

The product of two fractions is a fraction with the following properties:

1. The numerator is the product of the numerators of the given fractions.
2. The denominator is the product of the denominators of the given fractions.

In general, for any numbers a, b, x, and y, when $b \neq 0$ and $y \neq 0$:

$$\frac{a}{b} \cdot \frac{x}{y} = \frac{ax}{by}$$

We can find the product of $\frac{7}{27}$ and $\frac{9}{4}$ in lowest terms by using either of two methods.

METHOD 1. $\frac{7}{27} \times \frac{9}{4} = \frac{7 \times 9}{27 \times 4} = \frac{63}{108} = \frac{7 \times 9}{12 \times 9} = \frac{7}{12} \times \frac{9}{9} = \frac{7}{12} \times 1 = \frac{7}{12}$

METHOD 2. $\frac{7}{27} \times \frac{9}{4} = \frac{7}{\underset{3}{\cancel{27}}} \times \frac{\overset{1}{\cancel{9}}}{4} = \frac{7}{12}$

Notice that Method 2 requires less computation than Method 1 since the reduced form of the product was obtained by dividing the numerator and the denominator by a common factor *before* the product was found. This method may be called the **cancellation method**.

The properties that apply to the multiplication of arithmetic fractions also apply to the multiplication of algebraic fractions.

To multiply $\frac{5x^2}{7y}$ by $\frac{14y^2}{15x^3}$ and express the product in lowest terms, we may use either of the two methods. In this example, $x \neq 0$ and $y \neq 0$.

METHOD 1. $\frac{5x^2}{7y} \cdot \frac{14y^2}{15x^3} = \frac{5x^2 \cdot 14y^2}{7y \cdot 15x^3} = \frac{70x^2y^2}{105x^3y} = \frac{2y}{3x} \cdot \frac{35x^2y}{35x^2y} = \frac{2y}{3x} \cdot 1 = \frac{2y}{3x}$

METHOD 2. $\frac{5x^2}{7x} \cdot \frac{14y^2}{15x^3} = \frac{\overset{1}{\cancel{5x^2}}}{\underset{1}{\cancel{7y}}} \cdot \frac{\overset{2y}{\cancel{14y^2}}}{\underset{3x}{\cancel{15x^3}}} = \frac{2y}{3x}$ (the cancellation method)

While Method 1 is longer, it has the advantage of displaying each step as a property of fractions. This can be helpful for checking work.

Procedure

To multiply fractions:

METHOD 1

1. Multiply the numerators of the given fractions.
2. Multiply the denominators of the given fractions.
3. Reduce the resulting fraction, if possible, to lowest terms.

METHOD 2

1. Factor any polynomial that is not a monomial.
2. Use cancellation to divide a numerator and a denominator by each common factor.
3. Multiply the resulting numerators and the resulting denominators to write the product in lowest terms.

EXAMPLE 1

Multiply and express the product in reduced form: $\frac{5a^3}{9bx} \cdot \frac{6bx}{a^2}$

Solution **METHOD 1**

(1) Multiply the numerators and denominators of the given fractions:

$$\frac{5a^3}{9bx} \cdot \frac{6bx}{a^2} = \frac{5a^3 \cdot 6bx}{9bx \cdot a^2} = \frac{30a^3bx}{9a^2bx}$$

(2) Reduce the resulting fraction to lowest terms:

$$= \frac{10a}{3} \cdot \frac{3a^2bx}{3a^2bx} = \frac{10a}{3} \cdot 1 = \frac{10a}{3}$$

METHOD 2

(1) Divide the numerators and denominators by the common factors $3bx$ and a^2:

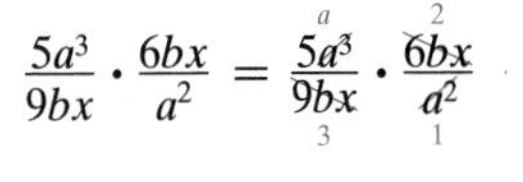

$$\frac{5a^3}{9bx} \cdot \frac{6bx}{a^2} = \frac{\overset{a}{5a^3}}{\underset{3}{9bx}} \cdot \frac{\overset{2}{6bx}}{\underset{1}{a^2}}$$

(2) Multiply the resulting numerators and the resulting denominators:

$$= \frac{10a}{3}$$

Answer $\frac{10a}{3}$ $(a \neq 0, b \neq 0, x \neq 0)$

EXAMPLE 2

Multiply and express the product in simplest form: $12a \cdot \frac{3}{8a}$

Solution Think of $12a$ as $\frac{12a}{1}$.

$$12a \cdot \frac{3}{8a} = \frac{12a}{1} \cdot \frac{3}{8a} = \frac{36a}{8a}$$
$$= \frac{4a}{4a} \cdot \frac{9}{2} = 1 \cdot \frac{9}{2} = \frac{9}{2}$$

Answer $\frac{9}{2}$ $(a \neq 0)$

EXAMPLE 3

Multiply and simplify the product: $\frac{x^2 - 5x + 6}{3x} \cdot \frac{2}{4x - 12}$

Solution $\frac{x^2 - 5x + 6}{3x} \cdot \frac{2}{4x - 12} = \frac{\overset{1}{(x-3)}(x-2)}{3x} \cdot \frac{\overset{1}{2}}{\underset{2}{4}\underset{1}{(x-3)}} = \frac{x-2}{6x}$

Answer $\frac{x-2}{6x}$ $(x \neq 0, 3)$

EXERCISES

Writing About Mathematics

1. When reduced to lowest terms, a fraction whose numerator is $x^2 - 3x + 2$ equals -1. What is the denominator of the fraction? Explain your answer.

2. Does $\frac{x^2}{xz + z^2} \cdot \frac{x^2 - z^2}{x^2 - xz} = \frac{x}{z}$ for all values of x and z? Explain your answer.

Developing Skills

In 3–41, find each product in lowest terms. In each case, list any values of the variable for which the fractions are not defined.

3. $\frac{8}{12} \cdot \frac{30a}{36}$
4. $36 \cdot \frac{5y}{9y}$
5. $\frac{1}{2} \cdot 20x$
6. $\frac{5}{d} \cdot d^2$
7. $\frac{x^2}{36} \cdot 20$
8. $mn \cdot \frac{8}{m^2n^2}$
9. $\frac{24x}{35y} \cdot \frac{14y}{8x}$
10. $\frac{12x}{5y} \cdot \frac{15y^2}{36x^2}$
11. $\frac{m^2}{8} \cdot \frac{32}{3m}$
12. $\frac{6r^2}{5s^2} \cdot \frac{10rs}{6r^3}$
13. $\frac{30m^2}{18n} \cdot \frac{6n}{5m}$
14. $\frac{24a^3b^2}{7c^3} \cdot \frac{21c^2}{12ab}$
15. $\frac{7}{8} \cdot \frac{2x+4}{21}$
16. $\frac{3a+9}{15a} \cdot \frac{a^3}{18}$
17. $\frac{5x-5y}{x^2y} \cdot \frac{xy^2}{25}$
18. $\frac{12a-4}{b} \cdot \frac{b^3}{12}$
19. $\frac{ab-a}{b^2} \cdot \frac{b^3-b^2}{a}$
20. $\frac{x^2-1}{x^2} \cdot \frac{3x^2-3x}{15}$
21. $\frac{2r}{r-1} \cdot \frac{r-1}{10}$
22. $\frac{7s}{s+2} \cdot \frac{2s+4}{21}$
23. $\frac{8x}{2x+6} \cdot \frac{x+3}{x^2}$
24. $\frac{1}{x^2-1} \cdot \frac{2x+2}{6}$
25. $\frac{a^2-9}{3} \cdot \frac{12}{2a-6}$
26. $\frac{x^2-x-2}{3} \cdot \frac{21}{x^2-4}$
27. $\frac{a(a-b)^2}{4b} \cdot \frac{4b}{a(a^2-b^2)}$
28. $\frac{(a-2)^2}{4b} \cdot \frac{16b^3}{4-a^2}$
29. $\frac{a^2-7a-8}{2a+2} \cdot \frac{5}{a-8}$
30. $\frac{x^2+6x+5}{9y^2} \cdot \frac{3y}{x+1}$
31. $\frac{y^2-2y-3}{2c^3} \cdot \frac{4c^2}{2y+2}$
32. $\frac{4a-6}{4a+8} \cdot \frac{6a+12}{5a-15}$
33. $\frac{x^2-25}{4x^2-9} \cdot \frac{2x+3}{x-5}$
34. $\frac{4x+8}{6x+18} \cdot \frac{5x+15}{x^2-4}$
35. $\frac{y^2-81}{(y+9)^2} \cdot \frac{10y+90}{5y-45}$
36. $\frac{8x}{2x^2-8} \cdot \frac{8x+16}{32x^2}$
37. $\frac{2-x}{2x} \cdot \frac{3x}{3x-6}$
38. $\frac{x^2-3x+2}{2x^2-2} \cdot \frac{2x}{x-2}$
39. $\frac{b^2+81}{b^2-81} \cdot \frac{81-b^2}{81+b^2}$
40. $\frac{d^2-25}{4-d^2} \cdot \frac{5d^2-20}{d+5}$
41. $\frac{a^2+12a+36}{a^2-36} \cdot \frac{36-a^2}{36+a^2}$

42. What is the value of $\frac{x^2-4}{6x+12} \cdot \frac{4x-12}{x^2-5x+6}$ when $x = 65{,}908$?

14-4 DIVIDING FRACTIONS

We know that the operation of division may be defined in terms of the multiplicative inverse, the reciprocal. A quotient can be expressed as the dividend times the reciprocal of the divisor. Thus:

$$8 \div 5 = \frac{8}{1} \times \frac{1}{5} = \frac{8 \times 1}{1 \times 5} = \frac{8}{5} \qquad \text{and} \qquad \frac{8}{7} \div \frac{5}{3} = \frac{8}{7} \times \frac{3}{5} = \frac{8 \times 3}{7 \times 5} = \frac{24}{35}$$

We use the same rule to divide algebraic fractions. In general, for any numbers a, b, c, and d, when $b \neq 0$, $c \neq 0$, and $d \neq 0$:

$$\frac{a}{b} \div \frac{c}{d} = \frac{a}{b} \cdot \frac{d}{c} = \frac{ad}{bc}$$

Procedure

To divide by a fraction, multiply the dividend by the reciprocal of the divisor.

EXAMPLE 1

Divide: $\frac{16c^3}{21d^2} \div \frac{24c^4}{14d^3}$

Solution *How to Proceed*

(1) Multiply the dividend by the reciprocal of the divisor:

$$\frac{16c^3}{21d^2} \div \frac{24c^4}{14d^3} = \frac{16c^3}{21d^2} \cdot \frac{14d^3}{24c^4}$$

(2) Divide the numerators and denominators by the common factors:

$$= \frac{\overset{2}{\cancel{16c^3}}}{\underset{3}{\cancel{21d^2}}} \cdot \frac{\overset{2d}{\cancel{14d^3}}}{\underset{3c}{\cancel{24c^4}}}$$

(3) Multiply the resulting numerators and the resulting denominators:

$$= \frac{4d}{9c}$$

Answer $\frac{4d}{9c}$ $(c \neq 0, d \neq 0)$

EXAMPLE 2

Divide: $\frac{8x + 24}{x^2 - 25} \div \frac{4x}{x^2 + 8x + 15}$

Solution *How to Proceed*

(1) Multiply the dividend by the reciprocal of the divisor:

$$\frac{8x + 24}{x^2 - 25} \div \frac{4x}{x^2 + 8x + 15} = \frac{8x + 24}{x^2 - 25} \cdot \frac{x^2 + 8x + 15}{4x}$$

(2) Factor the numerators and denominators, and divide by the common factors:

$$= \frac{\overset{2}{\cancel{8}}(x + 3)}{\underset{1}{\cancel{(x + 5)}}(x - 5)} \cdot \frac{\overset{1}{\cancel{(x + 5)}}(x + 3)}{\underset{1}{\cancel{4}}x}$$

(3) Multiply the resulting numerators and the resulting denominators:

$$= \frac{2(x + 3)^2}{x(x - 5)}$$

Answer $\frac{2(x + 3)^2}{x(x - 5)}$ $(x \neq 0, 5, -5, -3)$

Note: If $x = 5, -5$, or -3, the dividend and the divisor will not be defined. If $x = 0$, the reciprocal of the divisor will not be defined.

EXERCISES

Writing About Mathematics

1. Explain why the quotient $\frac{2}{x-2} \div \frac{x-3}{5}$ is undefined for $x = 2$ and for $x = 3$.

2. To find the quotient $\frac{3}{2(x-4)} \div \frac{x-4}{5}$, Ruth canceled $(x - 4)$ in the numerator and denominator and wrote $\frac{3}{2} \div \frac{1}{5} = \frac{3}{2} \cdot \frac{5}{1} = \frac{15}{2}$. Is Ruth's answer correct? Explain why or why not.

Developing Skills

In 3–27, find each quotient in lowest terms. In each case, list any values of the variables for which the quotient is not defined.

3. $\frac{7a}{10} \div \frac{21}{5}$

4. $\frac{12}{35} \div \frac{4b}{7}$

5. $8 \div \frac{x}{2y}$

6. $\frac{x}{9} \div \frac{x}{3}$

7. $\frac{3x}{5y} \div \frac{21x}{2y}$

8. $\frac{7ab^2}{10cd} \div \frac{14b^3}{5c^2d^2}$

9. $\frac{xy^2}{x^2y} \div \frac{x}{y^3}$

10. $\frac{6a^2b^2}{8c} \div 3ab$

11. $\frac{4x+4}{9} \div \frac{3}{8x}$

12. $\frac{3y^2+9y}{18} \div \frac{5y^2}{27}$

13. $\frac{a^3-a}{b} \div \frac{a^3}{4b^3}$

14. $\frac{x^2-1}{5} \div \frac{x-1}{10}$

15. $\frac{x^2-5x+4}{2x} \div \frac{2x-2}{8x^2}$

16. $\frac{4a^2-9}{10} \div \frac{10a+15}{25}$

17. $\frac{b^2-b-6}{2b} \div \frac{b^2-4}{b^2}$

18. $\frac{a^2-ab}{4a} \div (a^2 - b^2)$

19. $\frac{12y-6}{8} \div (2y^2 - 3y + 1)$

20. $\frac{(x-2)^2}{4x^2-16} \div \frac{21x}{3x+6}$

21. $\frac{x^2-2xy-8y^2}{x^2-16y^2} \div \frac{5x+10y}{3x+12y}$

22. $\frac{x^2-4x+4}{3x-6} \div (2 - x)$

23. $(9 - y^2) \div \frac{y^2+8y+15}{2y+10}$

24. $\frac{x-1}{x+1} \cdot \frac{2x+2}{x+2} \div \frac{4x-4}{x+2}$

25. $\frac{x+y}{x^2+y^2} \cdot \frac{x}{x-y} \div \frac{(x+y)^2}{x^4-y^4}$

26. $\frac{2a+6}{a^2-9} \div \frac{3+a}{3-a} \cdot \frac{a+3}{4}$

27. $\frac{(a+b)^2}{a^2-b^2} \div \frac{a+b}{b^2-a^2} \cdot \frac{a-b}{(a-b)^2}$

28. For what value(s) of a is $\frac{a^2-2a+1}{a^2} \div \frac{a^2-1}{a}$ undefined?

29. Find the value of $\frac{y^2-6y+9}{y^2-9} \div \frac{10y-30}{y^2+3y}$ when $y = 70$.

30. If $x \div y = a$ and $y \div z = \frac{1}{a}$, what is the value of $x \div z$?

14-5 ADDING OR SUBTRACTING ALGEBRAIC FRACTIONS

We know that the sum (or difference) of two arithmetic fractions that have the same denominator is another fraction whose numerator is the sum (or difference) of the numerators and whose denominator is the common denominator of the given fractions. We use the same rule to add algebraic fractions that have the same nonzero denominator. Thus:

Arithmetic fractions	*Algebraic fractions*
$\frac{5}{7} + \frac{1}{7} = \frac{5+1}{7} = \frac{6}{7}$	$\frac{a}{x} + \frac{b}{x} = \frac{a+b}{x}$
$\frac{5}{7} - \frac{1}{7} = \frac{5-1}{7} = \frac{4}{7}$	$\frac{a}{x} - \frac{b}{x} = \frac{a-b}{x}$

Procedure

To add (or subtract) fractions that have the same denominator:

1. Write a fraction whose numerator is the sum (or difference) of the numerators and whose denominator is the common denominator of the given fractions.

2. Reduce the resulting fraction to lowest terms.

EXAMPLE 1

Add and reduce the answer to lowest terms: $\frac{5}{4x} + \frac{9}{4x}$

Solution $\frac{5}{4x} + \frac{9}{4x} = \frac{5+9}{4x} = \frac{14}{4x} = \frac{7}{2x}$

Answer $\frac{7}{2x}$ $(x \neq 0)$

EXAMPLE 2

Subtract: $\frac{4x+7}{6x} - \frac{2x-4}{6x}$

Solution $\frac{4x+7}{6x} - \frac{2x-4}{6x} = \frac{(4x+7)-(2x-4)}{6x} = \frac{4x+7-2x+4}{6x} = \frac{2x+11}{6x}$

Answer $\frac{2x+11}{6x}$ $(x \neq 0)$

Note: In Example 2, since the fraction bar is a symbol of grouping, we enclose numerators in parentheses when the difference is written as a single fraction. In this way, we can see all the signs that need to be changed for the subtraction.

In arithmetic, in order to add (or subtract) fractions that have *different* denominators, we change these fractions to equivalent fractions that have the *same* denominator, called the **common denominator**. Then we add (or subtract) the equivalent fractions.

For example, to add $\frac{3}{4}$ and $\frac{1}{6}$, we use any common denominator that has 4 and 6 as factors.

METHOD 1. Use the product of the denominators as the common denominator. Here, a common denominator is 4×6, or 24.

$$\frac{3}{4} + \frac{1}{6} = \frac{3}{4} \times \frac{6}{6} + \frac{1}{6} \times \frac{4}{4} = \frac{18}{24} + \frac{4}{24} = \frac{22}{24} = \frac{11}{12}$$ *Answer*

METHOD 2. To simplify our work, we use the **least common denominator (LCD)**, that is, the least common multiple of the given denominators. The LCD of $\frac{3}{4}$ and $\frac{1}{6}$ is 12.

$$\frac{3}{4} + \frac{1}{6} = \frac{3}{4} \times \frac{3}{3} + \frac{1}{6} \times \frac{2}{2} = \frac{9}{12} + \frac{2}{12} = \frac{11}{12} \quad \textit{Answer}$$

To find the least common denominator of two fractions, we factor the denominators of the fractions completely. The LCD is the product of all of the factors of the first denominator times the factors of the second denominator that are not factors of the first.

$$\begin{aligned} 4 &= 2 \cdot 2 \\ 6 &= 2 \quad\;\; \cdot 3 \\ \text{LCD} &= 2 \cdot 2 \cdot 3 \end{aligned}$$

Then, to change each fraction to an equivalent form that has the LCD as the denominator, we multiply by $\frac{x}{x}$, where x is the number by which the original denominator must be multiplied to obtain the LCD.

$$\frac{3}{4}\left(\frac{x}{x}\right) = \frac{}{12} \qquad\qquad \frac{1}{6}\left(\frac{x}{x}\right) = \frac{}{12}$$

$$\frac{3}{4}\left(\frac{3}{3}\right) = \frac{9}{12} \qquad\qquad \frac{1}{6}\left(\frac{2}{2}\right) = \frac{2}{12}$$

Note that the LCD is the smallest possible common denominator.

Procedure

To add (or subtract) fractions that have different denominators:

1. Choose a common denominator for the fractions.
2. Change each fraction to an equivalent fraction with the chosen common denominator.
3. Write a fraction whose numerator is the sum (or difference) of the numerators of the new fractions and whose denominator is the common denominator.
4. Reduce the resulting fraction to lowest terms.

Algebraic fractions are added in the same manner as arithmetic fractions, as shown in the examples that follow.

EXAMPLE 3

Add: $\frac{5}{a^2b} + \frac{2}{ab^2}$

Solution *How to Proceed*

(1) Find the LCD of the fractions:

$$a^2b = a \cdot a \cdot b$$
$$ab^2 = a \cdot b \cdot b$$
$$\text{LCD} = a \cdot a \cdot b \cdot b = a^2b^2$$

(2) Change each fraction to an equivalent fraction with the least common denominator, a^2b^2:

$$\frac{5}{a^2b} + \frac{2}{ab^2} = \frac{5}{a^2b} \cdot \frac{b}{b} + \frac{2}{ab^2} \cdot \frac{a}{a}$$
$$= \frac{5b}{a^2b^2} + \frac{2a}{a^2b^2}$$

(3) Write a fraction whose numerator is the sum of the numerators of the new fractions and whose denominator is the common denominator:

$$= \frac{5b + 2a}{a^2b^2}$$

Answer $\frac{5b + 2a}{a^2b^2}$ $(a \neq 0, b \neq 0)$

EXAMPLE 4

Subtract: $\frac{2x + 5}{3} - \frac{x - 2}{4}$

Solution

$$\text{LCD} = 3 \cdot 4 = 12$$

$$\frac{2x + 5}{3} - \frac{x - 2}{4} = \frac{4}{4} \cdot \frac{2x + 5}{3} - \frac{3}{3} \cdot \frac{x - 2}{4}$$
$$= \frac{8x + 20}{12} - \frac{3x - 6}{12}$$
$$= \frac{(8x + 20) - (3x - 6)}{12}$$
$$= \frac{8x + 20 - 3x + 6}{12}$$
$$= \frac{5x + 26}{12} \text{ Answer}$$

EXAMPLE 5

Express as a fraction in simplest form: $y + 1 - \frac{1}{y - 1}$

Solution LCD $= y - 1$

$$y + 1 - \frac{1}{y - 1} = \frac{y + 1}{1} \cdot \left(\frac{y - 1}{y - 1}\right) - \frac{1}{y - 1} = \frac{y^2 - 1}{y - 1} - \frac{1}{y - 1} = \frac{y^2 - 1 - 1}{y - 1} = \frac{y^2 - 2}{y - 1}$$

Answer $\frac{y^2 - 2}{y - 1}$ $(y \neq 1)$

EXAMPLE 6

Subtract: $\frac{6x}{x^2 - 4} - \frac{3}{x - 2}$

Solution

$$x^2 - 4 = (x - 2)(x + 2)$$
$$x - 2 = (x - 2)$$
$$\text{LCD} = (x - 2)(x + 2)$$

$$\frac{6x}{x^2 - 4} - \frac{3}{x - 2} = \frac{6x}{(x - 2)(x + 2)} - \frac{3(x + 2)}{(x - 2)(x + 2)}$$
$$= \frac{6x - 3(x + 2)}{(x - 2)(x + 2)}$$
$$= \frac{6x - 3x - 6}{(x - 2)(x + 2)}$$
$$= \frac{3x - 6}{(x - 2)(x + 2)}$$
$$= \frac{3(x - 2)}{(x - 2)(x + 2)}$$
$$= \frac{3}{x + 2}$$

Answer $\frac{3}{x + 2}$ $(x \neq 2, -2)$

EXERCISES

Writing About Mathematics

1. In Example 2, the answer is $\frac{2x + 11}{6x}$. Can we divide $2x$ and $6x$ by 2 to write the answer in lowest terms as $\frac{x + 11}{3x}$? Explain why or why not.

2. Joey said that $2 - \frac{x - a}{a - x} = 3$ if $a \neq x$. Do you agree with Joey? Explain why or why not.

Developing Skills

In 3–43, add or subtract the fractions as indicated. Reduce each answer to lowest terms. In each case, list the values of the variables for which the fractions are not defined.

3. $\frac{11}{4c} + \frac{5}{4c} - \frac{6}{4c}$

4. $\frac{5r}{t} - \frac{2s}{t}$

5. $\frac{6}{10c} + \frac{9}{10c} - \frac{3}{10c}$

6. $\frac{x}{x + 1} + \frac{1}{x + 1}$

7. $\frac{6y - 4}{4y + 3} + \frac{7 - 2y}{4y + 3}$

8. $\frac{9d + 6}{2d + 1} - \frac{7d + 5}{2d + 1}$

9. $\frac{6x - 5}{x^2 - 1} - \frac{5x - 6}{x^2 - 1}$

10. $\frac{r^2 + 4r}{r^2 - r - 6} + \frac{8 - r^2}{r^2 - r - 6}$

11. $\frac{x}{3} + \frac{x}{2}$

12. $\frac{5x}{6} - \frac{2x}{3}$

13. $\frac{y}{6} + \frac{y}{5} - \frac{y}{2}$

14. $\frac{ab}{5} + \frac{ab}{4}$

15. $\frac{8x}{5} - \frac{3x}{4} + \frac{7x}{10}$

16. $\frac{5a}{6} - \frac{3a}{4}$

17. $\frac{a}{7} + \frac{b}{14}$

18. $\frac{9}{4x} + \frac{3}{2x}$

19. $\frac{1}{2x} - \frac{1}{x} + \frac{1}{8x}$

20. $\frac{9a}{8b} - \frac{3a}{4b}$

21. $d + \frac{7}{5d}$

22. $\frac{a - 3}{3} + \frac{a + 1}{6}$

23. $\frac{3y - 4}{5} - \frac{y - 2}{4}$

24. $\frac{b - 3}{5b} - \frac{b + 2}{10b}$

25. $\frac{y - 4}{4y^2} + \frac{3y - 5}{3y}$

26. $\frac{3c - 7}{2c} - \frac{3c - 3}{6c^2}$

27. $3 + \frac{5}{x + 1}$

28. $5 - \frac{2x}{x + y}$

29. $\frac{5}{x - 3} + \frac{7}{2x - 6}$

30. $\frac{9}{y + 1} - \frac{3}{4y + 4}$

31. $\frac{2}{3a - 1} + \frac{7}{15a - 5}$

32. $\frac{10}{3x - 6} + \frac{3}{2x - 4}$

33. $\frac{1x}{8x - 8} - \frac{3x}{4x - 4}$

34. $\frac{5}{y^2 - 9} - \frac{3}{y - 3}$

35. $\frac{6}{y^2 - 16} - \frac{5}{y + 4}$

36. $\frac{x}{x^2 - 36} - \frac{4}{3x + 18}$

37. $\frac{1}{y - 3} + \frac{2}{y + 4} + \frac{2}{3}$

38. $a + 1 + \frac{1}{a + 1}$

39. $x - 5 - \frac{x}{x + 3}$

40. $\frac{2x - 1}{x + 2} + 2x - 3$

41. $\frac{x + 2y}{3x + 12y} - \frac{6x - y}{x^2 + 3xy - 4y^2}$

42. $\frac{7a}{(a - 1)(a + 3)} + \frac{2a - 5}{(a + 3)(a + 2)}$

43. $\frac{2a + 7}{a^2 - 2a - 15} - \frac{3a - 4}{a^2 - 7a + 10}$

Applying Skills

In 44–46, represent the perimeter of each polygon in simplest form.

44. The lengths of the sides of a triangle are represented by $\frac{x}{2}$, $\frac{3x}{5}$, and $\frac{7x}{10}$.

45. The length of a rectangle is represented by $\frac{x + 3}{4}$, and its width is represented by $\frac{x - 4}{3}$.

46. Each leg of an isosceles triangle is represented by $\frac{2x - 3}{7}$, and its base is represented by $\frac{6x - 18}{21}$.

In 47 and 48, find, in each case, the simplest form of the indicated length.

47. The perimeter of a triangle is $\frac{17x}{24}$, and the lengths of two of the sides are $\frac{3x}{8}$ and $\frac{2x - 5}{12}$. Find the length of the third side.

48. The perimeter of a rectangle is $\frac{14x}{15}$, and the measure of each length is $\frac{x + 2}{3}$. Find the measure of each width.

49. The time t needed to travel a distance d at a rate of speed r can be found by using the formula $t = \frac{d}{r}$.

a. For the first part of a trip, a car travels x miles at 45 miles per hour. Represent the time that the car traveled at that speed in terms of x.

b. For the remainder of the trip, the car travels $2x + 20$ miles at 60 miles per hour. Represent the time that the car traveled at that speed in terms of x.

c. Express, in terms of x, the total time for the two parts of the trip.

50. Ernesto walked 2 miles at a miles per hour and then walked 3 miles at $(a - 1)$ miles per hour. Represent, in terms of a, the total time that he walked.

51. Fran rode her bicycle for x miles at 10 miles per hour and then rode $(x + 3)$ miles farther at 8 miles per hour. Represent, in terms of x, the total time that she rode.

14-6 SOLVING EQUATIONS WITH FRACTIONAL COEFFICIENTS

The following equations contain fractional coefficients:

$$\frac{1}{2}x = 10 \qquad \frac{x}{2} = 10 \qquad \frac{1}{3}x + 60 = \frac{5}{6}x \qquad \frac{x}{3} + 60 = \frac{5x}{6}$$

Each of these equations can be solved by finding an equivalent equation that does not contain fractional coefficients. This can be done by multiplying both sides of the equation by a common denominator for all the fractions present in the equation. We usually multiply by the least common denominator, the LCD.

Note that the equation $0.5x = 10$ can also be written as $\frac{1}{2}x = 10$, since a decimal fraction can be replaced by a common fraction.

Procedure

To solve an equation that contains fractional coefficients:

1. Find the LCD of all coefficients.
2. Multiply both sides of the equation by the LCD.
3. Solve the resulting equation using the usual methods.
4. Check in the original equation.

EXAMPLE 1

Solve and check: $\frac{x}{3} + \frac{x}{5} = 8$

Solution

How to Proceed		*Check*
(1) Write the equation:	$\frac{x}{3} + \frac{x}{5} = 8$	$\frac{x}{3} + \frac{x}{5} = 8$
(2) Find the LCD:	$\text{LCD} = 3 \cdot 5 = 15$	$\frac{15}{3} + \frac{15}{5} \stackrel{?}{=} 8$
(3) Multiply both sides of the equation by the LCD:	$15\left(\frac{x}{3} + \frac{x}{5}\right) = 15(8)$	$5 + 3 \stackrel{?}{=} 8$
(4) Use the distributive property:	$15\left(\frac{x}{3}\right) + 15\left(\frac{x}{5}\right) = 15(8)$	$8 = 8$ ✔
(5) Simplify:	$5x + 3x = 120$	
(6) Solve for x:	$8x = 120$	
	$x = 15$	

Answer $x = 15$

EXAMPLE 2

Solve:

a. $\frac{3x}{4} = 20 + \frac{x}{4}$ **b.** $\frac{2x + 7}{6} - \frac{2x - 9}{10} = 3$

Solution

a.

$$\begin{aligned} \frac{3x}{4} &= 20 + \frac{x}{4} \\ \text{LCD} &= 4 \\ 4\left(\frac{3x}{4}\right) &= 4\left(20 + \frac{x}{4}\right) \\ 4\left(\frac{3x}{4}\right) &= 4(20) + 4\left(\frac{x}{4}\right) \\ 3x &= 80 + x \\ 2x &= 80 \\ x &= 40 \quad \textit{Answer} \end{aligned}$$

b.

$$\begin{aligned} \frac{2x + 7}{6} - \frac{2x - 9}{10} &= 3 \\ \text{LCD} &= 30 \\ 30\left(\frac{2x + 7}{6} - \frac{2x - 9}{10}\right) &= 30(3) \\ 30\left(\frac{2x + 7}{6}\right) - 30\left(\frac{2x - 9}{10}\right) &= 30(3) \\ 5(2x + 7) - 3(2x - 9) &= 90 \\ 10x + 35 - 6x + 27 &= 90 \\ 4x + 62 &= 90 \\ 4x &= 28 \\ x &= 7 \quad \textit{Answer} \end{aligned}$$

In Example 2, the check is left to you.

EXAMPLE 3

A woman purchased stock in the PAX Company over 3 months. In the first month, she purchased one-half of her present number of shares. In the second month, she bought two-fifths of her present number of shares. In the third month, she purchased 14 shares. How many shares of PAX stock did the woman purchase?

Solution Let x = total number of shares of stock purchased.

Then $\frac{1}{2}x$ = number of shares purchased in month 1,

$\frac{2}{5}x$ = number of shares purchased in month 2,

14 = number of shares purchased in month 3.

The sum of the shares purchased over 3 months is the total number of shares.

month 1 + month 2 + month 3 = total

$$\begin{aligned} \frac{1}{2}x + \frac{2}{5}x + 14 &= x \\ 10\left(\frac{1}{2}x + \frac{2}{5}x + 14\right) &= 10(x) \\ 5x + 4x + 140 &= 10x \\ 9x + 140 &= 10x \\ 140 &= x \end{aligned}$$

Check month 1 = $\frac{1}{2}(140) = 70$, month 2 = $\frac{2}{5}(140) = 56$, month 3 = 14

$$70 + 56 + 14 = 140 \checkmark$$

Answer 140 shares

EXAMPLE 4

In a child's coin bank, there is a collection of nickels, dimes, and quarters that amounts to \$3.20. There are 3 times as many quarters as nickels, and 5 more dimes than nickels. How many coins of each kind are there?

Solution

Let x = the number of nickels.
Then $3x$ = the number of quarters,
and $x + 5$ = the number of dimes.
Also, $0.05x$ = the value of the nickels,
$0.25(3x)$ = the value of the quarters,
and $0.10(x + 5)$ = the value of the dimes.

Write the equation for the value of the coins. To simplify the equation, which contains coefficients that are decimal fractions with denominators of 100, multiply each side of the equation by 100.

The total value of the coins is \$3.20.

$$\begin{aligned} 0.05x + 0.25(3x) + 0.10(x + 5) &= 3.20 \\ 100[0.05x + 0.25(3x) + 0.10(x + 5)] &= 100(3.20) \\ 5x + 25(3x) + 10(x + 5) &= 320 \\ 5x + 75x + 10x + 50 &= 320 \\ 90x + 50 &= 320 \\ 90x &= 270 \\ x &= 3 \end{aligned}$$

Check There are 3 nickels, 3(3) = 9 quarters, and 3 + 5 = 8 dimes.

The value of 3 nickels is \$0.05(3) = \$0.15
The value of 9 quarters is \$0.25(9) = \$2.25
The value of 8 dimes is \$0.10(8) = \$0.80
\$3.20 ✔

Answer There are 3 nickels, 9 quarters, and 8 dimes.

Note: In a problem such as this, a chart such as the one shown below can be used to organize the information:

Coins	Number of Coins	Value of One Coin	Total Value
Nickels	x	0.05	$0.05x$
Quarters	$3x$	0.25	$0.25(3x)$
Dimes	$x + 5$	0.10	$0.10(x + 5)$

EXERCISES

Writing About Mathematics

1. Abby solved the equation $0.2x - 0.84 = 3x$ as follows:

$$\begin{array}{rcr} 0.2x - 0.84 & = & 3x \\ \underline{-0.2x } & & \underline{-0.2x} \\ -0.84 & = & 0.1x \\ \\ -8.4 & = & x \end{array}$$

Is Abby's solution correct? Explain why or why not.

2. In order to write the equation $0.2x - 0.84 = 3x$ as an equivalent equation with integral coefficients, Heidi multiplied both sides of the equation by 10. Will Heidi's method lead to a correct solution? Explain why or why not. Compare Heidi's method with multiplying by 100 or multiplying by 1,000.

Developing Skills

In 3–37, solve each equation and check.

3. $\frac{x}{7} = 3$

4. $\frac{1}{6}t = 18$

5. $\frac{3x}{5} = 15$

6. $\frac{x + 8}{4} = 6$

7. $\frac{m - 2}{9} = 3$

8. $\frac{2r + 6}{5} = -4$

9. $\frac{5y - 30}{7} = 0$

10. $\frac{5x}{2} = \frac{15}{4}$

11. $\frac{m - 5}{35} = \frac{5}{7}$

12. $\frac{2x + 1}{3} = \frac{6x - 9}{5}$

13. $\frac{3y + 1}{4} = \frac{44 - y}{5}$

14. $\frac{x}{5} + \frac{x}{3} = \frac{8}{15}$

15. $10 = \frac{x}{3} + \frac{x}{7}$

16. $\frac{r}{3} - \frac{r}{6} = 2$

17. $\frac{3t}{4} - 6 = \frac{t}{12}$

18. $\frac{a}{2} + \frac{a}{3} + \frac{a}{4} = 26$

19. $\frac{7y}{12} - \frac{1}{4} = 2y - \frac{5}{3}$

20. $\frac{y + 2}{4} - \frac{y - 3}{3} = \frac{1}{2}$

21. $\frac{t - 3}{6} - \frac{t - 25}{5} = 4$

22. $\frac{3m + 1}{4} = 2 - \frac{3 - 2m}{6}$

23. $0.03y - 1.2 = 8.7$

24. $0.4x + 0.08 = 4.24$

25. $2c + 0.5c = 50$

26. $0.08y - 0.9 = 0.02y$

27. $1.7x = 30 + 0.2x$

28. $0.02(x + 5) = 8$

29. $0.05(x - 8) = 0.07x$

30. $0.4(x - 9) = 0.3(x + 4)$

31. $0.06(x - 5) = 0.04(x + 8)$

32. $0.04x + 0.03(2{,}000 - x) = 75$

33. $0.02x + 0.04(1{,}500 - x) = 48$

34. $0.05x + 10 = 0.06(x + 50)$

35. $0.08x = 0.03(x + 200) - 4$

36. $\frac{0.4a}{3} + \frac{0.2a}{4} = 2$

37. $\frac{0.1a}{6} - \frac{0.3a}{4} = 3$

38. The sum of one-half of a number and one-third of that number is 25. Find the number.

39. The difference between one-fifth of a positive number and one-tenth of that number is 10. Find the number.

40. If one-half of a number is increased by 20, the result is 35. Find the number.

41. If two-thirds of a number is decreased by 30, the result is 10. Find the number.

42. If the sum of two consecutive integers is divided by 3, the quotient is 9. Find the integers.

43. If the sum of two consecutive odd integers is divided by 4, the quotient is 10. Find the integers.

44. In an isosceles triangle, each of the congruent sides is two-thirds of the base. The perimeter of the triangle is 42. Find the length of each side of the triangle.

45. The larger of two numbers is 12 less than 5 times the smaller. If the smaller number is equal to one-third of the larger number, find the numbers.

46. The larger of two numbers exceeds the smaller by 14. If the smaller number is equal to three-fifths of the larger, find the numbers.

47. Separate 90 into two parts such that one part is one-half of the other part.

48. Separate 150 into two parts such that one part is two-thirds of the other part.

Applying Skills

49. Four vegetable plots of unequal lengths and of equal widths are arranged as shown. The length of the third plot is one-fourth the length of the second plot.

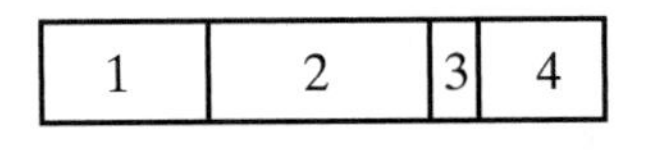

The length of the fourth plot is one-half the length of the second plot. The length of the first plot is 10 feet more than the length of the fourth plot. If the total length of the four plots is 100 feet, find the length of each plot.

50. Sam is now one-sixth as old as his father. In 4 years, Sam will be one-fourth as old as his father will be then. Find the ages of Sam and his father now.

51. Robert is one-half as old as his father. Twelve years ago, he was one-third as old as his father was then. Find their present ages.

52. A coach finds that, of the students who try out for track, 65% qualify for the team and 90% of those who qualify remain on the team throughout the season. What is the smallest number of students who must try out for track in order to have 30 on the team at the end of the season?

53. A bus that runs once daily between the villages of Alpaca and Down makes only two stops in between, at Billow and at Comfort. Today, the bus left Alpaca with some passengers. At Billow, one-half of the passengers got off, and six new ones got on. At Comfort, again one-half of the passengers got off, and, this time, five new ones got on. At Down, the last 13 passengers on the bus got off. How many passengers were aboard when the bus left Alpaca?

54. Sally spent half of her money on a present for her mother. Then she spent one-quarter of the cost of the present for her mother on a treat for herself. If Sally had $6.00 left after she bought her treat, how much money did she have originally?

55. Bob planted some lettuce seedlings in his garden. After a few days, one-tenth of these seedlings had been eaten by rabbits. A week later, one-fifth of the remaining seedlings had been eaten, leaving 36 seedlings unharmed. How many lettuce seedlings had Bob planted originally?

56. May has 3 times as many dimes as nickels. In all, she has $1.40. How many coins of each type does she have?

57. Mr. Jantzen bought some cans of soup at $0.39 per can, and some packages of frozen vegetables at $0.59 per package. He bought twice as many packages of vegetables as cans of soup. If the total bill was $9.42, how many cans of soup did he buy?

58. Roger has $2.30 in dimes and nickels. There are 5 more dimes than nickels. Find the number of each kind of coin that he has.

59. Bess has $2.80 in quarters and dimes. The number of dimes is 7 less than the number of quarters. Find the number of each kind of coin that she has.

60. A movie theater sold student tickets for $5.00 and full-price tickets for $7.00. On Saturday, the theater sold 16 more full-price tickets than student tickets. If the total sales on Saturday were $1,072, how many of each kind of ticket were sold?

61. Is it possible to have $4.50 in dimes and quarters, and have twice as many quarters as dimes? Explain.

62. Is it possible to have $6.00 in nickels, dimes, and quarters, and have the same number of each kind of coin? Explain.

63. Mr. Symms invested a sum of money in 7% bonds. He invested $400 more than this sum in 8% bonds. If the total annual interest from these two investments is $257, how much did he invest at each rate?

64. Mr. Charles borrowed a sum of money at 10% interest. He borrowed a second sum, which was $1,500 less than the first sum, at 11% interest. If the annual interest on these two loans is $202.50, how much did he borrow at each rate?

14-7 SOLVING INEQUALITIES WITH FRACTIONAL COEFFICIENTS

In our modern world, many problems involve inequalities. A potential buyer may offer *at most* one amount for a house, while the seller will accept *no less* than another amount. Inequalities that contain fractional coefficients are handled in much the same way as equations that contain fractional coefficients. The chart on the right helps us to translate words into algebraic symbols.

Words	Symbols
a is greater than b	$a > b$
a is less than b	$a < b$
a is at least b a is no less than b	$a \geq b$
a is at most b a is no greater than b	$a \leq b$

Procedure

To solve an inequality that contains fractional coefficients:

1. Find the LCD, a positive number.
2. Multiply both sides of the inequality by the LCD.
3. Solve the resulting inequality using the usual methods.

EXAMPLE 1

Solve the inequality, and graph the solution set on a number line:

a. $\frac{x}{3} - \frac{x}{6} > 2$

b. $\frac{3y}{2} + \frac{8 - 4y}{7} \leq 3$

Solution **a.**

$$\frac{x}{3} - \frac{x}{6} > 2$$
$$6\left(\frac{x}{3} - \frac{x}{6}\right) > 6(2)$$
$$6\left(\frac{x}{3}\right) - 6\left(\frac{x}{6}\right) > 12$$
$$2x - x > 12$$
$$x > 12$$

Since no domain was given, use the domain of real numbers.

Answer: $x > 12$

0 2 4 6 8 10 12 14 16

b.

$$\frac{3y}{2} + \frac{8 - 4y}{7} \leq 3$$
$$14\left(\frac{3y}{2} + \frac{8 - 4y}{7}\right) \leq 14(3)$$
$$14\left(\frac{3y}{2}\right) + 14\left(\frac{8 - 4y}{7}\right) \leq 42$$
$$21y + 16 - 8y \leq 42$$
$$13y \leq 26$$
$$y \leq 2$$

Since no domain was given, use the domain of real numbers.

Answer: $y \leq 2$

–1 0 1 2 3

EXAMPLE 2

Two boys want to pool their money to buy a comic book. The younger of the boys has one-third as much money as the older. Together they have more than $2.00. Find the smallest possible amount of money each can have.

Solution Let x = the number of cents that the older boy has.
Then $\frac{1}{3}x$ = the number of cents that the younger boy has.

The sum of their money in cents is greater than 200.

$$x + \tfrac{1}{3}x > 200$$
$$3\left(x + \tfrac{1}{3}x\right) > 3(200)$$
$$3x + x > 600$$
$$4x > 600$$
$$x > 150$$
$$\tfrac{1}{3}x > 50$$

The number of cents that the younger boy has must be an integer greater than 50. The number of cents that the older boy has must be a multiple of 3 that is greater than 150. The younger boy has at least 51 cents. The older boy has at least 153 cents. The sum of 51 and 153 is greater than 200.

Answer The younger boy has at least $0.51 and the older boy has at least $1.53.

EXERCISES

Writing About Mathematics

1. Explain the error in the following solution of an inequality.

$$\tfrac{x}{-3} > -2$$
$$-3\left(\tfrac{x}{-3}\right) > -3(-2)$$
$$x > 6$$

2. In Example 2, what is the domain for the variable?

Developing Skills

In 3–23, solve each inequality, and graph the solution set on a number line.

3. $\frac{1}{4}x - \frac{1}{5}x > \frac{9}{20}$

4. $y - \frac{2}{3}y < 5$

5. $\frac{5}{6}c > \frac{1}{3}c + 3$

6. $\frac{x}{4} - \frac{x}{8} \leq \frac{5}{8}$

7. $\frac{y}{6} \geq \frac{y}{12} + 1$

8. $\frac{y}{9} - \frac{y}{4} > \frac{5}{36}$

9. $\frac{t}{10} \leq 4 + \frac{t}{5}$

10. $1 + \frac{2x}{3} \geq \frac{x}{2}$

11. $2.5x - 1.7x > 4$

12. $2y + 3 \geq 0.2y$

13. $\frac{3x - 1}{7} > 5$

14. $\frac{5y - 30}{7} \leq 0$

15. $2d + \frac{1}{4} < \frac{7d}{12} + \frac{5}{3}$

16. $\frac{4c}{3} - \frac{7}{9} \geq \frac{c}{2} + \frac{7}{6}$

17. $\frac{2m}{3} \geq \frac{7 - m}{4} + 1$

18. $\frac{3x - 30}{6} < \frac{x}{3} - 2$

19. $\frac{6x - 3}{2} > \frac{37}{10} + \frac{x + 2}{5}$

20. $\frac{2y - 3}{3} + \frac{y + 1}{2} < 10$

21. $\frac{2r - 3}{5} - \frac{r - 3}{3} \leq 2$

22. $\frac{3t - 4}{3} \geq \frac{2t + 4}{6} + \frac{5t - 1}{9}$

23. $\frac{2 - a}{2} \leq \frac{a + 2}{5} - \frac{2a + 3}{6}$

24. If one-third of an integer is increased by 7, the result is at most 13. Find the largest possible integer.

25. If two-fifths of an integer is decreased by 11, the result is at least 4. Find the smallest possible integer.

26. The sum of one-fifth of an integer and one-tenth of that integer is less than 40. Find the greatest possible integer.

27. The difference between three-fourths of a positive integer and one-half of that integer is greater than 28. Find the smallest possible integer.

28. The smaller of two integers is two-fifths of the larger, and their sum is less than 40. Find the largest possible integers.

29. The smaller of two positive integers is five-sixths of the larger, and their difference is greater than 3. Find the smallest possible integers.

Applying Skills

30. Talk and Tell Answering Service offers customers two monthly options.

OPTION 1	*Measured Service*	OPTION 2	*Unmeasured Service*
	base rate is \$15		base rate is \$20
	each call costs \$0.10		no additional charge per call

Find the least number of calls for which unmeasured service is cheaper than measured service.

31. Paul earned some money mowing lawns. He spent one-half of this money for a book, and then one-third for a CD. If he had less than \$3 left, how much money did he earn?

32. Mary bought some cans of vegetables at \$0.89 per can, and some cans of soup at \$0.99 per can. If she bought twice as many cans of vegetables as cans of soup, and paid at least \$10, what is the least number of cans of vegetables she could have bought?

33. A coin bank contains nickels, dimes, and quarters. The number of dimes is 7 more than the number of nickels, and the number of quarters is twice the number of dimes. If the total value of the coins is no greater than \$7.20, what is the greatest possible number of nickels in the bank?

34. Rhoda is two-thirds as old as her sister Alice. Five years from now, the sum of their ages will be less than 60. What is the largest possible integral value for each sister's present age?

35. Four years ago, Bill was $1\frac{1}{4}$ times as old as his cousin Mary. The difference between their present ages is at least 3. What is the smallest possible integral value for each cousin's present age?

36. Mr. Drew invested a sum of money at $7\frac{1}{2}\%$ interest. He invested a second sum, which was \$200 less than the first, at 7% interest. If the total annual interest from these two investments is at least \$160, what is the smallest amount he could have invested at $7\frac{1}{2}\%$?

37. Mr. Lehtimaki wanted to sell his house. He advertised an asking price, but knew that he would accept, as a minimum, nine-tenths of the asking price. Mrs. Patel offered to buy the house, but her maximum offer was seven-eighths of the asking price. If the difference between the seller's lowest acceptance price and the buyer's maximum offer was at least \$3,000, find:

a. the minimum asking price for the house;

b. the minimum amount Mr. Lehtimaki, the seller, would accept;

c. the maximum amount offered by Mrs. Patel, the buyer.

38. When packing his books to move, Philip put the same number of books in each of 12 boxes. Once packed, the boxes were too heavy to lift so Philip removed one-fifth of the books from each box. If at least 100 books in total remain in the boxes, what is the minimum number of books that Philip originally packed in each box?

14-8 SOLVING FRACTIONAL EQUATIONS

An equation is called an **algebraic equation** when a variable appears in at least one of its sides. An algebraic equation is a **fractional equation** when a variable appears in the denominator of one, or more than one, of its terms. For example,

$$\frac{1}{3} + \frac{1}{x} = \frac{1}{2} \qquad \frac{2}{3d} + \frac{1}{3} = \frac{11}{6d} - \frac{1}{4} \qquad \frac{a^2 + 1}{a - 1} + \frac{a}{2} = a + 2 \qquad \frac{1}{y^2 + 2y - 3} + \frac{1}{y - 2} = 3$$

are all fractional equations. To simplify such an equation, clear it of fractions by multiplying both sides by the least common denominator of all fractions in the equation. Then, solve the simpler equation. As is true of all algebraic fractions, a fractional equation has meaning only when values of the variable do not lead to a denominator of 0.

KEEP IN MIND When both sides of an equation are multiplied by a variable expression that may represent 0, the resulting equation may not be equivalent to the given equation. Such equations will yield **extraneous solutions**, which are solutions that satisfy the *derived* equation but not the *given* equation. Each solution, therefore, must be checked in the original equation.

EXAMPLE 1

Solve and check: $\frac{1}{3} + \frac{1}{x} = \frac{1}{2}$

Solution Multiply both sides of the equation by the least common denominator, $6x$.

$$\frac{1}{3} + \frac{1}{x} = \frac{1}{2}$$
$$6x\left(\frac{1}{3} + \frac{1}{x}\right) = 6x\left(\frac{1}{2}\right)$$
$$6x\left(\frac{1}{3}\right) + 6x\left(\frac{1}{x}\right) = 6x\left(\frac{1}{2}\right)$$
$$2x + 6 = 3x$$
$$6 = x$$

Check

$$\frac{1}{3} + \frac{1}{x} = \frac{1}{2}$$
$$\frac{1}{3} + \frac{1}{\mathbf{6}} \stackrel{?}{=} \frac{1}{2}$$
$$\frac{2}{6} + \frac{1}{6} \stackrel{?}{=} \frac{1}{2}$$
$$\frac{3}{6} \stackrel{?}{=} \frac{1}{2}$$
$$\frac{1}{2} = \frac{1}{2} \checkmark$$

Answer $x = 6$

EXAMPLE 2

Solve and check: $\frac{5x + 10}{x + 2} = 7$

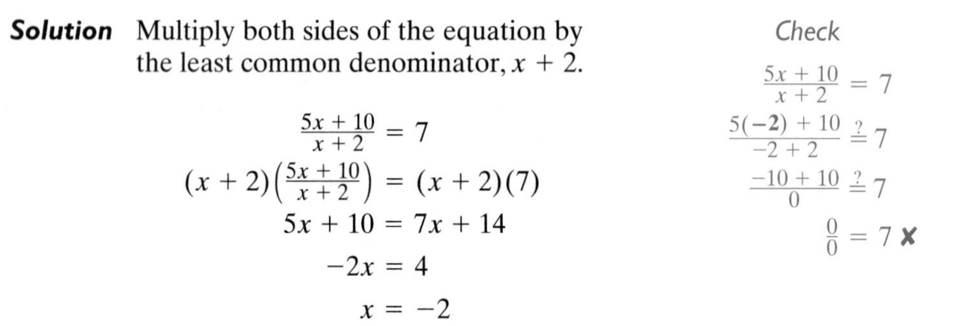

Solution Multiply both sides of the equation by the least common denominator, $x + 2$.

$$\frac{5x + 10}{x + 2} = 7$$
$$(x + 2)\left(\frac{5x + 10}{x + 2}\right) = (x + 2)(7)$$
$$5x + 10 = 7x + 14$$
$$-2x = 4$$
$$x = -2$$

Check

$$\frac{5x + 10}{x + 2} = 7$$
$$\frac{5(\mathbf{-2}) + 10}{\mathbf{-2} + 2} \stackrel{?}{=} 7$$
$$\frac{-10 + 10}{0} \stackrel{?}{=} 7$$
$$\frac{0}{0} = 7 \text{ ✗}$$

The only possible value of x is a value for which the equation has no meaning because it leads to a denominator of 0. Therefore, there is no solution for this equation.

Answer The solution set is the empty set, $\varnothing$ or $\{\ \}$.

EXAMPLE 3

Solve and check: $\frac{2}{x} = \frac{6 - x}{4}$

Solution

METHOD 1

Multiply both sides of the equation by the LCD, $4x$:

$$4x\left(\frac{2}{x}\right) = 4x\left(\frac{6-x}{4}\right)$$
$$8 = x(6-x)$$
$$8 = 6x - x^2$$
$$x^2 - 6x + 8 = 0$$
$$(x-2)(x-4) = 0$$

$x - 2 = 0$ | $x - 4 = 0$

$x = 2$ | $x = 4$

METHOD 2

Use the rule for proportion: *the product of the means equals the product of the extremes.*

$$\frac{2}{x} = \frac{6-x}{4}$$
$$x(6-x) = 8$$
$$6x - x^2 = 8$$
$$-x^2 + 6x - 8 = 0$$
$$x^2 - 6x + 8 = 0$$
$$(x-2)(x-4) = 0$$

$x - 2 = 0$ | $x - 4 = 0$

$x = 2$ | $x = 4$

Check

$x = 2$:

$$\frac{2}{x} = \frac{6-x}{4}$$
$$\frac{2}{\mathbf{2}} = \frac{6-\mathbf{2}}{4}$$
$$1 = 1 \checkmark$$

$x = 4$:

$$\frac{2}{x} = \frac{6-x}{4}$$
$$\frac{2}{\mathbf{4}} = \frac{6-\mathbf{4}}{4}$$
$$\frac{1}{2} = \frac{1}{2} \checkmark$$

Answer $x = 2$ or $x = 4$

EXAMPLE 4

Solve and check: $\frac{1}{x+2} + \frac{1}{x-1} = \frac{1}{2(x-1)}$

Solution Multiply both sides of the equation by the LCD, $2(x+2)(x-1)$:

$$2(x+2)(x-1)\left(\frac{1}{x+2} + \frac{1}{x-1}\right) = \left(\frac{1}{2(x-1)}\right)2(x+2)(x-1)$$
$$\frac{2(x+2)(x-1)}{x+2} + \frac{2(x+2)(x-1)}{x-1} = \frac{2(x+2)(x-1)}{2(x-1)}$$
$$\frac{2\cancel{(x+2)}(x-1)}{\cancel{x+2}} + \frac{2(x+2)\cancel{(x-1)}}{\cancel{x-1}} = \frac{2(x+2)\cancel{(x-1)}}{2\cancel{(x-1)}}$$
$$2(x-1) + 2(x+2) = x + 2$$
$$4x + 2 = x + 2$$
$$3x = 0$$
$$x = 0$$

Check

$$\frac{1}{\mathbf{0}+2} + \frac{1}{\mathbf{0}-1} \stackrel{?}{=} \frac{1}{2(\mathbf{0}-1)}$$
$$\frac{1}{2} + \frac{1}{-1} \stackrel{?}{=} \frac{1}{2(-1)}$$
$$-\frac{1}{2} = -\frac{1}{2} \checkmark$$

Answer $x = 0$

EXERCISES

Writing About Mathematics

1. Nathan said that the solution set of $\frac{2}{r-5} = \frac{10}{5r-25}$ is the set of all real numbers. Do you agree with Nathan? Explain why or why not.

2. Pam multiplied each side of the equation $\frac{y+5}{y^2-25} = \frac{3}{y+5}$ by $(y+5)(y-5)$ to obtain the equation $y + 5 = 3y - 15$, which has as its solution $y = 10$. Pru said that the equation $\frac{y+5}{y^2-25} = \frac{3}{y+5}$ is a proportion and can be solved by writing the product of the means equal to the product of the extremes. She obtained the equation $3(y^2 - 25) = (y + 5)^2$, which has as its solution 10 and -5. Both girls used a correct method of solution. Explain the difference in their answers.

Developing Skills

In 3–6, explain why each fractional equation has no solution.

3. $\frac{6x}{x} = 3$ **4.** $\frac{4a+4}{a+1} = 5$ **5.** $\frac{2}{x} = 4 + \frac{2}{x}$ **6.** $\frac{x}{x-1} + 2 = \frac{1}{x-1}$

In 7–45, solve each equation, and check.

7. $\frac{10}{x} = 5$ **8.** $\frac{15}{y} = 3$ **9.** $\frac{3}{2x} = \frac{1}{2}$

10. $\frac{15}{4x} = \frac{1}{8}$ **11.** $\frac{10}{x} + \frac{8}{x} = 9$ **12.** $\frac{15}{y} - \frac{3}{y} = 4$

13. $\frac{9}{2x} = \frac{7}{2x} + 2$ **14.** $\frac{30}{x} = 7 + \frac{18}{2x}$ **15.** $\frac{y-2}{2y} = \frac{3}{8}$

16. $\frac{x-5}{x} + \frac{3}{x} = \frac{2}{3}$ **17.** $\frac{y+9}{2y} + 3 = \frac{15}{y}$ **18.** $\frac{1}{3a} + \frac{5}{12} = \frac{2}{a}$

19. $\frac{b-6}{b} - \frac{1}{6} = \frac{4}{b}$ **20.** $\frac{7}{8} - \frac{x-3}{x} = \frac{1}{4}$ **21.** $\frac{5+x}{2x} - 1 = \frac{x+1}{x}$

22. $\frac{2+x}{6x} = \frac{3}{5x} + \frac{1}{30}$ **23.** $\frac{a}{a+2} - \frac{a-2}{a} = \frac{1}{a}$ **24.** $\frac{x+1}{2x} + \frac{2x+1}{3x} = 1$

25. $\frac{6}{3x-1} = \frac{3}{4}$ **26.** $\frac{2}{3x-4} = \frac{1}{4}$ **27.** $\frac{5x}{x+1} = 4$

28. $\frac{3}{5-3a} = \frac{1}{2}$ **29.** $\frac{4z}{7+5z} = \frac{1}{3}$ **30.** $\frac{1-r}{1+r} = \frac{2}{3}$

31. $\frac{3}{y} = \frac{2}{5-y}$ **32.** $\frac{5}{a} = \frac{7}{a-4}$ **33.** $\frac{2}{m} = \frac{5}{3m-1}$

34. $\frac{2}{a-4} = \frac{5}{a-1}$ **35.** $\frac{12}{2-x} = \frac{15}{7+x}$ **36.** $\frac{12y}{8y+5} = \frac{2}{3}$

37. $\frac{y}{y+1} - \frac{1}{y} = 1$ **38.** $\frac{x}{x^2-9} = \frac{1}{x+3}$ **39.** $\frac{2}{x} = \frac{x-3}{2}$

40. $\frac{a-1}{6} = \frac{1}{a}$ **41.** $\frac{1}{b} = \frac{b-1}{2}$ **42.** $\frac{3}{2b+1} = \frac{b}{2}$

43. $\frac{1}{b-1} + \frac{1}{6} = \frac{b+2}{12}$ **44.** $\frac{1}{x} - \frac{1}{9} = \frac{x-5}{18}$ **45.** $\frac{1}{x+2} + \frac{1}{2x+4} = -\frac{3}{2}x$

In 46–49, solve each equation for x in terms of the other variables.

46. $\frac{t}{x} - k = 0$ **47.** $\frac{t}{x} - k = 5k$ **48.** $\frac{a+b}{x} = c$ **49.** $\frac{d}{x} = \frac{d-1}{x-e}$

50. If $x = \frac{by}{c}$, $y = \frac{c^2}{a}$, $b = \frac{a}{c}$, $a \neq 0$, and $c \neq 0$, is it possible to know the numerical value of x without knowing numerical values of a, b, c, and y? Explain your answer

Applying Skills

51. If 24 is divided by a number, the result is 6. Find the number.

52. If 10 is divided by a number, the result is 30. Find the number.

53. The sum of 20 divided by a number, and 7 divided by the same number, is 9. Find the number.

54. When the reciprocal of a number is decreased by 2, the result is 5. Find the number.

55. The numerator of a fraction is 8 less than the denominator of the fraction. The value of the fraction is $\frac{3}{5}$. Find the fraction.

56. The numerator and denominator of a fraction are in the ratio 3 : 4. When the numerator is decreased by 4 and the denominator is increased by 2, the value of the new fraction, in simplest form, is $\frac{1}{2}$. Find the original fraction.

57. The ratio of boys to girls in the chess club is 4 to 5. After 2 boys leave the club and 2 girls join, the ratio is 1 to 2. How many members are in the club?

58. The length of Emily's rectangular garden is 4 feet greater than its width. The width of Sarah's rectangular garden is equal to the length of Emily's and its length is 18 feet. The two gardens are similar rectangles, that is, the ratio of the length to the width of Emily's garden equals the ratio of the length to the width of Sarah's garden. Find the possible dimensions of each garden. (Two answers are possible.)

CHAPTER SUMMARY

An **algebraic fraction** is the quotient of two algebraic expressions. If the algebraic expressions are polynomials, the fraction is called a **rational expression** or a **fractional expression**. An algebraic fraction is defined only if values of the variables do not result in a denominator of 0.

Fractions that are equal in value are called **equivalent fractions**. A fraction is **reduced to lowest terms** when an equivalent fraction is found such that its numerator and denominator have no common factor other than 1 or −1. This fraction is considered a **lowest terms fraction**.

Operations with algebraic fractions follow the same rules as operations with arithmetic fractions:

Multiplication

$$\frac{a}{x} \cdot \frac{b}{y} = \frac{ab}{xy}\ (x \neq 0, y \neq 0)$$

Division

$$\frac{a}{x} \div \frac{b}{y} = \frac{a}{x} \cdot \frac{y}{b} = \frac{ay}{bx}\ (x \neq 0, y \neq 0, b \neq 0)$$

Addition/subtraction with the same denominator

$$\frac{a}{c} + \frac{b}{c} = \frac{a+b}{c}\ (c \neq 0), \frac{a}{c} - \frac{b}{c} = \frac{a-b}{c}(c \neq 0)$$

Addition/subtraction with different denominators

(first, obtain the common denominator):

$$\frac{a}{b} + \frac{c}{d} = \frac{a}{b} \cdot \frac{d}{d} + \frac{c}{d} \cdot \frac{b}{b} = \frac{ad}{bd} + \frac{bc}{bd} = \frac{ad + bc}{bd}\ (b \neq 0, d \neq 0)$$

A **fractional equation** is an equation in which a variable appears in the denominator of one or more than one of its terms. To simplify a fractional equation, or any equation or inequality containing fractional coefficients, multiply both sides by the **least common denominator (LCD)** to eliminate the fractions. Then solve the simpler equation or inequality and check for **extraneous solutions**.

VOCABULARY

14-1 Algebraic fraction • Fractional expression • Rational expression

14-2 Reduced to lowest terms • Lowest terms fraction • Equivalent fractions • Division property of a fraction • Cancellation • Multiplication property of a fraction

14-3 Cancellation method

14-5 Common denominator • Least common denominator

14-8 Algebraic equation • Fractional equation • Extraneous solution

REVIEW EXERCISES

1. Explain the difference between an algebraic fraction and a fractional expression.
2. What fractional part of 1 centimeter is x millimeters?
3. For what value of y is the fraction $\frac{y-1}{y-4}$ undefined?
4. Factor completely: $12x^3 - 27x$

In 5–8, reduce each fraction to lowest terms.

5. $\frac{8bg}{12bg}$ **6.** $\frac{14d}{7d^2}$ **7.** $\frac{5x^2 - 60}{5}$ **8.** $\frac{8y^2 - 12y}{8y}$

In 9–23, in each case, perform the indicated operation and express the answer in lowest terms.

9. $\frac{3x^2}{4} \cdot \frac{8}{9x}$ **10.** $\frac{2x - 2}{3y} \cdot \frac{3xy}{2x}$ **11.** $6c^2 \div \frac{c}{2}$

12. $\frac{3a}{7b} \div \frac{18a}{35}$ **13.** $\frac{5m}{6} - \frac{m}{6}$ **14.** $\frac{9}{k} - \frac{3}{k} + \frac{4}{k}$

15. $\frac{ax}{3} + \frac{ax}{4}$ **16.** $\frac{5}{xy} - \frac{2}{yz}$ **17.** $\frac{4x + 5}{3x} + \frac{2x + 1}{2x}$

18. $\frac{x^2 - 5x}{x^2} \cdot \frac{x}{2x - 10}$ **19.** $\frac{2a}{a + b} + \frac{2b}{a + b}$ **20.** $\frac{y + 7}{5} - \frac{y + 3}{4}$

21. $\frac{x^2 - 25}{12} \div \frac{x^2 - 10x + 25}{4}$ **22.** $\frac{c - 3}{12} + \frac{c + 3}{8}$ **23.** $\frac{3a - 9a^2}{a} \div (1 - 9a^2)$

24. If the sides of a triangle are represented by $\frac{b}{2}$, $\frac{5b}{6}$, and $\frac{2b}{3}$, express the perimeter of the triangle in simplest form.

25. If $a = 2$, $b = 3$, and $c = 4$, what is the sum of $\frac{b}{a} + \frac{a}{c}$?

In 26–31, solve each equation and check.

26. $\frac{k}{20} = \frac{3}{4}$ **27.** $\frac{x - 3}{10} = \frac{4}{5}$ **28.** $\frac{y}{2} - \frac{y}{6} = 4$

29. $\frac{6}{m} = \frac{20}{m} - 2$ **30.** $\frac{2t}{5} - \frac{t - 2}{10} = 2$ **31.** $\frac{1}{a - 1} - \frac{1}{a} = \frac{1}{20}$

In 32–34, solve each equation for r in terms of the other variables.

32. $\frac{S}{h} = 2\pi r$ **33.** $\frac{c}{2r} = \pi$ **34.** $\frac{a}{r} - n = 0$

35. Mr. Vroman deposited a sum of money in the bank. After a few years, he found that the interest equaled one-fourth of his original deposit and he had a total sum, deposit plus interest, of $2,400 in the bank. What was the original deposit?

36. One-third of the result obtained by adding 5 to a certain number is equal to one-half of the result obtained when 5 is subtracted from the number. Find the number.

37. Of the total number of points scored by the winning team in a basketball game, one-fifth was scored in the first quarter, one-sixth was scored in the second quarter, one-third was scored in the third quarter, and 27 was scored in the fourth quarter. How many points did the winning team score?

38. Ross drove 300 miles at r miles per hour and 360 miles at $r + 10$ miles per hour. If the time needed to drive 300 miles was equal to the time needed to drive 360 miles, find the rates at which Ross drove. (Express the time needed for each part of the trip as $t = \frac{d}{r}$.)

39. The total cost, T, of n items that cost a dollars each is given by the equation $T = na$.

a. Solve the equation $T = na$ for n in terms of T and a.

b. Use your answer to **a** to express n_1, the number of cans of soda that cost $12.00 if each can of soda costs a dollars.

c. Use your answer to **a** to express n_2, the number of cans of soda that cost $15.00 if each can of soda costs a dollars.

d. If the number of cans of soda purchased for $12.00 is 4 less than the number purchased for $15.00, find the cost of a can of soda and the number of cans of soda purchased.

40. The cost of two cups of coffee and a bagel is $1.75. The cost of four cups of coffee and three bagels is $4.25. What is the cost of a cup of coffee and the cost of a bagel?

41. A piggybank contains nickels, dimes, and quarters. The number of nickels is 4 more than the number of dimes, and the number of quarters is 3 times the number of nickels. If the total value of the coins is no greater than $8.60, what is the greatest possible number of dimes in the bank?

Exploration

Some rational numbers can be written as terminating decimals and others as infinitely repeating decimals.

(1) Write each of the following fractions as a decimal:

$$\frac{1}{2}, \frac{1}{4}, \frac{1}{5}, \frac{1}{8}, \frac{1}{10}, \frac{1}{16}, \frac{1}{20}, \frac{1}{25}, \frac{1}{50}, \frac{1}{100}$$

(2) What do you observe about the decimals written in (1)?

(3) Write each denominator in factored form.

(4) What do you observe about the factors of the denominators?

(5) Write each of the following fractions as a decimal:

$$\frac{1}{3}, \frac{1}{6}, \frac{1}{9}, \frac{1}{11}, \frac{1}{12}, \frac{1}{15}, \frac{1}{18}, \frac{1}{22}, \frac{1}{24}, \frac{1}{30}$$

What do you observe about the decimals written in (5)?

(7) Write each denominator in factored form.

(8) What do you observe about the factors of the denominators?

(9) Write a statement about terminating and infinitely repeating decimals based on your observations.

CUMULATIVE REVIEW CHAPTERS 1–14

Part I

Answer all questions in this part. Each correct answer will receive 2 credits. No partial credit will be allowed.

1. The product of $3a^2$ and $5a^5$ is
(1) $15a^{10}$ (2) $15a^7$ (3) $8a^{10}$ (4) $8a^7$

2. In the coordinate plane, the point whose coordinates are $(-2, 1)$ is in quadrant
(1) I (2) II (3) III (4) IV

3. In decimal notation, 3.75×10^{-2} is
(1) 0.0375 (2) 0.00375 (3) 37.5 (4) 375

4. The slope of the line whose equation is $3x - y = 5$ is
(1) 5 (2) -5 (3) 3 (4) -3

5. Which of the following is an irrational number?
(1) 1.3 (2) $\frac{2}{3}$ (3) $\sqrt{9}$ (4) $\sqrt{5}$

6. The factors of $x^2 - 7x - 18$ are
(1) 9 and -2
(2) -9 and 2
(3) $(x - 9)$ and $(x + 2)$
(4) $(x + 9)$ and $(x - 2)$

7. The dimensions of a rectangular box are 8 by 5 by 9. The surface area is
(1) 360 cubic units
(2) 360 square units
(3) 157 square units
(4) 314 square units

8. The length of one leg of a right triangle is 8 and the length of the hypotenuse is 12. The length of the other leg is
(1) 4 (2) $4\sqrt{5}$ (3) $4\sqrt{13}$ (4) 80

9. The solution set of $\frac{a^2}{a+1} - 1 = \frac{1}{a+1}$ is
(1) $\{-1, 2\}$ (2) $\{2\}$ (3) $\{0, 1\}$ (4) $\{1, -2\}$

10. In the last n times a baseball player was up to bat, he got 3 hits and struck out the rest of the times. The ratio of hits to strike-outs is
(1) $\frac{3}{n}$ (2) $\frac{n-3}{n}$ (3) $\frac{3}{n-3}$ (4) $\frac{n-3}{3}$

Part II

Answer all questions in this part. Each correct answer will receive 2 credits. Clearly indicate the necessary steps, including appropriate formula substitutions, diagrams, graphs, charts, etc. For all questions in this part, a correct numerical answer with no work shown will receive only 1 credit.

11. Mrs. Kniger bought some stock on May 1 for \$3,500. By June 1, the value of the stock was \$3,640. What was the percent of increase of the cost of the stock?

12. A furlong is one-eighth of a mile. A horse ran 10 furlongs in 2.5 minutes. What was the speed of the horse in feet per second?

Part III

Answer all questions in this part. Each correct answer will receive 3 credits. Clearly indicate the necessary steps, including appropriate formula substitutions, diagrams, graphs, charts, etc. For all questions in this part, a correct numerical answer with no work shown will receive only 1 credit.

13. Two cans of soda and an order of fries cost \$2.60. One can of soda and two orders of fries cost \$2.80. What is the cost of a can of soda and of an order of fries?

14. a. Draw the graph of $y = x^2 - 2x$.

b. From the graph, determine the solution set of the equation $x^2 - 2x = 3$.

Part IV

Answer all questions in this part. Each correct answer will receive 4 credits. Clearly indicate the necessary steps, including appropriate formula substitutions, diagrams, graphs, charts, etc. For all questions in this part, a correct numerical answer with no work shown will receive only 1 credit.

15. If the measure of the smallest angle of a right triangle is 32° and the length of the shortest side is 36.5 centimeters, find the length of the hypotenuse of the triangle to the *nearest tenth of a centimeter.*

16. The area of a garden is 120 square feet. The length of the garden is 1 foot less than twice the width. What are the dimensions of the garden?

CHAPTER 15

PROBABILITY

Mathematicians first studied probability by looking at situations involving games of chance. Today, probability is used in a wide variety of fields. In medicine, it helps us to determine the chances of catching an infection or of controlling an epidemic, and the likelihood that a drug will be effective in curing a disease. In industry, probability tells us how long a manufactured product should last or how many defective items may be expected in a production run. In biology, the study of genes inherited from one's parents and grandparents is a direct application of probability. Probability helps us to predict when more tellers are needed at bank windows, when and where traffic jams are likely to occur, and what kind of weather we may expect for the next few days. While the list of applications is almost endless, all of them demand a strong knowledge of higher mathematics.

As you study this chapter, you will learn to solve problems such as the following: A doctor finds that, as winter approaches, 45% of her patients need flu shots, 20% need pneumonia shots, and 5% need both. What is the probability that the next patient that the doctor sees will need either a flu shot or a pneumonia shot?

Like the early mathematicians, we will begin a formal study of probability by looking at games and other rather simple applications.

Chapter Table of Contents

15-1 EMPIRICAL PROBABILITY

Probability is a branch of mathematics in which the chance of an event happening is assigned a numerical value that predicts how likely that event is to occur. Although this prediction tells us little about what may happen in an individual case, it can provide valuable information about what to expect in a large number of cases.

A decision is sometimes reached by the toss of a coin: "Heads, we'll go to the movies; tails, we'll go bowling." When we toss a coin, we don't know whether the coin will land with the head side or the tail side facing upward. However, we believe that heads and tails have equal chances of happening whenever we toss a fair coin. We can describe this situation by saying that the probability of heads is $\frac{1}{2}$ and the probability of tails is $\frac{1}{2}$, symbolized as:

$$P(\text{heads}) = \tfrac{1}{2} \text{ or } P(\text{H}) = \tfrac{1}{2} \qquad P(\text{tails}) = \tfrac{1}{2} \text{ or } P(\text{T}) = \tfrac{1}{2}$$

Before we define probability, let us consider two more situations.

1. Suppose we toss a coin and it lands heads up. If we were to toss the coin a second time, would the coin land tails up? Is your answer "I don't know"? Good! We cannot say that the coin must now be tails because we cannot predict the next result with certainty.

2. Suppose we take an index card and fold it down the center. If we then toss the card and let it fall, there are only three possible results. The card may land on its side, it may land on its edge, or it may form a tent when it lands. Can we say $P(\text{edge}) = \frac{1}{3}$, $P(\text{side}) = \frac{1}{3}$, and $P(\text{tent}) = \frac{1}{3}$?

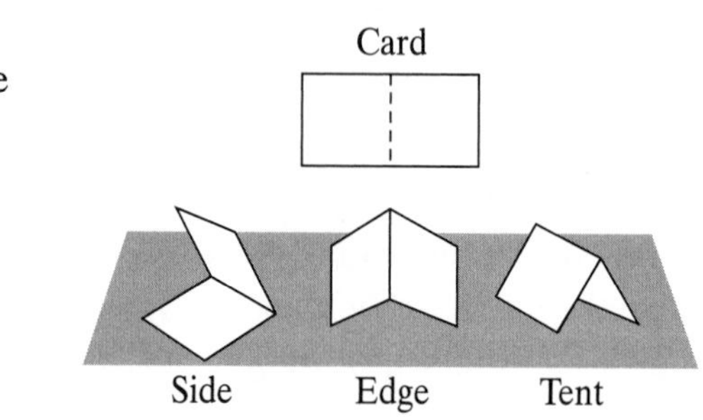

Again, your answer should be "I don't know." We cannot assign a number as a probability until we have some evidence to support our claim. In fact, if we were to gather evidence by tossing this card, we would find that the probabilities are *not* $\frac{1}{3}$, $\frac{1}{3}$, and $\frac{1}{3}$ because, unlike the result of tossing the coin, each result is not equally likely to occur.

Variables that might affect the experiment include the dimensions of the index card, the weight of the cardboard, and the angle measure of the fold. (An index card with a 10° opening would be much less likely to form a tent than an index card with a 110° opening.)

An Empirical Study

Let us go back to the problem of tossing a coin. While we cannot predict the result of one toss of a coin, we can still say that the probability of heads is $\frac{1}{2}$ based on observations made in an empirical study. In an **empirical study**, we perform an experiment many times, keep records of the results, and then analyze these results.

For example, ten students decided to take turns tossing a coin. Each student completed 20 tosses and the number of heads was recorded as shown at the right. If we look at the results and think of the probability of heads as a fraction comparing the number of heads to the total number of tosses, only Maria, with 10 heads out of 20 tosses, had results where the probability was $\frac{10}{20}$, or $\frac{1}{2}$. This fraction is called the **relative frequency**. Elizabeth had the lowest relative frequency of heads, $\frac{6}{20}$. Peter and Debbie tied for the highest relative frequency with $\frac{13}{20}$. Maria's relative frequency was $\frac{10}{20}$ or $\frac{1}{2}$. This does *not* mean that Maria had correct results while the other students were incorrect; the coins simply fell that way.

	Number of Heads	Number of Tosses
Albert	8	20
Peter	13	20
Thomas	12	20
Maria	10	20
Elizabeth	6	20
Joanna	12	20
Kathy	11	20
Jeanne	7	20
Debbie	13	20
James	9	20

The students decided to combine their results, by expanding the chart, to see what happened when all 200 tosses of the coin were considered. As shown in columns 3 and 4 of the table on the next page, each cumulative result is found by adding all the results up to that point. For example, in the second row, by adding the 8 heads that Albert tossed and the 13 heads that Peter tossed, we find the cumulative number of heads up to this point to be 21, the total number of heads tossed by Albert and Peter together. Similarly, when the 20 tosses that Albert made and the 20 tosses that Peter made are added, the cumulative number of tosses up to this point is 40.

Since each student completed 20 coin tosses, the cumulative number of tosses should increase by 20 for each row. The cumulative number of heads should increase by varying amounts for each row since each student experienced different results.

	(COL. 1) Number of Heads	(COL. 2) Number of Tosses	(COL. 3) Cumulative Number of Heads	(COL. 4) Cumulative Number of Tosses	(COL. 5) Cumulative Relative Frequency
Albert	8	20	8	20	$\frac{8}{20} = .400$
Peter	13	20	21	40	$\frac{21}{40} = .525$
Thomas	12	20	33	60	$\frac{33}{60} = .550$
Maria	10	20	43	80	$\frac{43}{80} = .538$
Elizabeth	6	20	49	100	$\frac{49}{100} = .490$
Joanna	12	20	61	120	$\frac{61}{120} = .508$
Kathy	11	20	72	140	$\frac{72}{140} = .514$
Jeanne	7	20	79	160	$\frac{79}{160} = .494$
Debbie	13	20	92	180	$\frac{92}{180} = .511$
James	9	20	101	200	$\frac{101}{200} = .505$

In column 5, the **cumulative relative frequency** is found by dividing the total number of heads at or above a row by the total number of tosses at or above that row. The cumulative relative frequency is shown as a fraction and then, for easy comparison, as a decimal. The decimal is given to the nearest thousandth.

While the relative frequency for individual students varied greatly, from $\frac{6}{20}$ for Elizabeth to $\frac{13}{20}$ for Peter and Debbie, the cumulative relative frequency, after all 200 tosses were combined, was a number very close to $\frac{1}{2}$.

A graph of the results of columns 4 and 5 will tell us even more. In the graph, the horizontal axis is labeled "Number of tosses" to show the cumulative results of column 4; the vertical axis is labeled "Cumulative relative frequency of heads" to show the results of column 5.

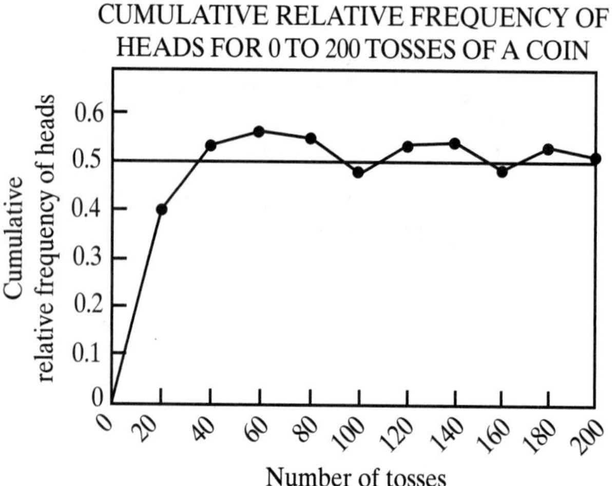

On the graph, we have plotted the points that represent the data in columns 4 and 5 of the preceding table, and we have connected these points to form a line graph. Notice how the line moves up and down around the relative frequency of 0.5, or $\frac{1}{2}$. The graph shows that the more times the coin is tossed, the closer the relative frequency comes to $\frac{1}{2}$. In other words, the line seems to level out at a relative frequency of $\frac{1}{2}$. We say that the cumulative relative frequency **converges** to the number $\frac{1}{2}$ and the coin will land heads up about one-half of the time.

Even though the cumulative relative frequency of $\frac{101}{200}$ is not exactly $\frac{1}{2}$, we sense that the line will approach the number $\frac{1}{2}$. When we use carefully collected evidence about tossing a fair coin to guess that the probability of heads is $\frac{1}{2}$, we have arrived at this conclusion empirically, that is, by experimentation and observation.

▶ ***Empirical probability*** **may be defined as the most accurate scientific estimate, based on a large number of trials, of the cumulative relative frequency of an event happening.**

Experiments in Probability

A single attempt at doing something, such as tossing a coin only once, is called a **trial**. We perform **experiments** in probability by repeating the same trial many times. Experiments are aimed at finding the probabilities to be assigned to the occurrences of an event, such as heads coming up on a coin. The objects used in an experiment may be classified into one of two categories:

1. **Fair and unbiased objects** have not been weighted or made unbalanced. An object is fair when the different possible results have equal chances of happening. Objects such as coins, cards, and spinners will always be treated in this book as fair objects, unless otherwise noted.
2. **Biased objects** have been tampered with or are weighted to give one result a better chance of happening than another. The folded index card described earlier in this section is a biased object because the probability of each of three results is not $\frac{1}{3}$. The card is weighted so that it will fall on its side more often than it will fall on its edge.

You have seen how to determine empirical probability by the tables and graph previously shown in this section. Sometimes, however, it is possible to guess the probability that should be assigned to the result described before you start an experiment.

In Examples 1–5, use common sense to guess the probability that might be assigned to each result. (The answers are given without comment here. You will learn how to determine these probabilities in the next section.)

EXAMPLE 1

A **die** is a six-sided solid object (a cube). Each side (or face) is a square. The sides are marked with 1, 2, 3, 4, 5, and 6 pips, respectively. (The plural of *die* is *dice*.) In rolling a fair die, getting a 4 means that the side showing four pips is facing up. Find the probability of getting a 4, or $P(4)$.

Die

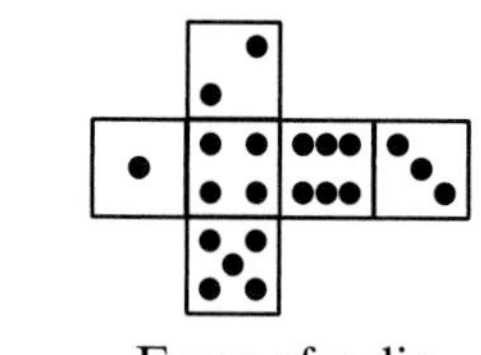

Faces of a die

Answer $P(4) = \frac{1}{6}$

EXAMPLE 2

A **standard deck of cards** contains 52 cards. There are four suits: hearts, diamonds, spades, and clubs. Each suit contains 13 cards: 2, 3, 4, 5, 6, 7, 8, 9, 10, jack, queen, king, and ace. The diamonds and hearts are red; the spades and clubs are black. In selecting a card from the deck without looking, find the probability of drawing:

a. the 7 of diamonds **b.** a 7 **c.** a diamond

Answers **a.** There is only one 7 of diamonds. $P(\text{7 of diamonds}) = \frac{1}{52}$

b. There are four 7s. $P(7) = \frac{4}{52}$ or $\frac{1}{13}$

c. There are 13 diamonds. $P(\text{diamond}) = \frac{13}{52}$ or $\frac{1}{4}$

EXAMPLE 3

There are 10 digits in our numeral system: 0, 1, 2, 3, 4, 5, 6, 7, 8, and 9. In selecting a digit without looking, what is the probability it will be:

a. the 8? **b.** an odd digit?

Answers **a.** $P(8) = \frac{1}{10}$ **b.** $P(\text{odd}) = \frac{5}{10}$ or $\frac{1}{2}$

EXAMPLE 4

A jar contains eight marbles: three are white and the remaining five are blue. In selecting a marble without looking, what is the probability it will be blue? (All marbles are the same size.)

Answer $P(\text{blue}) = \frac{5}{8}$

EXAMPLE 5

The English alphabet contains 26 letters. There are 5 vowels (A, E, I, O, U). The other 21 letters are consonants. If a person turns 26 tiles from a word game facedown and each tile represents a different letter of the alphabet, what is the probability of turning over:

a. the A? **b.** a vowel? **c.** a consonant?

Answers **a.** $P(\text{A}) = \frac{1}{26}$ **b.** $P(\text{vowel}) = \frac{5}{26}$ **c.** $P(\text{consonant}) = \frac{21}{26}$

EXERCISES

Writing About Mathematics

1. Alicia read that, in a given year, one out of four people will be involved in an automobile accident. There are four people in Alicia's family. Alicia concluded that this year, one of the people in her family will be involved in an automobile accident. Do you agree with Alicia's conclusion? Explain why or why not.

2. A library has a collection of 25,000 books. Is the probability that a particular book will be checked out $\frac{1}{25{,}000}$? Explain why or why not.

Developing Skills

In 3–8, in each case, a fair, unbiased object is involved. These questions should be answered without conducting an experiment; take a guess.

3. The six sides of a number cube are labeled 1, 2, 3, 4, 5, and 6. Find $P(5)$, the probability of getting a 5 when the die is rolled.

4. In drawing a card from a standard deck without looking, find $P(\text{any heart})$.

5. Each of 10 pieces of paper contains a different number from the set $\{0, 1, 2, 3, 4, 5, 6, 7, 8, 9\}$. The pieces of paper are folded and placed in a bag. In selecting a piece of paper without looking, find $P(7)$.

6. A jar contains five marbles, all the same size. Two marbles are black, and the other three are white. In selecting a marble without looking, find $P(\text{black})$.

7. Using the lettered tiles from a word game, a boy places 26 tiles, one tile for each letter of the alphabet, facedown on a table. After mixing up the tiles, he picks one. What is the probability that the tile contains one of the letters in the word MATH?

8. A *tetrahedron* is a four-sided object. Each side, or face, is an equilateral triangle. The numerals 1, 2, 3, and 4 are used to number the different faces, as shown in the figure. A trial consists of rolling the tetrahedron and reading the number that is face-down. Find $P(4)$.

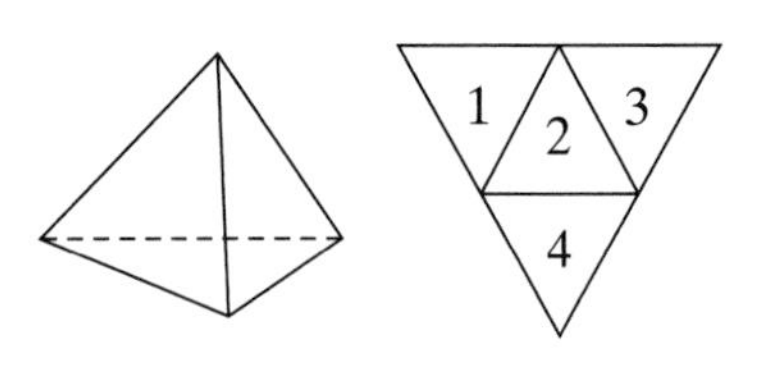

Tetrahedron Faces of a tetrahedron

9. By yourself, or with some classmates, conduct any of the experiments described in Exercises 3–8 to verify that you have assigned the correct probability to the event or result described. A good experiment should contain at least 100 trials.

10. The figure at the right shows a spinner that has an equal chance of landing on one of four sectors. The regions are equal in size and are numbered 1, 2, 3, and 4.

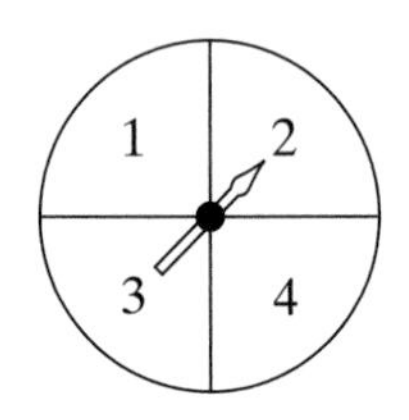

An experiment was conducted by five people to find the probability that the arrow will land on the 2. Each person spun the arrow 100 times. When the arrow landed on a line, the result did not count and the arrow was spun again.

a. Before doing the experiment, what probability would you assign to the arrow landing on the 2? (In symbols, $P(2) = ?$)

b. Copy and complete the table below to find the cumulative results of this experiment. In the last column, record the cumulative relative frequencies as fractions and as decimals to the *nearest thousandth*.

	Number of Times Arrow Landed on 2	Number of Spins	Cumulative Number of Times Arrow Landed on 2	Cumulative Number of Spins	Cumulative Relative Frequency
Barbara	29	100	29	100	$\frac{29}{100} = .290$
Tom	31	100	60	200	
Ann	19	100			
Eddie	23	100			
Cathy	24	100			

c. Did the experiment provide evidence that the probability you assigned in part **a** was correct?

d. Form a group of five people and repeat the experiment, making a table similar to the one shown above. Do your results provide evidence that the probability you assigned in **a** was correct?

In 11–15, a biased object is described. A probability can be assigned to a result only by conducting an experiment to determine the cumulative relative frequency of the event. While you may wish to guess at the probability of the event before starting the experiment, conduct at least 100 trials to determine the best probability to be assigned.

11. An index card is folded in half and tossed. As described earlier in this section, the card may land in one of three positions: on its side, on its edge, or in the form of a tent. In tossing the folded card, find *P*(tent), the probability that the card will form a tent when it lands.

12. A paper cup is tossed. It can land in one of three positions: on its top, on its bottom, or on its side, as shown in the figure. In tossing the cup, find *P*(top), the probability that the cup will land on its top.

Top Bottom Side

13. A nickel and a quarter are glued or taped together so that the two faces seen are the head of the quarter and the tail of the nickel. (This is a very crude model of a weighted coin.) In tossing the coin, find *P*(head).

14. A paper cup in the shape of a cone is tossed. It can land in one of two positions: on its base or on its side, as shown in the figure. In tossing this cup, find *P*(side), the probability that the cup will land on its side.

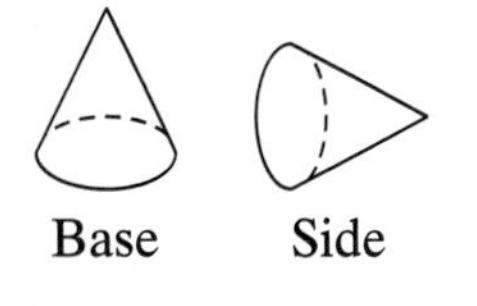

15. A thumbtack is tossed. It may land either with the pin up or with the pin down, as shown in the figure. In tossing the thumbtack, find *P*(pin up).

16. The first word is selected from a page of a book written in English.

a. What is the probability that the word contains at least one of the letters *a*, *e*, *i*, *o*, *u*, or *y*?

b. In general, what is the largest possible probability?

c. What is the probability that the word does *not* contain at least one of the letters *a*, *e*, *i*, *o*, *u*, or *y*?

d. In general, what is the smallest possible probability?

Applying Skills

17. An insurance company's records show that last year, of the 1,000 cars insured by the company, 210 were involved in accidents. What is the probability that an insurance policy, chosen at random from their files, is that of a car that was *not* involved in an accident?

18. A school's attendance records for last year show that of the 885 students enrolled, 15 had no absences for the year. What is the probability that a student, chosen at random, had no absences?

19. A chess club consists of 45 members of whom 24 are boys and 21 are girls. If a member of the club is chosen at random to represent the club at a tournament, what is the probability that the person chosen is a boy?

15-2 THEORETICAL PROBABILITY

An empirical approach to probability is necessary whenever we deal with biased objects. However, common sense tells us that there is a simple way to define the probability of an event when we deal with fair, unbiased objects.

For example, let us suppose that Alma is playing a game in which each player must roll a die. To win, Alma must roll a number greater than 4. What is the probability that Alma will win on her next turn? Common sense tells us that:

1. The die has an equal chance of falling in any one of *six* ways: 1, 2, 3, 4, 5, and 6.

2. There are *two* ways for Alma to win: rolling a 5 or a 6.

3. Therefore: $P(\text{Alma wins}) = \frac{\text{number of winning results}}{\text{number of possible results}} = \frac{2}{6} = \frac{1}{3}$.

Terms and Definitions

Using the details of the preceding example, let us examine the correct terminology to be used.

An **outcome** is a result of some activity or experiment. In rolling a die, 1 is an outcome, 2 is an outcome, 3 is an outcome, and so on. There are six outcomes when rolling a six-sided die.

A **sample space** is a set of all possible outcomes for the activity. When rolling a die, there are six possible outcomes in the sample space: 1, 2, 3, 4, 5, and 6. We say that the sample space is {1, 2, 3, 4, 5, 6}.

An **event** is a subset of the sample space. We use the term *event* in two ways. In ordinary conversation, it means a situation or happening. In the technical language of probability, it is the subset of the sample space that lists all of the outcomes for a given situation.

When we focus on a particular event, such as heads facing up when we toss a coin, we refer to it as the **favorable event**. The other event or events in the sample space, in this case tails, are called **unfavorable**.

When we roll a die, we may define many different situations. Each of these is called an event.

1. For Alma, the event of rolling a number *greater than* 4 contains only two outcomes: 5 and 6.

2. For Lee, a different event might be rolling a number *less than* 5. This event contains four outcomes: 1, 2, 3, and 4.

3. For Sandi, the event of rolling a 2 contains only one outcome: 2. When there is only one outcome, we call this a **singleton event**.

We can now define **theoretical probability** for fair, unbiased objects:

► **The theoretical probability of an event is the number of ways that the event can occur, divided by the total number of possibilities in the sample space.**

In symbolic form, we write:

$$P(E) = \frac{n(E)}{n(S)}$$

where

$P(E)$ represents the probability of event E;
$n(E)$ represents the number of ways event E can occur or the number of outcomes in event E;
$n(S)$ represents the total number of possibilities, or the number of outcomes in sample space S.

Since theoretical probability relies on calculation as opposed to experimentation, it is sometimes referred to as **calculated probability**.

Let us now use this formula to write the probabilities of the three events described above.

1. For Alma, there are two ways to roll a number greater than 4, and there are six possible ways that the die may fall. We say:

E = the set of numbers on a die that are greater than 4: $\{5, 6\}$.
$n(E)$ = 2 since there are 2 outcomes in this event.
S = the set of all possible outcomes: $\{1, 2, 3, 4, 5, 6\}$.
$n(S)$ = 6 since there are six outcomes in the sample space.

Therefore:

$$P(E) = \frac{n(E)}{n(S)} = \frac{2}{6} = \frac{1}{3}$$

2. For Lee, the probability of rolling a number less than 5 is

$$P(E) = \frac{n(E)}{n(S)} = \frac{4}{6} = \frac{2}{3}$$

3. For Sandi, the probability of rolling a 2 is

$$P(E) = \frac{n(E)}{n(S)} = \frac{1}{6}$$

Uniform Probability

A sample space is said to have **uniform probability**, or to contain **equally likely outcomes**, when each of the possible outcomes has an equal chance of occurring. In rolling a die, there are six possible outcomes in the sample space; each is equally likely to occur. Therefore:

$$P(1) = \frac{1}{6}; P(2) = \frac{1}{6}; P(3) = \frac{1}{6}; P(4) = \frac{1}{6}; P(5) = \frac{1}{6}; P(6) = \frac{1}{6}$$

We say that the die has uniform probability.

If, however, a die is weighted to make it biased, then one or more sides will have a probability greater than $\frac{1}{6}$, while one or more sides will have a probability less than $\frac{1}{6}$. A weighted die does not have uniform probability; the rule for theoretical probability does *not* apply to weighted objects.

Random Selection

When we select an object from a collection of objects without knowing any of the special characteristics of the object, we are making a **random selection**. Random selections are made when drawing a marble from a bag, when taking a card from a deck, or when picking a name out of a hat. In the same way, we may use the word *random* to describe the outcomes when tossing a coin or rolling a die; the outcomes happen without any special selection on our part.

Procedure

To find the simple probability of an event:

1. Count the total number of outcomes in the sample space S: $n(S)$.
2. Count all the possible outcomes of the event E: $n(E)$.
3. Substitute these values in the formula for the probability of event E:

$$P(E) = \frac{n(E)}{n(S)}$$

Note: The probability of an event is usually written as a fraction. A standard calculator display, however, is in decimal form. Therefore, if you use a calculator when working with probability, it is helpful to know fraction-decimal equivalents.

Recall that some common fractions have equivalent decimals that are terminating decimals:

$$\frac{1}{2} = 0.5 \qquad \frac{1}{4} = 0.25 \qquad \frac{1}{5} = 0.2 \qquad \frac{1}{8} = 0.125$$

Others have equivalent decimals that are repeating decimals:

$$\tfrac{1}{3} = 0.\overline{3} \qquad \tfrac{2}{3} = 0.\overline{6} \qquad \tfrac{1}{6} = 0.1\overline{6} \qquad \tfrac{1}{9} = 0.\overline{1}$$

The graphing calculator has a function that will change a decimal fraction to a common fraction in lowest terms. For example, the probability of a die showing a number greater than 4 can be displayed on a calculator as follows:

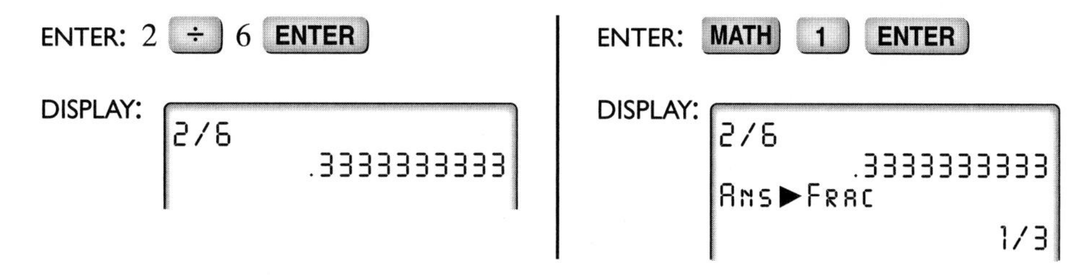

EXAMPLE 1

A standard deck of 52 cards is shuffled. Daniella draws a single card from the deck at random. What is the probability that the card is a jack?

Solution S = sample space of all possible outcomes, or 52 cards. Thus, $n(S) = 52$. J = event of selecting a jack. There are four jacks in the deck: jack of hearts, of diamonds, of spades, and of clubs. Thus, $n(J) = 4$.

$$P(J) = \frac{\text{number of possible jacks}}{\text{number of possible cards}} = \frac{n(J)}{n(S)} = \frac{4}{52} = \frac{1}{13} \textit{ Answer}$$

Calculator Solution ENTER: 4 ÷ 52 MATH 1 ENTER

DISPLAY:

```
4/52▶Frac
                 1/13
```

EXAMPLE 2

An aquarium at a pet store contains 8 goldfish, 6 angelfish, and 7 zebrafish. David randomly chooses a fish to take home for his fishbowl.

a. How many possible outcomes are in the sample space?

b. What is the probability that David takes home a zebrafish?

Solution **a.** Each fish represents a distinct outcome. Therefore, if S is the sample space of all possible outcomes, then $n(S) = 8 + 6 + 7 = 21$.

b. Since there are 7 zebrafish: $P(\text{zebrafish}) = \frac{7}{21} = \frac{1}{3}$.

Answers **a.** 21 **b.** $\frac{1}{3}$

EXAMPLE 3

A spinner contains eight regions, numbered 1 through 8, as shown in the figure. The arrow has an equally likely chance of landing on any of the eight regions. If the arrow lands on a line, the result is not counted and the arrow is spun again.

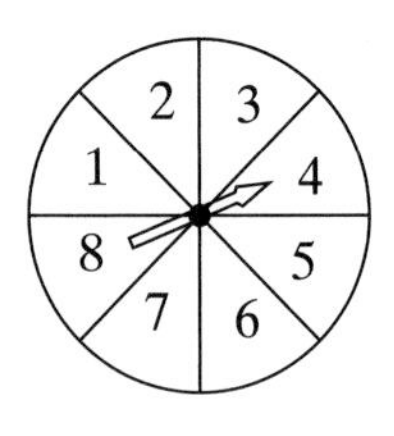

a. How many possible outcomes are in the sample space S?

b. What is the probability that the arrow lands on the 4? That is, what is $P(4)$?

c. List the set of possible outcomes for event O, in which the arrow lands on an odd number.

d. Find the probability that the arrow lands on an odd number.

Answers **a.** $n(S) = 8$

b. Since there is only one region numbered 4 out of eight regions: $P(4) = \frac{1}{8}$

c. Event $O = \{1, 3, 5, 7\}$

d. Since $O = \{1, 3, 5, 7\}$, $n(O) = 4$. $P(O) = \frac{n(O)}{n(S)} = \frac{4}{8} = \frac{1}{2}$

EXERCISES

Writing About Mathematics

1. A spinner is divided into three sections numbered 1, 2, and 3, as shown in the figure. Explain why the probability of the arrow landing on the region numbered 1 is not $\frac{1}{3}$.

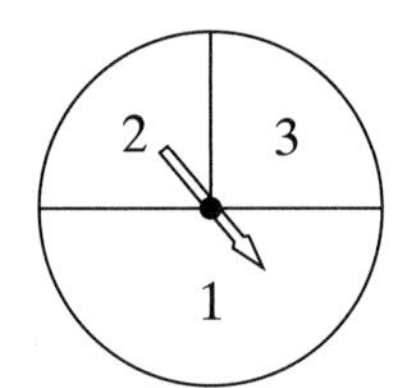

2. Mark said that since there are 50 states, the probability that the next baby born in the United States will be born in New Jersey is $\frac{1}{50}$. Do you agree with Mark? Explain why or why not.

Developing Skills

3. A fair coin is tossed.

a. List the sample space.

b. What is $P(\text{head})$, the probability that a head will appear?

c. What is $P(\text{tail})$?

In 4–9, a fair die is tossed. For each question: **a.** List the possible outcomes for the event. **b.** State the probability of the event.

4. The number 3 appears.

5. An even number appears.

6. A number less than 3 appears.

7. An odd number appears.

8. A number greater than 3 appears.

9. A number greater than or equal to 3 appears.

In 10–15, a spinner is divided into five equal regions, numbered 1 through 5, as shown below. The arrow is spun and lands in one of the regions. For each question:

a. List the outcomes for the event. **b.** State the probability of the event.

10. The arrow lands on number 3.

11. The arrow lands on an even number.

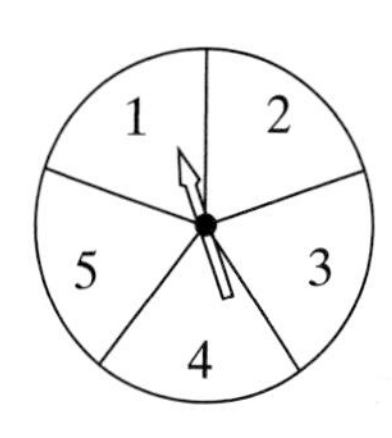

12. The arrow lands on a number less than 3.

13. The arrow lands on an odd number.

14. The arrow lands on a number greater than 3.

15. The arrow lands on a number greater than or equal to 3.

16. A standard deck of 52 cards is shuffled, and one card is drawn. What is the probability that the card is:

a. the queen of hearts? **b.** a queen? **c.** a heart?
d. a red card? **e.** the 7 of clubs? **f.** a club?
g. an ace? **h.** a red 7? **i.** a black 10?
j. a picture card (king, queen, jack)?

17. A person does not know the answer to a test question and takes a guess. Find the probability that the answer is correct if the question is:

a. a multiple-choice question with four choices

b. a true-false question

c. a question where the choices given are "sometimes, always, or never"

18. A marble is drawn at random from a bag. Find the probability that the marble is green if the bag contains marbles whose colors are:

a. 3 blue, 2 green **b.** 4 blue, 1 green **c.** 5 red, 2 green, 3 blue
d. 6 blue, 4 green **e.** 3 green, 9 blue **f.** 5 red, 2 green, 9 blue

19. The digits of the number 1,776 are written on disks and placed in a jar. What is the probability that the digit 7 will be chosen on a single random draw?

20. A letter is chosen at random from a given word. Find the probability that the letter is a vowel if the word is:

a. APPLE **b.** BANANA **c.** GEOMETRY **d.** MATHEMATICS

Applying Skills

21. There are 16 boys and 14 girls in a class. The teacher calls students at random to the chalkboard. What is the probability that the first person called is:

a. a boy? **b.** a girl?

22. There are 840 tickets sold in a raffle. Jay bought five tickets, and Lynn bought four tickets. What is the probability that:

a. Jay has the winning ticket? **b.** Lynn has the winning ticket?

23. The figures that follow are eight polygons: a square; a rectangle; a parallelogram that is not a rectangle; a right triangle; an isosceles triangle that does not contain a right angle; a trapezoid that does not contain a right angle; an equilateral triangle; a regular hexagon.

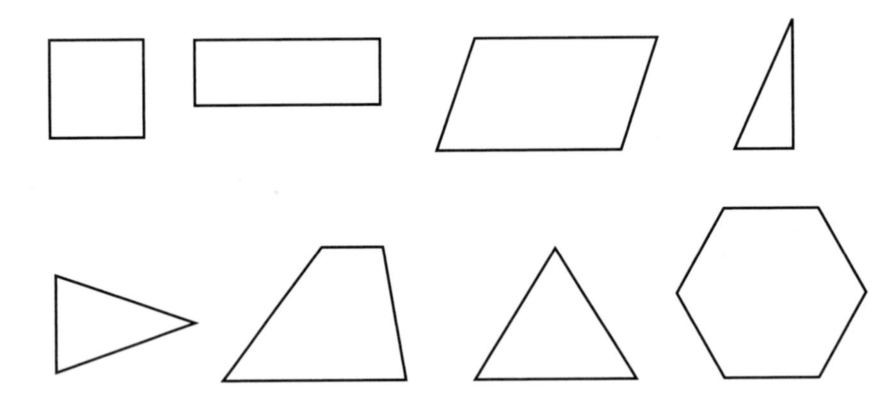

One of the figures is selected at random. What is the probability that this polygon:

a. contains a right angle? **b.** is a quadrilateral?

c. is a triangle? **d.** has at least one acute angle?

e. has all sides congruent? **f.** has at least two sides congruent?

g. has fewer than five sides? **h.** has an odd number of sides?

i. has four or more sides? **j.** has at least two obtuse angles?

15-3 EVALUATING SIMPLE PROBABILITIES

We have called an event for which there is only one outcome a singleton. For example, when rolling a die only once, getting a 3 is a singleton.

However, when rolling a die only once, some events are not singletons. For example:

1. The event of rolling an even number on a die is $\{2, 4, 6\}$.

2. The event of rolling a number less than 6 on a die is $\{1, 2, 3, 4, 5\}$.

The Impossible Case

On a single roll of a die, what is the probability that the number 7 will appear? We call this case an **impossibility** because there is no way in which this event can occur. In this example, event E = rolling a 7; so, $E = \{\,\}$ or $\varnothing$, and $n(E) = 0$. The sample space S for rolling a die contains six possible outcomes, and $n(S) = 6$. Therefore:

$$P(E) = \frac{\text{number of ways to roll a 7}}{\text{number of outcomes for the die}} = \frac{n(E)}{n(S)} = \frac{0}{6} = 0$$

In general, for any sample space S containing k possible outcomes, we say $n(S) = k$. For any impossible event E, which cannot occur in any way, we say $n(E) = 0$. Thus, the probability of an impossible event is:

$$P(E) = \frac{n(E)}{n(S)} = \frac{0}{k} = 0$$

and we say:

▶ **The probability of an impossible event is 0.**

There are many other impossibilities where the probability must equal zero. For example, the probability of selecting the letter E from the word PROBABILITY is 0. Also, selecting a coin worth 9 cents from a bank containing a nickel, a dime, and a quarter is an impossible event.

The Certain Case

On a single roll of a die, what is the probability that a whole number less than 7 will appear? We call this case a **certainty** because every one of the possible outcomes in the sample space is also an outcome for this event. In this example, the event E = rolling a whole number less than 7, so $n(E) = 6$. The sample space S for rolling a die contains six possible outcomes, so $n(S) = 6$. Therefore:

$$P(E) = \frac{\text{number of ways to roll a number less than 7}}{\text{number of outcomes for the die}} = \frac{n(E)}{n(S)} = \frac{6}{6} = 1$$

When an event E is certain, the event E is the same as the sample space S, that is, $E = S$ and $n(E) = n(S)$.

In general, for any sample space S containing k possible outcomes, $n(S) = k$. When the event E is certain, every possible outcome for the sample space is also an outcome for event E, or $n(E) = k$. Thus, the probability of a certainty is given as:

$$P(E) = \frac{n(E)}{n(S)} = \frac{k}{k} = 1$$

and we say:

▶ **The probability of an event that is certain to occur is 1.**

There are many other certainties where the probability must equal 1. Examples include the probability of selecting a consonant from the letters JFK or selecting a red sweater from a drawer containing only red sweaters.

The Probability of Any Event

The smallest possible probability is 0, for an impossible case; no probability can be less than 0. The largest possible probability is 1, for a certain event; no probability can be greater than 1. Many other events, as seen earlier, however, have probabilities that fall between 0 and 1. Therefore, we conclude:

► **The probability of any event E must be equal to or greater than 0, and less than or equal to 1:**

$$0 \leq P(E) \leq 1$$

Subscripts in Sample Spaces

A sample space may sometimes contain two or more objects that are exactly alike. To distinguish one object from another, we make use of subscripts. A **subscript** is a number, usually written in smaller size to the lower right of a term.

For example, a box contains six jellybeans: two red, three green, and one yellow. Using R, G, and Y to represent the colors red, green, and yellow, respectively, we can list this sample space in full, using subscripts:

$$\{R_1, R_2, G_1, G_2, G_3, Y_1\}$$

Since there is only one yellow jellybean, we could have listed the last outcome as Y instead of Y_1.

EXAMPLE 1

An arrow is spun once and lands on one of three equally likely regions, numbered 1, 2, and 3, as shown in the figure.

a. List the sample space for this experiment.

b. List all eight possible events for one spin of the arrow.

Solution **a.** The sample space $S = \{1, 2, 3\}$.

b. Since events are subsets of the sample space S, the eight possible events are the eight subsets of S:

{ }, the empty set for *impossible* events.
The arrow lands on a number other than 1, 2, or 3.

{1}, a *singleton*.
The arrow lands on 1 or the arrow lands on a number less than 2.

{2}, a *singleton*.
The arrow lands on 2 or the arrow lands on an even number.

{3}, a *singleton*.
The arrow lands on 3 or the arrow lands on a number greater than 2.

{1, 2}, an event with two possible outcomes.
The arrow does not land on 3 or does land on a number less than 3.

{1, 3}, an event with two possible outcomes.
The arrow lands on an odd number or does not land on 2.

{2, 3}, an event with two possible outcomes.
The arrow lands on a number greater than 1 or does not land on 1.

{1, 2, 3}, the sample space itself for events that are certain.
The arrow lands on a whole number less than 4 or greater than 0.

Answers **a.** $S = \{1, 2, 3\}$ **b.** { }, {1}, {2}, {3}, {1, 2}, {1, 3}, {2, 3}, {1, 2, 3}

EXAMPLE 2

A piggy bank contains a nickel, two dimes, and a quarter. A person selects one of the coins. What is the probability that the coin is worth:

a. exactly 10 cents? **b.** at least 10 cents?

c. exactly 3 cents? **d.** more than 3 cents?

Solution The sample space for this example is $\{N, D_1, D_2, Q\}$. Therefore, $n(S) = 4$.

a. There are two coins worth exactly 10 cents, D_1 and D_2. Therefore, $n(E) = 2$ and

$$P(\text{coin worth 10 cents}) = \frac{n(E)}{n(S)} = \frac{2}{4} = \frac{1}{2}.$$

b. There are three coins worth at least 10 cents, D_1, D_2, and Q. Therefore, $n(E) = 3$ and

$$P(\text{coin worth at least 10 cents}) = \frac{n(E)}{n(S)} = \frac{3}{4}.$$

c. There is no coin worth exactly 3 cents. This is an impossible event. Therefore,

$$P(\text{coin worth 3 cents}) = 0.$$

d. Each of the four coins is worth more than 3 cents. This is a certain event. Therefore,

$$P(\text{coin worth more than 3 cents}) = 1.$$

Answers **a.** $\frac{1}{2}$ **b.** $\frac{3}{4}$ **c.** 0 **d.** 1

EXAMPLE 3

In the Sullivan family, there are two more girls than boys. At random, Mrs. Sullivan asks one of her children to go to the store. If she is equally likely to ask any one of her children, and the probability that she asks a girl is $\frac{2}{3}$, how many boys and how many girls are there in the Sullivan family?

Solution Let $x =$ the number of boys

$x + 2 =$ the number of girls

$2x + 2 =$ the number of children.

Then:

$$P(\text{girl}) = \frac{\text{number of girls}}{\text{number of children}}$$

$$\frac{2}{3} = \frac{x+2}{2x+2}$$

$$2(2x + 2) = 3(x + 2)$$

$$4x + 4 = 3x + 6$$

$$x = 2$$

$$\text{Then } x + 2 = 4$$

$$\text{and } 2x + 2 = 6$$

Check

$$P(\text{girl}) = \frac{\text{number of girls}}{\text{number of children}}$$

$$\frac{2}{3} \stackrel{?}{=} \frac{4}{6}$$

$$\frac{2}{3} = \frac{2}{3} \checkmark$$

Answer There are two boys and four girls.

EXERCISES

Writing About Mathematics

1. Describe three events for which the probability is 0.

2. Describe three events for which the probability is 1.

Developing Skills

3. A fair coin is tossed, and its sample space is $S = \{H, T\}$.

a. List all four possible events for the toss of a fair coin.

b. Find the probability of each event named in part **a**.

In 4–11, a spinner is divided into seven equal sectors, numbered 1 through 7. An arrow is spun to fall into one of the regions. For each question, find the probability that the arrow lands on the number described.

4. the number 5

5. an even number

6. a number less than 5

7. an odd number

8. a number greater than 5

9. a number greater than 1

10. a number greater than 7

11. a number less than 8

12. A marble is drawn at random from a bag. Find the probability that the marble is black if the bag contains marbles whose colors are:

a. 5 black, 2 green **b.** 2 black, 1 green **c.** 3 black, 4 green, 1 red

d. 9 black **e.** 3 green, 4 red **f.** 3 green

13. Ted has two quarters, three dimes, and one nickel in his pocket. He pulls out a coin at random. Find the probability that the coin is worth:

a. exactly 5 cents **b.** exactly 10 cents **c.** exactly 25 cents

d. exactly 50 cents **e.** less than 25 cents **f.** less than 50 cents

g. more than 25 cents **h.** more than 1 cent **i.** less than 1 cent

14. A single fair die is rolled. Find the probability for each event.

a. The number 8 appears. **b.** A whole number appears.

c. The number is less than 5. **d.** The number is less than 1.

e. The number is less than 10. **f.** The number is negative.

15. A standard deck of 52 cards is shuffled, and a card is picked at random. Find the probability that the card is:

a. a jack **b.** a club **c.** a star

d. a red club **e.** a card from the deck **f.** a black club

g. the jack of stars **h.** a 17 **i.** a red 17

In 16–20, a letter is chosen at random from a given word. For each question: **a.** Write the sample space, using subscripts to designate events if needed. **b.** Find the probability of the event.

16. Selecting the letter E from the word EVENT

17. Selecting the letter S from the word MISSISSIPPI

18. Selecting a vowel from the word TRIANGLE

19. Selecting a vowel from the word RECEIVE

20. Selecting a consonant from the word SPRY

Applying Skills

21. There are 15 girls and 10 boys in a class. The teacher calls on a student in the class at random to answer a question. Express, *in decimal form*, the probability that the student called upon is:

a. a girl **b.** a boy **c.** a pupil in the class

d. a person who is not a student in the class

22. The last digit of a telephone number can be any of the following: 0, 1, 2, 3, 4, 5, 6, 7, 8, or 9. Express, *as a percent*, the probability that the last digit is:

a. 7 **b.** odd **c.** greater than 5 **d.** a whole number **e.** the letter R

23. A girl is holding five cards in her hand: the 3 of hearts, the 3 of diamonds, the 3 of clubs, the 4 of diamonds, the 7 of clubs. A player to her left takes one of these cards at random. Find the probability that the card selected from the five cards in the girl's hand is:

a. a 3 **b.** a diamond **c.** a 4
d. a black 4 **e.** a club **f.** the 4 of hearts
g. a 5 **h.** the 7 of clubs **i.** a red card
j. a number card **k.** a spade **l.** a number greater than 1 and less than 8.

24. A sack contains 20 marbles. The probability of drawing a green marble is $\frac{2}{5}$. How many green marbles are in the sack?

25. There are three more boys than girls in the chess club. A member of the club is to be chosen at random to play in a tournament. Each member is equally likely to be chosen. If the probability that a girl is chosen is $\frac{3}{7}$, how many boys and how many girls are members of the club?

26. A box of candy contains caramels and nut clusters. There are six more caramels than nut clusters. If a piece of candy is to be chosen at random, the probability that it will be a caramel is $\frac{3}{5}$. How many caramels and how many nut clusters are in the box?

27. At a fair, each ride costs one, two, or four tickets. The number of rides that cost two tickets is three times the number of rides that cost one ticket. Also, seven more rides cost four tickets than cost two tickets. Tycella, who has a book of tickets, goes on a ride at random. If the probability that the ride cost her four tickets is $\frac{4}{7}$, how many rides are there at the fair?

15-4 THE PROBABILITY OF (*A* AND *B*)

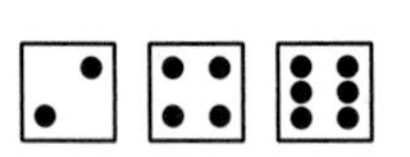

If a fair die is rolled, we can find simple probabilities, since we know that $S = \{1, 2, 3, 4, 5, 6\}$. For example, let event A be rolling an even number. Then $A = \{2, 4, 6\}$ and

$$P(A) = \frac{n(A)}{n(S)} = \frac{3}{6}$$

Let event B be rolling a number less than 3. Then $B = \{1, 2\}$ and

$$P(B) = \frac{n(B)}{n(S)} = \frac{2}{6}$$

Now, what is the probability of obtaining a number on the die that is even and less than 3? We may think of this as event (A *and* B), which consists of those elements of the sample space that are in A and also in B. In set notation we say,

$$(A \text{ and } B) = A \cap B = \{2\}$$

Only 2 is both even and less than 3. Notice that we use the symbol for intersection ($\cap$) to denote *and*. Since $n(A \text{ and } B) = 1$,

$$P(A \text{ and } B) = \frac{n(A \text{ and } B)}{n(S)} = \frac{1}{6}$$

Let us consider another example in which a fair die is rolled. What is the probability of rolling a number that is both odd *and* a 4?

Event $C = \{1, 3, 5\}$. Three numbers on a die are odd.

$$P(C) = \frac{n(C)}{n(S)} = \frac{3}{6}$$

Event $D = \{4\}$. One number on the die is 4.

$$P(D) = \frac{n(D)}{n(S)} = \frac{1}{6}$$

$(C \text{ and } D) = \{\text{numbers on a die that are odd } \textit{and } 4\} = C \cap D = \varnothing$, the empty set.

Since there is *no* outcome common to both C and D, $n(C \text{ and } D) = 0$. Therefore,

$$P(C \cap D) = \frac{n(C \cap D)}{n(S)} = \frac{0}{6} = 0$$

Conclusions

There is no simple rule or formula whereby the $n(A)$ and $n(B)$ can be used to find $n(A \text{ and } B)$. We must simply count the number of outcomes that are common in both events or write the intersection of the two sets and count the number of elements in that intersection.

KEEP IN MIND Event $(A \text{ and } B)$ consists of the outcomes that are in event A *and* in event B. Event $(A \text{ and } B)$ may be regarded as the *intersection* of sets, namely, $A \cap B$.

EXAMPLE 1

A fair die is rolled once. Find the probability of obtaining a number that is greater than 3 and less than 6.

Solution Event $A = \{\text{numbers greater than } 3\} = \{4, 5, 6\}$.

Event $B = \{\text{numbers less than } 6\} = \{1, 2, 3, 4, 5\}$.

Event $(A \text{ and } B) = \{\text{numbers greater than 3 } \textit{and} \text{ less than } 6\} = \{4, 5\}$.

Therefore: $P(A \text{ and } B) = \frac{n(A \text{ and } B)}{n(S)} = \frac{2}{6}$ or $P(A \cap B) = \frac{n(A \cap B)}{n(S)} = \frac{2}{6}$.

Answer $P(\text{number greater than 3 and less than } 6) = \frac{2}{6}$ or $\frac{1}{3}$

EXERCISES

Writing About Mathematics

1. Give an example of two events A and B, such that $P(A \text{ and } B) = P(A)$.

2. If $P(A \text{ and } B) = P(A)$, what must be the relationship between set A and set B?

Developing Skills

3. A fair die is rolled once. The sides are numbered 1, 2, 3, 4, 5, and 6. Find the probability that the number rolled is:

a. greater than 2 and odd
b. less than 4 and even
c. greater than 2 and less than 4
d. less than 2 and even
e. less than 6 and odd
f. less than 4 and greater than 3

4. From a standard deck of cards, one card is drawn. Find the probability that the card is:

a. the king of hearts
b. a red king
c. a king of clubs
d. a black jack
e. a 10 of diamonds
f. a red club
g. a 2 of spades
h. a black 2
i. a red picture card

Applying Skills

5. A set of polygons consists of an equilateral triangle, a square, a rhombus that is not a square, and a rectangle. One of the polygons is selected at random.

Find the probability that the polygon contains:

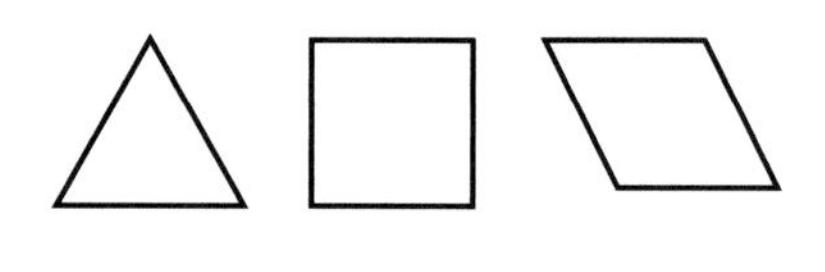

a. all sides congruent and all angles congruent
b. all sides congruent and all right angles
c. all sides congruent and two angles not congruent
d. at least two congruent sides and at least two congruent angles
e. at least three congruent sides and at least two congruent angles

6. In a class of 30 students, 23 take science, 28 take math, and all take either science or math.

a. How many students take both science *and* math?
b. A student from the class is selected at random. Find:

(1) P(takes science) (2) P(takes math) (3) P(takes science *and* math)

7. At a karaoke party, some of the boys and girls take turns singing songs. Of the five boys, Patrick and Terence are teenagers while Brendan, Drew, and Kevin are younger. Of the seven girls, Heather and Claudia are teenagers while Maureen, Elizabeth, Gwen, Caitlin, and Kelly are younger. Find the probability that the first song is sung by:

   a. a girl
   b. a boy
   c. a teenager
   d. someone under 13 years old
   e. a boy under 13
   f. a girl whose initial is C
   g. a teenage girl
   h. a girl under 13
   i. a boy whose initial is C
   j. a teenage boy

15-5 THE PROBABILITY OF (*A* OR *B*)

If a fair die is rolled, we can find simple probabilities. For example, let event A be rolling an even number. Then:

$$P(A) = \frac{n(A)}{n(S)} = \frac{3}{6}$$

Let event C be rolling a number less than 2. Then:

$$P(C) = \frac{n(C)}{n(S)} = \frac{1}{6}$$

Now, what is the probability of obtaining a number on the die that is even or less than 2? We may think of this as event (A *or* C).

For the example above, there are four outcomes in the event (A or C): 1, 2, 4, and 6. Each of these numbers is either even or less than 2 or both. Since $n(A \text{ or } C) = 4$ and there are six elements in the sample space:

$$P(A \text{ or } C) = \frac{n(A \text{ or } C)}{n(S)} = \frac{4}{6}$$

Observe that $P(A) = \frac{3}{6}$, $P(C) = \frac{1}{6}$ and $P(A \text{ or } C) = \frac{4}{6}$. In this case, it appears that

$$P(A) + P(C) = P(A \text{ or } C).$$

Will this simple addition rule hold true for all problems? Before you say "yes," consider the next example, in which a fair die is rolled.

Event $A = \{\text{even numbers on a die}\} = \{2, 4, 6\}$.

$$P(A) = \frac{n(A)}{n(S)} = \frac{3}{6}$$

Event $B = \{\text{numbers less than 3 on a die}\} = \{1, 2\}$.

$$P(B) = \frac{n(B)}{n(S)} = \frac{2}{6}$$

Then, event $(A \text{ or } B) = \{\text{numbers that are even or less than 3}\}$.

$$P(A \text{ or } B) = \frac{n(A \text{ or } B)}{n(S)} = \frac{4}{6}$$

Here $P(A) = \frac{3}{6}$, $P(B) = \frac{2}{6}$, and $P(A \text{ or } B) = \frac{4}{6}$. In this case, the simple rule of addition does not work: $P(A) + P(B) \neq P(A \text{ or } B)$. What makes this example different from $P(A \text{ or } C)$, shown previously?

A Rule for the Probability of (*A* or *B*)

Probability is based on the number of outcomes in a given event. For the event (A or B) in our example, we observe that the outcome 2 is found in event A *and* in event B. Therefore, we may describe rolling a 2 as the event (A and B).

The simple addition rule does not work for the event (A or B) because we have counted the event (A and B) twice: first in event A, then again in event B. In order to count the event (A and B) only once, we must subtract the number of shared elements, $n(A \text{ and } B)$, from the overall number of elements in (A or B).

Thus, the rule becomes: $n(A \text{ or } B) = n(A) + n(B) - n(A \text{ and } B)$

For this example: $n(A \text{ or } B) = 3 + 2 - 1 = 4$

Dividing each term by $n(S)$, we get an equivalent equation:

For this example:
$$\frac{n(A \text{ or } B)}{n(S)} = \frac{n(A)}{n(S)} + \frac{n(B)}{n(S)} - \frac{n(A \text{ and } B)}{n(S)}$$
$$\frac{n(A \text{ or } B)}{n(S)} = \frac{3}{6} + \frac{2}{6} - \frac{1}{6} = \frac{4}{6}$$

Since $P(A \text{ or } B) = \frac{n(A \text{ or } B)}{n(S)}$, we can write a general rule:

$$P(A \text{ or } B) = P(A) + P(B) - P(A \text{ and } B)$$

In set terminology, the rule for probability becomes:

$$P(A \text{ or } B) = P(A \cup B) = P(A) + P(B) - P(A \cap B)$$

Note the use of the union symbol ($\cup$) to indicate *or*.

Mutually Exclusive Events

We have been examining three different events that occur when a die is tossed.

Event A = {an even number} = $\{2, 4, 6\}$

Event B = {a number less than 3} = $\{1, 2\}$

Event C = {a number less than 2} = $\{1\}$

We found that

$$P(A \text{ or } C) = P(A) + P(C)$$
$$P(A \text{ or } B) = P(A) + P(B) - P(A \text{ and } B)$$

Why are these results different?

Of these sets, A and C are *disjoint*, that is they have no element in common. Events A and C are said to be **mutually exclusive events** because only one of the events can occur at any one throw of the dice. Events that are disjoint sets are mutually exclusive.

If two events A and C are mutually exclusive:

$$\mathbf{P(A \text{ or } C) = P(A) + P(C)}$$

Sets A and B are *not* disjoint sets. They have element 2 in common. Events A and B are *not* mutually exclusive events because both can occur at any one throw of the dice. Events that are not disjoint sets are not mutually exclusive.

If two events A and B are not mutually exclusive:

$$\mathbf{P(A \text{ or } B) = P(A) + P(B) - P(A \text{ and } B)}$$

Some examples of mutually exclusive events include the following.

- Drawing a spade or a red card when one card is drawn from a standard deck.
- Choosing a ninth-grade boy or a tenth-grade girl from a student body that consists of boys and girls in each grade 9 through 12.
- Drawing a quarter or a dime when one coin is drawn from a purse that contains three quarters, two dimes, and a nickel.
- Drawing a consonant or a vowel when a letter is drawn from the word PROBABILITY.

Some examples of events that are not mutually exclusive include the following.

- Drawing an ace or a red card when one card is drawn from a standard deck.
- Choosing a girl or a ninth-grade student from a student body that consists of boys and girls in each grade 9 through 12.
- Drawing a dime or a coin worth more than five cents when one coin is drawn from a purse that contains a quarter, two dimes, and a nickel.
- Drawing a Y or a letter that follows L in the alphabet when a letter is drawn from the word PROBABILITY.

EXAMPLE 1

A standard deck of 52 cards is shuffled, and one card is drawn at random. Find the probability that the card is:

a. a king or an ace **b.** red or an ace

Solution **a.** There are four kings in the deck, so $P(\text{king}) = \frac{4}{52}$.

There are four aces in the deck, so $P(\text{ace}) = \frac{4}{52}$.

These are mutually exclusive events. The set of kings and the set of aces are disjoint sets, having no elements in common.

$$P(\text{king or ace}) = P(\text{king}) + P(\text{ace}) = \frac{4}{52} + \frac{4}{52} = \frac{8}{52} \text{ or } \frac{2}{13} \quad \textit{Answer}$$

b. There are 26 red cards in the deck, so $P(\text{red}) = \frac{26}{52}$.

There are four aces in the deck, so $P(\text{ace}) = \frac{4}{52}$.

There are two red aces in the deck, so $P(\text{red and ace}) = \frac{2}{52}$.

These are not mutually exclusive events. Two cards are both red and ace.

$$\begin{aligned} P(\text{red or ace}) &= P(\text{red}) + P(\text{ace}) - P(\text{red and ace}) \\ &= \tfrac{26}{52} + \tfrac{4}{52} - \tfrac{2}{52} = \tfrac{28}{52} \text{ or } \tfrac{7}{13} \quad \textit{Answer} \end{aligned}$$

Alternative Solution There are 26 red cards and two more aces not already counted (the ace of spades, the ace of clubs). Therefore, there are 26 + 2, or 28, cards in this event. Then:

$$P(\text{red or ace}) = \frac{28}{52} \text{ or } \frac{7}{13} \quad \textit{Answer}$$

EXAMPLE 2

There are two events, A and B. Given that $P(A) = .3$, $P(B) = .5$, and $P(A \cap B) = .1$, find $P(A \cup B)$.

Solution Since the probability of $A \cap B$ is not 0, A and B are not mutually exclusive.

$$\begin{aligned} P(A \cup B) &= P(A) + P(B) - P(A \cap B) \\ &= .3 + .5 - .1 = .7 \end{aligned}$$

Answer $P(A \cup B) = .7$

EXAMPLE 3

A town has two newspapers, the *Times* and the *Chronicle*. One out of every two persons in the town subscribes to the *Times*, three out of every five persons in the town subscribe to the *Chronicle*, and three out of every ten persons in the town subscribe to both papers. What is the probability that a person in the town chosen at random subscribes to the *Times* or the *Chronicle*?

Solution Subscribing to the *Times* and subscribing to the *Chronicle* are not mutually exclusive events.

$$P(\textit{Times}) = \tfrac{1}{2} \qquad P(\textit{Chronicle}) = \tfrac{3}{5} \qquad P(\textit{Times} \text{ and } \textit{Chronicle}) = \tfrac{3}{10}$$

$$\begin{aligned} P(\textit{Times} \text{ or } \textit{Chronicle}) &= P(\textit{Times}) + P(\textit{Chronicle}) - P(\textit{Times} \text{ and } \textit{Chronicle}) \\ &= \tfrac{1}{2} + \tfrac{3}{5} - \tfrac{3}{10} \\ &= \tfrac{5}{10} + \tfrac{6}{10} - \tfrac{3}{10} = \tfrac{8}{10} \text{ or } \tfrac{4}{5} \text{ Answer} \end{aligned}$$

EXERCISES

Writing About Mathematics

1. Let A and B be two events. Is it possible for $P(A \text{ or } B)$ to be less that $P(A)$? Explain why or why not.

2. If B is a subset of A, which of the following is true: $P(A \text{ or } B) < P(A)$, $P(A \text{ or } B) = P(A)$, $P(A \text{ or } B) > P(A)$? Explain your answer.

Developing Skills

3. A spinner consists of five equal sectors of a circle. The sectors are numbered 1 through 5, and when an arrow is spun, it is equally likely to stop on any sector. For a single spin of the arrow, determine whether or not the events are mutually exclusive and find the probability that the number on the sector is:

a. 3 or 4 **b.** odd or 2 **c.** 4 or less **d.** 2 or 3 or 4 **e.** odd or 3

4. A fair die is rolled once. Determine whether or not the events are mutually exclusive and find the probability that the number rolled is:

a. 3 or 4
b. odd or 2
c. 4 or less than 4
d. 2, 3, or 4
e. odd or 3
f. less than 2 or more than 5
g. less than 5 or more than 2
h. even or more than 3

5. From a standard deck of cards, one card is drawn. Determine whether or not the events are mutually exclusive and find the probability that the card will be:

a. a queen or an ace
b. a queen or a 7
c. a heart or a spade
d. a queen or a spade
e. a queen or a red card
f. a jack or a queen or a king
g. a 7 or a diamond
h. a club or a red card
i. an ace or a picture card

6. A bank contains two quarters, six dimes, three nickels, and five pennies. A coin is drawn at random. Determine whether or not the events are mutually exclusive and find the probability that the coin is:

a. a quarter or a dime
b. a dime or a nickel
c. worth 10 cents or more than 10 cents
d. worth 10 cents or less
e. worth 1 cent or more
f. a quarter, a nickel, or a penny

In 7–12, in each case choose the numeral preceding the expression that best completes the statement or answers the question.

7. If a single card is drawn from a standard deck, what is the probability that it is a 4 or a 9?

(1) $\frac{2}{52}$ (2) $\frac{8}{52}$ (3) $\frac{13}{52}$ (4) $\frac{26}{52}$

8. If a single card is drawn from a standard deck, what is the probability that it is a 4 or a diamond?

(1) $\frac{8}{52}$ (2) $\frac{16}{52}$ (3) $\frac{17}{52}$ (4) $\frac{26}{52}$

9. If $P(A) = .2$, $P(B) = .5$, and $P(A \cap B) = .1$, then $P(A \cup B) =$

(1) .6 (2) .7 (3) .8 (4) .9

10. If $P(A) = \frac{1}{3}$, $P(B) = \frac{1}{2}$, and $P(A \text{ and } B) = \frac{1}{6}$, then $P(A \text{ or } B) =$

(1) $\frac{2}{5}$ (2) $\frac{2}{3}$ (3) $\frac{5}{6}$ (4) 1

11. If $P(A) = \frac{1}{4}$, $P(B) = \frac{1}{2}$, and $P(A \cap B) = \frac{1}{8}$, then $P(A \cup B) =$

(1) $\frac{1}{8}$ (2) $\frac{5}{8}$ (3) $\frac{3}{4}$ (4) $\frac{7}{8}$

12. If $P(A) = .30$, $P(B) = .35$, and $(A \cap B) = \varnothing$, then $P(A \cup B) =$

(1) .05 (2) .38 (3) .65 (4) 0

Applying Skills

13. Linda and Aaron recorded a CD together. Each sang some solos, and both of them sang some duets. Aaron recorded twice as many duets as solos, and Linda recorded six more solos than duets. If a CD player selects one of these songs at random, the probability that it will select a duet is $\frac{1}{4}$. Find the number of:

a. solos by Aaron **b.** solos by Linda **c.** duets

14. In a sophomore class of 340 students, some students study Spanish, some study French, some study both languages, and some study neither language. If $P(\text{Spanish}) = .70$, $P(\text{French}) = .40$, and $P(\text{Spanish } \textit{and} \text{ French}) = .25$, find:

a. the probability that a sophomore studies Spanish *or* French

b. the number of sophomores who study one or more of these languages

15. The Greenspace Company offers lawn care services and snow plowing in the appropriate seasons. Of the 600 property owners in town, 120 have contracts for lawn care, 90 for snow plowing, and 60 for both with the Greenspace Company. A new landscape company, the Earthpro Company offers the same services and begins a telephone campaign to attract customers, choosing telephone numbers of property owners at random. What is the probability that the Earthpro Company reaches someone who has a contract for lawn care or snow plowing with the Greenspace Company?

15-6 THE PROBABILITY OF (NOT *A*)

In rolling a fair die, we know that $P(4) = \frac{1}{6}$ since there is only one outcome for the event (rolling a 4). We can also say that $P(not\ 4) = \frac{5}{6}$ since there are five outcomes for the event (not rolling a 4): 1, 2, 3, 5, and 6.

We can think of these probabilities in another way. The event (4) and the event (not 4) are mutually exclusive events. Also, the event (4 or not 4) is a certainty whose probability is 1.

$$\begin{aligned} P(4 \text{ or not } 4) &= P(4) + P(\text{not } 4) \\ 1 &= P(4) + P(\text{not } 4) \\ 1 - P(4) &= P(\text{not } 4) \\ 1 - \tfrac{1}{6} &= P(\text{not } 4) \\ \tfrac{5}{6} &= P(\text{not } 4) \end{aligned}$$

The event (not A) is the **complement** of event A, when the universal set is the sample space, S.

In general, if $P(A)$ is the probability that some given result will occur, and $P(\text{not } A)$ is the probability that the given result will not occur, then:

1. $P(A) + P(\text{not } A) = 1$
2. $P(A) = 1 - P(\text{not } A)$
3. $P(\text{not } A) = 1 - P(A)$

Probability as a Sum

When sets are disjoint, we have seen that the probability of the union can be found by the rule $P(A \cup B) = P(A) + P(B)$. Since the possible outcomes that are singletons represent disjoint sets, we can say:

► **The probability of any event is equal to the sum of the probabilities of the singleton outcomes in the event.**

For example, when we draw a card from a standard deck, there are 52 singleton outcomes, each with a probability of $\frac{1}{52}$. Since all singleton events are disjoint, we can say:

$$\begin{aligned} P(\text{king}) &= P(\text{king of hearts}) + P(\text{king of diamonds}) + P(\text{king of spades}) \\ &\quad + P(\text{king of clubs}) \\ &= \tfrac{1}{52} + \tfrac{1}{52} + \tfrac{1}{52} + \tfrac{1}{52} \\ P(\text{king}) &= \tfrac{4}{52} \text{ or } \tfrac{1}{13} \end{aligned}$$

We also say:

▶ **The sum of the probabilities of all possible singleton outcomes for any sample space must always equal 1.**

For example, in tossing a coin,

$$P(S) = P(\text{heads}) + P\ (\text{tails}) = \tfrac{1}{2} + \tfrac{1}{2} = 1.$$

Also, in rolling a die,

$$\begin{aligned} P(S) &= P(1) + P(2) + P(3) + P(4) + P(5) + P(6) \\ &= \tfrac{1}{6} + \tfrac{1}{6} + \tfrac{1}{6} + \tfrac{1}{6} + \tfrac{1}{6} + \tfrac{1}{6} \\ &= 1 \end{aligned}$$

EXAMPLE 1

Dr. Van Brunt estimates that 4 out every 10 patients that he will see this week will need a flu shot. What is the probability that the next patient he sees will *not* need a flu shot?

Solution The probability that a patient will need a flu shot is $\frac{4}{10}$ or $\frac{2}{5}$.

The probability that a patient will not need a flu shot is $1 - \frac{2}{5} = \frac{3}{5}$. *Answer*

EXAMPLE 2

A letter is drawn at random from the word ERROR.

a. Find the probability of drawing each of the different letters in the word.

b. Demonstrate that the sum of these probabilities is 1.

Solution **a.** $P(E) = \frac{1}{5}$; $P(\text{R}) = \frac{3}{5}$; $P(\text{O}) = \frac{1}{5}$ *Answer*

b. $P(\text{E}) + P(\text{R}) + P(\text{O}) = \frac{1}{5} + \frac{3}{5} + \frac{1}{5} = \frac{5}{5} = 1$ *Answer*

EXERCISES

Writing About Mathematics

1. If event A is a certainty, $P(A) = 1$. What must be true about $P(\text{not } A)$? Explain your answer.

2. If A and B are disjoint sets, what is $P(\text{not } A \text{ or not } B)$? Explain your answer.

Developing Skills

3. A fair die is rolled once. Find each probability:

a. $P(3)$ **b.** $P(\text{not } 3)$ **c.** $P(\text{even})$
d. $P(\text{not even})$ **e.** $P(\text{less than } 3)$ **f.** $P(\text{not less than } 3)$
g. $P(\text{odd or even})$ **h.** $P[\text{not (odd or even)}]$ **i.** $P[\text{not } (2 \text{ or } 3)]$

4. From a standard deck of cards, one card is drawn. Find the probability that the card is:

a. a heart **b.** not a heart **c.** a picture card
d. not a picture card **e.** not an 8 **f.** not a red 6
g. not the queen of spades **h.** not an 8 or a 6

5. One letter is selected at random from the word PICNICKING.

a. Find the probability of drawing each of the different letters in the word.
b. Demonstrate that the sum of these probabilities is 1.

6. If the probability of an event happening is $\frac{1}{7}$, what is the probability of that event not happening?

7. If the probability of an event happening is .093, what is the probability of that event not happening?

8. A jar contains seven marbles, all the same size. Three are red and four are green. If a marble is chosen at random, find each probability:

a. $P(\text{red})$ **b.** $P(\text{green})$ **c.** $P(\text{not red})$
d. $P(\text{red or green})$ **e.** $P(\text{red and green})$ **f.** $P[\text{not (red or green)}]$

9. A box contains 3 times as many black marbles as green marbles, all the same size. If a marble is drawn at random, find the probability that it is:

a. black **b.** green **c.** not black **d.** black or green **e.** not green

10. A letter is chosen at random from the word PROBABILITY. Find each probability:

a. $P(\text{A})$ **b.** $P(\text{B})$ **c.** $P(\text{C})$
d. $P(\text{A or B})$ **e.** $P(\text{A or I})$ **f.** $P(\text{a vowel})$
g. $P(\text{not a vowel})$ **h.** $P(\text{A or B or L})$ **i.** $P(\text{A or not A})$

11. A single card is drawn at random from a well-shuffled deck of 52 cards. Find the probability that the card is:

a. a 6
b. a club
c. the 6 of clubs
d. a 6 or a club
e. not a club
f. not a 6
g. a 6 or a 7
h. not the 6 of clubs
i. a 6 and a 7
j. a black 6
k. a 6 or a black card
l. a black card or not a 6

Applying Skills

12. A bank contains three quarters, four dimes, and five nickels. A coin is drawn at random.

a. Find the probability of drawing:

(1) a quarter (2) a dime (3) a nickel

b. Demonstrate that the sum of the three probabilities given as answers in part **a** is 1.

c. Find the probability of not drawing:

(1) a quarter (2) a dime (3) a nickel

13. The weather bureau predicted a 30% chance of rain. Express *in fractional form*:

a. the probability that it will rain
b. the probability that it will not rain

14. Mr. Jacobsen's mail contains two letters, three bills, and five ads. He picks up the first piece of mail without looking at it. Express, *in decimal form*, the probability that this piece of mail is:

a. a letter
b. a bill
c. an ad
d. a letter or an ad
e. a bill or an ad
f. not a bill
g. not an ad
h. a bill and an ad

15. The square dartboard shown at the right, whose side measures 30 inches, has at its center a shaded square region whose side measures 10 inches. If darts directed at the board are equally likely to land anywhere on the board, what is the probability that a dart does *not* land in the shaded region?

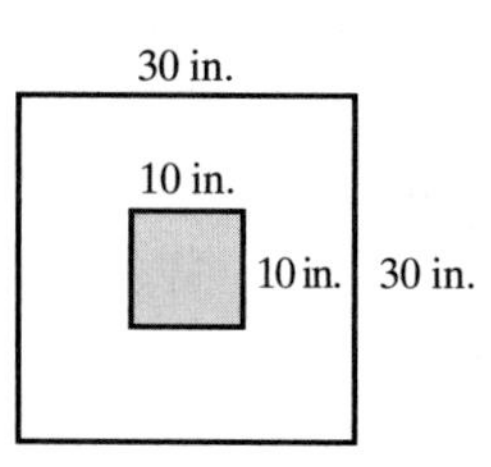

16. A telephone keypad contains the ten digits: 0, 1, 2, 3, 4, 5, 6, 7, 8, 9. Mabel is calling a friend. Find the probability that the last digit in her friend's telephone number is:

a. 6
b. 6 or a larger number
c. a number smaller than 6
d. 6 or an odd number
e. 6 or a number smaller than 6
f. not 6

g. 6 and an odd number

h. a number not larger than 6

i. a number smaller than 2 and larger than 6

j. a number smaller than 2 or larger than 6

k. a number smaller than 6 and larger than 2

l. a number smaller than 6 or larger than 2

15-7 THE COUNTING PRINCIPLE, SAMPLE SPACES, AND PROBABILITY

So far, we have looked at simple problems involving a single activity, such as rolling one die or choosing one card. More realistic problems occur when there are two or more activities, such as rolling two dice or dealing a hand of five cards. An event consisting of two or more activities is called a **compound event**.

Before studying the probability of events based on two or more activities, let us study ways to count the number of elements or outcomes in a sample space for two or more activities. For example:

A store offers five flavors of ice cream: vanilla, chocolate, strawberry, peach, and raspberry. A sundae can be made with either a hot fudge topping or a marshmallow topping. If a sundae consists of one flavor of ice cream and one topping, how many different sundaes are possible?

We will use letters to represent the five flavors of ice cream (V, C, S, P, R) and the two toppings (F, M). We can show the number of elements in the sample space in three ways:

1. **Tree Diagram.** The tree diagram at the right first branches out to show five flavors of ice cream. For each of these flavors the tree again branches out to show the two toppings. In all, there are 10 paths or branches to follow, each with one flavor of ice cream and one topping. These 10 branches show that the sample space consists of 10 possible outcomes, in this case, sundaes.

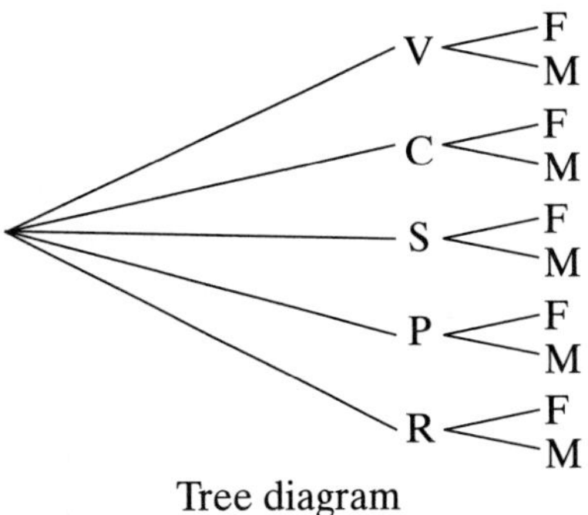

Tree diagram

2. **List of Ordered Pairs.** It is usual to order a sundae by telling the clerk the flavor of ice cream first and then the type of topping. This suggests a listing of ordered pairs. The first component of the ordered pair is the ice-cream flavor, and the second component is the type of topping. The set of pairs (ice cream, topping) is shown below.

   {(V, F), (C, F), (S, F), (P, F), (R, F), (V, M), (C, M), (S, M), (P, M), (R, M)}

   These 10 ordered pairs show that the sample space consists of 10 possible sundaes.

3. **Graph of Ordered Pairs.** Instead of listing pairs, we may construct a graph of the ordered pairs. At the right, the five flavors of ice cream appear on the horizontal scale or line, and the two toppings are on the vertical line. Each point in the graph represents an ordered pair. For example, the point circled shows the ordered pair (P, F), that is, (peach ice cream, fudge topping).

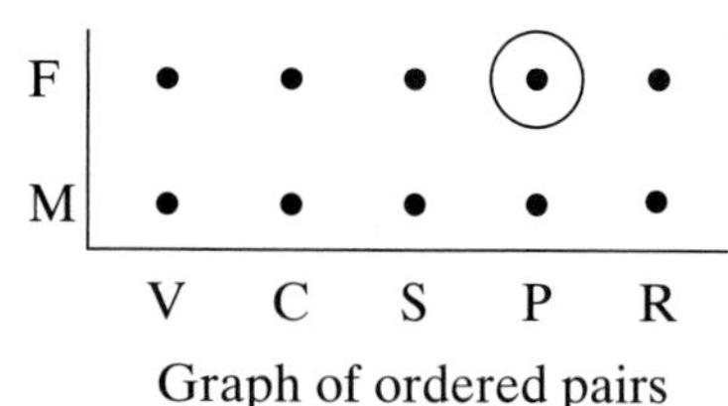

Graph of ordered pairs

This graph of 10 points, or 10 ordered pairs, shows that the sample space consists of 10 possible sundaes.

Whether we use a tree diagram, a list of ordered pairs, or a graph of ordered pairs, we recognize that the sample space consists of 10 sundaes. The number of elements in the sample space can be found by multiplication:

$$\underbrace{\text{number of flavors of ice cream}}_{5} \times \underbrace{\text{number of toppings}}_{2} = \underbrace{\text{number of possible sundaes}}_{10}$$

Suppose the store offered 30 flavors of ice cream and seven possible toppings. To find the number of elements in the sample space, we multiply:

$$30 \times 7 = 210 \text{ possible sundaes}$$

This simple multiplication procedure is known as the **counting principle**, because it enables us to count the number of elements in a sample space.

► ***The Counting Principle:*** **If one activity can occur in any of m ways and, following this, a second activity can occur in any of n ways, then both activities can occur in the order given in $m \times n$ ways.**

We can extend this rule to include three or more activities by extending the multiplication process. We can also display three or more activities by extending the branches on a tree diagram, or by listing ordered elements such as ordered triples and ordered quadruples.

For example, a coin is tossed three times in succession.

- On the first toss, the coin may fall in either of two ways: a head or a tail.
- On the second toss, the coin may also fall in either of two ways.
- On the third toss, the coin may still fall in either of two ways.

By the counting principle, the sample space contains $2 \times 2 \times 2$ or 8 possible outcomes. By letting H represent a head and T represent a tail, we can illustrate the sample space by a tree diagram or by a list of ordered triples:

We did *not* attempt to draw a graph of this sample space because we would need a horizontal scale, a vertical scale, and a third scale, making the graph

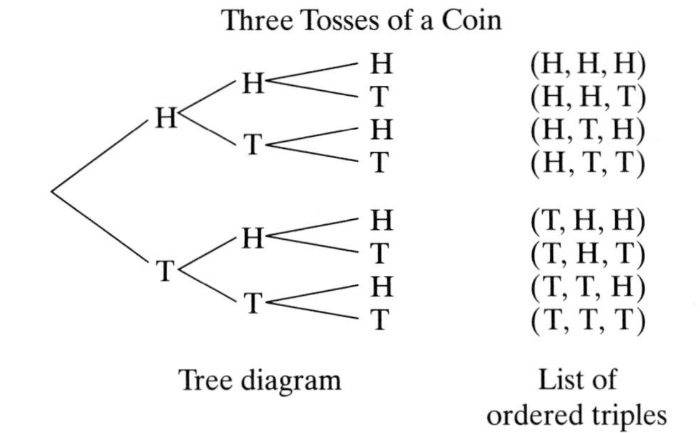

three-dimensional. Although such a graph can be drawn, it is too difficult at this time. We can conclude that:

1. Tree diagrams, or lists of ordered elements, are effective ways to indicate any compound event of two or more activities.
2. Graphs should be limited to ordered pairs, or to events consisting of exactly two activities.

EXAMPLE 1

The school cafeteria offers four types of salads, three types of beverages, and five types of desserts. If a lunch consists of one salad, one beverage, and one dessert, how many possible lunches can be chosen?

Solution By the counting principle, we multiply the number of possibilities for each choice:

$$4 \times 3 \times 5 = 12 \times 5 = 60 \text{ possible lunches } \textit{Answer}$$

Independent Events

The probability of rolling 5 on one toss of a die is $\frac{1}{6}$. What is the probability of rolling a pair of 5's when two dice are tossed?

When we roll two dice, the number obtained on the second die is in no way determined by of the result obtained on the first die. When we toss two coins, the face that shows on the second coin is in no way determined by the face that shows on the first coin.

When the result of one activity in no way influences the result of a second activity, the results of these activities are called **independent events**. In cases where two events are independent, we may extend the counting principle to find the probability that both independent events occur at the same time.

For instance, what is the probability that, when two dice are thrown, a 5 will appear on each of the dice? Let S represent the sample space and F represent the event (5 on both dice).

(1) Use the counting principle to find the number of elements in the sample space. There are 6 ways in which the first die can land and 6 ways in which the second die can land. Therefore, there are 6×6 or 36 pairs of numbers in the sample space, that is, $n(S) = 36$.

(2) There is only one face on each die that has a 5. Therefore there is 1×1 or 1 pair in the event F, that is $n(F) = 1$.

(3) $P(\text{5 on both dice}) = \frac{n(F)}{n(S)} = \frac{1}{36}$

The probability of 5 on both dice can also be determined by using the probability of each of the independent events.

$$P(\text{5 on first die}) = \frac{1}{6}$$
$$P(\text{5 on second die}) = \frac{1}{6}$$
$$P(\text{5 on both dice}) = P(\text{5 on first}) \times P(\text{5 on second}) = \frac{1}{6} \times \frac{1}{6} = \frac{1}{36}$$

We can extend the counting principle to help us find the probability of any two or more independent events.

► ***The Counting Principle for Probability:* E and F are independent events. The probability of event E is m $(0 \leq m \leq 1)$ and the probability of event F is n $(0 \leq n \leq 1)$. The probability of the event in which E and F occur jointly is the product $m \times n$.**

Note 1: The product $m \times n$ is within the range of values for a probability, namely, $0 \leq m \times n \leq 1$.

Note 2: Not all events are independent, and this simple product rule cannot be used to find the probability when events are not independent.

EXAMPLE 2

Mr. Gillen may take any of three buses, A or B or C, to get to the train station. He may then take the 6th Avenue train or the 8th Avenue train to get to work. The buses and trains arrive at random and are equally likely to arrive. What is the probability that Mr. Gillen takes the B bus and the 6th Avenue train to get to work?

Solution

$$P(\text{B bus}) = \frac{1}{3} \text{ and } P(\text{6th Ave. train}) = \frac{1}{2}$$

Since the train taken is independent of the bus taken:

$$P(B\text{ bus and 6th Ave. train}) = P(B\text{ bus}) \times P(\text{6th Ave. train})$$
$$= \frac{1}{3} \times \frac{1}{2} = \frac{1}{6} \textit{ Answer}$$

EXERCISES

Writing About Mathematics

1. Judy said that if a quarter and a nickel are tossed, there are three equally likely outcomes; two heads, two tails, or one head and one tail. Do you agree with Judy? Explain why or why not.

2. a. When a green die and a red die are rolled, is the probability of getting a 2 on the green die and a 3 on the red die the same as the probability of getting 3 on both dice? Explain why or why not.

b. When rolling two fair dice, is the probability of getting a 2 and a 3 the same as the probability of getting two 3's? Explain why or why not.

Developing Skills

3. A quarter and a penny are tossed simultaneously. Each coin may fall heads or tails. The tree diagram at the right shows the sample space involved.

H — H, T
T — H, T
Quarter Penny

a. List the sample space as a set of ordered pairs.

b. Use the counting principle to demonstrate that there are four outcomes in the sample space.

c. In how many outcomes do the coins both fall heads up?

d. In how many outcomes do the coins land showing one head and one tail?

4. Two dice are rolled simultaneously. Each die may land with any one of the six numbers faceup.

a. Use the counting principle to determine the number of outcomes in this sample space.

b. Display the sample space by drawing a graph of the set of ordered pairs.

5. A sack contains four marbles: one red, one blue, one white, one green. One marble is drawn and placed on the table. Then a second marble is drawn and placed to the right of the first.

a. How many possible marbles can be selected on the first draw?

b. How many possible marbles can be selected on the second draw?

c. How many possible ordered pairs of marbles can be drawn?

d. Draw a tree diagram to show all of the possible outcomes for this experiment. Represent the marbles by R, B, W, and G.

6. A sack contains four marbles: one red, one blue, one white, one green. One marble is drawn, its color noted and then it is replaced in the sack. A second marble is drawn and its color noted to the right of the first.

a. How many possible colors can be noted on the first draw?

b. How many possible colors can be noted on the second draw?

c. How many possible ordered pairs of colors can be noted?

d. Draw a tree diagram to show all of the possible outcomes for this experiment. Represent the marbles by R, B, W, and G.

7. A fair coin and a six-sided die are tossed simultaneously. What is the probability of obtaining:

a. a head on the coin?

b. a 4 on the die?

c. a head on the coin and a 4 on the die on a single throw?

8. A fair coin and a six-sided die are tossed simultaneously. What is the probability of obtaining on a single throw:

a. a head and a 3?

b. a head and an even number?

c. a tail and a number less than 5?

d. a tail and a number greater than 4?

9. Two fair coins are tossed. What is the probability that both land heads up?

10. Three fair coins are tossed. **a.** Find $P(\text{H, H, H})$. **b.** Find $P(\text{T, T, T})$.

11. A fair spinner contains congruent sectors, numbered 1 through 8. If the arrow is spun twice, find the probability that it lands:

a. $(7, 7)$ **b.** not $(7, 7)$ **c.** (not 7, not 7) **d.** (7, not 7) or (not 7, 7)

e. $(2, 8)$ **f.** not $(2, 8)$ **g.** (not 2, not 8) **h.** (2, not 8) or (not 2, 8)

12. A fair coin is tossed 50 times and lands heads up each time. What is the probability that it will land heads up on the next toss? Explain your answer.

Applying Skills

13. Tell how many possible outfits consisting of one shirt and one pair of pants Terry can choose if he owns:

a. 5 shirts and 2 pairs of pants

b. 10 shirts and 4 pairs of pants

c. 6 shirts and 6 pairs of pants

14. There are 10 doors into the school and eight staircases from the first floor to the second. How many possible ways are there for a student to go from outside the school to a classroom on the second floor?

15. A tennis club has 15 members: eight women and seven men. How many different teams may be formed consisting of one woman and one man on the team?

16. A dinner menu lists two soups, seven main courses, and three desserts. How many different meals consisting of one soup, one main course, and one dessert are possible?

17. The school cafeteria offers the menu shown.

Main Course	Dessert	Beverage
Pizza	Yogurt	Milk
Frankfurter	Fruit salad	Juice
Ham sandwich	Jello	
Tuna sandwich	Apple pie	
Veggie burger		

a. How many meals consisting of one main course, one dessert, and one beverage can be selected from this menu?

b. Joe does not like ham and tuna. How many meals (again, one main course, one dessert, and one beverage) can Joe select, not having ham and not having tuna?

c. If the pizza, frankfurters, yogurt, and fruit salad have been sold out, how many different meals can JoAnn select from the remaining menu?

18. A teacher gives a quiz consisting of three questions. Each question has as its answer either true (T) or false (F).

a. Using T and F, draw a tree diagram to show all possible ways the questions can be answered.

b. List this sample space as a set of ordered triples.

19. A test consists of multiple-choice questions. Each question has four choices. Tell how many possible ways there are to answer the questions on the test if the test consists of the following number of questions:

a. 1 question **b.** 3 questions **c.** 5 questions **d.** n questions

20. Options on a bicycle include two types of handlebars, two types of seats, and a choice of 15 colors. The bike may also be ordered in ten-speed, in three-speed, or standard. How many possible versions of a bicycle can a customer choose from, if he selects a specific type of handlebars, type of seat, color, and speed?

21. A state issues license plates consisting of letters and numbers. There are 26 letters, and the letters may be repeated on a plate; there are 10 digits, and the digits may be repeated. Tell how many possible license plates the state may issue when a license consists of each of the following:

a. 2 letters, followed by 3 numbers

b. 2 numbers, followed by 3 letters

c. 4 numbers, followed by 2 letters

22. In a school cafeteria, the menu rotates so that $P(\text{hamburger}) = \frac{1}{4}$, $P(\text{apple pie}) = \frac{2}{3}$, and $P(\text{soup}) = \frac{4}{5}$. The selection of menu items is random so that the appearance of hamburgers, apple pie, and soup are independent events. On any given day, what is the probability that the cafeteria offers hamburger, apple pie, and soup on the same menu?

23. A quiz consists of true-false questions only. Harry has not studied, and he guesses every answer. Find the probability that he will guess correctly to get a perfect score if the test consists of: **a.** 1 question **b.** 4 questions **c.** n questions

24. The probability of the Tigers beating the Cougars is $\frac{2}{3}$. The probability of the Tigers beating the Mustangs is $\frac{1}{4}$. If the Tigers play one game with the Cougars and one game with the Mustangs, find the probability that the Tigers: **a.** win both games **b.** lose both games

25. On Main Street in Pittsford, there are two intersections that have traffic lights. The lights are not timed to accommodate traffic. They are independent of one another. At each of the intersections, $P(\text{red light}) = .3$ and $P(\text{green light}) = .7$ for cars traveling along Main Street. Find the probability that a car traveling on Main Street will be faced with each set of given conditions at the two traffic lights shown.

a. Both lights are red.

b. Both lights are green.

c. The first light is red, and the second is green.

d. The first light is green, and the second is red.

e. At least one light is red, that is, not both lights are green.

f. Both lights are the same color.

26. A manufacturer of radios knows that the probability of a defect in any of his products is $\frac{1}{400}$. If 10,000 radios are manufactured in January, how many are likely to be defective?

27. Past records from the weather bureau indicate that the probability of rain in August on Cape Cod is $\frac{2}{7}$. If Joan goes to Cape Cod for 2 weeks in August, how many days will it probably rain if the records hold true?

28. A nationwide fast-food chain has a promotion, distributing to customers 2,000,000 coupons for the prizes shown below. Each coupon awards the customer one of the following prizes.

1 Grand Prize: $25,000 cash
2 Second-Place Prizes: New car
100 Third-Place Prizes: New TV
Fourth-Place Prizes: Free meal
Consolation Prizes: 25 cents off any purchase

a. Find the probability of winning:

(1) the grand prize (2) a new car (3) a new TV

b. If a customer has one coupon, what is the probability of winning one of the first three prizes (cash, a car, or a TV)?

c. If the probability of winning a free meal is $\frac{1}{400}$, how many coupons are marked as fourth-place prizes?

d. How many coupons are marked "25 cents off any purchase"?

e. If a customer has one coupon, what is the probability of not winning one of the first three prizes?

15-8 PROBABILITIES WITH TWO OR MORE ACTIVITIES

Without Replacement

Two cards are drawn at random from an ordinary pack of 52 cards. In this situation, a single card is drawn from a deck of 52 cards, and then a second card is drawn from the remaining 51 cards in the deck. What is the probability that both cards drawn are kings?

On the first draw, there are four kings in the deck of 52 cards, so

$$P(\text{first king}) = \tfrac{4}{52}$$

If a second card is drawn without replacing the first king selected, there are now only three kings in the 51 cards remaining. Therefore we are considering the probability of drawing a king, given that a king has already been drawn.

$$P(\text{second king}) = \tfrac{3}{51}$$

By the counting principle for probabilities:

$$P(\text{both kings}) = P(\text{first king}) \times P(\text{second king}) = \tfrac{4}{52} \times \tfrac{3}{51} = \tfrac{1}{13} \times \tfrac{1}{17} = \tfrac{1}{221}$$

This result could also have been obtained using the counting principle. The number of elements in the sample space is 52×51. The number of elements in the event (two kings) is 4×3. Then,

$$P(\text{both kings}) = \frac{4 \times 3}{52 \times 51} = \frac{12}{2{,}652} = \frac{1}{221}$$

This is called a problem **without replacement** because the first king drawn was not placed back into the deck. These are **dependent events** because the probability of a king on the second draw depends on whether or not a king appeared on the first draw.

In general, if A and B are two dependent events:

$$\boldsymbol{P(A \text{ and } B) = P(A) \times P(B \text{ given that } A \text{ has occurred})}$$

Earlier in the chapter we discussed $P(A \text{ and } B)$ where $(A \text{ and } B)$ is a single event that satisfies both conditions. Here $(A \text{ and } B)$ denotes two dependent events with A the outcome of one event and B the outcome of the other. The conditions of the problem will indicate which of these situations exists. When A and B are dependent events, $P(A \text{ and } B)$ can also be written as $P(A, B)$.

With Replacement

A card is drawn at random from an ordinary deck, the card is placed back into the deck, and a second card is then drawn and replaced. In this situation, it is clear the deck contains 52 cards each time that a card is drawn and that the same card could be drawn twice. What is the probability that the card drawn each time is a king?

On the first draw, there are four kings in the deck of 52 cards.

$$P(\text{first king}) = \frac{4}{52}$$

If the first king drawn is now placed back into the deck, then, on the second draw, there are again four kings in the deck of 52 cards.

$$P(\text{second king}) = \frac{4}{52}$$

By the counting principle for probabilities:

$$P(\text{both kings}) = P(\text{first king}) \times P(\text{second king}) = \frac{4}{52} \times \frac{4}{52} = \frac{1}{13} \times \frac{1}{13} = \frac{1}{169}$$

This is called a problem **with replacement** because the first card drawn was placed back into the deck. In this case, the events are independent because the probability of a king on the second draw does not depend on whether or not a king appeared on the first draw. Since the card drawn is replaced, the number of cards in the deck and the number of kings in the deck remain constant. In this case,

$$P(B \text{ given that } A \text{ has occurred}) = P(B).$$

In general, if A and B are two independent events,

$$\mathbf{P(A \text{ and } B) = P(A) \times P(B)}$$

Rolling two dice is similar to drawing two cards with replacement because the number of faces on each die remains constant, as did the number of cards in the deck. Typical problems with replacement include rolling dice, tossing coins (each coin always has two sides), and spinning arrows.

KEEP IN MIND

1. If the problem does not specifically mention *with replacement* or *without replacement*, ask yourself: "Is this problem with or without replacement?" or "Are the events dependent or independent?"
2. For many compound events, the probability can be determined most easily by using the counting principle.
3. Every probability problem can always be solved by:
   - counting the number of elements in the sample space, $n(S)$;
   - counting the number of outcomes in the event, $n(E)$;
   - substituting these numbers in the probability formula,

$$P(E) = \frac{n(E)}{n(S)}$$

Conditional Probability

The previous discussion involved the concept of **conditional probability**. For both dependent and independent events, in order to find the probability of A followed by B, it is necessary to calculate the probability that B occurs given that A has occurred.

Notation for conditional probability is $P(B$ given that A has occurred$) = P(B \mid A)$. Then the following statement is true for both dependent and independent events:

$$P(A \text{ and } B) = P(A) \times P(B \mid A)$$

If A and B independent events, $P(B \mid A) = P(B)$. Therefore, for independent events:

$$P(A \text{ and } B) = P(A) \times P(B \mid A) = P(A) \times P(B)$$

The general formula $P(A \text{ and } B) = P(A) \times P(B \mid A)$ can be solved for $P(B \mid A)$:

$$P(B \mid A) = \frac{P(A \text{ and } B)}{P(A)}$$

For example, suppose a box contains one red marble, one blue marble, one green marble, and one yellow marble. Two marbles are drawn without replacement. Let R be the event {red marble} and Y be the event {yellow marble}. The probability of R is $\frac{1}{4}$ and the probability of Y is $\frac{1}{4}$. If we want to find the probability of drawing a red marble *followed by* a yellow marble, R and Y are dependent events. We need the probability of Y given that R has occurred, which is $\frac{1}{3}$ since once the red marble has been drawn, only 3 marbles remain, one of which is yellow.

$$\begin{aligned} P(R \text{ followed by } Y) = P(R \text{ and } Y) &= P(R) \times P(Y \mid R) \\ &= \tfrac{1}{4} \times \tfrac{1}{3} \\ &= \tfrac{1}{12} \end{aligned}$$

This result can be shown by displaying the sample space.

$$\begin{array}{cccc} (R, B) & (B, R) & (G, R) & (Y, R) \\ (R, G) & (B, G) & (G, B) & (Y, B) \\ (R, Y) & (B, Y) & (G, Y) & (Y, G) \end{array}$$

There are 12 possible outcomes in the sample space and one of them is (R, Y). Therefore,

$$P(R, Y) = \tfrac{1}{12}.$$

EXAMPLE 1

Three fair dice are thrown. What is the probability that all three dice show a 5?

Solution These are independent events. There are six possible faces that can come up on each die.

On the first die, there is one way to obtain a 5 so $P(5) = \frac{1}{6}$.

On the second die, there is one way to obtain a 5 so $P(5) = \frac{1}{6}$.

On the third die, there is one way to obtain a 5 so $P(5) = \frac{1}{6}$.

By the counting principle:

$$\begin{aligned} P(5 \text{ on each die}) &= P(5 \text{ on first}) \times P(5 \text{ on second}) \times P(5 \text{ on third}) \\ &= \tfrac{1}{6} \times \tfrac{1}{6} \times \tfrac{1}{6} \\ &= \tfrac{1}{216} \textit{ Answer} \end{aligned}$$

EXAMPLE 2

If two cards are drawn from an ordinary deck without replacement, what is the probability that the cards form a pair (two cards of the same face value but different suits)?

Solution These are dependent events. On the first draw, any card at all may be chosen, so:

$$P(\text{any card}) = \tfrac{52}{52}$$

There are now 51 cards left in the deck. Of these 51, there are three that match the first card taken, to form a pair, so:

$$P(\text{second card forms a pair}) = \tfrac{3}{51}$$

Then:

$$\begin{aligned} P(\text{pair}) &= P(\text{any card}) \times P(\text{second card forms a pair}) \\ &= \tfrac{52}{52} \times \tfrac{3}{51} \\ &= 1 \times \tfrac{1}{17} = \tfrac{1}{17} \textit{ Answer} \end{aligned}$$

EXAMPLE 3

A jar contains four white marbles and two blue marbles, all the same size. A marble is drawn at random and not replaced. A second marble is then drawn from the jar. Find the probability that:

a. both marbles are white **b.** both marbles are blue **c.** both marbles are the same color

Solution These are dependent events.

a. On the first draw:

$$P(\text{white}) = \tfrac{4}{6}$$

Since the white marble drawn is not replaced, five marbles, of which three are white, are left in the jar. On the second draw:

$$P(\text{white given that the first was white}) = \tfrac{3}{5}$$

Then:

$$P(\text{both white}) = \tfrac{4}{6} \times \tfrac{3}{5} = \tfrac{12}{30} \text{ or } \tfrac{2}{5} \textit{ Answer}$$

b. Start with a full jar of six marbles of which two are blue.

On the first draw:

$$P(\text{blue}) = \tfrac{2}{6}$$

Since the blue marble drawn is not replaced, five marbles, of which only one is blue, are left in the jar. On the second draw:

$$P(\text{blue given that the first was blue}) = \tfrac{1}{5}$$

Then:

$$P(\text{both blue}) = \tfrac{2}{6} \times \tfrac{1}{5} = \tfrac{2}{30} \text{ or } \tfrac{1}{15} \textit{ Answer}$$

c. If both marbles are the same color, both are white or both are blue. These are disjoint or mutually exclusive events:

$$P(A \text{ or } B) = P(A) + P(B)$$

Therefore:

$$\begin{aligned} P(\text{both white or both blue}) &= P(\text{both white}) + P(\text{both blue}) \\ &= \tfrac{2}{5} + \tfrac{1}{15} \\ &= \tfrac{6}{15} + \tfrac{1}{15} \\ &= \tfrac{7}{15} \textit{ Answer} \end{aligned}$$

Note: By considering the complement of the set named in part **c**, we can easily determine the probability of drawing two marbles of *different* colors.

$$\begin{aligned} P[\text{not (both white or both blue)}] &= 1 - P(\text{both white or both blue}) \\ &= 1 - \tfrac{7}{15} \\ &= \tfrac{8}{15} \end{aligned}$$

EXAMPLE 4

A fair die is rolled.

a. Find the probability that the die shows a 4 given that the die shows an even number.

b. Find the probability that the die shows a 1 given that the die shows a number less than 5.

Solution **a.** Use the formula for conditional probability.

$$P(4 \text{ given an even number}) = P(4 \mid \text{even}) = \frac{P(4 \text{ and even})}{P(\text{even})}$$

The event "4 and even" occurs whenever the outcome is 4. Therefore, $P(4 \text{ and even}) = P(4)$.

$$P(4 \mid \text{even}) = \frac{P(4 \text{ and even})}{P(\text{even})} = \frac{P(4)}{P(\text{even})} = \frac{\frac{1}{6}}{\frac{3}{6}} = \frac{1}{3} \textit{ Answer}$$

b. The event "1 and less than 5" occurs whenever the outcome is 1. Therefore, $P(1 \text{ and less than } 5) = P(1)$.

$$P(1 \mid \text{less than } 5) = \frac{P(1 \text{ and less than } 5)}{P(\text{less than } 5)} = \frac{P(1)}{P(\text{less than } 5)} = \frac{\frac{1}{6}}{\frac{5}{6}} = \frac{1}{5} \textit{ Answer}$$

EXAMPLE 5

Fred has two quarters and one nickel in his pocket. The pocket has a hole in it, and a coin drops out. Fred picks up the coin and puts it back into his pocket. A few minutes later, a coin drops out of his pocket again.

a. Draw a tree diagram or list the sample space for all possible pairs that are outcomes to describe the coins that fell.

b. What is the probability that the same coin fell out of Fred's pocket both times?

a. What is the probability that the two coins that fell have a total value of 30 cents?

b. What is the probability that a quarter fell out at least once?

Solution **a.** Because there are two quarters, use subscripts.

The three coins are $\{Q_1, Q_2, N\}$, where Q represents a quarter and N represents a nickel. This is a problem with replacement. The events are independent.

$$\{(Q_1, Q_1), (Q_1, Q_2), (Q_1, N), (Q_2, Q_1), (Q_2, Q_2), (Q_2, N), (N, Q_1), (N, Q_2), (N, N)\}$$

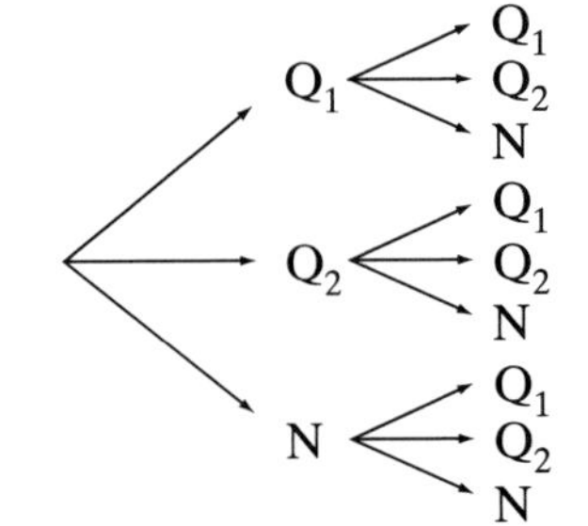

b. Of the nine outcomes, three name the same coin both times: (Q_1, Q_1), (Q_2, Q_2) and (N, N). Therefore:

$$P(\text{same coin}) = \tfrac{3}{9} \text{ or } \tfrac{1}{3} \text{ Answer}$$

c. Of the nine outcomes, the four that consist of a quarter and a nickel total 30 cents: (Q_1, N), (Q_2, N), (N, Q_1), (N, Q_2).

$$P(\text{total value of 30 cents}) = \tfrac{4}{9} \text{ Answer}$$

d. Of the nine outcomes, only (N, N) does not include a quarter. Eight contain at least one quarter, that is, one or more quarters.

$$P(\text{at least one quarter}) = \tfrac{8}{9} \text{ Answer}$$

EXAMPLE 6

Of Roosevelt High School's 1,000 students, 300 are athletes, 200 are in the Honor Roll, and 120 play sports and are in the Honor Roll. What is the probability that a randomly chosen student who plays a sport is also in the Honor Roll?

Solution Let A = the event that the student is an athlete, and

B = the event that the student is in the Honor Roll.

Then, A and B = the event that a student is an athlete *and* is in the Honor Roll.

Therefore, the conditional probability that the student is in the Honor Roll given that he or she is an athlete is:

$$P(B \mid A) = \frac{P(A \text{ and } B)}{P(A)}$$

$$= \frac{\frac{120}{1{,}000}}{\frac{300}{1{,}000}} = \frac{120}{300} = .4 \text{ Answer}$$

EXERCISES

Writing About Mathematics

1. The name of each person who attends a charity luncheon is placed in a box and names are drawn for first, second, and third prize. No one can win more than one prize. Is the probability of winning first prize greater than, equal to, or less than the probability of winning second prize? Explain your answer.

2. Six tiles numbered 1 through 6 are placed in a sack. Two tiles are drawn. Is the probability of drawing a pair of tiles whose numbers have a sum of 8 greater than, equal to, or less than the probability of obtaining a sum of 8 when rolling a pair of dice? Explain your answer.

Developing Skills

3. A jar contains two red and five yellow marbles. If one marble is drawn at random, what is the probability that the marble drawn is: **a.** red? **b.** yellow?

4. A jar contains two red and five yellow marbles. A marble is drawn at random and then replaced. A second draw is made at random. Find the probability that:

   a. both marbles are red
   b. both marbles are yellow
   c. both marbles are the same color
   d. the marbles are different in color

5. A jar contains two red and five yellow marbles. A marble is drawn at random. Then without replacement, a second marble is drawn at random. Find the probability that:

   a. both marbles are red
   b. both marbles are yellow
   c. both marbles are the same color
   d. the marbles are different in color
   e. the second marble is red given that the first is yellow

6. In an experiment, an arrow is spun twice on a circular board containing four congruent sectors numbered 1 through 4. The arrow is equally likely to land on any one of the sectors.

   a. Indicate the sample space by drawing a tree diagram or writing a set of ordered pairs.
   b. Find the probability of spinning the digits 2 and 3 in that order.
   c. Find the probability that the same digit is spun both times.
   d. What is the probability that the first digit spun is larger than the second?

7. A jar contains nine orange disks and three blue disks. A girl chooses one at random and then, without replacing it, chooses another. Let O represent orange and B represent blue.

   a. Find the probability of each of the following outcomes:
   (1) (O, O) (2) (O, B) (3) (B, O) (4) (B, B)

   b. Now, find for the disks chosen, the probability that:
   (1) neither was orange
   (2) only one was blue
   (3) at least one was orange
   (4) they were the same color
   (5) at most one was orange
   (6) they were the same disk
   (7) the second is orange given that the first was blue

8. A card is drawn at random from a deck of 52 cards. Given that it is a red card, what is the probability that it is: **a.** a heart? **b.** a king?

9. A fair coin is tossed three times.

   a. What is the probability of getting all heads, given that the first toss is heads?
   b. What is the probability of getting all heads, given that the first two tosses are heads?

Applying Skills

10. Sal has a bag of hard candies: three are lemon (L) and two are grape (G). He eats two of the candies while waiting for a bus, selecting them at random one after another.

a. Using subscripts, draw a tree diagram or list the sample space of all possible outcomes showing which candies are eaten.

b. Find the probability of each of the following outcomes:

(1) both candies are lemon (2) neither candy is lemon

(3) the candies are the same flavor (4) at least one candy is lemon

c. What is the probability that the second candy that Sal ate was lemon given that the first was grape?

11. Carol has five children: three girls and two boys. One of her children was late for lunch. Later that day, one of her children was late for dinner.

a. Indicate the sample space by a tree diagram or list of ordered pairs showing which children were late.

b. If each child was equally likely to be late, find the probability of each outcome below.

(1) Both children who were late were girls.

(2) Both children who were late were boys.

(3) The same child was late both times.

(4) At least one of the children who was late was a boy.

c. What is the probability that the child who was late for dinner was a boy given that the child who was late for lunch was a girl?

12. Several players start playing a game, each with a full deck of 52 cards. Each player draws two cards at random, one at a time without replacement. Find the probability that:

a. Flo draws two jacks **b.** Frances draws two hearts

c. Jerry draws two red cards **d.** Mary draws two picture cards

e. Ann does not draw a pair **f.** Stephen draws two black kings

g. Carrie draws a 10 on her second draw given that the first was a 5

h. Bill draws a heart and a club in that order

i. Ann draws a king on her second draw given that the first was not a king

13. Saverio has four coins: a half dollar, a quarter, a dime, and a nickel. He chooses one of the coins at random and puts it in a bank. Later he chooses another coin and also puts that in the bank.

a. Indicate the sample space of coins saved.

b. If each coin is equally likely to be saved, find the probability that:

(1) the coins saved will be worth a total of 35 cents

(2) the coins saved will add to an even amount

(3) the coins saved will include the half-dollar

(4) the coins saved will be worth a total of less than 30 cents

(5) the second coin saved will be worth more than 20 cents given that the first coin saved was worth more than 20 cents.

(6) the second coin saved will be worth more than 20 cents given that the first coin saved was worth less than 20 cents.

14. Farmer Brown must wake up before sunrise to start his chores. Dressing in the dark, he reaches into a drawer and pulls out two loose socks. There are eight white socks and six red socks in the drawer.

a. Find the probability that both socks are:

(1) white (2) red (3) the same color (4) not the same color

b. Find the minimum number of socks Farmer Brown must pull out of the drawer to guarantee that he will get a matching pair.

15. Tillie is approaching the toll booth on an expressway. She has three quarters and four dimes in her purse. She takes out two coins at random from her purse. Find the probability of each outcome.

a. Both coins are quarters.

b. Both coins are dimes.

c. The coins are a dime and a quarter, in any order.

d. The value of the coins is enough to pay the 35-cent toll.

e. The value of the coins is not enough to pay the 35-cent toll given that one of the coins is a dime.

f. The value of the coins was 35 cents given that one coin was a quarter.

g. At least one of the coins picked was a quarter.

16. One hundred boys and one hundred girls were asked to name the current Secretary of State. Thirty boys and sixty girls knew the correct name. One of these boys and girls is selected at random.

a. What is the probability that the person selected knew the correct name?

b. What is the probability that the person selected is a girl, given that that person knew the correct name?

c. What is the probability that the person selected knew the correct name, given that the person is a boy?

17. In a graduating class of 400 seniors, 200 were male and 200 were female. The students were asked if they had ever downloaded music from an online music store. 75 of the male students and 70 of the female students said that they had downloaded music.

a. What is the probability that a randomly chosen senior has downloaded music?

b. What is the probability that a randomly chosen senior has downloaded music given that the senior is male?

18. Gracie baked one dozen sugar cookies and two dozen brownies. She then topped two-thirds of the cookies and half the brownies with chocolate frosting.

a. What is the probability that a randomly chosen treat is an unfrosted brownie?

b. What is the probability that a baked good chosen at random is a cookie given that it is frosted?

19. In a game of Tic-Tac-Toe, the first player can put an X in any of the four corner squares, four edge squares, or the center square of the grid. The second player can then put an O in any of the eight remaining open squares.

a. What is the probability that the second player will put an O in the center square given the first player has put an X in an edge square?

b. If the first player puts an X in a corner square, what is the probability that the second player will put an O in a corner square?

c. In a game of Tic-Tac-Toe, the first player puts an X in the center square and the second player puts an O in a corner square. What is the probability that the first player will put her next X in an edge square?

20. Of the 150 members of the high school marching band, 30 play the trumpet, 40 are in the jazz band, and 18 play the trumpet and are also in the jazz band.

a. What is the probability that a randomly chosen member of the marching band plays the trumpet but is not in the jazz band?

b. What is the probability that a randomly chosen member of the marching band is also in the jazz band but does not play the trumpet?

15-9 PERMUTATIONS

A teacher has announced that Al, Betty, and Chris, three students in her class, will each give an oral report today. How many possible ways are there for the teacher to choose the order in which these students will give their reports?

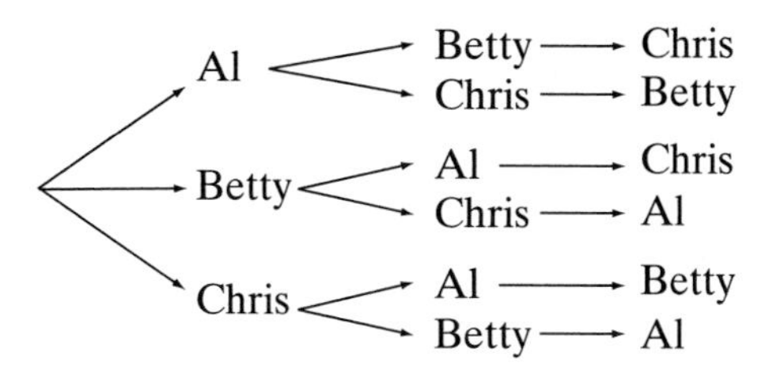

A tree diagram shows that there are six possible orders or arrangements. For example, Al, Betty, Chris is one possible arrangement; Al, Chris, Betty is another. Each of these arrangements is called a permutation. A **permutation** is an arrangement of objects or things in some specific order. (In discussing permutations, the words "objects" or "things" are used in a mathematical sense to include all elements in question, whether they are people, numbers, or inanimate objects.)

The six possible permutations in this case may also be shown as a set of ordered triples. Here, we let A represent Al, B represent Betty, and C represent Chris:

$$\{(A, B, C), (A, C, B), (B, A, C), (B, C, A), (C, A, B), (C, B, A)\}$$

Let us see, from another point of view, why there are six possible orders. We know that any one of three students can be called on to give the first report. Once the first report is given, the teacher may call on any one of the two remaining students. After the second report is given, the teacher must call on the one remaining student. Using the counting principle, we see that there are $3 \times 2 \times 1$ or 6 possible orders.

Consider another situation. A chef is preparing a recipe with 10 ingredients. He puts all of one ingredient in a bowl, followed by all of another ingredient, and so on. How many possible orders are there for placing the 10 ingredients in a bowl? Using the counting principle, we have:

$$10 \times 9 \times 8 \times 7 \times 6 \times 5 \times 4 \times 3 \times 2 \times 1 = 3{,}628{,}800 \text{ possible ways}$$

If there are more than 3 million possible ways of placing 10 ingredients in a bowl, can you imagine in how many ways 300 people who want to buy tickets for a football game can be arranged in a line? Using the counting principle, we find the number of possible orders to be $300 \times 299 \times 298 \times \cdots \times 3 \times 2 \times 1$. To symbolize such a product, we make use of the **factorial symbol**, !. We represent the product of these 300 numbers by the symbol 300!, read as "three hundred factorial" or "factorial 300."

Factorials

In general, for any natural number n, we define **n factorial** or **factorial n** as follows:

DEFINITION

$$n! = n(n-1)(n-2)(n-3) \times \cdots \times 3 \times 2 \times 1$$

Note that 1! is the natural number 1.

Calculators can be used to evaluate factorials. On a graphing calculator, the factorial function is found by first pressing MATH and then using the left arrow key to highlight the PRB menu. For example, to evaluate 5!, use the following sequence of keys:

ENTER: 5 MATH ◄ 4 ENTER

DISPLAY:

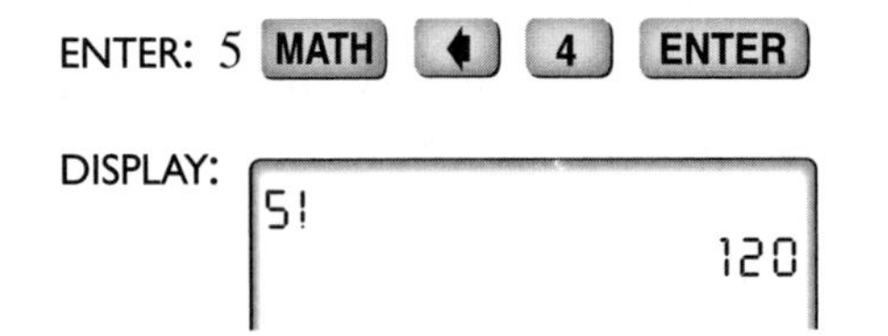

Of course, whether a calculator does or does not have a factorial function, we can always use repeated multiplication to evaluate a factorial:

$$5! = 5 \times 4 \times 3 \times 2 \times 1$$

When using a calculator, we must keep in mind that factorial numbers are usually very large. If the number of digits in a factorial exceeds the number of places in the display of the calculator, the calculator will shift from standard decimal notation to scientific notation. For example, evaluate 15! on a calculator:

ENTER: 15 MATH ◀ 4 ENTER

DISPLAY:
```
15!
        1.307674368E12
```

The number in the display can be written in scientific notation or in decimal notation.

$$1.307674368 \text{ E } 12 = 1.307674368 \times 10^{12} = 1{,}307{,}674{,}368{,}000$$

Representing Permutations

We have said that permutations are arrangements of objects in different orders. For example, the number of different orders in which four people can board a bus is 4! or $4 \times 3 \times 2 \times 1$ or 24. There are 24 permutations, that is, 24 different orders or arrangements, of these four people, in which all four of them get on the bus.

We may also represent this number of permutations by the symbol ${}_4P_4$. The symbol ${}_4P_4$ is read as: "the number of permutations of four objects taken four at a time." Here, the letter P represents the word *permutation*.

The small ${}_4$ written to the lower left of P tells us that four objects are available to be used in an arrangement, (four people are waiting for a bus).

The small ${}_4$ written to the lower right of P tells us how many of these objects are to be used in each arrangement, (four people getting on the bus).

Thus, ${}_4P_4 = 4! = 4 \times 3 \times 2 \times 1 = 24$.

Similarly, ${}_5P_5 = 5! = 5 \times 4 \times 3 \times 2 \times 1 = 120$.

In the next section, we will study examples where not all the objects are used in the arrangement. We will also examine a calculator key used with permutations. For now, we make the following observation:

► **For any natural number n, the number of permutations of n objects taken n at a time can be represented as:**

$$_nP_n = n! = n(n - 1)(n - 2) \times \cdots \times 3 \times 2 \times 1$$

EXAMPLE 1

Compute the value of each expression.

a. $6!$ **b.** ${}_2P_2$ **c.** $\frac{7!}{3!}$

Solution **a.** $6! = 6 \times 5 \times 4 \times 3 \times 2 \times 1 = 720$

b. ${}_2P_2 = 2! = 2 \times 1 = 2$

c. $\frac{7!}{3!} = \frac{7 \times 6 \times 5 \times 4 \times 3 \times 2 \times 1}{3 \times 2 \times 1} = \frac{7 \times 6 \times 5 \times 4}{1} \times \frac{3 \times 2 \times 1}{3 \times 2 \times 1} = 840$

Calculator Solution

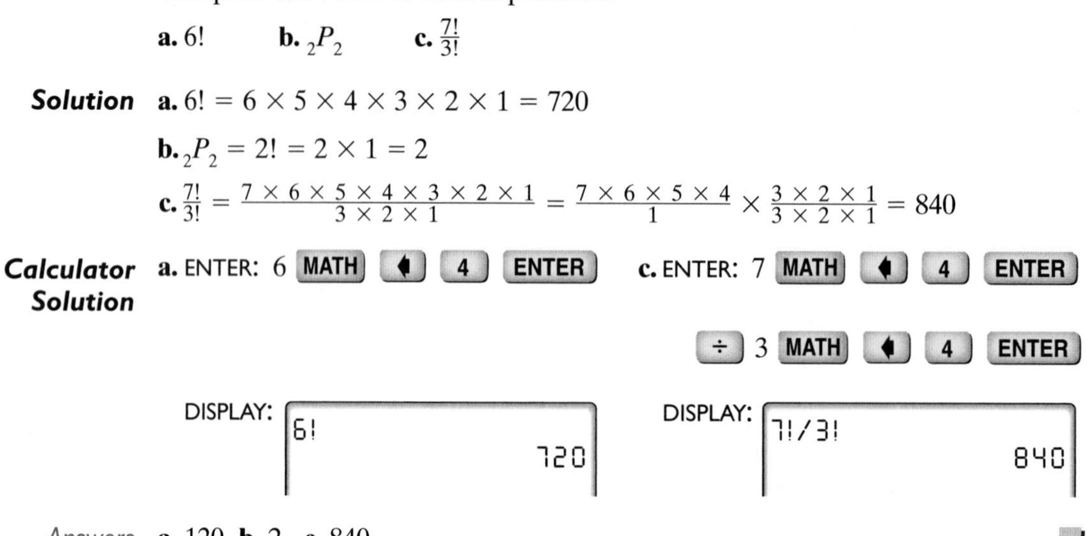

Answers **a.** 120 **b.** 2 **c.** 840

EXAMPLE 2

Paul wishes to call Virginia, but he has forgotten her unlisted telephone number. He knows that the exchange (the first three digits) is 555, and that the last four digits are 1, 4, 7, and 9, but he cannot remember their order. What is the maximum number of telephone calls that Paul may have to make in order to dial the correct number?

Solution The telephone number is 555- . Since the last four digits will be an arrangement of 1, 4, 7, and 9, this is a permutation of four numbers, taken four at a time.

$${}_4P_4 = 4! = 4 \times 3 \times 2 \times 1 = 24 \text{ possible orders}$$

Answer The maximum number of calls that Paul may have to make is 24.

Permutations That Use Some of the Elements

At times, we deal with situations involving permutations in which we are given n objects, but we use fewer than n objects in each arrangement. For example, a teacher has announced that he will call students from the first row to explain homework problems at the board. The students in the first row are George, Helene, Jay, Karla, Lou, and Marta. If there are only two homework problems, and each problem is to be explained by a different student, in how many orders may the teacher select students to go to the board?

We know that the first problem can be assigned to any of six students. Once this problem is explained, the second problem can be assigned to any of the five remaining students. We use the counting principle to find the number of possible orders in which the selection can be made.

$$6 \times 5 = 30 \text{ possible orders}$$

If there are three problems, then after the first two students have been selected, there are four students who could be selected to explain the third problem. Extend the counting principle to find the number of possible orders in which the selection can be made.

$$6 \times 5 \times 4 = 120 \text{ possible orders}$$

Note that the starting number is the number of persons in the group from which the selection is made. Each of the factors is one less than the preceding factor. The number of factors is the number of choices to be made.

Using the language of permutations, we say that the number of permutations of six objects taken three at a time is 120.

The Symbols for Permutations

In general, if we have a set of n different objects, and we make arrangements of r objects from this set, we represent the number of arrangements by the symbol ${}_nP_r$. The subscript, r, representing the number of factors being used, must be less than or equal to n, the total number of objects in the set. Thus:

► **For numbers n and r, where $r \leq n$, the permutation of n objects, taken r at a time, is found by the formula:**

$$ {}_nP_r = \underbrace{n(n-1)(n-2)\cdots}_{r \text{ factors}} $$

This formula can also be written as:

$$ {}_nP_r = n(n-1)(n-2)\cdots(n-r+1) $$

Note that when there are r factors, the last factor is $(n - r + 1)$.

In the example given above, in which three students were selected from a group of six, $n = 6$, $r = 3$, and the last factor is $n - r + 1 = 6 - 3 + 1 = 4$.

Permutations and the Calculator

There are many ways to use a calculator to evaluate a permutation. In the three solutions presented here, we will evaluate the permutation ${}_8P_3$, the order in which, from a set of 8 elements, 3 can be selected.

METHOD 1. *Use the Multiplication Key,* ×.

The number permutation is simply the product of factors. In ${}_8P_3$, the first factor, 8, is multiplied by (8 − 1), or 7, and then by (8 − 2), or 6. Here, exactly three factors have been multiplied.

ENTER: 8 × 7 × 6 ENTER

DISPLAY:

```
8*7*6
                336
```

This approach can be used for any permutation. In the general permutation ${}_nP_r$, where $r \leq n$, the first factor n is multiplied by $(n - 1)$, and then by $(n - 2)$, and so on, until exactly r factors have been multiplied.

METHOD 2. *Use the Factorial Function.*

When evaluating ${}_8P_3$, it appears as if we start to evaluate 8! but then we stop after multiplying only three factors. There is a way to write the product $8 \times 7 \times 6$ using factorials. As shown below, when we divide 8! by 5!, all but the first three factors will cancel, leaving $8 \times 7 \times 6$ or 336. Since the factorial in the denominator uses all but three factors, $8 - 3 = 5$ can be used to find that factorial:

$$\begin{aligned} {}_8P_3 &= 8 \times 7 \times 6 \\ &= 8 \times 7 \times 6 \times \frac{5 \times 4 \times 3 \times 2 \times 1}{5 \times 4 \times 3 \times 2 \times 1} \\ &= \frac{8 \times 7 \times 6 \times 5 \times 4 \times 3 \times 2 \times 1}{5 \times 4 \times 3 \times 2 \times 1} \\ &= \frac{8!}{5!} \\ &= \frac{8!}{(8-3)!} \end{aligned}$$

We now evaluate this fraction on a calculator.

ENTER: 8 MATH ◄ 4 ÷ 5 MATH ◄ 4 ENTER

DISPLAY:

```
8!/5!
                336
```

This approach shows us that there is another formula that can be used for the general permutation ${}_nP_r$, where $r < n$:

$$\boldsymbol{{}_nP_r = \frac{n!}{(n-r)!}}$$

METHOD 3. *Use the Permutation Function.*

Calculators have a special function to evaluate permutations. On a graphing calculator, nPr is found by first pressing MATH and then using the left arrow key to highlight the PRB menu. The value of n is entered first, then the nPr symbol is entered followed by the value of r.

ENTER: 8 MATH ◀ 2 3 ENTER

DISPLAY:

```
8 nPr 3
            336
```

EXAMPLE 3

Evaluate ${}_6P_2$.

Solution This is a permutation of six objects, taken two at a time. There are two possible formulas to use.

$${}_6P_2 = 6 \times 5 = 30 \qquad {}_6P_2 = \frac{6!}{4!} = \frac{6 \times 5 \times 4 \times 3 \times 2 \times 1}{4 \times 3 \times 2 \times 1} = 6 \times 5 = 30$$

Calculator Solution ENTER: 6 MATH ◀ 2 2 ENTER

DISPLAY:

```
6 nPr 2
             30
```

Answer ${}_6P_2 = 30$

EXAMPLE 4

How many three-letter "words" (arrangements of letters) can be formed from the letters L, O, G, I, C if each letter is used only once in a word?

Solution Forming three-letter arrangements from a set of five letters is a permutation of five, taken three at a time. Thus:

$${}_5P_3 = 5 \times 4 \times 3 = 60$$

or

$${}_5P_3 = \frac{5!}{(5-3)!} = \frac{5!}{2!} = \frac{5 \times 4 \times 3 \times 2 \times 1}{2 \times 1} = 5 \times 4 \times 3 = 60$$

Answer 60 words

EXAMPLE 5

A lottery ticket contains a four-digit number. How many possible four-digit numbers are there when:

a. a digit may appear only once in the number?

b. a digit may appear more than once in the number?

Solution **a.** If a digit appears only once in a four-digit number, this is a permutation of 10 digits, taken four at a time. Thus:

$$_{10}P_4 = 10 \times 9 \times 8 \times 7 = 5{,}040$$

b. If a digit may appear more than once, we can choose any of 10 digits for the first position, then any of 10 digits for the second position, and so forth. By the counting principle:

$$10 \times 10 \times 10 \times 10 = 10{,}000$$

Answers **a.** 5,040 **b.** 10,000

EXERCISES

Writing About Mathematics

1. Show that $n! = n(n - 1)!$.

2. Which of these two values, if either, is larger: $_9P_9$ or $_9P_8$? Explain your answer.

Developing Skills

In 3–17, compute the value of each expression.

3. $4!$ **4.** $6!$ **5.** $7!$ **6.** $3! + 2!$ **7.** $(3 + 2)!$

8. $_3P_3$ **9.** $_8P_8$ **10.** $\frac{8!}{5!}$ **11.** $_6P_3$ **12.** $_{10}P_2$

13. $_{20}P_2$ **14.** $_{11}P_4$ **15.** $_7P_6$ **16.** $_{255}P_2$ **17.** $\frac{999!}{(999 - 5)!}$

In 18–21, in each case, how many three-letter arrangements can be formed if a letter is used only once?

18. LION **19.** TIGER **20.** MONKEY **21.** LEOPARD

22. Write the following expressions in the order of their values, beginning with the smallest: $_{60}P_5$, $_{45}P_6$, $_{24}P_7$, $_{19}P_7$.

Applying Skills

23. Using the letters E, M, I, T: **a.** How many arrangements of four letters can be found if each letter is used only once in the "word"? **b.** List these "words."

24. In how many different ways can five students be arranged in a row?

25. In a game of cards, Gary held exactly one club, one diamond, one heart, and one spade. In how many different ways can Gary arrange these four cards in his hand?

26. There are nine players on a baseball team. The manager must establish a batting order for the players at each game. The pitcher will bat last. How many different batting orders are possible for the eight remaining players on the team?

27. There are 30 students in a class. Every day the teacher calls on different students to write homework problems on the board, with each problem done by only one student. In how many ways can the teacher call students to the board if the homework consists of:

a. only 1 problem? **b.** 2 problems? **c.** 3 problems?

28. At the Olympics, three medals are given for each competition: gold, silver, and bronze. Tell how many possible winning orders there are for the gymnastic competition if the number of competitors is:

a. 7 **b.** 9 **c.** 11 **d.** n

29. How many different ways are there to label the three vertices of a scalene triangle, using no letter more than once, when:

a. we use the letters R, S, T?

b. we use all the letters of the English alphabet?

30. A class of 31 students elects four people to office, namely, a president, vice president, secretary, and treasurer. In how many possible ways can four people be elected from this class?

31. How many possible ways are there to write two initials, using the letters of the English alphabet, if:

a. an initial may appear only once in each pair?

b. the same initial may be used twice?

In 32–35: **a.** Write each answer in factorial form. **b.** Write each answer, after using a calculator, in scientific notation.

32. In how many different orders can 60 people line up to buy tickets at a theater?

33. We learn the alphabet in a certain order, starting with A, B, C, and ending with Z. How many possible orders are there for listing the letters of the English alphabet?

34. Twenty-five people are waiting for a bus. When the bus arrives, there is room for 18 people to board. In how many ways could 18 of the people who are waiting board the bus?

35. Forty people attend a party at which eight door prizes are to be awarded. In how many orders can the names of the winners be announced?

15-10 PERMUTATIONS WITH REPETITION

How many different "words" or arrangements of four letters can be formed using each letter of the word PEAK?

This is the number of permutations of four things, taken four at a time. Since ${}_4P_4 = 4! = 4 \times 3 \times 2 \times 1 = 24$, there are 24 possible words or arrangements.

PEAK	PAEK	EAPK	AEPK	EPAK	APEK
PAKE	PEKA	APKE	EPKA	AEKP	EAKP
PKAE	PKEA	AKPE	EKPA	AKEP	EKAP
KPAE	KPEA	KAEP	KEAP	KAPE	KEPA

Now consider a related question. How many different words or arrangements of four letters can be formed using each letter of the word PEEK?

This is an example of a permutation with repetition because the letter E is repeated in the word. We can try to list the different arrangements by simply replacing the A in each of the arrangements given for the word PEAK. Let the E from PEAK be E_1 and the E that replaces A be E_2. The arrangements can be written as follows:

PE_1E_2K	PE_2E_1K	E_1E_2PK	E_2E_1PK	E_1PE_2K	E_2PE_1K
PE_2KE_1	PE_1KE_2	E_2PKE_1	E_1PKE_2	E_2E_1KP	E_1E_2KP
PKE_2E_1	PKE_1E_2	E_2KPE_1	E_1KPE_2	E_2KE_1P	E_1KE_2P
KPE_2E_1	KPE_1E_2	KE_2E_1P	KE_1E_2P	KE_2PE_1	KE_1PE_2

We have 24 different arrangements if we consider E_1 to be different from E_2. But they are not really different. Notice that if we consider the E's to be the same, every word in the first column is the same as a word in the second column, every word in the third column is the same as a word in the fourth column, and every word in the fifth column is the same as a word in the sixth column. Therefore, only the first, third, and fifth columns are different and there are $\frac{24}{2}$ or 12 arrangements of four letters when two of them are the same. This is the number of arrangements of four letters divided by the number of arrangements of two letters.

Now consider a third word. In how many ways can the letters of EEEK be arranged? We will use the 24 arrangements of the letters of PE_1E_2K and write E_3 in place of P.

$E_3E_1E_2K$	$E_3E_2E_1K$	$E_1E_2E_3K$	$E_2E_1E_3K$	$E_1E_3E_2K$	$E_2E_3E_1K$
$E_3E_2KE_1$	$E_3E_1KE_2$	$E_2E_3KE_1$	$E_1E_3KE_2$	$E_2E_1KE_3$	$E_1E_2KE_3$
$E_3KE_2E_1$	$E_3KE_1E_2$	$E_2KE_3E_1$	$E_1K\ E_3E_2$	$E_2KE_1E_3$	$E_1KE_2E_3$
$KE_3E_2E_1$	$K\ E_3E_1E_2$	$KE_2E_1E_3$	$KE_1E_2E_3$	$KE_2E_3E_1$	$KE_1E_3E_2$

Notice that each row is the same arrangement if we consider all the E's to be the same letter. The 4! arrangements of four letters are in groups of 3! or 6, the number of different orders in which the 3 E's can be arranged among themselves.

Therefore the number of possible arrangements is:

$$\frac{4!}{3!} = \frac{4 \times 3 \times 2 \times 1}{3 \times 2 \times 1} = \frac{24}{6} = 4$$

If we consider all E's to be the same, the four arrangements are:

EEEK EEKE EKEE KEEE

In general, the number of permutations of n things, taken n at a time, with r of these things identical, is:

$$\frac{n!}{r!}$$

EXAMPLE 1

How many six-digit numerals can be written using all of the following digits: 2, 2, 2, 2, 3, and 5?

Solution This is the number of permutations of six things taken six at a time, with 4 of the digits 2, 2, 2, 2 identical. Therefore:

$$\frac{6!}{4!} = \frac{6 \times 5 \times 4 \times 3 \times 2 \times 1}{4 \times 3 \times 2 \times 1} = 6 \times 5 = 30$$

Calculator Solution ENTER: 6 MATH ◀ 4 ÷ 4 MATH ◀ 4 ENTER

DISPLAY:

```
6!/4!
                30
```

Answer 30 six-digit numerals

EXAMPLE 2

Three children, Rita, Ann, and Marie, take turns doing the dishes each night of the week. At the beginning of each week they make a schedule. If Rita does the dishes three times, and Ann and Marie each do them twice, how many different schedules are possible?

Solution We can think of this as an arrangement of the letters RRRAAMM, that is, an arrangement of seven letters (for the seven days of the week) with R appearing three times and A and M each appearing twice. Therefore we will divide 7! by 3!, the number of arrangements of Rita's days; then by 2!, the number of arrangements of Ann's days; and finally by 2! again, the number of arrangements of Marie's days.

$$\text{Number of arrangements} = \frac{7!}{3! \times 2! \times 2!} = \frac{7 \times 6 \times 5 \times 4 \times 3 \times 2 \times 1}{3 \times 2 \times 1 \times 2 \times 1 \times 2 \times 1} = 210$$

Answer 210 possible schedules

EXERCISES

Writing About Mathematics

1. **a.** List the six different arrangements or permutations of the letters in the word TAR.
   b. Explain why exactly six arrangements are possible.
2. **a.** List the three different arrangements or permutations of the letters in the word TOT.
   b. Explain why exactly three arrangements are possible.

Developing Skills

In 3–6, how many five-letter permutations are there of the letters of each given word?

3. APPLE **4.** ADDED **5.** VIVID **6.** TESTS

In 7–14: **a.** How many different six-letter arrangements can be written using the letters in each given word? **b.** How many different arrangements begin with E? **c.** If an arrangement is chosen at random, what is the probability that it begins with E?

7. SIMPLE **8.** FREEZE **9.** SYSTEM **10.** BETTER
11. SEEDED **12.** DEEDED **13.** TATTOO **14.** ELEVEN

In 15–22, find the number of distinct arrangements of the letters in each word.

15. STREETS **16.** INSISTS **17.** ESTEEMED **18.** DESERVED
19. TENNESSEE **20.** BOOKKEEPER **21.** MISSISSIPPI **22.** UNUSUALLY

In 23–26, in each case find: **a.** How many different five-digit numerals can be written using all five digits listed? **b.** How many of the numerals formed from the given digits are greater than 12,000 and less than 13,000? **c.** If a numeral formed from the given digits is chosen at random, what is the probability that it is greater than 12,000 and less than 13,000?

23. 1, 2, 3, 4, 5 **24.** 1, 2, 2, 2, 2 **25.** 1, 1, 2, 2, 2 **26.** 2, 2, 2, 2, 3

In 27–31, when written without using exponents, a^2x can be written as aax, axa, or xaa. How many different arrangements of letters are possible for each given expression when written without exponents?

27. b^3y **28.** a^2b^5 **29.** abx^6 **30.** a^2by^7 **31.** a^4b^8

Applying Skills

32. A bookseller has 7 copies of a novel and 3 copies of a biography. In how many ways can these 10 books be arranged on a shelf?

33. In how many ways can 6 white flags and 3 blue flags be arranged one above another on a single rope on a flagstaff?

34. Florence has 6 blue beads, 8 white beads, and 4 green beads, all the same size. In how many ways can she string these beads on a chain to make a necklace?

35. Frances has 8 tulip bulbs, 10 daffodil bulbs, and 7 crocus bulbs. In how many ways can Frances plant these bulbs in a border along the edge of her garden?

36. Anna has 2 dozen Rollo bars and 1 dozen apples as treats for Halloween. In how many ways can Anna hand out 1 treat to each of 36 children who come to her door?

37. A dish of mixed nuts contains 7 almonds, 5 cashews, 3 filberts, and 4 peanuts. In how many different orders can Jerry eat the contents of the dish, one nut at a time?

38. Print your first and last names using capital letters. How many different arrangements of the letters in your full name are possible?

15-11 COMBINATIONS

Comparing Permutations and Combinations

A **combination** is a collection of things in which order is not important. Before we discuss combinations, let us start with a problem we know how to solve.

Ann, Beth, Carlos, and Dava are the only members of a school club. In how many ways can they elect a president and a treasurer for the club?

Any one of the 4 students can be elected as president. After this happens, any one of the 3 remaining students can be elected treasurer. Thus there are 4×3 or 12 possible outcomes. Using the initials to represent the students involved, we can write these 12 arrangements:

$$(A, B)\ (B, A)\ (B, C)\ (C, B)$$
$$(A, C)\ (C, A)\ (B, D)\ (D, B)$$
$$(A, D)\ (D, A)\ (C, D)\ (D, C)$$

These are the permutations. We could have found that there are 12 permutations by using the formula:

$${}_4P_2 = 4 \times 3 = 12$$

Answer: 12 permutations

Now let us consider two problems of a different type involving the members of the same club.

Ann, Beth, Carlos, and Dava are the only members of a school club. In how many ways can they choose two members to represent the club at a student council meeting?

If we look carefully at the list of 12 possible selections given in the answer to problem 1, we can see that while (A, B) and (B, A) are two different choices for president and treasurer, sending Ann and Beth to the student council meeting is exactly the same as sending Beth and Ann. For this problem let us match up answers that consist of the same two persons.

$$(A, B) \leftrightarrow (B, A) \qquad (B, C) \leftrightarrow (C, B)$$
$$(A, C) \leftrightarrow (C, A) \qquad (B, D) \leftrightarrow (D, B)$$
$$(A, D) \leftrightarrow (D, A) \qquad (C, D) \leftrightarrow (D, C)$$

Although order is important in listing slates of officers in problem 1, there is no reason to consider the order of elements in this problem. In fact, if we think of two representatives to the student council as a *set* of two club members, we can list the sets of representatives as follows:

$$\{A, B\} \qquad \{B, C\}$$
$$\{A, C\} \qquad \{B, D\}$$
$$\{A, D\} \qquad \{C, D\}$$

From this list we can find the number of *combinations* of 4 things, taken 2 at a time, written in symbols as ${}_4C_2$. The answer to this problem is found by dividing the number of permutations of 4 things taken 2 at a time by 2!. Thus:

$${}_4C_2 = \frac{{}_4P_2}{2!} = \frac{4 \times 3}{2 \times 1} = \frac{12}{2} = 6$$

Answer: 6 combinations

Ann, Beth, Carlos, and Dava are the only members of a school club. In how many ways can they choose a three-member committee to work on the club's next project? Is order important to this answer? If 3 officers were to be elected, such as a president, a treasurer, and a secretary, then order would be important and the number of permutations would be needed. However, a committee is a set of people. In listing the elements of a set, order is not important. Compare the permutations and combinations of 4 persons taken 3 at a time:

Permutations	*Combinations*
(A, B, C) (A, C, B) (B, A, C) (B, C, A) (C, A, B) (C, B, A)	{A, B, C}
(A, B, D) (A, D, B) (B, A, D) (B, D, A) (D, A, B) (D, B, A)	{A, B, D}
(A, C, D) (A, D, C) (C, A, D) (C, D, A) (D, A, C) (D, C, A)	{A, C, D}
(B, C, D) (B, D, C) (C, B, D) (C, D, B) (D, B, C) (D, C, B)	{B, C, D}

While there are 24 permutations, written as ordered triples, there are only 4 combinations, written as sets. For example, in the first row of permutations, there are 3! or $3 \times 2 \times 1$ or 6 ordered triples (slates of officers) including Ann, Beth, and Carlos.

However, there is only one set (committee) that includes these three persons. Therefore, the number of ways to select a three-person committee from a group of four persons is the number of *combinations* of 4 things taken 3 at a time. This number is found by dividing the number of permutations of 4 things taken 3 at a time by 3!, the number of arrangements of the three things.

$$_4C_3 = \frac{_4P_3}{3!} = \frac{4 \times 3 \times 2}{3 \times 2 \times 1} = \frac{24}{6} = 4$$

Answer: 4 combinations

In general, for counting numbers n and r, where $r \leq n$, the number of combinations of n things taken r at a time is found by using the formula:

$$_nC_r = \frac{_nP_r}{r!}$$

On a graphing calculator, the sequence of keys needed to find $_nC_r$ is similar to that for $_nP_r$. The combination symbol is entry 3 in the PRB menu.

ENTER: 4 MATH ◄ 3 3 ENTER

DISPLAY:

```
4 nCr 3
                4
```

Note: The notation $\binom{n}{r}$ also represents the number of combinations of n things taken r at a time. Thus:

$$\binom{n}{r} = {_nC_r} \quad \text{or} \quad \binom{n}{r} = \frac{_nP_r}{r!}$$

Some Relationships Involving Combinations

Given a group of 5 people, how many different 5-person committees can be formed? Common sense tells us that there is only 1 such committee, namely, the committee consisting of all 5 people. Using combinations, we see that

$$_5C_5 = \frac{_5P_5}{5!} = \frac{5 \times 4 \times 3 \times 2 \times 1}{5 \times 4 \times 3 \times 2 \times 1} = 1$$

Also,

$$_3C_3 = \frac{_3P_3}{3!} = \frac{3 \times 2 \times 1}{3 \times 2 \times 1} = 1 \quad \text{and} \quad _4C_4 = \frac{_4P_4}{4!} = \frac{4 \times 3 \times 2 \times 1}{4 \times 3 \times 2 \times 1} = 1$$

► **For any counting number n, $_nC_n = 1$.**

Given a group of 5 people, in how many different ways can we select a committee consisting of no people, or 0 people? Common sense tells us that there is only 1 way to select no one. Thus, using combinations, $_5C_0 = 1$. Let us agree to the following generalization:

► **For any counting number n, $_nC_0 = 1$.**

In how many ways can we select a committee of 2 people from a group of 7 people? Since a committee is a combination,

$$_7C_2 = \frac{_7P_2}{2!} = \frac{7 \times 6}{2 \times 1} = 21.$$

Now, given a group of 7 people, in how many ways can 5 people *not* be appointed to the committee? Since each set of people not appointed is a combination:

$$_7C_5 = \frac{_7P_5}{5!} = \frac{7 \times 6 \times 5 \times 4 \times 3}{5 \times 4 \times 3 \times 2 \times 1} = \frac{7 \times 6}{2 \times 1} \times \frac{5 \times 4 \times 3}{5 \times 4 \times 3} = \frac{7 \times 6}{2 \times 1} \times 1 = 21$$

Notice that $_7C_2 = {}_7C_5$. In other words, starting with a group of 7 people, the number of sets of 2 people that can be selected is equal to the number of sets of 5 people that can be *not* selected. In the same way it can be shown that $_7C_3 = {}_7C_4$, that $_7C_6 = {}_7C_1$, and that $_7C_7 = {}_7C_0$.

In general, starting with n objects, the number of ways to choose r objects for a combination is equal to the number of ways to *not* choose $(n - r)$ objects for the combination.

► **For whole numbers n and r, where $r \leq n$,**

$$_nC_r = {}_nC_{n-r}$$

KEEP IN MIND

PERMUTATIONS

1. *Order is important.*

   Think of ordered elements such as ordered pairs and ordered triples.

2. An *arrangement* or *a slate of officers* indicates a permutation.

COMBINATIONS

1. *Order is not important.*

   Think of sets.

2. A *committee*, or a *selection of a group*, indicates a combination.

EXAMPLE 1

Evaluate: $_{10}C_3$

Solution This is the number of combinations of 10 things, taken 3 at a time. Thus

$$_{10}C_3 = \frac{_{10}P_3}{3!} = \frac{10 \times 9 \times 8}{3 \times 2 \times 1} = \frac{720}{6} = 120$$

Calculator Solution

ENTER: 10 MATH ◄ 3 3 ENTER

DISPLAY:

```
10 nCr 3
             120
```

Answer 120

EXAMPLE 2

Evaluate: $\binom{25}{23}$

Solution (1) This is an alternative form for the number of combinations of 25 things, taken 23 at a time: $\binom{25}{23} = {}_{25}C_{23}$.

(2) Since ${}_nC_r = {}_nC_{n-r}$, ${}_{25}C_{23} = {}_{25}C_2$.

(3) Using ${}_{25}C_2$, perform the shorter computation. Thus:

$$\binom{25}{23} = {}_{25}C_{23} = {}_{25}C_2 = \frac{25 \times 24}{2 \times 1} = 300$$

Answer 300

EXAMPLE 3

There are 10 teachers in the science department. How many 4-person committees can be formed in the department if Mrs. Martens and Dr. Blumenthal, 2 of the teachers, must be on each committee?

Solution Since Mrs. Martens and Dr. Blumenthal must be on each committee, the problem becomes one of filling 2 positions on a committee from the remaining 8 teachers.

$${}_8C_2 = \frac{{}_8P_2}{2!} = \frac{8 \times 7}{2 \times 1} = 28$$

Answer 28 committees

EXAMPLE 4

There are six points in a plane, no three of which are collinear. How many straight lines can be drawn using pairs of these three points?

Solution Whether joining points A and B, or points B and A, only 1 line exists, namely, $\overleftrightarrow{AB}$. Since order is not important here, this is a combination of 6 points, taken 2 at a time.

$${}_6C_2 = \frac{{}_6P_2}{2!} = \frac{6 \times 5}{2 \times 1} = 15$$

Answer 15 lines

EXAMPLE 5

Lisa Dwyer is a teacher at a local high school. In her class, there are 10 boys and 20 girls. Find the number of ways in which Ms. Dwyer can select a team of 3 students from the class to work on a group project if the team consists of:

a. any 3 students **b.** 1 boy and 2 girls **c.** 3 girls **d.** at least 2 girls

Solution **a.** The class contains 10 boys and 20 girls, for a total of 30 students. Since order is not important on a team, this is a combination of 30 students, taken 3 at a time.

$$_{30}C_3 = \frac{_{30}P_3}{3!} = \frac{30 \times 29 \times 28}{3 \times 2 \times 1} = 4{,}060$$

b. This is a compound event. To find the number of ways to select 1 boy out of 10 boys for a team, use $_{10}C_1$. To find the number of ways to select 2 girls out of 20 for the team, use $_{20}C_2$. Then, by the counting principle, multiply the results.

$$_{10}C_1 \times {}_{20}C_2 = \frac{10}{1} \times \frac{20 \times 19}{2 \times 1} = 10 \times 190 = 1{,}900$$

c. This is another compound event, in which 0 boys out of 10 boys and 3 girls out of 20 girls are selected. Recall that $_{10}C_0 = 1$. Thus:

$$_{10}C_0 \times {}_{20}C_3 = 1 \times \frac{20 \times 19 \times 18}{3 \times 2 \times 1} = 1{,}140$$

Note that this could also have been thought of as the simple event of selecting 3 girls out of 20 girls:

$$_{20}C_3 = \frac{20 \times 19 \times 18}{3 \times 2 \times 1} = 1{,}140$$

d. A team of at least 2 girls can consist of exactly 2 girls (see part **b**) or exactly 3 girls (see part **c**). Since these events are disjoint, add the solutions to parts **b** and **c**:

$$1{,}900 + 1{,}140 = 3{,}040$$

Answers **a.** 4,060 teams **b.** 1,900 teams **c.** 1,140 teams **d.** 3,040 teams

EXERCISES

Writing About Mathematics

1. Explain the difference between a permutation and a combination.
2. A set of r letters is to be selected from the 26 letters of the English alphabet. For what value of r is the number of possible sets of numbers greatest?

Developing Skills

In 3–14, evaluate each expression.

3. ${}_{15}C_2$ **4.** ${}_{12}C_3$ **5.** ${}_{10}C_4$ **6.** ${}_{25}C_1$

7. ${}_{13}C_0$ **8.** ${}_{14}C_{14}$ **9.** ${}_{9}C_8$ **10.** ${}_{200}C_{198}$

11. $\binom{7}{3}$ **12.** $\binom{9}{4}$ **13.** $\binom{17}{17}$ **14.** $\binom{499}{2}$

15. Find the number of combinations of 6 things, taken 3 at a time.

16. How many different committees of 3 people can be chosen from a group of 9 people?

17. How many different subsets of exactly 7 elements can be formed from the set

$$\{0, 1, 2, 3, 4, 5, 6, 7, 8, 9\}?$$

18. For each given number of non-collinear points in a plane, how many straight lines can be drawn?

a. 3 **b.** 4 **c.** 5 **d.** 7 **e.** 8 **f.** n

19. Consider the following formulas:

(1) ${}_nC_r = \frac{{}_nP_r}{r!}$ (2) ${}_nC_r = \frac{n!}{(n-r)!r!}$ (3) ${}_nC_r = \frac{n(n-1)(n-2)\cdots(n-r+1)}{r!}$

a. Evaluate ${}_8C_3$ using each of the three formulas.

b. Evaluate ${}_{11}C_7$ using each of the three formulas.

c. Can all three formulas be used to find the combination of n things, taken r at a time?

Applying Skills

20. A coach selects players for a team. If, while making this first selection, the coach pays no attention to the positions that individuals will play, how many teams are possible?

a. Of 14 candidates, Coach Richko needs 5 for a basketball team.

b. Of 16 candidates, Coach Jones needs 11 for a football team.

c. Of 13 candidates, Coach Greves needs 9 for a baseball team.

21. A disc jockey has 25 recordings at hand, but has time to play only 22 on the air. How many sets of 22 recordings can be selected?

22. There are 14 teachers in a mathematics department.

a. How many 4-person committees can be formed in the department?

b. How many 4-person committees can be formed if Mr. McDonough, 1 of the 14, must be on the committee?

c. How many 4-person committees can be formed if Mr. Goldstein and Mrs. Friedel, 2 of the 14, must be on the committee?

23. There are 12 Republicans and 10 Democrats on a senate committee. From this group, a 3-person subcommittee is to be formed. Find the number of 3-person subcommittees that consist of:

a. any members of the senate committee
b. Democrats only
c. 1 Republican and 2 Democrats
d. at least 2 Democrats
e. John Clark, who is a Democrat, and any 2 Republicans

24. Sue Bartling loves to read mystery books and car-repair manuals. On a visit to the library, Sue finds 9 new mystery books and 3 car-repair manuals. She borrows 4 of these books. Find the number of different sets of 4 books Sue can borrow if:

a. all are mystery books
b. exactly 2 are mystery books
c. only 1 is a mystery book
d. all are car-repair manuals

25. Cards are drawn at random from a 52-card deck. Find the number of different 5-card poker hands possible consisting of:

a. any 5 cards from the deck
b. 3 aces and 2 kings
c. 4 queens and any other card
d. 5 spades
e. 2 aces and 3 picture cards
f. 5 jacks

26. How many committees consisting of 7 people or more can be formed from a group of 10 people?

27. There are 12 roses growing in Heather's garden. How many different ways can Heather choose roses for a bouquet consisting of more than 8 roses?

15-12 PERMUTATIONS, COMBINATIONS, AND PROBABILITY

In this section, a variety of probability questions are presented. In some cases, permutations should be used. In other cases, combinations should be used. When answering these questions, the following should be kept in mind:

1. If the question asks "How many?" or "In how many ways?" the answer will be a whole number.
2. If the question asks "What is the probability?" the answer will be a value between 0 and 1 inclusive.
3. If order is important, use permutations.
4. If order is not important, use combinations.

In Examples 1–4, Ms. Fenstermacher must select 4 students to represent the class in a spelling bee. Her best students include 3 girls (Callie, Daretta, and Jessica) and 4 boys (Bandu, Carlos, Sanjit, and Uri).

EXAMPLE 1

Ms. Fenstermacher decides to select the 4 students from the 7 best by drawing names from a hat. How many different groups of 4 are possible?

Solution In choosing a group of 4 students out of 7, order is not important. Therefore, use combinations.

$$_7C_4 = \frac{_7P_4}{4!} = \frac{7 \times 6 \times 5 \times 4}{4 \times 3 \times 2 \times 1} = 35$$

Answer 35 groups

EXAMPLE 2

What is the probability that the 4 students selected for the spelling bee will consist of 2 girls and 2 boys?

Solution (1) The answer to Example 1 shows 35 possible groups, or $n(S) = 35$.

(2) Since order is not important in choosing groups, use combinations.

The number of ways to choose 2 girls out of 3 is $_3C_2 = \frac{3 \times 2}{2 \times 1} = 3$.

The number of ways to choose 2 boys out of 4 is $_4C_2 = \frac{4 \times 3}{2 \times 1} = 6$.

Event E is the compound event of choosing 2 girls and 2 boys, or

$$n(E) = {_3C_2} \times {_4C_2} = 3 \times 6 = 18.$$

(3) Thus $P(E) = \frac{_3C_2 \times {_4C_2}}{_7C_4} = \frac{18}{35}$.

Answer $P(\text{2 girls, 2 boys}) = \frac{18}{35}$

EXAMPLE 3

The students chosen are Callie, Jessica, Carlos, and Sanjit. In how many orders can these 4 students be called upon in the spelling bee?

Solution The number of ways in which 4 students can be called upon in a spelling bee means that someone is first, someone else is second, and so on. Since this is a problem about order, use permutations.

$$_4P_4 = 4! = 4 \times 3 \times 2 \times 1 = 24$$

Answer 24 orders

EXAMPLE 4

For the group consisting of Callie, Jessica, Carlos, and Sanjit, what is the probability that the first 2 students called upon will be girls?

Solution This question may be answered using various methods, two of which are shown below.

METHOD 1. *Counting Principle*

There are 2 girls out of 4 students who may be called first. Once a girl is called upon, only 1 girl remains in the 3 students not yet called upon. Apply the counting principle of probability.

$$P(\text{first 2 are girls}) = P(\text{girl first}) \times P(\text{girl second}) = \tfrac{2}{4} \times \tfrac{1}{3} = \tfrac{2}{12} = \tfrac{1}{6}$$

METHOD 2. *Permutations*

Let $n(S)$ equal the number of ways to call 2 of the 4 students in order, and let $n(E)$ equal the number of ways to call 2 of the 2 girls in order.

$$P(E) = \frac{n(E)}{n(S)} = \frac{{}_2P_2}{{}_4P_2} = \frac{2 \times 1}{4 \times 3} = \frac{2}{12} = \frac{1}{6}$$

Answer $P(\text{first 2 are girls}) = \frac{1}{6}$

EXERCISES

Writing About Mathematics

1. A committee of three persons is to be chosen from a group of eight persons. Olivia is one of the persons in that group. Olivia said that since ${}_8C_3 = {}_8C_5$, the probability that she will be chosen for that committee is equal to the probability that she will not be chosen. Do you agree with Olivia? Explain why or why not.

2. Four letters are to be selected at random from the alphabet. Jenna found the probability that the four letters followed S in alphabetical order by using permutations. Colin found the probability that the four letters followed S in alphabetical order by using combinations. Who was correct? Explain your answer.

Applying Skills

3. The Art Club consists of 4 girls (Jennifer, Anna, Gloria, Teresa) and 2 boys (Mark and Dan).
   a. In how many ways can the club elect a president and a treasurer?
   b. Find the probability that the 2 officers elected are both girls.
   c. How many 2-person teams can be selected to work on a project?
   d. Find the probability that a 2-person team consists of:
   (1) 2 girls (2) 2 boys (3) 1 girl and 1 boy (4) Anna and Mark

4. A committee of 4 is to be chosen at random from 4 men and 3 women.

a. How many different 4-member committees are possible?

b. How many 4-member committees contain 3 men and 1 woman?

c. What is the probability that a 4-member committee will contain exactly 1 woman?

d. What is the probability that a man will be on the committee?

e. What is the probability that the fourth person chosen is a woman given that 3 men have already been chosen?

5. A committee of 6 people is to be chosen from 9 available people.

a. How many 6-person committees can be chosen?

b. The committee, when chosen, has 4 students and 2 teachers. Find the probability that a 3-person subcommittee from this group includes:

(1) students only

(2) exactly 1 teacher

(3) at least 2 students

(4) a teacher given that 2 students have been chosen

6. A box of chocolate-covered candies contains 7 caramels and 3 creams all exactly the same in appearance. Jim selects 4 pieces of candy.

a. Find the number of selections possible of 4 pieces of candy that include:

(1) 4 caramels

(2) 1 caramel and 3 creams

(3) 2 caramels and 2 creams

(4) any 4 pieces

b. Find the probability that Jim's selection included:

(1) 4 caramels

(2) 1 caramel and 3 creams

(3) 2 caramels and 2 creams

(4) no caramels

(5) a second cream given that 2 caramels and a cream have been selected

7. Two cards are drawn at random from a 52-card deck without replacement. Find the probability of drawing:

a. the ace of clubs and jack of clubs in either order

b. a red ace and a black jack in either order

c. 2 jacks

d. 2 clubs

e. an ace and a jack in either order

f. an ace and a jack in that order

g. a heart given that the king of hearts was drawn

h. a king given that a queen was drawn

8. Mrs. Carberry has 4 quarters and 3 nickels in her purse. If she takes 3 coins out of her purse without looking at them, find the probability that the 3 coins are worth:

a. exactly 75 cents

b. exactly 15 cents

c. exactly 35 cents

d. exactly 55 cents

e. more than 10 cents

f. less than 40 cents

9. A 3-digit numeral is formed by selecting digits at random from $\{2, 4, 6, 7\}$ without repetition. Find the probability that the number formed:

a. is less than 700 **b.** is greater than 600

c. contains only even digits **d.** is an even number

10. a. There are 10 runners on the track team. If 4 runners are needed for a relay race, how many different relay teams are possible?

b. Once the relay team is chosen, in how many different orders can the 4 runners run the race?

c. If Nicolette is on the relay team, what is the probability that she will lead off the race?

11. Lou Grant is an editor at a newspaper employing 10 reporters and 3 photographers.

a. If Lou selects 2 reporters and 1 photographer to cover a story, from how many possible 3-person teams can he choose?

b. If Lou hands out 1 assignment per reporter, in how many ways can he assign the first 3 stories to his 10 reporters?

c. If Lou plans to give the first story to Rossi, a reporter, in how many ways can he now assign the first 3 stories?

d. If 3 out of 10 reporters are chosen at random to cover a story, what is the probability that Rossi is on this team?

12. Chris, Willie, Tim, Matt, Juan, Bob, and Steven audition for roles in the school play.

a. If 2 male roles in the play are those of the hero and the clown, in how many ways can the director select 2 of the 7 boys for these roles?

b. Chris and Willie got the 2 leading male roles.

(1) In how many ways can the director select a group of 3 of the remaining 5 boys to work in a crowd scene?

(2) How many of these groups of 3 will include Tim?

(3) Find the probability that Tim is in the crowd scene.

13. There are 8 candidates for 3 seats in the student government. The candidates include 3 boys (Alberto, Peter, Thomas) and 5 girls (Elizabeth, Maria, Joanna, Rosa, Danielle). If all candidates have an equal chance of winning, find the probability that the winners include:

a. 3 boys **b.** 3 girls

c. 1 boy and 2 girls **d.** at least 2 girls

e. Maria, Peter, and anyone else **f.** Danielle and any other two candidates

g. Alberto, Elizabeth, and Rosa **h.** Rosa, Peter, and Thomas

14. A gumball machine contains 6 lemon-, 4 lime-, 3 cherry-, and 2 orange-flavored gumballs. Five coins are put into the machine, and 5 gumballs are obtained.

a. How many different sets of 5 gumballs are possible?

b. How many of these will contain 2 lemon and 3 lime gumballs?

c. Find the probability that the 5 gumballs dispensed by the machine include:

(1) 2 lemon and 3 lime
(2) 3 cherry and 2 orange
(3) 2 lemon, 2 lime, and 1 orange
(4) lemon only
(5) lime only
(6) no lemon

15. The letters in the word HOLIDAY are rearranged at random.

a. How many 7-letter words can be formed?

b. Find the probability that the first 2 letters are vowels.

c. Find the probability of no vowels in the first 3 letters.

16. At a bus stop, 5 people enter a bus that has only 3 empty seats.

a. In how many different ways can 3 of the 5 people occupy these empty seats?

b. If Mrs. Costa is 1 of the 5 people, what is the probability that she will *not* get a seat?

c. If Ann and Bill are 2 of the 5 people, what is the probability that they both will get seats?

CHAPTER SUMMARY

Probability is a branch of mathematics in which the chance of an event happening is assigned a numerical value that predicts how likely that event is to occur.

Empirical probability may be defined as the most accurate scientific estimate, based on a large number of trials, of the cumulative relative frequency of an event happening.

An **outcome** is a result of some activity or experiment. A **sample space** is a set of all possible outcomes for the activity. An **event** is a subset of the sample space.

The **theoretical probability** of an event is the number of ways that the event can occur, divided by the total number of possibilities in the sample space. If $P(E)$ represents the probability of event E, $n(E)$ represents the number of outcomes in event E and $n(S)$ represents the number of outcomes in the sample space S. The formula, which applies to **fair and unbiased** objects and situations is:

$$P(E) = \frac{n(E)}{n(S)}$$

The probability of an **impossible** event is 0.

The probability of an event that is **certain** to occur is 1.

The probability of any event E must be equal to or greater than 0, and less than or equal to 1:

$$0 \leq P(E) \leq 1$$

For the shared outcomes of events A and B:

$$P(A \text{ and } B) = \frac{n(A \cap B)}{n(S)}$$

If two events A and C are **mutually exclusive**:

$$P(A \text{ or } C) = P(A) + P(C)$$

If two events A and B are not mutually exclusive:

$$P(A \text{ or } B) = P(A) + P(B) - P(A \text{ and } B)$$

For any event A:

$$P(A) + P(\text{not } A) = 1$$

The sum of the probabilities of all possible singleton outcomes for any sample space must always equal 1.

The Counting Principle: If one activity can occur in any of m ways and, following this, a second activity can occur in any of n ways, then both activities can occur in the order given in $m \times n$ ways.

The Counting Principle for Probability: E and F are independent events. The probability of event E is m $(0 \leq m \leq 1)$ and the probability of event F is n $(0 \leq n \leq 1)$, the probability of the event in which E and F occur jointly is the product $m \times n$.

When the result of one activity in no way influences the result of a second activity, the results of these activities are called **independent events**. Two events are called **dependent events** when the result of one activity influences the result of a second activity.

If A and B are two events, **conditional probability** is the probability of B given that A has occurred. The notation for conditional probability is $P(B \mid A)$.

$$P(B \mid A) = \frac{P(A \text{ and } B)}{P(A)}$$

The general result $P(A \text{ and } B) = P(A) \times P(B \mid A)$ is true for both dependent and independent events since for independent events, $P(B \mid A) = P(B)$.

A **permutation** is an arrangement of objects in some specific order. The number of permutations of n objects taken n at a time, ${}_nP_n$, is equal to **n factorial:**

$${}_nP_n = n! = n \times (n-1) \times (n-2) \times \cdots \times 3 \times 2 \times 1$$

The number of permutations of n objects taken r at a time, where $r \leq n$, is equal to:

$${}_nP_r = \underbrace{n(n-1)(n-2)\cdots}_{r \text{ factors}} \quad \text{or} \quad {}_nP_r = \frac{n!}{(n-r)!}$$

A permutation of n objects taken n at a time in which r are identical is equal to $\frac{n!}{r!}$.

A **combination** is a set of objects in which order is not important, as in a committee. The formula for the number of combinations of n objects taken r at a time ($r \leq n$) is

$$_nC_r = \frac{_nP_r}{r!}$$

From this formula, it can be shown that $_nC_n = 1$, $_nC_0 = 1$, and $_nC_r = {_nC_{n-r}}$. Permutations and combinations are used to evaluate $n(S)$ and $n(E)$ in probability problems.

VOCABULARY

15-1 Probability • Empirical study • Relative frequency • Cumulative relative frequency • Converge • Empirical probability • Trial • Experiment • Fair and unbiased objects • Biased objects • Die • Standard deck of cards

15-2 Outcome • Sample space • Event • Favorable event • Unfavorable event • Singleton event • Theoretical probability • Calculated probability • Uniform probability • Equally likely outcomes • Random selection

15-3 Impossibility • Certainty • Subscript

15-5 Mutually exclusive events

15-7 Compound event • Tree diagram • List of ordered pairs • Graph of ordered pairs • Counting Principle • Independent events

15-8 Without replacement • Dependent events • With replacement • Conditional probability

15-9 Permutation • Factorial symbol (!) • n factorial

15-11 Combination

REVIEW EXERCISES

1. In a dish of jellybeans, some are black. Aaron takes a jellybean from the dish at random without looking at what color he is taking. Jake chooses the same color jellybean that Aaron takes. Is the probability that Aaron takes a black jellybean the same as the probability that Jake takes a black jellybean? Explain your answer.

2. The probability that Aaron takes a black jellybean from the dish is $\frac{8}{25}$. Can we conclude that there are 25 jellybeans in the dish and 8 of them are black? Explain why or why not.

3. The numerical value of ${}_nP_r$ is the product of r factors. Express the smallest of those factors in terms of n and r.

4. If $P(A) = .4$, $P(B) = .3$, and $P(A \cap B) = .2$, find $P(A \cup B)$.

5. If $P(A) = .5$, $P(B) = .2$, and $P(A \cup B) = .5$, find $P(A \cap B)$.

In 6–11, evaluate each expression:

6. 8! **7.** ${}_5P_5$ **8.** ${}_{12}P_2$ **9.** ${}_5C_5$ **10.** ${}_{12}C_3$ **11.** ${}_{40}C_{38}$

12. How many 7-letter words can be formed from the letters in UNUSUAL if each letter is used only once in a word?

13. A SYZYGY is a nearly straight-line configuration of three celestial bodies such as the Earth, Moon, and Sun during an eclipse.

   a. How many different 6-letter arrangements can be made using the letters in the word SYZYGY?

   b. Find the probability that the first letter in an arrangement of SYZYGY is

   (1) G (2) Y (3) a vowel (4) not a vowel

14. From a list of 10 books, Gwen selects 4 to read over the summer. In how many ways can Gwen make a selection of 4 books?

15. Mrs. Moskowitz, the librarian, checks out the 4 books Gwen has chosen. In how many different orders can the librarian stamp these 4 books?

16. A coach lists all possible teams of 5 that could be chosen from 8 candidates. How many different teams can he list?

17. The probability that Greg will get a hit the next time at bat is 35%. What is the probability that Greg will *not* get a hit?

18. **a.** How many 3-digit numerals can be formed using the digits 2, 3, 4, 5, 6, 7, and 8 if repetition is not allowed?

   b. What is the probability that such a 3-digit numeral is greater than 400?

19. **a.** If a 4-member committee is formed from 3 girls and 6 boys in a club, how many committees can be formed?

   b. If the members of the committee are chosen at random, find the probability that the committee consists of:

   (1) 2 girls and 2 boys (2) 1 girl and 3 boys (3) 4 boys

   c. What is the probability that the fourth member chosen is a girl if the first three are 2 boys and a girl?

   d. What is the probability that a boy is on the committee?

20. From a 52-card deck, 2 cards are drawn at random without replacement. Find the probability of selecting:

a. a ten and a king, in that order
b. 2 tens
c. a ten and a king, in either order
d. 2 spades
e. a king as the second card if the king of hearts is the first card selected.

21. If a card is drawn at random from a 52-card deck, find the probability that the card is:

a. an eight or a queen
b. an eight or a club
c. red or an eight
d. not a club
e. red and a club
f. not an eight or not a club

22. From an urn, 3 marbles are drawn at random with no replacement. Find the probability that the 3 marbles are the same color if the urn contains, in each case, the given marbles:

a. 3 red and 2 white
b. 2 red and 2 blue
c. 3 red and 4 blue
d. 4 white and 5 blue
e. 10 blue
f. 2 red, 1 white, 3 blue

23. From a class of girls and boys, the probability that one student chosen at random to answer a question will be a girl is $\frac{1}{3}$. If four boys leave the class, the probability that a student chosen at random to answer a question will be a girl is $\frac{2}{5}$. How many boys and girls are there in the class before the four boys leave?

24. A committee of 5 is to be chosen from 4 men and 3 women.

a. How many different 5-person committees are possible?

b. Find the probability that the committee includes:

(1) 2 men and 3 women
(2) 3 men and 2 women
(3) at least 2 women
(4) all women

c. In how many ways can this 5-person committee select a chairperson and a secretary?

d. If Hilary and Helene are on the committee, what is the probability that one is selected as chairperson and the other as secretary?

25. Assume that $P(\text{male}) = P(\text{female})$. In a family of three children, what is the probability that all three children are of the same gender?

26. a. Let $n(S)$ = the number of five-letter words that can be formed using the letters in the word RADIO. Find $n(S)$.

b. Let $n(E)$ = the number of five-letter words that can be formed using the letters in the word RADIO, in which the first letter is a vowel Find $n(E)$.

c. If the letters in RADIO are rearranged at random, find the probability that the first letter is a vowel by using the answers to parts **a** and **b**,

d. Find the answer to part **c** by a more direct method.

In 27–32, if the letters in each given word are rearranged at random, use two different methods to find the probability that the first letter in the word is A.

27. RADAR **28.** CANVAS **29.** AZALEA

30. AA **31.** DEFINE **32.** CANOE

33. Twenty-four women and eighteen men are standing in a ticket line. What is the probability that the first five persons in line are women? (The answer need not be simplified.)

34. A four-digit code consists of numbers selected from the set of even digits: 0, 2, 4, 6, 8. No digit is used more than once in any code and the code can begin with 0. What is the probability that the code is a number less than 4,000?

35. In a class there are 4 more boys than girls. A student from the class is chosen at random. The probability that the student is a boy is $\frac{3}{5}$. How many girls and how many boys are there in the class?

36. The number of seniors in the chess club is 2 less than twice the number of juniors, and the number of sophomores is 7 more than the number of juniors. If one person is selected at random to represent the chess club at a tournament, the probability that a senior is chosen is $\frac{2}{5}$. Find the number of students from each class who are members of the club.

37. In a dish, Annie has 16 plain chocolates and 34 candy-coated chocolates, of which 4 are blue, 12 are purple, 15 are green, and 3 are gray.

a. What is the probability that Annie will randomly choose a plain chocolate followed by a blue chocolate?

b. Annie eats 2 green chocolates and then passes the dish to Bob. What is the probability that he will randomly choose a purple chocolate?

38. Find the probability that the next patient of the doctor described in the chapter opener on page 575 will need either a flu shot or a pneumonia shot.

Exploration

Rachael had a box of disks, all the same size and shape. She removed 20 disks from the box, marked them, and then returned them to the box. After mixing the marked disks with the others in the box, she removed a handful of disks and recorded the total number of disks and the number of marked disks. Then she returned the disks to the box, mixed them, and removed another handful. She repeated this last step eight times. The chart below shows her results.

Disks	12`	10	15	11	8	10	7	12	9	13
Marked Disks	3	4	5	4	3	2	2	5	3	5

a. Use the data to estimate the number of disks in the box.

b. Repeat Rachael's experiment using a box containing an unknown number of disks. Compare your estimate with the actual number of disks in the box.

c. Explain how this procedure could be used to estimate the number of fish in a pond.

CUMULATIVE REVIEW CHAPTERS 1–15

Part I

Answer all questions in this part. Each correct answer will receive 2 credits. No partial credit will be allowed.

1. Which of the following is a rational number?

(1) $\sqrt{169}$ (2) $\sqrt{12}$ (3) $\sqrt{9+4}$ (4) π

2. The area of a trapezoid is 21 square inches and the measure of its height is 6.0 inches. The sum of the lengths of the bases is

(1) 3.5 inches (2) 5.3 inches (3) 7.0 inches (4) 14 inches

3. A point on the line whose equation is $y = -3x + 1$ is

(1) $(4, -1)$ (2) $(-1, -2)$ (3) $(-2, 1)$ (4) $(1, -2)$

4. Which of the following is the graph of an exponential function?

(1)
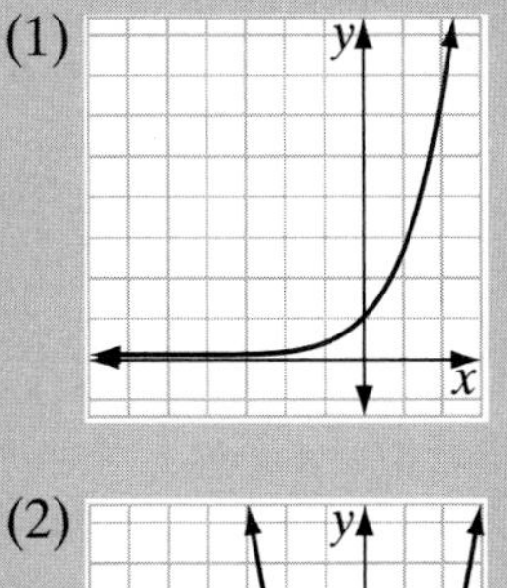

(3)
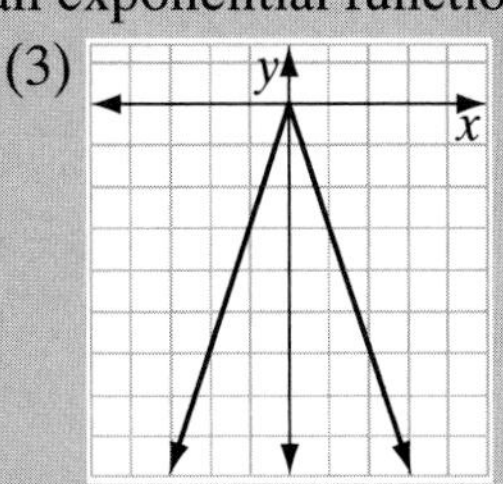

(2)
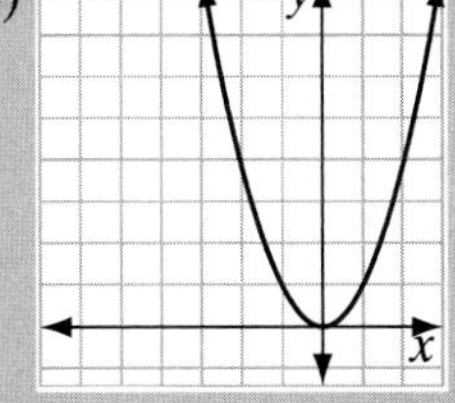

(4)
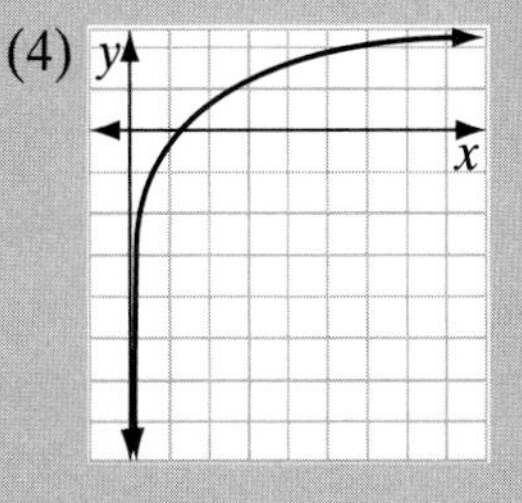

5. The height of a cone is 12.0 centimeters and the radius of its base is 2.30 centimeters. What is the volume of the cone to three significant digits?
(1) 52.2 square centimeters (3) 66.5 square centimeters
(2) 52.3 square centimeters (4) 66.6 square centimeters

6. If $12 - 3(x - 1) = 7x$, the value of x is
(1) 1.5 (2) 4.5 (3) 0.9 (4) 3.75

7. The product of $4x^{-2}$ and $5x^3$ is
(1) $20x^{-6}$ (2) $20x$ (3) $9x^{-6}$ (4) $9x$

8. There are 12 members of the basketball team. At each game, the coach selects a group of five team members to start the game. For how many games could the coach make different selections?
(1) 12! (2) 5! (3) $\frac{12!}{5!}$ (4) $\frac{12!}{7! \cdot 5!}$

9. Which of the following is an example of direct variation?
(1) the area of any square compared to the length of its side
(2) the perimeter of any square compared to the length of its side
(3) the time it takes to drive 500 miles compared to the speed
(4) the temperature compared to the time of day

10. Which of the following intervals represents the solution set of the inequality $-3 \leq 2x+1 < 7$?
(1) $(-3, 7)$ (2) $(-2, 3)$ (3) $[-3, 7)$ (4) $[-2, 3)$

Part II

Answer all questions in this part. Each correct answer will receive 2 credits. Clearly indicate the necessary steps, including appropriate formula substitutions, diagrams, graphs, charts, etc. For all questions in this part, a correct numerical answer with no work shown will receive only 1 credit.

11. The vertices of parallelogram $ABCD$ are $A(0, 0)$, $B(7, 0)$, $(12, 8)$, and $D(5, 8)$. Find, to the nearest degree the measure of $\angle DAB$.

12. Marty bought 5 pounds of apples for 98 cents a pound. A week later, she bought 7 pounds of apples for 74 cents a pound. What was the average price per pound that Marty paid for apples?

Part III

Answer all questions in this part. Each correct answer will receive 3 credits. Clearly indicate the necessary steps, including appropriate formula substitutions, diagrams, graphs, charts, etc. For all questions in this part, a correct numerical answer with no work shown will receive only 1 credit.

13. Each student in an international study group speaks English, Japanese, or Spanish. Of these students, 100 speak English, 50 speak Japanese, 100 speak Spanish, 45 speak English and Spanish, 10 speak English and

Japanese, 13 speak Spanish and Japanese, and 5 speak all three languages. How many students do not speak English?

14. Abe, Brian, and Carmela share the responsibility of caring for the family pets. During a seven-day week, Abe and Brian each take three days and Carmela the other one. In how many different orders can the days of a week be assigned?

Part IV

Answer all questions in this part. Each correct answer will receive 4 credits. Clearly indicate the necessary steps, including appropriate formula substitutions, diagrams, graphs, charts, etc. For all questions in this part, a correct numerical answer with no work shown will receive only 1 credit.

15. Mrs. Martinez is buying a sweater that is on sale for 25% off of the original price. She has a coupon that gives her an additional 20% off of the sale price. Her final purchase price is what percent of the original price?

16. The area of a rectangular parcel of land is 720 square meters. The length of the land is 4 meters less than twice the width.

a. Write an equation that can be used to find the dimensions of the land.

b. Solve the equation written in part **a** to find the dimensions of the land.

CHAPTER

16

STATISTICS

Chapter Table of Contents

Every four years, each major political party in the United States holds a convention to select the party's nominee for President of the United States. Before these conventions are held, each candidate assembles a staff whose job is to plan a successful campaign. This plan relies heavily on statistics: on the collection and organization of data, on the results of opinion polls, and on information about the factors that influence the way people vote. At the same time, newspaper reporters and television commentators assemble other data to keep the public informed on the progress of the candidates.

Election campaigns are just one example of the use of statistics to organize data in a way that enables us to use available information to evaluate the current situation and to plan for the future.

16-1 COLLECTING DATA

In our daily lives, we often deal with problems that involve many related items of numerical information called **data**. For example, in the daily newspaper we can find data dealing with sports, with business, with politics, or with the weather.

Statistics is the study of numerical data. There are three typical steps in a statistical study:

STEP 1. The collection of data.

STEP 2. The organization of these data into tables, charts, and graphs.

STEP 3. The drawing of conclusions from an analysis of these data.

When these three steps, which describe and summarize the formation and use of a set of data, are included in a statistical study, the study is often called **descriptive statistics**. You will study these steps in this first course. In some cases, a fourth step, in which the analyzed data are used to predict trends and future events, is added.

Data can be either **qualitative** or **quantitative**. For example, a restaurant may ask customers to rate the meal that was served as excellent, very good, good, fair, or poor. This is a qualitative evaluation. Or the restaurant may wish to make a record of each customer tip at different times of the day. This is a quantitative evaluation, which lends itself more readily to further statistical analysis.

Data can be collected in a number of ways, including the following:

1. A written *questionnaire* or list of questions that a person can answer by checking one of several categories or supplying written responses. Categories to be checked may be either qualitative or quantitative. Written responses are usually qualitative.

2. An *interview*, either in person or by telephone, in which answers are given verbally and responses are recorded by the person asking the questions. Verbal answers are usually qualitative.

3. A *log* or a diary, such as a hospital chart or an hourly recording of the outdoor temperature, in which a person records information on a regular basis. This type of information is usually quantitative.

Note: Not all numerical data are quantitative data. For instance, a researcher wishes to investigate the eye color of the population of a certain island. The researcher assigns “blue” to 0, “black” to 1, “brown” to 2, and so on. The resulting data, although numerical, are qualitative since it represents eye color and the assignment was arbitrary.

Sampling

A statistical study may be useful in situations such as the following:

1. A doctor wants to know how effective a new medicine will be in curing a disease.
2. A quality-control team wants to know the expected life span of flashlight batteries made by its company.
3. A company advertising on television wants to know the most frequently watched TV shows so that its ads will be seen by the greatest number of people.

When a statistical study is conducted, it is not always possible to obtain information about every person, object, or situation to which the study applies. Unlike a **census**, in which every person is counted, some statistical studies use only a **sample**, or portion, of the items being investigated.

To find effective medicines, pharmaceutical companies usually conduct tests in which a sample, or small group, of the patients having the disease under study receive the medicine. If the manufacturer of flashlight batteries tested the life span of every battery made, the warehouse would soon be filled with dead batteries. The manufacturer tests only a sample of the batteries to determine their average life span. An advertiser cannot contact every person owning a TV set to determine which shows are being watched. Instead, the advertiser studies TV ratings released by a firm that conducts polls based on a small sample of TV viewers.

For any statistical study, whether based on a census or a sample, to be useful, data must be collected carefully and correctly. Poorly designed sampling techniques result in **bias**, that is, the tendency to favor a selection of certain members of the population which, in turn, produces unreliable conclusions.

Techniques of Sampling

We must be careful when choosing samples:

1. The sample must be fair or unbiased, to reflect the entire population being studied. To know what an apple pie tastes like, it is not necessary to eat the entire pie. Eating a sample, such as a piece of the apple pie, would be a fair way of knowing how the pie tastes. However, eating only the crust or only the apples would be an unfair sample that would not tell us what the entire pie tastes like.
2. The sample must contain a reasonable number of the items being tested or counted. If a medicine is generally effective, it must work for many people. The sample tested cannot include only one or two patients. Similarly, the manufacturer of flashlight batteries cannot make claims based on testing five or 10 batteries. A better sample might include 100 batteries.

3. Patterns of sampling or random selection should be employed in a study. The manufacturer of flashlight batteries might test every 1,000th battery to come off the assembly line. Or, the batteries to be tested might be selected at random.

These techniques will help to make the sample, or the small group, representative of the entire population of items being studied. From the study of the small group, reasonable conclusions can be drawn about the entire group.

EXAMPLE 1

To determine which television programs are the most popular in a large city, a poll is conducted by selecting people at random at a street corner and interviewing them. Outside of which location would the interviewer be most likely to find an unbiased sample?

(1) a ball park (2) a concert hall (3) a supermarket

Solution People outside a ball park may be going to a game or purchasing tickets for a game in the future; this sample may be biased in favor of sports programs. Similarly, those outside a concert hall may favor musical or cultural programs. The best (that is, the fairest) sample or cross section of people for the three choices given would probably be found outside a supermarket.

Answer (3)

Experimental Design

So far we have focused on data collection. In an **experiment**, a researcher imposes a treatment on one or more groups. The **treatment group** receives the treatment, while the **control group** does not.

For instance, consider an experiment of a new medicine for weight loss. Only the treatment group is given the medicine, and conditions are kept as similar as possible for both groups. In particular, both groups are given the same diet and exercise. Also, both groups are of large enough size and are chosen such that they are comprised of representative samples of the general population.

However, it is often not enough to have just a control group and a treatment group. The researcher must keep in mind that people often tend to respond to *any* treatment. This is called the **placebo effect**. In such cases, subjects would report that the treatment worked even when it is ineffective. To account for the placebo effect, researchers use a group that is given a **placebo** or a dummy treatment.

Of course, subjects in the experimental and placebo groups should not know which group they are in (otherwise, psychology will again confound the results). The practice of not letting people know whether or not they have been given the real treatment is called **blinding**, and experiments using blinding are said to be **single-blind experiments**. When the variable of interest is hard to measure or define, **double-blind experiments** are needed. For example, consider an experiment measuring the effectiveness of a drug for attention deficit disorder. The problem is that "attention deficiency" is difficult to define, and so a researcher with a bias towards a particular conclusion may interpret the results of the placebo and treatment groups differently. To avoid such problems, the researchers working directly with the test subjects are not told which group a subject belongs to.

Interpreting Graphs of Data

Oftentimes embellishments to graphs distort the perception of the data, and so you must exercise care when interpreting graphs of data.

1. Two- and three-dimensional figures.

As the graph on the right shows, graphs using two- or three-dimensional figures can distort small changes in the data. The *lengths* show the decrease in crime, but since our eyes tend to focus on the *areas*, the total decrease appears greater than it really is. The reason is because linear changes are increased in higher dimensions. For instance, if a length doubles in value, say from x to $2x$, the area of a square with sides of length x will increase by

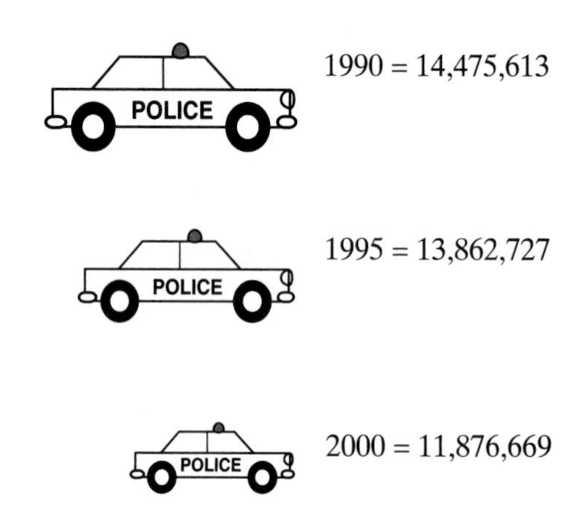

$$x^2 \to (2x)^2 = 4x^2,$$

a four-fold increase. Similarly, the volume of a cube with edges of length x will increase by

$$x^3 \to (2x)^3 = 8x^3,$$

an eight-fold increase!

2. Horizontal and vertical scales.

The scales used on the vertical and horizontal axes can exaggerate, diminish, and/or distort the nature of the change in the data. For instance, in the graph on the left of the following page, the total change in weight is less than a pound, which is negligible for an adult human. However, the scale used apparently amplifies this amount. While on the right, the unequal horizontal scale makes the population growth appear linear.

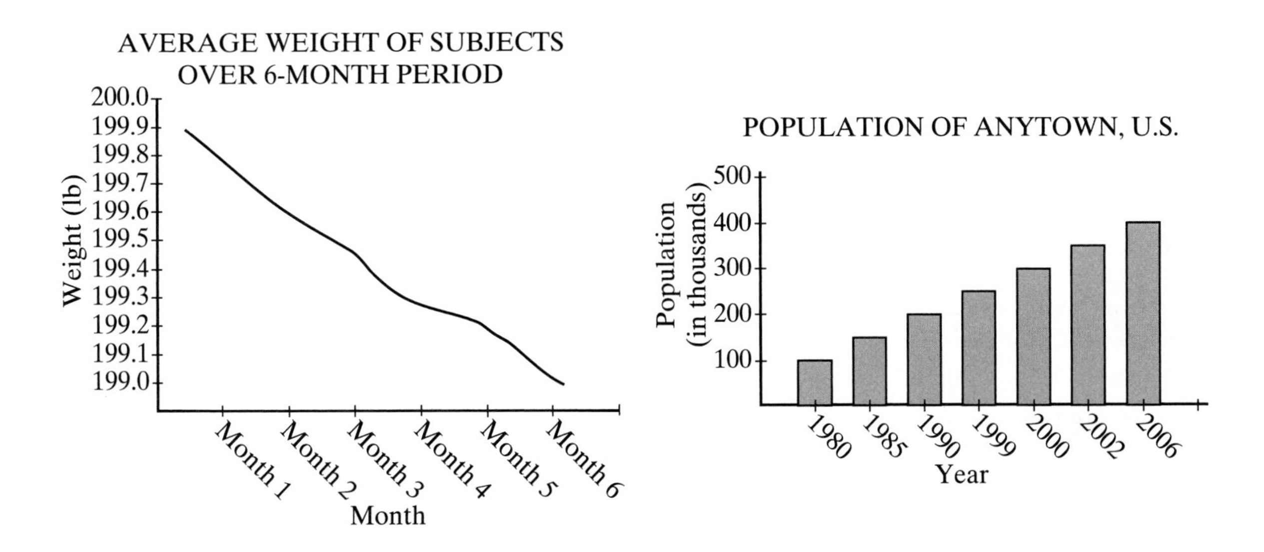

EXERCISES

Writing About Mathematics

1. A census attempts to count every person. Explain why a census may be unreliable.

2. A sample of a new soap powder was left at each home in a small town. The occupants were asked to try the powder and return a questionnaire evaluating the product. To encourage the return of the questionnaire, the company promised to send a coupon for a free box of the soap powder to each person who responded. Do you think that the questionnaires that were returned represent a fair sample of all of the persons who tried the soap? Explain why or why not.

Developing Skills

In 3–10, determine if each variable is quantitative or qualitative.

3. Political affiliation

4. Opinions of students on a new music album

5. SAT scores

6. Nationality

7. Cholesterol level

8. Class membership (freshman, sophomore, etc.)

9. Height

10. Number of times the word "alligator" is used in an essay.

In 11–18, in each case a sample of students is to be selected and the height of each student is to be measured to determine the average height of a student in high school. For each sample:

a. Tell whether the sample is biased or unbiased.

b. If the sample is biased, explain how this might affect the outcome of the survey.

11. The basketball team

12. The senior class

13. All 14-year-old students

14. All girls

15. Every tenth person selected from an alphabetical list of all students

16. Every fifth person selected from an alphabetical list of all boys

17. The first three students who report to the nurse on Monday

18. The first three students who enter each homeroom on Tuesday

In 19–24, in each case the Student Organization wishes to interview a sample of students to determine the general interests of the student body. Two questions will be asked: "Do you want more pep rallies for sports events? Do you want more dances?" For each location, tell whether the Student Organization would find an unbiased sample at that place. If the sample is biased, explain how this might influence the result of the survey.

19. The gym, after a game

20. The library

21. The lunchroom

22. The cheerleaders' meeting

23. The next meeting of the Junior Prom committee

24. A homeroom section chosen at random

25. A statistical study is useful when reliable data are collected. At times, however, people may exaggerate or lie when answering a question. Of the six questions that follow, find the *three* questions that will most probably produce the largest number of *unreliable* answers.
(1) What is your height?
(2) What is your weight?
(3) What is your age?
(4) In which state do you live?
(5) What is your income?
(6) How many people are in your family?

26. List the three steps necessary to conduct a statistical study.

27. Explain why the graph below is misleading.

SUMMER OLYMPIC GAMES CHAMPIONS
100-METER RACE

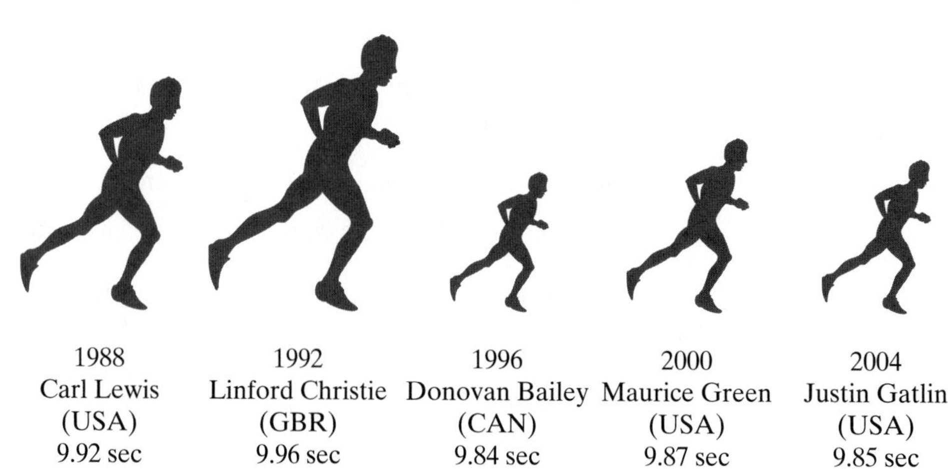

28. Investigators at the University of Kalamazoo were interested in determining whether or not women can determine a man's preference for children based on the way that he looks. Researchers asked a group of 20 male volunteers whether or not they liked children. The researchers then showed photographs of the faces of the men to a group of 10 female volunteers and asked them to pick out which men they thought liked children. The women correctly identified over 90% of the men who said they liked children. The researchers concluded that women could identify a man's preference for children based on the way that he looks. Identify potential problems with this experiment.

Hands-On Activity

Collect *quantitative* data for a statistical study.

1. Decide the topic of the study. What data will you collect?

2. Decide how the data will be collected. What will be the source(s) of that data?

- **a.** Questionnaires
- **b.** Personal interviews
- **c.** Telephone interviews
- **d.** Published materials from sources such as almanacs or newspapers.

3. Collect the data. How many values are necessary to obtain reliable information?

Keep the data that you collect to use as you learn more about statistical studies.

16-2 ORGANIZING DATA

Data are often collected in an unorganized and random manner. For example, a teacher recorded the number of days each of 25 students in her class was absent last month. These absences were as follows:

$$0, 3, 1, 0, 4, 2, 1, 3, 5, 0, 2, 0, 0, 0, 4, 0, 1, 1, 2, 1, 0, 7, 3, 1, 0$$

How many students were absent fewer than 2 days? What was the number of days for which the most students were absent? How many students were absent more than 5 days? To answer questions such as these, we find it helpful to organize the data.

One method of organizing data is to write it as an ordered list. In order from least to greatest, the absences become:

$$0, 0, 0, 0, 0, 0, 0, 0, 0, 1, 1, 1, 1, 1, 1, 2, 2, 2, 3, 3, 3, 4, 4, 5, 7$$

We can immediately observe certain facts from this ordered list: more students were absent 0 days than any other number of days, the same number of students were absent 5 and 7 days. However, for more a quantitative analysis, it is useful to make a table.

Preparing a Table

In the left column of the accompanying table, we list the data values (in this case the number of absences) in order. We start with the largest number, 7, at the top and go down to the smallest number, 0.

For each occurrence of a data value, we place a **tally** mark, |, in the row for that number. For example, the first data value in the teacher's list is 0, so we place a tally in the 0 row; the second value is 3, so we place a tally in the 3 row. We follow this procedure until a tally for each data value is recorded in the proper row. To simplify counting, we write every fifth tally as a diagonal mark passing through the first four tallies: ~~||||~~.

Absences	Tally
7	|
6	
5	|
4	||
3	|||
2	|||
1	~~||||~~ |
0	~~||||~~ ||||

Once the data have been organized, we can count the number of tally marks in each row and add a column for the **frequency**, that is, the number of times that a value occurs in the set of data. When there are no tally marks in a row, as for the row showing 6 absences, the frequency is 0. The sum of all of the frequencies is called the **total frequency**. In this case, the total frequency is 25. (It is always wise to check the total frequency to be sure that no data value was overlooked or duplicated in tallying.) From the table, called a **frequency distribution table**, it is now easy to see that 15 students were absent fewer than 2 days, that more students were absent 0 days (9) than any other number of days, and that 1 student was absent more than 5 days.

Absences	Tally	Frequency
7	|	1
6		0
5	|	1
4	||	2
3	|||	3
2	|||	3
1	~~||||~~ |	6
0	~~||||~~ ||||	9
Total frequency		25

Grouped Data

A teacher marked a set of 32 test papers. The grades or scores earned by the students were as follows:

90, 85, 74, 86, 65, 62, 100, 95, 77, 82, 50, 83, 77, 93, 73, 72,
98, 66, 45, 100, 50, 89, 78, 70, 75, 95, 80, 78, 83, 81, 72, 75

Because of the large number of different scores, it is convenient to organize these data into **groups** or **intervals**, which must be equal in size. Here we will use six intervals: 41–50, 51–60, 61–70, 71–80, 81–90, 91–100. Each interval has a length of 10, found by subtracting the starting point of an interval from the starting point of the next higher interval.

For each test score, we now place a tally mark in the row for the interval that includes that score. For example, the first two scores in the list above are 90 and 85, so we place two tally marks in the interval 81–90. The next score is 74, so we place a tally mark in the interval 71–80. When all of the scores have been tallied, we write the frequency for each interval.

Interval	Tally	Frequency
91–100	~~||||~~ |	6
81–90	~~||||~~ |||	8
71–80	~~||||~~ ~~||||~~ |	11
61–70	||||	4
51–60		0
41–50	|||	3

This table, containing a set of intervals and the corresponding frequency for each interval, is an example of **grouped data**.

When unorganized data are grouped into intervals, we must follow certain rules in setting up the intervals:

1. The intervals must cover the complete range of values. The **range** is the difference between the highest and lowest values.

2. The intervals must be equal in size.

3. The number of intervals should be between 5 and 15. The use of too many or too few intervals does not make for effective grouping of data. We usually use a large number of intervals, for example, 15, only when we have a very large set of data, such as hundreds of test scores.

4. Every data value to be tallied must fall into one and only one interval. Thus, the intervals should not overlap. When an interval ends with a counting number, the following interval begins with the next counting number.

5. The intervals must be listed in order, either highest to lowest or lowest to highest.

These rules tell us that there are many ways to set up tables, all of them correct, for the same set of data. For example, here is another correct way to group the 32 unorganized test scores given at the beginning of this section. Note that the length of the interval here is 8.

Interval	Tally	Frequency
93–100	卌 |	6
85–92	||||	4
77–84	卌 ||||	9
69–76	卌 ||	7
61–68	|||	3
53–60		0
45–52	|||	3

Constructing a Stem-and-Leaf Diagram

Another method of displaying data is called a **stem-and-leaf diagram**. The stem-and-leaf diagram groups the data without losing the individual data values.

A group of 30 students were asked to record the length of time, in minutes, spent on math homework yesterday. They reported the following data:

$$38, 15, 22, 20, 25, 44, 5, 40, 38, 22, 20, 35, 20, 0, 36,$$
$$27, 37, 26, 33, 25, 17, 45, 22, 30, 18, 48, 12, 10, 24, 27$$

To construct a stem-and-leaf diagram for the lengths of time given, we begin by choosing part of the data values to be the stem. Since every score is a one- or two-digit number, we will choose the tens digit as a convenient stem. For the one-digit numbers, 0 and 5, the stem is 0; for the other data values, the stem is 1, 2, 3, or 4. Then the units digit will be the leaf. We construct the diagram as follows:

STEP 1. List the stems, starting with 4, under one another to the left of a vertical line beneath a crossbar.

Stem	Leaf
4	
3	
2	
1	
0	

STEP 2. Enter each score by writing its leaf (the units digit) to the right of the vertical line, following the appropriate stem (its tens value). For example, enter 38 by writing 8 to the right of the vertical line, after stem 3.

Stem	Leaf
4	
3	8
2	
1	
0	

STEP 3. Add the other scores to the diagram until all are entered.

Stem	Leaf
4	4 0 5 8
3	8 8 5 6 7 3 0
2	2 0 5 2 0 0 7 6 5 2 4 7
1	5 7 8 2 0
0	5 0

STEP 4. Arrange the leaves in order after each stem.

STEP 5. Add a key to demonstrate the meaning of each value in the diagram.

Stem	Leaf
4	0 4 5 8
3	0 3 5 6 7 8 8
2	0 0 0 2 2 2 4 5 5 6 7 7
1	0 2 5 7 8
0	0 5

Key: 3 | 0 = 30

EXAMPLE 1

The following data consist of the weights, in kilograms, of a group of 30 students:

70, 43, 48, 72, 53, 81, 76, 54, 58, 64, 51, 53, 75, 62, 84,
67, 72, 80, 88, 65, 60, 43, 53, 42, 57, 61, 55, 75, 82, 71

a. Organize the data in a table. Use five intervals starting with 40–49.

b. Based on the grouped data, which interval contains the greatest number of students?

c. How many students weigh less than 70 kilograms?

Solution **a.**

Interval	Tally	Frequency (number)
80–89	𝍸	5
70–79	𝍸 II	7
60–69	𝍸 I	6
50–59	𝍸 III	8
40–49	IIII	4

b. The interval 50–59 contains the greatest number of students, 8. *Answer*

c. The three lowest intervals, namely 40–49, 50–59, and 60–69, show weights less than 70 kilograms. Add the frequencies in these three intervals:
$4 + 8 + 6 = 18$ *Answer*

EXAMPLE 2

Draw a stem-and-leaf diagram for the data in Example 1.

Solution Let the tens digit be the stem and the units digit the leaf.

(1) Enter the data values in the given order:

Stem	Leaf
8	1 4 0 8 2
7	0 2 6 5 2 5 1
6	4 2 7 5 0 1
5	3 4 8 1 3 3 7 5
4	3 8 3 2

(2) Arrange the leaves in numerical order after each stem:

Stem	Leaf
8	0 1 2 4 8
7	0 1 2 2 5 5 6
6	0 1 2 4 5 7
5	1 3 3 3 4 5 7 8
4	2 3 3 8

(3) Add a key indicating unit of measure: Key: 5 | 1 = 51 kg

EXERCISES

Writing About Mathematics

1. Of the examples given above, which gives more information about the data: the table or the stem-and-leaf diagram? Explain your answer.

2. A set of data ranges from 2 to 654. What stem can be used for this set of data when drawing a stem-and-leaf diagram? What leaves would be used with this stem? Explain your choices.

Developing Skills

3. a. Copy and complete the table to group the data, which represent the heights, in centimeters, of 36 students:

162, 173, 178, 181, 155, 162, 168, 147, 180,
171, 168, 183, 157, 158, 180, 164, 160, 171,
183, 174, 166, 175, 169, 180, 149, 170, 150,
158, 162, 175, 171, 163, 158, 163, 164, 177

Interval	Tally	Frequency
180–189		
170–179		
160–169		
150–159		
140–149		

b. Use the grouped data to answer the following questions:

(1) How many students are less than 160 centimeters in height?

(2) How many students are 160 centimeters or more in height?

(3) Which interval contains the greatest number of students?

(4) Which interval contains the least number of students?

c. Display the data in a stem-and-leaf diagram. Use the first two digits of the numbers as the stems.

d. What is the range of the data?

e. How many students are taller than 175 centimeters?

4. a. Copy and complete the table to group the data, which gives the lifespan, in hours, of 50 flashlight batteries:

73, 81, 92, 80, 108, 76, 84, 102, 58, 72,
82, 100, 70, 72, 95, 105, 75, 84, 101, 62,
63, 104, 97, 85, 106, 72, 57, 85, 82, 90,
54, 75, 80, 52, 87, 91, 85, 103, 78, 79,
91, 70, 88, 73, 67, 101, 96, 84, 53, 86

Interval	Tally	Frequency
50–59		
60–69		
70–79		
80–89		
90–99		
100–109		

b. Use the grouped data to answer the following questions:

(1) How many flashlight batteries lasted for 80 or more hours?

(2) How many flashlight batteries lasted fewer than 80 hours?

(3) Which interval contains the greatest number of batteries?

(4) Which interval contains the least number of batteries?

c. Display the data in a stem-and-leaf diagram. Use the digits from 5 through 10 as the stems.

d. What is the range of the data?

e. What is the probability that a battery selected at random lasted more than 100 hours?

5. The following data consist of the hours spent each week watching television, as reported by a group of 38 teenagers:

13, 20, 17, 36, 25, 21, 9, 32, 20, 17, 12, 19, 5, 8, 11, 28, 25, 18,
19, 22, 4, 6, 0, 10, 16, 3, 27, 31, 15, 18, 20, 17, 3, 6, 19, 25, 4, 7

a. Construct a table to group these data, using intervals of 0–4, 5–9, 10–14, 15–19, 20–24, 25–29, 30–34, and 35–39.

b. Construct a table to group these data, using intervals of 0–7, 8–15, 16–23, 24–31, and 32–39.

c. Display the data in a stem-and-leaf diagram.

d. What is the range of the data?

e. What is the probability that a teenager, selected at random from this group, spends less than 4 hours watching television each week?

6. The following data show test scores for 30 students:

90, 83, 87, 71, 62, 46, 67, 72, 75, 100, 93, 81, 74, 75, 82,
83, 83, 84, 92, 58, 95, 98, 81, 88, 72, 59, 95, 50, 73, 93

a. Construct a table, using intervals of length 10 starting with 91–100.

b. Construct a table, using intervals of length 12 starting with 89–100.

c. For the grouped data in part **a**, which interval contains the greatest number of students?

d. For the grouped data in part **b**, which interval contains the greatest number of students?

e. Do the answers for parts **c** and **d** indicate the same general region of test scores, such as "scores in the eighties"? Explain your answer.

7. For the ungrouped data from Exercise 5, tell why each of the following sets of intervals is not correct for grouping the data.

a.

Interval
25–38
13–24
0–12

b.

Interval
30–39
20–29
10–19
5–9
0–4

c.

Interval
32–40
24–32
16–24
8–16
0–8

d.

Interval
33–40
25–32
17–24
9–16
1–8

Hands-On Activity

Organize the data that you collected in the Hands-On Activity for Section 16-1.

1. Use a stem-and-leaf diagram.

a. Decide what will be used as stems.

b. Decide what will be used as leaves.

c. Construct the diagram.

d. Check that the number of leaves in the diagram equals the number of values in the data collected.

2. Use a frequency table.

a. How many intervals will be used?

b. What will be the length of each interval?

c. What will be the starting and ending points of each interval? Check that the intervals do not overlap, are equal in size, and that every value falls into only one interval.

d. Tally the data.

e. List the frequency for each interval.

f. Check that the total frequency equals the number of values in the data collected.

3. Decide which method of organization is better for your data. Explain your choice.

Keep your organized data to work with as you learn more about statistics.

16-3 THE HISTOGRAM

In Section 16-2 we organized data by grouping them into intervals of equal length. After the data have been organized, a graph can be used to visualize the intervals and their frequencies.

The table below shows the distribution of test scores for 32 students in a class. The data have been organized into six intervals of length 10.

Test Scores (Intervals)	Frequency (Number of Scores)
91–100	6
81–90	8
71–80	11
61–70	4
51–60	0
41–50	3

We can use a histogram to display the data graphically. A **histogram** is a vertical bar graph in which each interval is represented by the width of the bar and the frequency of the interval is represented by the height of the bar. The bars are placed next to each other to show that, as one interval ends, the next interval begins.

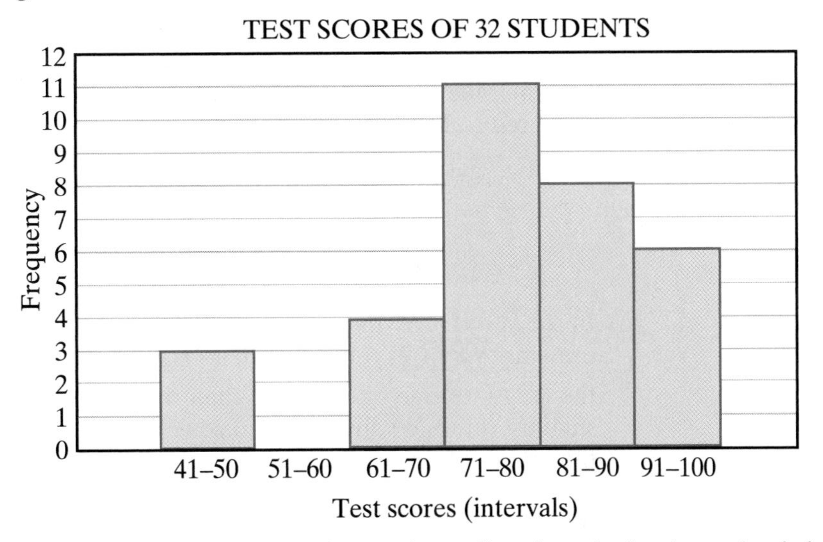

In the above histogram, the intervals are listed on the horizontal axis in the order of increasing scores, and the frequency scale is shown on the vertical axis. The first bar shows that 3 students had test scores in the interval 41–50. Since no student scored in the interval 51–60, there is no bar for this interval. Then, 4 students scored between 61 and 70; 11 between 71 and 80; 8 between 81 and 90; and 6 between 91 and 100.

Except for an interval having a frequency of 0, the interval 51–60 in this example, there are no gaps between the bars drawn in a histogram. Since the histogram displays the frequency, or number of data values, in each interval, we sometimes call this graph a **frequency histogram**.

A graphing calculator can display a frequency histogram from the data on a frequency distribution table.

(1) Clear L_1 and L_2 with the ClrList function by pressing STAT 4 2nd L1 , 2nd L2 ENTER.

(2) Press STAT 1 to edit the lists. L_1 will contain the minimum value of each interval. Move the cursor to the first entry position in L_1. Type the value and then press ENTER. Type the next value and then press ENTER. Repeat this process until all the minimum values of the intervals have been entered.

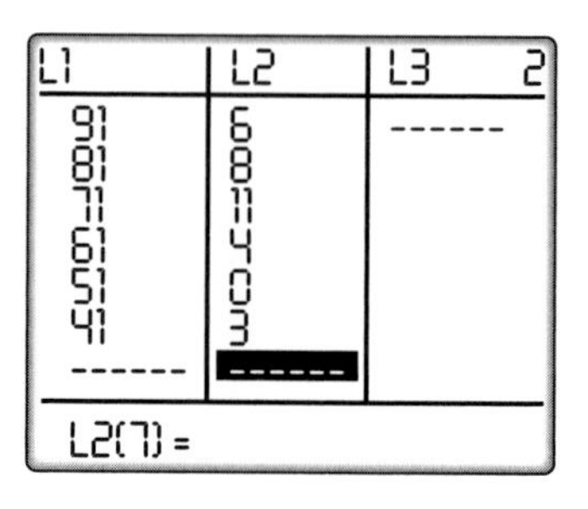

(3) Repeat the process to enter the frequencies that correspond to each interval in L_2.

(4) Clear any functions in the Y= menu.

(5) Turn on Plot1 from the STAT PLOT menu, and configure it to graph a histogram. Make sure to also set Xlist to L_1 and Freq to L_2.

ENTER: 2nd STAT PLOT 1 ENTER ▼ ► ► ENTER ► 2nd L1 ▼ 2nd L2

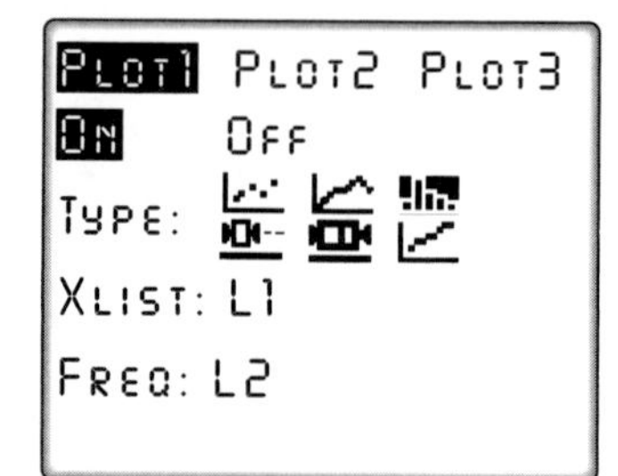

(6) In the WINDOW menu, accessed by pressing WINDOW, enter Xmin as 31, the length of one interval less than the smallest interval value and Xmax as 110, the length of one interval more than the largest interval value. Enter Xscl as 10, the length of the interval. The Ymin is 0 and Ymax is 12 to be greater than the largest frequency.

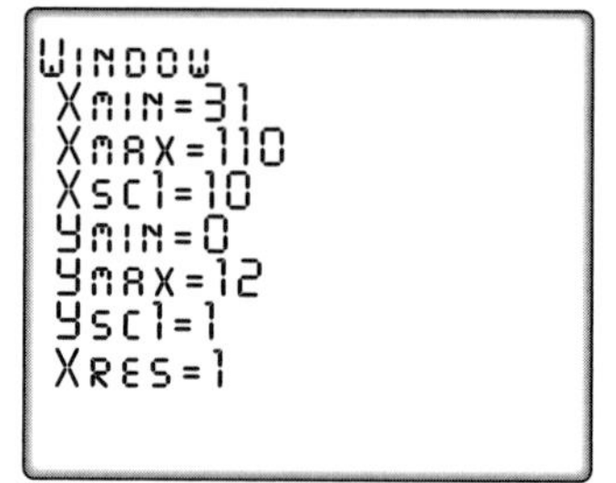

(7) Press GRAPH to draw the graph. We can view the frequency (n) associated with each interval by pressing TRACE. Use the left and right arrow keys to move between intervals.

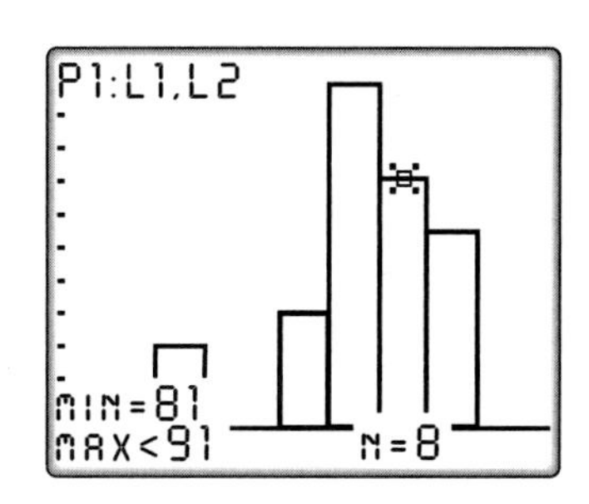

EXAMPLE 1

The table on the right represents the number of miles per gallon of gasoline obtained by 40 drivers of compact cars. Construct a frequency histogram based on the data.

Interval	Frequency
16–19	5
20–23	11
24–27	8
28–31	5
32–35	7
36–39	3
40–43	1

Solution (1) Draw and label a vertical scale to show frequencies. The scale starts at 0 and increases to include the highest frequency in any one interval (here, it is 11).

(2) Draw and label intervals of equal length on a horizontal scale. Label the horizontal scale, telling what the numbers represent.

(3) Draw the bars vertically, leaving no gaps between the intervals.

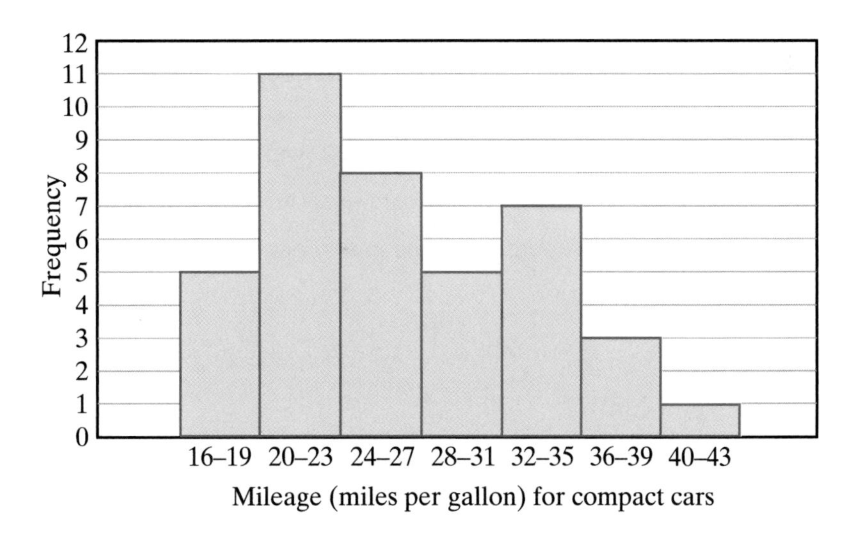

Calculator Solution

(1) Press **STAT** **1** to edit the lists and enter the minimum value of each interval into L_1: 16, 20, 24, 28, 32, 36, 40. Use the arrow key to move into L_2, and enter the corresponding frequencies: 5, 11, 8, 5, 7, 3, 1.

(2) Go to the STAT PLOT menu and choose Plot1 by pressing **2nd** **STAT PLOT** **1**. Move the cursor with the arrow keys, then press **ENTER** to select On and the histogram. Type **2nd** **L1** into Xlist and **2nd** **L2** into Freq.

(3) Set the Window. Each interval has length 4, so set Xmin to 12 (4 less than the smallest interval value), Xmax to 44 (4 more than the largest interval value), and Xscl to 4. Make Ymin 0 and Ymax 12 to be greater than the largest frequency.

(4) Draw the graph by pressing **GRAPH**. Press **TRACE** and use the right and left arrow keys to show the frequencies, the heights of the vertical bars.

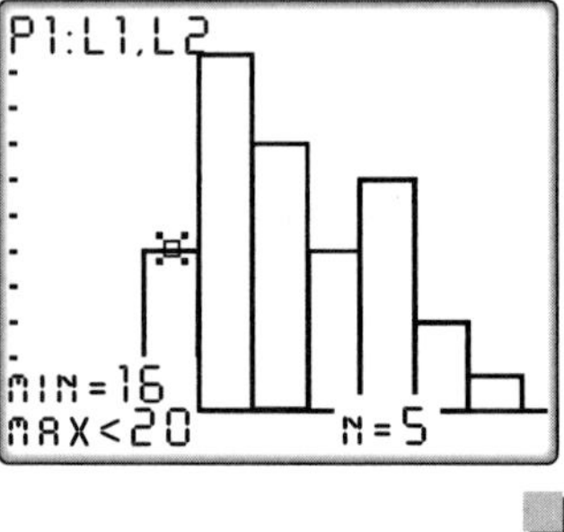

EXAMPLE 2

Use the histogram constructed in Example 1 to answer the following questions:

a. In what interval is the greatest frequency found?

b. What is the number (or frequency) of cars reporting mileages between 28 and 31 miles per gallon?

c. For what interval are the fewest cars reported?

d. How many of the cars reported mileage greater than 31 miles per gallon?

e. What percent of the cars reported mileage from 24 to 27 miles per gallon?

Solution

a. 20–23

b. 5

c. 40–43

d. Add the frequencies for the three highest intervals. The interval 32–35 has a frequency of 7; 36–39 a frequency of 3; 40–43 a frequency of 1: $7 + 3 + 1 = 11$.

e. The interval 24–27 has a frequency of 8. The total frequency for this survey is 40. $\frac{8}{40} = \frac{1}{5} = 20\%$.

Answers **a.** 20–23 **b.** 5 **c.** 40–43 **d.** 11 **e.** 20%

EXERCISES

Writing About Mathematics

1. Compare a stem-and-leaf diagram with a frequency histogram. In what ways are they alike and in what ways are they different?

2. If the data in Example 1 had been grouped into intervals with a lowest interval of 16–20, what would be the endpoints for the other intervals? Would you be able to determine the frequency for each new interval? Explain why or why not.

Developing Skills

In 3–5, in each case, construct a frequency histogram for the grouped data. Use graph paper or a graphing calculator.

3.

Interval	Frequency
91–100	5
81–90	9
71–80	7
61–70	2
51–60	4

4.

Interval	Frequency
30–34	5
25–29	10
20–24	10
15–19	12
10–14	0
5–9	2

5.

Interval	Frequency
1–3	24
4–6	30
7–9	28
10–12	41
13–15	19
16–18	8

6. For the table of grouped data given in Exercise 5, answer the following questions:

a. What is the total frequency in the table?

b. What interval contains the greatest frequency?

c. The number of data values reported for the interval 4–6 is what percent of the total number of data values?

d. How many data values from 10 through 18 were reported?

Applying Skills

7. Towering Ted McGurn is the star of the school's basketball team. The number of points scored by Ted in his last 20 games are as follows:

36, 32, 28, 30, 33, 36, 24, 33, 29, 30, 30, 25, 34, 36, 34, 31, 36, 29, 30, 34

a. Copy and complete the table to find the frequency for each interval.

b. Construct a frequency histogram based on the data found in part **a**.

c. Which interval contains the greatest frequency?

d. In how many games did Ted score 32 or more points?

e. In what percent of these 20 games did Ted score fewer than 26 points?

Interval	Tally	Frequency
35–37		
32–34		
29–31		
26–28		
23–25		

8. Thirty students on the track team were timed in the 200-meter dash. Each student's time was recorded to the *nearest tenth* of a second. Their times are as follows:

29.3, 31.2, 28.5, 37.6, 30.9, 26.0, 32.4, 31.8, 36.6, 35.0,
38.0, 37.0, 22.8, 35.2, 35.8, 37.7, 38.1, 34.0, 34.1, 28.8,
29.6, 26.9, 36.9, 39.6, 29.9, 30.0, 36.0, 36.1, 38.2, 37.8

a. Copy and complete the table to find the frequency in each interval.

b. Construct a frequency histogram for the given data.

c. Determine the number of students who ran the 200-meter dash in under 29 seconds.

d. If a student on the track team is chosen at random, what is the probability that he or she ran the 200-meter dash in fewer than 29 seconds?

Interval	Tally	Frequency
37.0–40.9		
33.0–36.9		
29.0–32.9		
25.0–28.9		
21.0–24.9		

Hands-On-Activity

Construct a histogram to display the data that you collected and organized in the Hands-On Activities for Sections 16-1 and 16-2.

1. Draw the histogram on graph paper.

2. Follow the steps in this section to display the histogram on a graphing calculator.

16-4 THE MEAN, THE MEDIAN, AND THE MODE

In a statistical study, after we have collected the data, organized them, and presented them graphically, we then analyze the data and summarize our findings. To do this, we often look for a representative, or typical, score.

Averages in Arithmetic

In your previous study of arithmetic, you learned how to find the average of two or more numbers. For example, to find the average of 17, 25, and 30:

STEP 1. Add these three numbers: $17 + 25 + 30 = 72$.

STEP 2. Divide this sum by 3 since there are three numbers: $72 \div 3 = 24$.
The average of the three numbers is 24.

Averages in Statistics

The word **average** has many different meanings. For example, there is an *average* of test scores, a batting *average*, the *average* television viewer, an *average* intelligence, and the *average* size of a family. These averages are *not* found by the same rule or procedure. Because of this confusion, in statistics we speak of **measures of central tendency**. These measures are numbers that usually fall somewhere in the center of a set of organized data.

We will discuss three measures of central tendency: the mean, the median, and the mode.

The Mean

In statistics, the arithmetic average previously studied is called the **mean** of a set of numbers. It is also called the **arithmetic mean** or the **numerical average**. The mean is found in the same way as the arithmetic average is found.

Procedure

To find the mean of a set of *n* numbers, add the numbers and divide the sum by *n*. The symbol used for the mean is $\bar{x}$.

For example, if Ralph's grades on five tests in science during this marking period are 93, 80, 86, 72, and 94, he can find the mean of his test grades as follows:

STEP 1. Add the five data values: $93 + 80 + 86 + 72 + 94 = 425$.

STEP 2. Divide this sum by 5, the number of tests: $425 \div 5 = 85$.

The mean (arithmetic average) is 85.

Let us consider another example. In a car wash, there are seven employees whose ages are 17, 19, 20, 17, 46, 17, and 18. What is the mean of the ages of these employees?

Here, we add the seven ages to get a sum of 154. Then, $154 \div 7 = 22$. While the mean age of 22 is the correct answer, this measure does *not* truly represent the data. Only one person is older than 22, while six people are under 22. For this reason, we will look at another measure of central tendency that will eliminate the extreme case (the employee aged 46) that is *distorting* the data.

The Median

The **median** is the middle value for a set of data arranged in numerical order. For example, the median of the ages 17, 19, 20, 17, 46, 17, and 18 for the car-wash employees can be found in the following manner:

STEP 1. Arrange the ages in numerical order: 17, 17, 17, 18, 19, 20, 46

STEP 2. Find the middle number: 17, 17, 17, 18, 19, 20, 46
↑ (18)

The median is 18 because there are three ages less than 18 and three ages greater than 18. The median, 18, is a better indication of the typical age of the employees than the mean, 22, because there are so many younger people working at the car wash.

Now, let us suppose that one of the car-wash employees has a birthday, and her age changes from 17 to 18. What is now the median age?

STEP 1. Arrange the ages in numerical order: 17, 17, 18, 18, 19, 20, 46

STEP 2. Find the middle number: 17, 17, 18, 18, 19, 20, 46
↑ (18)

The median, or middle value, is again 18. We can no longer say that there are three ages less than 18 because one of the three youngest employees is now 18.

We can say, however, that:

1. the median is 18 because there are three ages less than or equal to 18 and three ages greater than or equal to 18; or
2. the median is 18 because, when the data values are arranged in numerical order, there are three values below this median, or middle number, and three values above it.

Recently, the car wash hired a new employee whose age is 21. The data now include eight ages, an even number, so there is no middle value. What is now the median age?

STEP 1. Arrange the ages in numerical order: 17, 17, 18, 18, 19, 20, 21, 46

STEP 2. There is no single middle number. Find the *two* middle numbers: 17, 17, 18, 18, 19, 20, 21, 46
↑ ↑ (18, 19)

STEP 3. Find the mean (arithmetic average) of the two middle numbers: $\frac{18 + 19}{2} = 18\frac{1}{2}$

The median is now $18\frac{1}{2}$. There are four ages less than this center value of $18\frac{1}{2}$ and four ages greater than $18\frac{1}{2}$.

Procedure

To find the median of a set of *n* numbers:

1. Arrange the numbers in numerical order.
2. If n is odd, find the middle number. This number is the median.
3. If n is even, find the mean (arithmetic average) of the *two* middle numbers. This average is the median.

The Mode

The **mode** is the data value that appears most often in a given set of data. It is usually best to arrange the data in numerical order before finding the mode.

Let us consider some examples of finding the mode:

1. The ages of employees in a car wash are 17, 17, 17, 18, 19, 20, 46. The mode, which is the number appearing most often, is 17.
2. The number of hours each of six students spent reading a book are 6, 6, 8, 11, 14, 21. The mode, or number appearing most frequently, is 6. In this case, however, the mode is not a useful measure of central tendency. A better indication is given by the mean or the median.
3. The number of photographs printed from each of Renee's last six rolls of film are 8, 8, 9, 11, 11, and 12. Since 8 appears twice and 11 appears twice, we say that there are two modes: 8 and 11. We do not take the average of these two numbers since the mode tells us where most of the scores appear. We simply report both numbers. When *two* modes appear within a set of data, we say that the data are **bimodal**.
4. The number of people living in each house on Meryl's street are 2, 2, 3, 3, 4, 5, 5, 6, 8. These data have *three* modes: 2, 3, and 5.
5. Ralph's test scores in science are 72, 80, 86, 93, and 94. Here, every number appears the same number of times, once. Since *no* number appears more often than the others, we define such data as having no mode.

Procedure

To find the mode for a set of data, find the number or numbers that occur most often.

1. If one number appears most often in the data, that number is the mode.
2. If two or more numbers appear more often than all other data values, and these numbers appear with the same frequency, then each of these numbers is a mode.
3. If each number in a set of data occurs with the same frequency, there is no mode.

KEEP IN MIND Three measures of central tendency are:

1. The mean, or mean average, found by adding n data values and then dividing the sum by n.
2. The median, or middle score, found when the data are arranged in numerical order.
3. The mode, or the value that appears most often.

A graphing calculator can be used to arrange the data in numerical order and to find the mean and the median. The calculator solution in the following example lists the keystrokes needed to do this.

EXAMPLE 1

The weights, in pounds, of five players on the basketball team are 195, 168, 174, 182, and 181. Find the average weight of a player on this team.

Solution The word *average*, by itself, indicates the *mean*. Therefore:

(1) Add the five weights: 195 + 168 + 174 + 182 + 181 = 900.

(2) Divide the sum by 5, the number of players: 900 ÷ 5 = 180.

Calculator Solution Enter the data into list L_1. Then use 1-Var Stats from the STAT CALC menu to display information about this set of data.

ENTER: STAT ▸ ENTER ENTER

DISPLAY:

```
1-Var Stats
 x̄=180
 Σx=900
 Σx²=162410
 Sx=10.12422837
 σx=9.055385138
↓n=5
```

The first value given is $\overline{x}$, the mean.

Answer 180 pounds

The second value given is Σx = 900. The symbol Σ represents a sum and Σx = 900 can be read as "The sum of the values of x is 900." The list shows other values related to this set of data. The arrow at the bottom of the display indicates that more entries follow what appears on the screen. These can be displayed by pressing the down arrow. One of these is the median (Med = 181). The display also shows that there are 5 data values (n = 5). Others we will use in later sections in this chapter and in more advanced courses.

EXAMPLE 2

Renaldo has marks of 75, 82, and 90 on three mathematics tests. What mark must he obtain on the next test to have an average of exactly 85 for the four math tests?

Solution The word *average*, by itself, indicates the *mean*.

Let x = Renaldo's mark on the fourth test.

The sum of the four test marks divided by 4 is 85.

$$\frac{75 + 82 + 90 + x}{4} = 85$$
$$\frac{247 + x}{4} = 85$$
$$247 + x = 340$$
$$x = 93$$

Check

$$\frac{75 + 82 + 90 + 93}{4} \stackrel{?}{=} 85$$
$$\frac{340}{4} \stackrel{?}{=} 85$$
$$85 \stackrel{?}{=} 85 \checkmark$$

Answer Renaldo must obtain a mark of 93 on his fourth math test.

EXAMPLE 3

Find the median for each distribution.

a. 4, 2, 5, 5, 1 **b.** 9, 8, 8, 7, 4, 3, 3, 2, 0, 0

Solution **a.** Arrange the data in numerical order: 1, 2, 4, 5, 5

The median is the middle value: 1, 2, 4, 5, 5 (↑ at 4)

Answer median = 4

b. Since there is an even number of values, there are two middle values. Find the mean (average) of these two middle values:

9, 8, 8, 7, 4, 3, 3, 2, 0, 0 (↑ ↑ at 4 and 3)

$$\frac{4 + 3}{2} = \frac{7}{2} = 3\frac{1}{2}$$

Answer median = $3\frac{1}{2}$ or 3.5

EXAMPLE 4

Find the mode for each distribution.

a. 2, 9, 3, 7, 3 **b.** 3, 4, 5, 4, 3, 7, 2 **c.** 1, 2, 3, 4, 5, 6, 7

Solution **a.** Arrange the data in numerical order: 2, 3, 3, 7, 9.

The mode, or most frequent value, is 3.

b. Arrange the data in numerical order: 2, 3, 3, 4, 4, 5, 7. Both 3 and 4 appear twice. There are two modes.

c. Every value occurs the same number of times in the data set 1, 2, 3, 4, 5, 6, 7.

There is no mode.

Answers **a.** The mode is 3. **b.** The modes are 3 and 4. **c.** There is no mode.

Linear Transformations of Data

Multiplying each data value by the same constant or adding the same constant to each data value is an example of a **linear transformation of a set of data.**

Let us start by examining additive transformations. For instance, consider the data 2, 2, 3, 4, 5. If 10 is added to each data value, the data set becomes:

12, 12, 13, 14, 15

Notice that every measure of central tendency has been shifted to the right by 10 units:

old mean $= \frac{2 + 2 + 3 + 4 + 5}{5} = 3.2$	new mean $= \frac{12 + 12 + 13 + 14 + 15}{5} = 13.2$
old median $= 3$	new median $= 13$
old mode $= 2$	new mode $= 12$

In fact, this result is valid for any additive transformation of a data set. In general:

▶ **If $\bar{x}$, d, and o are the mean, median, and mode of a set of data and the constant c is added to each data value, then $\bar{x} + c$, $d + c$, and $o + c$ are the mean, median, and mode of the transformed data.**

It can be also shown that a similar result holds for multiplicative transformations, that is:

▶ **If $\bar{x}$, d, and o are the mean, median, and mode of a set of data and each data value is multiplied by the *nonzero* constant c, then $c\bar{x}$, cd, and co are the mean, median, and mode of the transformed data.**

EXAMPLE 5

In Ms. Huan's Algebra class, the average score on the most recent quiz was 65. Being in a generous mood, Ms. Huan decided to curve the quiz by adding 10 points to each quiz score. What will be the new average score for the class?

Answer $65 + 10 = 75$ points

EXERCISES

Writing About Mathematics

1. On her first two math tests, Rene received grades of 67 and 79. Her mean (average) grade for these two tests was 73. On her third test she received a grade of 91. Rene found the mean of 73 and 91 and said that her mean for the three tests was 82. Do you agree with Rene? Explain why or why not.

2. Carlos said that when there are n numbers in a set of data and n is an odd number, the median is the $\frac{n+1}{2}$th number when the data are arranged in order. Do you agree with Carlos? Explain why or why not.

Developing Skills

3. For each set of data, find the mean.

a. 7, 3, 5, 11, 9

b. 22, 38, 18, 14, 22, 30

c. $5\frac{1}{2}, 2\frac{3}{4}, 7\frac{1}{2}, 5\frac{3}{4}, 4\frac{1}{2}$

d. 1.00, 0.01, 1.10, 0.12, 1.00, 1.03

4. Find the median for each set of data.

a. 1, 2, 5, 3, 4

b. 2, 9, 2, 9, 7

c. 3, 8, 12, 7, 1, 0, 4

d. 80, 83, 97, 79, 25

e. 3.2, 8.7, 1.4

f. 2.00, 0.20, 2.20, 0.02, 2.02

g. 21, 24, 23, 22, 20, 24, 23, 21, 22, 23

h. 5, 7, 9, 3, 8, 7, 5, 6

5. What is the median for the digits 1, 2, 3, . . . , 9?

6. What is the median for the counting numbers from 1 through 100?

7. Find the mode for each distribution.

a. 2, 2, 3, 4, 8

b. 2, 2, 3, 8, 8

c. 2, 2, 8, 8, 8

d. 2, 3, 4, 7, 8

e. 2, 2, 3, 8, 8, 9, 9

f. 1, 2, 1, 2, 1, 2, 1

g. 1, 2, 3, 2, 1, 2, 3, 2, 1

h. 3, 19, 21, 75, 0, 6

i. 3, 2, 7, 6, 2, 7, 3, 1, 4, 2, 7, 5

j. 19, 21, 18, 23, 19, 22, 18, 19, 20

8. A set of data consists of six numbers: 7, 8, 8, 9, 9, and x. Find the mode for these six numbers when:

a. $x = 9$

b. $x = 8$

c. $x = 7$

d. $x = 6$

9. A set of data consists of the values 2, 4, 5, x, 5, 4. Find a possible value of x such that:

a. there is no mode because all scores appear an equal number of times

b. there is only one mode

c. there are two modes

10. For the set of data 5, 5, 6, 7, 7, which statement is true?

(1) mean = mode

(2) median = mode

(3) mean = median

(4) mean $<$ median

11. For the set of data 8, 8, 9, 10, 15, which statement is true?

(1) mean $<$ median

(2) mean $>$ mode

(3) median $<$ mode

(4) mean = median

12. When the data consists of 3, 4, 5, 4, 3, 4, 5, which statement is true?
(1) mean > median
(2) mean > mode
(3) median < mode
(4) mean = median

13. For which set of data is there no mode?
(1) 2, 1, 3, 1, 2
(2) 1, 2, 3, 3, 3
(3) 1, 2, 4, 3, 5
(4) 2, 2, 3, 3, 3

14. For which set of data is there more than one mode?
(1) 8, 7, 7, 8, 7
(2) 8, 7, 4, 5, 6
(3) 8, 7, 5, 7, 6, 5
(4) 1, 2, 2, 3, 3, 3

15. For which set of data does the median equal the mode?
(1) 3, 3, 4, 5, 6
(2) 3, 3, 4, 5
(3) 3, 3, 4
(4) 3, 4

16. For which set of data will the mean, median, and mode all be equal?
(1) 1, 2, 5, 5, 7
(2) 1, 2, 5, 5, 8, 9
(3) 1, 1, 1, 2, 5
(4) 1, 1, 2

17. The median of the following data is 11:

2, 5, 9, 11, 40, 3, 4, 5, 10, 45, 32, 40, 67, 7, 11, 9, 20, 34, 5, 1, 8, 15, 16, 19, 39

a. If 4 is subtracted from each data value, what is the median of the transformed data set?

b. If the largest data value is doubled and the smallest data value is halved, what is the median of the new data set?

18. The mean of the following data is 37.625:

3, 0, 1, 7, 8, 11, 31, 15, 99, 98, 92, 81, 85, 87, 55, 54, 34, 27, 26, 21, 14, 17, 19, 18

If each data value is multiplied by 2 and increased by 5, what is the mean of the transformed data set?

19. Three consecutive integers can be represented by x, $x + 1$, and $x + 2$. The average of these consecutive integers is 32. What are the three integers?

20. Three consecutive even integers can be represented by x, $x + 2$, and $x + 4$. The average of these consecutive even integers is 20. Find the integers.

21. The mean of three numbers is 31. The second is 1 more than twice the first. The third is 4 less than 3 times the first. Find the numbers.

Applying Skills

22. Sid received grades of 92, 84, and 70 on three tests. Find his test average.

23. Sarah's grades were 80 on each of two of her tests and 90 on each of three other tests. Find her test average.

24. Louise received a grade of x on each of two of her tests and of y on each of three other tests. Represent her average for all the tests in terms of x and y.

25. Andy has grades of 84, 65, and 76 on three social studies tests. What grade must he obtain on the next test to have an average of exactly 80 for the four tests?

26. Rosemary has grades of 90, 90, 92, and 78 on four English tests. What grade must she obtain on the next test so that her average for the five tests will be 90?

27. The first three test scores are shown below for each of four students. A fourth test will be given and averages taken for all four tests. Each student hopes to maintain an average of 85. Find the score needed by each student on the fourth test to have an 85 average, or explain why such an average is not possible.

a. Pat: 78, 80, 100

b. Bernice: 79, 80, 81

c. Helen: 90, 92, 95

d. Al: 65, 80, 80

28. The average weight of Sue, Pam, and Nancy is 55 kilograms.

a. What is the total weight of the three girls?

b. Agnes weighs 60 kilograms. What is the average weight of the four girls: Sue, Pam, Nancy, and Agnes?

29. For the first 6 days of a week, the average rainfall in Chicago was 1.2 inches. On the last day of the week, 1.9 inches of rain fell. What was the average rainfall for the week?

30. If the heights, in centimeters, of a group of students are 180, 180, 173, 170, and 167, what is the mean height of these students?

31. What is the median age of a family whose members are 42, 38, 14, 13, 10, and 8 years old?

32. What is the median age of a class in which 14 students are 14 years old and 16 students are 15 years old?

33. In a charity collection, ten people gave amounts of $1, $2, $1, $1, $3, $1, $2, $1, $1, and $1.50. What was the median donation?

34. The test scores for an examination were 62, 67, 67, 70, 90, 93, and 98. What is the median test score?

35. The weekly salaries of six employees in a small firm are $440, $445, $445, $450, $450, and $620.

a. For these six salaries, find: (1) the mean (2) the median (3) the mode

b. If negotiations for new salaries are in session and you represent management, which measure of central tendency will you use as the average salary? Explain your answer.

c. If negotiations are in session and you represent the labor union, which measure of central tendency will you use as an average salary? Explain your answer.

36. In a certain school district, bus service is provided for students living at least $1\frac{1}{2}$ miles from school. The distances, rounded to the nearest half mile, from school to home for ten students are $0, \frac{1}{2}, \frac{1}{2}, 1, 1, 1, 1, 1\frac{1}{2}, 3\frac{1}{2}$, and 10 miles.

a. For these data, find: (1) the mean (2) the median (3) the mode

b. How many of the ten students are entitled to bus service?

c. Explain why the mean is not a good measure of central tendency to describe the average distance between home and school for these students.

37. Last month, a carpenter used 12 boxes of nails each of which contained nails of only one size. The sizes marked on the boxes were:

$$\tfrac{3}{4}\text{ in.}, \tfrac{3}{4}\text{ in.}, \tfrac{3}{4}\text{ in.}, \tfrac{3}{4}\text{ in.}, \tfrac{3}{4}\text{ in.}, \tfrac{3}{4}\text{ in.}, \tfrac{3}{4}\text{ in.}, \tfrac{3}{4}\text{ in.}, 1\text{ in.}, 1\text{ in.}, 2\text{ in.}, 2\text{ in.}$$

a. For these data, find: (1) the mean (2) the median (3) the mode

b. Describe the average-size nail used by the carpenter, using at least one of these measures of central tendency. Explain your answer.

Hands-On Activity

Find the mean, the median, and the mode for the data that you collected in the Hands-On Activity for Section 16-1. It may be necessary to go back to your original data to do this.

16-5 MEASURES OF CENTRAL TENDENCY AND GROUPED DATA

Intervals of Length 1

In a statistical study, when the range is small, we can use intervals of length 1 to group the data. For example, each member of a class of 25 students reported the number of books he or she read during the first half of the school year. The data are as follows:

5, 3, 5, 3, 1, 8, 2, 4, 2, 6, 3, 8, 8, 5, 3, 4, 5, 8, 5, 3, 3, 5, 6, 2, 3

These data, for which the values range from 1 to 8, can be organized into a table such as the one shown at the right, with each value representing an interval.

Since 25 students were included in this study, the total frequency, N, is 25. We can use this table, with intervals of length 1, to find the mode, median, and mean for these data.

Interval	Frequency
8	4
7	0
6	2
5	6
4	2
3	7
2	3
1	1
	$N = 25$

Mode of a Set of Grouped Data

Since the greatest frequency, 7, appears for interval 3, the mode for the data is 3. In general:

▶ **For a set of grouped data, the mode is the value of the interval that contains the greatest frequency.**

Median of a Set of Grouped Data

We have learned that the median for a set of data in numerical order is the middle value.

For these 25 numbers, there are 12 numbers greater than or equal to the median, and 12 numbers less than or equal to the median. Therefore, when the numbers are written in numerical order, the median is the 13th number from either end.

1, 2, 2, 2, 3, 3, 3, 3, 3, 3, 3, 4, 4, 5, 5, 5, 5, 5, 5, 6, 6, 8, 8, 8, 8

↑ The median is 4.

When the data are grouped in the table shown earlier, a simple counting procedure can be used to find the median, the 13th number. When we add the frequencies of the first four intervals, starting at the top, we find that these intervals include data for:

$$4 + 0 + 2 + 6 = 12 \text{ students}$$

Therefore, the next lower interval (with frequency greater than 0) must include the median, the value for the 13th student. This is the interval for the data value 4.

When we add the frequencies of the first three intervals, starting at the bottom, we find that these intervals include data for:

$$1 + 3 + 7 = 11 \text{ students}$$

The next higher interval contains two scores, one for the 12th student and one that is the median, or the value for the 13th student. Again this is the interval for the data value 4.

In general:

► **For a set of grouped data, the median is the value of the interval that contains the middle data value.**

Mean of a Set of Grouped Data

By adding the four 8's in the ungrouped data, we see that four students, reading eight books each, have read 8 + 8 + 8 + 8 or 32 books. We can arrive at this same number by using the grouped intervals in the table: we multiply the four 8's by the frequency 4. Thus, (4)(8) = 32. Applying this multiplication shortcut to each row of the table, we obtain the third column of the following table:

Interval	Frequency	(Interval) × (Frequency)
8	4	$8 \times 4 = 32$
7	0	$7 \times 0 = 0$
6	2	$6 \times 2 = 12$
5	6	$5 \times 6 = 30$
4	2	$4 \times 2 = 8$
3	7	$3 \times 7 = 21$
2	3	$2 \times 3 = 6$
1	1	$1 \times 1 = 1$
	$N = 25$	Total = 110

The total (110) represents the sum of all 25 pieces of data. We can check this by adding the 25 scores in the unorganized data.

Finally, to find the mean, we divide the total number, 110, by the number of items, 25. Thus, the mean for the data is: $110 \div 25 = 4.4$.

Procedure

To find the mean for *N* values in a table of grouped data when the length of each interval is 1:

1. For each interval, multiply the interval value by its corresponding frequency.
2. Find the sum of these products.
3. Divide this sum by the total frequency, *N*.

Calculator Solution for Grouped Data

The calculator can be used to find the mean and median for the grouped data shown above. Enter the number of books read by each student into L_1 and the frequency for each number of books into L_2. Then use the 1-Var Stats from the STAT CALC menu to display information about the data.

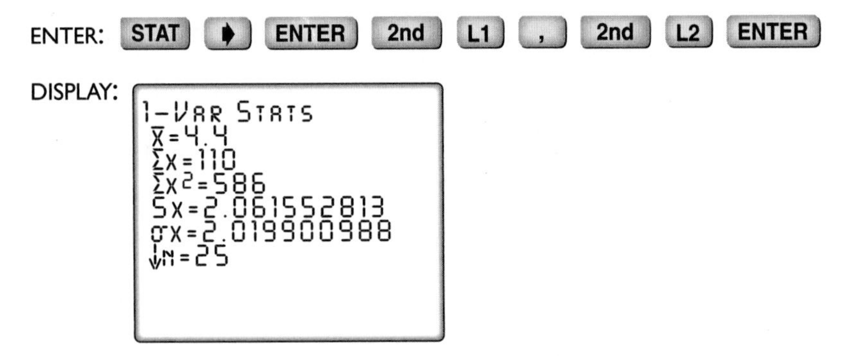

The display shows that the mean, $\bar{x}$, is 4.4, the sum of the number of books read is 110, and the number of students, the total frequency, N, is 25. Use the down arrow to display the median, Med = 4.

Intervals Other Than Length 1

There are specific mathematical procedures to find the mean, median, and mode for grouped data with intervals other than length 1, but we will not study them at this time. Instead, we will simply identify the intervals that contain some of these measures of central tendency. For example, a small industrial plant surveyed 50 workers to find the number of miles each person commuted to work. The commuting distances were reported, to the nearest mile, as follows:

0, 0, 1, 1, 2, 2, 2, 3, 3, 4, 4, 4, 5, 5, 6, 6, 6, 7, 7, 7, 9,

10, 10, 10, 10, 10, 10, 10, 10, 12, 12, 14, 15, 17, 17,

18, 22, 23, 25, 28, 30, 32, 32, 33, 34, 34, 36, 37, 37, 52

These data are organized into a table with intervals of length 10, as follows:

Interval (commuting distance)	Frequency (number of workers)
50–59	1
40–49	0
30–39	9
20–29	4
10–19	15
0–9	21
	N = 50

Modal Interval

In the table, interval 0–9 contains the greatest frequency, 21. We say that interval 0–9 is the **group mode**, or **modal interval**, because this group of numbers has the greatest frequency. The modal interval is *not* the same as the mode. The modal interval is a group of numbers; the mode is usually a single number. For this example, the original data (before being placed into the table) show that the number appearing most often is 10. Hence, the mode is 10. The modal interval, which is 0–9, tells us that, of the six intervals in the table, the most frequently occurring commuting distance is 0 to 9 miles.

Both the mode and the modal interval depend on the concept of greatest frequency. For the mode, we look for a single number that has the greatest frequency. For the modal interval, we look for the interval that has the greatest frequency.

Interval Containing the Median

To find the interval containing the median, we follow the procedure described earlier in this section. For 50 numbers, the median, or middle number, will be at a point where 25 numbers are at or above the median and 25 are at or below it.

Count the frequencies in the table from the uppermost interval and move downward. We add $1 + 0 + 9 + 4 = 14$. Since there are 15 numbers in the next lower interval, and $14 + 15 = 29$, we see that the 25th number will be reached somewhere in that interval, 10–19.

Count from the bottom interval and move up. We have 21 numbers in the first interval. Since there are 15 numbers in the next higher interval, and $21 + 15 = 36$, we see that 25th number will be reached somewhere in that interval, 10–19. This is the same result that we obtained when we moved downward. The interval containing the median for this grouping is 10–19.

In this course, we will not deal with problems in which the median is not found in any interval.

Interval Containing the Mean

When data are grouped using intervals of length other than 1, there is no simple procedure to identify the interval containing the mean. However, the mean can be *approximated* by assuming that the data are equally distributed throughout each interval. The mean is then found by using the midpoint of each interval as the value of each entry in the interval. This problem is studied in higher-level courses.

EXAMPLE 1

In the table, the data indicate the heights, in inches, of 17 basketball players. For these data find:
a. the mode **b.** the median **c.** the mean

Height (inches)	Frequency (number)
77	2
76	0
75	5
74	3
73	4
72	2
71	1

Solution **a.** The greatest frequency, 5, occurs for the height of 75 inches. The mode, or height appearing most often, is 75.

b. For 17 players, the median is the 9th number, so there are 8 heights greater than or equal to the median and 8 heights less than or equal to the median. Counting the frequencies going down, we have $2 + 0 + 5 = 7$. Since the frequency of the next interval is 3, the 8th, 9th, and 10th heights are in this interval, 74.

Counting the frequencies going up, we have $1 + 2 + 4 = 7$. Again, the frequency of the next interval is 3, and the 8th, 9th, and 10th heights are in this interval. The 9th height, the median, is 74.

c. (1) Multiply each height by its corresponding frequency:

$77 \times 2 = 154$ $\quad 76 \times 0 = 0$ $\quad 75 \times 5 = 375$ $\quad 74 \times 3 = 222$

$73 \times 4 = 292$ $\quad 72 \times 2 = 144$ $\quad 71 \times 1 = 71$

(2) Find the total of these products:

$$154 + 0 + 375 + 222 + 292 + 144 + 71 = 1{,}258$$

(3) Divide this total, 1,258, by the total frequency, 17 to obtain the mean:

$$1258 \div 17 = 74$$

Calculator Solution Clear any previous data that may be stored in L_1 and L_2. Enter the heights of the players into L_1 and the frequencies into L_2. Then use 1-Var Stats from the STAT CALC menu to display information about the data. The screen will show the mean, $\bar{x}$. Press the down arrow key to display the median.

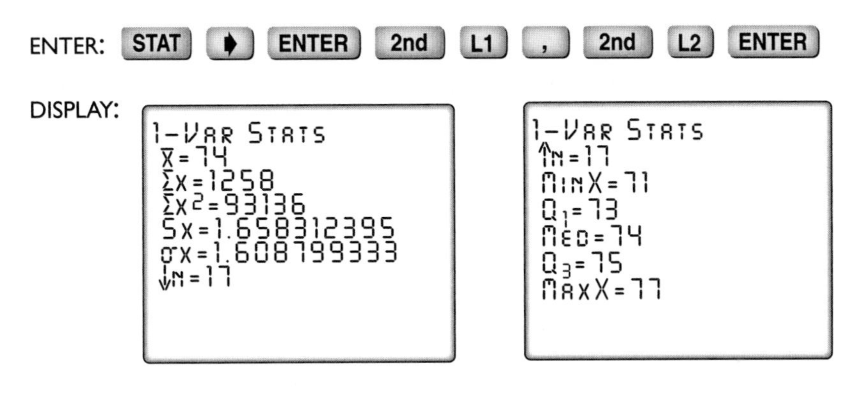

Answers **a.** mode = 75 **b.** median = 74 **c.** mean = 74

EXERCISES

Writing About Mathematics

1. The median for a set of 50 data values is the average of the 25th and 26th data values when the data is in numerical order. What must be true if the median is equal to one of the data values? Explain your answer.

2. What must be true about a set of data if the median is *not* one of the data values? Explain your answer.

Developing Skills

In 3–5, the data are grouped in each table in intervals of length 1.
Find: **a.** the total frequency **b.** the mean **c.** the median **d.** the mode

3.

Interval	Frequency
10	1
9	2
8	3
7	3
6	4
5	3

4.

Interval	Frequency
15	3
16	2
17	4
18	1
19	5
20	6

5.

Interval	Frequency
25	4
24	0
23	3
22	2
21	4
20	5
19	2

In 6–8, the data are grouped in each table in intervals other than length 1. Find: **a.** the total frequency **b.** the interval that contains the median **c.** the modal interval

6.

Interval	Frequency
55–64	3
45–54	8
35–44	7
25–34	6
15–24	2

7.

Interval	Frequency
4–9	12
10–15	13
16–21	9
22–27	12
28–33	15
34–39	10

8.

Interval	Frequency
126–150	4
101–125	6
76–100	6
51–75	3
26–50	7
1–25	2

Applying Skills

9. On a test consisting of 20 questions, 15 students received the following scores:

17, 14, 16, 18, 17, 19, 15, 15, 16, 13, 17, 12, 18, 16, 17

a. Make a frequency table for these students listing scores from 12 to 20.

b. Find the median score.

c. Find the mode.

d. Find the mean.

10. A questionnaire was distributed to 100 people. The table shows the time taken, in minutes, to complete the questionnaire.

a. For this set of data, find: (1) the mean (2) the median (3) the mode

b. How are the three measures found in part **a** related for these data?

Interval	Frequency
6	12
5	20
4	36
3	20
2	12

11. A storeowner kept a tally of the sizes of suits purchased in the store, as shown in the table.

a. For this set of data, find:

(1) the total frequency (2) the mean

(3) the median (4) the mode

b. Which measure of central tendency should the storeowner use to describe the average suit sold?

Size of Suit (interval)	Number Sold (frequency)
48	1
46	1
44	3
42	5
40	3
38	8
36	2
34	2

12. Test scores for a class of 20 students are as follows:

93, 84, 97, 98, 100, 78, 86, 100, 85, 92, 72, 55, 91, 90, 75, 94, 83, 60, 81, 95

a. Organize the data in a table using 51–60 as the smallest interval.

b. Find the modal interval.

c. Find the interval that contains the median.

13. The following data consist of the weights, in pounds, of 35 adults:

176, 154, 161, 125, 138, 142, 108, 115, 187, 158, 168, 162
135, 120, 134, 190, 195, 117, 142, 133, 138, 151, 150, 168
172, 115, 148, 112, 123, 137, 186, 171, 166, 166, 179

a. Organize the data in a table, using 100–119 as the smallest interval.

b. Construct a frequency histogram based on the grouped data.

c. In what interval is the median for these grouped data?

d. What is the modal interval?

16-6 QUARTILES, PERCENTILES, AND CUMULATIVE FREQUENCY

Quartiles

When the values in a set of data are listed in numerical order, the median separates the values into two equal parts. The numbers that separate the set into four equal parts are called **quartiles**.

To find the quartile values, we first divide the set of data into two equal parts and then divide each of these parts into two equal parts.

The heights, in inches, of 20 students are shown in the following list. The median, which is the average of the 10th and 11th data values, is shown here enclosed in a box.

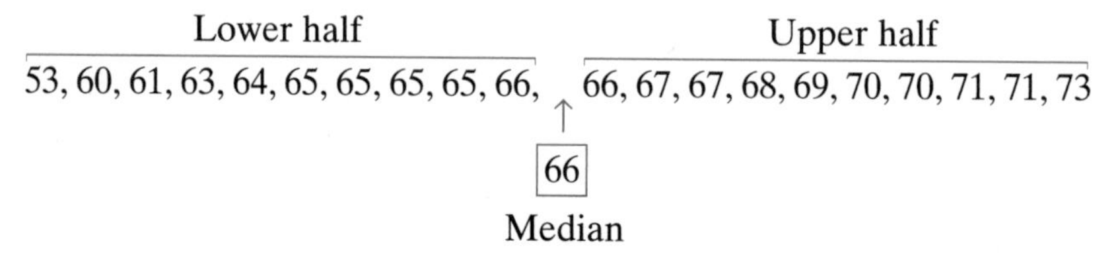

Ten heights are listed in the lower half, 53–66. The middle value for these 10 heights is the average of the 5th and 6th values from the lower end, or 64.5. This value separates the lower half into two equal parts.

Ten heights are also listed in the upper half, 66–73. The middle value for these 10 heights is the average of the 5th and 6th values from the upper end, or 69.5. This value separates the upper half into two equal parts.

The 20 data values are now separated into four equal parts, or quarters.

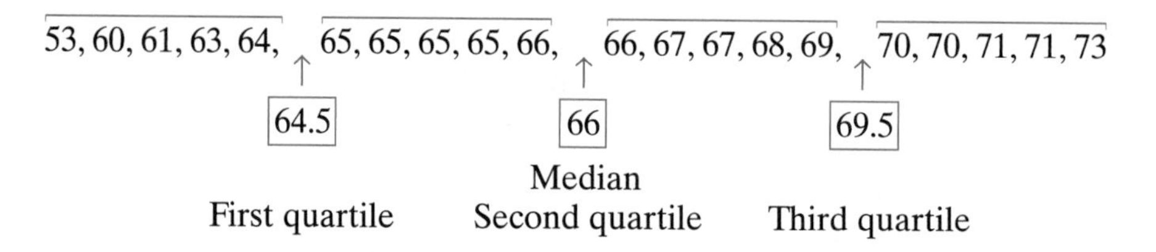

The numbers that separate the data into four equal parts are the quartiles. For this set of data:

1. Since one quarter of the heights are less than or equal to 64.5 inches, 64.5 is the **lower quartile**, or **first quartile**.

2. Since two quarters of the heights are less than or equal to 66 inches, 66 is the **second quartile**. The second quartile is always the same as the median.

3. Since three quarters of the heights are less than or equal to 69.5 inches, 69.5 is the **upper quartile**, or **third quartile**.

Note: The quartiles are sometimes denoted Q_1, Q_2, and Q_3.

Procedure

To find the quartile values for a set of data:

1. Arrange the data in ascending order from left to right.
2. Find the median for the set of data. The median is the second quartile value.
3. Find the middle value for the lower half of the data. This number is the first, or lower, quartile value.
4. Find the middle value for the upper half of the data. This number is the third, or upper, quartile value.

Note that when finding the first quartile, use all of the data values less than or equal to the median, but do not include the median in the calculation. Similarly, when finding the third quartile, use all of the data values greater than or equal to the median, but do not include the median in the calculation.

Constructing a Box-and-Whisker Plot

A **box-and-whisker plot** is a diagram that uses the quartile values, together with the maximum and minimum values, to display information about a set of data. To draw a box-and-whisker plot, we use the following steps.

STEP 1. Draw a scale with numbers from the minimum to the maximum value of a set of data. For example, for the set of heights of the 20 students, the scale should include the numbers from 53 to 73.

STEP 2. Above the scale, place dots to represent the five numbers that are the **statistical summary** for this set of data: the minimum value, the first quartile, the median, the third quartile, and the maximum value. For the heights of the 20 students, these numbers are 53, 64.5, 66, 69.5 and 73.

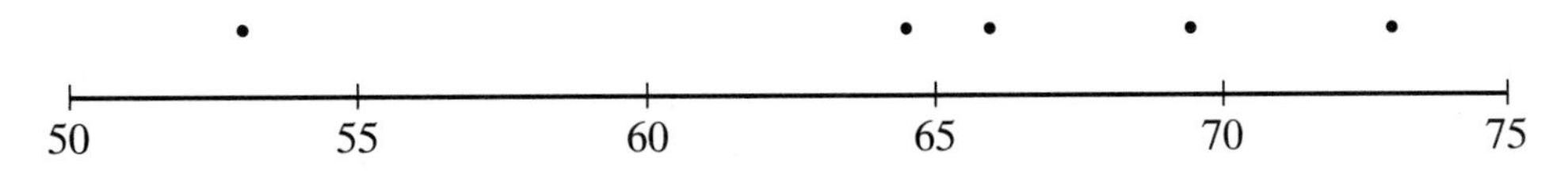

STEP 3. Draw a box between the dots that represent the lower and upper quartiles, and a vertical line in the box through the point that represents the median.

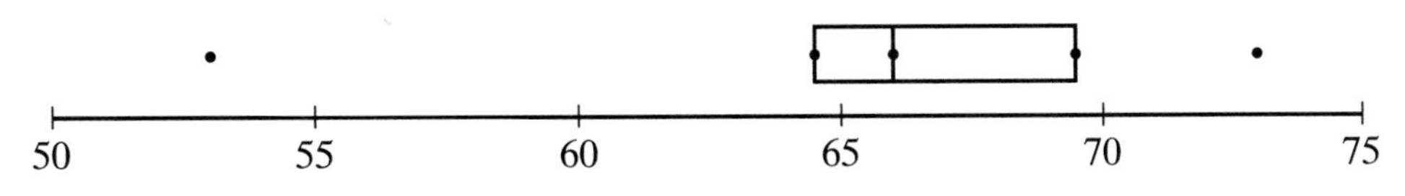

STEP 4. Add the whiskers by drawing a line segment joining the dots that represent the minimum data value and the lower quartile, and a second line segment joining the dots that represent the maximum data value and the upper quartile.

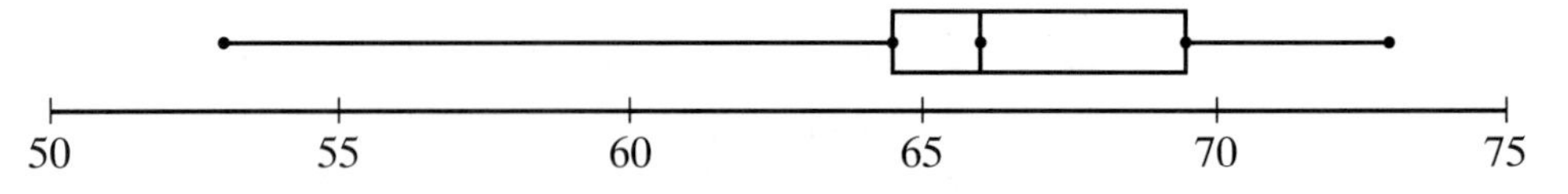

The box indicates the ranges of the middle half of the set of data. The long whisker at the left shows us that the data are more scattered at the lower than at the higher end.

A graphing calculator can display a box-and-whisker plot. Enter the data in L_1, then go to the STAT PLOT menu to select the type of graph to draw.

ENTER: **2nd** **STAT PLOT** **1** **ENTER**

▼ **▶** **▶** **▶** **▶** **ENTER** **▼**

2nd **L1** **▼** **ALPHA** **1**

PLOT1 PLOT2 PLOT3
ON OFF
TYPE:
XLIST: L1
FREQ: 1

Now display the box-and-whisker plot by entering **ZOOM** **9**.

We can press **TRACE** and the right and left arrow keys to display the minimum value, first quartile, median, third quartile, and maximum value.

P1:L1
MED=66

The five statistical summary can also be displayed in 1-Var Stats. Scroll down to the last five values.

ENTER: **STAT** **▶** **ENTER** **ENTER**

1-VAR STATS
↑N
MINX=53
Q_1=64.5
MED=66
Q_3=69.5
MAXX=73

EXAMPLE 1

Find the five statistical summary for the following set of data:

8, 5, 12, 9, 6, 2, 14, 7, 10, 17, 11, 8, 14, 5

Solution (1) Arrange the data in numerical order:

2, 5, 5, 6, 7, 8, 8, 9,10, 11, 12, 14, 14, 17

We can see that 2 is the minimum value and 17 is the maximum value.

(2) Find the median. Since there are 14 data values in the set, the median is the average of the 7th and 8th values.

$$\text{Median} = \frac{8 + 9}{2} = 8.5$$

Therefore, 8.5 is the second quartile.

(3) Find the first quartile. There are seven values less than 8.5. The middle value is the 4th value from the lower end of the set of data, 6. Therefore, 6 is the first, or lower, quartile.

(4) Find the third quartile. There are seven values greater than 8.5. The middle value is the 4th value from the upper end of the set of data, 12. Therefore, 12 is the third, or upper, quartile.

Answer The minimum is 2, first quartile is 6, the second quartile is 8.5, the third quartile is 12, and the maximum is 17.

Note: The quartiles 6, 8.5, and 12 separate the data values into four equal parts even though the original number of data values, 14, is not divisible by 4:

2, 5, 5, [6,] 7, 8, 8, ↑ 9, 10, 11, [12,] 14, 14, 17

[8.5]

The first and third quartile values, 6 and 12, are data values. If we think of each of these as a half data value in the groups that they separate, each group contains $3\frac{1}{2}$ data values, which is 25% of the total.

Percentiles

A **percentile** is a number that tells us what percent of the total number of data values lies at or below a given measure.

Let us consider again the set of data values representing the heights of 20 students. What is the percentile rank of 65? To find out, we separate the data into the values that are less than or equal to 65 and those that are greater than or equal to 65, so that the four 65's in the set are divided equally between the two groups:

53, 60, 61, 63, 64, 65, 65,	65, 65, 66, 66, 67, 67, 68, 69, 70, 70, 71, 71, 73

Half of 4, or 2, of the 65's are in the lower group and half are in the upper group.

Since there are seven data values in the lower group, we find what percent 7 is of 20, the total number of values:

$$\frac{7}{20} = 0.35 = 35\%$$

Therefore, 65 is at the 35th percentile.

To find the percentile rank of 69, we separate the data into the values that are less than or equal to 69 and those that are greater or equal to 69:

53, 60, 61, 63, 64, 65, 65, 65, 65, 66, 66, 67, 67, 68,	69,	70, 70, 71, 71, 73

Because 69 occurs only once, we will include it as half of a data value in the lower group and half of a data value in the upper group. Therefore, there are $14\frac{1}{2}$ or 14.5 data values in the lower group.

$$\frac{14.5}{20} = 0.725 = 72.5\%$$

Because percentiles are usually not written using fractions, we say that 69 is at the 73rd percentile.

EXAMPLE 2

Find the percentile rank of 87 in the following set of 30 marks:

56, 65, 65, 67, 72, 73, 75, 77, 77, 78, 78, 78, 80, 80, 80,
82, 83, 85, 85, 85, 86, 87, 87, 87, 88, 90, 92, 93, 95, 98

Solution (1) Find the sum of the number of marks less than 87 and half of the number of 87's:

$$\begin{aligned} \text{Number of marks less than } 87 &= 21 \\ \text{Half of the number of 87's } (0.5 \times 3) &= \underline{\ 1.5} \\ & \ 22.5 \end{aligned}$$

(2) Divide the sum by the total number of marks:

$$\frac{22.5}{30} = 0.75$$

(3) Change the decimal value to a percent: $0.75 = 75\%$.

Answer: A mark of 87 is at the 75th percentile.

Note: 87 is also the upper quartile mark.

Cumulative Frequency

In a school, a final examination was given to all 240 students taking biology. The test grades of these students were then grouped into a table. At the same time, a histogram of the results was constructed, as shown below.

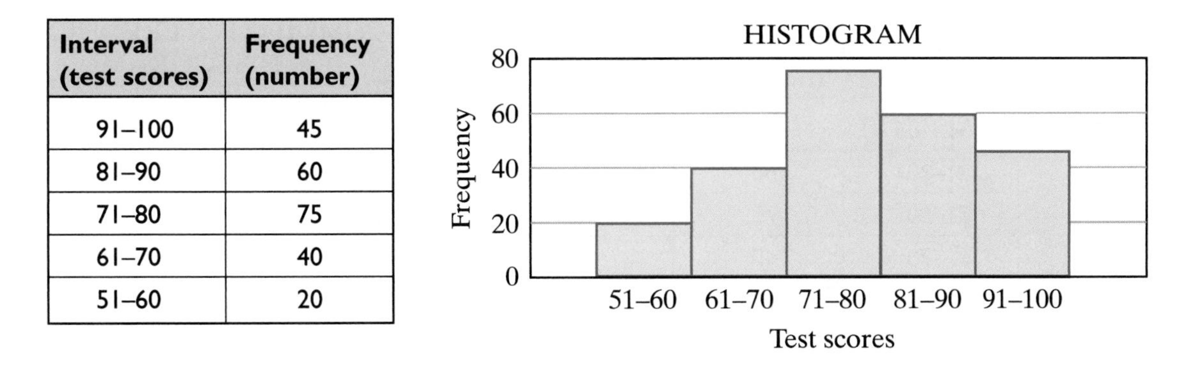

Interval (test scores)	Frequency (number)
91–100	45
81–90	60
71–80	75
61–70	40
51–60	20

From the table and the histogram, we can see that 20 students scored in the interval 51–60, 40 students scored in the interval 61–70, and so forth. We can use these data to construct a new type of histogram that will answer the question, "How many students scored *below* a certain grade?"

By answering the following questions, we will gather some information before constructing the new histogram:

1. How many students scored 60 or less on the test?

From the lowest interval, 51–60, we know that 20 students scored 60 or less.

2. How many students scored 70 or less on the test?

By adding the frequencies for the two lowest intervals, 51–60 and 61–70, we see that 20 + 40, or 60, students scored 70 or less.

3. How many students scored 80 or less on the test?

By adding the frequencies for the three lowest intervals, 51–60, 61–70, and 71–80, we see that 20 + 40 + 75, or 135, students scored 80 or less.

4. How many students scored 90 or less on the test?

Here, we add the frequencies in the four lowest intervals. Thus, 20 + 40 + 75 + 60, or 195, students scored 90 or less.

5. How many students scored 100 or less on the test?

By adding the five lowest frequencies, 20 + 40 + 75 + 60 + 45, we see that 240 students scored 100 or less. This result makes sense because 240 students took the test and all of them scored 100 or less.

Constructing a Cumulative Frequency Histogram

The answers to the five questions we have just asked were found by adding, or *accumulating*, the frequencies for the intervals in the grouped data to find the **cumulative frequency**. The accumulation of data starts with the lowest interval of data values, in this case, the lowest test scores. The histogram that displays these accumulated figures is called a **cumulative frequency histogram**.

Interval (test scores)	Frequency (number)	Cumulative Frequency
91–100	45	240
81–90	60	195
71–80	75	135
61–70	40	60
51–60	20	20

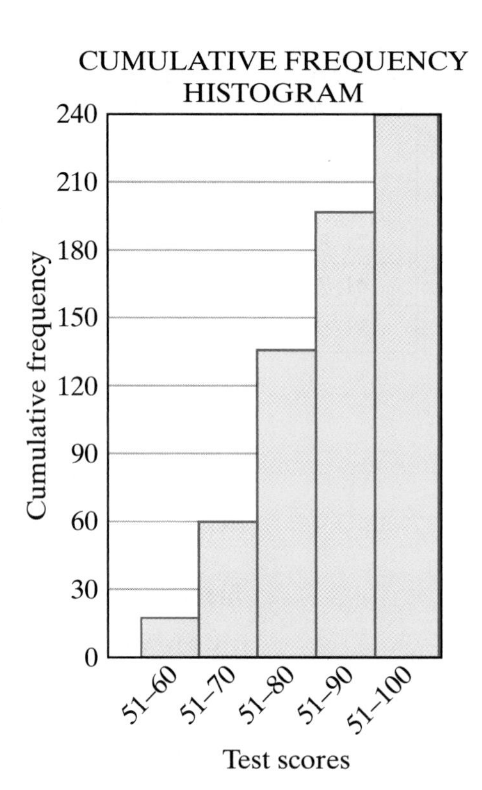

To find the cumulative frequency for each interval, we add the frequency for that interval to the frequencies for the intervals with lower values. To draw a cumulative frequency histogram, we use the cumulative frequencies to determine the heights of the bars.

For our example of the 240 biology students and their scores, the frequency scale for the cumulative frequency histogram goes from 0 to 240 (the total frequency for all of the data). We can replace the scale of the cumulative frequency histogram shown above with a different one that expresses the cumulative frequency in percents. Since 240 students represent 100% of the students taking the biology test, we write 100% to correspond to a cumulative frequency of 240. Similarly, since 0 students represent 0% of the students taking the biology test, we write 0% to correspond to a cumulative frequency of 0.

If we divide the percent scale into four equal parts, we can label the three added divisions as 25%, 50%, and 75%.

Thus the graph relates each cumulative frequency to a percent of the total number of biology students. For example, 120 students (half of the total number) corresponds to 50%.

Let us use the percent scale to answer the question, "What percent of the students scored 70 or below on the test?" The height of each bar represents both the number of students and the percent of the students who had scores at or below the largest number in the interval represented by that bar. Since 25%, or a quarter, of the scores were 70 or below, we say that 70 is an approximate value for the lower quartile, or the 25th percentile.

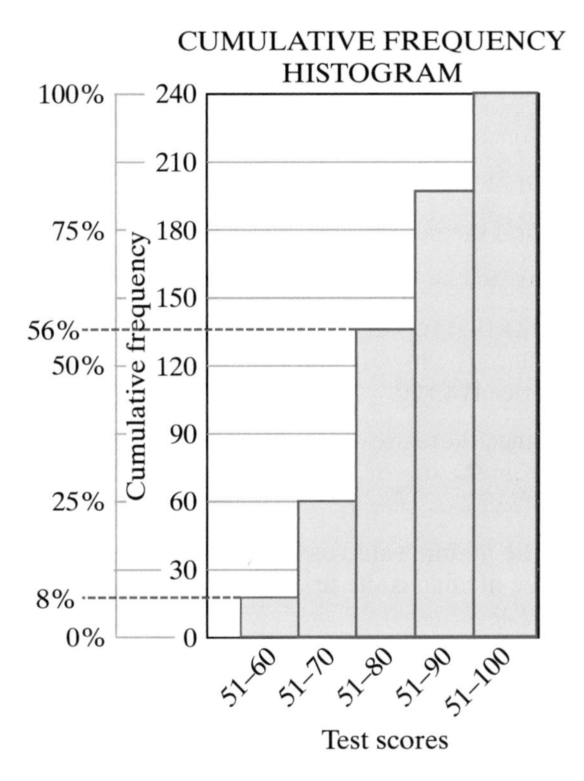

From the histogram, we can see that about 56% of the students had scores at or below 80. Thus, the second quartile, the median, is in the 51–80 interval. For these data, the upper quartile is in the 51–90 interval.

From the histogram, we can also conveniently read the approximate percentiles for the scores that are the end values of the intervals. For example, to find the percentile for a score of 60, the right-end score of the first interval, we draw a horizontal line segment from the height of the first interval to the percent scale, as shown by the dashed line in the histogram above. The fact that the horizontal line crosses the percent scale at about one-third the distance between 0% and 25% tells us that approximately 8% of the students scored 60 or below 60. Thus, the 8th percentile is a good estimate for a score of 60.

EXAMPLE 3

A reporter for the local newspaper is preparing an article on the ice cream storés in the area. She listed the following prices for a two-scoop cone at 15 stores.

$2.48, $2.57, $2.30, $2.79, $2.25, $3.00, $2.82, $2.75,
$2.55, $2.98, $2.53, $2.40, $2.80, $2.50, $2.65

a. List the data in a stem-and-leaf diagram.

b. Find the median.

c. Find the first and third quartiles.

d. Construct a box-and-whisker plot.

e. Draw a cumulative frequency histogram.

f. Find the percentile rank of a price of $2.75.

Solution **a.** The first two digits in each price will be the stem. The lowest price is $2.25 and the highest price is $3.00.

Stem	Leaf
3.0	0
2.9	8
2.8	0 2
2.7	5 9
2.6	5
2.5	0 3 5 7
2.4	0 8
2.3	0
2.2	5

Key: 2.9 | 8 = $2.98

b. Since there are 15 prices, the median is the 8th from the top or from the bottom. The median is $2.57.

c. The middle value of the set of numbers below the median is the first quartile. That price is $2.48.

The middle value of the set of numbers above the median is the third quartile. That price is $2.80.

d. Use a scale from $2.25 to $3.00. Place dots at $2.48, $2.57, and $2.80 for the first quartile, the median, and the third quartile. Draw the box around the quartiles with a vertical line through the median. Add the whiskers.

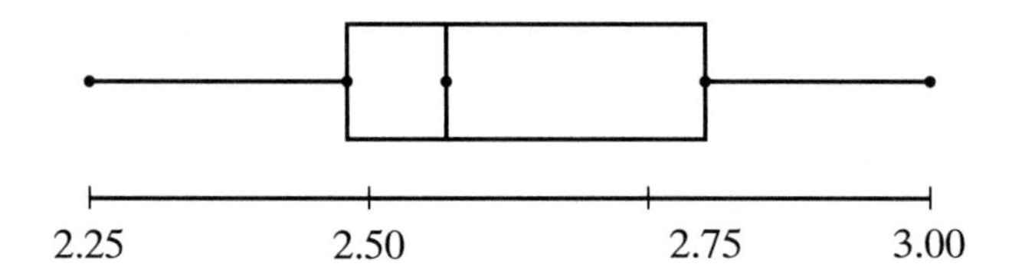

Interval	Frequency	Cumulative Frequency
3.00–3.09	1	15
2.90–2.99	1	14
2.80–2.89	2	13
2.70–2.79	2	11
2.60–2.69	1	9
2.50–2.59	4	8
2.40–2.49	2	4
2.30–2.39	1	2
2.20–2.29	1	1

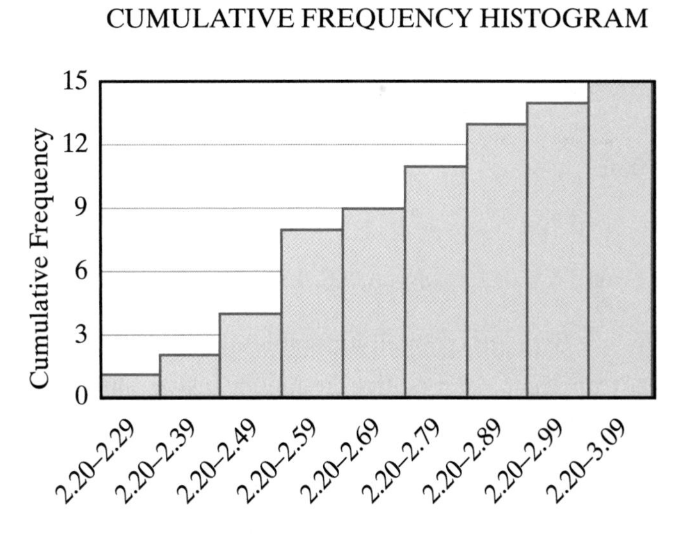

e. Make a cumulative frequency table and draw the histogram. Use 2.20–2.29 as the smallest interval.

f. There are 9 data values below \$2.75. Add $\frac{1}{2}$ for the data value \$2.75.

$$\text{Percentile rank: } \frac{9\frac{1}{2}}{15} = 0.6\overline{3} \approx 63\%$$

Answers **a.** Diagram **b.** median = \$2.57 **c.** first quartile = \$2.48; third quartile = \$2.80 **d.** Diagram **e.** Diagram **f.** 63rd percentile

Note: A cumulative frequency histogram can be drawn on a calculator just like a regular histogram. In list L_2, where we previously entered the frequencies for each individual interval, we now enter each cumulative frequency.

EXERCISES

Writing About Mathematics

1. a. Is it possible to determine the percentile rank of a given score if the set of scores is arranged in a stem-and-leaf diagram? Explain why or why not.

b. Is it possible to determine the percentile rank of a given score if the set of scores is shown on a cumulative frequency histogram? Explain why or why not.

2. A set of data consisting of 23 consecutive numbers is written in numerical order from left to right.

a. The number that is the first quartile is in which position from the left?

b. The number that is the third quartile is in which position from the left?

Developing Skills

In 3–6, for each set of data: **a.** Find the five numbers of the statistical summary **b.** Draw a box-and-whisker plot.

3. 12, 17, 20, 21, 25, 27, 29, 30, 32, 33, 33, 37, 40, 42, 44

4. 67, 70, 72, 77, 78, 78, 80, 84, 86, 88, 90, 92

5. 0, 0, 1, 1, 1, 1, 2, 2, 2, 2, 3, 3, 3, 3, 5, 7, 9, 9

6. 3.6, 4.0, 4.2, 4.3, 4.5, 4.8, 4.9, 5.0

In 7–9, data are grouped into tables. For each set of data:

a. Construct a cumulative frequency histogram.

b. Find the interval in which the lower quartile lies.

c. Find the interval in which the median lies.

d. Find the interval in which the upper quartile lies.

7.

Interval	Frequency
41–50	8
31–40	5
21–30	2
11–20	5
1–10	4

8.

Interval	Frequency
25–29	3
20–24	1
15–19	3
10–14	9
5–9	9

9.

Interval	Frequency
1–4	4
5–8	3
9–12	7
13–16	2
17–20	2

10. For the data given in the table:

a. Construct a cumulative frequency histogram.

b. In what interval is the median?

c. The value 10 occurs twice in the data. What is the percentile rank of 10?

Interval	Frequency
21–25	5
16–20	4
11–15	6
6–10	3
1–5	2

11. For the data given in the table:

a. Construct a cumulative frequency histogram.

b. In what interval is the median?

c. In what interval is the upper quartile?

d. What percent of scores are 17 or less?

e. In what interval is the 25th percentile?

Interval	Frequency
33–37	4
28–32	3
23–27	7
18–22	12
13–17	8
8–12	5
3–7	1

Applying Skills

12. A group of 400 students were asked to state the number of minutes that each spends watching television in 1 day. The cumulative frequency histogram shown below summarizes the responses as percents.

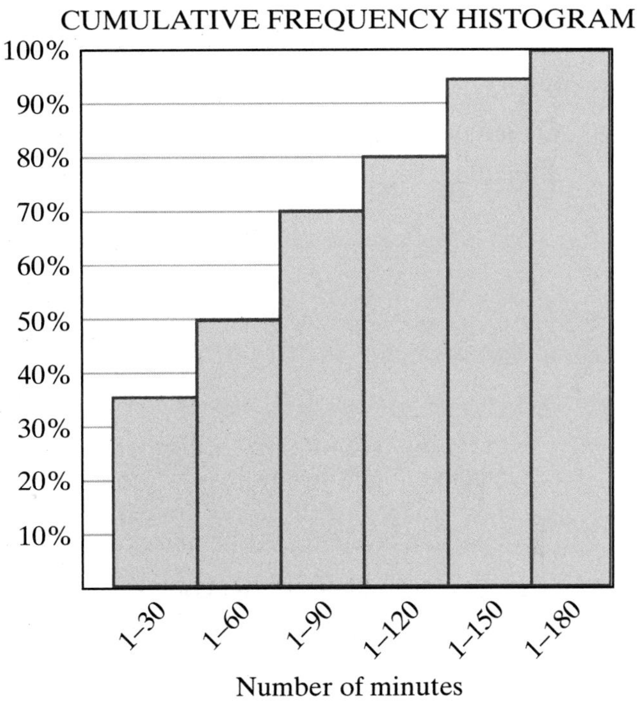

a. What percent of the students questioned watch television for 90 minutes or less each day?

b. How many of the students watch television for 90 minutes or less each day?

c. In what interval is the upper quartile?

d. In what interval is the lower quartile?

e. If one of these students is picked at random, what is the probability that he or she watches 30 minutes or less of television each day?

CUMULATIVE FREQUENCY HISTOGRAM

13. A journalism student was doing a study of the readability of the daily newspaper. She chose several paragraphs at random and listed the number of letters in each of 88 words. She prepared the following chart.

a. Copy the chart, adding a column that lists the cumulative frequency

b. Find the median.

c. Find the first and third quartiles.

d. Construct a box-and whisker plot.

e. Draw a cumulative frequency histogram.

f. Find the percentile rank of a word with 7 letters.

Number of letters	Frequency
1	4
2	14
3	20
4	20
5	3
6	18
7	5
8	2
9	1
10	1

14. Cecilia's average for 4 years is 86. Her average is the upper quartile for her class of 250 students. At most, how many students in her class have averages that are less than Cecilia's?

15. In the table at the right, data are given for the heights, in inches, of 22 football players.

a. Copy and complete the table.

b. Draw a cumulative frequency histogram.

c. Find the height that is the lower quartile.

d. Find the height that is the upper quartile.

Height (inches)	Frequency	Cumulative Frequency
77	2	
76	2	
75	7	
74	5	
73	3	
72	2	
71	1	

16. The lower quartile for a set of data was 40. These data consisted of the heights, in inches, of 680 children. At most, how many of these children measured more than 40 inches?

In 17 and 18, select, in each case, the numeral preceding the correct answer.

17. On a standardized test, Sally scored at the 80th percentile. This means that

(1) Sally answered 80 questions correctly.

(2) Sally answered 80% of the questions correctly.

(3) Of the students who took the test, about 80% had the same score as Sally.

(4) Of the students who took the test, at least 80% had scores that were less than or equal to Sally's score.

18. For a set of data consisting of test scores, the 50th percentile is 87. Which of the following could be *false*?

(1) 50% of the scores are 87.

(2) 50% of the scores are 87 or less.

(3) Half of the scores are at least 87.

(4) The median is 87.

16-7 BIVARIATE STATISTICS

We have been studying **univariate statistics** or statistics that involve a single set of numbers. Statistics are often used to study the relationship between two different sets of values. For example, a dietician may want to study the relationship between the number of calories from fat in a person's diet and the level of cholesterol in that person's blood, or a merchant may want to study the relationship between the amount spent on advertising and gross sales. Although these examples involving **two-valued statistics** or **bivariate statistics** require complex statistical methods, we can investigate some of the properties of similar but simpler problems by looking at graphs and by using a graphing calculator. A graph that shows the pairs of values in the data as points in the plane is called a **scatter plot**.

Correlation

We will consider five cases of two-valued statistics to investigate the relationship or **correlation** between the variables based on their scatter plots.

CASE 1 *The data has positive linear correlation. The points in the scatter plot approximate a straight line that has a positive slope.*

A driver recorded the number of gallons of gasoline used and the number of miles driven each time she filled the tank. In this example, there is both correlation and **causation** since the increase in the number of miles driven causes the number of gallons of gasoline needed to increase.

Gallons	7.2	5.8	7.0	5.5	5.6	7.1	6.0	4.4	5.0	6.2	4.7	5.7
Miles	240	188	226	193	187	235	202	145	167	212	154	188

This scatter plot can be duplicated on your graphing calculator. Enter the number of miles as L_1 and the number of gallons of gasoline as corresponding entries in L_2. The miles will be graphed as x-values and the gallons of gasoline as y-values. First, turn on Plot 1:

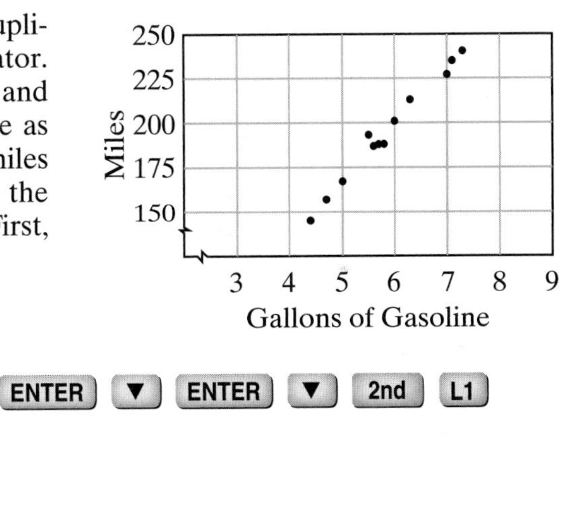

ENTER: 2nd STAT PLOT 1 ENTER ▼ ENTER ▼ 2nd L1 ▼ 2nd L2

DISPLAY:

```
Plot1 Plot2 Plot3
On    Off
Type: 
Xlist: L1
Ylist: L2
Mark: □ + ·
```

Now use ZoomStat from the ZOOM menu to construct a window that will include all values of x and y.

ENTER: 

DISPLAY: 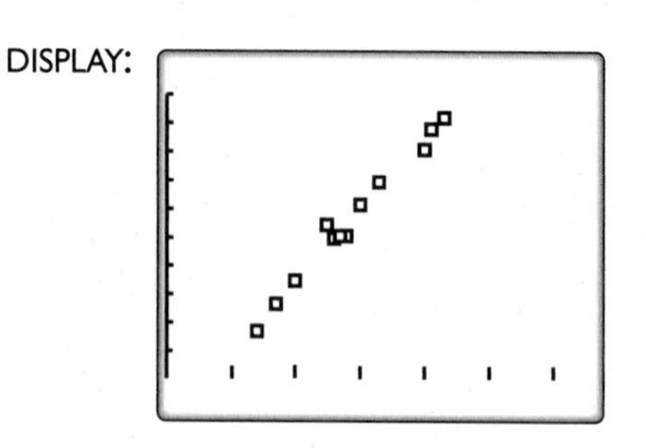

CASE 2 *The data has moderate positive correlation. The points in a scatter plot do not lie in a straight line but there is a general tendency for the values of y to increase as the values of x increase.*

Last month, each student in an English class was required to choose a book to read. The teacher recorded, for each student in the class, the number of days spent reading the book and the number of pages in the book.

Days	8	14	12	26	9	17	28	13	15	30	18	20
Pages	225	300	298	356	200	412	205	215	310	357	209	250
Days	29	22	17	14	11	14	22	19	16	7	18	30
Pages	314	288	256	225	232	256	300	305	276	172	318	480

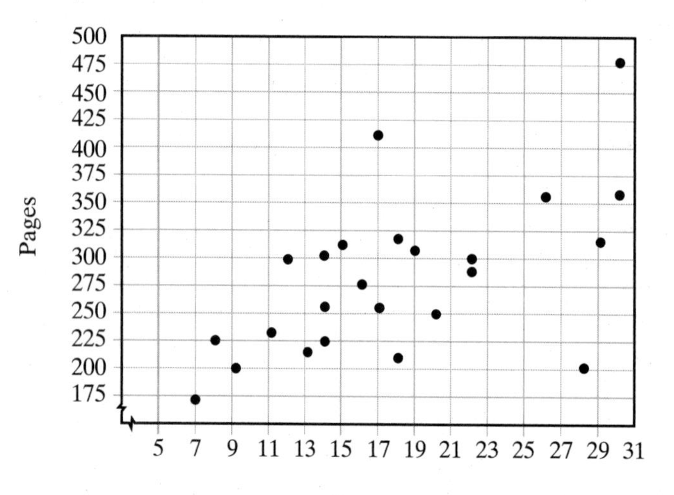

While the books with more pages may have required more time, some students read more rapidly and some spent more time each day reading. The graph shows that, in general, as the number of days needed to read a book increased, the number of pages that were read also increased.

CASE 3 *The data has no correlation.*

Before giving a test, a teacher asked each student how many minutes each had spent the night before preparing for the test. After correcting the test, she prepared the table below which compares the number of minutes of study to the number of correct answers.

Minutes of Study	20	15	40	5	10	25	30	12	5	20	35	40
Correct Answers	15	10	3	19	16	6	12	3	5	8	16	14

The graph shows that there is *no* correlation between the time spent studying just before the test and the number of correct answers on the test.

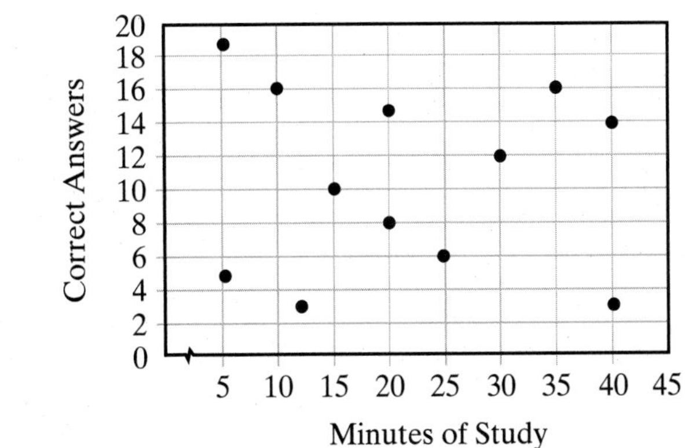

CASE 4 *The data has moderate negative correlation. The points in a scatter plot do not lie in a straight line but there is a general tendency for the values of y to decrease as the values of x increase.*

A group of children go to an after-school program at a local youth club. The director of the program keeps a record, shown below, of the time, in minutes, each student spends playing video games and doing homework.

Games	20	30	90	60	30	50	70	40	80	60
Homework	50	60	10	40	40	35	15	30	30	10

In this instance, the unit of measure, minutes, is found in the problem rather than in the table. To create meaningful graphs, always include a unit of measure on the horizontal and vertical axes.

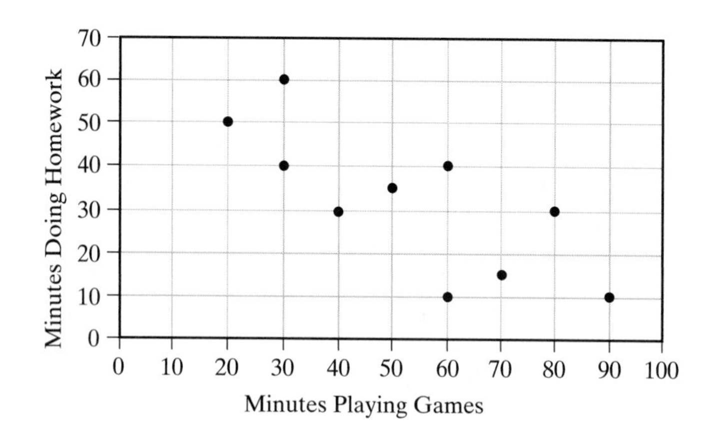

The graph shows that, in general, as the number of minutes spent playing video games increases, the number of minutes spent doing homework decreases.

CASE 5 *The data has negative linear correlation. The points in the scatter plot approximate a straight line that has a negative slope.*

A long-distance truck driver travels 500 miles each day. As he passes through different areas on his trip, his average speed and the length of time he drives each day vary. The chart below shows a record of average speed and time for a 10-day period.

Speed	50	64	68	60	54	66	70	62	64	58
Time	10	7.9	7.5	8.5	9.0	7.0	7.1	8.0	8.2	9.0

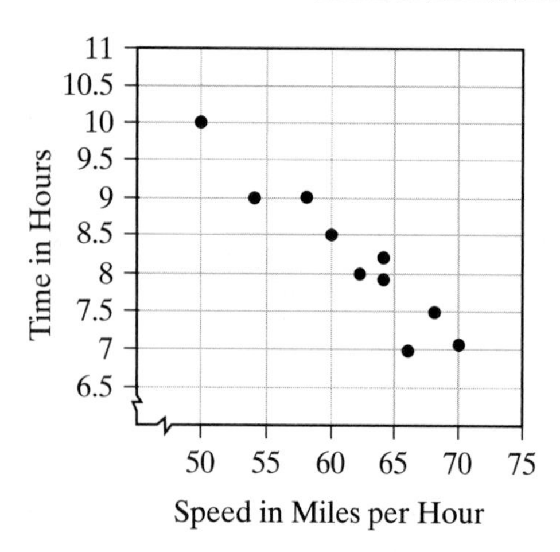

In this case, the increase in the average speed causes the time required to drive a fixed distance to decrease. This example indicates both negative correlation and causation.

It is important to note that correlation is not the same as causation. Correlation is an indication of the strength of the linear relationship or association between the variables, but it does not mean that changes in one variable are the cause of changes in the other. For example, suppose a study found there was a strong positive correlation between the number of pages in the daily newspaper and the number of voters who turn out for an election. One would not be correct in concluding that a greater number of pages causes a greater turnout. Rather, it is likely that the urgency of the issues is reflected in the increase of both the size of the newspaper and the size of the turnout.

Another example where there is no causation occurs in **time series** or data that is collected at regular intervals over time. For instance, the population of the U.S. recorded every ten years is an example of a time series. In this case, we cannot say that time causes a change in the population. All we can do is note a general trend, if any.

Line of Best Fit

When it makes sense to consider one variable as the independent variable and the other as the dependent variable, and the data has a linear correlation (even if it is only moderate correlation), the data can be represented by a **line of best fit**. For example, we can write an equation for the data in Case 1. Enter the data into L_1 and L_2 if it is not already there. Find the mean values for x, the number of miles driven, and for y, the number of gallons of gasoline used. Then use 2-Var Stats from the STAT CALC menu:

ENTER: STAT ▶ 2 ENTER

DISPLAY:

```
2-Var Stats
x̄=5.85
Σx=70.2
Σx²=419.88
Sx=.9150260801
σx=.8760707734
↓n=12
```

The calculator gives $\bar{x} = 5.85$ and, by pressing the down arrow key, $\bar{y} = 194.75$. We will use these mean values, (5.85, 194.75), as one of the points on our line. We will choose one other data point, for example (7.1, 235), as a second point and write the equation of the line using the slope-intercept form $y = mx + b$. First find the slope:

$$m = \frac{y_2 - y_1}{x_2 - x_1} = \frac{194.75 - 235}{5.85 - 7.1} = \frac{-40.25}{-1.25} = 32.2$$

Now use one of the points to find the y-intercept:

$$194.75 = 32.2(5.85) + b$$

$$194.75 = 188.37 + b$$

$$6.38 = b$$

Round the values to three significant digits. A possible equation for a line of best fit is $y = 32.2x + 6.38$.

The calculator can also be used to find a line called the **regression line** to fit a bivariate set of data. Use the LinReg(ax+b) function in the STAT CALC menu.

ENTER: STAT ▶ 4 ENTER

DISPLAY:

```
LinReg
y=ax+b
a=32.67643865
b=3.592833876
```

If we round the values to three significant digits, the equation of the regression line is $y = 32.7x + 3.59$. In this case, the difference between these two equations is negligible. *However, this is not always the case.* The regression line is a special line of best fit that minimizes the square of the vertical distances to each data point.

We can compare these two equations with the actual data. Graph the scatter plot of the data using ZoomStat. Then write the two equations in the Y= menu.

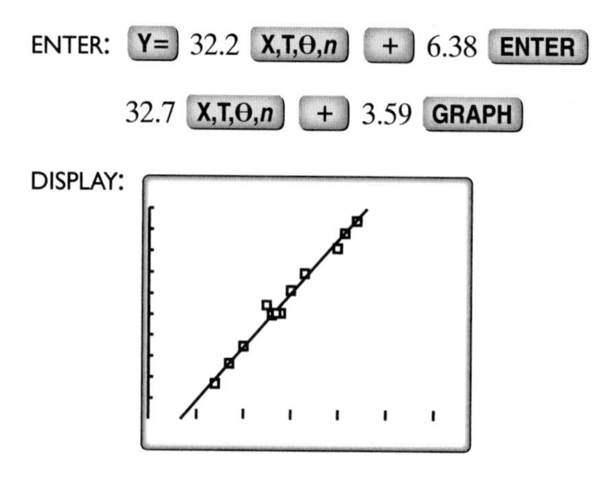

Notice that the lines are very close and do approximate the data.

Note 1: The equation of the line of best fit is very sensitive to rounding. Try to round the coefficients of the line of best fit to at least three significant digits or to whatever the test question asks.

Note 2: A line of best fit is appropriate only for data that exhibit a linear pattern. In more advanced courses, you will learn how to deal with nonlinear patterns.

These equations can be used to predict values. For example, if the driver has driven 250 miles before filling the tank, how many gallons of gasoline should be needed? We will use the equation from the calculator.

$$y = 32.7x + 3.59$$
$$250 = 32.7x + 3.59$$
$$246.41 = 32.7x$$
$$7.535474006 = x$$

It is reasonable to say that the driver can expect to need about 7.5 gallons of gasoline.

What we just did is called **extrapolation**, that is, using the line of best fit to make a prediction outside of the range of data values. Using the line of best fit to make a prediction *within* the given range of data values is called **interpolation**.

In general, interpolation is usually safe, while care should be taken when extrapolating. The observed correlation pattern may not be valid outside of the given range of data values. For example, consider the scatter plot of the population of a town shown below. The population grew at a constant rate during the years in which the data was gathered. However, we do not expect the population to continue to grow forever, and thus, it may not be possible to extrapolate far into the future.

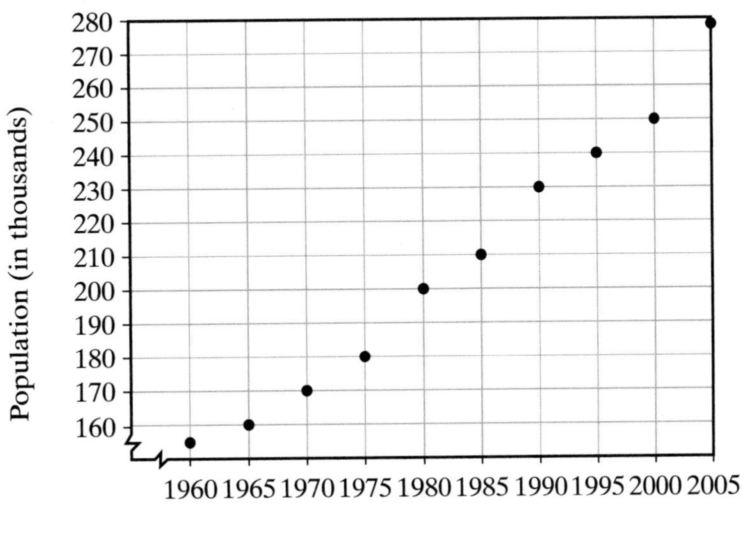

Keep in Mind In general, when a given relationship involves two sets of data:

1. In some cases a straight line, a line of best fit, can be drawn to approximate the relationship between the data sets.
2. If a line of best fit has a positive slope, the data has positive linear correlation.
3. If the line of best fit has a negative slope, the data has negative linear correlation.
4. A line of best fit can be drawn through $(\bar{x}, \bar{y})$, the point whose coordinates are the means of the given data. Any data point that appears to lie on or near the line of best fit can be used as a second point to write the equation.

5. A calculator can be used to find the regression line as the line of best fit.
6. When the graphed data points are so scattered that it is not possible to draw a straight line that approximates the given relationship, the data has no correlation.

To study bivariate data without using a graphing calculator:

1. Make a table that lists the data.
2. Plot the data as points on a graph.
3. Find the mean of each set of data and locate the point $(\bar{x}, \bar{y})$ on the graph.
4. Draw a line that best approximates the data.
5. Choose the point $(\bar{x}, \bar{y})$ and one other point or any two points that are on or close to the line that you drew. Use these points to write an equation of the line.
6. Use the equation of the line to predict related outcomes.

To study bivariate data using a graphing calculator:

1. Enter the data into L_1 and L_2 or any two lists in the memory of the calculator.
2. Use STAT PLOT to turn on a plot and to choose the type of plot needed. Enter the names of the lists in which the data is stored and choose the mark to be used for each data point.
3. Use ZoomStat to choose a viewing window that shows all of the data points.
4. Find the regression line using LinReg(ax+b) from the STAT CALC menu.
5. Enter the equation of the regression line in the Y= menu and use GRAPH to show the relationship between the data and the regression line.
6. Use the equation of the line to predict related outcomes.

In this course, we have found a line of best fit by finding a line that seems to represent the data or by using a calculator. In more advanced courses in statistics, you will learn detailed methods for finding the line of best fit.

EXAMPLE 1

The table below shows the number of calories and the number of grams of carbohydrates in a half-cup serving of ten different canned or frozen vegetables.

Carbohydrates	9	23	4	5	19	8	12	7	13	17
Calories	45	100	20	25	110	35	50	30	70	80

a. Draw a scatter plot on graph paper. Let the horizontal axis represent grams of carbohydrates and the vertical axis represent the number of calories.

b. Find the mean number of grams of carbohydrates in a serving of vegetables and the mean number of calories in a serving of vegetables.

c. On the graph, draw a line that approximates the data in the table, and determine its equation.

d. Enter the data in L_1 and L_2 on your calculator and find the linear regression equation, LinReg(ax+b).

e. Use each equation to find the expected number of calories in a serving of vegetables with 20 grams of carbohydrates. Compare the answers.

Solution **a.**

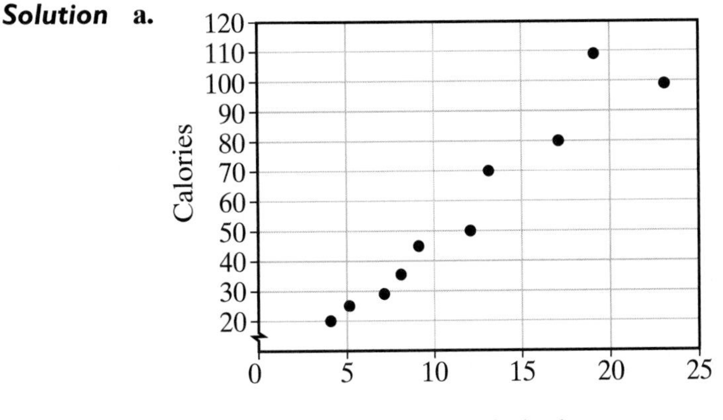

b. Enter the number of grams of carbohydrates in L_1 and the number of calories in L_2. Find $\bar{x}$ and $\bar{y}$, using 2-Var Stats.

ENTER: 

The means of the x- and y-coordinates are $\bar{x} = 11.7$ and $\bar{y} = 56.5$.

Locate the point (11.7, 56.5) on the graph.

c.

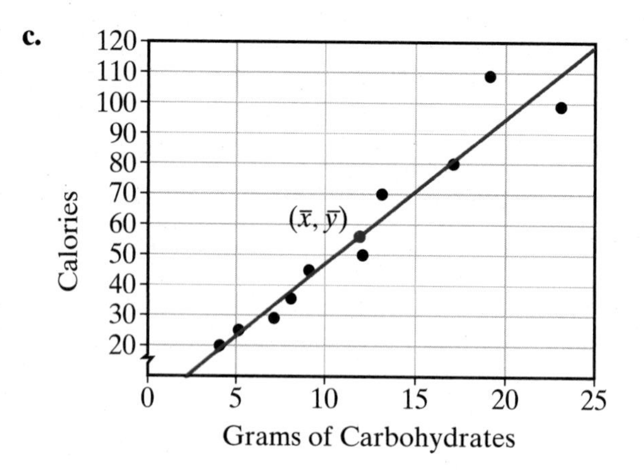

The line we have drawn seems to go through the point (4, 20). We will use this point and the point with the mean values, (11.7, 56.5), to write an equation of a line of best fit.

$$m = \frac{56.5 - 20}{11.7 - 4} = \frac{36.5}{7.7} \approx 4.74$$

$$y = mx + b$$
$$20 = \tfrac{36.5}{7.7}(4) + b$$
$$1.05 \approx b$$

An equation of a best fit line is $y = 4.74x + 1.05$.

d. The data are in L_1 and L_2.

ENTER: STAT ▶ 4 ENTER

DISPLAY:

```
LinReg
 y=ax+b
 a=4.885506842
 b=-.6604300475
```

The equation of the regression line is $y = 4.89x - 0.660$.

e. Let $x = 20$.

Use the equation from **c.**

$y = 4.74x + 1.05$

$y = 4.74(20) + 1.05$

$y = 95.85$

Use the equation from **d.**

$y = 4.89x - 0.660$

$y = 4.89(20) - 0.660$

$y = 97.14$

The two equations give very similar results. It would be reasonable to say that we could expect the number of calories to be about 96 or close to 100.

EXERCISES

Writing About Mathematics

1. **a.** Give an example of a set of bivariate data that has negative correlation.

   b. Do you think that the change in the independent variable in your example causes the change in the dependent variable?

2. Explain the purpose of finding a line of best fit.

Applying Skills

3. When Gina bought a new car, she decided to keep a record of how much gas she uses. Each time she puts gas in the car, she records the number of gallons of gas purchased and the number of miles driven since the last fill-up. Her record for the first 2 months is as follows:

Gallons of gas	10	12	9	6	11	10	8	12	10	7
Miles driven	324	375	290	190	345	336	250	375	330	225

   a. Draw a scatter plot of the data. Let the horizontal axis represent the number of gallons of gas and the vertical axis represent the number of miles driven.

   b. Does the data have positive, negative, or no correlation?

   c. Is this a causal relationship?

   d. Find the mean number of gallons of gasoline per fill-up.

   e. Find the mean number of miles driven between fill-ups.

   f. Locate the point that represents the mean number of gallons of gasoline and the mean number of miles driven. Use (0, 0) as a second point. Draw a line through these two points to approximate the data in the table.

   g. Use the line drawn in part **d** to approximate the number of miles Gina could drive on 3 gallons of gasoline.

4. Gemma made a record of the cost and length of each of the 14 long-distance telephone calls that she made in the past month. Her record is given below.

Minutes	3.7	1.0	19.6	0.8	4.3	34.8	2.9
Cost	$0.35	$0.11	$2.12	$0.09	$0.47	$3.78	$0.24
Minutes	2.5	7.1	10.9	5.8	1.5	1.4	8.0
Cost	$0.27	$0.79	$1.21	$0.65	$0.20	$0.17	$0.89

   a. Draw a scatter plot of the data on graph paper. Let the horizontal axis represent the number of minutes, and the vertical axis represent the cost of the call.

   b. Does the data have positive, negative, or no correlation?

   c. Is this a causal relationship?

d. Find the mean number of minutes per call.

e. Find the mean cost of the calls.

f. On the graph, draw a line of best fit for the data in the table and write its equation.

g. Use a calculator to find the equation of the regression line.

h. Approximate the cost of a call that lasted 14 minutes using the equation written in **d.**

i. Approximate the cost of a call that lasted 14 minutes using the equation written in **e.**

5. A local store did a study comparing the cost of a head of lettuce with the number of heads sold in one day. Each week, for five weeks, the price was changed and the average number of heads of lettuce sold per day was recorded. The data is shown in the chart below.

Cost per Head of Lettuce	$1.50	$1.25	$0.90	$1.75	$0.50
No. of Heads Sold	48	52	70	42	88

a. Draw a scatter plot of the data. Let the horizontal axis represent the cost of a head of lettuce and the vertical axis represent the number of heads sold.

b. Does the data have positive, negative, or no correlation?

c. Is this a causal relationship?

d. Find the mean cost per head.

e. Find the mean number of heads sold.

f. On the graph, draw a line that approximates the data in the table.

g. What appears to be the result of raising the price of a head of lettuce?

6. The chart below shows the recorded heights in inches and weights in pounds for the last 24 persons who enrolled in a health club.

Height	Weight	Height	Weight	Height	Weight	Height	Weight
69	160	75	180	66	145	71	165
67	160	76	155	66	130	66	155
63	135	70	175	68	160	67	140
73	185	73	170	68	140	78	210
71	215	68	190	72	170	72	160
79	225	74	190	77	195	69	145

a. Draw a scatter plot on graph paper to display the data.

b. Does the data have positive, negative, or no linear correlation?

c. Is this a causal relationship?

d. Draw and find the equation of a line of best fit. Use (77, 195) as a second point.

e. Use a calculator to find the linear regression line.

f. According to the equation written in **d.**, if the next person who enrolls in the health club is 62 inches tall, what would be the expected weight of that person?

g. According to the equation written in **e.**, if the next person who enrolls in the health club weighs 200 pounds, what would be the expected height of that person?

7. The chart below shows the number of millions of cellular telephones in use in the United States by year from 1994 to 2003.

Year	'94	'95	'96	'97	'98	'99	'00	'01	'02	'03
Phones	24.1	33.8	44.0	55.3	69.2	86.0	109.5	128.3	140.8	158.7

Let L_1 be the number of years after 1990: 4, 5, 6, 7, 8, 9, 10, 11, 12, 13.

a. Draw a scatter plot on graph paper to display the data.

b. Does the data have positive, negative, or no linear correlation?

c. Is this a causal relationship?

d. Draw and find the equation of a line of best fit.

e. On the graph, draw a line that approximates the data in the table, and determine its equation. Use (6, 44.0) as a second point.

f. If the line of best fit is approximately correct for years beyond 2003, estimate how many cellular phones will be in use in 2007.

8. The chart below shows, for the last 20 Supreme Court Justices to have left the court before 2000, the age at which the judge was nominated and the number of years as a Supreme Court judge.

Age	47	64	62	62	59	55	54	45	43	56
Years	15	16	24	17	24	4	3	31	23	5
Age	50	56	62	59	50	56	57	49	49	62
Years	33	16	16	7	18	7	13	6	12	1

a. Draw a scatter plot on graph paper to display the data.

b. Does the data have positive, negative, or no linear correlation?

c. Is there a causal relationship?

d. Draw and find an equation of a line of best fit.

e. Use a calculator to find the linear regression line.

f. Do you think that the data, the line of best fit, the regression line, or none of these could be used to approximate the number of years as a Supreme Court justice for the next person to retire from that office?

9. A cook was trying different recipes for potato salad and comparing the amount of dressing with the number of potatoes given in the recipe. The following data was recorded.

Number of Potatoes	7	4	2	8	6	7	5	4
Cups of Dressing	$1\frac{1}{2}$	$\frac{7}{8}$	$\frac{3}{4}$	$1\frac{1}{4}$	1	$1\frac{3}{4}$	$1\frac{1}{8}$	$\frac{3}{4}$

a. Draw a scatter plot on graph paper to display the data.

b. Does the data have positive, negative, or no linear correlation?

c. Draw and find the equation of a line of best fit. Use $\left(4, \frac{7}{8}\right)$ as a second point.

d. Use a calculator to find the linear regression line.

e. According to the equation written in **c.**, if the cook needs to use 10 potatoes to have enough salad, approximately how many cups of dressing are needed?

CHAPTER SUMMARY

Statistics is the study of numerical data. In a statistical study, data are collected, organized into tables and graphs, and analyzed to draw conclusions.

Data can either be quantitative or qualitative. **Quantitative data** represents counts or measurements. **Qualitative data** represents categories or qualities.

In an **experiment**, a researcher imposes a treatment on one or more groups. The **treatment group** receives the treatment, while the **control group** does not.

Tables and **stem-and-leaf diagrams** are used to organize data. A table should have between five and fifteen **intervals** that include all data values, are of equal size, and do not overlap.

A **histogram** is a bar graph in which the height of a bar represents the **frequency** of the data values represented by that bar.

A **cumulative frequency histogram** is a bar graph in which the height of the bar represents the total frequency of the data values that are less than or equal to the upper endpoint of that bar.

The mean, median, and mode are three **measures of central tendency**. The **mean** is the sum of the data values divided by the total frequency. The **median** is the middle value when the data values are placed in numerical order. The **mode** is the data value that has the largest frequency.

Quartile values separate the data into four equal parts. A **box-and-whisker plot** displays a set of data values using the **minimum**, the **first quartile**, the median, the **third quartile**, and the **maximum** as significant measures. The **percentile** rank tells what percent of the data values lie at or below a given measure.

In **two-valued statistics** or **bivariate statistics**, a relation between two different sets of data is studied. The data can be graphed on a **scatter plot**. The data may have positive, negative, or no correlation. Data that has positive or negative linear correlation can be represented by a **line of best fit**.

The line of best fit can be used to predict values not in the included data set. **Interpolation** is predicting within the given data range. **Extrapolation** is predicting outside of the given data range.

VOCABULARY

16-1 Data • Statistics • Descriptive statistics • Qualitative data • Quantitative data • Census • Sample • Bias • Experiment • Treatment group • Control group • Placebo effect • Placebo • Blinding • Single-blind experiment • Double-blind experiment

16-2 Tally • Frequency • Total frequency • Frequency distribution table • Group • Interval • Grouped data • Range • Stem-and-leaf diagram •

16-3 Histogram • Frequency histogram

16-4 Average • Measures of central tendency • Mean • Arithmetic mean • Numerical average • Median • Mode • Bimodal • Linear transformation of a data set

16-5 Group mode • Modal interval

16-6 Quartile • Lower quartile • First quartile • Second quartile • Upper quartile • Third quartile • Box-and-whisker plot • Five statistical summary • Percentile • Cumulative frequency • Cumulative frequency histogram

16-7 Univariate statistics • Two-valued statistics • Bivariate statistics • Scatter plot • Correlation • Causation • Time series • Line of best fit • Regression line • Extrapolation • Interpolation

REVIEW EXERCISES

1. Courtney said that the mean of a set of consecutive integers is the same as the median and that the mean can be found by adding the smallest and the largest numbers and dividing the sum by 2. Do you agree with Courtney? Explain why or why not.

2. A set of data contains N numbers arranged in numerical order.

a. When is the median one of the numbers in the set of data?

b. When is the median not one of the numbers in the set of data?

3. For each of the following sets of data, find: **a.** the mean **b.** the median **c.** the mode (if one exists)

(1) 3, 4, 3, 4, 3, 5

(2) 1, 3, 5, 7, 1, 2, 4

(3) 9, 3, 2, 8, 3, 3

(4) 9, 3, 2, 3, 8, 2, 7

4. Express, in terms of y, the mean of $3y - 2$ and $7y + 18$.

5. For the following data:

78, 91, 60, 65, 81, 72, 78, 80, 65, 63, 59, 78, 78, 54, 87, 75, 77

a. Use a stem-and-leaf diagram to organize the data.

b. Draw a histogram, using 50–59 as the lowest interval.

c. Draw a cumulative frequency histogram.

d. Draw a box-and-whisker plot.

6. The weights, in kilograms, of five adults are 53, 72, 68, 70, and 72.

a. Find: (1) the mean (2) the median (3) the mode

b. If each of the adults lost 5 kilograms, find, for the new set of weights: (1) the mean (2) the median (3) the mode

7. Steve's test scores are 82, 94, and 91. What grade must Steve earn on a fourth test so that the mean of his four scores will be exactly 90?

8. From Monday to Saturday of a week in May, the recorded high temperature readings were 72°, 75°, 79°, 83°, 83°, and 88°. For these data, find:

a. the mean b. the median c. the mode

9. Paul worked the following numbers of hours each week over a 20-week period:

15, 3, 7, 6, 2, 14, 9, 25, 8, 12, 8, 8, 15, 0, 8, 12, 28, 10, 14, 10

a. Organize the data in a frequency table, using 0–5 as the lowest interval.

b. Draw a frequency histogram of the data.

c. In what interval does the median lie?

d. Which interval contains the lower quartile?

10. The table shows the scores of 25 test papers.

Score	Frequency	Cumulative Frequency
60	1	
70	9	
80	8	
90	2	
100	5	

a. Is the data univariate or bivariate?

b. Find the mean score.

c. Find the median score.

d. Find the mode.

e. Copy and complete the table.

f. Draw a cumulative frequency histogram.

g. Find the percentile rank of 90.

h. What is the probability that a paper chosen at random has a score of 80?

11. The electoral votes cast for the winning presidential candidate in elections from 1900 to 2004 are as follows:

292, 336, 321, 435, 277, 404, 382, 444, 472, 523, 449, 432, 303, 442,
457, 303, 486, 301, 520, 297, 489, 525, 426, 370, 379, 271, 286

a. Organize the data in a stem-and-leaf diagram. (Use the first digit as the stems, and the last two digits as the leaves.)

b. Find the median number of electoral votes cast for the winning candidate.

c. Find the first-quartile and third-quartile values.

d. Draw a box-and-whisker plot to display the data.

12. The ages of 21 high school students are shown in the table at the right.

a. What is the median age?

b. What is the percentile rank of age 15?

c. When the ages of these 21 students are combined with the ages of 20 additional students, the median age remains unchanged. What is the smallest possible number of students under 16 in the second group?

Age	Frequency
18	1
17	4
16	2
15	7
14	2
13	5

13. For each variable, determine if it is qualitative or quantitative.

a. Major in college

b. GPA in college

c. Wind speed of a hurricane

d. Temperature of a rodent

e. Yearly profit of a corporation

f. Number of students late to class

g. Zip code

h. Employment status

14. Researchers looked into a possible relationship between alcoholism and pneumonia. They conducted a study of 100 current alcoholics, 50 former alcoholics, and 1,000 non-alcoholics who were hospitalized for a mild form of pneumonia. The researchers found that 30% of alcoholics and 30% of former alcoholics, versus only 15% of the non-alcoholics developed a more dangerous form of pneumonia. The researchers concluded that alcoholism raises the risk for developing pneumonia.

Discuss possible problems with this study.

15. Aurora buys oranges every week. The accompanying table lists the weights and the costs of her last 10 purchases of oranges.

Weight (lb)	2.2	1.2	3.6	4.5	1.0	2.5	1.8	5.0	3.5	1.7
Cost ($)	1.22	0.60	1.04	1.58	0.50	0.89	0.95	1.88	1.46	0.70

a. Is the data univariate or bivariate?

b. Draw a scatter plot of the data on graph paper. Let the horizontal axis represent the weights of the oranges and the vertical axis the costs.

c. Is there a correlation between the weight and the cost of the oranges? If so, is it positive or negative?

d. If the price is determined by the number of oranges purchased, do the variables have a causal relationship? Explain your answer.

e. On the graph, draw a line of best fit that approximates the data in the table and write its equation.

f. Use the equation written in **d** to approximate the cost of 4 pounds of oranges.

16. Explain why the graph on the right is misleading. (*Hint:* In accounting, numbers enclosed by parentheses denote negative numbers.)

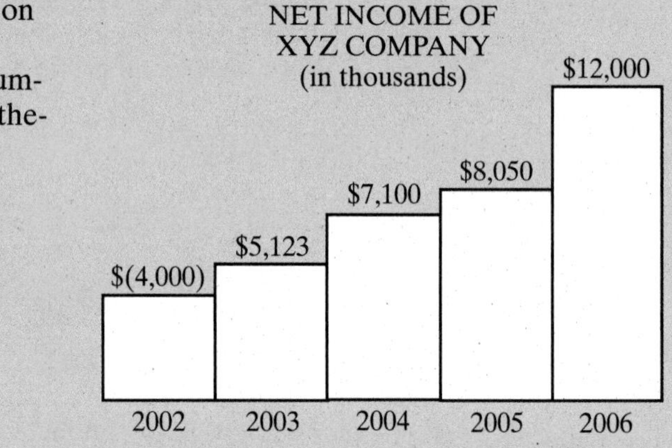

Exploration

a. Marny took the SAT in 2004 and scored a 1370. She was in the 94th percentile. Jordan took the SAT in 2000 and scored 1370. He was in the 95th percentile. Explain how this is possible.

b. Taylor's class rank stayed the same even though he had a cumulative grade point average of 3.4 one semester and 3.8 the next semester. Explain how this is possible.

CUMULATIVE REVIEW CHAPTERS 1–16

Part I

Answer all questions in this part. Each correct answer will receive 2 credits. No partial credit will be allowed.

1. When the domain is the set of integers, the solution set of the inequality $0 < 0.1x - 0.4 \leq 0.2$ is

(1) { } (2) {4, 5} (3) {4, 5, 6} (4) {5, 6}

2. The product $(2a + 3)(2a - 3)$ can be written as

(1) $2a^2 - 9$
(2) $4a^2 - 9$
(3) $4a^2 + 9$
(4) $4a^2 - 12a + 9$

3. When 0.00034 is written in the form 3.4×10^n, the value of n is

(1) −3 (2) −4 (3) 3 (4) 4

4. When $\frac{x}{3} + \frac{1}{2} = \frac{x-2}{6}$, x equals

(1) −5 (2) −1 (3) $\frac{1}{2}$ (4) 1

5. The mean of the set of even integers from 2 to 100 is

(1) 49 (2) 50 (3) 51 (4) 52

6. The probability that 9 is the sum of the numbers that appear when two dice are rolled is

(1) $\frac{4}{6}$ (2) $\frac{2}{36}$ (3) $\frac{2}{6}$ (4) $\frac{4}{36}$

7. If the circumference of a circle is 12 centimeters, then the area of the circle is

(1) 36 square centimeters
(2) 144 square centimeters
(3) $\frac{36}{\pi}$ square centimeters
(4) $\frac{144}{\pi}$ square centimeters

8. Which of the following is not an equation of a function?

(1) $y = 3x + 2$
(2) $y = x^2 + 3x + 1$
(3) $y^2 = x$
(4) $y = |x|$

9. The value of ${}_{10}P_8$ is

(1) 80 (2) 90 (3) 1,814,400 (4) 3,628,800

10. Which of the following is an equation of a line parallel to the line whose equation is $y = -2x + 4$?

(1) $2x + y = 7$ (2) $y - 2x = 7$ (3) $2x - y = 7$ (4) $y = 2x + 7$

Part II

Answer all questions in this part. Each correct answer will receive 2 credits. Clearly indicate the necessary steps, including appropriate formula substitutions, diagrams, graphs, charts, etc. For all questions in this part, a correct numerical answer with no work shown will receive only 1 credit.

11. In a bridge club, there are three more women than men. How many persons are members of the club if the probability that a member chosen at random, is a woman is $\frac{3}{5}$?

12. Find to the nearest degree the measure of the smallest angle in a right triangle whose sides measure 12, 35, and 37 inches.

Part III

Answer all questions in this part. Each correct answer will receive 3 credits. Clearly indicate the necessary steps, including appropriate formula substitutions, diagrams, graphs, charts, etc. For all questions in this part, a correct numerical answer with no work shown will receive only 1 credit.

13. The lengths of the sides of a triangle are in the ratio 3 : 5 : 6. The perimeter of the triangle is 49.0 meters. What is the length of each side of the triangle?

14. Huy worked on an assignment for four days. Each day he worked half as long as he worked the day before and spent a total of 3.75 hours on the assignment.

a. How long did Huy work on the assignment each day?

b. Find the mean number of hours that Huy worked each day.

Part IV

Answer all questions in this part. Each correct answer will receive 4 credits. Clearly indicate the necessary steps, including appropriate formula substitutions, diagrams, graphs, charts, etc. For all questions in this part, a correct numerical answer with no work shown will receive only 1 credit.

15. The perimeter of a garden is 16 feet. Let x represent the width of the garden.

a. Write an equation for the area of the land, y, in terms of x.

b. Sketch the graph of the equation that you wrote in **a**.

c. What is the maximum area of the land?

16. Each morning, Malcolm leaves for school at 8:00 o'clock. His brother Marvin leaves for the same school at 8:15. Malcolm walks at 2 miles an hour and Marvin rides his bicycle at 8 miles an hour. They follow the same route to school and arrive at the same time.

a. At what time do Malcolm and Marvin arrive at school?

b. How far is the school from their home?

INDEX

C

D

Q

R

X

Y

Z